Lecture Notes in Networks and Systems

Volume 37

Series editor

Janusz Kacprzyk, Polish Academy of Sciences, Warsaw, Poland
e-mail: kacprzyk@ibspan.waw.pl

The series "Lecture Notes in Networks and Systems" publishes the latest developments in Networks and Systems—quickly, informally and with high quality. Original research reported in proceedings and post-proceedings represents the core of LNNS.

Volumes published in LNNS embrace all aspects and subfields of, as well as new challenges in, Networks and Systems.

The series contains proceedings and edited volumes in systems and networks, spanning the areas of Cyber-Physical Systems, Autonomous Systems, Sensor Networks, Control Systems, Energy Systems, Automotive Systems, Biological Systems, Vehicular Networking and Connected Vehicles, Aerospace Systems, Automation, Manufacturing, Smart Grids, Nonlinear Systems, Power Systems, Robotics, Social Systems, Economic Systems and other. Of particular value to both the contributors and the readership are the short publication timeframe and the world-wide distribution and exposure which enable both a wide and rapid dissemination of research output.

The series covers the theory, applications, and perspectives on the state of the art and future developments relevant to systems and networks, decision making, control, complex processes and related areas, as embedded in the fields of interdisciplinary and applied sciences, engineering, computer science, physics, economics, social, and life sciences, as well as the paradigms and methodologies behind them.

More information about this series at http://www.springer.com/series/15179

Mohamed Ben Ahmed
Anouar Abdelhakim Boudhir

Editors

Innovations in Smart Cities and Applications

Proceedings of the 2nd Mediterranean Symposium
on Smart City Applications

 Springer

Editors
Mohamed Ben Ahmed
Computer Sciences Department,
  Faculty of Sciences and Techniques
Abdelmalek Essaadi University
Tangier
Morocco

Anouar Abdelhakim Boudhir
Computer Sciences Department,
  Faculty of Sciences and Techniques
Abdelmalek Essaadi University
Tangier
Morocco

ISSN 2367-3370   ISSN 2367-3389   (electronic)
Lecture Notes in Networks and Systems
ISBN 978-3-319-74499-5   ISBN 978-3-319-74500-8   (eBook)
https://doi.org/10.1007/978-3-319-74500-8

Library of Congress Control Number: 2018931878

Printed on acid-free paper

This Springer imprint is published by the registered company Springer International Publishing AG
part of Springer Nature
The registered company address is: Gewerbestrasse 11, 6330 Cham, Switzerland

Preface

The volume contains latest research work presented in the Second Edition of the Mediterranean Symposium on Smart City Applications (SCAMS2017) held on October 25–27, 2017, in Tangier, Morocco. This book presents original research results, new ideas, and practical development experiences. It includes papers from all areas of Smart City Applications.

Topics related to smart mobility, big data, smart grids, smart homes, smart buildings, smart environment, cloud, social networks and security issues will be discussed in SCAMS2017.

The conference stimulates the cutting-edge research discussions among many academic researchers, scientists, industrial engineers, and students from all around the world. The topics covered in this book also focus on innovative issues at international level by bringing together the experts from different countries.

The scope of SCAMS2017 includes methods and practices which bring various emerging Internetworking and data technologies together to capture, integrate, analyze, mine, annotate, and visualize data in a meaningful and collaborative manner. A series of international workshops are organized as invited sessions in the SCAMS2017:

- The 2nd International Workshop on Smart Learning and Innovative Educations,
- The 1st International Workshop on Smart Healthcare,
- The 1st International Workshop on Mathematics for Smart City,
- The 1st International Workshop on Industry 4.0 and Smart Manufacturing.

We would like as well to thank the Local Arrangement Chairs for making excellent local arrangements for the conference. We also would like to thank the Workshop Chairs for their delicate work.

Many thanks to the Springer staff for their support and guidance. In particular, our special thanks to Dr. Thomas Ditzinger and Ms. Varsha Prabakaran for their kind support.

Mohamed Ben Ahmed
Anouar Abdelhakim Boudhir

Organization

SCAMS2017 General Chairs

Mohamed Ben Ahmed   UAE University, Faculty of Sciences
and Techniques, Tangier, Morocco

Anouar Abdelhakim Boudhir   UAE University, Faculty of Sciences
and Techniques, Tangier, Morocco

SCAMS2017 Local Chairs

Mohamed Bouhorma   UAE University, Faculty of Sciences
and Techniques, Tangier, Morocco

Chaker El Amrani   UAE University, Faculty of Sciences
and Techniques, Tangier, Morocco

Workshop Chairs

SLIED'17 Chairs

El Bouhdidi Jaber   ENSATE, UAE University, Morocco

Ghailani Mohamed   ENSAT, UAE University, Morocco

Ait Kbir M'hamed   FSTT, UAE University, Morocco

SHEALTH'17 Chairs

El Hfid Mohamed   Faculty of Medicine and Pharmacy,
                                             UAE University, Morocco
Boudhir Anouar Abdelhakim   FSTT, UAE University, Morocco
El Ouaai Fatiha   FSTT, UAE University, Morocco

MSC'17 Chairs

Naji Ahmed   FSTT, UAE University, Morocco
Er-Riani Mustapha   FSTT, UAE University, Morocco
El Merzguioui M'hamed   FSTT, UAE University, Morocco

WI4SM'17 Chair

El Mustapha Ouardouz   FSTT, UAE University, Morocco

Organizing Committee

Astitou Abdelali   FSTT, Abdelmalek Essaadi University, Morocco
Aknin Noura   FS, Abdelmalek Essaadi University, Morocco
Bri Seddik   EST, Moulay Ismail University, Morocco
Boulmalf Mohammed   UIR, Morocco
El Brak Mohamed   FSTT, Abdelmalek Essaadi University, Morocco
El Jarroudi Mustapha   FSTT, Abdelmalek Essaadi University, Morocco
En-Naimi El Mokhtar   FSTT, Abdelmalek Essaadi University, Morocco
Ezziyyani Mostafa   FSTT, Abdelmalek Essaadi University, Morocco
Fennan Abdelhadi   FSTT, Abdelmalek Essaadi University, Morocco
Ghadi Abderrahim   FSTT, Abdelmalek Essaadi University, Morocco
Khattabi Mohamed Aissam   FSJEST, Abdelmalek Essaadi University,
                                             Morocco
Moussanif Hajar   FSS, Cadi AYYAD University, Morocco
Ziani Ahmed   FS, Abdelmalek Essaadi University, Morocco
Zili Hassan   FSTT, Abdelmalek Essaadi University, Morocco

PhD Committee

Allach Samir   FSTT, Abdelmalek Essaadi University, Morocco
Mahboub Aziz   FSTT, Abdelmalek Essaadi University, Morocco
Benchigra Younes   FSTT, Abdelmalek Essaadi University, Morocco

Program Committee Members

Ahmad S. Almogren	King Saud University, Saudi Arabia
Abdel-Badeeh M. Salem	Ain shams University, Egypt
Alabdulkarim Lamya	King Saud University, Saudi Arabia
Aknin Noura	FS UAE, Morocco
Accorsi Riccardo	Bologna University, Italy
Alghamdi Jarallah	Prince Sultan University, Saudi Arabia
Ahmed Kadhim Hussein	Babylon University, Iraq
Anabtawi Mahasen	Al-Quds University, Palestine
Arioua Mounir	UAE, Morocco
Astitou Abdelali	UAE, Morocco
Assaghir Zainab	Lebanese University, Lebanon
Bahraoui Fatima	UAE, Morocco
Bessai-Mechmach Fatma Zohra	CERIST, Algeria
Ben Ahmed Mohamed	FSTT, UAE, Morocco
Benaouicha Said	UAE, Morocco
Ben Yahya Sadok	Faculty of Sciences of Tunis, Tunisia
Boulmalf Mohammed	UIR, Morocco
Boutejdar Ahmed	German Research Foundation, Bonn, Germany
Britel Reda	UAE, Morocco
Chadli Lala Saadia	University Sultan Moulay Slimane, Beni Mellal, Morocco
Damir Žarko	Zagreb University, Croatia
Dousset Bernard	UPS, Toulouse, France
Dominique Groux	UPJV, France
Elhaddadi Anass	Mohammed Premier University, Morocco
El-Hebeary Mohamed Rashad	Cairo University, Egypt
El Kafhali Said	Hassan 1st University, Settat, Morocco
El Yassini Khalid	Moulay Ismail University, Morocco
El Yadari Mourad	FP, Errachidia, Morocco
Ensari Tolga	Istanbul University, Turkey
Enrique Arias	Castilla-La Mancha University, Spain
En-Naimi El Mokhtar	UAE, Morocco
Haddadi Kamel	IEMN Lille University, France
Jaime Lioret Mauri	Polytechnic University of Valencia, Spain
Jus Kocijan	Nova Gorica University, Slovenia
Khoudeir Majdi	IUT, Poitiers University, France
Labib Arafeh	Al-Quds University, Palestine
Lalam Mustapha	Mouloud Mammeri University of Tizi-Ouzou, Algeria
Loncaric Sven	Zagreb University, Croatia
Mademlis Christos	Aristotle University of Thessaloniki, Greece

Hazim Tawfik	Cairo University, Egypt
Miranda Serge	Nice University, France
Mousannif Hajar	Cadi Ayyad University, Morocco
Ouederni Meriem	INP-ENSEEIHT Toulouse, France
Sagahyroon Assim	American University of Sharjah, United Arab Emirates
Senthil Kumar	Hindustan College of Arts and Science, India
Senan Sibel	Istanbul University, Turkey
Sonja Grgić	Zagreb University, Croatia
Tebibel Bouabana Thouraya	ESI, Alger, Algeria

Workshops' Program Committee Members

SLIED'17

Abdellatif Medouri	ENSATE UAE, Morocco
M'hamed Ait Kbir	FSTT, UAE, Morocco
Aknin Noura	FS, UAE, Morocco
Kamal eddine El Kadiri	ENSATE UAE, Morocco
Amal Nejaari	ENSATE UAE, Morocco
Mounia Abik	ENSIAS, UM V, Rabat
Anouar Abtoy	ENSA, Tetuan
Abdellah Azmani	FST, Tangier
Samir Bennani	EMI, UM V, Rabat
Mohamed El Haddad	ENSA, Tangier
Mohamed Chrayah	ENSA, Tetuan
Bernard Dousset	UPS, Toulouse, France
Kunalè Florent Kudagba	Suptem, Tangier
Anass El Haddadi	ENSA El Hoceima
El Mokhtar En-Naimi	FST, Tangier
Abdelhadi Fennan	FST, Tangier
Soufiane Rouissi	BORDEAUX Montaigne University, France
Sahbi Sidhom	Lorraine University, Nancy, France
Mariam Tanana	ENSA, Tangier, Morocco
Mourad El Yadari	FP, Errachidia, Morocco
Abdelhamid Zouhair	ENSA, El Hoceima
Abderrahim Tahiri	ENSA, Tetuan
Mohammed El Achab	ENSA, Tetuan
Mohamed Lazaar	ENSA, Tetuan
Yassine Zaoui Seghroucheni	FS, Tetuan

Mohamed Yassine Chkouri   ENSA, Tetuan
Dominique Groux-Leclet   UPJV, France
Abderrahim El Mhouti   FSTH, Mohamed Premier University
Zakaria Itahriouan   FSI, UPF, Fez, Morocco

SHEALTH'17

Bernadetta Kwintiana Ane   University of Stuttgart, Germany
Farhat Fnaiech   ENSIT, Tunisia
Mousa Al-Akhras   Saudi Electronic University,
                                   Kingdom of Saudi Arabia
Zilong Ye   California State University, Los Angeles, USA
Adel Alti   University of SETIF-1, Algeria
Boulmalf Mohamed   University Internationale de Rabat, Morocco
Arsalane Zarghili   USMBA, Fez, Morocco
Joel Rodrigues   National Institute of Telecommunications, Brazil
Latif Rachid   ENSAA, Ibn Zohr University, Morocco
Lamya Alabdulkarim   King Saud University, Saudi Arabia
El Achhab Mohamed   University Abdelmalek Essaadi, Morocco
Khalid NAFIl   University Mohamed V of Rabat, Morocco
Mahasweta Sarkar   San Diego State University, California, USA
Abdelkrim Haqiq   University Hassan I of Settat, Morocco
Kashif Saleem   King Saud University, Kingdom of Saudi Arabia
Ben Ahmed Mohamed   FSTT, UAE, Morocco
Harroud Hamid   University AlAkhawayn of Ifrane, Morocco
Otman Abdoun   University Abdelmalek Essaadi, Morocco
Abeer Mohamed ElKorany   Faculty of Computers and Information Cairo
                                   University, Egypt
Josep Lluis de la Rosa   Girone Université, Spain
My Lahcen Hasnaoui   University Moulay Ismail Meknes, Morocco

MSC'17

Mohamed El Ghami   Nord Nesna University, Nesna, Norway
Chia-Ming Fan   National Taiwan Ocean University, Taiwan
Khalid El Yassini   Moulay Ismail University, Meknes, Morocco
Rhoudaf Mohamed   FS, Moulay Ismail University, Meknes, Morocco
Er-Riani Mustapha   FSTT, Abdelmalek Essaadi University, Tangier,
                                   Morocco
Imlahi Abdelouahid   FSTT, Abdelmalek Essaadi University, Tangier,
                                   Morocco
Naji Ahmed   FSTT, Abdelmalek Essaadi University, Tangier,
                                   Morocco

Amrani Mofdi   FSTT, Abdelmalek Essaadi University, Tangier,
                               Morocco
Serghini Abdelhafid   ESTO, Mohammed Premier University, Oujda,
                               Morocco
El Hajaji Abdelmajid   ENCG, Chouaib Doukkali University, El Jadida,
                               Morocco
Louartiti Khalid   FS, HASSAN II University, Casablanca,
                               Morocco
Akharif Abdelhadi   FSTT, Abdelmalek Essaadi University, Tangier,
                               Morocco
Bernoussi Abdessamd   FSTT, Abdelmalek Essaadi University, Tangier,
                               Morocco
Hamdoun Said   FSTT, Abdelmalek Essaadi University, Tangier,
                               Morocco
El Halimi Rachid   FSTT, Abdelmalek Essaadi University, Tangier,
                               Morocco
Settati Abdel   FSTT, Abdelmalek Essaadi University, Tangier,
                               Morocco
Belhadj Hassan   FSTT, Abdelmalek Essaadi University, Tangier,
                               Morocco
Chadli Lala Saadia   FST, University Sultan Moulay Slimane,
                               Beni Mellal, Morocco

WI4SM'17

Mustapha Ouardouz   FST of Tangier, Morocco
Abdes-samed Bernoussi   FST of Tangier, Morocco
Mustapha Zekraoui   FST of Beni Mellal, Morocco
Abdelkarim Chouaf   ENSEM of Casablanca, Morocco
Kamal Reklaoui   FST of Tangier, Morocco
Naoufal Sefiani   FST of Tangier, Morocco
Ibrahim Hadj Baraka   FST of Tangier, Morocco
Driss Sarsri   ENSAT of Tangier, Morocco
Mohammed Bsiss   FST of Tangier, Morocco
Abdelali Astito   FST of Tangier, Morocco
Ahmed Rachid   UPJV, France
Bernadetta Kwintiana Ane   University of Stuttgart, Germany

SCAMS2017 Keynote Talks

Smart Cities Promised Technological and Social Revolution

Mohamed Essaaidi

ENSIAS, Mohammed V University, Rabat, Morocco

Abstract. Many cities around the world are currently transitioning toward smart cities to attain several key objectives such as a low-carbon environment, high quality of living, and resource-efficient economy. Urban performance depends not only on the city's endowment of hard infrastructure, but also on the availability and quality of knowledge communication and social infrastructure. There is a growing importance of information and communication technologies (ICTs) and social and environmental capital in profiling the competitiveness of cities. Information and communication technologies play a critical role in building smart cities and supporting comprehensive urban information systems.

This brings together citizens and integrates technologies and services such as transportation, broadband communications, buildings, health care, and other utilities. Advanced communication and computing techniques can facilitate a participatory approach for achieving integrated solutions and creating novel applications to improve urban life and build a sustainable society. Extensive research is taking place on a wide range of enabling information and communication technologies, including cloud, network infrastructure, wireless and sensing technologies, mobile crowdsourcing, social networking, and big data analytics.

The main purpose of this keynote talk is to present an overview of smart city technologies, applications, opportunities, and research challenges.

Inclusive Smart City for Specially Abled People

Parthasarathy Subashini

Avinashilingam University, India

Abstract. The concept of smart city itself is still emerging, and the work of defining and conceptualizing it is in progress. This concept is used all over the world with different nomenclatures, context, and meanings stressing the need for public and private stakeholders to put the citizen at the heart of any smart city project. Citizens, the inhabitants of the intelligent cities become agents of change, fully aware of the city challenges and play a qualified role in the civic network, characterized by participation, civic engagement, territorial commitment, and the will of sharing knowledge of creativity. The challenges in the cities and societies are mostly related to the fact that disability causes specific barriers, such as limitations of mobility, visual and hearing impairments, and a high disease susceptibility. ICT solution implemented in the cities can help overcome mobility and visual and cognitive movements. Innovations in areas such as remote sensors, embedded systems, robotics, or wireless mobile networks provide building blocks for intelligent ambient systems that can support especially abled people and allow them to move around the city without any assistance.

The current market tends to expand products designed to serve large portions of the population, providing tools, objects, and furnishings with a design that meets the requirements or needs of people with or without disabilities. Regarding that perspective, there has been an increase in the variability of these products marketed to the most diverse purposes, including for daily life activities.

Assistive technologies are a key tool to any physically challenged person when checking their ability toward accessibility. Understanding of these technologies goes hand in hand with understanding the requirements of their accessibility also. Assistive technology can be defined as "any item, piece of equipment, or product system, whether acquired commercially or off the shelf, modified, or customized, that is used to increase, maintain, or improve functional capabilities of individuals with disabilities." Assistive technology includes assistive, adaptive, and rehabilitation devices for people with disabilities. Especially, the ability to communicate is fundamental to a basic quality of life, yet for many people effective communication is difficult because of a physical impairment, language disorder, or learning disability. The proposed topic is toward various assistive technology systems that facilitate people who face just such challenges.

Assistive technology service includes any service that directly assists an individual with a disability in the selection, acquisition, or use of an assistive technology device.

The use of ATs can improve the physical and social functioning of persons with disabilities and elderly people.

The proposed keynote speech describes about various opportunities and challenges related to "inclusive smart city," expanding the possibilities for persons with special needs in the urban space.

Cognitive IoT for Smart Urban Sensing

Mounir Ghogho[1,2]

[1] University of Leeds, UK
[2] UIR, Morocco

Abstract. In this talk, I will briefly introduce the Internet of things (IoT) and its enabling technologies. Then, I will give a brief overview of the different wireless connectivity solutions (cellular and non-cellular) that compete to have a share of the IoT market. Fundamental limits on coverage/throughput and their relation with network density will be presented and used to explain the advantages and disadvantages of the different wireless solutions. I will then present the concept of cognitive IoT, its components, and its challenges. The focus will be on cognitive sensing, communication, and energy harvesting. Finally, I will briefly present my ongoing IoT projects on urban sensing.

Data for Smart Spaces

Hajar Mousannif

Cadi Ayyad University, Morocco

Abstract. With the significant advances in information and communication technology over the last half-century, cloud computing, big data, Internet of things, and mobile technology are rapidly emerging as new pillars of the next generation of IT.

With the cloud, we have infinite power of computation right in our pockets. Big data takes advantage of the technically "unlimited" storage and computing capabilities of the cloud to make predictive and prescriptive analytics to extract insights about every aspect of our lives, while the Internet of things connects everything that can be connected. They also generate huge amounts of data waiting to be processed and analyzed. Mobile technology makes the picture even more beautiful by making information available anywhere anytime!

Cloud computing, big data, Internet of things, and mobile technology blur the traditional boundaries and profoundly revolutionize the value chain of many vital sectors including health care, transportation, education, and industry. They also raise many pressing ethical issues waiting to be addressed. The aim of this talk is to highlight the potential of all afore-mentioned technologies through four real-world deployment projects we carried out in Morocco, which fall under four categories: smart health, smart transportation, smart education, and smart monitoring. Interesting proto-types will be demonstrated.

New Modeling Approaches for Micro-Grids Enabling Frugal Social Solar Smart City

Aawatif Hayar

University Hassan II Casablanca Morocco, ENSEM

Abstract. The Frugal Smart City concept we recommend for Casablanca and, in general, for developing countries puts citizens at the center of the transformation process, creating a public–private–people partnership where citizens are actors in and builders of their smart city. This concept is also aiming at limiting investment risks by adopting data-driven cost-effective or "frugal" bottom-up approach. It is based on the use of mobiquitous devices, such as smartphones, crowdsourcing tools, and open data analysis techniques, to develop IT-driven innovation cycle and e-services that track and answer citizens' economic cultural, social, and ecological needs.

This participatory-oriented social innovation approach will allow, step-by-step, to build a set of interconnected pilot projects to set up gradually a sustainable smart city and collaborative innovation ecosystem creating at the end inclusive sustainable economy which turns societal and economic challenges into business opportunities.

This talk will present the Frugal Smart City concept and some applications based on data-assisted green energy integration and management in smart micro-grids.

We will also present some recent theoretical advances and practical results in smart grids modeling and implementation from pilot projects set up in the context of Casablanca Frugal Smart City.

Intelligent Transport Systems

Abdelhakim Senhaji Hafid

University of Montreal, Canada

Abstract. Intelligent Transportation Systems (ITS) make use of advanced communications and information technology to improve the safety and efficiency (e.g., in terms of mobility and energy consumption) of transportation. Mobile vehicular networks represent the key component of ITS; indeed, they are the foundation of a wide spectrum of novel safety, traffic control, and entertainment applications which are realized mainly through vehicle to vehicle (V2V) and vehicle to infrastructure (V2I) communications. To enable these applications, novel schemes and protocols need to be developed to support their requirements in terms of performance and reliability. In this talk, we define mobile vehicular networks and we present a number of applications that can be supported by this type of networks. Then, we present key challenges facing the realization of the potential of vehicular networks. We will also briefly overview related contributions produced by the Network Research Laboratory at the University of Montreal. We conclude by presenting our view on the future of vehicular networks.

Contribution of IoT Applications to Enhance Authenticating Individual's Geo-location

Hassan Zili

Abdelmalek Essaadi University, Morocco

Abstract. This talk is about the invention that relates to geo-locations software management utility. Specifically, it provides a method and system for authenticating an individual's geo-location via a communication network and applications using the same. The method comprises the following steps:

- Providing an individual with a smartphone having a GPS receiving unit associated with a communications network;
- Providing the individual with a biometric user identification technology that may effectively utilize a wristband worn by the individual for passive biometric user identification and wherein the wristband is wirelessly coupled to the smartphone to obtain the individual's electrocardiogram;
- Obtaining via the communications network the geo-location of the smartphone utilizing the GPS receiving unit;
- Identifying the user with the biometric user identification technology by obtaining biometric characteristics that are unique to each human (in our case, it's the heartbeat);
- Verifying the biometric user identification technology is within a preset proximity to the smartphone to authenticate the individual's geo-location.

This invention is directed to a cost-effective, efficient, method and system for authenticating, accessing, acquiring, storing, and managing each individual's geo-location position data via a communication network.

This method includes recording the individual's authenticated geo-locations for a period of time, wherein during recording of the individual's authenticated geo-locations.

Reference

Patent number: 9801058
Patent Publication Number: 20160021535
Inventors: Tarik Tali (Lakeway, TX), Hassan Zili (Tangier), Abdelhak Tali (Tangier)
Application Number: 14/699,460

Big Data in the Smart City:
The Big Bridge Example

Gaetan Robert Lescouflair[1] and Serge Miranda[2]

[1] LSIS, University of Marseille and Aix, France
[2] MBDS, University of Nice Sophia Antipolis, France

Abstract. We entered a new data-centric economy era with the widespread use of supporting big data infrastructures to deliver predictive real-time analysis and augmented intelligence in the three "P" sectors (Public, Private, and Professional). Every economic sector will be drastically impacted. Such big data infrastructure involves two major scientific fields: computer science for *data integration* (i.e., building a virtual or real data lake) and mathematics for *machine learning*—ML—(AI for *deep learning*—DL). This concept of data lake was first introduced in 1999 by Pyle.

Three types of data are involved in a big data architecture: *structured data* (with predefined schema), *semi-structured data* (around XML with metadata), and *unstructured data* (no schema, no metadata). A data lake is a generalization of data warehouse to semi-structured and non structured data. The data lake could be real (with pumping systems like in most data warehouses) or virtual with distributed large data sets. Today, there is no SQL standard to manage a data lake with many proprietary proposals encompassing new key features like "external tables".

Expected use of a data lake is predictive real-time analysis by data scientists using a large variety of ML and DL methods generally in supervised, unsupervised ou reinforcement modes; no interactivity exists among these methods.

This conference encompasses two parts:

- A state of the art of SQL extensions to manage NO SQL data bases and a real or virtual data lake.
- A presentation of the BIG BRIDGE project with the creation of a data lake within Nice smart city, with historical SQL data (air pollution and heart emergencies in the hospital) and NO SQL data coming from smart watch (monitoring heart beat for people at risk).

SDN for Smart City

Noureddine Idboufker

Cadi Ayyad University, Morocco

Abstract. In spite of the increasing popularity that the concept of smart city has gained over the past few years, there is still no universal definition for it. Smart city concept is mainly about a city that provides ICT-based services in different sectors of activity, in order to mitigate urban challenges, increase efficiency, reduce costs, and enhance the quality of life. Generally, Smart city definition varies in function of city resources, development, and its ability of changing.

The role of ICT for enabling smart city features should be highlighted; especially with the advent of new technologies provided by new features and characteristics (high throughput, low latency, high QoS.).

It is inevitable to mention WSN as a basic layer that should be present in each smart city initiative. Indeed, the WSN layer allows service provider to design, to develop, and to provide a rich service catalogue with a real added value. However, WSN is based on the use of sensors with limited energy efficiency, processing, and storage capabilities.

We propose the Software Defined Networks (SDNs) approach, as one of the potential solutions to face the challenge associated with the use of sensors in smart city. SDN is based on separating control and data planes, and centralizing network intelligence within SDN controllers, which are directly programmable. Thus, this new paradigm provides a deep knowledge of the state of the network resources, in order to alleviate the load and add agility and flexibility to network. By separating control and data planes, SDN makes possible the use of a wireless layer based on sensor with a high level of energy efficiency and low level of processing and storage, leading to a real improvement of the Network Scalability.

New Technology for Effective E-learning and Smart Campus

Noura Aknin

Abdelmalek Essaadi University, Morocco

Abstract. Technology is one of the main factors that regularly disrupts the world of education in terms of educational resources and learning environment. Its implementation facilitates the transition from a traditional knowledge transfer model to a more efficient system based on collaboration, autonomy, and involvement.

Today, there is a growing importance of information and communication technologies (ICTs) and their applications offer momentous opportunities for development of new approaches that allow the learner to deepen his knowledge and develop the necessary skills to succeed his "cognitive project." In addition to these approaches, the actors and educational leaders are obliged to reconsider the necessary spaces for an efficient and intelligent learning.

Among the latest technologies, the Internet of things (IoT) used to sense its surrounding environments plays a key role for the future development of smart campus. Indeed, its implementation in campus using e-learning can enable interaction between all components of the educational system and physical spaces for learning purposes or communication. Thus, with the proliferation of connected objects, campuses can collect data more easily to interpret learner behaviors and activities for effective e–learning application toward a smart campus.

This talk focuses on new advanced technologies for effective learning and challenges of a smart campus.

Partial Differential Equation (PDE) Models for Ocean Modeling

J. Rafael Rodríguez Galván

Universidad de Cádiz, Spain

Abstract. We focus on Partial Differential Equation (PDE) models for ocean modeling. In particular, we consider the Boussinesq equations (with variable density depending on temperature and salinity) and the well-known incompressible Navier–Stokes equations. Different simplifications which are usually introduced in oceanography are considered. Then we focus on one of them, where the anisotropy between vertical and horizontal scales in maritime domains leads to anisotropic Navier–Stokes equations (a generalization of Primitive Equations of the ocean).

The additional difficulties which are introduced by anisotropic Navier–Stokes equations are analyzed, and some recent techniques for the numerical approximation of these equations by means of the finite element method are introduced. Some numerical simulations are shown, focusing in water flow along the Gibraltar Strait.

Toward Efficient Numerical Models for Assessment and Management of Sea-Pollution Risks

Mohammed Seaid

Durham University, UK

Abstract. Nowadays risk management is a key success for designing smart cities and sustainable development. In our contribution, we present recent trends in modeling sea-pollution and its management. Modeling hydrodynamics for water free-surface along with dispersion of contaminants are part of this study. Here we consider both the single-layer and multi-layer depth-averaged Navier–Stokes equations for flow modeling and advection–diffusion models for the concentration of pollutant. To solve the flow equations, we consider a robust finite volume method with mesh adaptation, and for the transport and dispersion of contaminants, we propose a new particle method. Stochastic and turbulent effects are accounted for in these models, and efficient numerical tools are presented for their treatment. Inverse problems to localize the pollution source on the sea surface are also presented in the current work. Installing alert systems along the coast areas are also examined, and optimal solutions will be provided. As a real-life application, we present results obtained for oil spills on the North Sea.

Use of 3D Laser Scanning Technology in Plant Virtual Planning

Younes Gouaiti

Business Development and Marketing Director of Enigma, Morocco

Abstract. In recent years, 3D laser scanning has revolutionized the areas of digital plant planning, documentation of industrial facilities, architecture, monument protection, landscape, and virtual reality. Visualization of buildings or complex geometries in three dimensions allows a more precise, safer, and faster work.

Laser scanners scan their environment by sending a laser beam. The reflected signal creates a very dense cloud of points. These points, with the help of appropriate algorithms, make it possible to reconstruct an image of the environment. The scanner associates these points with photos of the real environment. The result is a reproduction of the surfaces in a three-dimensional space.

The applications are diverse we quote:

- Transformations and Extensions: Precise 3D documentation of the real state of buildings as a planning basis for transformations and extensions;
- Off-Site Manufacturing: Off-site manufacturing and assembly capability with high accuracy made possible by accurate 3D CAD data retrieval and dimensional control;
- Asset Management: Simplification of facilities management, maintenance, training, etc., thanks to complete 3D data, simulations, and virtual trainings;
- Supervision of Works: Improved coordination of different trades, complete documentation, and supervision of all works;
- 3D Digitization of Crimes Scenes or Accidents: capturing essential details for the subsequent reconstruction of the crime or accident;
- Deformation Control: determine if the object or structure studied changes shape or moves;
- Quality Control: ensure that the final state of the as-built conforms to the design plans.

The research areas associated with 3D laser scanning are diverse and cover the whole process: from acquisition to restitution through identification and recognition techniques, classification, analysis, and processing.

Contents

Contents

**The 1st International Workshop on Mathematics for Smart City:
MCS'17**

The 2nd Mediterranean Symposium on Smart City Applications

Logical Model of AS Implementing IPv6 Addressing

Acia Izem[1]([✉]), Mohamed Wakrim[1], and Abderrahim Ghadi[1,2]

[1] LabSI, Department of Mathematics and Informatics, FSA,
Ibn Zohr University, Agadir, Morocco
acia8891@gmail.com, m.wakrim@uiz.ac.ma, ghadi05@gmail.com
[2] LIST, Department of Computer Science, FST,
Abdelmalek Essaadi University, Tangier, Morocco

Abstract. IPv6 address space is a public resource that must be managed in a prudent manner. Wherever possible, address space should be distributed in a hierarchical manner. This is necessary to permit the aggregation of the routing information by ISPs (Internet Service Providers) and to limit the expansion of internet routing tables.

The aim of this paper is to propose a model that may help to resolve the massive growth in the size of routing tables. Inspired from the tree of Pythagor, we get to distribute the prefixes in a hierarchical manner that minimizes and divides up the routing tables. The process of building the logical model, presented in this paper, is a grouping tool that can be used by network architects of autonomous systems or companies to build their networks with minimal routing tables.

Keywords: Model · Network · Logical · Autonomous systems
Routing table · IPv6 · Tree of Pythagor

1 Introduction

1.1 Related Work

There has in recent years been a surge of interest within the research community in modeling the dynamic and the structure of complex networks including internet, we cite for example [1–7]: Erdos-Renyi random graph model, Barabasi-Albert model, Watts-Strogatz small world model and Kleinberg model. All these models are elegant and simple but all of them have some limitations such as: The number of nodes N is fixed, the internet is the result of a growth process that continuously increases N. The size of the network influence the readability of the graph: the bigger the size of the network the less we can see the nodes and the connections between them.

More importantly, the graphs presented by these models are significantly different from the structure of internet. They lack of what we call in this paper the representative nodes. For example: The computers are nodes, they are connected

© Springer International Publishing AG, part of Springer Nature 2018
M. Ben Ahmed and A. A. Boudhir (Eds.): SCAMS 2017, LNNS 37, pp. 3–16, 2018.
https://doi.org/10.1007/978-3-319-74500-8_1

to internet through routers therefore the routers are special nodes which we call the representative nodes they can represent a group of computers as well as a group of routers by using the summarization technique.

1.2   The Internet Structure and the Growth of Routing Tables

The structure of internet is determined by the routed protocols such as IPv4 and IPv6. In other words, how the IP addresses are assigned or distributed.

With the rapid development of the internet applications and the explosion of the end users, the IANA announced in February 2011 the exhaustion of the available public IPv4 addresses. Besides the exhaustion problem, the random distribution of public IPv4 addresses enormously grows the routing tables [8].

The summarization technique that was used to minimize the routing table was useless because this technique should go hand in hand with the geographic distribution. For example: According to the topology presented in Fig. 1a, the routing table of the router R contains 9 records as shown in Table 1.

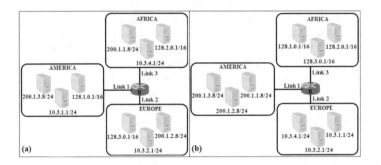

Fig. 1. **(a)** Random distribution of IPv4 addresses: We have 3 servers in Africa (200.1.1.8/24, 10.3.4.1/24 and 128.2.0.1/16), 3 servers in Europe (200.1.2.8/24, 10.3.2.1/24 and 128.3.0.1/16) and 3 servers in America (200.1.3.8/24, 10.3.1.1/24 and 128.1.0.1/16). **(b)** Geographic distribution of IPv4 addresses: All the IPs that begin with 128 (Resp. 200.1 (Resp. 10.3)) will be found in Africa (Resp. America (Resp. Europe)).

The problem of the random distribution arises when the routing table of R contains hundreds records, in this case if the packet destination address was 200.1.1.8/24, the router R has to search in all its long routing table to decide which link (1, 2 or 3) will be used to deliver the given packet.

Applying the summarization technique, in the topology presented in Fig. 1a, will not resize the routing table of R. However, by reassigning the IPs (See Fig. 1b) then using the summarization technique, the routing table of R will contain just 3 records instead of 9 records (See Table 2).

Finally, by comparing Tables 1 and 2, it is obvious that using the geographic distribution of IPs and the summarization technique may resolve the problem of enormous growth of global routing tables. On one hand, this solution cannot be applied in the IPv4 addressing because the IPs that are dispatched randomly could not be retrieved and the available public IPv4 addresses are already

Table 1. The routing table of R: Whenever R receives a packet whose destination address is 200.1.1.8/24, R uses link 3 to deliver it.

Destination	Next hop
200.1.1.8/24	link 3
10.3.4.1/24	link 3
128.2.0.1/16	link 3
200.1.2.8/24	link 2
10.3.2.1/24	link 2
128.3.0.1/1	link 2
200.1.3.8/24	link 1
10.3.1.1/24	link 1
128.1.0.1/16	link 1

Table 2. The new routing table of R: Whenever R receives a packet whose destination address starts with 200.1, R uses link 1 to deliver it whatever the remaining address bits are (The routers in America will deliver it to the right server or destination).

Destination	Next hop
200.1.0.0/16	link 1
10.3.0.0/16	link 2
128.0.0.0/8	link 3

exhausted. On the other hand, it can be used in the IPv6 addressing. However, the tools used by network architects to design networks such as GNS3, Microsoft Visio and Packet tracer give them total freedom to assign IPs and connect subnets together without controlling the sizes of the routing tables.

In this paper, we do not try to model the whole structure of internet, but we focus on modeling the structure of its components. Therefore, we propose a logical model of an autonomous system (or a network) based on summary routes. This logical model may help network architects and network managers, implementing IPv6 in their organizations, to design their networks with optimal routing tables.

The structure of the paper is organized as follows. In Sect. 2, we review the networking and mathematical concepts used in the proposed method. In Sect. 3, we discuss how we get inspired. In Sect. 4, we illustrate the proposed model.

2 Definitions and Concepts

2.1 IPv6 Summary Routes

In OSPFv3, the network is subdivided to areas. An area is a collection of routers where each router interface belongs to an area.

The Area Border Router (ABR) is a kind of router whose interfaces belong to multiple areas and it (ABR) is responsible for the delivery of the summaries. For example: In the topology presented in Fig. 2, the Area Border Router R3 receives from R1 and R2 the prefixes: $2001 : 0DB8 : 6783 : 045A :: /64$ and $2001 : 0DB8 : 6783 : 045B :: /64$. Subsequently, they are added to its routing table.

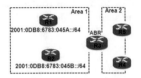

Fig. 2. The summary route.

As being ABR, R3 delivers their summary ($2001 : 0DB8 : 6783 : 045A :: /63$) to R4 and R5. Thus, whenever R4 or R5 receives a packet whose first 63 bits of the destination address is $2001 : 0DB8 : 6783 : 045A$, they deliver it to R3 without seeing the remaining bits. By seeing the 64th bit of this address, R3 can decide to deliver the given packet to R1 or R2.

2.2   The Basic Address Plan Structure

The illustration shown in Fig. 3 represents the basic address plan structure.

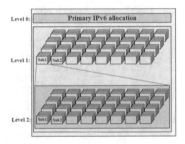

Fig. 3. Basic address plan structure: Level 0 represents an organization's primary IPv6 allocation from the ISP or RIR. Level 1 represents blocks of subnets defined and assigned to some network or attribute within the organization. Level 2 would be the next block of assigned subnets within each level 1 subnet. Level 3 would likely be the smallest assignable subnet (again, with each instance of level 3 fitting into each level 2 subnet) [9].

2.3   Pythagorean Tree

The construction of the Pythagorean tree begins with a square. In the top edge of this square, we construct a right isosceles triangle whose hypotenuse is the edge of the square. Along each of the other two sides of this isosceles triangle we construct squares. The same procedure is then applied recursively to the two smaller squares. The limit of this construction is called the Pythagorean tree (See Fig. 4).

Fig. 4. The Pythagorean tree

3   Synthesis: Comparison, Inspiration and Assumption

3.1   The Representative Node

By returning to Fig. 2, the routers R4 and R5 do not know the prefixes: $2001 : 0DB8 : 6783 : 045A :: /64$ and $2001 : 0DB8 : 6783 : 045B :: /64$. But they know their summary: $2001 : 0DB8 : 6783 : 045A :: /63$.

Hence, we define the summary $2001 : 0DB8 : 6783 : 045A :: /63$ as the representative of $2001 : 0DB8 : 6783 : 045A :: /64$ and $2001 : 0DB8 : 6783 : 045B :: /64$. In this paper, we schematize this kind of relation between prefixes with the graph presented in Fig. 5.

Fig. 5. The representative node (the black node): The leaf nodes represent the prefixes $2001 : 0DB8 : 6783 : 045A :: /64$ and $2001 : 0DB8 : 6783 : 045B :: /64$. The both leaves are connected to a new node (The representative node) that has their summary $(2001 : 0DB8 : 6783 : 045A :: /63)$ as a label. A node in this representation can be a router or a representative of a group of routers

3.2   Pythagorean Tree versus Autonomous System (AS) Structure

We assume that the Pythagorean tree represents nearly the logical structure of an AS. The reasons why we came up with this assumption are explained as follows.

On one hand, the summarization technique consists of grouping small prefixes to get one large prefix, however the construction of the Pythagorean tree consists of creating small squares from large square. Thus, the construction processes go in the opposite direction but the final shapes are similar: Large entity (square or prefix) connected to two smallest entities (Squares or prefixes).

On the other hand, in order to limit the expansion of internet routing tables, the AS has to optimize its internal routing tables and represent the generated subnets with the generating prefixes by using the summarization technique. By looking at the tree presented in Fig. 4, we notice that all the leaf squares (the smallest squares) are originally created from the root square (the largest square). Furthermore, suppose that the construction of the Pythagorean tree (Fig. 4) was done from the leaf squares to the root square (the opposite direction of the real construction), then in each level of aggregation every pair of consecutive small squares is represented by a new large square as shown in Fig. 6.

Despite the difference in the number of aggregated elements, the aggregation process presented in Fig. 6 is similar to the summarization technique: By looking at Fig. 6, we notice that we always gather just two small squares to get one large square, however in the summarization technique we can aggregate two or more

Fig. 6. Pythagorean tree versus summarization technique: In the first level of aggregation, the squares 1 and 2 (Resp. the squares 3 and 4) are represented by the square 5 (Resp. the square 6). In the second level of aggregation, the squares 5 and 6 are represented by the square 7.

small prefixes to get one large prefix. In Sect. 3, we illustrate how we preserve the hierarchical idea of the Pythagorean tree to get the logical model of an autonomous system.

4   The Proposed Method

Generally speaking, the proposed method is actually a grouping process that minimizes the size of the routing tables. To illustrate our algorithm, suppose that an ISP is given the prefix $2C0F : FB20 :: /32$.

The ISPs usually classify their clients into 3 categories: The first and the second client categories are large-sized clients and medium-sized clients. They are, usually, respectively given prefixes with /48 and /56 as length. The third client category is the small-sized clients which usually has prefixes with /60 or /64 as length.

Table 3. The generated prefixes

2C0F : FB20 :: /36	2C0F : FB20 : 8000 :: /36
2C0F : FB20 : 1000 :: /36	2C0F : FB20 : 9000 :: /36
2C0F : FB20 : 2000 :: /36	2C0F : FB20 : A000 :: /36
2C0F : FB20 : 3000 :: /36	2C0F : FB20 : B000 :: /36
2C0F : FB20 : 4000 :: /36	2C0F : FB20 : C000 :: /36
2C0F : FB20 : 5000 :: /36	2C0F : FB20 : D000 :: /36
2C0F : FB20 : 6000 :: /36	2C0F : FB20 : E000 :: /36
2C0F : FB20 : 7000 :: /36	2C0F : FB20 : F000 :: /36

To aggregate the prefixes or the subnets generated from the prefix $2C0F : FB20 :: /32$, we propose the following process:

Let k and n be integers where k is the number of client categories and $n \in \{2, 4, 8, 16\}$.

Step 1: Determine k and the prefix length of each category:

– 1st case: $k = 1$.
  In this case, the ISP is using the prefix $2C0F : FB20 :: /32$ to generate prefixes for one client category. For illustration purposes, we take /36 as the prefix length of this category.
– 2nd case: $k > 1$.
  In this case, the ISP is using the prefix $2C0F : FB20 :: /32$ to generate prefixes that will be offered to clients from different categories. For illustration purposes, we take k = 2, /36 and /40 as the prefix lengths of these categories.

Step 2: If $k = 1$ then we follow the following substeps:

1. Generate all the subnets related to $2C0F : FB20 :: /36$. We get the prefixes listed in Table 3.
2. Sort the generated prefixes in ascending order (The prefixes in Table 3 are already sorted). We consider that: A000 < B000 and 1000 < 2000.
3. Represent each prefix in Table 3 with a node (See Fig. 7):
4. From the left to the right we begin the aggregation, in every aggregation level:
   - We specify the value of n.
   - We arrange the nodes into groups of n nodes. Each group will be represented by a new node.
   - The new node label is the, nodes labels, summary of its corresponding group.
   - We aggregate the new nodes in the next aggregation level.

We repeat the aggregation process until we get one node. By applying this process we get:

Fig. 7. Representation of prefixes with nodes: The nodes are sorted in ascending order.

Level 0: In this level (See Fig. 8), we have the same nodes as in Fig. 7.

Level 1: In this level, we aggregate the nodes of Level 0. Let $n = 2$ (It can also be 4, 8 or 16), as shown in Fig. 8, the first group of n nodes is

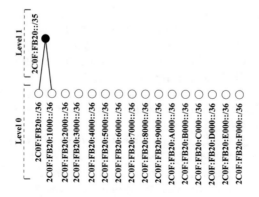

Fig. 8. The aggregation process

formed from the nodes labeled with the prefixes $2C0F : FB20 :: /36$ and $2C0F : FB20 : 1000 :: /36$, this group is represented in Level 1 with a black node (The representative node) labeled with the summary $2C0F : FB20 :: /35$ (The summary of $2C0F : FB20 :: /36$ and $2C0F : FB20 : 1000 :: /36$). In the same way, we create the representative nodes (The black nodes) for the remaining groups as shown in Fig. 9.

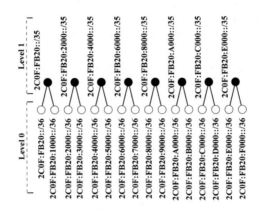

Fig. 9. The aggregation process

Level 2: In this level, we aggregate the nodes of Level 1. Let n = 2 (It can also be 4 or 8, but it cannot be 16 because the number of nodes to aggregate is not multiple of 16). As shown in Fig. 10, the first group of n nodes is formed from the nodes labeled with the prefixes $2C0F : FB20 :: /35$ and $2C0F : FB20 : 2000 :: /35$, this group is represented in Level 2 with a black node (The representative node) labeled with the summary $2C0F : FB20 :: /34$ (The summary of $2C0F : FB20 :: /35$ and $2C0F : FB20 : 2000 :: /35$). In the same way, we create the representative nodes (The black nodes) for the remaining groups.

Level 3: In this level, we aggregate the nodes of Level 2. Let $n = 2$ (n can also be 4 but it cannot be 8 (Resp. 16) because the number of the nodes to aggregate is not multiple of 8 (Resp. 16)). As shown in Fig. 11, the first group of n nodes is formed from the nodes labeled with the prefixes $2C0F : FB20 :: /34$ and $2C0F : FB20 : 4000 :: /34$, this group is represented in Level 3 with a node labeled with the summary $2C0F : FB20 :: /33$ (The summary of $2C0F : FB20 :: /34$ and $2C0F : FB20 : 4000 :: /34$). Likewise, we create the representative nodes for the remaining groups as shown in Fig. 11.

Level 4: In this level, we aggregate the nodes of Level 3. Let $n = 2$, it cannot be 4 (Resp. 8 (Resp. 16)) because the number of the nodes to aggregate is not multiple of 4 (Resp. 8 (Resp. 16)). As shown in Fig. 11, we can create just one group of n nodes which is formed from the nodes labeled with the prefixes $2C0F : FB20 :: /33$ and $2C0F : FB20 : 8000 :: /33$, this group is

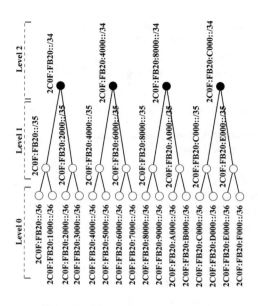

Fig. 10. The aggregation process

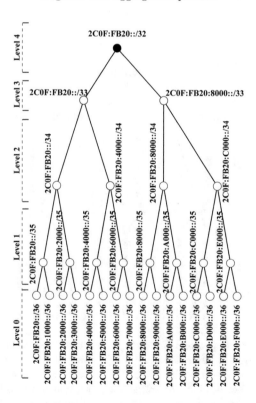

Fig. 11. The aggregation process: The basic logical model

represented in the level 4 with a black node (The representative node) labeled with the summary $2C0F : FB20 :: /32$ (The summary of $2C0F : FB20 :: /33$ and $2C0F : FB20 : 8000 :: /33$).

Level 4 (See Fig. 11) contains just one node, therefore the aggregation process is ended.

Step 3: If $k > 1$ (the 2nd case) then we proceed as follows.

1. Select the client category that has the smallest prefix length.
2. Apply the substeps of **Step 2**.
3. From the remaining client categories, select the client category that has the smallest prefix length.
4. Use at least one of the leaves labels, of the tree gotten in (2), as a prefix and apply the substeps of **Step 2**.
5. We recursively reapply the substeps (3) and (4) as long as we have remaining client categories.

In other words, for illustrative purposes, in the 2nd case we took $k = 2$, /36 and /40 as prefix lengths of the client categories. According to the basic address plan structure shown in Fig. 3, we generate the prefixes related to the category /36 then we use at least one of them to generate prefixes related to the category /40. Therefore, to generate the corresponding logical model, we do the following substeps:

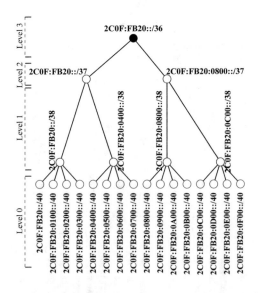

Fig. 12. The aggregation process: **Level 0:** We have the generated prefixes. **Level 1:** We aggregate the nodes of Level 0 with n = 4 (It can also be 2, 8 or 16). **Level 2:** We aggregate the nodes of Level 1 with n = 2 (It can also be 4 but it cannot be 8 (Resp. 16) because the number of nodes to aggregate is not multiple of 8 (Resp. 16)). **Level 3:** We aggregate the nodes of Level 2 with n = 2, n cannot be 4 (Resp. 8 (Resp. 16)) because the number of nodes to aggregate is not multiple of 4 (Resp. 8 (Resp. 16))

1. We select /36: it is the category that has the smallest prefix length.
2. By using $2C0F : FB20 :: /36$ as a prefix we apply the substeps of **Step 2**. In the end we get the tree seen in Fig. 11.
3. We have only one remaining category: /40, thus we select it.
4. From the leaves labels of the graph presented in Fig. 11, we choose one: let be the label 2C0F : FB20 :: /36. Afterwards, we consider the prefix $2C0F : FB20 :: /36$ as the generating prefix, instead of $2C0F : FB20 :: /32$, then we reapply the substeps of **Step 2**:
   (a) From $2C0F : FB20 :: /36$, we generate all the subnets related to the category /40.
   (b) We sort the generated prefixes. We consider that: $0A00 < 0B00$.
   (c) We represent the prefixes with nodes (See Fig. 12-Level 0).
   (d) We aggregate the nodes (See Fig. 12):

In Fig. 12, the root of the tree (The black node) is already a leaf in the tree presented in Fig. 11. Thus, by merging the two graphs we get the tree presented in Fig. 13.

Finally, the AS is given in general more than one Global Routing Prefix therefore its logical structure is a forest where the Global Routing Prefixes are the roots of its trees.

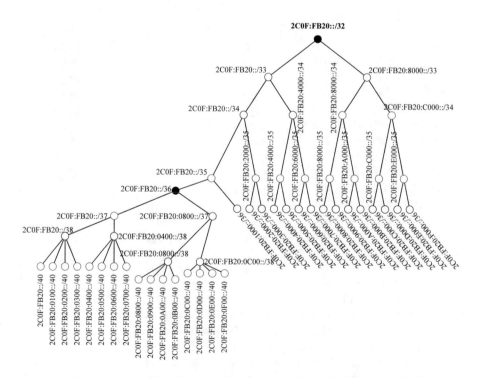

Fig. 13. Aggregation process

5   Discussion

The basic address plan structure (Fig. 3) depicts how to generate subnets. However the logical model proposed in this paper shows how to group these subnets. Furthermore, the hierarchical structure of the proposed model allows to minimize and divide up the routing tables and consequently increase the speed of the routing process: On one hand, by looking at the logical model presented in Fig. 11, the network architects may know the order of connecting the subnets together, and the order of connecting the flows or the traffic coming from them. The more we mess the order presented in Fig. 11, the more the routing tables sizes of this network become bigger. On the other hand, as already mentioned the node X is the node that deliver the summary X, in Fig. 11 whenever the node $2C0F : FB20 :: /32$ receives a packet whose destination is one of the leaf nodes, the node $2C0F : FB20 :: /32$ has two options whether give it to the node $2C0F : FB20 :: /33$ or the node $2C0F : FB20 : 8000 :: /33$. The both will have also two options. Furthermore, suppose the leaf nodes are directly connected to the root, in this case the node $2C0F : FB20 :: /32$ has 16 options in its routing list. It is obvious that searching in a list of two options is faster than searching in a list of 16 options.

5.1   The Value of n

By looking at the binary format of the hexadecimal digits (1, 2, 3, 4, 5, 6, 7, 8, 9, A, B, C, D, E, F) in ascending order, we notice that according to the value of the first bit in the left, the hexadecimal digits are divided into two groups, each group has 8 elements as shown in the first column of Table 4. Likewise, according to the value of the second (Resp. the third) bit in the left, each group presented in the first (Resp. the second) column of Table 4 is divided into two groups, each group has 4 (Resp. 2) elements as shown in the second (Resp. the third) column of Table 4.

Table 4. The arrangement

1st arrangement	2nd arrangement	3th arrangement
0000	0000	0000
0001	0001	0001
0010	0010	0010
0011	0011	0011
0100	0100	0100
0101	0101	0101
0110	0110	0110
0111	0111	0111
1000	1000	1000
1001	1001	1001
1010	1010	1010
1011	1011	1011
1100	1100	1100
1101	1101	1101
1110	1110	1110
1111	1111	1111

By looking at the prefixes presented in Table 3, we notice that their 9th hexadecimal digits are different but their first 8 hexadecimal digits are the same, therefore these prefixes can be arranged into one group of 16 elements.

Hence we may assume that $n \in \{2, 4, 8, 16\}$.

5.2   Flexibelity of the Proposed Logical Model

The proposed logical model is totally flexible. Whenever we apply the following rule, we get a new logical model.

In Fig. 11, we call the node $2C0F : FB20 : E000 :: /35$ the 1st order representative of the node $2C0F : FB20 : F000 :: /36$ and the node $2C0F : FB20 : C000 :: /34$ the 2nd order representative of the node $2C0F : FB20 : F000 :: /36$.

Rule: Let X, Y and Z be nodes, where Y (Resp. Z) is the 1st (Resp. 2nd) order representative of X. If Y is deleted X will be directly connected to Z.

For example, by deleting the nodes $2C0F : FB20 :: /35$ and $2C0F : FB20 : C000 :: /34$ the logical model seen in Fig. 11 becomes as shown in the simulation presented in Fig. 14.

Last and not least, the choice of the logical model depends on the features of the routers that will be used in the

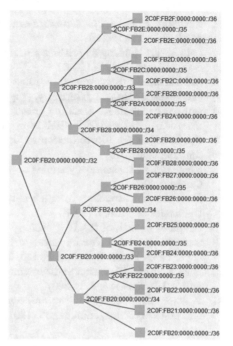

Fig. 14. Flexibility of the logical model

network: The more the router has good features the more its routing table can be long.

6   Conclusion

We believe that the great purpose of modeling is solving problems. The hierarchy of the model presented in this paper uses effectively the summarization technique and its flexibility permits to divide up the routing tables between routers according to their features. Consequently, implementing networks based on the proposed logical model may help to resolve the problem of the enormous growth of routing tables and increase the speed of the routing process.

Acknowledgment. Financial support provided by CNRST, funded by Moroccan government, is gratefully acknowledged. The authors also wish to thank an anonymous reviewer for his helpful comments on an earlier draft of the manuscript.

References

1. Newman, M.E.J.: The structure and function of complex networks. SIAM Rev. **45**(2), 167–256 (2003)
2. Lebhar, E.: Routing algorithms and random models for small world graphs. Thesis, ENS of Lyon, France (2005)
3. Reichardt, J.: Structure and function of complex modular networks. Thesis, University of Bremen, Germany, August 2006
4. van der Hofstad, R.: Random graphs and complex networks (2016). http://www.win.tue.nl/~hofstad/NotesRGCN.html
5. van Steen, M.: An Introduction to Graph Theory and Complex Networks. Lexington, January 2010
6. Noel, P.A.: Stochastic dynamics on complex networks. Thesis, Laval University, Quebec (2012)
7. Levorato, V.: Modeling complex networks: pretopology and Application. Thesis, University of Paris 8 (2008)
8. Graph of growth of the BGP Table - 1994 to Present. http://bgp.potaroo.net/
9. Grazian, R.: IPv6 Fundamentals a Straightforward Approach to Understanding IPv6. Cisco Press, Indianapolis (2013)
10. IPv6 Address Allocation and Assignment Policy. APNIC, ARIN, RIPE NCC, Document ID: ripe-412, July 2007
11. Wallace, K.: CCNP Routing and Switching Route 300–101 Official Certification Guide. Cisco Press, Indianapolis (2015)

Model Driven Modernization and Cloud Migration Framework with Smart Use Case

Khadija Sabiri[(✉)], Faouzia Benabbou[(✉)], and Adil Khammal[(✉)]

Information Technology and Modeling Laboratory, Science Faculty of Ben M'sik,
Casablanca, Morocco
Khadija.sabiry@gmail.com, faouziabenabbou@yahoo.fr,
infoadil@gmail.com

Abstract. With the fast evolving cloud market, many enterprises have attempted to move their legacy application in this new environment to take its full advantage. However, this move is strongly associated with the fact of packaging the legacy application in an image or in a container encapsulated into a virtual machine and deployed in a single instance.

So, the need of making the legacy application more agile and flexible is becoming a must on moving to a cloud environment, yet to fully benefit from it. It is important to change the architecture of the legacy application before deploying it in a cloud environment. In consequence, there is a need for an approach that assure two important things: how the architecture of legacy application has to be changed to transform it into a cloud native application architecture and then to be deployed it in a cloud environment.

In this work, we propose a modernization iterative and incremental process to overcome the listed issues above. This process is based on the concept of smart use case combining with the architecture driven modernization (ADM) approach. The last consists on followed the steps: Reverse Engineering, Transformation/ Upgrade, and Forward Engineering. This process aims to not only raise the IT agility, but moreover its business agility.

Keywords: Cloud computing · ADM · Smart use case · Native cloud application
Legacy application

1 Introduction

Cloud computing has changed the way that the application is developed, delivered, consumed and managed. The goal is to take advantage of the numerous cloud benefits such as agility, continuity, cost reduction, autonomy, and easy management of resources [1]. Though, enterprise adoption of the cloud has been lately grown up [2].

However, a migrated application may not benefit from the cloud as long as a migration concept is simply considered as packaging the legacy architecture into a virtual machine instance. although, this migrated application is tended to be considered as a native cloud application, focusing on pay as you go model [3, 4]. This type of migration generally aims to involve some legacy technologies of the application.

© Springer International Publishing AG, part of Springer Nature 2018
M. Ben Ahmed and A. A. Boudhir (Eds.): SCAMS 2017, LNNS 37, pp. 17–27, 2018.
https://doi.org/10.1007/978-3-319-74500-8_2

Hence, as a consequence of this, the migrated application keeps the same complexity as before, even if it may be built modularly. That complexity may be appeared in the following ways: the application is deployed as a single unit, scaled as a single application, needed to redeploy the whole application in case either updated, maintained or added new features, and the entire application will be felt down if just one service is out of order.

However, it's not just putting the application into the cloud or splitting it into smaller containers or images. It has to be redesigned, re-architected in a way that exploits full benefits of cloud.

The reason behind the cloud migration, is to transform into a native cloud application to ensure business continuity requirements as well as the longevity of IT. As a consequence, there is a need for such a transformation to truly gain agility in running on a cloud.

This paper investigates the process of modernizing the legacy application by means of transforming into a cloud-native application and by transiting to a cloud environment. To tackle this problem, an Architecture Driven Modernization (ADM) approach has been adopted.

The remainder of this paper is structured as follows. Section 2 describes related research to this work. Section 3 explains the background used. Section 4 presents a proposed approach. Finally, Sect. 5 concludes the paper.

2   Cloud Migration Methods

The direction of this work is for the modernization legacy application and for the migration to a cloud application. In the literature, cloud migration may be classified into the followed three categories [5]:

2.1   Replace Component(s) with Cloud Offering

This is the least invasive type of migration, where one or more (architectural) components are replaced by cloud services [6]. This type of migration is considered as a cloud, enhancing treatment around the legacy application, rather than a pure cloud enabling [7]. This type of migration requires a series of reconfigurations to adjust incompatibilities to use functionalities of the ported layer [8] and may be more expensive than rewriting.

Although Fehling et al. [2] identify two mainly challenges of the replacement strategy: the maintenance of the new system, which will not be as familiar with the old system; and the lack of a guarantee that the new system will be as functional as the original one.

2.2   Outsourcing and Wrapped

The application functionality which is based on exposing them as services, hosting and running in a cloud environment. It requires the modernization of the legacy system by presenting them as autonomous components Service Oriented Architecture (SOA) in order to take many advantages offered by SOA technology. Although the migration of a legacy system towards SOA is not an easy task, it must answer two major questions such as: what can be migrated from the legacy system [8], and how can be executed the migration process [9]. As a result, not all legacy systems have matured enough to take up this transformation, and two studies are required for moving to the Cloud:

- Establishing a study to migrate to SOA
- Establishing a study to migrate to SOA application to the cloud

2.3   Cloudified

This technique is focused on at the level of architectural models by redesigning the architecture of a legacy system within a service oriented architecture.

To sum up, the most of the presented cloud migration methods are mainly based on migrating to SOA autonomous components. This concept has potential advantages, but has also some significant drawbacks. As examples, not all applications are enough mature to transform it into services and not all legacy application's functionalities are able to be reused by exposing them as services.

3   Background Theory

3.1   Architecture Driven Modernization

Concept

Architecture Driven Modernization (ADM) is an initiative's Object Management Group (OMG) launched in 2003 [10], which has been promoted as the process of understanding and transforming a legacy system following model driven development principles [11]. The goal of the ADM is to conduct the reengineering processes following the model driven architecture (MDA) by representing all the artifacts as models, i.e., platform independent model (PIM), platform specific model (PSM) and computational-independent model (CIM) [12].

ADM is concerned with the modernization of all aspects of the current architecture system and transformation to a new target architecture based on user's perspectives. ADM copes with all model harmoniously, which allows formalizing a model to model transformation between them by means of all the principles of model-driven engineering (MDE), i.e., models, metamodels, models-transformation. The models produced in the process of the ADM may be able to be reused.

The ADM process is divided into three major phases [13]:

- Reverse engineering: it is the phase of analyzing the legacy system to identify its components and their relationships. This phase aims to represent the system at a higher level of abstraction through obtaining models.
- Transformation/Upgrade: this phase aims to transform the source models, resulted from the previous phase, in target models at the same abstraction level.
- Forward engineering: this last phase is about to generate the new system from the target model at a low abstraction level.

Standard: KDM

As stated before, ADM is the conduction of engineering process considering MDA techniques. To carry out a software modernization, ADM has provided Knowledge Discovery Metamodel (KDM). KDM is an OMG specification which represents information related to existing software systems. The goal of the KDM is to define a meta-model to represent all artifacts of the legacy system such as user interfaces, source code, data-bases, configuration files, etc. [11].

KDM assists in the whole process modernization, by building models at different abstraction levels from different legacy system artifacts along ADM phases, i.e., reverse engineering, transformation/upgrade and forward engineering. In the first phase, KDM is used to represent the legacy system as models by obtaining KDM models. In the second and last phase, KDM is considered as a common repository which has the ability to exchange information between all legacy system artifacts.

The meta model of the KDM [13] is divided into four layers: infrastructure layer, program element layer, runtime resource layer, and abstraction layer.

Each layer is represented as packages. In turn, each package has a set of metamodel elements which aim to represent a certain independent facet of knowledge related to existing software systems.

3.2 Native Cloud Application

Cloud native application is purpose built and ran to fully exploit the advantages of the cloud computing model.

National Institute of Standards and Technology (NIST) [14] defined cloud computing as a model for enabling ubiquitous, convenient, on-demand (ability to adjust computing capabilities as needed) network access (availability computing capabilities over the network) to a shared pool (multi-tenant service) of configurable computing resources (e.g. networks, servers, storage, applications, and services) that can be rapidly provisioned and released with minimal management effort (scale rapidly with demand) or service provider interaction.

We provide a definition of a native cloud applications by specifying their properties. According to [15], the cloud native application should ensure IDEAL properties, by being:

- Isolated state: it means the application doesn't hold a state.
- Distributed by nature: components are distributed among resources in the environment.
- Elastic by scaling out: ability to add the independent IT resources to address increasing workload the number of resources assigned to an application.
- Operated via an Automated management system: the ability to manage, automatically adding and removing the resources during runtime.
- And its components should be Loosely coupled: it's about reduce the dependencies between application components.

In this paper [16] the authors present the common motivations behind moving to cloud-native application architecture, such as:

- Deliver software-based solutions more quickly (speed): The ability to innovate, experiment, and deliver value more quickly.
- Fault isolating: limit the scope of a failure in any one component to just that component; it reflects the ability to limit a bugger program from affecting another program.
- Fault tolerating: it prevents a failure in one of the components, if an application crashes, some detection will see it's a problem, destroy the broken instance, and replace it with a working new one.
- Automatic recovering way (safety): engaging in the automated process of identification and recovery.
- Enabling horizontal instead application scaling.
- And finally, client diversity: to handle a huge diversity of (mobile) platforms and legacy systems.

4   Proposed Method

This section presents the concept of the smart use cases, its definition and its use. Then, the modernization chain is explained with its process steps.

4.1   Smart Use Case Concept

According to [17], this concept comes to answer many issues (various stakeholders, heterogeneous technologies used…) in a software development project, which is characterized by functional and technical highly complex. In order to mitigate this complexity, the project would be expressed using a single of unit work, which in turn is expressed in the smart use case. By and large, the functional requirement of the project is modeled using traditional use cases, from defining the users' objectives to execute the steps. However, the traditional use case may vary considerably in its size and its complexity in terms of identification and realization of services. This makes the traditional use cases hard to specify, so in a result, it will be hard to implement and hard to test too. As an example, a use case may be required a hundred pages, a hundred scenario and many services to be described.

So for designing the smart use cases, the use cases are split up into different levels of granularity in terms of size and complexity, such as:

- Cloud: regroups clusters of the different business process that belong together.
- Kite: Individual business processes are generally placed.
- Sea: In this level, each single use case describes a single elementary business process and achieves a single goal.
- Fish: This level is used to model autonomous functionality supporting the sea level.
- Claim: the processes often need to be deeply modeled appeared as sub steps in its up level.

4.2   The Modernization Chain

This method is mainly based on ADM combined with the smart use cases concept.
The following Fig. 1 shows the whole process steps:

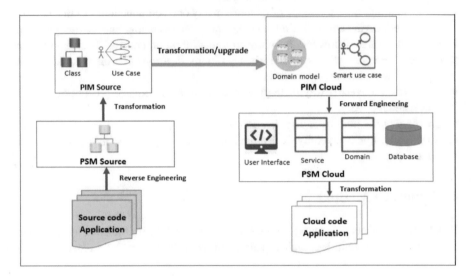

Fig. 1.   Modernization and cloud migration framework

Reverse Engineering

It's the stage of analyzing a system to identify components and relationships of the system, and to design new representations of the system in another form or at a higher level of abstraction.

The reverse engineering stage consists of two sub-stages:

Source code to KDM

The first sub-stage is about the code, reverse engineering, which aims to recover the model code of the application, and to create PSM conformed to KDM metamodel. This sub-stage is supported by means of code-to-model transformation. The PSM resulted is a model tailored to specify a system in terms of a specific platform, i.e., JAVAEE, expressed in Unified Modeling Language (UML). The subset of UML diagrams that are useful for PSM, includes the class diagram and the state diagram. This sub-stage is automated with MoDisco tools [18] to extract models from such applications. MoDisco proposes a generic and extensible metamodel-driven approach to model discovery. This tool is taking as an input source code, and producing as an output, a KDM model conformed to the KDM metamodel.

For a given example, MoDisco takes as an input Java code source discovers to Java code model, which represents every artifact in the entered code source, and as an output KDM model. This resulted by means of Java-to-KDM model transformation between Java model and KDM model passing by an intermediate Java source code model.

KDM to PIM

In the second sub-stage is about to transform the KDM into PIM-UML models by means of a model to model transformation, that takes as an input a model conforming to the KDM metamodel and produces as outputs models conforming to the UML metamodel.

In order to be achieved, we propose two kinds of model to model transformations, in a way the source metamodel corresponds to the KDM metamodel and the both targets metamodels correspond respectively to the UML metamodel of use cases and UML metamodel of a class diagram. Those two model-to-model transformations specified are represented as follows:

- The first model to model transformation, calledKDM2UseCases transformation, produces use cases diagram as the first target model from the KDM source model.
- The second one, called KDM2ClassDiagram, produces a class diagram as the second target model from the KDM source model.

Transformation/Upgrade

In this second stage, PIM source will be the subject of transforming to PIM cloud. PIM cloud is defined as a model to design a native cloud application as discussed in the previous section.

PIM cloud will be designed in a manner that each business functionality does one thing independently. PIM cloud should conform to this cloud application metamodel [19, 20].

The aim of this stage is, to end up with PIM cloud, which will be mapped to any cloud platforms given. For this purpose, this stage has two inputs generated in the previous stage: class diagram and use cases of the application.

Class diagram describes the state of the application by attributes, and its behavior through operations as well as the relationships (association, aggregation...) which represent the logical connections between classes. The class diagram is used to define the conceptual model for a given domain. The domain is decomposed into a set of conceptual classes; he subject area to which the user applies a program is the domain of the software.

Understanding the domain models of the application is by means of the PIM source represented at the highest level of abstraction regardless of any implementation complexity.

In order to transform the PIM source into PIM cloud, two sub-stages should have taken:

- Extract different domain models of the application by means of UML class diagram. Each package will be represented as a set of classes describing all entities in a given domain. The aim of this sub-stage is to mitigate the complexity of the application by structuring it into separate domains with well defined bounded context- where a domain model is defined and applicable.
- A use case describes the functionalities of the application and the processes to be performed. Each use case overviews a piece of functionality and describes a set of actions to be done. The smart use case will be identified as sub-functions of its use case mother. Each smart use case represents an elementary business process.

This use case will be decomposed into two levels of granularity: a sea level and a fish level use case.

The sea level use case describes a single use case that represents the elementary business process, and the fish level use case represents zero, one, or more use cases that support the execution of the sea level use case.

So, with the concept of smart use cases, PIM cloud will be made up of granularity of use cases, which make it easier to be implemented, tested, deployed, scaled and so on. A single elementary business process is modeled in a smart use case.

Additionally, this decomposition of PIM cloud makes the business application more agile that makes easier to add new business functionalities to the application.

In the end, each smart use case will be tied to its business process, to which is belongs. In turn, its business process mother will be assigned to its own domain model, in order to keep the smart use related to the domain together.

Since PIM is transformed into PIM Cloud, with the granularity of use cases and well defined separated domains, so from now on, PIM Cloud will be able to be implemented into any given provider cloud platform.

Forward Engineering Stage: Transformation PIM Cloud to PSM Cloud

After transforming the PIM to PIM cloud, the latter may be represented by means of smart use cases, which will be the subject of the target cloud platform, by means of a PIM cloud of PSM cloud transformation.

It is important to note that this PIM cloud may be transformed to any given platform clouds.

The smart use case is treated as input models that can then be transformed down to code.

In this stage, we present how to transform this PIM cloud to cloud platform specific model. Each smart use in the PIM cloud will be implemented in the following way:

- A smart use case is implemented independently, which enable the application to be speedy scalable and deployed.
- Each smart use case handles, navigation separately, which ensures supporting a huge diversity of user interfaces.
- Each smart use case executes individual tasks, addressing a very specific business scope.
- Each smart use case has its own domain which ensures the relation between the smart use of its domain (package).
- And finally, each smart use case has its own database which enables to have different types of cloud storage.

5   Comparaison Cloud Migration Methods

The spite the fact that the following methods, studied in this paper [5], are based on the models paradigm, the application resulted, migrated and deployed to the cloud, doesn't conform to cloud native application properties cited in the third section. The following

Table 1 illustrates how the output of the following methods doesn't respect those criteria to be a native cloud application.

Table 1. Comparative study cloud migration methods

Methods	Criterion						
	Speed	Resiliency	Scale	Loosely coopled	Add new features	Deploy, update, maintain modularly	Distributed
Cloud-MIG [21]	No	No	Yes	No	No	No	No
Remics [22]	No	No	Yes	Yes	No	No	Yes
Artist [23]	No	No	Yes	No	No	No	No
Based on service [24]	No	No	Yes	No	No	No	No
PaaS migration [25]	No	No	Yes	No	No	No	No

The key of interpretation this tabular, is in the following way: each table cell value determines whether those cloud migration methods have been included the native cloud application properties. For instance, CloudMIG doesn't respect any cloud application proprieties marked by the value "No". As a consequence, we have noticed that the present methods aren't effectively realized resiliency, speed, adding new features... objectives behind moving forward to cloud environment.

The presented methods have particularly focused on the way to adapt and to adjust the legacy application to be deployed in a cloud environment. However, this way may hold the same complexity as before, even if it may be built modularly. That complexity may be appeared in the following ways: the application is deployed as a single unit, scaled as a single application, needed to redeploy the whole application in case either updated, maintained or added new features, and the entire application will be felt down if just one service is out of order. Nevertheless, our proposed method mitigates this complexity through transforming PIM into PIM cloud, which the latter is composed of granularity of smart use case, that will be deployed, scaled, updated and maintained modularly. This PIM cloud conformed to native cloud application architecture.

6 Conclusion

This work has presented a modernization iterative and incremental process for supporting a transition to a cloud environment and a transformation legacy application architecture to a cloud native application. In concrete, the goal of the modernization process presented has consisted on adopting ADM combined with the smart use case. This process has aimed to make the application more agile to ensure business continuity requirements as well as the longevity of IT. Thanks to this process, the application may be scaled modularly, and the components of the application, which means the smart use cases, were implemented, deployed, configured, distributed and maintain independently. It is important to note that this modernization iterative and incremental process is semi-automatically supported at the moment, notably transformation/upgrade and forward engineering stages, which are developed and applied by tailored transformation chain for a particular application. Future direction includes the approve this approach by using case study and make our process automated.

References

1. Zhao, J.F., Zhou, J.T.: Strategies and methods for cloud migration. Int. J. Autom. Comput. **11**(2), 143–152 (2014). https://doi.org/10.1007/s11633-014-0776-7
2. Fehling, C., Leymann, F., Ruehl, S.T., Rudek, M., Verclas, S.: Service migration patterns decision support and best practices for the migration of existing service-based applications to cloud environments. In: Proceedings of the 6th IEEE International Conference on Service Oriented Computing and Applications (SOCA) (2013)
3. Fehling, C., Konrad, R., Leymann, F., Mietzner, R., Pauly, M., Schumm, D.: Flexible process-based applications in hybrid clouds. In: Proceedings of the 2011 IEEE International Conference on Cloud Computing (CLOUD) (2011)
4. Sabiri, K., Benabbou, F., Hain, M., Moutachouik, H., Akodadi, K.: A survey of cloud migration methods: a comparaison and proposition. IJACSA **7**(5), 598–604 (2016). https://doi.org/10.14569/IJACSA.2016.070579
5. Buyya, R., Yeo, C.S., Venugopal, S.: Market-oriented cloud computing: vision, hype, and reality for delivering it services as computing utilities. In: Proceedings of the 10th IEEE International Conference on High Performance Computing and Communications, Dalian, pp. 5–13. IEEE (2008)
6. Low, L.C.Y., Chen, Y., Wu, M.C.: Understanding the determinants of Cloud computing adoption. Ind. Manag. Data Syst. **111**, 1006–1023 (2008)
7. Watson, P., Lord, P., Gibson, F., Periorellis, P., Pitsilis, G.: Cloud computing for e-science with Carmen. In: Proceedings of the 2nd Iberian Grid Infrastructure Conference, Porto, Portugal, pp. 3–14 (2008)
8. Baiardi, F., Sgandurra, D.: Securing a community cloud. In: IEEE 30th International Conference on Distributed Computing Systems Workshops, pp. 32–41. IEEE Computer Society (2010)
9. Barbacci, M.R., et al.: Quality Attribute Workshops (QAWs), Technical report CMU/SEI-2003-TR-016, SEI, Carnegie Mellon University, USA (2003)
10. OMG ADM Force Task: Why do we need standards for the modernization of existing systems? (2012). http://adm.omg.org/legacy/ADM_whitepaper.pdf
11. Perez-Castillo, R., de Guzman, I.G.-R., Avila-Garcia, O., Piattini, M.: On the use of ADM to contextualize data on legacy source code for software modernization. In: Proceedings of the 2009 16th Working Conference on Reverse Engineering, WCRE 2009, pp. 128–132. IEEE Computer Society, Washington, D.C. (2009)
12. Ulrich, W.M., Newcomb, P.: Information Systems Transformation: Architecture-Driven Modernization Case Studies. Morgan Kaufmann Publishers Inc., San Francisco (2010)
13. OMG Object Management Group (OMG) architecture-driven modernisation
14. Mell, P., Grace, T.: The NIST Definition of Cloud Computing. National Institute of Standards and Technology (NIST), Gaithersburg (2011)

15. Fehling, C., Leymann, F., Retter, R., Schupeck, W., Arbitter, P.: Cloud Computing Patterns: Fundamentals to Design, Build, and Manage Cloud Applications. Springer, Heidelberg (2014)
16. Stine, M.: Migrating to Cloud-Native Application Architectures. O'Reilly Media Inc., Sebastopol (2015)
17. Hoogendoorn, S., (Capgemini): Pragmatic model driven development using smart use cases and domain driven design, 7 April 2009
18. MoDisco. https://eclipse.org/gmt/modisco/
19. Hamdaqa, M., Livogiannis, T., Tahvildari, L.: A reference model for developing cloud applications. In: Proceedings of the 1st International Conference on Cloud Computing and Services Science, CLOSER 2011, pp. 98–103 (2011). https://doi.org/10.5220/0003393800 980103
20. Sabiri, K., Benabbou, F., Moutachaouik, H., Hain, M.: Towards a cloud migration framework. In: Third World Conference on Complex Systems (WCCS) (2015). https://doi.org/10.1109/ICoCS.2015.7483315
21. Frey, S., Hasselbring, W.: Model-based migration of legacy software systems to scalable and resource-efficient cloud-based applications: the CloudMIG approach. In: The First International Conference on Cloud Computing, GRIDs, and Virtualization, CLOUD COMPUTING 2010 (2010)
22. Parastoo, M., Berre, A.J., Henry, A., Barbier, F., Sadovykh, A.: REMICS-REuse and migration of legacy applications to interoperable cloud services. In: Third European Conference, ServiceWave, pp. 195–196 (2010)
23. Menychtas, A., Santzaridou, C., Echevarria, L.O., Alonso, J.: ARTIST methodology and framework: a novel approach for the migration of legacy software on the cloud. In: 15th International Symposium on Symbolic and Numeric Algorithms for Scientific Computing (2013)
24. Zhang, W.Q., Berre, A.J., Roman, D., Huru, H.A.: Migrating legacy applications to the service Cloud. In: 14th Conference Companion on Object Oriented Programming Systems Languages and Applications (2009)
25. Beslic, A., Bendraou, R., Sopena, J., Rigolet, J.-Y.: Towards a solution avoiding vendor lock-in to enable migration between cloud platforms. In: 2nd International Workshop on Model-Driven Engineering for High Performance and Cloud computing (MDHPCL) (2013)

SDL Modeling and Validation of Home Area Network in Smart Grid Systems

Zahid Soufiane[1]([✉]) [iD], En-Nouaary Abdeslam[1], and Bah Slimane[2]

[1] Institut National des Postes et Télécommunications (INPT), Rabat, Morocco
zahidsoufiane@gmail.com, abdeslam@inpt.ac.ma
[2] Ecole Mohammadia d'Ingénieurs (EMI), Rabat, Morocco
slimane.bah@emi.ac.ma

Abstract. The Smart Grid is an intelligent power network featured by its two-way flows of electricity and information. The integrated communication infrastructure allows Smart Grid systems to manage the operation of all connected components to provide reliable and sustainable electricity supplies. The Home Area Network (HAN) is a dedicated network connecting devices in the home, as well as electrical vehicles. The HAN market is now emerging within the smart grid sector to serve home with different solutions. Such solutions should be devised with rigor to avoid any possible errors or anomalies along the process life cycle, from inception until deployment and operation. Modeling and validation is one of the powerful techniques used to achieve such goals. This paper presents an approach to modeling and validating the HAN network and its four services. We use SDL (Specification and Description Language) as a standardized language to describe especially the "demand response" service and its two programs types. The resulting design is a generic model that focuses on the main functions of the HAN network. We then validate the resulting SDL model using the reachability analysis technique with the support of IBM Rational SDL suite. The final validated model can be used to generate code for concrete or virtual Smart Grid solutions.

Keywords: Smart Grid · Communication protocols and services
Formal modeling · V&V · Reachability analysis · SDL · MSC · HAN

1 Introduction

Smart Grid, as shown in Fig. 1, is an integration of power delivery systems with communication and information technology (CIT) to provide better services and improve the traditional electrical grid to be more reliable, cooperative, responsive, and economical. It is a complex system, made of a large number of heterogeneous entities with local interactions, multiple levels of structure and organization, which forms a whole system that is both hard to predict and describe. In order to analyze Smart Grid systems, the network architecture must be defined with all the communication protocols and interfaces between its composing entities. One of the main components of Smart Grid systems is the Home Area Network (HAN). It has commanded a great attention among

© Springer International Publishing AG, part of Springer Nature 2018
M. Ben Ahmed and A. A. Boudhir (Eds.): SCAMS 2017, LNNS 37, pp. 28–43, 2018.
https://doi.org/10.1007/978-3-319-74500-8_3

researchers and solutions' providers because it has many facets that must be examined by utilities with regard to its use and potential benefits for both customers and providers.

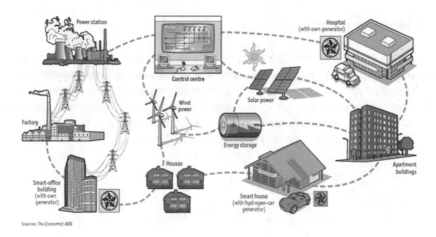

Fig. 1. Smart Grid

In a previous work [1], we have shown that there are several conceptual models proposed by international organizations, such as NIST, ITU and IEEE; and each country or utility defines its own network based on these models. Also the technologies involved in the implementation are widely different because the smart grid supports both wired and wireless technologies. To cope with the complexity and the diversity of SG systems, modeling is one of the powerful techniques that can be used to visualize the system, simulate it and validate it before proceeding to its development.

This paper proposes a generic model for the HAN network in SG systems. This model describes the structure and the behavior of the HAN, especially in terms of offered services according to the U.S. Department of Energy (DOE) report [2]: Advanced Metering Infrastructure (AMI), Demand Response (DR), Electric Vehicles (EV) and Distributed Energy Resources and Storage (DER). It takes into account the energy generation and management, the in-home devices administration and the connection with utility grid. We use SDL (Specification and Description Language) as a standardized language to describe especially the DR service and its two programs types. The resulting design is a generic model that focuses on the main functions of a HAN network. We then validate the resulting SDL model using the reachability analysis technique with the support of IBM Rational SDL suite. The final validated model can be used to generate code for concrete or virtual Smart Grid solutions.

The remainder of this paper is organized as follows. Section 2 gives an overview of the HAN network and describes its main components and functionalities, especially the DR service and its two programs. Section 3 introduces the modeling process of the Smart Grid using SDL. It presents our design approach in terms of UML use case diagram, an MSC (Message Sequence Charts) for the main scenario, and the SDL diagrams (the structure, the data, and the behavior aspects) for the most important components of the

HAN. Section 4 addresses the verification and validation of the model. Section 5 concludes the paper and presents future work.

2   Overview of HAN

A Home Area Network (HAN) is a network contained within a user's home that connects user's appliances and electrical vehicles. It also contains software applications to monitor and control these devices, as well as other resources such as renewable resources and energy storage equipments. In our previous work [1], we presented a smart grid communication infrastructure model based on international roadmaps and guides, and in particular, we described the HAN architecture as shown in Fig. 2.

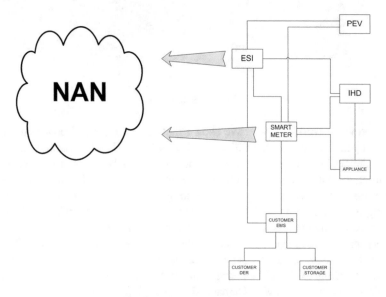

Fig. 2.   The HAN architecture

The main component of the HAN is the Home (or Customer) Energy Management System (HEMS). It utilizes renewable energy effectively by visualizing load equipment information in the home (such as air conditioner, storage battery and EV) and controlling it properly. So, it optimizes the performance of energy generation, consumption and storage in the HAN. This component is generally integrated with the In-Home Display (IHD) but could theoretically be considered separately. This IHD is an interface, between the customer and the home, connected to all devices in HAN. It is designed to deliver information related to energy such as consumption, pricing, or service messages from utility. Other components are the Energy Services Interface (ESI) and the Smart Meter (SM). The SM collects information about energy usage in the HAN, as well as manages control services such as circuit disconnection. It can store the metering data internally, and send it to utility via a two-way communications. On the other hand, the ESI is the gateway, mostly provided by utilities, that routes data between the HAN and the neighborhood area network (NAN). It can be considered as the HAN gateway through which a customer's HAN communicates with the utility companies or any other entity that provides energy management services. Generally, The HAN gateway is embedded in the meter device physically, but, it is logically separated from the meter. The last two components are the appliance and the Plug-in Electrical Vehicle (PEV). Appliances are all devices in the home that can be connected to the electricity network, and they may include a technology known as Smart Plug (SP). To establish a secure communication connection between utilities and the HAN, all HAN customer devices must register themselves to HEMS. Then, the SPs collect the power consumption data for home appliances, and send accumulated power consumption data for each appliance to HEMS in each period of time. As to PEV, it is connected to grid and registered to HEMS like other devices. It can charge its battery or send energy stored back to grid if necessary.

Note that the architecture of Fig. 2 requires a scalable, sophisticated, reliable and fast communication infrastructure. Generally, the HAN appliances are equipped with wireless or wired short range communication technologies, such as, ZigBee, TCP/IP and PLC. The choice between these technologies depends either on the constraints imposed by the utilities network, or on the user needs in case the network supports more than one technology. This communication architecture allows customer to benefit from smart grid services, the most important of which is the DR service. It reduces peak loads, when the system is under stress, by minimizing the consumption of electric energy in response to an increase in the price of electricity or heavy burdens on the system [2]. DR is achieved through the application of a variety of DR resource types, including distributed generation, dispatchable load, storage and Plug-in Electric Vehicles. So, anything that may contribute to modify the power supplied by the main grid, or somehow affects electricity usage can be considered as a DR resource. In general, the DR resources can be classified into two categories: reactive resources and proactive resources. Reactive resources include customers that receive DR signals from a third party to reduce, shift or shut down their demand in order to adjust their consumption voluntarily. On the other hand, proactive resources refer to components which initiate a DR action by sending requests or bids to the utility to either curtail their consumption in exchange for payments, or to negotiate a price for buying energy.

Every customer may participate in what is called DR programs. The main reasons for encouraging customers to participate in such programs vary from monetary savings, to understanding the obligation to help avoiding blackouts, to a sense of responsibility. DR programs are classified into two major categories, namely Incentive Based Programs and Time-based Programs. Each of these categories contains several programs illustrated in Table 1 [4].

In the Incentive Based Programs the activities are initiated by the utility or the DR Service Provider. Customers, participating in one of these programs, receive DR signals from the provider in order to motivate them to reduce their electricity consumption in exchange for an incentive payment, bill credit or contractual arrangements between electricity suppliers and customers. Generally, DR signals are sent in times of high electricity consumption and may be voluntary demand reduction requests or mandatory commands. In voluntary programs like DLC or EDRP, customers are not penalized if they do not curtail consumption. But in mandatory programs such as I/C, they suffer penalties if they do not curtail when directed. The adoption of Incentive Based DR would bring benefits to both customers and utilities. In fact, the demand is changed to follow available supply so that the amount of power generated is minimized significantly. This reduces or even eliminates the overloads in distribution system. However, with the Time-based Programs the electricity price changes for different periods. So, they rely on customer's choice to decrease or change their consumption in response to changes of electricity's price during a period. In the TOU program, the day is divided into blocks of hours and the kWh price varies between blocks; the price for each period is prede-termined and constant. In the RTP program, the price may vary as often as hourly, and in the CPP program, very high prices are applied for certain peak days; the utility informs the customer one day in advance of the critical peak pricing day. For the rest of this paper, we assume that the customer is registered in two programs, one from each cate-gory of Table 1.

Table 1. DR programs classification

Incentive based programs	Time-based programs
Direct Load Control (DLC)	Time-of-Use (TOU) program
Interruptible/curtail able service (I/C)	Real Time Pricing (RTP) program
Emergency Demand Response Program (EDRP)	Critical Peak Pricing (CPP) program
Capacity Market Program (CMP)	
Demand Bidding/Buy Back	
Ancillary Service Markets (A/S)	

3   The Modeling Process of Smart Grid

The modeling process of a complex system such as smart grid goes through several steps. Generally, we start with defining the formal specification from an informal description presented in RFC or other supports. Then, we use a tool for exhaustive reachability analysis in the V&V (Validation and Verification) step. After that, we use analytical or empirical methods to analyze the performance of the system. Then, we move to the implementation step by generating an executable code using the validation model. Finally, the conformance testing is applied to confirm that the implementation conforms to the specification. Figure 3 illustrates the whole modeling process using SDL (Specification and Description Language) and related support tools [8]. This paper presents only our contributions with regard to the first two steps, namely the formal specification and the validation and verification. The rest of the steps are left to future work.

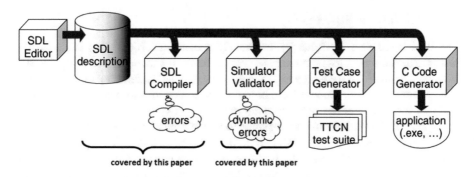

Fig. 3. An overview of the SDL modeling process

3.1 Specification and Description Language (SDL)

The SDL language [5] is a standardized language used for the description of the architecture, the behavior, and the data of telecommunication systems. It has been developed and standardized by the ITU (International Telecommunication Union) in the Z.100 Recommendation. The choice of this language is based on its several advantages and characteristics. SDL is a graphical and open source language that does not depend on operating systems. It includes traditional object-oriented features, such as encapsulation and polymorphism, and provides a high degree of reuse.

In SDL, the architecture is modeled as a system containing one or more blocks. Each block may contain either other blocks or processes. Each process contains an extended finite state machine. State machines are interconnected with each other and with the environment via channels and signal routes, which serve to transmit and exchange signals. When a signal arrives to a state machine, it is saved in the queue. And when a state machine consumes a signal from its queue, it executes a transition from one state to another state. Figure 4 resumes this description [6].

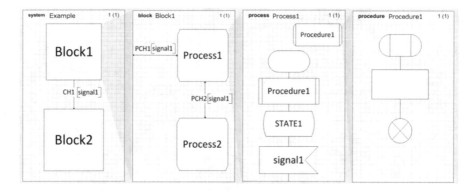

Fig. 4. An overview of SDL model

3.2   IBM Rational SDL Suite

The IBM Rational SDL Suite [7] (previously named Tau SDL Suite) is a real-time software development tool for complex systems and software described using the SDL language standard. The tool provides the following graphical interfaces: a drag-and-drop SDL editor to draw the model from high-level specifications to detailed behavior, an analyzer for errors and warning checking in the drawn system, a simulator for testing the behavior of SDL systems, an explorer to validate this behavior using reachability analysis and several other interfaces. With this tool, we can describe typical scenarios of the communication behavior between system components and their environment by means of message interchange with a visual trace language using the Message Sequence Chart (MSC). And after the model simulation and validation, SDL Suite enables the automatic generation of a C executable code.

3.3   Our Design Approach of the HAN

We begin our modeling process with a simple use case analysis. The system consists of three actors: the customer, the utility and the environment which refers to renewable energy sources. Figure 5 shows the use case diagram. The "control device" use case represents the act of controlling either the appliance, by powering on/off, or the EV by plugging or unplugging it in the network and launch the charging process. This is represented by the "generalization" relationship between this use case and the "control appliance" and "control EV" use cases. The customer actor can control the devices or define parameters to authorize the storage change (allowing the sale of energy). The "change

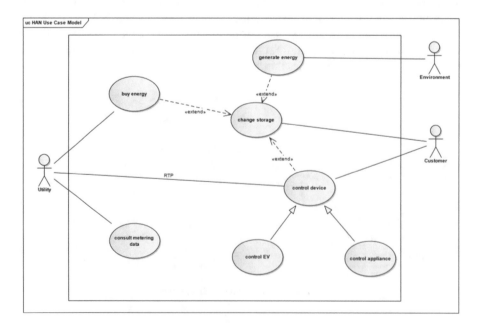

Fig. 5.   The use case diagram of the HAN

storage" use case extends three use cases: "control device", "buy energy" and "generate energy". This last is associated with the environment actor which represents all possible sources of renewable generation (wind, solar...) as we mentioned before. The third actor is the utility, which can either "consult the metering data" of a customer, send a bid to "buy energy" or "control device" in the HAN by a DR process.

In order to provide a more visible interaction diagram, we can translate this use case diagram into message sequence charts. We present here only the interaction between the environment ENV (that includes the three actors: customer, environment and utility) and the HAN system as shown in the Fig. 6. This is just an example of possible MSC diagram, and we can even propose other MSC diagrams as many as possible scenarios that our system can achieve. In Fig. 6, we start with the energy generation, and then we allow the use home energy storage by devices or by selling it to utility. The user powers on five appliances, plugs the EV and starts charging it. Next, the utility consults the metering data of the customer, sends a bid to buy energy and finally request the user to curtail the consumption. In the end, the customer shutdowns appliances and unplugs the EV.

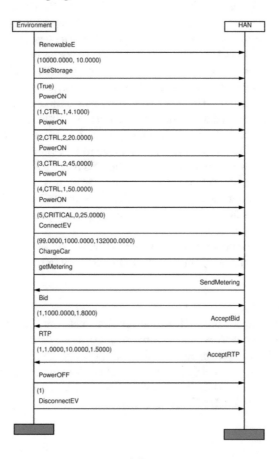

Fig. 6. A HAN MSC example

3.4 SDL Model Description

Our SDL model consists of six entities. Each entity represents one of the components in Fig. 1, except for customer DER and customer storage, which are grouped together into a single entity. When a customer turns on an appliance, this latter sends a registration request to IHD. The database is stored in the HEMS; it is permanently updated and can be used when a RTP signal is received to know which device to turn off. The same process is activated when an EV is plugged in the network; the only difference is that customer has the choice to charge the car immediately or in an upcoming time. We suppose that only one EV can be connected to the network at a specific time; however, we can connect infinity appliances. The devices consumption is calculated by the smart meter, and is sent to utility to calculate the billing tariff. But the user can decide to use energy from utility grid or use its own energy generated by renewable sources and stored in the home. In this case, the smart meter displayed value does not change.

The HAN receive RTP and Bid signal to, respectively, curtail its consumption and sale energy to utility. We suppose that the HAN processes only one request from the same type at a time. When the RTP request is received, the HEMS decides to accept the curtailment or not depending on algorithm that we will explain in the next section. We regroup the appliances in two types and three priorities. The types are CTRL and CRITICAL. The first one is controllable appliances; this type can be turned off if the RTP request is accepted. For the critical type, the appliances are not affected by the signal. Regarding the priorities, we chose three classes: the priority 0 for the CRITICAL; the priorities 1 and 2 related to CTRL type. The difference between them is that the appliances of type 2 can be turned off if the RTP signal requires this; however the appliances of type 1 are crucial for customer, so that they cannot be affected by this signal. This is only a simple method to choose the DR resources to turn off during the event. Other complex methods and algorithms can be used, but this is out the scope of this paper. On the other hand, in the process of selling energy, we compare the amount of energy requested by utility and the energy stored in the home as well as the EV battery. If the available power covers the utility needs, then the Bid is accepted. Else, it is refused and the process stops.

The complete SDL model consists of over 20 pages of diagrams, and is available online[1]. For the sake of space, only selected diagrams are introduced and explained in the next section.

Behavior Specification

In this section, we present our SDL model for HAN network. The system level of the model is presented in Fig. 7. We have six main blocks as we described previously. They interact with ENV via 5 channels. Channels C1, C4 and C9 are related to customer interaction with HAN; channel C11 is linked to the renewable energy resources and the last one is C14 that connects the HAN with the NAN. The variables and signals of our model are grouped into the HANmessages package.

---

[1] https://github.com/zahidsoufiane/HANmodel/.

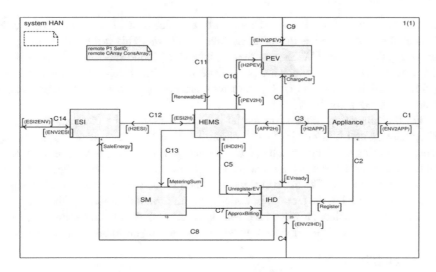

Fig. 7. The system view of the HAN SDL model

The Package HANmessages

We regrouped all declarations (variables, signals) in one package and we imported it in the system level to be visible for all blocks and process. This package is called "HANmessages" and is shown in Fig. 8. To make the scheme clearer, we regroup signals of the same channel in SIGNALLIST. The "powerON" signal has four parameters, namely, the ID, the type and the priority of the appliance; in addition to the consumption. For this last parameter, it represents the device consumption in a specific time interval.

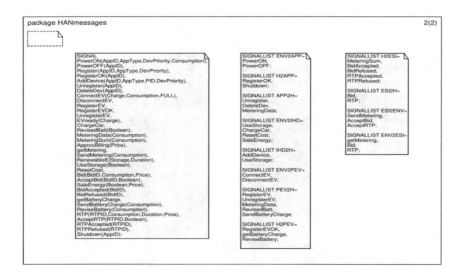

Fig. 8. Signal declaration of the HAN model

In our case, this interval is 60 s. This is only a choice to avoid sending a very large number of signals to update the metering data in unit time. This value can be change by modifying the value of TIMER. Another signal is the "ConnectEV". It has three parameters: the vehicle charge, which is a percentage value of the battery level, the consumption like appliance, and the FULL parameter representing the amount of energy needed to achieve a charge of 100%.

For the generation process, the signal is "RenewableE", with two parameters: one of type storage and the second of type duration. The storage variable stands for the energy generated during the duration value. It can be changed and take into consideration other parameters (seasons, temperature, wind movement...). The last two signals that we describe here are the RTP and Bid signals, related to the DR process. The RTP signal is identified by an ID, the amount of curtailment, the event duration and the new price if the event is accepted. However, the Bid signal is defined by three parameters; an ID as the RTP, the utility's energy needs to be purchased from customer, and the sale price.

The Block HEMS

The block "HEMS" is the main block in our model. It contains three processes, namely, DeviceDB, EnergyUse and DemandResponse, as shown in Fig. 9. The process DeviceDB is a database of all devices and EV plugged in the home. It stores information in three array indexed by DevID and contain, respectively, the PID, the type and the priority of devices; as well as a Boolean EVexist which indicates that an EV is connected to home or not. These four variables are "revealed", permanently and in real time, to other processes in the same block. The second process, EnergyUse, is responsible for the generation and metering operations. It contains a "revealed" variable "Store" showing

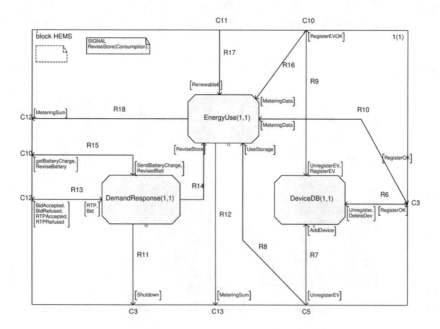

Fig. 9. The process view of the HEMS block

the amount of energy stored at home. This variable value increases when the process receives the signal RenewableE from ENV; and decreases if the energy is sold to the utility or if the customer chooses to use internal storage instead of energy from the grid.

The DemandResponse process manages the DR signals. Figure 10 shows the state machine corresponding to Bid event. To do this, the process uses the value of two revealed variables: Store and EVexist. When the Bid signal is consumed, the process checks the Store value and compares it with the utility need. In case the Store covers this need, the Bid is accepted. Otherwise, the process use a copy of EVexist to know if an EV is plugged; if the value of this Boolean is false, the Bid is refused; else, the process requests the EV charge of battery and goes to state waitBC. Upon receiving the Battery charge, the process reads the real time value of Store, adds it to receiving EV charge value and compares the sum to needed energy. The Bid is refused if the sum cannot cover the utility request; otherwise, a revision requests are sent to Store and EV to update their values. When a confirmation of revision is received (because the customer may unplug the EV before the revision takes place), a confirmation of Bid accepted is transferred to the utility.

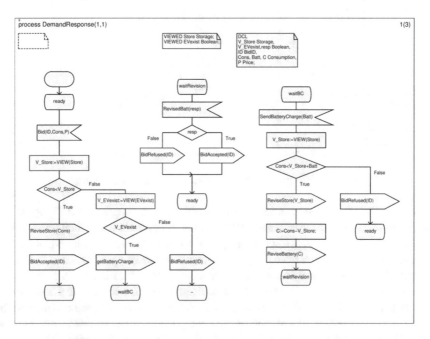

Fig. 10. The process DemandResponse: Bid event handling

On the other hand, the RTP event is presented in Figs. 11 and 12. We will discuss the operation in two steps: the first step is the choice of DR resources concerned by this event (Fig. 11); and the second one is to either accept or refuse this RTP (Fig. 12). The process in this algorithm uses three "revealed" variables from other processes in the same block: three arrays indexed by DevID and store the PID, type and priority of devices. The process uses also two other variables imported from "appliance" block: the powerset, of created devices, that we mentioned earlier; and the consumption array of these devices. The difference between a "revealed" and an "exported" variable is that the first one is used to exchange variable value between processes in the same block and in real time; however, the "exported" mechanism shares a copy of variable value between processes in two different blocks, so, the value of imported variable may be different from the original variable.

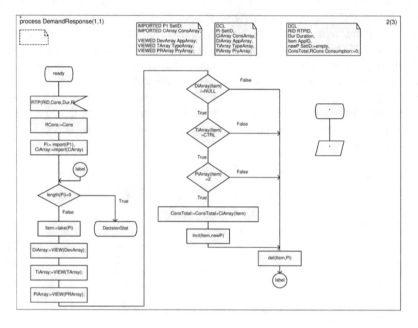

Fig. 11. The process DemandResponse: RTP event handling

The step of DR resources selection starts with the consumption of RTP signal. The process imports the two variables from "appliance" block and assigns their value to local variables: Pi and CiArray. It tests the length of variable Pi; and if this length is not null, we get an arbitrary element from this powerset with the "take" operator. Then, we "view" the three "revealed" arrays and do three consecutive tests for the chosen element. The objective of these tests is to know if the device is still ON, its type is CTRL and its priority is 2. If the item validates these conditions, we add its consumption to the variable ConsTotal (initialized to 0) and include the item to a new Powerset newP. After that we delete the item from Pi and we go back to the start of the loop. We repeat this operation until Pi becomes empty. The process goes then to state DecisionStat.

From this state, we can start the second step. When the input NONE is performed (spontaneous transition), we compare the value of ConsTotal to the curtailment value multiplied by a coefficient 1.1 to avoid turn off multiple devices. This is only a choice, so we can use a coefficient 1 or any value higher than 1. If ConsTotal is less than the other terminal, the RTP is refused. Else, we use a variable CurP of type Powerset to

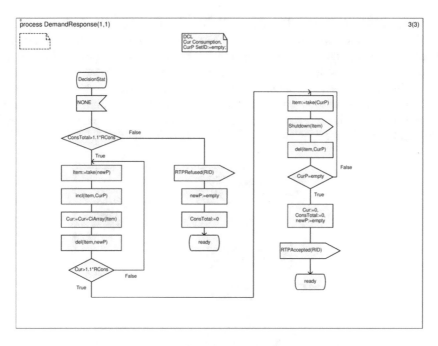

Fig. 12. The process DemandResponse: RTP event handling (continued)

indicate the device ID to shutdown. We start by choosing an arbitrary item from newP, add it to CurP and delete it from newP. We repeat it until the sum of consumptions of devices in CurP exceeds the threshold. Then, we send a signal Shutdown to all devices in CurP; we reset the variables and send a signal RTPAccepted to the utility.

4 Verification and Validation

The IBM Rational SDL explorer offers many automatic state space exploration and verification features to automatically find errors and inconsistencies in a system, or to verify the system against requirements. These techniques are: exhaustive, bit-state, random walk bit state exploration and verification against a given MSC [7]. Exhaustive exploration is an algorithm suited only for small system. It executes all the SDL symbols at least once and also all the behaviors of the model. This method is not adapted for our complex model (we ran this exploration but the explorer has crashed). Bit-state exploration is an efficient algorithm for reasonably large SDL systems. It is useful in checking deadlocks by storing the hash-code of each model state instead of the whole state itself. Another method, but this time it is useful for very large model, is the random walk exploration. It explores the behavior tree by repeatedly choosing a random path down the tree. The last one is the verification of a system level MSC to check if there is a possible execution path for the SDL system that satisfies the MSC. In our case, we will use the last three techniques.

The steps of a bit-state exploration (the same as exhaustive exploration) are presented in Fig. 13 [8]. According to the configuration used, the validation is finished when all reachable states of the model are explored. If errors reports are generated during the validation, we correct the model and start the procedure again. However, several states cannot be explored because of the large number of global states. In this case the model configuration must be reduced by changing the queue length, limiting the exploration depth and so on. Otherwise, if we cannot reduce this configuration, the validation is finished in a partial manner because some states remain unexplored.

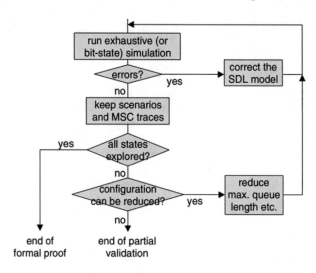

Fig. 13. The bit-state simulation method

In our case, we defined the test values for signals and ran two bit-state explorations: the first with the default events priorities; and the second with the same priority for all events. In fact, the events in SDL can be divided into five classes [7] and for each class a priority of 1, 2, 3, 4 or 5 are assigned:

- Internal events: Events local to the processes in the system, e.g., tasks, decisions, inputs, outputs.
- Input from ENV: Reception of signals from the environment.
- Timeout events: Expiration of SDL timers.
- Channel outputs: A signal is removed from a channel queue and put into another channel queue or the input port of a process instance.
- Spontaneous transitions: A transition in a process caused by input of NONE.

By default, internal events and channel outputs are assigned priority 1, and the other events are assigned priority 2. In the second scenario, we consider that all event classes are assigned priority 1.

In the first scenario, and after we corrected all errors, the validation ends without exploring all states. The transition that interests us is the transition that is executed when the SmartPlug process is in the state "ready" and receives a signal RegisterOK.

Normally, this transition must be executed if we send a PowerON followed by a PowerOFF signal for the same ID. It remains unexplored because the internal events (priority 1) are considered before the input from ENV (priority 2). As to the second scenario, the exploration took several hours (about 30 h) to finish. The transition not executed with scenario 1 was executed in this scenario, as well as other symbols.

The exploration using the random walk leads to the same results, but in a very short time compared to bit-state exploration. The last method that we used is the verification against a given MSC. For this, we used the MSC that we have presented earlier in this paper (Fig. 4). It defines only signals to and from the environment. In this case the MSC was verified, i.e., the behavior described in the MSC was indeed possible. We can use other MSC behaviors, with signals exchanged between blocks.

5   Conclusion and Future Works

We presented in this paper a generic model of HAN network and its services using SDL language and IBM Rational SDL Suite. After explaining the different elements of the model and related aspects, we proceeded to the verification and validation of the resulting model using the bit-state and random walk explorations as well as the verification using MSC. The result is a validated SDL model that can be used for automatic generation of codes for concrete or virtual Smart Grid solutions.

As part of our future work, we aim to verify this model using model checking approach. We intend to generate the SDL/PR (textual Phrase Representation) from the SDL/GR (Graphic Representation) described in this paper. Then, we will convert the model into Promela language and use the SPIN model checker to verify the system against specific safety and liveness properties.

References

1. Soufiane, Z., Slimane, B., Abdeslam, E.-N.: A synthesis of communication architectures and services of smart grid systems. In: International Conference on Systems of Collaboration (SysCo). IEEE (2016)
2. Department of Energy: Communications Requirements of Smart Grid Technologies, October 2010
3. Kuzlu, M., Pipattanasomporn, M., Rahman, S.: Communication network requirements for major smart grid applications in HAN, NAN and WAN. Comput. Netw. **67**, 74–88 (2014)
4. Khajavi, P., Abniki, H., Arani, A.B.: The role of incentive based demand response programs in smart grid. In: 2011 10th International Conference on Environment and Electrical Engineering (EEEIC). IEEE (2011)
5. Belina, F., Hogrefe, D.: The CCITT-specification and description language SDL. Comput. Netw. ISDN Syst. **16**(4), 311–341 (1989)
6. Casanovas, J., Figueras, J., Guasch, A.: Formalizing geographical models using specification and description language: the wildfire example. In: Proceedings of the 2013 Winter Simulation Conference: Simulation: Making Decisions in a Complex World. IEEE Press (2013)
7. IBM Support. http://www-01.ibm.com/support/docview.wss?uid=swg27041363. Accessed 01 Apr 2017
8. Doldi, L.: Validation of Communications Systems with SDL: The Art of SDL Simulation and Reachability Analysis. Wiley, Chichester (2003)

Automatics Tools and Methods
for Patents Analysis: Efficient Methodology
for Patent Document Clustering

Ayoub El Khammal[1], El Mokhtar En-Naimi[1(✉)] , Mohamed Kanas[1],
Jaber El Bouhdidi[2], and Anass El Haddadi[3]

[1] LIST Laboratory, Department of Computer Science, The Faculty of Sciences
and Technologies, UAE, Tangier, Morocco
Ayoub.elkhammal@gmail.com, ennaimi@gmail.com
[2] The National School of Applied Sciences, Tétouan, Morocco
jaber.f15@gmail.com
[3] The National School of Applied Sciences, Al-Hoceima, Morocco
anass.elhaddadi@gmail.com

Abstract. Patents have become a potentially powerful data sources and a wealth
of information for companies and organizations that tend to analyze and exploit
them for a variety of purposes and interests. However, with the ever-increasing
volume of patents filed year after year and the multiplicity of patent global data
bases, the task of patent research and analysis is becoming increasingly complicated and traditional analytical approaches have shown their limitations and have
become costly in terms of time and labor. Thus, various techniques and
approaches have been proposed to help specialists in their tasks of patent collection, analysis and results visualization. Document Clustering is one of these
common technics widely used in patent analysis. Over time, several powerful
clustering techniques and algorithms have emerged, but they often require modifications and adaptations depending on the fields of application and the target
data. In view of this, we propose in this paper a methodology for obtaining an
efficient clustering for patent documents based on the k-means, k-means ++
algorithm and various data-mining and text-mining techniques. Commons issues
often faced during the analysis of patents such as the manipulation and representation of textual data or the curse of dimension will also be addressed in this study.

Keywords: Patent · Patent metadata · IPC · Unstructured data · Clustering
K-means · K-means ++ · High dimensional problem · PCA · Data extraction
Text extraction

1 Introduction

Patent information is often exploited to study technological developments, trends and
technological potential [1] as well as decision-making process in research and development [2]. Technology and innovation trends can be analyzed using patents [3].

© Springer International Publishing AG, part of Springer Nature 2018
M. Ben Ahmed and A. A. Boudhir (Eds.): SCAMS 2017, LNNS 37, pp. 44–53, 2018.
https://doi.org/10.1007/978-3-319-74500-8_4

Data. Information from patent documents can prevent firms from investing in obsolete technologies [4] and improves strategic planning [5]. Patent analysis provides key information about the technological environment [7] and deals with component technologies [6]. Therefore, patent analysis is appropriate for detecting the relationships between different technologies. Details of approaches to patent analysis are provided in [7]. However, with the ever-growing volume of patents lodged each year and the multiplicity of databases in the world, the task of collecting and analyzing patent document is becoming increasingly complicated, especially for non-specialists, because patent information is enormous and rich in technical and legal terms. Therefore, information about patents needs to be transformed into something simpler and easier to understand.

Patent documents have a rigidly fixed structure, containing standardized fields such as patent number, applicant, inventors, assignee, classification of technological fields, description, claims, etc. Most of this information may be found on the home page of a patent document and are called the patent metadata. All these special and specific features of patent documents make it a valuable source of knowledge. In this paper we will show how can we use this knowledge in combination with data extracted from patent text in order to increase the accuracy of patents clustering.

Grouping or clustering is commonly used technic researchers often use classifications or groupings in patent analysis.

The grouping consists in partitioning a set $O = \{O1, O2,..., On\}$ of objects in homogeneous clusters maximizing intra-cluster similarity while minimizing inter-cluster similarity. Groups are formed without any prior information about the objects that are grouped together. All labels associated with objects are obtained only from the data. Grouping enable to identifies important topics or concepts from a set of documents and contributes to highlighting patterns of undetected or unexpected concentrations in databases. This is one of the most popular and frequently used approaches to form meaningful groups from unlabeled objects.

Clustering is a basic technique in many disciplines, including machine learning. It has been widely used in various decision-making processes. As the number of patents increases and the volume of data increases, it is not possible to successfully analyze any patent set without grouping. Therefore, grouping is the essential function that any patent analysis tool should provide.

Over time, many special clustering algorithms have been designed to be applied to patent databases. An example of this patent analysis platform is Patent iNSIGHT Pro [8] or the Patent Cluster search engine [9]. The clustering usually used in these platforms is often based on the content of the patent text, in particular summaries and claims parts, but ignores the information contained in the patents metadata, which is not the case for the k-clustering used in our methodology. The idea behind the k-clustering is to try to detect k optimal clusters by an iterative reinstallation method based on an optimization function. The most popular k-clustering is the k-means algorithm [10]. The purpose of k-means is to group the segments of N given observations into k clusters in which each observation belongs to the cluster with the nearest means. It uses the means (centroid) as the representative of a cluster. One of the main strengths of k-means clustering is its scalability.

These basic clustering algorithms require modifications and adaptations according to the fields of application and the target data in order to take maximum advantage of them. In this context, we propose a clustering methodology to efficiently group together patent documents based on unstructured data, mostly derived from the free text contained in patent documents, in particular the abstract part, title and description. Also we exploit the knowledge extracted from the patent metadata fields such as the IPC to increase the clustering accuracy.

The following section describes the main phases of the proposed methodology.

2   Methodology

The proposed methodology consists of 4 main phases. First a search equation is defined and the patent documents are extracted from the international patent bases. Then the documents are processed and the keywords are extracted. After that, a representation in vectors of the patent document collection is performed. Given the low structure density of the matrix obtained.

The PCA analysis is conducted in order to reduce the dimension size of the data model.

Once the final matrix is obtained we enrich it by adding additional dimensions that are fed by relevant data from patent documents metadata such as IPC or the inventor field etc. This will help increasing the final clustering accuracy.

The next step is the identify the number of clusters and, the k-means ++ algorithm is executed for a better initialization of the k centers.

The final step is the k-means launch and analysis of the results. All these steps are shown in Fig. 1.

2.1   Definition of the Search Function and Patent Collections

The very first step is to target the areas of technologies that interest analysts, and then relying on experts a search equation is defined which is nothing but a logical combination of carefully selected keywords.

Subsequently the search equation is used to extract the corresponding patents directly from the patent databases. All the patents documents that matches the equation are retrieved. Thus, documents belonging to a wide range of industries are collected from patent global databases like the United States Patent, Trademark Office (EPO), or WIPS (www.wipson.com).

2.2   Document Pre-processing and Keywords Extraction

The second step is to extract relevant keywords from each document. To do so, we are interested only in the title, abstract and description. These parts contain mainly natural language, therefore, we use text mining and NLP techniques to extract, potentially useful keywords from each document. Before this a this pre-processing step is needed.

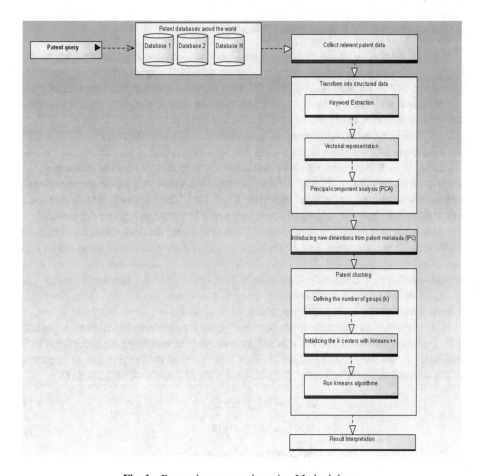

Fig. 1. Patent documents clustering Methodology.

The objective of the text-preprocessing is to remove unnecessary information such as tags, stop words and also to find a standard representation of the words present in the documents.

Since different keywords may be used to designate the same technique or functionality, depending on the technology filed. We define a kind of mapping table in order to replace these different keywords by a standard one a single term based on experts opinion.

In addition to the keywords obtained after this step, the experts are also consulted in order to enrich the extracted keywords by multi-word or phrasal sentences because sometimes they are more significant than a single word. At the end of this step each document is represented by a set of terms as follow:

Doc1 = {word1, word2,…. wordm}
Doc2 = {word2, word4,… wordm,… wordn}.

2.3   Data Model for Representing Patent Documents

The list of standardized keywords of each document obtained in the previous step is merged to form a matrix whose columns represent the keywords and rows represent the patent documents as proposed by [13]. Where a dimension (an axis) of the vector space is associated with each term or keyword. As for the matrix weights, several representations are possible. A simple representation has been proposed by Yoon, B., & Park, where the data model is represented as an existence matrix with the columns representing the keywords indexes $(1,\dots, j,\dots, n)$ and the rows representing the documents indexes $(1,\dots, i,\dots, m)$. If the j-term exists in the text of the i-document then the element (i, j) is supplied with the value 1 otherwise the value 0 is filled. In the end, we obtain an existence matrix whose values are fulfilled by either 1 or 0 [11]. The drawback of this method is that it does not show the degree of importance of a keyword in a given document.

Another representation was proposed by Jun et al. using the frequency of a term in the document. This ultimately give importance to terms that occur most frequently in the document than others [12]. However, it may happen that some less interesting terms appear much more frequently than others in a document like the word "today" for example. In the proposed methodology we have considered the TF-IDF (term frequency x inverse document frequency) weighting method which has been proposed by Salton [14]. This method has been shown to be effective for the similarity calculation, especially with the cosine distance. The basic principle is simple, the weighting is proportional to its frequency in the document (number of times it appears) as well as the inverse frequency of the occurrence in the reference corpus. According to the following formula (3):

$$w(i, j) = tf(i, j) \times idf(i) \tag{1}$$

$$tf(i, j) = n(i, j)/||dj|| \tag{2}$$

$$idf(i) = idf(i) = \log(m/mi) \tag{3}$$

- $w(i, j)$: the score tf-idf of the term j in document j
- $idf(i)$: inverse documentary frequency
- $tf(i, j)$: frequency of the term in the document
- $n(i, j)$: the number of occurrences of the term i in the document j
- dj: is the length of the document
- mi: the number of documents where the term i appears
- M: Is the total number of documents.

The logarithm function is applied for the IDF calculation to avoid overweighting very rare terms [13–15]. The Table 1 shows document-term matrix with the TF-IDF weighting. We refer to this stricture as PKM matrix.

Table 1. PKM: Patent-keyword matrix

	word 1	word 2	...	keyword n
Patent 1	w11	w12	...	w1n
Patent 2	w21	w22	...	w2n
...	...	....	...	...
Patent M	wM1	...	...	wMn

Generally, a given document contains only a small part of the total set of keywords since, as a matter of principle, most terms are present only in a reduced set of documents. as a consequence, the document-term structure is likely to be very sparse and above all its dimension is huge. This will make the task of similarity calculation less effective while it is a crucial step to get better clustering.

To solve this problem, we have considered conducting a Principal component analysis PCA as defined by Jun et al. [12]. The Table 2 shows the new structure after applying PCA analysis. We call this new structure the PPM matrix.

- pi, j represents the value of the jth principal component and the number of column k of the new structure is much smaller than the number of columns of the original PKM matrix.

Table 2. PPM: Patent principal components matrix

	axe1	axe2	...	axe k
Patent1	p11	p12	...	p1n
Patent2	p21	p22	...	p2n
...	...		...	...
PatentM	pm1	pm2	...	pMn

2.4   Enrich the Model with the Metadata

The data contained on patent metadata can provide us with additional information that can improve significantly the quality of the clustering if we use them correctly, Therefore, at this stage of our methodology we propose to add a new dimension to the PPM (Patent Principal components Matrix). This new dimension will be filled by the international Patent Classification IPC which classifies patents in a hierarchical way according to the domain of technology to which they belong [16]. The IPC codes are assigned carefully by the experts of the patent offices. As such, it represents a vital information that can increase accuracy of the clustering. The Table 3 shows the new structure after the introduction of the dimension relating to IPC. Where ipci is a normalized value of the IPC code corresponding to the patent document i. We refer to this new structure as the matrix PICPM.

Table 3. PICPM: Patent IPC principal components matrix

	IPC	Var1	...	varm
Patent1	ipc1	p11	...	p1n
Patent2	ipc2	p21	...	p2n
...	...	...	...	...
PatentM	ipcm	pm1	pm2	pmn

2.5   Clustering of Patents

2.5.1   Clustering Algorithm

Many algorithms are used for grouping. K-means clustering is one of the popular terms for cluster analysis. The real advantage of K-means is that it works with a large number of variables, which is often the case specially when we deal with unstructured data, as it is the case for the analysis we are conducting. K-means is faster than hierarchical clustering, and it can produces tighter clusters than hierarchical clusters, especially for globular clusters [18].

The main objective of the algorithm is to Partition the dataset in k-clusters according to some computational value.

k-means is widely used for clustering patent documents but most of the approaches that exists have one main limitation. Indeed the k-mean is applied either on the structured or unstructured data, but not on the combined data.

In this paper, we take advantage of both structured and unstructured data contained in patent document to form relevant clusters by combining structured and unstructured data for the purpose of obtaining an accurate clustering result.

2.5.2   K-Means Initialization

One of the major drawbacks of the K-means algorithm is that doesn't perform well with non-globular clusters. Actually, a different initial partition can result in different final clusters [18]. The initialization of the centers is a critical step of K-means which show that for different initializations of random centers we can obtain different k-groups.

We can try to run the k-means several times from different initialization. But there is no guarantee to find out the an optimal clustering.

We have understood that the best mitigation for this issue, is to start the algorithm with a good initialization of the centers. This means the K centers should be selected well at the beginning. In order to address this requirement we have decided to use the K-means ++ initialization algorithm proposed by Arthur and Vassilvitskii [17] which introduces three intermediate steps replacing the initialization step of the classical algorithm (Fig. 2):

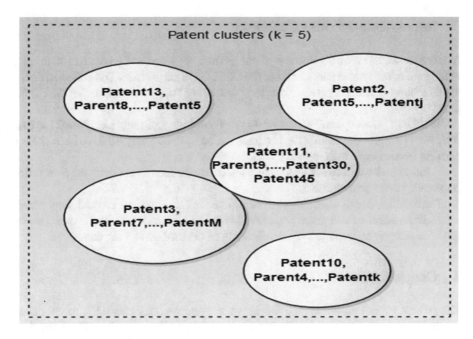

Fig. 2.   Patent groups after running the clustering algorithm.

ALGORITHM 1: K-means with Kmeans++ initialization

Begin

1) arbitrary chose the first center;

While (nember of centers is not yet reached) do

> *2) Sequential selection of the remaining centers while promoting the candidates furthest from the centers already selected;*
> *3) assignment of the new element to all the centers;*

End while

While (non-stabilized centers) do

> *1.select a document (randomly);*
> *2. assign to the closest center;*
> *3. readjust the centers;*

End while

end

---

3   Implementation and Evaluation

In order to validate the effectiveness of our methodology we have retrieved 100 patents in the field of renewable energies from the USPTO US patent base. The collected documents mainly contained patents that deal with either "the wind turbine" or "the solar panel".

Thanks to java implementation of this methodology and tools like Word2vec that where very helpful on making the implementation easiest. Thus, we have been able to perform several experiments with $k = 2$.

In most cases we have obtained a very good distribution of data with and F-measure coefficients average closer to 1.

The conducted tests were able to demonstrate the validity of our methodology which seems to be effective and gives good results on a small data sets as is the case for our test. The next objective is to test the efficiency of the method at a big data scale.

4   Conclusion

The present paper introduces an efficient generic approach for clustering patent documents using the k-means ++ & k-means algorithm, while using the ACP method to reduce the data model dimension obtained after extracting relevant keywords from patent documents through the use of text-mining techniques. In the proposed methodology, we have also demonstrate how to take advantage of patents metadata that contains valuable structured data which can be easily extracted and used to improve the clustering accuracy. Thus, we have used the IPC code to improve the clustering quality. In the same way, other fields can be exploited as the inventor, date of publication, country of origin and many others. The choice of including or excluding metadata fields essentially depends on the purpose of the analysis to be carried out.

We were able to test the effectiveness of our methodology on a set of patents in the field of renewable energies retrieved from the USPTO patent database. The tests carried out were aimed at grouping the patents into two groups, patents in which the "Wind Turbine" on one side and the "solar panel" on the other side were concerned. Indeed, this test written in java languages has validated the effectiveness of the proposed methodology but we want to push the tests even further by implementing this methodology in a bigdata environment. With a significant amount of patent documents.

The proposed approach seems encouraging but it still has some limitations that we will try to solve in our future works. We have identified several areas of possible improvement in order to limit as much as possible the intervention of experts during a patent analysis process, such as:

1. make the tasks of extracting, selecting and filtering keywords completely automatic.
2. find an efficient way to detect the number of clusters k.
3. develop indices to measure the quality of the obtained clustering and calculate its relevance.

References

1. Zhang, L.: Identifying key technologies in Saskatchewan, Canada: evidence from patent information. World Pat. Inf. **33**(4), 364–370 (2011). https://doi.org/10.1016/j.snb.2011.07.062

2. Pilkington, A., Lee, L.L., Chan, C.K., Ramakrishna, S.: Defining keyinventors: a comparison of fuel cell and nanotechnology industries. Technol. Forecast. Soc. Change **76**(1), 118–127 (2009). https://doi.org/10.1016/j.techfore.2008.03.015

3. Thorleuchter, D., Poel, D.V., Prinzie, A.: A compared R&D-based and patent-based cross impact analysis for identifying relationships between technologies. Technol. Forecast. Soc. Change **77**(7), 1037–1050 (2010). https://doi.org/10.1016/j.techfore.2010.03.002

4. Wang, Y.L., Huang, S., Wu, Y.C.J.: Information technology innovation in India: the top 100 IT firms. Technol. Forecast. Soc. Change **79**(4), 700–708 (2012). https://doi.org/10.1016/j.techfore.2011.10.009

5. Abraham, B.P., Moitra, S.D.: Innovation assessment through patent analysis. Technovation **21**(4), 245–252 (2001). https://doi.org/10.1016/j.techfore.2011.10.009

6. Trappey, A.J.C., Trappey, C.V., Wu, C.Y., Lin, C.W.: A patent quality analysis for innovative technology and product development. Adv. Eng. Inform. **26**(1), 26–34 (2012). https://doi.org/10.1016/j.aei.2011.06.005

7. Porter, A.L., Cunningham, S.W.: Tech Mining: Exploiting New Technologies for Competitive Advantage. Wiley, Hoboken (2005). ISBN 978-0-471-47567-5

8. Allison, J., Lemley, M., Moore, K., Trunkey, R.: Valuable Patents. Berkeley Olin Program in Law & Economics Working Paper Series (2003)

9. Allison, J., Lemley, M., Walker, J.: Extreme value or trolls on top? The characteristics of the most litigated patents. Univ. Pa. Law Rev. **158**(1), 1–37 (2009)

10. Kukolj, D., Tekic, Z., Nikolic, L.J., Panjkov, Z., Pokric M., Drazic, M., Vitas, M., Nemet, D.: Comparison of algorithms for patent documents clusterization. In: Proceedings of 35th MIPRO, Opatija, MIPRO 2012, pp. 1176–1178 (2012). ISBN: 978-953-233-068-7

11. Kim, Y.G., Suh, J.H., Park, S.C.: Visualization of patent analysis for emerging technology. Expert Syst. Appl. Int. J. **34**, 1804–1812 (2008)

12. Jun, S., Park, S.-S., Jang, D.-S.: Document clustering method using dimension reduction and support vector clustering to overcome sparseness. Expert Syst. Appl. **41**, 3204–3212 (2014)

13. Salton, G., McGill, M.J.: Introduction to Modern Information Retrieval. McGraw-Hill Inc., New York (1986)

14. Aizawa, A.: An information-theoretic perspective of tf–idf measures. Inf. Process. Manage. Int. J. **39**, 45–65 (2003)

15. Salton, G., Buckley, C.: Weighting approaches in automatic text retrieval. Inf. Process. Manage. **24**(5), 513–523 (1988)

16. WIPO - International Patent Classification: http://www.wipo.int/classifications/ipc/en/

17. Arthur, D., Vassilvitskii, S.: K-means ++: the advantages of careful seeding. In: Proceedings of the 18th Annual ACM-SIAM Symposium on Discrete Algorithms, SODA 2007, pp. 1027–1035. Society for Industrial and Applied Mathematics, Philadelphia (2007)

18. Huang, Z.: Clustering Large datasets with mixed Numeric and Categorical Values. CSIRO Mathematical and Information Sciences, Australia (1997)

Impact of Aggregation and Deterrence Function Choice on the Parameters of Gravity Model

Asma Sbai$^{(\boxtimes)}$ ⓘ and Fattehallah Ghadi

Laboratory of Science Engineering, Faculty of Science,
Ibn Zohr University, Agadir, Morocco
{a.sbai,f.ghadi}@uiz.ac.ma

Abstract. Mobility is part of our everyday lives. Modeling transport must take in consideration economical, social and environmental aspects of a city. The main purpose of this paper is to present a comprehensive presentation of different formulas of gravity models to estimate origin destination matrix (ODM) using the most used deterrence functions to provide parameters of calibration of trip distribution model. Network managers need an accurate ODM to operate their activities such as failure management, anomaly detection, design and traffic engineering. Thus, to improve the network management, it's a prerequisite to model the traffic between different zones through the estimation of OD matrix. We will discuss in detail the gravity-entropy model and the generation of this model using entropy maximization approach and we will focus on the calibration process using Hyman methods for three different deterrence functions using a practical application on Moroccan national mobility. We also demonstrate that changing the level of aggregation of data is significantly influencing the parameters values of ODM estimation.

Keywords: Origin destination matrix · Gravity model · Estimation
Interurban mobility · Deterrence function · Calibration

1 Introduction

A smart city is a qualification of rich environment of communication network that support different kinds of digital applications. Those applications produce rich datasets that improve significantly the management and decision making system of a city. One of the most important uses is the Intelligent Transportation System ITS which consists on interconnecting instruments for a quick and efficient decision making. Thanks to the empowerment of cities with different sensor technologies, collecting data to estimate ODM became easier but raises new issues on what model is more adequate.

Estimation of ODM is the second sub model of the classical four model for the Urban Transportation Planning Process (UTPP). As illustrated in Fig. 1, this model is composed of four sub models. The first one is the trip generation sub model, which consists on determining the traffic attracted and produced from each zone. The second one is the trip distribution; called also zonal interchange analysis which matches origin with destination, the result is a trip table with n lines and n colons where n represents

© Springer International Publishing AG, part of Springer Nature 2018
M. Ben Ahmed and A. A. Boudhir (Eds.): SCAMS 2017, LNNS 37, pp. 54–66, 2018.
https://doi.org/10.1007/978-3-319-74500-8_5

the number of zones. This matrix indicates the number of trips between areas. The third sub-model is the modal split. The result of this step is a range of Origin-Destination matrix, each one refers to a singular transportation mode and the share of each mode. Finally, trip assignment concerns the selection of route between origin and destination.

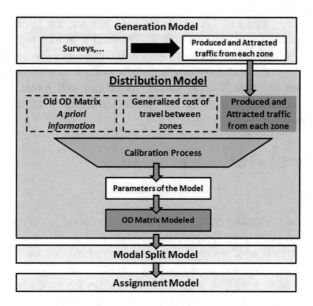

Fig. 1. The four model of UTPP

According to past studies, we can categorize ODM estimation to static and dynamic method based on time dependency. Static approach is time-independent and aimed for long-time transportation planning, while dynamic approach is meant for short-term strategies like route guidance or traffic control. Many studies have also point the factors influencing the reliability of ODM estimation: a-priori matrix and the optimum traffic count location [18]. The need for sufficient a-priori information and the position of sensors will save consequent work for data collection.

The evolution of technology allowed cities to be instrumented which allowed the collection of richer data than before and faster measurement and decision making.

Smart cities are equipped with a huge amount of sensors that differ on the type of data collected, their costs and reliability. For that Antonio [16] have distinguish three types of sensors: Point sensors: the most widely use are inductive loop detectors and Weight In Motion (WIM) systems, point to point sensors: for example the Automated Vehicle Identification system (AVI) which is based on identifying vehicles on various points of the network. The CCTV cameras records license plates and send this information to a center system to process it. Finally, area wide sensors: researchers are paying attention to the use of Smartphone and GPS to investigate dynamic estimation of traffic. Other researchers are more focused on using multiple data source and the fusion of all resources to achieve better results.

In this article, we will relay on a-priori ODM from a survey conducted in the National Project of Mobility Master Plan (also known by: SMDN 2020-2035). We will process the calibration of the gravity model using three different deterrence functions and two level of aggregation.

The paper is organized as follow: In the next part we will briefly present some researches on the gravity model and how we obtained it from entropy maximization approach. Then we will list different deterrence functions and their authors. We will find the parameters of the model using a Hyman calibration process solved using a three level Furness approach. Then in the third section, we will present the data and the level of aggregation of the tests. A forth section is illustrating the results. Finally a fifth section as a conclusion.

2 Related Works and State of the Art

2.1 Related Works

The calibration of the gravity model and the impact of level of aggregation has been recently presented in 2016 by Delgado and Bonnel [12] they demonstrated using the case of Lyon that the level of zoning which is selected when constructing O–D matrices and calibrating the parameters of the gravity model has a very strong influence on parameter transferability. In [19] authors are comparing the goodness-of-fit of aggregate and disaggregate gravity modeling using automated passengers counting (APC) data of Seoul metropolitan and demonstrates that the variation in goodness-of-fit and forecasting power largely depends on the estimation method and selected variables. In that study, the disaggregate modeling approach outperforms that of the aggregate. In the next references we will see in detail the state of the art on gravity model, deterrence functions and the process of calibration.

2.2 Gravity Model Formulation

Trip distribution models are supposed to produce the best prediction of traveler's destination choices on the basis of trip generation and attraction model for each travel zone.

One of the most frequently tool in the field of spatial interaction is the gravity model. Transportation planning seems to mostly use this model to analyze and describe mobility between zones of a study area. In the basis of multiple empirical studies, gravity model is an analogy from the Newton's law. It assumes that the interaction between any two zones is proportional to their magnitudes and inversely proportional to distance between them.

Many improvements have been tested over the years and trip distribution is usually performed using doubly constrained gravity model in the following formula:

$$T_{ij} = A_i O_i B_j D_j f(C_{ij}) \tag{1}$$

$$A_i = 1 / \sum B_j D_j f(C_{ij}) \tag{2}$$

$$B_j = 1 / \sum A_i O_i f(C_{ij}) \tag{3}$$

Trips produced and attracted by each zone is determined a priori. The sums of rows correspond to the produced trips, while the sums of columns correspond to the attracted trips.

T_{ij}	Number of trips modeled from zone origin i to zone destination j
O_i	Trip produced from zone i
D_j	Trips attracted to zone j
C_{ij}	Distance between zone i and zone j
f	Deterrence function
A_i, B_j	Coefficient that are set to meet the margin constraints of the Origin Destination Matrix
t_{ij}	Number of trips observed from zone origin i to zone destination j
n	Number of zones

The same distribution is obtained from entropy maximization distribution by solving the following optimization problem:

$$minZ = \sum T_{ij} \ln(T_{ij} - 1) \tag{4}$$

s.t:

$$\sum_j T_{ij} = O_i \tag{5}$$

$$\sum_i T_{ij} = D_j \tag{6}$$

$$\sum_{ij} C_{ij} T_{ij} = C. \tag{7}$$

2.3 Deterrence Functions

Deterrence function has also take attention of researchers and has applied different formulation on real and hypothetical cases which lead all along to the development of calibration techniques. Table 1 represents some of the most known function and their authors.

Table 1. Different deterrence functions

Author	Function name	Function formulation
Wilson [5]	Exponential function	$f(C_{ij}) = e^{-\beta C_{ij}}$
	Power function	$f(C_{ij}) = C_{ij}^{\alpha}$
	Tanner function	$f(C_{ij}) = C_{ij}^{\alpha} \cdot e^{-\beta C_{ij}}$
	Log normal distribution	$f(C_{ij}) = \dfrac{1}{C_{ij} \cdot \sigma \cdot \sqrt{2\pi}} \cdot e^{-\frac{(lnC_{ij}-\mu)^2}{2\sigma^2}}$ σ: the standard deviation of the log-normal distribution μ: the mean
Fotheringham [6]	Competing Destination Model CDM	$f(C_{ij}) = (S_{ij})^\rho e^{-\beta C_{ij}}$ Where: $S_{ij} = \sum_{k=1}^{n} \quad D_k e^{-\sigma c_{ik}}$ $k \neq i, k \neq j$
Thorsen and Gillsen [8]	Competing Destination Model $\sigma = 1$	$f(C_{ij}) = (S_{ij})^\rho e^{-\beta C_{ij}}$ Where: $S_{ij} = \sum_{k=1}^{n} \quad D_k e^{-c_{ik}}$ $k \neq i, k \neq j$
Fang and Tsao [9]	Self Deterrence Model with Quadratic Cost SDMQC	$f(C_{ij}) = e^{-\beta C_{ij}-\mu T_{ij}C_{ij}}$
de Grange et al. [7]	Consolidated model SDMQC + CDM	$f(C_{ij}) = S_{ij}^\rho e^{-\beta C_{ij}-\mu T_{ij}C_{ij}}$

Exponential Function

Only one parameter β needs to be calibrated and is widely used in research and case study.

$$f(C_{ij}) = e^{-\beta C_{ij}} \tag{8}$$

Power Function

A power function is used to find a better estimate of trips. It could be considered as a special case of exponential deterrence function:

$$f(C_{ij}) = C_{ij}^{\alpha} \tag{9}$$

This can be expressed in this form:

$$f(C_{ij}) = e^{-\alpha log(C_{ij})} \tag{10}$$

The exponent α will be positive to give decreasing travel with increasing cost.

The exponential model reproduces the average cost of travel while maximizing entropy, the power model reproduces the average of the logarithm of cost of travel.

Tanner Function

Introduced by Tanner (1961), he combines the exponential and the power functions in the following formula:

$$f\left(C_{ij}\right) = C_{ij}^{\alpha}.e^{-\beta C_{ij}} \tag{11}$$

This can be written in an exponential form with a generalized cost that is least function of cost and log cost:

$$f\left(C_{ij}\right) = e^{-\alpha C_{ij} - \beta log\left(C_{ij}\right)} \tag{12}$$

The distribution modeled is constrained to reproduce the means of both cost and its logarithms.

2.4   Calibration

Calibration is a complex process. Although methods to find the best fit parameters exists but in general they are not implemented in all modeling software.

The calibration is done using the maximum likelihood approach. In general, calibration aims to synthesize distribution from observed trip ends plus a variety of deterrence parameters then comparing the results with the observed trip matrix.

The following constraints must be respected:

$$\sum_{j} T_{ij} = O_{i} \tag{13}$$

$$\sum_{i} T_{ij} = D_{j} \tag{14}$$

$$\sum_{ij} C_{ij} T_{ij} = \sum_{ij} C_{ij} t_{ij} \tag{15}$$

$$\sum_{ij} log(C_{ij}) T_{ij} = \sum_{ij} log(C_{ij}) t_{ij} \tag{16}$$

This can be done by extending the Furness balancing process into a third dimension if a one parameter deterrence function is used. However, advanced deterrence functions with multiple parameters such as tanner function are difficult to calibrate using this method.

2.5   The Goodness of Fit of Calibration

It's necessary to define indicators to evaluate the reproduction of matrices from observation. Several indicators are used in the literature to measure the distance between observed and simulated matrices. We choose to use:

- Correlation R^2 between observed and synthesizes trips
- Root Mean Square Error between observed and synthesize trips

$$RMSE = \left[\frac{\sum (T_{ij} - t_{ij})^2}{n} \right]^{1/2} \tag{17}$$

- Standardized Root Mean Square Error

$$SRMSE = \frac{\sqrt{\frac{\sum (t_{ij} - T_{ij})^2}{n^2}}}{\frac{\sum t_{ij}}{n^2}} \tag{18}$$

According to Knudsen and Fotheringham (1986), SRMSE is the most accurate measure for analyzing the performance of two or more models in replicating the same data set, or for comparing a model's performance in different spatial systems. It's lowest bound is zero which mean that we have a perfect reproduction of observation. Generally its value is less than one. But when it's greater than one it means that the mean error is greater than the mean value.

Then we will graphically compare the observed and synthesized trip cost distribution.

3   Data and Study Area

In order to verify the gravity model, we carried out several test using a real study case of mobility in Morocco and compared the results with case study in the literature.

Morocco is located in the Maghreb region of North Africa and count over then 33.8 million population and an area of 446.550 km^2. The study includes all major cities such as: Casablanca, Marrakesh, Tanger, Tetouan, Agadir, Oujda, and Nador.

Data were gathered in the context of a National Project of Mobility Master Plan (also known by: SMDN 2020-2035) which aim to implement a global strategy ensuring an adjusted development of the sector of transport and find solutions for a rational use of space and transport while insuring the respect of environment. The survey was conducted on 152 zones represented in Fig. 2. In this study, we also aggregated the data on 12 zones that represents the 12 Moroccan regions (Fig. 3).

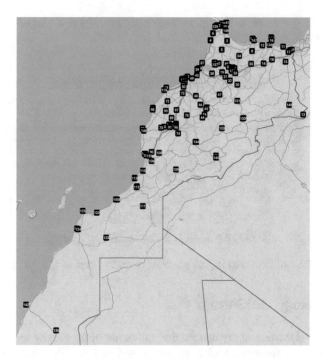

Fig. 2. Study zones aggregation level of 152 zones

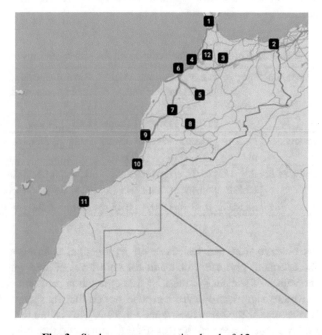

Fig. 3. Study zones aggregation level of 12 zones

Data will be processes as in Fig. 1 which highlights the first two sub-models of the classical four step models (Fig. 4).

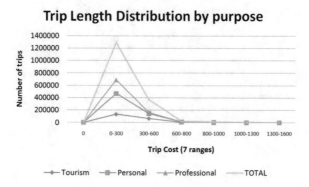

Fig. 4. Observed trip length distribution by purpose 152 * 152

4   Methodology and Results

We applied the algorithm of Hyman for the calibration of the gravity model using three different deterrence functions which are mentioned in Eqs. (8), (9) and (11) then we applied the gravity model and evaluate the goodness of fit for each aggregation level (12 zones and 152 zones). This was implemented using Matlab R2014a under a 32-bit machine (Table 2).

Table 2.   Results of Hyman calibration for gravity-entropy

Deterrence function		Gravity entropic doubly constraint 12 * 12			Gravity entropic doubly constraint 152 * 152		
		Exp	Power	Tanner	Exp	Power	Tanner
Parameters	Beta	0.0063	0.4151	0.1059	0.006	0.369	0.109
	mu	–	–	0.0017	–	–	0.0015
Goodness of fit	R^2	0.97	0.94	0.98	0.34	0.12	0.39
	RMSE	48.46	90.35	45.15	758	953	751
	SRMSE	0.49	0.80	0.45	10.11	12.72	10.04

As you can observe from our test case, the exponential deterrence function and Tanner function seems to perform a better fit for both level of aggregation. It is also obvious that the values of statistics in the 12 * 12 aggregation reconstruct the observed trip distribution more significantly. We can also notice that the parameters changes between the two levels for the same deterrence function. This is due to the fact that the less aggregated levels the fewer variables needs estimation and the greater aggregation the more zones pairs had to be estimated (Figs. 5, 6, 7, 8, 9 and 10).

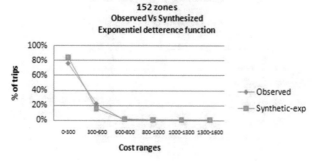

Fig. 5. Observed and synthetic trip cost distribution generated from doubly constraint gravity modeling and an exponential deterrence function using the parameter β resulting from calibration process (Zoning 152 * 152)

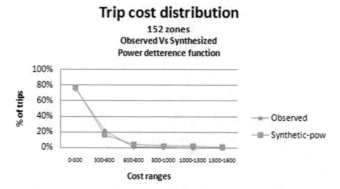

Fig. 6. Observed and synthetic trip cost distribution generated from doubly constraint gravity modeling and a power deterrence function using the parameter μ resulting from calibration process (Zoning 152 * 152)

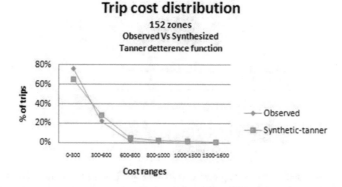

Fig. 7. Observed and synthetic trip cost distribution generated from doubly constraint gravity modeling and a Tanner deterrence function using the parameters β and μ from calibration process (Zoning 152 * 152)

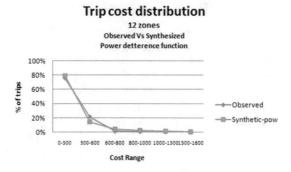

Fig. 8. Observed and synthetic trip cost distribution generated from doubly constraint gravity modeling and an exponential deterrence function using the parameter β resulting from calibration process (Zoning 12 * 12)

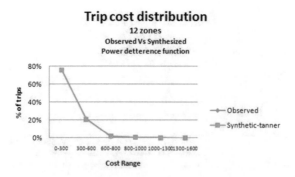

Fig. 9. Observed and synthetic trip cost distribution generated from doubly constraint gravity modeling and a power deterrence function using the parameter μ resulting from calibration process (Zoning 12 * 12)

Fig. 10. Observed and synthetic trip cost distribution generated from doubly constraint gravity modeling and a Tanner deterrence function using the parameters β and μ from calibration process (Zoning 12 * 12)

5 Conclusion

From the research that has been conduct, it is possible to conclude that the accuracy of the gravity-entropy model is highly influenced by the level of aggregation and the choice of deterrence function. And thus, data aggregation level used in calibration is an important factor to choose appropriately. Also increasing the aggregation makes significant changes in the parameters. The findings are of direct practical relevance, since we used a real case study of Morocco.

For future research, we aim to validate the gravity model with more complicated deterrence functions for different case study of different aggregation. Then we aim for the next stage of our research to include trip assignment steps in the calibration process and evaluate its influence.

References

1. Mathew, T.V., Krishna Rao, K.V.: Introduction to Transportation Engineering. NPTEL, 7 May 2007
2. Hensher, D.A., Button, K.J.: Handbook of Transport Modelling. Pergamon, Oxford (2000)
3. Wilson, A.G.A.: Statistical theory of spatial distribution models. Transp. Res. 1, 253–269 (1967)
4. Casey, H.J.: Applications to traffic engineering of the law of retail gravitation. Quarterly 100, 23–35 (1955)
5. Wilson, A.G.: Entropy in Urban and Regional Modelling. Pion Limited, London (1970)
6. Fotheringham, A.S.: A new set of spatial-interaction models: the theory of competing destinations. Environment and Planning A 15, 15–36 (1983)
7. de Grange, L., Fernández, E., de Cea, J.: A consolidated model of trip distribution. Transp. Res. Part E Logist. Transp. Rev. 46(1), 61–75 (2010)
8. Thorsen, I., Gitlesen, J.P.: Empirical evaluation of alternative model specifications to predict commuting flows. J. Regional Sci. 38(2), 273–292 (1998)
9. Fang, S.C., Tsao, S.J.: Linearly-constrained entropy maximization problem with quadratic cost and its applications to transportation planning problems. Transp. Sci. 29(4), 353–365 (1995)
10. Gonçalves, M.B., Bez, E.T.: A study about calibration methods of some trip distribution models
11. Abdel-Aal, M.M.M.: Calibrating a trip distribution gravity model stratified by the trip purposes for the city of Alexandria. Alex. Eng. J. 53(3), 677–689 (2014)
12. Delgado, J.C., Bonnel, P.: Level of aggregation of zoning and temporal transferability of the gravity distribution model: the case of Lyon. J. Transp. Geogr. 51, 17–26 (2016)
13. Williams, I.: A comparison of some calibration techniques for doubly constrained models with an exponential cost function. Transp. Res. 10(2), 91–104 (1976)
14. Hyman, G.M.: The calibration of trip distribution models. Environ. Plan. A 1(1), 105–112 (1969)
15. Smith, D.P., Hutchinson, B.G.: Goodness of fit statistics for trip distribution models. Transp. Res. Part A Gen. 15(4), 295–303 (1981)
16. Constantinos, A., Balakrishna, R., Koutsopoulos, H.N.: A synthesis of emerging data collection technologies and their impact on traffic management applications. Eur. Transp. Res. Rev. 3(3), 139–148 (2011)

17. Morimura, T., Kato, S.: Statistical origin-destination generation with multiple sources. In: 2012 21st International Conference on Pattern Recognition (ICPR), pp. 3443–3446. IEEE, November 2012
18. Ye, P., Wen, D.: Optimal traffic sensor location for origin-destination estimation using a compressed sensing framework. IEEE Trans. Intell. Transp. Syst. **18**, 1857–1866 (2016)
19. Kim, C., Choi, C.G., Cho, S., Kim, D.: A comparative study of aggregate and disaggregate gravity models using Seoul metropolitan subway trip data. Transp. Plan. Technol. **32**(1), 59–70 (2009)

A Semantic Web Architecture for Context Recommendation System in E-learning Applications

Bouchra Bouihi$^{(\boxtimes)}$ ⓘ and Mohamed Bahaj

Department of Mathematics and Computer Science, Faculty of Science and Technology,
University Hassan 1st, Km 3, B.P.: 577 Route de Casablanca, Settat, Morocco
bouchrabouihi@gmail.com, mohamedbahaj@gmail.com

Abstract. The widespread use of e-learning applications has put emphasis on the importance of having applications more personalized and adaptable to every learner needs. The one size fits all is no more working. Every learner should be delivered the right learning material that suits its learning context at the right time. The challenge is to incorporate the recommendation system in e-learning platforms in order to offer to learners a successful learning experience. In response to this challenge, in this paper, we propose a semantic web architecture of a context recommendation system in e-learning by means of which the learners will be offered learning content based on their profiles, activities and social interactions. The proposed architecture is a re-engineering of classical web architecture of current e-learning platforms. It's based on semantic web technologies. It comprises an ontology that guarantees a shareable and reusable modeling of the learning context and OWL Rules filtering that will be used as a recommendation technique.

Keywords: Semantic web · E-learning · OWL ontology
Recommendation system · Context-aware · SWRL

1 Introduction

Nowadays, e-learning platforms are widely used in education for both universities and companies. Because learners are given the opportunity to access electronic learning courses through the network. This access allows developing learners' skills, while making the process of learning independent of time and place. However, the continuous development of e-learning platforms has led to a huge amount of learning materials available on the network. It is time-consuming for learners to find the learning materials that they really need. "The one size fits all" is no more working. The challenge is to deliver to the learners the right learning materials at the right time.

To lead a successful learning experience, the learning materials should be recommended for the learners in coherence with their learning context. The contextual information such as; prior knowledge, activity history, interests, social interactions; should be taken into account in order to deliver to learners the learning materials suitable to what they really want [6]. This new learning pattern is specified as context-aware [7].

© Springer International Publishing AG, part of Springer Nature 2018
M. Ben Ahmed and A. A. Boudhir (Eds.): SCAMS 2017, LNNS 37, pp. 67–73, 2018.
https://doi.org/10.1007/978-3-319-74500-8_6

Context-aware applications should be developed with suitable context modelling and reasoning techniques. Therefore, ontology could be the suitable model to represent the context since this latter is considered as specific domain of knowledge. Complex context knowledge provided with formal semantics could be represented by ontology-based models. This representation allows to share and integrate context information [8].

Most of current e-learning platforms are based on a layered architecture which encapsulates the three levels of abstraction: data, application and presentation. In this paper, we propose a re-engineering of this architecture to integrate a semantic layer that holds an ontology and rule based approach for semantic recommendation. We aim to use the ontology as a domain knowledge for gathering the learning context information and OWL Rules filtering will be used as recommendation technique. The remainder of the paper is structured as follows. Section 2 summarizes the state of the art on semantic recommendation systems. Section 3 describes the proposed semantic architecture for context-aware applications. Finally, Sect. 4 concludes and shows some future lines of work.

2   Related Works

There have been many researches about personalized learning using semantic web technologies, mainly ontologies [9, 12]. Several ontologies-based approaches for context-aware e-learning platforms were proposed. Authors of [6] propose to make recommendation to realize context-awareness in learning content provisioning by exploiting knowledge about the learner (user context), knowledge about the content, and knowledge about the learning domain. For this purpose, they designed three ontologies with a focus on learner's prior knowledge and his learning goal in the recommendation process. But the social learner interactions are not taken into consideration.

The work presented in [10] proposes to recommend learning content based on the expert learning object knowledge base and personal learning progress where sequencing rules were used to connect learning objects. The rules were created from the knowledge base and competency gap. However, all the focus is on the learning content; the authors do not study the learner profile and its social interactions that are important contextual information.

[11] Proposed a framework to observe personalization in e-learning system based on ontology. They created user ontology, domain ontology and observation ontology. They also used reasoning mechanism over distributed Resource Description Framework (RDF) annotation. The query rule language used in this system is Triple. However, OWL has more powerful expressive capability than RDF.

[12] proposed a semantic recommender system for e-learning. It comprises ontology and web ontology language (OWL) rules. The proposed system is consisted of two subsystems; Semantic Based System and Rule Based System. In this work, the authors do not explain how they built the ontology and which Rule language they work with.

Our work differs from these researches by proposing a re-engineering of the layered architecture of current e-learning platform by integrating a semantic layer that will hold the semantic recommender system. Our proposed semantic recommender consists of two subsystems that are: E-learning Ontology Subsystem and the Semantic Rules Subsystem.

3 Semantic Web Architecture

Most of the current e-learning platforms are based on a layered architecture which encapsulates the three levels of abstraction: data, application and presentation. The first layer is concerned with the storage of the data. The second handles the requests of the user interface by querying the storage media, after performing the various treatments and returning the results to the third layer this last layer then manages their display. These solutions are not sufficiently aware of the learner context. The context-awareness is highly recommended to deliver to the learner the learning material relevant to the current situation of the learner. To achieve this, we adopt the ontological approach to define a model to represent and manage context information. In this work, we want to perform a re-engineering of this architecture, with a view to incorporate the technology of the Semantic Web. To this end, we are proposing to insert a Semantic Layer between the layer of data and that of the application. Figure 1 resumed schematically this architecture.

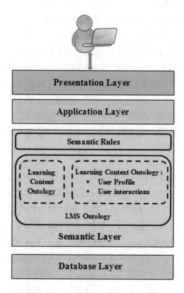

Fig. 1. Semantic web architecture for recommendation system in e-learning

In the following, we focus on the semantic Layer of our proposed architecture. It's organized in two main parts:

- E-learning Ontology
- Semantic Rules

3.1 E-learning Ontology

Our approach considerably relies on semantic modeling of the learner's context and environment. For this purpose, we make use of ontologies and Semantic Web technologies. The e-learning ontology is the ontology of the whole Learning Management System. Since we are working on current e-learning platforms that are already running and designed with UML diagrams and in order to limit the amount of effort required to build such a consistent ontology, we propose to build this ontology by adopting the approach UML-To-OWL proposed in our previous work [1]. Then the resulted OWL ontology will have some changes to be able to model the learner context. In order to keep the modeling task manageable, we divide this ontology into two sub-ontologies that are: Learning Content Ontology and Learning Context Ontology.

Learning Content Ontology: A learning content is an instantiation of Learning Objects-abbreviated LO. The LO are a digital small size components of a learning course which can be reused several times in different learning contexts. However, these Learning Objects are often designed and developed by different organizations and authors which make the learning content semantically heterogeneous. This heterogeneity affects its reusability. So it is essential to think of a shared modeling of LO in order to make them easily accessible, usable, reusable and semantically interoperable.

Different standards have been defined to help the development of learning systems and the representation of their joined LOs. Making use of these standards, not only guarantees the interoperability, but also the quality of the system [3]. Among these standards, we can cite LOM, SCORM and the IMS-LD. LOM is interested in learning content description, SCORM in content, structure, and the IMS-LD in learning scenario. In our work, we are interested in LOM standard. LOM (Learning Object Metadata) is a standard developed by IEEE consortium. It defines the structure of an instance of metadata for a LO. It is composed of a set of 80 elements divided in 9 categories performing each a different function [2]. To capture the semantics of LOs, we present this standard in an ontological way (Fig. 2).

Learning Context Ontology: The context of learning is a crucial aspect in e-learning. Therefore, it is important to determine according to the learner current context what are the relevant learning materials to deliver, how, and at what time. All the learning process must be adapted to context changes. To take into account the context in an e-learning system, it is necessary to find a way to represent it. This representation must provide a coherent model to store and process the context information in order to respond to the environment changes. At the semantic level, we define context information using a Learning Context Ontology that includes two interrelated sub-ontologies: Learner-Social ontology and Learner Activity ontology. This ontology will represent and store every learning context's information.

Figure 3 shows the Learner Social Ontology that is built from FOAF ontology. FOAF Ontology is an ontology that is built on the Resource Description Framework RDF2. It's conceived to represent people's personal information and their social relationships among a social network. People are represented as nodes and relationships by edges [5].

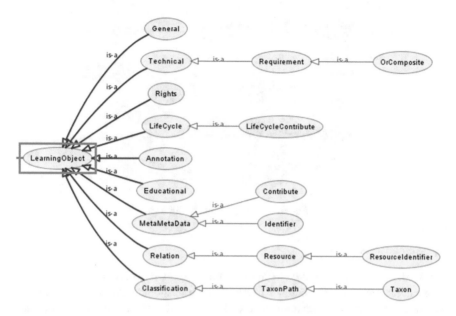

Fig. 2. Learning content ontology designed with Protégé according to LOM standard

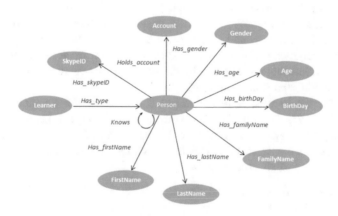

Fig. 3. Learner-social ontology

This ontology is widely used, [4] claims that the class foaf:Person has nearly one million instances spread over about 45,000 web documents. So it's relevant to reuse it to represent the context information about the learner profile and its social interactions.

Figure 4 shows the Learner Activity Ontology. This ontology represents and stores the different information about the learner's pedagogical interest and behavior. It shows in which topic the learner is interested, in which courses is enrolled and what are the specific activities he did.

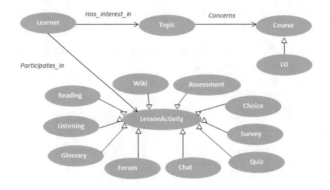

Fig. 4. Learner Activity Ontology

3.2 Semantic Rules

To take into account the context in an e-learning system, it is necessary to find a way to represent the context in the latter. This representation should provide a coherent model to store and process the context information in order to react to the environment changes. The e-learning ontology is the context model. After building this ontology, it's time now to apply techniques of refinement and adaptation of the LOs on it to deliver to the learner the learning object relevant to its context. A method to filter the LOs is to apply a set of business rules, indicating what LO to use in what context. These filtering rules will be used as our recommendation technique. They synthesize the domain knowledge and business constraints that must be met by the system. Business rules are translated into SWRL (Semantic Web Rule Language).

SWRL (Semantic Web Rule Language) is a language for Semantic Web rules, combining the OWL - DL and OWL-Lite with the unary/binary sub-language of RuleML (Rule Markup Language). The structure of SWRL rules consists of an antecedent and a consequent. A rule means «if the antecedent conditions are maintained, then the consequent conditions must also be applied».

If Antecedent Then Consequent Antecedent (Body) → Consequent (Head)
Example of a SWRL Rule:
If a Learner x knows (FOAF Object Property) another Learner y who is interested in a Topic T, then the learner x may also be interested in the same Topic T. So in this case, it's relevant to recommend to the Learner x the topic T. This recommendation rule is expressed with SWRL language as:
Learner (?x) ∧ Knows(?x, ?y) ∧ has_interest_in_topic(?y, ?T) → has_interest_in_topic(?x, ?T).

4   Conclusion and Future Works

Context-aware applications play an important role in education, especially e-learning. Context recommendation systems give the opportunity to learners to lead a successful learning experience by getting the right learning materials that suit their needs at the right time. Semantic web technologies make these systems more performant and relevant. In this paper, we have proposed our semantic web architecture for current e-learning platforms. This ontology and rule based architecture is proposed to make use of ontology as a domain knowledge for gathering the learning context information and OWL Rules filtering as a recommendation technique.

For future work, we will extend our semantic architecture that will offer a context recommendation system for mobile learning.

References

1. Bouihi, B., Bahaj, M.: Building an e-learning system's Owl ontology by exploring the UML model. J. Theor. Appl. Inf. Technol. **87**(3), 380 (2016)
2. Soualah-Alila, F., Nicolle, C., Mendes, F.: Une approche Web sémantique et combinatoire pour un système de recommandation sensible au contexte appliqué à l'apprentissage mobile. In: 11 ème édition de l'atelier Fouille de Données Complexes, No. 47–58 (2014)
3. Grandbastien, M., Huynh-Kim-Bang, B., Monceaux, A.: Les ontologies du prototype LUISA, une architecture fondée sur des Web Services Sémantiques pour les ressources de formation. In: 19es Journées Francophones d'Ingénierie des Connaissances, pp. 61–72 (2008)
4. Ding, L., Finin, T., Joshi, A.: Analyzing social networks on the semantic web. IEEE Intell. Syst. (Trends & Controversies) **8**(6), 815–820 (2004)
5. Buriano, L.: Exploiting social context information in context-aware mobile tourism guides. In: Proceedings of Mobile Guide (2006)
6. Yu, Z., Nakamura, Y., Jang, S., Kajita, S., Mase, K.: Ontology-based semantic recommendation for context-aware e-learning. In: Ubiquitous Intelligence and Computing, pp. 898–907 (2007)
7. Schmidt, A., Winterhalter, C.: User context aware delivery of e-learning material: approach and architecture. J. Univ. Comput. Sci. **10**(1), 28–36 (2004)
8. Villalon, M.P., Suárez-Figueroa, M.C., García-Castro, R., Gómez-Pérez, A.: A context ontology for mobile environments (2010)
9. Benlamri, R., Zhang, X.: Context-aware recommender for mobile learners. Hum.-centric Comput. Inf. Sci. **4**(1), 1–34 (2014)
10. Shen, L.P., Shen, R.M.: Ontology-based learning content recommendation. Int. J. Contin. Eng. Educ. Life Long Learn. **15**(3–6), 308–317 (2005)
11. Henze, N., Dolog, P., Nejdl, W.: Reasoning and ontologies for personalized e-learning in the semantic web. Educ. Technol. Soc. **7**(4), 82–97 (2004)
12. Shishehchi, S., Banihashem, S.Y., Zin, N.A.M.: A proposed semantic recommendation system for e-learning: a rule and ontology based e-learning recommendation system. In: 2010 International Symposium in Information Technology (ITSim), vol. 1, pp. 1–5. IEEE, June 2010

BECAMEDA: A Customizable Method to Specify and Verify the Behavior of Multi-agent Systems

Abdelhay Haqiq$^{(\boxtimes)}$ ⓘ and Bouchaib Bounabat

ENSIAS, Mohammed V University, Rabat, Morocco
{abdelhay.haqiq,b.bounabat}@um5s.net.ma

Abstract. Multi-Agent paradigm offers a viable solution to the increasing needs for smart and crisis system that reacts accordingly to the environment changes. The researches have largely focused on studying the development of the various approaches that deals with designing and implementing the multi-agent system. However, there is a lack in approaches that treat in depth the specification aspect of the Multi-Agent systems engineering process, which is important to better define and describe the behavior of the agents. This paper is interested in studying the specification and the verification of Multi-Agent behavior, it proposes in consequence an effective method called BECAMEDA. It is based on an iterative process that is useful to understand the system starting from goals identification that the system expects to achieve through the formal verification of the system properties. The method is illustrated by an example of a Crisis Management system.

Keywords: Multi-agent · Specification · Verification · Behavioral model
Model checking

1 Introduction

The most challenging thing in software engineering for Multi-Agent is to ensure that the requirement of the system is well identified and verified [1, 2], especially when the system is complex and appeals with many stakeholders. In this paper, we are interested to study the Multi-Agent system to cope with the dynamic, distributed and cooperative systems. A multi-agent system is a system made up of several agents which are in inter-action among them, this interaction is very important to define the overall behavior of the system. The behavior of the agent is defined to be a set of properties that an agent outlines to give responses to its environment [3]. Basically, the agent reacts to events based on the received decisions and gives responses in the form of actions according to the experienced situation.

To specify MAS, it is important to have processes and methods that will help engineers to understand how to build up customized system [4]. In this paper, we are interested of studying the specification of five essential properties that the agents are supposed to exhibit: reactivity, adaptability, conflict management, and reusability. The reactivity refers to the ability to respond to the environment, the adaptability is the capacity to cope with changes and disruptions of the environment, the conflict management is the ability to manage resource sharing, and the reusability is the capability to reuse agent

© Springer International Publishing AG, part of Springer Nature 2018
M. Ben Ahmed and A. A. Boudhir (Eds.): SCAMS 2017, LNNS 37, pp. 74–86, 2018.
https://doi.org/10.1007/978-3-319-74500-8_7

components. In literature, there are approaches [10–14] that study some of these properties, but there is still a lack of an all-in-one approach that considers all of the mentioned properties.

To deal with the complexity of building up a multi-agent system, this paper proposes a method that helps to make the specification and the verification of such a system to avoid errors happening within the implementation phase. Since MAS can be applied in different domains and systems, this paper proposes to have demonstration of this method through a Crisis Management system case study.

Thus, this paper intends to present the difficulty of how to define the specification of Crisis Management system for both individual and collective behavior. We propose a method called BECAMEDA (Behavioral spEcification and verifiCAtion of RMAS based on E-MDRA) for the specification and the verification of Reactive Multi-Agent System based on the E-MDRA (Extended Multi-Decisional Reactive Agent) model. The E-MDRA is an extension of The MDRA, which is developed by [5], The MDRA is being deployed in various domains such as wireless sensor networks [6], mobile systems [7] and organizational systems [8], this model helps on modeling reactive agents by putting forward decisional aspect facilitating the specification of the agent according to their behavior and their intelligence. Throughout this paper, the Crisis Management system is used as an example to demonstrate our approach [9]. The overall goals of the Crisis Management System (CMS) are: facilitating the rescue mission of the police by offering detailed information about the location of the crash; managing the dispatch of ambulances or other alternate emergency vehicles to transport victims from the crisis scene to hospitals; facilitating the first-aid missions by providing relevant medical history of identified victims to the first-aid workers by querying databases of local hospitals.

The rest of the paper is organized as follows. Section 2 presents a related work. Section 3 presents a background related to the Extended Multi Decisional Reactive Agent. Section 4 describes our approach through an example of a Crisis Management System. Section 5 concludes the paper and draws future perspectives.

2   Related Work

Many approaches have been proposed for the specification and the verification of MAS. [10] presents a formal method for specifying and verifying a Multi-Agent system based on design patterns, refinement, and event-B. The objective of the method is to satisfy the global properties of the system based on the agent's local behavior. [11] proposes an approach to specify by refinement in event-B the collaboration between agents in a critical system where the consideration of the fault tolerance property is essential. The authors define the SMA through four applets (A, M, E, R). A represents a collection of different classes of agents. M represents a middleware system that has the ability to ensure the communication flow between agents and resolve the communication disconnect issues, E represents a collection of system events, R represents a set of dynamic relationships between agents allowing making connections between active agents belonging either to the same classes or to different classes. [12] proposes High Level Multi-Agent Petri-Net (HMAP) to describe, model and analyze the dynamic behavior of MAS based on the Petri Net. HMAP is based on a formal logical method describing the behavior of the agent in nine sequential steps. [13] proposes an approach called ForMAAD (Formal Method for Agent-based Application Design) to specify the agent's behavior at the individual and the collective levels. The behavior is specified through a formal language called Temporal Z that integrates the first-order temporal logic with the Z notation. The approach proposes to divide the refinement process into four levels: cooperation, organizational, collective behavior and individual behavior. [14] proposes a translation of the AUML (Agent UML) diagrams; which is an extension of the UML language to model Multi-Agent system; into the RT-Maude formal language. RT-Maude supports the specification and the analysis of real-time systems, and more specifically object-oriented real-time systems. It is based on the programming environment based on the rewriting logic.

The Table 1 presents a comparison between the different approaches according to the reactivity, adaptability, conflict management, and reusability properties. We note that there is no specific approach that considers all the properties mentioned. Besides of this, there is a lack of a real process that helps engineer to model the specification phase, which is the most important phase on the software engineer process.

Table 1. Comparison of approaches to specify MAS.

	Reactivity	Adaptability	Conflict management	Reusability	Method support
[10]	**	–	*	*	**
[11]	**	*	*	–	–
[12]	**	*	**	*	–
[13]	**	–	*	*	**
[14]	**	**	–	*	–
[15]	***	**	**	*	–

Caption: ***: strongly; **: moderately; *: Weakly; –: not supported or not described

In our previous works [15], we have proposed an extension of MDRA model called E-MDRA to deal with the properties mentioned. However, we are still in need to have a method that helps engineers to correctly identify and define the specification. Thus, this paper focuses on the proposed method based on E-MDRA. The next section gives an overview of the E-MDRA specification model.

3   E-MDRA Model Overview

3.1   E-DRA

The E-DRA model is based on an external specification & an internal specification [15]. The external specification denotes a set of information sent to or received from the environment. It is characterized by:

- A: Set of actions exerted on the agent, each action represents a possible operation to be performed on this agent

- D: Set of decisions generated by the agent to provide solutions concerning the system behavior. A decision d is characterized by its decision horizon DurDec(d), that indicates the time during which this decision remains valid
- S: Set of Signal received by the agent. The signal represents the acknowledgement response to confirm the execution of a decision
- E': Set of external states delivered by the agent. Each external states represents the object state emitted to the environment

The internal specification represents a set of information generated and exchanged inside the agent. It is characterized by:

- E: Set of agent's internal states. Each one indicates the current state of the agent. An agent may exhibits parallel operations
- O: Set of agent's internal objectives. Each internal objective denotes the expected state after the execution of a decision
- O': Set of agent's external objectives, which can be achieved. These objectives represent the agent interpretation of each action, and define therefore its different behaviors

From a dynamic perspective, these sets determine the events received from the environment (A, S), events sent to the environment (D, E) and the internal events (E, O, O').

3.2 E-MDRA

The organization of Extended Multi Decisional Reactive (E-MDRA) is defined by a set of agents connected together with communication interfaces, thus forming a hierarchical structure based on two-level tree: an E-DRA Supervisor (E-DRAS) and two or several sub-agent components (E-MDRASi). The connection between supervisor and its sub-agents is realized through two communication interfaces: Decisional Interface (DI) and Signaling Interface (SI). Such a system interacts with its environment with Actions exerted by the environment and External States emitted to the environment.

The description of a MDRA is defined as a quadruplet S: < E-DRAS, DI, SI, S-MDRA >, with:

- E-DRAS: represents the E-DRA type that supervises S.
- DI: Decisional Interface of S, implements a translation function of a decision into several parallel actions, each of these actions belongs to a low-level sub-agent.
- SI: Signaling Interface of S, implements a translation functions of several external states into one and only signal.
- S-MDRA: characterizes a set of 2 or more E-MDRA components.

4   BECAMEDA Method

BECAMEDA (Behavioral spEcification and verifiCAtion of RMAS based on E-MDRA) method provides support for the specification and verification of multi-agent with the involvement of the reactivity, the adaptability, the conflicts management and the reusability properties. It captures the requirements and goals of the system, identifies the goal's actions, determines the system's organizational structure and models the internal and external behavioral specification of the agents.

BECAMEDA is made up of two layers: specification layer and verification layer. The specification layer is consisted of three phases: identification phase, definition phase and modeling phase. The identification phase is the phase where the system identifies its goals. A goal represents a desirable objective of the system. Goals are organized as a hierarchical structure made up through the refinement of higher-level goals towards lower-level goals. The lower-level goal is achieved through actions. The definition phase consists of two steps: the first step represents the agents and their organizational structure which is also classified into levels. Whereas, the second step determines the actions to be assigned to agents under each level. During this phase, the actions are a priori assigned to agents of level 1, then, according to the modeling phase, they are assigned to the following level. Basically, the modeling phase is made up through three steps for each agent under a level i greater than or equal 1. The First step, models the internal behavior actions of level i. the Second step, assigns actions to agents of the following level (level i + 1), and finally, the third step, models the external behavior actions of level i.

After having establishing the models of the agent for both internal and external behavior, the verification layer indicates the system properties to be verified, and performs afterwards a formal verification with the use of model-checking technique. Figure 1 illustrates the BECAMEDA phases.

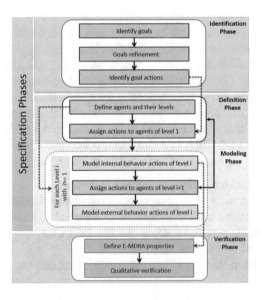

Fig. 1. BECAMEDA phases.

4.1 Identification Phase

The identification phase captures the system requirements and identifies the goals of the system. A goal defines the global objectives that the system desires to fulfill. This phase is made up following three steps: identifying goals, refining goals and identify goal actions. The two first goals are represented with two models: the goals model and the refined goals model. The goal model facilitates the understanding of a system with the generation of a set of abstract goals that offer a high-level understanding of the defined objective of the system, as for the refined goals model, it identifies the concrete goals that highlight the goal activities. The goal-actions model identifies the actions that realize the concrete goal. The entities forming this phase are built up following the meta-model of Fig. 2 where is defined the concept of goal, action and their different relationships.

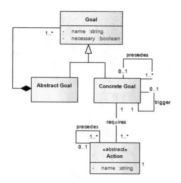

Fig. 2. Identification phase metamodel.

In the following, we present the BECAMEDA method throughout an example of a Crisis Management system.

Identify goals. In this step, we represent the system requirements as a set of structured goals consisted of sub-goals. The goals called "abstract goals" express the high-level objective of the system; they are connected by AND/OR/XOR logic gates. The AND indicates that all of the sub-goals must be completed to satisfy the parent goal, the OR indicates that there is an alternative way to represent the sub-goals, and the XOR indicates that the goal parent can be satisfied with one of its sub-goals.

Goal refinement. We decompose in this step the abstract goal into concrete goals to capture the dynamic aspects of the goals model and defines each goal through "precede" and "trigger" attributes. The "precedes" determines that the goal X must be completed before the goal Y is pursued. Whereas "triggers", means the instantiation of a new goal when an event occurs while another goal is still in activity.

The Fig. 3 presents the step 1 and 2 of the identification phase. The Goal "capture info Crisis (Goal 1)" depends on the achievement of three concrete goals: "receive information from witnesses (Goal 1.1)", "receive information from video surveillance (Goal 1.2)" or "receive information from the phone company (Goal 1.3)". We note that the three sub-goals are successive goals, since they are defined by the key word "preceded" and are alternative as well. Indeed, when a crisis event is reported, the first step to do is to get the maximum information from witnesses, in case of doubt, it is possibility to verify the statements of witnesses via either the surveillance system, if installed on the incident scene, or by checking the eligibility of witnesses by contacting the phone company. For the goal 2.1, it depends on the achievement of several sub-goals. Basically, depending on the context of the environment, we note different missions: "Establish communication with IR (Internal Resource) (Goal 2.1.1)" mission, "IR confirms mission (Goal 2.1.2)" mission and "Select IR (Goal 2.1.3)" mission.

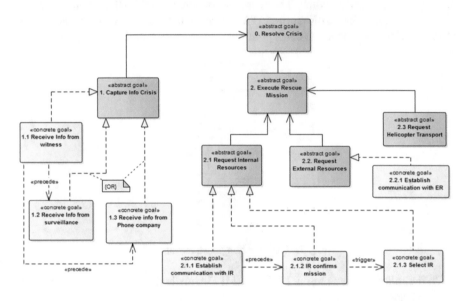

Fig. 3. Abstract & concrete goals crisis system.

Identify goal actions. In this step, we select the actions to carry out to satisfy and realize a specific goal. The figure below shows the model that associates the concrete goals to actions (Fig. 4).

Fig. 4. Goal actions model.

4.2 Definition Phase

The definition phase represents the agents and their organizational structure classified with levels, it also presents the actions to be assigned to agents under each level. This phase is intrinsically dependent on the modeling phase, since there is a back-and-forth process between the two phases. This process helps at recognizing actions associated to each agent level. The entities of this phase are described in the meta-model of Fig. 5.

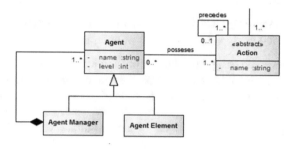

Fig. 5. Definition phase metamodel.

Define agents and their level. In this step, agents are defined and are organized under the specified hierarchical structure of the expected system. We note two kinds of agents: Agent Manager and Elementary Agent. The Agent Manager is the top-level agent who is responsible of taking decisions and maintaining the relationships between agents. Whereas, the Elementary Agent is the lower-level agent who interacts directly with resources. In some cases, an agent manager may supervise other managers and elementary agents.

Referring to the example mentioned above. To initiate the crisis management process, we need different agents: Coordinator agent, Surveillance system, Phone Company, Internal resource manager that supervises First Aid worker and the super Observer. The Coordinator and Internal resource manager represent the "agent manager" that coordinates the sub-agents. These agents denote the "Elementary Agent" since they are directly interacting with the resources to get information (Fig. 6).

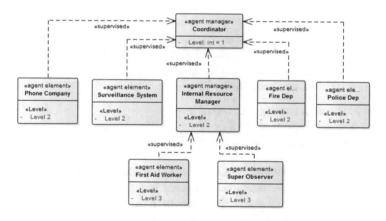

Fig. 6. Agent structure.

Assign actions to agents. In this step, we note the different actions that an agent will need to perform to accomplish its goal. An action is assigned to an agent if only it meets the following rules:

- If an action belongs to a level, then there has to be an agent associated to it
- An agent cannot comply with two actions of different levels

Figure 7 displays the assigning of the actions to agents of level 2 according to the CMS case study. The agent manger "Internal Resource Manager" possess two actions: "Mission analysis" and "Assign Internal Resource". Indeed, it has as role to evaluate the crisis, and then sends the appropriate resource to scene. For the other agents, there are elementary agent that possess each one of them a unique action to be accomplished.

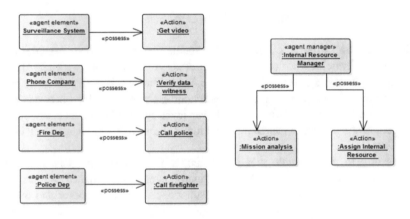

Fig. 7. Assign actions to agents of level 2.

4.3   Modeling Phase

The modeling phase models the high-level description of the internal behavior of the agent's action as well as the external behavior that defines the interaction between agents. There is an iterative process of three steps to model the behavior for each level i (where i >= 1) identified in the definition phase:

1. Model the internal behavior actions of level i
2. Assign actions to agents of level i + 1
3. Model the external behavior actions of level i

Model the internal behavior actions. Modeling the internal behaviour is mainly based on the representation of the decisions and states in terms of transition state diagram, which is based on the profile that we have proposed in [16]. Let us take the example of "Assign Internal Resource" action described in Fig. 8.

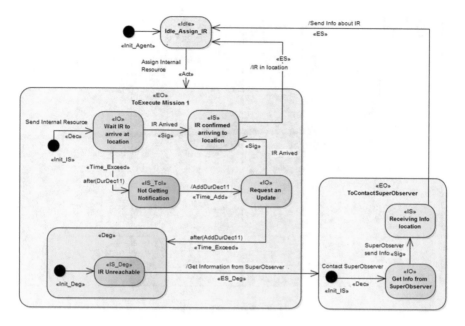

Fig. 8. Internal behavior.

Once the agent receives action "Assign Internal Resource", he makes the decision to "send the resource to the scene crisis". If he does not get any notification form IR (Internal Resource), he adds an extra time waiting for any information, if after that time he does not receive anything, he changes his behavior by contacting the super Observer agent. When all information gathered, he sends a report about IR.

Model the external behavior actions. In the external behavior, we describe the inter-action between an agent manager and its sub-agents across the communication interface. This interaction is represented by the activity diagram, which is designed to describe the organizational behavior. The Fig. 9 describes the interaction between the internal resource manager and its sub-agents: the first Aid worker and the Super Observer agent.

4.4   Verification Phase

In the verification phase, we verify the behavioral properties of the system throughout the model checking technique. The model checking technique starts with making a comparison between two descriptions of a system behavior, one is considered as the requirement and the other is considered as the actual design. Then it verifies the system by exploring the state space to determine if the system has satisfied a series of properties. This model provides a counter-example in case the model does not meet the specification [17].

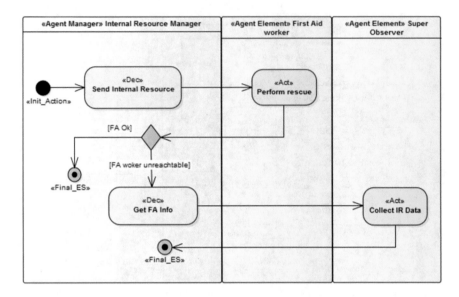

Fig. 9. External behavior.

The BECAMEDA's verification phase proposes to translate the models designed in the modeling phase to the SMV model checker [18, 19], the system's properties is then expressed with temporal logic formulas.

We have used the gNuSMV tool [20] to verify the temporal properties expressed in CTL (Computational Tree Logic) formulas [21]. The Fig. 10 presents an example of a formal verification of the following system properties:

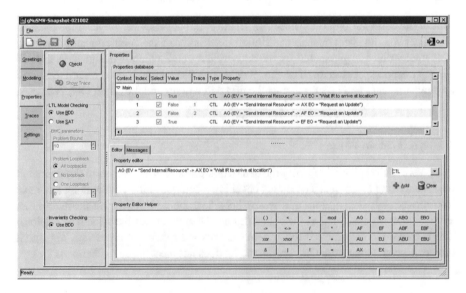

Fig. 10. Properties expressed in CTL.

SPEC AG ((EV = "Send Internal Resource") - > AX EO = "Wait IR to arrive at location")

When the agent takes the decision "Send Internal Resource", he will be in the state to wait for internal resource to arrive at location.

SPEC AG ((EV = "Send Internal Resource") - > AX EO = "Request an Update")

When the agent takes the decision "Send Internal Resource", he will be in the state to request an update

SPEC AG ((EV = "Send Internal Resource") - > AF EO = "Request an Update")

When the agent takes the decision "Send Internal Resource", he will always in the future be in the state to request an update

SPEC AG ((EV = "Send Internal Resource") - > EF EO = "Request an Update")

When the agent takes the decision "Send Internal Resource", he will in some point in the future be in the state to request an update.

5 Conclusion

In this paper, we have introduced a method called BECAMEDA for the specification and the verification of Multi-Agent System. This method is based on different processes that help understanding the system starting from goals identification that captures the system needs, and ending with a formal verification of the system properties. The process specifies four phases: identification, definition, modeling and verification. This approach uses model-checking technique to perform formal verification of the agents' behavior.

In order to illustrate the specification and the verification aspect of the method, we have used a Crisis Management System, the case study was kept simple to make the concepts of the method clear enough. Thus, the method has been effectively applied to a realistic example, which allows understanding more the system and being able to prove its accuracy. For more details about the verification phase, the reader can refer to our previous works [18, 19].

As future work, we are currently investigating to combine the BECAMEDA method with an existing approach that considers in depth the designing phase. This will be an important extension of the BECAMDA method to build up a complete method that deals with the different MAS engineering phases.

References

1. Mattei, S., Bisgambiglia, P.A., Delhom, M., Vittori, E.: Towards discrete event multi agent platform specification. In: Third International Conference on Computational Logics, Algebras, Programming, Tools, and Benchmarking, pp. 14–21 (2012)
2. Chuanjun, R., Hongbing, H., Shiyao, J.: Specification of agent in complex adaptive system. In: Computer Science and Computational Technology, ISCSCT 2008, vol. 2, pp. 210–216 (2008)
3. Sharma, M., Firdaus, M., Chatterjee, R.K., Sarkar, A.: Constraint specification in multi-agent system. In: Region 10 Conference (TENCON), pp. 2404–2409. IEEE (2016)

4. Subburajand, V.H., Urban, J.E.: Issues and challenges in building a framework for reactive agent systems. In: Complex, Intelligent and Software Intensive Systems (CISIS), pp. 600–605. IEEE (2010)

5. Bounabat, B., Romadi, R., Labhalla, S.: Designing multiagent reactive systems: a specification method based on reactive decisional agents. In: Pacific Rim International Workshop on Multi-Agents. LNAI, vol. 1733, pp. 197–210. Springer, Heidelberg (1999)

6. Aaroud, A., Labhalla, S.E., Bounabat, B.: Modelling the handover function of global system for mobile communication. Int. J. Model. Simul. 25(2), 99–105 (2005)

7. Romadi, R., Berbia, H., Bounabat, B.: Wireless sensor network simulation of the energy consumption by a multi-agents system. J. Theor. Appl. Inf. Technol. 25(1), 50–56 (2011)

8. Berrada, M.: Qualitative verification of multi-agents reactive decisional system using business process modeling notation. In: Intelligent Agent Technology, pp. 747–751. IEEE (2006)

9. Kienzle, J., Guelfi, N., Mustafiz, S.: Crisis management systems: a case study for aspect-oriented modeling. In: Transactions on Aspect-Oriented Software Development VII, pp. 1–22. Springer, Heidelberg (2010)

10. Graja, Z., Migeon, F., Maurel, C., Gleizes, M.P., Kacem, A.H.: A stepwise refinement based development of self-organizing multi-agent systems: application to the foraging ants. In: International Workshop on Engineering Multi-Agent Systems, pp. 40–57. Springer, Heidelberg (2014)

11. Pereverzeva, I., Troubitsyna, E., Laibinis, L.: Formal development of critical multi-agent systems: a refinement approach. In: Ninth European Dependable Computing Conference, pp. 156–161 (2012)

12. Chatterjee, R.K., Neha, N., Sarkar, A.: Behavioral modeling of multi agent system: high level petri net based approach. Int. J. Agent Technol. Syst. 7(1), 55–78 (2015)

13. Hadj-Kacem, A., Regayeg, A., Jmaiel, M.: ForMAAD: a formal method for agent-based application design. Int. J. Web Intell. Agent Syst. 5(4), 435–454 (2007)

14. Laouadi, M.A., Mokhati, F., Seridi-Bouchelaghem, H.: A novel organizational model for real time mas: towards a formal specification. In: Intelligent Systems for Science and Information, pp. 171–180. Springer International Publishing, Cham (2014)

15. Haqiq, A., Bounabat, B.: An extended approach for the behavioral and temporal constraints specification of reactive agent. In: 15th International Conference on Intelligent Systems Design and Applications, pp. 329–334. IEEE (2015)

16. Haqiq, A., Bounabat, B.: UML profile for modeling multi decisional reactive agent system. J. Lect. Notes Softw. Eng. 1(3), 224 (2013). ISSN:2301-3559

17. Clarke, E.M.: The birth of model checking. In: 25 Years of Model Checking, pp. 1–26. Springer, Heidelberg (2008)

18. Haqiq, A., Bounabat, B.: Model checking of multi decisional reactive agent system. In: 9th International Conference on Intelligent Systems: Theories and Application, Rabat, Morocco, vol. 1, pp. 133–140 (2014)

19. Haqiq, A., Bounabat, B.: Verification of multi decisional reactive agent using SMV model checker. In: 8th IEEE International Design and Test Symposium, Marrakesh, Morocco, pp. 1–6 (2013)

20. Cimatti, A., Clarke, E., Giunchiglia, E., Giunchiglia, F., Pistore, M., Roveri, M., Sebastiani, R., Tacchella, A.: NuSMV 2: an open source tool for symbolic model checking. In: 14th Conference on Computer Aided Verification, LNCS, vol. 2404. Springer, Heidelberg (2002)

21. Bolotov, A.: A clausal resolution method for CTL branching-time temporal logic. J. Exp. Theor. Artif. Intell. 11(1), 77–93 (1999)

Ontology of the CBA: Towards Operationalization and Implementation

Malika Sedra[(✉)] ⓘ and Samir Bennani ⓘ

Université Mohammed-V Ecole Mohammadia des Ingénieurs, Agdal, Rabat, Morocco
malikasedra@gmail.com

Abstract. To remedy inefficiency and increase the quality of learning, training systems have adopted a pedagogical approach based on competence bases. It is a concept that is common to the academic and professional communities; which imposes the use of ontology of the CBA to conceptualize a dynamic repository, determining the terminological semantics of skills, which is sectorial with pedagogical norms. We propose a CBA ontology that focuses on explicit modeling of competency in order to provide a shared formal representation that promotes exchange, interoperability and collaboration among the various users in the learning systems. Our goal is not just to build an appropriate ontology. Our ambition is to implement this solution in an authoring system to assist tutors with didactic intentions based on skills to produce CBA based educational content.

Keywords: Ontology · Competencies-based approach · Learning systems
Ontological engineering · System author

1 Introduction

An appropriate production approach of pedagogical content in e-Learning systems must offer an effective operating support to minimize the efforts deployed by the tutors in the technical part to allow them to concentrate more on didactic and pedagogical aspects than on technology. The objective of our research is to propose, in the near future, a pedagogical production tool based to the Competency-Based Approach (CBA), capable of appropriately reducing the efforts of tutors with didactic objectives based on skills. This approach is based on the concept of competencies that is currently of great importance to the academic and professional communities. First, we tried to have a clearer vision and a deeper understanding of the CBA. This has enabled us to discover approaching learning through skills requires a change in perspectives and a reorganization of learning. It involves adopting a progression approach w is to hic improve the expected learning outcomes and to establish the level of knowledge and attitudes that learners must achieve at the end of their training.

In this paper, we focus on the ontology concept and highlight the importance of the use of CBA ontology in e-Learning systems. This ontology can be characterized by the structuring of the concepts related to the CBA which will be gathered to enable us to express the knowledge available in this field.

© Springer International Publishing AG, part of Springer Nature 2018
M. Ben Ahmed and A. A. Boudhir (Eds.): SCAMS 2017, LNNS 37, pp. 87–94, 2018.
https://doi.org/10.1007/978-3-319-74500-8_8

Finally, we propose our approach for the implementation of an ontology of the CBA, as well as the difficulties encountered during ontological modeling and semantic normalization.

2 CBA: Evolution of Pedagogical Practices

2.1 What Is Competency?

The notion of competency has been the subject of several interpretations because of its mobilization in various disciplinary fields. In the field of the sciences of education, this notion is used under the following definition: "a competency refers to a set of elements that the subject can mobilize to deal with a situation successfully" [1]. There are also other definitions that characterize a competency as a skill that combines cognitive, social, emotional or psychomotor skills in a set of situations [2]. Indeed, it represents a set of resources (knowledge, know-how and attitudes) that the learner is able to combine to accomplish a specific task. In general, a competency in various educational activities is oriented towards action and know-how, which both emphasize an aspect of competency which is knowledge how to act. This knowledge involves capacities and power to act which are resources similar to knowledge and attitudes that the subject mobilizes while aiming for success, in complex situations. Through the analysis of the literature, we have noticed that all the definitions of the concept competencies converge to the capacity to achieve an efficient combination of knowledge, know-how, know-how to be and knowledge to act in a precise context. We have also noted three characteristics of this concept:

1. competency is the capacity to act «competency to act» or to be able to act which assumes the aptitude to combine and to mobilize pertinent resources;
2. Competency is contextual and is closely related to a task, context, situation or a series of situations;
3. Competency is a potential acquired of "knowledge, know-how, and know how to be" (see Fig. 1) and it is only through training and/or practice that a person can increase this potential and develop or even acquire other skills.

In our research, we have focused on a definition of competency that emphasizes "know how to act". Although there are several definitions that correspond to our concerns, we have chosen the definition of Romainville [4], which we recall below, as a reference of our reflections. "Competency is a complex knowledge to act that builds on the effective mobilization and combination of a variety of internal and external resources within a family of situations".

In short, competence implies an ability to act effectively in the face of a family of situations, which the pupil alone can master by having the necessary knowledge, abilities, resources and skills. Knowledge refers to a domain (knowledge domain). The abilities, according to paquette [6], describe processes or generic "meta-processes" independent of the domain of application which make it possible to memorize, assimilate, perceive, analyze, synthesize and evaluate the knowledge of a specific domain. The resources include: knowledge (theoretical, environmental…), know-how (cognitive,

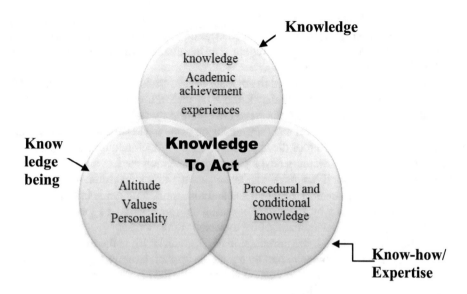

Fig. 1. The notion of "competencies"

operational, social…) and resource networks (emotional, physiological, documentary, human, software…) [7].

2.2 Competency-Based Approach

The competency based approach is a pedagogical orientation based on the principle of dealing with learning through skills. The aim is to introduce a didactic methodology that allows learners to create, develop and acquire competencies. Unlike traditional approaches, with the skill-based approach, the concept of success in a training process becomes closely tied to the mastery of knowledge and skills that the learner must prove to pass. We are talking about the pedagogy of success.

Currently, the CBA is the world's most widely advocated approach; [3] It cares about the students by integrating them into the heart of the pedagogical concern and focusing on their development and their needs imposed by social and economic considerations. The aim of the CBA is therefore to modernize the aims of education, in order to better adapt them to personal, social and professional needs. [14] Following the perspective of the CBA, several changes are observed: on the one hand, curricula are written in terms of expected competencies. In order to constitute courses, teachers have to change their didactics. They are called upon to deeply analyze each competency, to decompose, if necessary, a competency or skills merged combined into course sequences, in order to establish a progressing learning order. On the other hand, the way to consider and prepare the evaluation is also affected. In a traditional approach, evaluation often involves the taught knowledge. If this knowledge is mastered then competence is also mastered. In the CBA, however, there are two modes of formative and summative evaluation. The teacher adopts a strategy of frequent formative evaluation appropriate to the nature of each targeted skill (a project to be realized, a problem to solve, a creation, an artwork…). At the end of the learning process, the tutor carries out the summative assessment, which focuses on the final competency and requires the mobilization of all types of knowledge and skills acquired beforehand.

The introduction of the CBA in the domain of face-to-face or distance training has given rise to several constraints that hamper its realization. Indeed, the majority of education stakeholders who did not have the opportunity to have an overview of the approach in its application feel that they are losing content and are confused by the lack of a thorough understanding of the approach and lack of methodological support tools for both teachers and curriculum developers. Not to mention the fact that tutors express themselves in the natural language, can be subject to various interpretations that have their turns generating several terminological and conceptual ambiguities and accentuates the concerns expressed by the teachers during the preparation of the programs on expected skills and the assessment of the expected competencies.

To this end, we propose the introduction of a formalism [5] in the form of ontology, which will make it possible to explicitly use all the implicit terminology of the "competencies" concept in the field of education and facilitate the implementation of the CBA, specifically, via an e-Learning platform.

3   Ontology of the CBA

Ontology is a formal structuring of concepts that express the knowledge of a domain. It also makes it possible to create classes (general concept), instances (specific concept), relationships between these concepts, properties that describe these concepts, functions, constraints and rules, which make knowledge reusable and shareable. The use of ontology makes it possible to define each concept in a unique and explicit way and to extend its use, by using correct semantic relations (synonymy, antonym…). In our approach, we opted for a domain ontology "CBA ontology" as an infrastructure to eliminate ambiguity and support the extensive requirements of the "skills" concept since their description provides a rich set of modeling elements capable of expressing details of competencies.

3.1   Ontological Engineering

In the literature there are many methodologies of ontology construction but there is no consensus around a practical and effective procedure to adopt [9, 10]. Regardless of the methodology chosen, the development must go through an engineering phase that determines its life cycle. Since ontology is considered as a software component, its development process must be based on these main steps [11]:

- Specification of needs and definition of a domain and its scope;
- Conceptualization by identifying concepts, attributes, and values as well as Relations and Attribute Instances;

- Formalization, which consists of translating into a formal formalism the conceptual model resulting from the preceding stage. As a result, definitions of concepts become explicit and more precise;
- Implementation of ontology on a language. The chosen language must take into account the formalization model.

The tools for constructing ontology, also called modeling tools, are evolving rapidly following the progress made in this field in order to systematize the ontological engineering. During the last years several ontology languages have been developed. In the following we give a succinct presentation of some of them:

- Ontolingua; It is a server that offers a set of tools and services that support the cooperative construction of ontology, between geographically separated groups.
- KAON (Karlsuhe Ontology and Semantic Web), formerly known as ONTOEDIT, is an open source environment for the design, development and management of ontology.
- Protege 2000 of the Department of Medical Informatics of Stanford University; the successor of ProtegeWin, is a tool that allows: (1) to build an ontology of the domain, (2) to customize forms of knowledge acquisition and (3) to transfer knowledge of the domain.
- WebOnto of the Open University Knowledge Media Institute; WebOnto supports collaborative browsing, creating and editing ontologies on the Web.

3.2   The Construction of CBA Ontology

In our case, the ontology that we built to support the CBA in the education system was to make this approach a shared object between the different actors of learning environments to facilitate their action, allowing software agents to manipulate semantic information. To do this, we have, in the first step, identified the conceptual classes consisting of concepts and relations from a corpus [5], then modeled the primitives and determined the properties of the classes and the semantic approximations between the conceptual primitives of the domain. In the second step, we have structured and formalized these classes under a conceptual description describing the internal structure of the constituted concepts. The corpus we have analyzed [7, 8] is a set of documents expressed in an informal language, which covers the whole domain of knowledge on the CBA, in order to remove possible semantic ambiguities. The domain-specific knowledge has been identified in terms of conceptual primitives and axioms. Our ontology was created on this basis with a simple structure using semantic interrelations to define the basic semantic structure of ontology (see Fig. 2). A competency is composed of other competencies; a competency also implies other competencies.

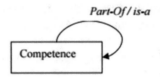

Fig. 2. The concept of competence

Our approach is based on an ontological modeling of a skills repository. This repository is constructed of several documents that offer an overall description of the expected skills of a student at the end of training. One can distinguish Preparatory competencies: Required competencies that the tutor deems mandatory to initiate a specific educational activity (Courses, Practical Works, Evaluation…).

3.3   Conceptualization and Formalism

The conceptualization phase consists of a set of steps leading to a set of semi-formal intermediate representations. It identifies and defines the vocabulary of the domain, starting from the information sources and knowledge of the domain, independently of the languages of formalizations to be used to represent the ontology. The Fig. 3 illustrates the outcome of the conceptualization and representation of the relationships between the different concepts. Competencies are decomposed into sub-skills. A competency is the combination of abilities and knowledge. Educational resources require competence to be used and to develop new skills. In this model, we can distinguish the competencies required (that the tutor deems mandatory to initiate a determined pedagogical resources) and competency acquired targeted by a resource.

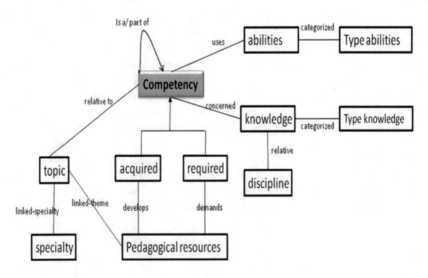

Fig. 3. Relationships and specializations between concepts.

We represent the binary relations between classes by a diagram in which the classes are represented by rectangles and the relations by arrows with the name of the relation to represent the hierarchy of the concepts. The conceptual ontology obtained in the preceding step must be formalized in logic of description, in order to facilitate its subsequent representation in a completely formal and operational language. The result of this step is a formal ontology consisting of a terminological part describing the concepts and relations and of a final part describing the instances. Our ontology is composed essentially of concepts (competency, acquired and required competencies), to which we can integrate other sub ontologies; in our case we have integrated the sub ontology "pedagogical resources" to identify the resources that reach a competency. At this level of research we were content to integrate an ontology of pedagogical resources. So in future work, we intend to integrate other necessary ontologies for the realization of our solution.

4   Conclusion

Systems to share and design educational content are little used and do not correspond to the expectations of most teachers. There are many barriers to sharing and reusing. In this article, we propose a CBA ontology that will allow tutors and educational designers to share a common repository to produce the content of their documents according to the competency-based approach and in an unambiguous, semantic and formal way. This repository will undoubtedly facilitate the task for the users and allow enrichment with new knowledge, inspired by the concepts and the relations of this ontology. In the course of our work, we will be able to focus on the completion of the implementation phase by choosing an adequate language of transcription. Thus, we will start another section which will engender more questions and request more reflections.

References

1. Minet, F.: Compétence: de la définition à l'utilisation. In: Jouvenot, C., Parlier, M. (eds.) Elaborer des référentiels de compétences, Editions du réseau ANACT, pp. 332–362 (2005)
2. Jonnaert, P.: Compétences et socioconstructivisme. Un cadre théorique. De Boeck, Bruxelles (2002)
3. Zarifian, P.: Le modèle de la compétence, trajectoire historique, enjeux actuels et propositions, Éditions Liaisons (2004)
4. Romainville, M.: What if we stopped drawing on skills? In: DIRECT, no 10, pp. 31–44 (2008)
5. Tardif, J.: L'évaluation des compétences - Documenter le parcours de développement. Chenelière Education, Montréal (2006)
6. Gilbert, P.: Modélisation des connaissances et des compétences. Presses de l'Université du Québec, Sainte-Foy, Québec (2002)
7. Georges, F., Poumay, M., Tardif, J.: Comment appréhender la complexité inhérente aux compétences? Recherche exploratoire sur les preuves de la compétence et sur les critères de son évaluation. In: Conférence présentée au 28ème congrès de l'AIPU, Mons (2014)
8. Assadi, H.: Construction d'ontologies à partir de textes techniques, these doctorat, Université Paris 6 (1998)

9. Aussenac–Gilles, N., Biebow, B., Szulman, S.: Revisiting ontology design: a method based on corpus analysis. In: porc. Ekaw 2000, juan–les-pins, pp. 172–188 (2000)
10. De Nicola, A., Missikoff, M., Navigli, R.: A software engineering approach to ontology building. Inf. Syst. **34**, 258–275 (2009)
11. Fernandez-Lopez, M., Gomez, A., Juristo, N.: METHONTOLOGY: from ontological art towards ontological engineering, pp. 33–40 (1997)
12. Berio, G., Harzallah, M.: De l'Ingénierie des Connaissances àla Gestion des Compétences Dans les actes de la conférenceIC 2005. 16éme journées francophones d'Ingénierie des connaissances (2005)

IT Collaboration Based on Actor Network Theory: Actors Identification Through Data Quality

Mohammed Salim Benqatla[(⊠)] [iD], Meryam Belhiah[(⊠)],
Dikra Chikhaoui[(⊠)], and Bounabat Bouchaib[(⊠)]

Alquasadi Team, Ensias Admir Laboratory, Rabat IT Center,
Mohammed V University, Rabat, Morocco
{salim.benqatla,meryam.belhiah,dikra.chikhaoui,
b.bounabat}@um5s.net.ma

Abstract. IT collaboration involves exchanging information and data within a network with several actors in order to achieve business objectives. Such cooperation is generally ensured by building a collaborative network. This work presents an approach of actors identification through data quality in Actor-Network mode of collaboration. Indeed, data quality is one of the important characteristics which expose the actor importance in the network. We investigate the translation process of ANT (Actor Network Theory), while focusing on the problematization phase in which actor-networks are identified according to the data quality level provided, and then translating this level into cost and analyzing all possible coalitions using cooperative game. The findings will allow identifying which coalitions enhance data quality. The build of such actor-network depends therefore on both data quality and the operating cost of these data between systems.

Keywords: Actor Network Theory · Data quality · Business collaboration network
Cooperative game theory · Shapley value

1 Introduction

Business Collaboration refers to the process where several organizations work together in an intersection of common goals. This organization manages to reach a set of strategic objectives through collaboration with partners and through the pooling of resources and the exchange of information and services with them. Objectives vary according to several criteria. However, restructuring resources, improving quality and efficiency of operations are among the main operational objectives. To define adequately the objectives and the context of the collaboration, partners must specify the needs and the goals to be reached as well as all the aspects likely to influence the choices and the mode of operation.

Data quality is among the major objectives of the collaboration, It is one of the important characteristics which exposes the actor importance in the network. It's a clue on how the actor will collaborate efficiently. Even if the dimensions of data quality are

© Springer International Publishing AG, part of Springer Nature 2018
M. Ben Ahmed and A. A. Boudhir (Eds.): SCAMS 2017, LNNS 37, pp. 95–106, 2018.
https://doi.org/10.1007/978-3-319-74500-8_9

not universally agreed, we assume in this article that data quality level is determined by actors according to business objectives.

In this paper, we introduce the different concepts and theories used in this work, mainly Actor-Network Theory (ANT) [4, 5] as the framework of collaboration, Data quality as a characteristic for the identification of actors, Cooperative Game and Shapley Value as mechanism of cost and coalition analyzing. Then we describe the proposed approach of building an Actor-Network for data quality objective. We illustrate this work by an experimental setup applied in a realistic Actor-network context. Finally, we conclude this paper by summarizing aim points and the perspective works.

2   Concepts

2.1   Actor Network Theory

In general, ANT conceptualizes social interactions in networks. Networks integrate both the material environment and the semiotic environment [24]. In this theory, there is no difference between the human and non-human parts of a technological system. ANT mentions that the world is full of hybrid entities [25]. The core of ANT analysis is to examine the process of translation where actors align their interests of others with their own. The actors translate their interests by constructing a network and breaking the resistance of other actors and their network [24]. These actors can be an authority that either influence and use others or have no motivation and will be under the control of other actors.

In order to reach a step of the construction of a network, Callon and Latour defined an approach, inspired by ethnomethodology [6], which bears on a sequence of steps called the translation sequence. To translate is to "express in his own language what others say and want, to set up as spokesman" [4], but translate it is also, negotiate, perform a series of movements of all kinds and thus to each sequence of the process, which can be defined in four main steps:

1. Problematization
   "The problematization or how to become essential?", "The problematization, as its name indicates asking at first a problem. This is to raise awareness to a number of actors that are concerned with this problem, and that everyone can find satisfaction through a solution that translators are able to offer" [8], so problematization is the effort made by the actors to convince that they have the right solution [3]. It "describes a system of alliances or associations between entities, defining this, their identity and what they want" [18].
2. Interessement
   "The incentive devices or how to seal alliances", the incentive is in fact for Callon "all actions through which an entity is trying to impose and stabilize the identity of the other players who is defined in problematization" [8] incentive is the second phase, consists of "deployment speeches, objects and devices intended to attract and attach different players to the Network" [9].

It is building the interface between the interests of different actors and the strengthening of the relationship between these interests. In the area of strategy, it can be a system of alliances to ensure that the different members of the organization are involved in the strategic process.

The main thing is to translate the interests of other actors in order to get them to take part in the network. To translate the interests of others, we can either convince them that there are common interests and that the proposed solution also serves their interests or manipulate their interests and objectives or finally become unavoidable.

3. Enrollment

   "How to define and coordinate the roles", Enrollment is "the set of multilateral negotiations, beatings forces or tricks that come with sharing and allow it to succeed" [8].

   For enrollment, each actor in the network is assigned a role. This role is related to the translation of their interests. For Callon, "the enrollment is to describe the set of multilateral negotiations, coups or intelligence accompanying sharing and allow it to succeed" [9]. The enrollment can thus be regarded as stabilizing the system of alliances set during the phase of the incentive. This system is the result of multilateral negotiations, trials of strength and stratagems [9]. It is during this phase to confront showdowns integrating new actors to the networks or by strengthening links between network members.

   The enrollment phase is the key to the success or failure of innovation [15], but this phase is not studied formally in the literature on control.

4. Mobilization

   The last phase of translation, the mobilization is to gather its allies. It is the cockpit of the various interests in a way that they remain more or less stable [10], it raises the question of the representation of stakeholders and enrolled in the project which is then established as spokespersons of the groups they represent [11]. However, "everyone can act very differently to the solution proposed: the abandon, accept it as it is, change the modalities which accompany or statement that it contains, or even they will be appropriated in the transferring in a completely different context" [9].

In a particular way, incentive phase of ANT can be analyzed from a cooperative game with transferable utility point of view [9]. Our objective is to identify the actor-network through data quality. For that, we use the Shapley Value to identify a better translation of the operating cost for improving data quality in an actor-network context (Fig. 1).

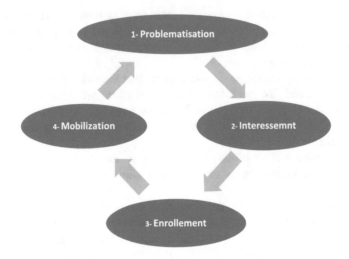

Fig. 1. Phases composing Actor Network Theory

2.2   Business Collaborative Mode

The collaboration represents one of the main relations between organizations partners. The most part of the definition of this concept agrees to mean joint job between several entities hoping for mutual benefits [12]. The same concept is used as well for the individuals as for organizations working together at common objectives. The collaboration connotes a relation lasting and made general for a complete commitment in favor of a common mission. Also, the collaboration consists in the commitment, mutual and coordinated by the partners to reach common.

The notion of network of collaboration is represented a group collaborating of entities, autonomous and heterogeneous, being recovered from different domains of governance and working jointly with a view to attaining a series of common objectives or even supplementary [13]. The entities of network collaboration are broadly autonomous and heterogeneous in terms of their environments of working, of their organizational cultures, and of their issued capitals. Correlations are supported by group of means of support implicating the systems of information of stakeholders. The collaboration job takes a multitude of forms; it is used in the field of private service as part of a network of public administrations. This situation takes the existence of a contract of collaboration representing in a definite manner the responsibilities of every member of the Network collaboration.

2.3   Data Quality

Data quality may be defined as "the degree to which information consistently meets the requirements and expectations of all knowledge workers who require it to perform their processes" [14], which can be summarized by the expression "fitness for use" [1]. The term data quality dimension is widely used to describe the measurement of the quality of data. Even if the key DQ dimensions are not universally agreed amongst the academic community, we can refer to Batini et al. [15] who have identified 15 dimensions:

Intrinsic: accuracy, believability, reputation and objectivity;
Contextual: value-added, relevance, completeness, timeliness and appropriate amount;

Representational and accessibility: understandability, interoperability, concise representation, accessibility, ease of operations and security.

All case studies that aimed at assessing and improving data quality have chosen a subset of data quality dimensions, depending on the objectives of the study [16–19]. Measurable metrics were then defined to score each dimension. While it is difficult to agree on the dimensions that will determine the data quality, it is however possible, when taking users' perspective into account [20], to define a basic subset of key dimensions, including: accuracy, completeness and timeliness.

Accuracy is defined as "the closeness of the results of observations of the true values or values accepted as being true" [15]. Benqatla et al. [1] Define accuracy as "the extent to which data are correct, reliable and certified". The associated metric is as follows:

$$\frac{number\ of\ accurate\ values}{Total\ number\ of\ all\ values} \tag{1}$$

Completeness. Completeness specifies how "data is not missing and is sufficient to the task at hand" [19]. As completeness has often to deal with the meaning of null values, it may be expressed in terms of the "ratio between the number of non-null values in a source and the size of the universal relation" [18]. Completeness is usually associated with the metric below [20]:

$$\frac{number\ of\ non-null\ values}{Total\ number\ of\ all\ values} \tag{2}$$

Depending on the context, both accuracy and completeness may be calculated for: a relation attribute, a database or a data warehouse [20].

Timeliness. Timeliness is a time-related dimension. It expresses "how current data are for the task at hand" [19].

As a matter of fact, even if a data is accurate and complete, it may be useless if not up-to-date.

$$\frac{number\ of\ values\ that\ are\ up-to-date}{Total\ number\ of\ all\ values} \tag{3}$$

2.4 Cooperative Game

Game theory is a mathematical tool which is used to analysis the strategic interaction between multiple decision makers [21]. Initially it was used in economics for understanding the concept of economic behavior. But now it is used in various fields such as communication, biology, psychology for modeling the decision making situation where the outcomes depend upon the interacting strategies of two or more agents [3].

The cooperative game theory can be applied to the case where actors can achieve more benefit by cooperating than staying alone, it consists of two elements: (i) a set of players, and (ii) a characteristic function specifying the value created by different subsets of the players in the game [22]. The coalition formation problem is one of the important issues of game theory, both in cooperative and non-cooperative games. There are several attempts to analyze this problem. Many papers tried to find stable coalition structures in a cooperative game theoretic fashion. If we suppose that forming the grand coalition generates the largest total surplus, it is natural to assume that the grand coalition structure will eventually form after some negotiations [23]. Then, the worth of the grand coalition has to be allocated to the individual players, according to the contribution of each player [22]. We are interested in this work in cost-sharing between coalition members likely to form using Game theory as a device for ANT interessement phase.

The Shapley value is a very common cost-sharing procedure in cooperative game theory essentially based on the so-called incremental costs [22]. The Shapley value of player i in the game given by the characteristic function V is the share of the surplus should be assigned. It's a weighted average of the contributions of a player i to reach of the possible coalition.

For example, consider a game with three players, i1, i2 and i3. Assume that player i1 is the first player of the game, i2 is the second player to join the game and player i3 is the last one. Player i1 is allocated a cost C ({i1}), player i2 is allocated a cost C ({i1, i2}) – C ({i1}), and player i3 a cost C ({i1, i2, i3}) – C ({i1, i2}). The Shapley value assumes that the order of arrival is random and the probability that a player joins first, second, third, etc. a coalition is the same for all players. Assume that force of each coalition is known in the form of the characteristic function V. The cost allocated to a player i in a game, including a set N of players is given by:

$$\phi i(N) = \left(\sum_{S \subseteq N : i \in S} \left(\frac{(|S| - 1)!(|N| - |S|)!}{|N|!} \right) \left([C(S) - C(S \setminus \{i\})] \right) \right) \tag{4}$$

|N| and |S| respectively, the total number of players and the one belonging to the coalition S.

An alternative equivalent formula for the Shapley value is:

$$\phi i(N) = \left(\frac{1}{|N|!} \sum_{R} (v(\text{PR}i \cup \{i\}) - v(\text{PR}i)) \right) \tag{5}$$

Where the sum ranges over all |N| orders R of the players and PRi is the set of players in N which precede i in the order R.

Choosing a method of cost allocation is not an easy thing. According to the literature Shapley value seems to be suitable to this context of actor-Network building game. In fact, Shapley imposes four axioms to be satisfied (Efficiency, Symmetry, Dummy and Additivity).

1. Efficiency: players precisely distribute among themselves the resources available to the grand coalition. Namely, Efficiency: $\sum i \in N \; \varphi_i(v) = v(N)$.
2. Symmetry: Players i, j $\in$ N are said to be symmetric with respect to game v if they make the same marginal contribution to any coalition, i.e., for each S $\subset$ N with i, j $\notin$ S, v(S $\cup$ i) = v(S $\cup$ j). In another way, if players i and j are symmetric with respect to game v, then $\varphi_i(v) = \varphi_j(v)$.
3. Dummy: If i is a dummy player, i.e., v (S $\cup$ i)- v (S) = 0 for every S $\subset$ N, then $\varphi_i(v) = 0$.
4. Additivity: $\varphi(v + w) = \varphi(v) + \varphi(w)$, where the game v + w is defined by (v + w) (S) = v (S) + w (S) for all S.

The dummy, symmetry (meaning that two players have the same strength Strategy will receive the same gain) and efficiency make the Shapley value, particularly attractive for treating the problem of equitable sharing of resources common to several economic agents.

3   Related Works

In this section, we will deal with a representative set of existing studies that work on collaboration network and theories used to build the network of collaboration in the context of IT Governance.

Collaborative Networks (CNs) are complex systems that can be described or modeled from multiple perspectives. The collaborative network was presented and studied by IT researchers as a virtual network. Camarinha Matos is one of the widely recognized researchers working in the field whose works are basically concerned about IT perspectives and requirements of Collaborative Networks, despite the fact that they bear solid similitudes to alternate approaches of Collaborative Network too. During the later years and as a typical result of the difficulties confronted by both scientific and business terms, it has watched an abundance in the sorts of rising Collaborative Networks [25]. Some research tries to recognize the particular impacts of the properties of network structure on the execution of firms (particularly, the quantity of licenses) [26]. Thought, the change of between firm connections and following impacts after some time were not considered. [27] Many researchers have examined this field by the approach of Camarinha Matos and expounded the IT tools and its necessities to move forward. What's more, different researchers have lead studies on a similar field of learning, however this time from various viewpoints, for example, organization together or arranges in development organizations.

For example, Ahuja assesses the effects of firm's network of relation on innovation and elaborates a theoretical framework that relates the aspects of firm's ego network-direct ties, indirect ties and structural holes (disconnections between a firm's partners). Chinowsky [7], studied the Construction Company's networks by the approach of Social Network Analysis in company level and project level. Heedae Park studied collaboration effects on the profit amount of projects in Korean international contractors.

In this context, there is no single modeling formalism or "universal language" that can cover all perspectives of interest. Since CNs have a clear multidisciplinary nature, it is natural that we search for applicable modeling tools and approaches originated in other disciplines. In fact, Computer Science, Engineering, and Management, among other fields have developed plenty of modeling tools that might have some applicability in CNs [24].

There are also many developments in other disciplines that can contribute to the start of a foundation for collaborative networks, e.g. in complexity theories, game theory, multi-agent systems, graph theory, formal engineering methods, federated systems, self-organizing systems, swarm intelligence, and social networks. The theoretical foundation work in the ECOLEAD project took the mentioned early works as a baseline.

Game theory can provide the concepts for the analysis of decision-making in cases involving multiple decision-makers who interact with each other. In the case of CNs, game theory could offer: tools to manage cost, risk and profit sharing among the network participants, and tools to design optimal incentives for the VBE, VO, etc.

4   Data Quality in Network Collaboration

4.1   Actors Identification Through Data Quality

In our approach data quality is determined by taking into account users' perspective and objectives. An actor should decide on a subset of dimensions as mentioned above, the data quality level is then deducted from accuracy, completeness, and timeliness scores.

Data exchange 'Relationship' (respecting ANT terminology) represents actor's interactions, which allows to spot future alliances and coalitions respectively conflicts and dissension.

We use the problematization step of the translation process to identify and characterize actors, as well as to analyze the scenario in which the network will operate. First, we identify actors with a height score of data quality, and then measuring data quality as proposed into [23] values and measuring the costs of operating this data in all possible coalitions. The choice of actor-network depends therefore on both data quality and the cost of using these data in each system, assessing the collaborative value of an actor can naturally be seen in terms of the cooperative game theory with a transferable utility, by means of Shapley value [2]. In fact the allocation of budgets depends on the contribution of each actor, in term of data quality translated to costs in our context.

4.2   Illustration

In order to experiment our approach of actors identification based on ANT, we work on the inter-organizational data exchange project for the administration of customs of Morocco (see Fig. 2).

- Information System of Administration of Customs (S1)
- Information System of Treasury Department (S2)
- Integrated Tax systems Governed by the Administration of Tax (S3)

- Information System of Public Enterprises and Privatization Department (S4)

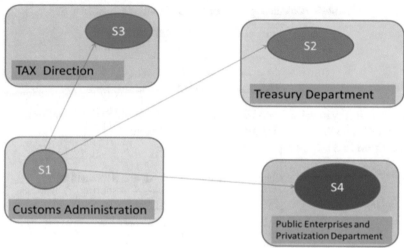

Fig. 2. Extract of a Public Financial Authority process interaction

In our case, each system is managed by administration, S1, S2, S3, S4 and each system has its level of data quality.

Assuming that the result of the data quality analysis is presented as follows (Table 1).

Table 1. Data quality of actors network

Administrations	Actors	Data quality level
Customs Administration	S1	3
Treasury Department	S2	3
Administration of Tax	S3	5
Department Public Enterprises and Privatization	S4	2

A budget is allocated for this project between this four public administration.

We supposed that administration S1 must collaborate with the S2 in view of the functional dependency between the two institutions and that S1 has the choice between cooperating with S3 or S4 for the implementation of the exchange project. We will measure costs in each coalition:

Coalition 1: S1, S2 (Table 2)

Table 2. Calculating Shapley Value of coalition S1 and S2

Administrations	Shapley value
S1	2.0
S2	5.0

Coalition 2: S1, S2, S4 (Table 3)

Table 3. Calculating Shapley Value of coalition S1, S2 and S4

Administrations	Shapley value
S1	4.0
S2	5.0
S4	8.0

The possibility of getting good data quality from S4, motivates the coalition building [S1, S2, S4]. The Shapley value gives the following results:

Coalition 3: S1, S2, S3 (Table 4)

Table 4. Calculating Shapley Value of coalition S1,S2 and S3

Administrations	Shapley value
S1	2.0
S2	5.0
S3	0.5

S3 has a similar situation as S4, and has the ability to produce high data quality. The best strategy of S1 is collaborating with S3 as it is the best source in terms of data quality, therefore a reduced cost, we get:

We have implemented a Web application developed in JAVA/J2EE, ANT MANGER is composed of several modules developed in JAVA/J2EE technology, with an architecture separating the presentation part (Front-end) from the Business part (Business, Backend), the aim of this architecture is to allow ANT Manager to interface with other platforms carried out by the research teams.

The goal of this platform is to provide a decision-making tool for information systems managers, architects, in order to build the most appropriate collaborative network for the administrative agency context.

ANT MANAGER is currently composed of the following modules:

- Management of user authorizations.
- Workflow theory of network actors.
- Managing collaboration networks (Adding an actor, Linking a player to another).
- Simulation management, according to the algorithm of the game theory (Perform several simulations, simulation backup, simulation suppression export the simulation result in CSV format in order to analyze the various simulations carried out by the architect or the person in charge of the information systems.

5   Conclusion

Displayed equations should be numbered consecutively in the paper, with the number set flush right and enclosed in parentheses.

Globalization leads to increased competition and higher customer expectations. At the same time, companies are stressed to reduce production costs while fronting the challenges of increasing product complexity, environmental concerns. The collaboration of organizations with networks is not a new phenomenon. However, permanent progress of IT in terms of new, reliable, and cheaper.

Information and communication technologies are a catalyst for collaboration in networks. Such collaboration is ensured by building a collaborative network. Our approach highlights the identification of actors through data quality in Actor-Network mode of collaboration, this operation by improving the level of data quality translates to cost that is analyzed to all possible coalitions using cooperative game Shapley value.

The proposed work is supported by a software tool which enables to design networks and calculate actors Shapley Value. The main contributions of this work can therefore be summarized as follows:

- As Data quality criteria to identify the selected actors to build the collaboration network.
- Translating data quality objective into cost to analyze coalition.
- Implementation of a web application in order to design and simulate the actor-network evolution based on the cost calculation approach.

After building network collaboration, we are particularly interested in applying social network analysis to analyze this established network.

References

1. Benqatla, M., Chikhaoui, D., Bounabat, B.: Actor network theory as a collaborative mode: the contribution of game theory in the interessement phase (2016)
2. Aisopos, F., Tserpes, K., Kardara, M., Panousopoulos, G., Phillips, S., Salamouras, S.: Information exchange in business collaboration using grid technologies. Identity Inf. Soc. 2(2), 189–204 (2009)
3. Callon, M.: Some elements of a sociology of translation: domestication of the scallops and fishermen of St. Brieuc Bay. In: Law, J. (ed.) Power Action and Belief: A New Sociology of Knowledge? pp. 196–233. Routledge, London (1986)
4. Dameri, R.P.: From IT compliance cost to IT governance benefits: an Italian business case. In: Management of the Interconnected World (2010)
5. Shapley, L.S.: A value for n-person games. In: Kuhn, H., Tucker, A.W. (eds.) Contributions to the Theory of Games, II. Princeton University Press, Princeton (1953)
6. Goulet, F., Vinck, D.: Innovation through withdrawal contribution to a sociology of detachment. Revue française de sociologie (English) 53, 117–146 (2012)
7. Chinowsky, P.S., Diekmann, J., O'Brien, J.: Project organizations as social networks. J. Constr. Eng. Manag. 136, 452–458 (2010)
8. Callon, M.: Sociologie de la traduction. In: Akrich, M., Callon, M., Latour, B. (eds.) Sociologie de la traduction- textes fondateur. Mines Paris (2006)
9. Mouritsen, J., Larsen, H.T., Bukh, P.N.: Intellectual Capital and the "capable firm": narrating, visualizing and numbering for managing knowledge. Account. Organizations Soc. 26(7–8), 735–762 (2001)
10. Parung, J., Bititci, U.S.: A conceptual metric for managing collaborative networks. J. Model. Manag. 1(2), 116–136 (2006)

11. Camarinha-Matos, L.M., Afsarmanesh, H.: J. Intell. Manuf. **16**, 439 (2005). https://doi.org/10.1007/s10845-005-1656-3
12. International Association for Information and Data Quality (2015). http://iaidq.org/main/glossary.shtml
13. Pipino, L.L., Lee, Y.W., Wang, R.Y.: Data quality assessment. Commun. ACM **45**, 211–218 (2002)
14. Aladwani, A.M., Palvia, P.C.: Developing and validating an instrument for measuring user-perceived web quality. Inf. Manag. **39**(6), 467–476 (2002)
15. Batini, C., Comerio, M., Viscusi, G.: Managing quality of large set of conceptual schemas in public administration: methods and experiences. In: Model and Data Engineering, pp. 31–42. Springer, Heidelberg (2012)
16. Scannapieco, M., Catarci, T.: Data quality under a computer science perspective. Archivi Comput. **2**, 1–15 (2002)
17. Närman, P., Johnson, P., Ekstedt, M., Chenine, M., König, J.: Enterprise architecture analysis for data accuracy assessments. In: Enterprise Distributed Object Computing Conference (2009)
18. Agndal, H., Nilsson, U.: Interorganizational cost management in the exchange process. Manag. Account. Res. **20**, 85–101 (2009)
19. Grabisch, M., Funaki, Y.: A coalition formation value for games with externalities. Version revise - Documents de travail du Centre d'Economie de la Sorbonne 2008. 76 - ISSN: 1955-611X, <halshs-00344797v2> (2011)
20. Williamson, Oe: Transaction-cost economics: the governance of contractual relations. J. Law Econ. **22**(20), 233–261 (1979)
21. Shin, D.H.: Convergence and divergence: Policy making about the convergence of technology in Korea. Government Inf. Q. **27**, 147–160 (2010)
22. Latour, B.: We have never been modern. Harvest Wheatsheaf, Hemel Hempstead (1993)
23. Belhiah, M., Benqatla, M.S., Bounabat, B.: Decision support system for implementing data quality projects. In: Helfert, M., Holzinger, A., Belo, O., Francalanci, C. (eds.) Data Management Technologies and Applications, DATA 2015 (2016)
24. Camarinha-Matos, L.M., Afsarmanesh, H. (eds.): Collaborative Networks: Reference Modeling. Springer Science & Business Media (2008)
25. Camarinha-Matos, L.M., Afsarmanesh, H.: Collaborative networks: a new scientific discipline. J. Intell. Manuf. **16**, 439–452 (2005)
26. Ahuja, G.: Collaboration networks, structural holes, and innovation: a longitudinal study. Adm. Sci. Q. **45**, 425–455 (2000)
27. Park, H., Han, S.H.: Impact of inter-firm collaboration networks in international construction projects: a longitudinal study. In: Construction Research Congress 2012 (2012)

Allocation of Static and Dynamic Wireless Power Transmitters Within the Port of Le Havre

Nisrine Mouhrim[1,2(✉)], Ahmed El Hilali Alaoui[1],
Jaouad Boukachour[2], and Dalila Boudebous[2]

[1] Modeling and Scientific Computing Laboratory,
Faculty of Science and Technology, Route d'Imouzzer, B.P. 2202, Fes, Morocco
nisrine.mouhrim@usmba.ac.ma
[2] Normandie Univ, UNIHAVRE, 76600 Le Havre, France

Abstract. The port of Le Havre, "Grand Port Maritime du Havre (GPMH)", is the first port in France and the fifth in the Europe's top port list in terms of container volume. This massification in container traffic has generated a large use of trucks that are a source of diesel pollution. To address the greenhouse gas emissions related to port's last mile logistic, a collaborative relationships are established between different parties of the port. This collaboration aims to create projects that can improve the air quality such as replacing conventional trucks by electric ones. In this context, our study aims to propose a strategic allocation of the infrastructure of charge for electric trucks. To this end, we adapt the technology Wireless Power Transfer that permit to an electric truck to charge its battery statically in a set of fixed nodes (breakpoints) or dynamically in a set of segments of the route during the electric trucks mobility. To model this problem, we propose an integer non-linear programming formulation. Afterward, we investigate the effectiveness of the population based algorithm particle swarm optimization to determine the efficient allocation.

Keywords: Electric trucks · Wireless power transmitters
Mathematic programming · Particle swarm optimization · Optimization

1 Introduction

Accordingly, to the European Environment Agency (EEA), the amount of energy used by transport account more than 33.2% of the total energy produced. The main transport source of energy is fuels gasoline and diesel. Burning this energy by road freight transportation releases almost one quarter 22.6% of the global greenhouse gas (GHG) emissions. Environmental implications of GHG emissions are in increase namely environmental degradation, climate change, global warming, species elimination, natural ecosystems destruction and human health risks.

In this regard, different sectors have enforced their concern at environmental issues to meet climate goals, one of them is the port sector. In this paper, we study one of the Port of Le Havre's actions to address greenhouse gas emissions, particularly those emitted from last mile delivery. The action aims to adopt a new clean technology, the

© Springer International Publishing AG, part of Springer Nature 2018
M. Ben Ahmed and A. A. Boudhir (Eds.): SCAMS 2017, LNNS 37, pp. 107–121, 2018.
https://doi.org/10.1007/978-3-319-74500-8_10

objective is to replace diesel trucks by electric ones. Electric trucks (ET) seems to be a green solution to build a clean last mile transport network. Besides been zero emissions, ET are silent, required less regular maintenance and regenerative braking. However, ET have short range, limited speed, long recharge time and high initial cost. The ET's battery is the origin of the high initial purchase price of ET, moreover ET's battery is power to weight ratio. To overcome the majority of this drawbacks, we have chosen the technology wireless electric trucks. This technology allows: (i) a quick charge to ET that can extend battery range, (ii) the dynamic mode permit to charge while driving, subsequently drivers benefit from quick charge without waste of time and with less range anxiety. Moreover, frequent recharge allowed by dynamic system permit small capacity of battery and then cheaper ET. (iii) No need to human intervention, drivers benefit from easy and safe charging process without need to use of dangerous and dirty cable (especially in bad weather).

The major inconvenient of wireless power transmitters is the high cost of infrastructure of recharge. In this regard, our work aims to present a strategic allocation of wireless power transmitters with both static and dynamic mode. Our objective is to determine a compromise between the cost of infrastructure and the cost of battery. In fact, the determination of the allocation and the number of static and dynamic segments of recharge is impacted by the capacity of the battery, this tradeoff is explained as follow: numerous number of static and dynamic segments of recharge permit a frequent charge for ET and subsequently, we can choose a smaller battery capacity. In contrast, few number of segments of recharge requires a battery with higher capacity.

To determine the best compromise, we have abstract the transport network of GPMH into a graph, then we have provided a mathematical formulation as a nonlinear integer program model that express our objective namely minimizing the cost of infrastructure of recharge so as battery cost with regard to defining a strategic allocation that respect the battery's capacity as well as avoiding the allocation into prohibited segments of route such as traffic circle and on bridges. To solve this problem, we have adapted our mathematical model to Particle Swarm Optimization since we have a nonlinear program.

The reminder of this paper is organized as follows: Sect. 2 presents scientific works related to the subject of our study. Section 3 describes the process of wireless system for ET. Section 4 contains the problem formulation and the mathematical model. Section 5 introduces the method of resolution. Finally, Sect. 6 displays and comments numerical results.

2   Literature Review

During the last years, Electric vehicles are beginning to catch the attention of more researchers. The variation of electric vehicles (EV) models and mode of charge has made this field of research quite rich. These researches can be subdivided into two areas:

Electric Vehicle Routing Problem (EVRP): is an extension of the well-known VRP, introduced first by Dantzig and Ramser in 1959 [1]. The EVRP aims to determine the least cost generated by an electric vehicle fleet during its visit to a set of customers. The objective can be minimizing energy consumption, distance, vehicle fleet or a combination of a subset of those objectives. The routing of each electric vehicle includes additional detours generated by visiting recharging stations. Each recharge station has a known location and can be visited by the same or different vehicles multiple times as needed [2]. There are other real-world logistics restrictions was added to the classical EVRP. For example, the vehicle freight capacity is limited and the service at each customer must occur within an associated time window, this variant of EVRP is commonly called (CEVRP-TW) [3]. Some works consider that the stay time of an electric vehicle at a charging stations is a function of battery level when it reach a station. Others, consider that time of charge is a variable to be determined, so they allowed to a vehicle to be partially charged [4]. Hiermann [5] considered a mixed fleet, the vehicles differ in their capacity of load charge, battery capacity and acquisition cost. In [6], Sassi et al. presented a case study of la poste, where the vehicle fleet is composed of conventional and heterogeneous electric vehicles. Felipe et al. [7] proposed an EVRP where both the amount of energy recharged and the technology used are a decision variable. In [8], authors propose a schedule of EV recharge taking into account the battery degradation cost so as the energy consumption. The case study in this work consider the service of carrying passengers from an airport to a hotel using electric vehicle.

Electric Vehicle Charging Station Allocation: in this area, researchers aim to optimally allocating/locating electric vehicles charging station. The widespread adoption of EV is related to the degree of ease of access to the infrastructure of recharge whence it came the sensitivity of this problem. In [9] authors presented an allocation based on driver's convenience, their model aims to enhance the trip ratio trip (ratio of successful trips). This allocation optimization problem was formulated as the Maximum Coverage Location Problem (MCLP). In [10], authors considered in their model of allocation the uncertainty information of urban traffic as it affects the vehicles energy consumption. The candidate solutions were simulated and the discrete event simulation was built in Arena. Baouch et al. [11] presented a model that aims to minimize the charging station fixed charge as well as the electric vehicle travel cost, the application was validated within the metropolitan area of Lyon. Chen et al. [12] developed a vehicle charging location model that minimize the station operating costs and maximize the satisfied demand, the model validation was applied to a case study to parking of Seattle, Washington. Besides previous cited works that describe the problem of Charging Station Allocation for a plug-in electric vehicle, researchers have opted for another promising technology that permit to charge a vehicle in motion namely on-road electric vehicle. This technology has strengthened research on wireless power transfer (WPT). This technologies was developed first by the Korean Advanced Institute of Science and Technology (KAIST) [13] and named On Line Electric Vehicles (OLEV). The KAIST was the first to test this system on its own campus. In [14] a detailed description of the system design of OLEV was introduced, nevertheless an overview of the system was presented in [15]. An optimal placement of WPT proposed by means of a mathematical program in the work of Jang et al. [16], their model considered just a unique path. In [17] authors present a mathematical framework that locates optimally the WPT with regard to the maximum traffic flow, since the traffic congestion affects the time travel. In [18] authors studied different scenarios by introducing an integer programming model that decides the suitable road segments to locate WPT in regard to sustain the battery range. In [19], authors provided the feasibility and the potential of WPT.

The above cited researches are limited on the installation of WPT in a circle network composed of a unique path, where the origin correspond to the destination. Thus, our contribution aims to develop a model that can deal with a transport network composed with a set of nodes where vehicles can benefit from static WPT during their service operation and from dynamic WPT installed on arcs connecting the set of nodes. The objective is to determine a strategic allocation of the two types of WPT that minimize the infrastructure cost as well as the battery cost.

3 Zero Emissions Trucks Challenge

The port of Le Havre in collaboration with logistic companies had recognized the need to introduce specific programs and new green policies to deal with pollution within port area caused by trucks that operates last mile freight transportation activities. The GPMH is undergoing a pre-project that meet the air quality goals. The purpose is to study the opportunity for new system implementation such as electric trucks.

To transport containers from the terminal to the final destination, two steps are operated: (i) A local transport from the terminal to the container transport company, this last mile transport operates mainly on short distance. (ii) An interurban transport from container transport company to the final destination and inversely.

This study concerns the local transport. Our contribution is focused on the strategic part of replacing current trucks with wireless electric trucks. The adopted technology combines two wireless recharging techniques:

Static WPT: the shutdown times; generated during the recovery or the deposit of containers or for the verification of documents in the entrance to the port makes the static WPT a convenient solution to recharge de electric truck's battery.

Dynamic WPT: the rationale of the choice of this technology is the possibility that offer by decreasing the weight of the batteries as well as reducing the charging time.

In this paper, we aim to determine the number of each type of WPT together with their strategic allocation and the battery capacity. The solution must take into account the feasibility of the set of all truck routing, regardless of the origin and the destination.

4 System Description

In this work, we deal with a new wireless power transfer technology, called On-Line Electric Vehicle (OLEV). This innovative technology allow to a vehicle to charge while in motion which permit to reduce the battery volume and weight.

A power-receiving installed under a vehicle pick up the power transmitted delivered the underground coil (see Fig. 1). The efficiency of the electric power transfer has reached the limit of 72%. In other word, power transmitters are composed of an inductive cable and inverter. The inverter is responsible to convert a 60 Hz AC to

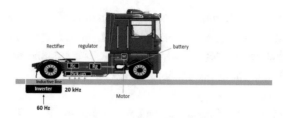

Fig. 1. System operation of OLEV

20 kHz Dc current. Afterword, the inductive cable generate a magnetic flux and transfer the power to the pic up device. In this study, the decision that we make about WPT concerns both inverters and inductive cable.

5 Model Definition and Formulation

Let a set of paths L, each of these paths connects an origin to a destination, a set of N vehicles operates on those paths. Our objective is to find a good compromise between the cost of static and dynamic WPT segments of charge and the cost of the battery while maintaining the quality of the ET. To achieve this goal the WPT must be strategically allocated. The objectives are conflicting because, for example, the minimization of the cost of the battery reduces its autonomy, which leads to a greater number of active segments and subsequently a higher installation cost and inversely.

To model decisions to be made with regards to electric battery specificities, we formally describe the problem by abstracting the transport network within the port of Le Havre into a graph $G = (V, A)$, where $V = \{v_0, v_1, \ldots, v_n\}$, in our graph we call by a vertex: transport companies, terminals, traffic circles and empty container depots. Each vertex v_i has a service time t_{v_i} (with $t_{v_i} = 0$ for traffic circles) and characterized by a required energy R_{v_i}, this amount represents the energy needed to make the internal displacement within the node v_i ($R_{v_i} = 0$ when v_i is a traffic circle). $A = \{(v_i, v_j) : i \neq j\}$ is the set of arcs. Each arc $(v_i, v_j) :\in A$ is partitioned to a set of segments with the same length. Thus, the number of segments that composes an arc is equal to its length divided by the length of segment (the length of a segment is the same for all arcs). The aim of this segmentation is transforming the problem of location of dynamic WPT to a problem of allocation. In such a way, the problem of allocation becomes a 0–1 problem. That is to say, for each segment we have to decide if it will be equipped with an inductive transmitter cable plus an inverter, only transmitter cable or be it just inactive segment. It is noted that a single inverter can power a successive series (which does not exceed its capacity) of active segment. $J_{v_i v_j}$ denotes The set of segments of the arc (v_i, v_j) indexed by j.

We denote by $t^l_{v_i v_j, k}$ the instant of arrival at the kth segment of the arc (v_i, v_j), passing by the path l, we assume that $t^l_{v_i v_j, k}$ is a data as the velocity is predefined as well as the distance from the origin to each kth segment of a given arc (v_i, v_j). The autonomy of the battery when the ET arrives to the kth segment of the arc (v_i, v_j) passing by path l at

$t^l_{v_iv_j,k}$ is denoted by $I(t^l_{v_iv_j,k})$, thus $I(t^l_{v_i})$ express the amount of energy in the battery when arrived to the node v_i at $t^l_{v_i}$ while traveling trough the path l. The maximum length of a series of WPT that an inverter can supply is L_{max}. The required power when the ET arrived to the kth segment of the arc (v_i, v_j) is denoted by $P_{bat}\left(t^l_{v_iv_j,k}\right)$. Let P_{cs} indicate the charging rate of the battery.

Practically, on some sections of the road, we cannot install the WPT because one of the follow reasons: the structure of route, the potion coincides with a traffic circle or bridge, not safe, not enough reliable). Thus, we define $s_{v_iv_j,k}$ as a binary data that takes the value 1 if the allocation of a recharge segment on the kth segment of the arc (v_i, v_j) is prohibited and 0 otherwise. Another binary data is $s^l_{v_iv_j,v_jv_k}$ that takes 1 if (v_i, v_j) and (v_j, v_k) are two adjacent arcs in the path l and 0 otherwise.

To quantify the cost of infrastructure as well as the battery cost, we introduce C_{cab} as the cost of unit cable of power transmitter, C_{inv} the cost of an inverter and C_{st} the cost of static WPT. The number of ET is N. To formulate the mathematical model, we need as a data to identify the first segment of each arc (v_i, v_j) denoted $O_{(v_i,v_j)}$ and its last one denoted $f_{(v_i,v_j)}$.

5.1 Decision Variables

As mentioned above the decision we make around WPT concerns inverters as well as inductive cable. For this reason, we introduce the decision variable $y_{v_iv_j,k}$ that is a binary variable equals to 1 if the kth segment of the arc (v_i, v_j) is active, and 0 otherwise as well as $z_{v_iv_j,k}$ that represents a binary variable equal to 1 if the kth segment of the arc (v_i, v_j) has inverter, and 0 otherwise; St_{v_i} is a 0–1 variable that takes the value 1 if the node v_i contains a static recharge segment, and 0 otherwise; I_{bat} is an integer variable representing the battery capacity, this variable is comprised between I_{min} and I_{max}.

5.2 Objectives

$$min(N * C_{bat} * I_{bat}; C_{cab} * \sum_{(v_i,v_j)\in A} \sum_{k\in J_{(v_i,v_j)}} y_{v_iv_j,k} +$$
$$C_{ond} * \sum_{(v_i,v_j)\in A} \sum_{k\in J_{(v_i,v_j)}} z_{v_iv_j,k} + C_{st} * \sum_{v_i\in V} St_{v_i})$$
(1)

The objective function (1) aims to minimize the total cost of battery, static WPT and dynamic WPT. The cost of dynamic WPT is composed of the cost of active segments of charge and the cost of inverters.

5.3   Constraints

$$y_{v_i v_j,k} \geq z_{v_i v_j,k} \qquad \forall (v_i, v_j) \in A, \forall k \in J_{v_i v_j}$$

$$I\left(t^l_{v_i v_j,k}\right) - \int_{t^l_{v_i v_j,k}}^{t^l_{v_i v_j,k+1}} P_{bat}(t)dt + P_{cs} * (t^l_{v_i v_j,k+1} - t^l_{v_i v_j,k}) * y_{v_i v_j,k} \geq I_{low} \tag{2}$$

$$\forall (v_i, v_j) \in A, \forall k, k+1 \in J_{v_i v_j} \tag{3}$$

$$I\left(t^l_{v_i v_j,k+1}\right) = \min\{I_{hight}, I\left(t^l_{v_i v_j,k}\right) -$$
$$\int_{t^l_{v_i v_j,k}}^{t^l_{v_i v_j,k+1}} P_{bat}(t)dt + P_{cs} * (t^l_{v_i v_j,k+1} - t^l_{v_i v_j,k}) * y_{v_i v_j,k}\} \tag{4}$$
$$\forall (v_i, v_j) \in A, \forall k, k+1 \in J_{v_i v_j}$$

if $k = f_{v_i v_j}$ et $k+1 = v_j$, constraints (2) and (3) become: $(*)$

$$I\left(t^l_{v_j}\right) - \int_{t^l_{v_i v_j,f_{v_i v_j}}}^{t^l_{v_j}} P_{bat}(t)dt + P_{cs} * (t^l_{v_j} - t^l_{v_i v_j,f_{v_i v_j}}) * y_{v_i v_j,f_{v_i v_j}} \geq I_{low} \tag{5}$$
$$\forall (v_i, v_j) \in A, \forall k \in J_{v_i v_j}$$

$$I\left(t^l_{v_j}\right) = \min\{I_{hight}, I\left(t^l_{v_i v_j,f_{v_i v_j}}\right)$$
$$- \int_{t^l_{v_i v_j,f_{v_i v_j}}}^{t^l_{v_j}} P_{bat}(t)dt + P_{cs} * (t^l_{v_j} - t^l_{v_i v_j,f_{v_i v_j}}) * y_{v_i v_j,f_{v_i v_j}}\} \tag{6}$$
$$\forall (v_i, v_j) \in A, \forall k, k+1 \in J_{v_i v_j}$$

$$I\left(t^l_{v_i v_j,0_{v_i v_j}}\right) = I\left(t^l_{v_i}\right) + P_{cs} * t_{v_i} * St_{v_i} - R_{v_i} \tag{7}$$

$$\sum_{k=i}^{i+L_{max}+1} y_{v_i v_j,k} \leq L_{max} \tag{8}$$
$$\forall (v_i, v_j) \in A, \forall k \in J_{v_i v_j}, \forall i \, with \, i \leq f_{v_i v_j} - L_{max}$$

$$\sum_{k=i}^{i+f_{v_i v_j}} y_{v_i v_j,k} + \sum_{j=0_{v_j v_l}}^{L_{max}+1-f_{v_i v_j}-k} y_{v_j v_l,k} \leq L_{max} \tag{9}$$
$$\forall (v_i, v_j), (v_j, v_l) \in A, \forall k, \in J_{v_i v_j} \quad \forall i \, with \, f_{v_i v_j} - L_{max} < i \leq f_{v_i v_j}$$

$$z_{v_i v_j,k} + z_{v_i v_j,k+1} \leq 1 \quad \forall (v_i, v_j) \in A, \forall j \in J_{v_i v_j} - f_{v_i v_j} \tag{10}$$

$$z_{v_i v_j,k} = (y_{v_i v_j,k} - y_{v_i v_j,k-1}) * y_{v_i v_j,k}$$
$$\forall (v_i, v_j) \in A, \forall k, k-1 \in J_{v_i v_j} \tag{11}$$

$$z_{v_i v_j, 0_{v_i v_j}} = \left(y_{v_i v_j, 0_{v_i v_j}} - \prod_{l=1}^{L} y_{v_i v_j, f_{v_i v_j}}\right) * y_{v_i v_j, 0_{v_i v_j}}$$
$$\forall (v_i, v_j) \in A, \forall k \in J_{v_i v_j} \tag{12}$$

$$y_{v_i v_j, k} \leq s_{v_i v_j, k} \quad \forall (v_i, v_j) \in A, \forall k, \in J_{v_i v_j} \tag{13}$$

$$St_{v_i}, y_{v_i v_j, k}, z_{v_i v_j, k} \in \{0, 1\} \tag{14}$$

$$I_{hight} \in [I_{min}, I_{max}] \tag{15}$$

Constraint (2) enforces inverters to not be linked with an inactive segment. Constraint (3) ensures that the amount of remained energy in the battery at the beginning of each segment is up to critical amount I_{low}. When an arc $(v_i, v_j) \in A$ is not a part of the path l, we have $t_{v_i v_j, k}^l = 0 \forall k \in J_{v_i v_j}$. Constraint (4) defines the remained energy at the beginning of each segment. This amount is the minimum value between the remained energy in the beginning of the previous segment minus the charge consumed at the previous segment plus the amount of energy added to the battery if the previous segment is active, and battery's maximum storage capacity. Constraint (5) ensures that each vehicle reaches its destination with an amount of energy greater than the critical value. Constraint (6) updates the state of charge of the battery in case we have (*). Constraint (7) updates the state of charge after leaving a node. Constraint (8) guarantees that at most L_{max} successive segments are active. Constraint (8) replaces constraint (9) when the series of studied segments spreads over two successive arcs (v_i, v_j) and (v_j, v_l) in the path l. Constraint (10) guarantees that two successive active segments have at most one inverter. Constraint (11) and (12) ensure that we must have an inverter in a series of active segments. Constraint (13) interdict the allocation of an active segment in a prohibited zone. Expressions (14) and (15) are the variables of the model, constraint (15) imposes that the battery capacity must be chosen in a specific interval.

6   Solution Approach

This model is a nonlinear integer program. Thus, heuristics are needed to obtain the good compromise between different costs. In this study, we use the Particle Swarm Optimization (PSO) approach as it has shown outstanding performance in solving nonlinear problems. In the next section, we will describe the PSO approach, then we extend this description to adopt the PSO to find a good solution to our model.

Particle swarm optimization (PSO) approach is relatively a new metaheuristic that was developed in 1995 by Kennedy and Eberhart [20]. PSO is derived from the concept of social interaction of flocks of bird and fish schooling. A swarm of bird cooperate with the swarm leader that have a best position relatively to the source of food to get closer to the target (food resource). The algorithm was widely used axing to its effectiveness, requires few parameter settings, computational memory and easy implemented.

6.1 Standard PSO

Like genetic algorithms, PSO is population based technique, where the swarm account for the population. Each particle i is characterized by a position $x_i = [x_{i,1}, x_{i,2}, \ldots, x_{i,D}]$ and a velocity $v_i = [v_{i,1}, v_{i,2}, \ldots, v_{i,D}]$ in D-dimensional space. The displacement of each particle i is relative to its best historical position extracted from its individual experience $Pbest_i = [Pbest_{i,1}, Pbest_{i,2}, \ldots, Pbest_{i,D}]$ and the global best position which the leader position $Gbest = [Gbest_1, Gbest_2, \ldots, Gbest_D]$. This process is expressed as follow:

$$v_{i,j}(t+1) = wv_{i,j}(t) + c_1 * r_1 * \left(Pbest_{i,j} - x_{i,j}(t)\right) + c_2 * r_2 * \left(Gbest_j - x_{i,j}(t)\right) \quad (16)$$

$$x_{i,j}(t+1) = x_{i,j}(t) + v_{i,j}(t+1) \quad (17)$$

Where r_1 and r_2 in (16) are two vectors of random numbers in the range of [0, 1], c_1 and c_2 are the acceleration coefficients, they describe the relative influence of the social and cognitive learning, mostly c_1 is equal to c_2. w is the inertia weight which determine the degree of resist change in velocity.

The pseudo code of The PSO algorithm is:

For each particle
    Initialize particles with random position and velocity
End
Do
    For each particle
        Evaluate the particle according to the objective function
        If the performance of the current particle is better than the best one $Pbest_i$
            Set current value as the new $Pbest_i$
    End
    Select the particle with the best performance $Gbest$
    For each particle
        Adjust the particle velocity
        Adjust the particle position
    End
While stopping condition is not met

6.2 Binary PSO

In binary problems, the decision variables can only take the values 0 or 1. In our mathematical model, we have three 0–1 decisions $\left(y_{v_i v_j, k}, z_{v_i v_j, k}, St_{v_i}\right)$, thus we have adopted the discrete binary PSO (BPSO) to our variables. The major difference between PSO and BPSO is that the update is made only for velocity, which expresses the probability that a bit of solution changes to its opposite.

$$v_{i,j}(t+1) = \begin{cases} 0, & if\ r \geq Seg(v_{i,j}(t+1)) \\ 1, & if\ r < Seg(v_{i,j}(t+1)) \end{cases} \tag{18}$$

Where r in (17) is a random in the range [0.1, 1.0] and Seg is a logistic function that transform the $v_{i,j}$ to a probability. This transformation is defined as follow:

$$Seg\left(v_{i,j}(t+1)\right) = \frac{1}{1+e^{-v_{i,j}(t+1)}} \tag{19}$$

$v_{i,j}$ must be taken between $[-v_{mac}, v_{max}]$. Mostly v_{max} is equal to 6, which limits de probability in the range of [0.0025 0.9975] (see [21]).

7   Numerical Experiments

7.1   Data

As mentioned above, our case study is the port of Le Havre. Thus, we have investigated the set of possible tours achieved by conventional trucks. The objective is to make the same tours achievable by an electric truck by determining a strategic allocation of static and dynamic infrastructure of charge. Three tours are studied:

- Duboc-Roulier-Duboc
- Duboc-Atlantique-Roseliere-Duboc
- DUBOC-Roulier-TDF-Duboc

Figures 2, 3 and 4 draw the three tours within the port of Le Havre.

Fig. 2. Duboc-Roulier-Duboc

The transport network is composed of transport companies, empty container depots, traffic circles and terminals. Figure 5 describes the graph representing the transport network.

Fig. 3. Duboc-Roulier-TDF-Duboc

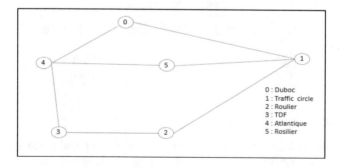

Fig. 4. Duboc-Atlantique-Roseliere-Duboc

Fig. 5. Graph representing the transport network

As previously mentioned, each arc is subdivided into several segments with the same length, we have considered that the length of a segment is 50 m. Table 1 presents the number of segment on each arc. Besides, Table 2 introduces other data.

Table 1. Number of segment on each arc

Arcs	Number of segments
Duboc-Traffic circle	111
Traffic circle-Roulier	174
Roulier-TDF	78
TDF-Roulier	70
Roulier-Traffic circle	191
Traffic circle-Duboc	122
Duboc-Atlantique	176
Atlantique-Traffic circle	95
Roseliere-Traffic circle	117

Table 2. Other data

$\vec{P}_{\vec{cs}}$	800
$\vec{C}_{\vec{cab}}$	60
$\vec{C}_{\vec{inv}}$	5000
$\vec{L}_{\vec{max}}$	250
$\vec{\alpha}$	0.2
$\vec{\beta}$	0.8
$\vec{C}_{\vec{bat}}$	400
N	10
$\vec{\eta}$	0.8

Table 3. Solutions costs

Solution	Battery capacity	Cost of the battery	Cost of infrastructure
1	16	6400	934200
2	20	8000	828800
3	14	5600	1402200

7.2   Results

Multiobjective PSO has allowed to decision maker three possible solutions. In each solution, the static WPT are recommended to be installed at Duboc and TDF, while the allocation of dynamic WPT differ within the battery capacity. Table 3 expose the three non-dominant solutions. Tables 4, 5 and 6 determine the distance of each active segment from the origin of the arc (m). We mentioned that the number of inverters in each solution is equal to the number of series of active segments.

The effectiveness of our results is proved by the strategic emplacement of active segments regarding to the speed profile of trucks. We find that dynamic segments are placed in zones where the truck is moving slowly.

Table 4. Location of active segments of solution 1

Arcs	Distance of each active segment from the origin of the arc (m)					
Duboc-Traffic circle	4100–4150	4200–4250				
Traffic circle-Roulier						
Roulier-TDF	1150–1200	1550–1600				
TDF-Roulier	400–450	1000–1050	2750–2800			
Roulier-Traffic circle	450–500	850–900				
Traffic circle-Duboc	0–50	2400–2450	3400–3450			
Duboc-Atlantique	250–300	450–500	3050–3100	4850–4900	6250–6300	8200–8250
Atlantique-Roselière	850–900					
Roselière-Traffic circle						

Table 5. Location of active segments of solution 2

Arcs	Distance of each active segment from the origin of the arc (m)					
Duboc-Traffic circle	1350–1400					
Traffic circle-Roulier	750–800	2050–2100	2350–2400	3550–3600	5050–5100	5500–5600
Roulier-TDF						
TDF-Roulier	2000–2050					
Roulier-Traffic circle	4250–4300	5000–5500				
Traffic circle-Duboc	4150–4200	5300–5350	5450–5500			
Duboc-Atlantique	2100–2150	5200–5250				
Atlantique-Roselière	2200–2250	3400–3450				
Roselière-Traffic circle						

Table 6. Location of active segments of solution 3

Arcs	Distance of each active segment from the origin of the arc (m)					
Duboc-Traffic circle	1200–1250	3300–3350	4500–4550			
Traffic circle-Roulier	350–400	2150–2200	2350–2400	4500–4550	6150–6200	7200–7250
Roulier-TDF	1950–2000	2450–2500				
TDF-Roulier	2200–2250	2800–2850				
Roulier-Traffic circle	5350–5400	7300–7350				
Traffic circle-Duboc	850–900	900–950	1350–1400	3250–3300	3650–3700	5700–5750
Duboc-Atlantique	5700–5750	3600–3650	4200–4250	7150–7200	8050–8100	8250–8300
Atlantique-Roselière	300–350	1900–1950	3400–3450			
Roselière-Traffic circle						

8 Conclusion

In this paper we have studied the problem of allocation of wireless power transmitters whiten the port of Le Havre. Our aim is in one side finding a set of solution that presents a strategic allocation of static and dynamic WPT, in other side minimizing the total costs of the infrastructure as well as the cost of the battery. Firstly, we have modeled this problem as an integer non-linear programming. Afterward, we have adapted our problem to the multiobjective particle swarm optimization approach. As results, we present to the decider-maker three different solutions, the decider must take into account other issues such as battery degradation and the maintenance cost of infrastructure. We note that our study is on the development and implementation of addition resolution method in order to compare it with the current one, such as to our knowledge, we are the first authors that study the problem of allocation of dynamic and static wireless transfer transmitter.

Acknowledgments. This research work was conducted as part of the Green Truck project. This project has received funding from Normandy region of France.

References

1. Dantzig, G., Ramser, J.: The truc dispatching problem. Manag. Sci. **6**(1), 80–91 (1959). https://doi.org/10.1287/mnsc.6.1.80
2. Lin, J., Zhou, W., Wolfson, O.: Electric vehicle routing problem. In: The 9th International Conference on City Logistics, Tenerife, Canary Islands, Spain, 17–19 June 2015 (2015)
3. Schneider, M., Stenger, A., Goeke, D.: The electric vehicle-routing problem with time windows and recharging stations. Transp. Sci. **48**(4), 500–520 (2014)
4. Keskin, M., Çatay, B.: Partial recharge strategies for the electric vehicle routing problem with time windows. Transp. Res. Part C: Emerg. Technol. **65**, 111–127 (2016). https://doi.org/10.1016/j.trc.2016.01.013, Transp. Rev. **71**(1), 111–128 (2016)
5. Hiermann, G., Puchinger, J., Hartl, R.F.: The electric fleet size and mix vehicle routing problem with time windows and recharging stations (working paper) (2014). http://prolog.univie.ac.at/research/publications/downloads/Hie_2014638.pdf. Accessed 17 July 2014
6. Sassi, O., Cherif-Khettaf, W.R., Oulamara, A.: Vehicle routing problem with mixed fleet of conventional and heterogeneous electric vehicles and time dependent charging costs. World Acad. Sci. Eng. Tech. Internat. J. Math. Comput. Phys. Electr. Comput. Eng. **9**(3), 163–173 (2016)
7. Felipe, A., Ortuno, M.T., Righini, G., Tirado, G.: A heuristic approach for the green vehicle routing problem with multiple technologies and partial recharges. Transp. Res. Part E: Logist. Transp. Rev. **71**(1), 111–128 (2014)
8. Barco, J., Guerra, A., Muñoz, L., Quijano, N.: Optimal routing and scheduling of charge for electric vehicles: case study. arXiv preprint arXiv:1310.0145 (2013)
9. Alhazmi, Y.A., Mostafa, H.A., Salama, M.M.: Optimal allocation for electric vehicle charging stations using Trip Success Ratio. Int. J. Electr. Power Energy Syst. **91**, 101–116 (2017)
10. Sebastiani, M.T., Lüders, R., Fonseca, K.V.O.: Allocation of charging stations in an electric vehicle network using simulation optimization. In: Proceedings of the 2014 Winter Simulation Conference (2014)
11. Baouchea, F., Billota, R., El Faouzi, N.-E., Trigui, R.: Electric vehicle charging stations allocation models. In: Proceedings of the Transport Research Arena Conference, pp. 1–10 (2014)
12. Chen, T.D., Kockelman, K.M., Khan, M.: Locating electric vehicle charging stations. Transp. Res. Rec. J. Transp. Res. Board **2385**(1), 28–36 (2013)

13. Nagatsuka, Y., Ehara, N., Kaneko, Y., Abe, S., Yasuda, T.: Compact contactless power transfer system for electric vehicles. In: The 2010 International Power Electronics Conference, 21–24 June 2010, pp. 807–813 (2010)
14. Shin, J., Kim, Y., Ahn, S., Lee, S., Jung, G., Jeon, S.-J., Cho, D.-H.: Design and implementation of shaped magnetic resonance based wireless power transfer system for roadway-powered moving electric vehicles. IEEE Trans. Ind. Electron. **61**, 1179–1192 (2013)
15. Qiu, C., Chau, K.T., Liu, C., Chan, C.: Overview of wireless power transfer for electric vehicle charging. In: 2013 World Electric Vehicle Symposium and Exhibition (EVS27), pp. 1–9. IEEE (2013)
16. Jang, Y.J., Ko, Y.D., Jeong, S.: Optimal design of the wireless charging electric vehicle. In: Proceedings of the IEEE IEVC, pp. 1–5 (2012)
17. Riemann, R., Wang, D.Z.W., Busch, F.: Optimal location of wireless charging facilities for electric vehicles: flow-capturing location model with stochastic user equilibrium. Transp. Res. Part C: Emerg. Technol. **58**, 1–12 (2015)
18. Ushijima, H.-M., Khan, M.Z., Chowdhury, M., Safro, I.: Optimal installation for electric vehicle wireless charging lanes. Transp. Res. Part C. arXiv:1704.01022 (2017)
19. Limb, B.J., Zane, R., Quinn, J.C., Bradley, T.H.: Infrastructure optimization and economic feasibility of in-motion wireless power transfer. In: IEEE Transportation Electrification Conference and Expo (ITEC), pp. 1–4. IEEE (2016)
20. Kennedy, J., Eberhart, R.: Particle swarm optimization. In: Proceedings of the IEEE International Conference on Neural Networks, pp. 1942–1948. IEEE Press (1995)
21. Nezamabadi-pour, H., Rostami Shahrbabaki, M., Maghfoori-Farsangi, M.: Binary particle swarm optimization: challenges and new solutions. CSI J. Comput. Sci. Eng. Persian **6**(1), 21–32 (2008)

Using the CBR Dynamic Method
to Correct the Generates Learning Path
in the Adaptive Learning System

Nihad El Ghouch[1], El Mokhtar En-Naimi[1(✉)] ,
Abdelhamid Zouhair[2], and Mohammed Al Achhab[3]

[1] LIST Laboratory, The Faculty of Sciences and Technologies, UAE, Tangier, Morocco
nihad_elghouch@hotmail.fr, ennaimi@gmail.com
[2] The National School of Applied Sciences, Al-Hoceima, Morocco
zouhair07@gmail.com
[3] The National School of Applied Sciences, Tetuan, Morocco
alachhab@gmail.com

Abstract. The adaptive learning systems have the capacity to adapt the learning process to the needs/the rhythms of each learner, the learning styles and the preferences, but they do not ensure an individualized follow-up in real time. In this article, we will present our architecture of an Adaptive Learning System using Dynamic Case-Based Reasoning. This architecture is based on the learning styles of Felder-Silverman and the Bayesian Network to propose the learning path according to the adaptive style and on the other hand on the approach of the Dynamic Case-Based Reasoning to ensure a prediction of the dynamic situation during the learning process, when the learner has difficulty learning. This approach is based on the reuse of past similar experiences of learning (learning path) by analyzing learners' traces.

Keywords: Adaptive learning systems · Learning style
Dynamic Case-Based Reasoning · Learning paths

1 First Section

The use of Web technologies in the educational domain has made it possible to consider new approaches and learning contexts. Several research projects have proposed the integration of pedagogical aspects into E-learning platforms that depend on the capacity of these approaches to provide learners with contents and educational paths adapted to their needs. The development of adaptive educational systems meets this objective. In this paper, we will discuss adaptive learning systems based on the detection of knowledge in relation to the teaching resources, the preferences of the learners and the learning processes. These systems allow for personalized learning and individualized follow-up. This follow-up makes it possible to study the behavior of the learner through his traces of interactions during the learning.

We propose an architecture that performs adaptive learning with individualized monitoring. This architecture is divided into two parts: the static part and the dynamic

© Springer International Publishing AG, part of Springer Nature 2018
M. Ben Ahmed and A. A. Boudhir (Eds.): SCAMS 2017, LNNS 37, pp. 122–128, 2018.
https://doi.org/10.1007/978-3-319-74500-8_11

part. The first part allows to detect the initial learning style and to propose the path of learning according to the style. The second part ensures an individualized and continuous follow-up of the learner during the learning process, using the Dynamic Case Based Reasoning approach that relies on the sharing and the re-use of successfully passed experiments (learning paths). The rest of this article will be organized as follows: In the next section, we will introduce adaptive learning systems. Then, in the third section, we present the Case-Based Reasoning approach. In the fourth section, we detail our proposed architecture and describe the different parts. Finally, in the last section, we come to a conclusion.

2   Adaptive Learning System

Adaptive learning systems are an important class of e-learning systems; they customize the learning process according to the needs, prerequisites, objectives, etc. of each learner, and then create a specific learning path. These systems can be categorized in two categories, according to the strategic methods according to which these systems work to ensure adaptability:

- Systems using explicit methods for the collection of information to build the learning profile based on forms and questionnaires in which the learners express their learning preferences.
- Systems using implicit methods which consist in collecting information, through the navigation of the learner within the learning platform. The objective of these two strategies is to provide the learning system, with the necessary information to be able to assign the learning objects most adapted to the characteristics of the learner by creating a learning path.

The problem of the Adaptive Learning Systems is that they assume the generated learning path is systematically the leading one, which is not necessarily the case, since we can always detect several negative results during the learning process [1] and they do not allow an individualized, continuous follow-up of the learner. It allowed us to conceive the architecture of an Adaptive Learning System using a decision support system capable of following the learner in real time, using his or her learning style and traces recorded in the learning platform, in order to predict and decrease the number of abandonments. This architecture allows for:

- Knowing the learning style;
- Proposing the learning path;
- Analyze the traces of learning;
- Take into account the dynamic change of the learning path;
- Providing solutions for change based on past experience (learning paths stored in a base of learning paths).

This is why we choose the approach of Case-Based Reasoning (CBR). It is a distributed artificial intelligence approach, considered as the most privileged method of modeling past experience of users.

3 Case-Based Reasoning

3.1 Definition

The CBR is an approach that solves new problems based on past experiences or solved problems available in system memory [2]. The Case-Based Reasoning CBR is a reasoning approach to solve new problems by adapting past cases already solved. The CBR consists in re-using a new problem, the solution of a similar old problem already dealt with and resolved. It is based on a large number of problems solved in the past instead of relying on a deep knowledge of the domain [3]. In the release of a new situation (a problem), a search for the most similar situation is made in the entire past experience, and the chosen solution will be reused and adapted (if necessary) to the new case. The advantages of CBR system:

- It does not require an acquisition of deep knowledge of the domain to find a solution for a problem. Indeed, the knowledge consists in establishing a description of a problem and its solution.
- It is relatively simple and easy to compare with other techniques of the IA.
- It facilitates the learning, by inserting new cases into the base of the cases [4].

3.2 Cycle of CBR

In general, Case-Based Reasoning (CBR) is an approach to the resolution of problems based on the re-use of past experience called "case". A case represents in particular a problem and a solution that has been applied (or a method to generate it). The CBR cycle generally consists of five steps (Fig. 1):

- **Elaboration:** allows collecting all the necessary information of a problem so as to build a new case called target case,
- **Retrieve:** consists in looking into the base of cases for one or several similar solved cases in the target case. This step is based on the measures of similarity; the correct selection of the most similar case in the base of cases depends on this measure;
- **Reuse:** allows obtaining a solution to the new case from the solution selected in the previous step. The knowledge of adaptation sometimes depends on the domains of application;
- **Revise:** allows estimating the proposed solution. In certain cases, it is possible that the proposed solution does not succeed in solving the problem, which allows for correcting or refusing the solution;
- **Retain:** allows adding the new case with its solution in the case base.

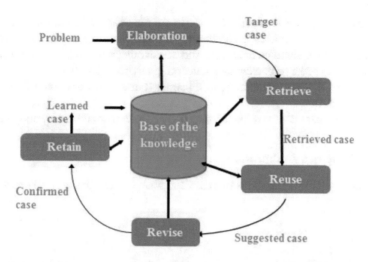

Fig. 1. Cycle of CBR

The Case-Based Reasoning allows the interpretation of a situation (target case), the retrieve of a similar situation (case sources), the proposal of an adaptation possible for the current situation with some possible repair and the saving of the result as a new experience of learning. The target cases of a CBR can be [6]:

- **Static target cases:** all the information of the case must first be presented before the search in the base of cases [7, 8].
- **Dynamic target cases:** the target case evolves dynamically in time; the CBR has to take into account this continual evolution. A dynamic case is described by one or several records which presents the evolution of one or several parameters, which is relevant for the prediction of the situation [7, 8]. The systems of static CBR suffer from limitations in the management of the dynamic parameters, and they are incapable of detecting automatically the evolution of their parameters as well as of adapting to the changes of the current situation [7, 8].

4   Proposed Approach

4.1   Description of the Approach

Our approach is to adapt the learning and to offer a personalized follow-up for diverse profiles of learners. This follow-up is based on utilizing other experiences successfully lived with other learners. When use the Case-Based Reasoning, the case is dynamic because the path of the learner evolves and changes dynamically with time (analysis of traces in real time). We consider that:

- The learner has an initial learning style through this style by assigning learning objects by constructing a learning path;

- The learning process is observed by the traces recorded in the platform. The exploitation of the traces makes it possible on the one hand to analyze the activity of a learner and understand its behavior. And on the other hand to extract information or knowledge in order to personalize its learning environment [9];
- The change of a way of learning of a learner (target case) triggers a CBR cycle to adapt the learning;
- The base of cases contains the learning style of a learner and his learning path.

4.2   Our Proposed Architecture

Our architecture (Fig. 2) contains two essential phases: static phase and dynamic phase.

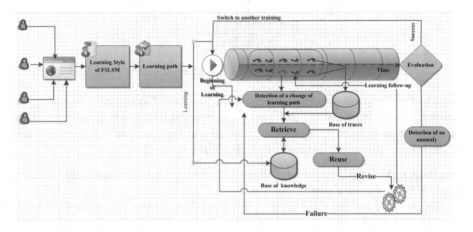

Fig. 2.   Proposed architecture

4.2.1   Static Phase

During this phase, two steps are executed:

- **Learning style detection:** This step allows us to detect the initial learning style, we used the Felder and Silverman FSLSM learning style model [10], there are several models, the choice of the FSLSM test is that this model returns that this model fills most of the criteria required by hypermedia systems [11]. This test defines learning styles in 4 dimensions: Active/Reflective, Sensory/Intuitive, Visual/Verbal, Sequential/Global. Learners who belong to each dimension prefer to use specific learning objects. The results of the FSLSM Test determine the nature and type of the learning object.
- **Proposition of a learning path:** After the learning style of each learner is detected, the learning objects are assigned according to the style. This assignment is carried out by the Bayesian network. The Bayesian Network [12, 13] allows calculating the probability to assign learning object according to a description of the learning style of the learner. The result of this phase is the modeling and determination of the learning style of the learner.

4.2.2   Dynamic Phase

This phase starts with the beginning of learning training/a course. It comprises:

- Learner behavior modeling: collects information from the traces of interaction between the learner and the platform.
- Detection of a change in the proposed learning path: launch of the CBR cycle.

During this phase, the CBR cycle will be applied with each one of its steps:

- **Elaboration:** allows to collect information such as learning styles and to observe the behavior of the learner by analyzing his traces in the system in a continuous way. Considering that the learner who does not follow the proposed learning path or who has a failure in the evaluation: a target case;
- **Retrieve:** allows to detect the source case(s) most similar to the target case, by looking for the learning path(s) most similar to the target case in the base of the source cases (past learning paths) based on the measure of similarity with learners with the same style or with all learners. The comparisons between the target case and the source cases are based on indicators, which are collected from the observation of the analysis of learners' traces during the learning: The number of connections to the platform, the number of visits to each unit of the course, the total time spent for each unit, the number of examples, the number of exercises, the duration allocated to the theoretical part, the duration allocated to the practical part, etc. This step will be activated each time a change in the traces of the learners leads to changing a learning path, i.e. an update of the target case with time;
- **Reuse:** Based on the retrieved case, the adaptation solves the target problem:
  - Applying the proposed solution without change.
  - Applying the proposed solution with change.
  - Applying human intervention.
- **Revise:** This step will be executed:
  - If the case chosen is the most similar source case of the target case
  - Otherwise, if the source case does not match the target case. In this case, the system must return to the Recall step.
- **Retain:** Memorize the target case as a new case in the system knowledge base (case basis). During this phase, the system intervenes by offering a base of the learning paths of learning indexed to build or adapt paths by relying on the traces of the learner interacting with the platform and learning style, in order to have an individualized follow-up in real time.

5   Conclusion and Perspectives

Our contribution consists in proposing architecture allowing an adaptive learning and an individualized follow-up of the learner. Our architecture bases on the test of FSLSM to detect the initial learning style of the learner (the preferences), the Bayesian Network to affect the learning objects according to the detected style by creating of learning path and a decision system allowing to adapt the learning, if the learner does not follow the proposed learning path or he suffers difficulties to learn.

In Our architecture, we used a Dynamic Case-Based Reasoning based on the traces by the learner during the learning process. These traces change in real time. Our future work consists in implementing the proposed architecture, developing the different steps of the cycle of the Dynamic Case-Based Reasoning.

References

1. Lenz, B.: Failure is essential to learning (2015). http://www.edutopia.org/blog/failure-essential-learning-bob-lenz
2. Kolodner, J.L.: Case-Based Reasoning. Morgan Kaufmann (1993)
3. Bichindaritz, I., Marling, C.: Case-based reasoning in the health sciences: what's next? Artif. Intel. Med. **36**, 127–135 (2006)
4. Watson, I., Marir, F.: Case-Based Reasoning: A Review, AI-CBR, Dept. of Computer Science, University of Auckland, New Zealand. http://www.aicbr.org/classroom/cbr-review.html. Accessed Aug 2009
5. Mille, A.: Traces based reasoning (TBR) definition, illustration and echoes with story telling. Rapport Technique RR-LIRIS-2006-002, LIRIS UMR 5205 CNRS/INSA de Lyon/Université Claude Bernard Lyon 1/Université Lumière Lyon2/Ecole Centrale de Lyon, January 2006
6. Loriette-Rougegrez, S.: Raisonnement à partir de cas pour desévolutions spatiotemporelles deprocessus. revue internationale degéomatique **8**(1–2), 207–227 (1998)
7. Zouhair, A.: Raisonnement à partir de cas dynamique multi-agents: application à un système de tuteur intelligent, Ph.D. in computer science, in Cotutelle between the Faculty of Sciences and Technologies of Tangier (Morocco) and the University of Le Havre (France), supported in October 2014
8. En-Naimi, E.M., Zouhair, A.: Intelligent dynamic case-based reasoning using multiagents system in adaptive e-service, e-commerce and elearning systems. Int. J. Knowl. Learn. **11**(1), 42–57 (2016)
9. Settouti, L.S., Prié, Y., Mille, A., Marty, J.-C.: Vers des systèmes à base de traces modélisées pour les eiah. LIRIS Research Report (2007)
10. Felder, R.M., Silverman, L.K.: Learning Styles and Teaching Styles in Engineering Education, November 1987
11. Popescu, E., Badica, C., Trigano, P.: Description and organization of instructional resources in an adaptive educational system focused on learning styles. In: Advances in Intelligent and Distributed Computing, pp. 177–186. Springer, Heidelberg (2008)
12. Elghouch, N., Seghroucheni, Y.Z., En-Naimi, E.M., El Mohajir, B.E., Al Achhab, M.: An application to index the didactic resources in an adaptive learning system. In: The Fifth International Conference on Information and Communication Technology and Accessibility (ICTA 2015), Marrakech, Morocco, 21–23 December 2015, pp. 1–3. IEEE Proceedings (2015)
13. Elghouch, N., En-Naimi, E.M., Seghroucheni, Y.Z., El Mohajir, B.E., Al Achhab, M.: ALS_CORR[LP]: an adaptive learning system based on the learning styles of Felder-Silverman and a Bayesian network. In: 4th IEEE International Colloquium on Information Science and Technology (CiSt) (2016)

Noisy Satellite Image Segmentation
Using Statistical Features

Salma El Fellah, Salwa Lagdali[✉], Mohammed Rziza, and Mohamed El Haziti

LRIT, Rabat IT Center, Faculty of Sciences, Mohammed V University in Rabat, Rabat, Morocco
salwalagdali@gmail.com

Abstract. Satellite image segmentation is a principal task in many applications of remote sensing such as natural disaster monitoring and residential area detection and especially for Smart cities, which make demands on Satellite image analysis systems. This type of image (satellite image) is rich and various in content however it suffers from noise that affects the image in the acquisition. The most of methods retrieve the textural features from various methods but they do not produce an exact descriptor features from the image and they do not consider the effect of noise. Therefore, there is a requirement of an effective and efficient method for features extraction from the noisy image. This paper presents an approach for satellite image segmentation that automatically segments image using a supervised learning algorithm into urban and non-urban area. The entire image is divided into blocks where fixed size sub-image blocks are adopted as sub-units. We have proposed a statistical feature including local feature computed by using the probability distribution of the phase congruency computed on each block. The results are provided and demonstrate the good detection of urban area with high accuracy in absence of noise but a low accuracy when noise is added which yields as to present a novel features based on higher order spectra known by their robustness against noise.

Keywords: Computer vision · Segmentation · Classification · Satellite image
Statistical feature · Phase gradient · Higher Order Statistics

1 Introduction

Remotely sensed images of the Earth that we can acquire through satellites are very large in number. The classification of data has long attracted the attention of the remote sensing community because classification results are the basis for many environmental and socioeconomic applications [1]. Scientists and practitioners have made much effort in developing advanced classification approaches and techniques for improving classification accuracy. Each image contains a lot of information inside it and can have a number of objects with characteristics related to the nature, shape, color, density, texture or structure. It is very difficult for any human to go through each image, extract, and store useful patterns. An automatic mechanism is needed to extract objects from the image and then do classification. The problem of urban object recognition on satellite images is rather complicated because of the huge amount of variability in the shape and

© Springer International Publishing AG, part of Springer Nature 2018
M. Ben Ahmed and A. A. Boudhir (Eds.): SCAMS 2017, LNNS 37, pp. 129–135, 2018.
https://doi.org/10.1007/978-3-319-74500-8_12

layout of an urban area, in addition to that, the occlusion effects, illumination, view angle, scaling, are uncontrolled [2]. Therefore, more robust methods are necessary for good detection of the objects in remotely sensed images. Generally, most of existing segmentation approaches are based on frequency features and use images in gray levels. Texture based methods partition an image into several homogenous regions in terms of texture similarity. Most of the work has concentrated on pixel-based techniques, [3]. The result of pixel-level segmentation is a thematic map in which each pixel is assigned a predefined label from a finite set. However, remote sensing images are often multi-spectral and of high resolution which makes its detailed semantic segmentation excessively computationally demanding task. This is the reason why some researchers decided to classify image blocks in-stead of individual pixels [4]. We also adopt this approach by automatically dividing the image into a single sub-image (block), and then evaluate classifiers based on support vector machines, which have shown good results in image classification. The process of generating descriptions represents the visual content of images. In this paper, we focus on the emerging image segmentation method that use statistical feature; in order to model local feature and we analyze the effect of noise on the classification accuracy.

The rest of the paper is organized as follows. Section 2 presents a review of some works related to our work, Sect. 3 describes the general framework of the proposed approach, while Sect. 4 shows the experimental results and finally Sect. 5 concludes the paper

2   Related Work

There are many techniques for classification of satellite images. We briefly review here some of the methods that are related to our work: Mehralian and Palhang, [5] separate urban terrains from non-urban terrains using a supervised learning algorithm. Extracted feature for image description is based on principal components analysis of gradient distribution. Ilea and Whelan [6] considered the adaptive integration of the color and texture attributes and proposed a color-texture-based approach for SAR image segmentation. Fauquer et al. [7] classify aerial images based on color, texture and structure features; the authors tested their algorithm on a dataset of 1040 aerial images from 8 categories. Ma and Manjunath [8] use Gabor descriptors for representing aerial images. Their work is centered on efficient content-based retrieval from the database of aerial images and they did not try to automatically classify images to semantic categories. In this work, we use local features, believing that is beneficial on identifying image and more suitable to represent complex and noisy scenes and events categories, our feature vector extracted will be the result of combination of the statistical local feature.

3   Proposed Approach of Urban Terrain Recognition

In this paper, a new approach for satellite images segmentation is presented. The method comprises three major steps:

Firstly, we start by splitting the image into blocks with size of (20 × 20).

The second step consists in feature extraction, we calculate the computation of phase congruency map of each sub image (block) then, the statistical local features (mean, variance and skewness) are calculated from the probability distribution of the phase congruency. These features are combined to construct a feature vector for each block. These vectors are used to characterize each sub image.

Finally, these vectors are used as training and testing where Gaussian noise is added to the test images. Classification is performed using SVM to distinguish urban classes from non-urban classes. The tests show that the proposed method can segment images with high accuracy in absence of noise but when the images are corrupted by noise the accuracy become progressively worse.

In what follows, it is assumed that satellite images are being analyzed for segmentation to urban and non-urban terrain. Below, we describe each of these steps.

3.1 Partitioning

For evaluation of the classifiers we used 800×800 pixel (RGB) image taken from Google Earth related to Larache city, Morocco, satellite images are sometimes very large and handling. We split this image into smaller blocks of 20×20 pixels, with an overlaps of 4 pixels at the borders. So we have in total 4493 blocks in our experiments. We classified all images into 2 categories, namely: urban and non-urban. Examples of sub-images from each class are shown in Fig. 1.

Fig. 1. Some sample blocks from satellite images; first row: Non-urban areas, second row urban areas.

3.2 Feature Extraction

To describe each sub-image, the statistical features of the 2D phase congruency histogram applied on each block and obtained values as principals used for features, so we have a 1D feature vector for each block, which will be used in training process.

The statistical features [9] provide information about the properties of the probability distribution. We use statistical features of the phase congruency histogram (PCH) as mean, variance and skewness that are computed by using the probability distribution of the different levels in the histograms of PCH. Let λi be a discrete random variable that represents different levels in a map and let $p(\lambda_i)$ be the respective probability density function. A histogram is an estimation of the probability of occurrence of values λj as measured by $p(\lambda_i)$. We content with three statistical feature of histogram:

- **Mean (m):** computes the average value. It is the standardized first central moment of the probability distribution in image.

$$m = \sum_{i=0}^{L-1} \lambda_i p(\lambda_i)$$

- **Variance (σ):** It's second central moment of the probability distribution, the expected value of the squared deviation from the mean.

$$\sigma = \sum_{i=0}^{L-1} p(\lambda_i) \lambda_i^2 - m^2$$

- **Skewness (k):** computes the symmetry of distribution. S gives zero value for a symmetric histogram about the mean and otherwise gives either positive or negative value depending on whether histogram has been skewed right or left to the mean.

$$k = \sum_{i=0}^{L-1} (\gamma_i - m)^3 p(\gamma_i)$$

After the calculation of these statistical features for each PCH, the feature vectors *fPC* of each block are constructed as:

$$fPC = \{mPCH, \sigma PCH, kPCH\}$$

The feature vectors of all the blocks images including urban and non-urban sub-image are constructed and stored to create a feature database.

After feature extraction, we use this vector feature to train and test SVM. Given a set of training examples, each marked as belonging to one of two categories. We used half of the images for training and the other half for testing.

In order to evaluate the effect of noise on the classification of satellite images, we add Gaussian noise with different noise levels (SNR = 20 dB, 10 dB and 5 dB) on the test images. The results of the classification experiment with and without noise are reported in Table 1.

4   Experimental Result of the Proposed Method

4.1   Phase Congruency Features

We experiment the approach on two samples of satellite image, taken both from Google Earth, related to Larache city, Morocco [10]. There sizes are 800 × 800 pixel (RGB) image. We have split (with an overlaps of 4 pixels at the borders) these images into smaller sub-images of 20 × 20 pixels. In total, it occur 4493 sub images.

We characterize each sub-image by 1D statistical feature vector, which is based on the statistical local features (mean, variance and skewness) calculated from the probability distribution of the phase congruency (Fig. 2).

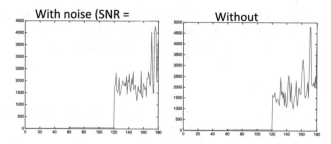

Fig. 2. Local feature plot of a satellite image with a level noise (5 dB) and without noise

Figure shows the plot of local feature of a satellite image block (see Fig. 3) with a very high level noise (5 dB) and without noise, it can be observed that some picks have despaired when adding noise; which means that the feature have been affected by the noise.

Fig. 3. Satellite image block without noise (first line) and with a level noise (5 dB) (second line)

We use half of the features for training and the other half for testing, we label the training data manually with 1 and −1, where label 1 refers to urban category and the −1 refers to non-urban category, and the obtained model will be tested on the test data. To evaluate the robustness of the features against noise, we add Gaussian noise with different levels noise on the test data. In the sections below the results of the test will be discussed.

To examine the ability of proposed approach, we have used accuracy and precision statistical measures:

Table 1. Table captions should be placed above the tables.

Metric	Without noise	SNR = 5 dB	SNR = 10 dB	SNR = 20 dB
Accuracy	0.94	0.74	0.85	0.93
Precision	0.96	0.76	0.87	0.95

From Table 1 we observe that in the absence of noise, the phase congruency features give good results (94%). However, when the images are corrupted by Gaussian noise, the correct accuracy becomes to deteriorate from 93% at SNR = 20 dB to 74% at SNR = 5 dB.

This degradation yields us to think about exploiting Higher order spectra in the feature extraction procedure. Higher order spectra and especially the third order namely the bispectrum are known by their ability to nullify Gaussian noise.

4.2   Bispectrum Features

Bispectrum is the third order spectrum known by its ability to nullify Gaussian noise where the bispectrum of Gaussian noise is zero [11, 12].

Mathematically the bispectrum $B(f1, f2)$ of 1D signal x is expressed as:

$$B(f1, f2) = X(f1) \cdot X(f2) \cdot X(f1 + f2)$$

Where $X(f1)$ and $X(f2)$ are respectively the fourier transform of x at frequencies $f1$ and $f2$ and $X(f1 + f2)$ the conjugate.

The bispectrum is a complex value and it can be expressed as:

$$B(f1, f2) = B(f1, f2) \vee exp(i\pi)$$

Where $B(f1, f2) \vee$ is the magnitude and π is the phase of bispectrum.

Features extracted from bispectrum are very robust against noise where every Gaussian noise added to the image is eliminated by the bispectrum. For this reason, we are interested in the bispectrum magnitude features.

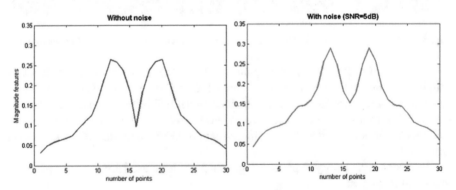

Fig. 4. Mean magnitude of bispectrum of a satellite image with a level noise (5 dB) and without noise.

Figure 4 shows the mean magnitude of bispectrum of a satellite image with a very high level noise (5 dB) and without noise, it can be observed that the feature do not affected by the noise. Hence it can be used as a feature for classifying satellite images.

5 Conclusion and Perspectives

In this work, we have focused on the type of Feature Extraction Technique, and we have proposed a statistical feature including local features in which we compute the probability of distribution of 2D phase congruency. We distinguish urban from non-urban terrain, the algorithm makes decision about image block (not a pixel) in both size (20×20), so each block are described by 1D vector features, then SVM are used for classification. The results of the approach yield good performance in absence of noise. However, when the images are corrupted by Gaussian noise the accuracy deteriorate which yields as to analyzing bispectrum features that do not change with noise. As a future work, we are interested in exploiting the presented magnitude feature to classify noisy satellite images.

References

1. Blaschke, T.: Object based image analysis for remote sensing. ISPRS J. **65**, 2–16 (2010)
2. Sirmacek, B., Unsalan, C.: A probabilistic approach to detect urban regions from remotely sensed images based on combination of local features. In: 5th RAST 2011 Conference (2011)
3. Szummer, M., Picard, R.W.: Indoor-outdoor image classification. In: Proceedings of the IEEE ICCV Workshop, Bombay, India, pp. 42–51, January 1998
4. Pagare, R., Shinde, A.: A study on image annotation techniques. Int. J. Comput. Appl. **37**(6), 42–45 (2012)
5. Mehralian, S., Palhang, M.: Principal components of gradient distribution for aerial images segmentation. In: 11th Intelligent Systems Conference (2013)
6. Ilea, D.E., Whelan, P.F.: Image segmentation based on the integration of color texture descriptors - a review. Patt. Recogn. **44**, 2479–2501 (2011)
7. Fauqueur, J., Kingsbury, G., Anderson, R.: Semantic discriminant mapping for classification and browsing of remote sensing textures and objects. In: Proceedings of IEEE ICIP 2005 (2005)
8. Ma, W.Y., Manjunath, B.S.: A texture thesaurus for browsing large aerial photographs. J. Am. Soc. Inf. Sci. **49**(7), 633–648 (1998)
9. Tiwari, S., Shukla, V.P., Biradar, S.R., Singh, A.K.: A blind blur detection scheme using statistical features of phase congruency and gradient magnitude. Adv. Elect. Eng. **2014,** 10 (2014). Article ID 521027. Lang. Syst. **15**(5), 795–825 (1993). http://doi.acm.org/10.1145/161468.16147
10. Salma E.F., Mohammed E.H., Mohamed R., Mohamed M.: A hybrid feature extraction for satellite image segmentation using statistical global and local feature. In: Lecture Notes in Electrical Engineering (LNEE), vol. 380, pp. 247–255, April 2016. https://doi.org/10.1007/978-3-319-30301-7_26
11. Nikias, C.L., Mendel, J.M.: Signal processing with higher-order spectra. IEEE Sig. Process. Mag. **10**(3), 10–37 (1993)
12. Petropulu, A.: Higher-order spectral analysis. In: Madisetti, V.K., Williams, D.B. (eds.) Digital Signal Processing Handbook. Chapman & Hall/CRCnetBASE (1999)

Including Personality Traits, Inferred from Social Networks, in Building Next Generation of AEHS

Kenza Sakout Andaloussi[1(✉)], Laurence Capus[1], and Ismail Berrada[2]

[1] Department of Computer Science and Software Engineering, Laval University, Quebec City, QC, Canada
kenza.sakout-andaloussi.1@ulaval.ca,
laurence.capus@ift.ulaval.ca
[2] FSDM FES, LIMS LAB,
Université Sidi Mohamed Ben Abdellah, Fez, Morocco
ismail.berrada@univ-lr.fr

Abstract. User profile inference on online social networks is a promising way for building recommender and adaptive systems. In the context of adaptive learning systems, user models are still constructed by means of classical techniques such as questionnaires. Those are too time-consuming and present a risk of dissuading learners to use the system. This paper explores the feasibility of learner modeling based on a proposed set of features extracted and inferred from social networks, according to the IMS-LIP specification. A suitable general architecture of an AEHS is presented, whose adaptation combines three distinct aspects: Felder and Silverman learning style, knowledge level and personality traits. This latter is a novel adaptation criterion, it is an interesting user feature to be incorporated in user models, a feature that is not yet considered by existing AEHS. However, adapting such systems to personality traits contributes to achieving a better adaptation by varying learning approaches, integrating collaboration and adapting feedback. The aim of this paper is to show how this contribution is doable through the proposed framework.

Keywords: Educational hypermedia system · Adaptation · Learner model
FSLM · Big five personality traits · Social networks

1 Introduction

E-learning environments are aspiring to respond to the growing need for personalized on-line learning by providing more support for adaptability and on-demand learning object generation [1]. Adaptive Educational Hypermedia systems (AEHS) are considered as one of the key areas for delivering personalized e-learning. The benefit of such learning is that it can be dynamically tailored to the individual student's abilities and skill attainment. This empowers the learner engagement and his learning outcomes. Although, some studies [2, 3] report that AEHS are not widely used since they are still challenging various issues. Particularly, in the earlier work [4], we have showed, through our analysis of 50 current AEHS, that for learner modeling, designers need to

© Springer International Publishing AG, part of Springer Nature 2018
M. Ben Ahmed and A. A. Boudhir (Eds.): SCAMS 2017, LNNS 37, pp. 136–148, 2018.
https://doi.org/10.1007/978-3-319-74500-8_13

consider the maximum amount of relevant data without overloading the user by questionnaires that may dissuade him/her. Since the quality of the provided person-alized learning depends largely on the characteristics and richness of the learner model, adaptive learning systems would benefit from improving their learner models. We have also noticed that there is a lack of standards use, which could have insured the inter-operability, the reusability and the scalability of the learner model. Finally, concerning the adaptation model, we have noticed that all AEHS adapt their features to one or two of the following aspects: content, navigation or presentation and very few systems adapt to all the three at the same time. We wonder whether there are other adaptation aspects that could boost learner motivation.

To cope with these issues, we investigate, in this paper, how social networks can help building AEHS. We propose a system architecture that acquires wider knowledge about a user, from his/her interaction with social networks. This will allow reaching a better adaptation and avoiding disadvantages of questionnaires, such as additional time that students need to spend and the influence of lack of motivation to fill out loaded questionnaires. In fact, on one hand, our literature review concerning social networks analysis [5, 6] shows that we can use such networks to analyze user behavior, extract his/her preferences and predict his/her personality traits. On the other hand, several published papers acknowledged that personality traits, in particular, influence learning and academic behaviors (how the learner likes to proceed, what motivates him and if he/she likes to collaborate with others) [7, 8]. That explain why we should include personality traits in our learner model as an adaptation criteria together with the learning style and the knowledge level. Effectively, adapting our system to the learner personality traits imply varying the learning approach for each type of learners, encouraging the use of collaborative tools when needed and providing the most con-venient feedback for each trait. This latter is a novel adaptation aspect that can be added to the three ones used to date: content, navigation and presentation.

In addition, this paper shows how we could use educational specifications (IMS-LIP [9] for user modeling) and standards (IEEE-LOM [10] and SCORM [11] for domain modeling) to support dynamic modeling and collaboration during the per-sonalized learning process that corresponds to the expected objectives.

The remainder of the paper is structured as follows. In Sect. 2, we present the big five personality model and the Felder and Silverman learning style model and how we can differentiate learning according to their characteristics. In Sect. 3 we describe our system architecture with its detailed models. Finally, we conclude with directions for future work in Sect. 4.

2 Related Work

Providing accurate personalized learning to users requires modeling their preferences, interests and needs. This is referred to learner modeling. Therefore, most of the researchers on adaptive learning systems have focused on the learner profile based on his knowledge level and/or learning style. We agree that including these data in an adaptive system is essential, since knowledge level serves to define the appropriable learning object difficulty and the learning style represents how the student like processing information. However, we think that if we can consider also the learner personality traits, this will extend the learner model and provide more adaptability. So we suggest an association between personality type, learning style and knowledge level as criteria for generating customized learning objects. In this work, we adopt the big five personality model [12] since it is one conceptualization of personality that has been increasingly studied and validated in the scientific literature [13] and the Felder and Silverman learning style model for its simplicity and well acceptance [4].

2.1 Big Five Personality Traits

Personality could be defined as the set of an individual's characteristics and behaviors that guide him/her to make decisions and act accordingly under specific conditions [14].

According to the big five personality model, most human personality traits can be described in five wide-ranging dimensions which are: Openness, Consciousness, Extraversion, Agreeableness and Neuroticism (OCEAN) [12]. This model has emerged for understanding the relationship between personality and academic behaviors [15–17]. Table 1 summarizes these relationships.

Applying the big five personality model in this work will allow us to determine which learners are thoughtful (deep approach), which ones process information more superficially (surface approach) and which ones focus on the product (achieving approach). Furthermore, this model will help us to determine if learners like to collaborate or not and which learners need an extrinsic motivation for learning. In fact learners with high scores (greater than 50%) in:

- Openness, Consciousness or Extraversion are more likely to proceed with a deep approach, so they make sense of what they are learning, they can relate it to their previous knowledge, they have positive emotion about learning and they like discuss their thoughts with others [18].
- Extraversion are more interested in obtaining high grades, they follow up all suggested material and exercises. So, it's more advantageous to remember them the objectives of courses and how those can help them to success in their career. This is called the achieving approach [15].
- Neuroticism are likely to limit their study to the minimal fundamentals, they don't make connections between pieces of information but can memorize what they learn in an atomistic way. This is called the surface approach [15].

Learners' personality traits can be measured in different ways. The explicit way uses questionnaires such as the International Personality Item Pool of the NEO (IPIP-NEO) [19], the mini-IPIP scale [20], the Big Five Inventory (BFI) [12] or the NEO Five Factor Inventory (NEO-FFI) [21]. The implicit way, which particularly interests us in this work, consists of predicting the learner personality traits from his digital footprints of behavior in social networks especially Facebook or Twitter[1] [22].

---

[1] https://applymagicsauce.com/demo.html.

Table 1. Correlation between personality traits and learning characteristics

Personality trait	Characteristics	Learning approach	Collaboration type	Motivation type
Openness (high +50%)	Asking questions, analyzing arguments, critical, logical, relating learning to previous knowledge	Deep approach	Collaborative	Intrinsic
Consciousness (high +50%)	Concentration, autonomy, organization, caring about learning conditions, clear goals	Deep and strategic approach	Individual	Intrinsic
Extraversion (high +50%)	Perceiving studying as a means of getting hold of a degree or finding a well-paid job	Deep approach	Collaborative	Academic success motivation
Agreeableness (high +50%)	Friendliness, Trustworthiness, and cooperativeness	Achieving approach	Collaborative	Extrinsic
Neuroticism (high +50%)	Lack of concentration, fear of failure, problems in understanding how things relate to each other	Surface approach	Individual	Extrinsic

2.2 Felder and Silverman Learning Style Model

Felder and Silverman, in their inventory of learning style [23], outline various dimensions regarding how people process information, and each dimension has two possible values:

- Information processing: Active (A)/Reflective (R)
- Perception: Sensing (S)/Intuitive (I)
- Input: Visual (Vi)/Verbal (Ve)
- Understanding: Sequential (Seq)/Global (G).

The combinations of these preferences result in total of 16 learning styles types and are typically denoted by four letters to represent a person, for example one learner can have as learning style: (A, I, Ve, G).

The relation between the learning style and learning strategy has attracted many researchers. The results of their works showed that learner, tend to favor a particular teaching strategy enabling him to better assimilate the course. Some authors [24, 25] suggest an association between learning style and learning objects in E-learning context. Table 2 summarizes the preferences of each style according to the Felder and Silverman model.

The Felder and Silverman learning style is generally determined within the administration of the Index of Learning Styles Questionnaire (ILSQ) which is an online

Table 2. Felder & Silverman scale and its implications for learning preferences

Learning style dimension	Type	Preferences
Information processing	Active	Applied exercises, experimentation, simulations, role-play, project, group work
	Reflective	Less exercises, situations problems, summaries, case study, individual work
Perception	Sensing	Applied exercise, experiences, concrete facts, first examples then theory
	Intuitive	Theoretical data, theory before examples, abstract problems
Entry channel	Visual	Graphs, photos, diagrams, charts, videos, multimedia
	Verbal	Text, audio, hypertext, conferences, lecturing, verbal information
Understanding	Sequential	Exercises after theory, summaries after course, logical fixed order
	Global	Holistic approach: overview, exercises and summary before course

form composed of 44 questions [26]. However, it has been noticed that questionnaire is not reliable since given answers could not accord with the real behavior the questions aim to investigate (either the user deforms his answers intentionally or not) [27]. To come up with this issue, some authors proposed probabilistic methods in order to detect the style of the learner by investigating his behavior while using the system. These methods include neural networks [28], KNN [29], Bayesian networks [30], etc.

3   Our Proposed System Description

In this section we propose a framework for our adaptive learning system. This framework consists of methods and mechanisms to provide a customized educational experience which meets the educational interests and needs specific to each learner. Our system is distinguished from the existing ones by the fact of adopting social networks APIs to initiate the learner model. That implies several modifications to the classical models used to date:

- for the learner model, it will consider the personality traits as a data and an adaptation factor;
- for the domain model, it will take into account collaborative tools as learning objects, so the system can show them when needed;
- for the adaptation model, it will add feedback as an adaptation aspect, in a way that it can be adapted to the learner personality traits.

Our proposed architecture is conforming to the LAOS theoretical framework [31] as shown in Fig. 1. We have chosen this model since it is comparatively the most

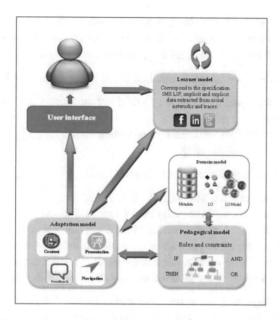

Fig. 1. Macro architecture of our adaptive learning system.

recent one, it provides a clear separation of the major parts of an AEHS and it contributes to modeling the pedagogical process of a course.

The learner model refers to the user model (UM) in LAOS, the pedagogical model is the equivalent of the goals and constraints model (GM), the domain model and the adaptation model maintain the same nomination (DM) and (AM) and the user interface represents the presentation model (PM).

3.1 Learner Model

Our learner model is based on the IMS-LIP specification model [9]. This specification will ensure the interoperability and the richness of the learner model, an aspect that was rarely considered by the existing AEHS. The proposed learner model includes implicitly and explicitly acquired data. The adaptation to this model essentially considers: the personality traits, the learning style according to the Felder & Silverman model and the knowledge level.

A novel way to respond to the IMS-LIP specification without overloading the user by questionnaires is the use of social networks. In fact, domain-independent data that could be retrieved via the Facebook, Twitter and LinkedIn APIs are:

- Identification: Name, e-mail, age, gender, mother tongue.
- Affiliation: names of the groups user is affiliated.
- QCL: qualifications, certifications and licenses.
- Accessibility: language skills.
- Interests: hobbies, entertainment.
- Goal: in terms of career.

- Accessibility: personality traits which are detected from navigation traces and behavior of the user on Facebook or Twitter. Once those traits are defined, we can conclude the motivation type and the preference to collaborative work.

The system uses tests and/or questionnaires to fill in the following fields:

- Competency: skills and knowledge acquired before starting a course and at the end of each learning sequence.
- Activity: other activities initiated by the learner.
- Goal: specific to the field study.

The system automatically detects the following fields:

- Relationship: between the system and other data structures.
- Security key: password and security codes assigned to the learner.
- Transcript: a summary of the results obtained when using the system. We will enrich this field with a complete description of the learner's navigation and his/her connection time. These data which will be used to predict the learning style of the user according to the Felder & Silverman model (trait of the Accessibility field).

Once the user registered, the system stores his personal data, that is mostly extracted from his social networks if he owns one at least, especially Facebook. Otherwise, the user is asked to fill in a form that allows to respond to the IMS LIP specifications and to answer the big five personality traits questionnaire. Concerning the learning style, the user has the choice to answer the ILS questionnaire or to let the system detecting it automatically from his interaction and navigation traces while using the system.

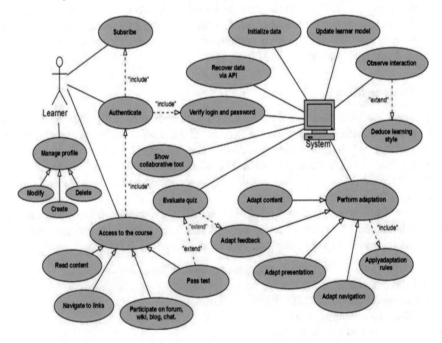

Fig. 2. Use case diagram of the learner and the system

Mostly, the learner, as a principal actor of the adaptive system, must subscribe if he/she is a new user, either by providing his/her login/password in order to access to his/her social networks or by answering a questionnaire. If the learner has already used the system before, he/she just need to authenticate his/herself.

The learner can always manage his/her profile by adding, removing or changing some data. Finally, the learner can access to the courses available within the system, he/she can: read their contents, navigate to the links proposed by the system, participate on forums, wikis, blogs and chats and pass tests/quizzes, so the system can evaluate him/her.

Figure 2 outlines the prospective use cases of the learner in his/her interaction with the system, according to the modeling language UML (Unified Modeling Language).

3.2 Domain Model

The proposed domain model is composed of assembled learning objects. This new tendency plays an important role in the development of e-learning systems by allowing the use and reuse of digital courses.

To ensure the reuse of learning objects by virtual platforms, their information structure is standardized. The most common standards are: LOM (Learning Object Metadata) [10], SCORM (Sharable Content Object Reference Metadata) [11], IMS-LD (IMS Learning Design) [32] and ISO/IEC 19788 – MLR (Metadata Learning Resources) [33].

For our domain model, we have chosen to use the standards LOM and SCORM. Since LOM, standardized by the IEEE organization, provides a metadata model for describing learning objects, which facilitates their indexing and reuse. While SCORM proposes a mechanism for exploiting various learning objects in a system and controlling their use.

Multiple versions of the same learning object are created in agreement to the principle of differentiated pedagogy [34], and are indexed by the means of the LOM content metadata.

The LOM standard indexes each learning object within nine descriptive elements as follows: 1. General, 2. Lifecycle, 3. Meta metadata, 4. Technical, 5. Educational, 6. Rights, 7. Relation, 8. Annotation and 9. Classification.

Table 3 shows our list of descriptors that are essential for adapting resources to the learner profile.

Table 3. Our list of descriptors used for learning object adaptation.

Identifier	Element	Description
1	**General**	
1.2	Title	Title of the resource
1.3	Language	Language(s) of the resource
1.4	Description	Description of the resource content
1.5	Keyword	keywords giving information about the theme carried by the resource content
1.7	Structure	Basic structured organization of the resource (collection, linear, hierarchical, etc.)
4	**Technical**	
4.1	Format	Sound, textual
5	**Educational**	
5.1	Interactivity type	Description of the predominant learning mode promoted by the resource
5.2	Learning resource type	Free activity, narrative text, auto evaluation, case study, summary, demonstration, formative evaluation, exercise, experience, exploration, reading text/presentation, educational game, role-play, project, educational scenario, simulation
5.3	Interactivity level	High (1) or low (0)
5.4	Semantic density	Very low to very high
5.6	Context	Description of the pedagogical use of the resource
5.7	Typical age range	Age of users
5.8	Difficulty	Resource difficulty: very easy, easy, medium, difficult, very difficult
5.9	Typical learning time	Approximate learning time
7	**Relation**	
7.1	Kind	Kind of relations between learning objects (prerequisite, part of, based on, etc.)
9	**Classification**	
9.1	Purpose	Purpose of the resource

3.3   Pedagogical Model

The purpose of our pedagogical model is to describe the learning strategies. It consists of rules and constraints that synthesize the domain knowledge hierarchy (prerequisites, parts, equivalents, etc.) and the constraints that have to be respected by the system for a better adaptation of learning objects. We propose a set of rules, on the basis of the works of [24, 25], that allows adaptation to the knowledge level, to the personality traits and to the Felder and Silverman learning style:

Knowledge level

- **Rule1**: IF "test result" <50% THEN "LOM.Educational.Difficulty (5.8)" = very easy.
- **Rule2**: IF 50% <= "test result" <60% THEN "LOM.Educational.Difficulty (5.8)" = easy.
- **Rule3**: IF 60% <= "test result" <70% THEN "LOM.Educational.Difficulty (5.8)" = medium.
- **Rule4**: IF 70% <= "test result" <90% THEN "LOM.Educational.Difficulty (5.8)" = difficult.
- **Rule5**: IF "test result"> = 90% THEN "LOM.Educational.Difficulty (5.8)" = very difficult.

Personality traits

- **Rule1**: IF "personality type" = Openness, Extraversion or Agreeableness (+50%) THEN "LOM.Educational.Interactivity level (5.3)" =1.
- **Rule2**: IF "personality type" = Consciousness or Neuroticism (+50%) THEN "LOM.Educational.Interactivity level (5.3)" =0.
- **Rule3**: IF "personality type" = Openness, Consciousness or Extraversion (+50%) THEN "LOM.Educational.Semantic density (5.4)"> 2.
- **Rule4**: IF "personality type" = Agreeableness or Extraversion (+50%) THEN display LOM.Classifcation.Purpose (9.1).
- **Rule5**: IF "personality type" = Neuroticism (+50%) THEN display LOM.Classifcation.Purpose (9.1) and "LOM.educational.Semantic density (5.4)" = 0 or 1.

Felder & Silverman learning style

- **Rule1**: IF "learning style" = Active THEN "LOM.Educational.Learning resource type (5.2)" = "exercise or experience or simulation" AND "LOM.Educational.Interactivity level (5.3)" =1.
- **Rule2**: IF "learning style" = Reflective THEN "LOM.Educational.Learning resource type (5.2)" = "case study or exploration or summary".
- **Rule3**: IF "learning style" = Sensing THEN "LOM.Educational.Learning resource type (5.2)" = "exercise or experience or simulation" and "LOM.General.Structure (1.7)" = examples before theory.
- **Rule4**: IF "learning style" = Intuitive THEN "LOM. Educational.Learning resource type (5.2)" = "case study or narrative text" and "LOM.General.Structure (1.7)" = theory before examples.
- **Rule5**: IF "learning style" = Visual THEN "LOM.technical.format (4.1)"= "textual or video".
- **Rule6**: IF "learning style" = Verbal THEN "LOM.technical.format (4.1)"= "sound or video".
- **Rule7''**: IF "learning style" = Sequential THEN "LOM.General.Structure (1.7)" = theory before exercises.
- **Rule8**: IF "learning style" = Global THEN "LOM.General.Structure (1.7)" = summary and exercises before theory.

3.4   Adaptation Model

We propose a probabilistic adaptation model, such us Bayesian Networks [30], which dynamically calculates the probability that a learning object is the most suitable to the learner's learning style, personality traits and knowledge level. It allows also giving feedback to learners based on their test results and their personality traits following to the recommendations of the authors [35].

The adaptation model is, all the time, on communication with the learner model, the domain model and the pedagogical model. It first checks if the title of the resource, its purpose, description, context, language and typical age range correspond to the requested course by the learner and his profile. After what, and on basis of the learner prerequisite test results, the adaptation model defines the difficulty of the corresponding resource (following the rules of the pedagogical model). Then, it applies the rules corresponding to the learner's personality type. Finally, once the learner's learning style is defined, the adaptation model selects the most probable convenient learning objects.

In virtue of the probabilistic and dynamic aspects enclosed in automatic detection of learning styles, our system gradually and constantly adjusts the learner model. Consequently, this inquires the adjustment of the learning objects generation, in a way that those concord with the updated learner model. It's another advantage of our system, that was not widely applied in the current AEHS.

4   Conclusion and Future Work

Within this paper, we have proposed an initial formalization of our adaptive learning system architecture. We have investigated a new way to initialize the learner model by using social networks, while meeting the specifications of the IMS LIP. The compliance to such specification will grants the interoperability and the reusability of the learner model, in contrast to the existing AEHS learner models. Also the extracted data from social networks will enrich our learner model and will reduce the use of questionnaires.

We introduced adaptation rules corresponding to the learner's personality traits, on the basis of the literature review concerning the subject. Such traits have never been used by the current AEHS. However, we have noticed that the adaptation to personality traits is quite important. It involves considering collaborative learning objects, adapting the learning approach and providing adequate feedback to learners. This will increase their motivation to use the system and will promote interaction and collaboration among users, whose personality shows that they like to interact. That is, in fact, one of the principles of the connectivism learning theory [36].

We discussed also how we can model the domain, in a way that it corresponds to the most common standards and it allows dynamic adaptation of the adaptation model.

The proposal is under implementation in a prototype system, once done we should test it on a sample of learners and discuss the results.

References

1. Brusilovsky, P.: Adaptive hypermedia. User Model. User Adap. Inter. **11**(1–2), 87–110 (2001)
2. Baki, A., Güven, B., Karal, H., Özyurt, Ö., Özyurt, H.: Evaluation of an adaptive and intelligent educational hypermedia for enhanced individual learning of mathematics: a qualitative study. Exp. Syst. Appl. **39**, 12092–12104 (2012)
3. Wilson, C., Scott, B.: Adaptive systems in education: a review and conceptual unification. Int. J. Inf. Learn. Technol. **34**(1), 2–19 (2017)
4. Sakout, A.K., Capus, L., Berrada, I.: Adaptive educational hypermedia systems: current developments and challenges. In: Proceedings of the 2nd International Conference on Big Data, Cloud and Applications, BDCA 2017, Tetouan, Morocco, 29–30 March 2017. ACM (2017)
5. Liu, Y., Wang, J., Jiang, Y.: PT-LDA: a latent variable model to predict personality traits of social network users. Neurocomputing **210**, 155–163 (2016). SI:Behavior Analysis In SN
6. Staiano, J., Lepri, B., Aharony, N., Pianesi, F., Sebe, N., Pentland, A.: Friends don't lie: inferring personality traits from social network structure. In: Proceedings of the 2012 ACM Conference on Ubiquitous Computing, UbiComp 2012, pp. 321–330. ACM, New York (2012)
7. Faria, A.R., Almeida, A., Martins, C., Gonçalves, R., Figueiredo, L.: Personality traits, learning preferences and emotions. In: Proceedings of the Eighth International C* Conference on Computer Science and Software Engineering, C3S2E 2015, pp. 63–69. ACM, New York (2008)
8. Hazrati-Viari, A., Rad, A.T., Torabi, S.S.: The effect of personality traits on academic performance: the mediating role of academic motivation. Procedia - Soc. Behav. Sci. **32**, 367–371 (2012). The 4th International Conference of Cognitive Science
9. IMS: IMS Learner Information Package Information Model v1 (2001). http://www.imsglobal.org/profiles/lipinfo01.html. Accessed 20 Apr 2017
10. IMS: IMS Meta-data Best Practice Guide for IEEE 1484.12.1-2002 Standard for Learning Object Metadata (2006). http://www.imsglobal.org/metadata/mdv1p3/imsmd_bestv1p3.html. Accessed 20 Apr 2017
11. ADL: Sharable Content Object Reference Model (SCORM) 2004, 4th Edition Content Aggregation Model (CAM) Version 1.1 (2009)
12. John, O.P., Naumann, L.P., Soto, C.J.: Paradigm shift to the integrative big-five trait taxonomy: history, measurement, and conceptual issues. In: John, O.P., Robins, R.W., Pervin, L.A. (eds.) Handbook of Personality: Theory and Research, pp. 114–158. Guilford Press, New York (2008)
13. Cobb-Clark, D.A., Schurer, S.: The stability of big-five personality traits. Econ. Lett. **115**(1), 11–15 (2012)
14. Larsen, R.J., Buss, D.M.: Personality Psychology: Domains of Knowledge About Human Nature, 2nd edn. McGraw Hill, New York (2005)
15. Chamorro-Premuzic, T., Furnahm, A., Lewis, M.: Personality and approaches to learning predict preferences for different teaching methods. Learn. Individ. Differ. **17**, 241–250 (2007)
16. Entwistle, N.: Motivational factors in students' approaches to learning. In: Schmeck, R.R. (ed.) Learning Strategies and Learning Styles, pp. 21–49. Plenum Press, New York (1988)
17. Marcela, V.: Learning strategy, personality traits and academic achievement of university students. Procedia - Soc. Behav. Sci. **174**, 3473–3478 (2015). International Conference on New Horizons in Education, INTE 2014, 25–27 June 2014, Paris, France

18. Heinström, J.: The impact of personality and approaches to learning on information behavior. Inf. Res. **5**(3) (2000)
19. Johnson, J.A.: Measuring thirty facets of the Five Factor Model with a 120-item public domain inventory: development of the IPIP-NEO-120. J. Res. Pers. **51**, 78–89 (2014)
20. Donnellan, M.B., Oswald, F.L., Baird, B.M., Lucas, R.E.: The mini-IPIP scales: tiny-yet-effective measures of the Big Five factors of personality. Psychol. Assess. **18**, 192–203 (2006)
21. Costa, P.T., McCrae, R.R.: Revised NEO Personality Inventory (NEO-PI-R) and NEO Five-Factor Inventory (NEO-FFI) Manual, Odessa, FL. Psychological Assessment Resources (1992)
22. Kosinski, M., Stillwell, D., Graepel, T.: Private traits and attributes are predictable from digital records of human behavior. Proc. Nat. Acad. Sci. **110**(15), 5802–5805 (2013)
23. Felder, R.M., Silverman, L.K.: Learning styles and teaching styles in engineering education. Eng. Educ. **78**(7), 674–681 (1988)
24. Franzoni, A.L., Assar, S.: Student learning styles adaptation method based on teaching strategies and electronic media. Educ. Technol. Soc. **12**(4), 15–29 (2009)
25. Karagiannidis, C., Sampson, D.: Adaptation rules relating learning styles research and learning objects meta-data. In: Workshop on Individual Differences in Adaptive Hypermedia, 3rd International Conference on Adaptive Hypermedia and Adaptive Web-Based Systems, Eindhoven, The Netherlands (2004)
26. Index of Learning Styles Questionnaire. https://www.engr.ncsu.edu/learningstyles/ilsweb.html. Accessed 23 Jan 2017
27. Draper, S.: Observing, measuring and evaluating a courseware: a conceptual introduction. In: Implementing Learning Technologies, Learning Technology Dissemination Initiative, pp. 58–65 (1996). http://www.icbl.hw.ac.uk/ltdi/implementing-it/measure.pdf. Accessed 18 Jan 2017
28. Zatarain-Cabada, R., Barrón-Estrada, M.L., Angulo, V.P., García, A.J., García, C.A.R.: Identification of Felder-Silverman learning styles with a supervised neural network. In: Advanced Intelligent Computing Theories and Applications. With Aspects of Artificial Intelligence, pp. 479–486. Springer, Heidelberg (2010)
29. Zatarain-Cabada, R., Barrón-Estrada, M., Zepeda-Sánchez, L., Sandoval, G., OsorioVelazquez, J., Urias-Barrientos, J.: A Kohonen network for modeling students' learning styles in Web 2.0 collaborative learning systems. In: Advances in Artificial Intelligence, MICAI 2009, pp. 512–520 (2009)
30. Carmona, C., Castillo, G., Millán, E.: Designing a Dynamic Bayesian Network for modeling student's learning styles. In: Díaz, P., Kinshuk, A.I., Mora, E. (eds.) ICALT 2008, pp. 346–350. IEEE Computer Society, Los Alamitos (2008)
31. Cristea, A., de Mooij, A.: LAOS: layered WWW AHS authoring model and their corresponding algebraic operators. In: WWW 2003 Proceedings of World Wide Web International Conference. ACM, New York (2003)
32. IMS: IMS Learning Design Information Model Revision, 20 January 2003. http://www.imsglobal.org/learningdesign/ldv1p0/imsld_infov1p0.html. Accessed 20 Apr 2017
33. ISO/IEC 19788-1: Information technology – Learning, education and training – Metadata for learning resources – Part 1: Framework (2011)
34. Grenier, N., Moldoveanu, M.: Differentiated pedagogy: a new teaching model in multiethnic elementary school settings in Quebec, Canada. In: EDULEARN11 Proceedings, pp. 758–765 (2011)
35. Dennis, M., Masthoff, J., Mellish, C.: Adapting progress feedback and emotional support to learner personality. Int. J. Artif. Intell. Educ. **26**, 877–931 (2016)
36. Siemens, G.: Connectivism: a learning theory for the digital age. Int. J. Instr. Technol. Distance Learn. **2**(1), 3–10 (2005)

Embedded Systems HW/SW Partitioning Based on Lagrangian Relaxation Method

Adil Iguider[(✉)] ⬤, Mouhcine Chami, Oussama Elissati,
and Abdeslam En-Nouaary

Institut National des Postes et Télécomunications, Lab. STRS,
Av. Allal El Fassi, Madinat Al Irfane, Rabat, Morocco
{iguider, chami, elissati, abdeslam}@inpt.ac.ma

Abstract. Embedded systems (ES) are nowadays, in the heart of every complex electronic device. An ES is a system that combines both hardware blocks and software blocks in a single chip. The necessity to decrease the cost and the development time of the design flow of the ES and to keep the overall performance of the system require the development of new design approaches for such systems. The compound design (co-design) is a very interesting approach used to fulfill the latter requirements. The partitioning of blocks between hardware and software is one of the most important steps in this process of co-design. In this paper, we present a novel method (heuristic) based on optimal path optimization technique (lagrangian relaxation method) to deal with the partitioning problem. The solution aims to optimize the hardware area (cost) of the ES while respecting a given constraint time of execution. To validate the effectiveness of our approach, we give a comparison with the results obtained with the Genetic Algorithm (GA).

Keywords: Embedded systems · HW/SW partitioning · Lagrangian relaxation
Heuristic algorithms · Co-design

1 Introduction

The Hardware/Software co-design plays a major role in designing modern embedded systems application. In fact, it facilitates the integration of embedded systems in many critical and important sections. Particularly, in smart cars, smart building, home automation, smart grid and many other sections which compose modern smart cities. The Hardware/Software co-design as defined in [1] is the design of cooperating hardware components and software components in a single design effort. Choosing a good balance between hardware implementation and software implementation is driving by several factors such as performance, energy efficiency, power density, design complexity, design cost and design schedules. In this article, we make a focus on the performance factor which is related to the execution time and the cost factor which is related to the hardware area. Several approaches had been proposed in the objective to optimize the partitioning while dealing with those two factors. There are mainly two approaches' families, the exact algorithms and the heuristic algorithms. In the **exact algorithms** family, we find especially Branch and Bound method (BB),

© Springer International Publishing AG, part of Springer Nature 2018
M. Ben Ahmed and A. A. Boudhir (Eds.): SCAMS 2017, LNNS 37, pp. 149–160, 2018.
https://doi.org/10.1007/978-3-319-74500-8_14

Integer Linear Programming (ILP) and Dynamic Programming (DP). BB is defined in [2] the algorithm is based on binary tree, the objective is to find the path from the top to the bottom of the tree. An example of its application to Hardware Software Partitioning (HSP) problem is presented in [3]. ILP formulation consists of a set of variables, a set of linear inequalities, and a single linear function of the variables that serves as an objective function, ILP was used in HSP problem in [4]. DP is a method, in which large problems are broken down into smaller problems, and through solving the individual smaller problems, the solution to the larger problem is discovered, an example of using dynamic programming in HSP is described in [5].

The **heuristic algorithms** family contains Simulated Annealing (SA), Genetic Algorithm (GA), Tabu Search (TS), Greedy Algorithm (GR), Hill Climbing Algorithm (HC) and Particle Swarm Optimization (PSO). SA algorithm is based on the analogy between the solid annealing and the combinatorial optimization problem, the algorithm is explained in [6], an enhancement of the SA algorithm is presented in [7]. Genetic Algorithm (GA) mimics the process of natural evolution and is based on the survival-of the fitness principle, the steps are: (1) Population Initialization, (2) Parents selection, (3) Crossover, (4) Mutation, and the process is repeated from step 2 until the termination condition is met, the algorithm is explained in [8], in which the authors proposed an heuristic algorithm based on a combination of simulated annealing algorithm and genetic algorithm. Many other researches were proposed based on genetic algorithm as in [9–15]. Tabu Search algorithm employs local search methods to a problem and checks its immediate neighbors in the hope of finding an improved solution, an example of tabu search implementation is presented in [6]. Other studies were based on tabu search algorithm, as in [16, 17]. Greedy Algorithm constructs a solution in iterative way, it starts by a candidate set, and at each step it adds the element that gives the best optimization (optimize the objective function under a set of constraints), an implementation of greedy algorithm for HSP problem is described in [18]. In [19], the authors propose an enhanced greedy algorithm that escapes local minima and leads to the globally optimal solution. Hill Climbing Algorithm consists of starting with a sub-optimal solution to a problem, and then repeatedly improves the solution until some condition is maximized. Unlike the Greedy Algorithm, Hill Climbing Algorithm has the ability to avoid local minima. In [20], the authors propose a novel technique for the neighbors search. Particle Swarm Optimization (PSO) is defined in [21]. PSO consists of a swarm of particles, where particle represent a potential solution (better condition). Particle will move through a multidimensional search space to find the best position in that space. An example of using PSO in HSP problem is presented in [22]. In [23], the authors propose a method based on the shortest path algorithm. Table 1 summarizes and gives taxonomy of cited algorithms. This paper treats the HSP problem in the same objective. A novel heuristic approach is proposed to minimize the global cost under a given temporal constraint. As in [23], the Data Flow Graph (DFG) is used to model the system. In our approach, each block is supposed to communicate directly with the next one. The idea of our proposal is to construct a double value directed graph with all possible implementations (hardware or software) by duplicating each block. In this constructed graph, each node represents the block's implementation (hardware or software), and each edge has an execution time and has a cost which are related to the current block's implementation and its successor. The objective is to find the best path which has the minimal cost while respecting the global time constraint. The algorithm can be applied to optimize the cost and the execution time interchangeably. Also, it can be applied when multiple types of hardware and software are used.

Table 1. Algorithms in HW/SW partitioning

Reference	Based algorithms			Optimized metrics	
	Algorithm	Exact	Heuristic	Execution time	HW area
[3]	Branch & Bound	–			–
[4]	ILP	–			–
[5]	Dynamic Programming	–		–	–
[6, 7]	SA		–	–	
[9–15]	GA		–	–	–
[16]	GA+TS		–	–	
[18, 19]	Greedy Algorithm		–		–
[20]	Hill Climbing		–		–
[22]	PSO		–		–
[23]	Shortest path		–	–	
Proposed	Shortest path		–	–	–

This paper is organized as follow. After the introduction, in Sect. 2, we present the problematic and the proposed solution based on Lagrangian Relaxation method. In Sect. 3, we give the results of tests and the comparison with the Genetic Algorithm. Finally, Sect. 4 gives the conclusion and the future works.

2 Optimal Path Optimization for HW/SW Partitioning Problem

2.1 Problem Formalization

A hardware/software partition is defined using two sets H and S, where H is the set of blocks designed in hardware and S is the set of blocks designed in software. The system ES is composed on N blocks $B = \{B1, B2, B3, ..., Bn\}$, the system is represented as DFG model, and each block is supposed to communicate directly with the adjacent block as shown in the example in Fig. 1. The blocks are executed in parallel and each block is firing upon the required tokens are presents at its inputs. It is assumed that the system is not influenced by the external systems, in this case even if the blocks are executed in parallel, the data must traverse the blocks in a sequential manner, and therefore the execution time of the system is the sum of the execution time of each block within the system. The partitioning problem consists of finding the optimal sets H and S that optimize the cost (hardware area) of the system within a global time of execution constraint, where $H \cap B = \varnothing$ and $H \cup S = B$.

Fig. 1. DFG model of four blocks

As described in the articles of the latter section, the objective is minimize the $\sum C(B_i)$ under a constraint on $\sum T(B_i)$ where $C(B_i)$ and $T(B_i)$ are respectively the cost and the execution time of the block B_i. The idea of our approach is as follow, from the DFG graph (Fig. 1) we construct a directed graph by duplicating each block as represented in Fig. 2, each node in the constructed graph represents the nature of the block's implementation (hardware or software). We also add two fictional blocks which are Entry block and Exit block which represents respectively the entry point and exit point of the data that will traverse the system. Each edge in the graph has a cost and a time, the cost represents the needed cost for the data to use the edge, and the time represents the execution time needed for the data to reach the next node by using this edge. The cost and the time depend on the nature of the current block and the nature of its successor (hardware or software). The data will then traverse the graph from the entry point to the exit point by using a possible path. The objective is to find the best path that optimizes the global cost and that respects a given global time of execution constraint.

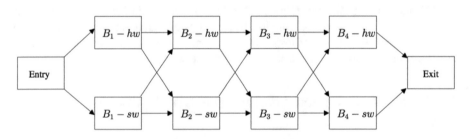

Fig. 2. Graph representation with four blocks

The formula for individual cost (C_i) calculation and individual execution time (T_i) calculation of each edge are as follow (Fig. 3):

- $C_i(hw - hw) = A(B_i)$
- $C_i(hw - sw) = A(B_i) +$ communication cost $(hw - sw)$
- $C_i(sw - sw) = $ Size $(.elf\ B_i)$
- $C_i(sw - hw) = $ Size $(.elf\ B_i) +$ comm. cost $(sw - hw)$
- $T_i(hw - hw) = $ Texec $(B_i hw)$
- $T_i(hw - sw) = $ Texec $(B_i hw) +$ Tcomm $(hw - sw)$
- $T_i(sw - sw) = $ Texec $(B_i sw)$
- $T_i(sw - hw) = $ Texec $(B_i sw) +$ Tcomm $(sw - hw)$

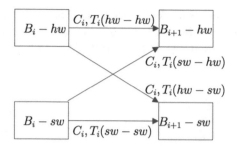

Fig. 3. Two adjacent blocs of the graph

With:

- $A(B_i)$: hardware area of block B_i
- Size (.elf B_i): size of binary file of block B_i
- Texec (B_i): execution time of block B_i

The cost of a hardware block is related to its area and the cost of a software block is related to the size of its binary file (.elf file). The cost of communication between a hardware block and a software block is related to bus usage. Therefore, the cost of an edge which is between two hardware nodes is the cost of the hardware area of the first node (B_i in hardware). The cost of an edge which is between two software nodes is the cost of the size of the binary file of the first node (B_i in software). The cost of an edge which is between a hardware node and a software node is the cost of the hardware area of the first node plus the communication cost between the first node (hardware) and the second node (software). Finally, the cost of an edge which is between a software node and a hardware node is the cost of the binary file of the first node plus the communication cost between the first node (software) and the second node (hardware). The communication time between a hardware block and a software block is related to the time used by the bus, and then the communication time between a hardware block and a hardware block or between a software block and a software block is neglected. In this case, an edge between two hardware nodes has an execution time equal to the execution time of the first node (B_i in hardware). An edge between two software nodes has an execution time equal to the execution time of the first node (B_i in software). An edge between a hardware node and a software node has an execution time equal to the execution time of the first node (hardware) plus the time needed to communicate with the second node (software). Lastly, an edge between a software node and a hardware node has an execution time equal to the execution time of the first node (software) plus the time needed to communicate with the second node (hardware). For simplicity, we don't consider the impact of the external systems in term of cost and execution time at the entry point and at the exit point. The edge between the entry node and the first block (hardware or software node) has a cost of zero and an execution time equal to zero:

- $C_0 = 0$
- $T_0 = 0$

The edge between the last block in hardware (software) and the exit node has a cost equal to the cost of the hardware area (size of binary file) of the last block in hardware (software), and its execution time is equal to the execution time of the last block in hardware (software):

- $C_n(hw) = A(B_n)$
- $C_n(sw) = Size (.elf B_n)$
- $T_n(hw) = Texec (B_n hw)$
- $T_n(sw) = Texec (B_n sw)$.

2.2   Lagrangian Relaxation Method

The general representation of the Lagrangian relaxation method is given in [24]. The method is widely used to solve integer programming problems in different domains. In [25], the lagrangian method is used to solve a partitioning problem of class formation for training sessions. The langrangian relaxation method can also be applied to resolve the optimal path optimization problem of a double value graph.

The problem (P) described in the last section, is an optimization problem of a double value directed graph. The graph has the following properties:

1. Named (X, U, c, t)
2. Oriented and each edge u has two values c and t
3. $c: U \to R^+$: is the cost for using the edge u
4. $t: U \to R^+$: the necessary time (execution time) to traverse the edge u

For each path p (from entry node to the exit node), we associate a function x: $U \to \{0, 1\}$, where $x(u) = 1$ if the edge u is traversed by the path p and $x(u) = 0$ if the edge u is not traversed by p. The goal of the problem (P) is to find the optimal path for which the cost is minimal and in the same time it respects a global temporal constraint as described below:

(P): $min (f(x))$ where $f(x) = \sum c(u).x(u)$ under the constraints:

- $x \in S$: S set of functions associated to possible paths
- $g(x) \le 0$, where $g(x) = \sum t(u).x(u) - T$
  (*T is the global execution time constraint*)

The lagrangian relaxation method doesn't give the exact solution of min(f(x)), in fact, it consists of finding the biggest minor of this minimum. This minor is named ω^* and we have:

$$\omega^* \le min(f(x))\{x \in S; g(x) \le 0\} \tag{1}$$

We call the Lagrange function, the function L: S. $R^+ \to R$ defined by:

$$L(x, \lambda) = f(x) + \lambda.g(x) \tag{2}$$

λ is a positive coefficient called Lagrange multiplier. We call the dual function, the function ω: $R^+ \rightarrow R$ defined by:

$$\omega(\lambda) = min(L(x, \lambda))\{x \in S\} \tag{3}$$

For each fixed λ, $\omega(\lambda)$ is calculated using the same graph (Fig. 2), and each edge is now one-value, the value is: $c(u) + \lambda.t(u)$. On the latter graph, the value of the optimized path is calculated using Dijkstra's algorithm, which is $min(\sum(c(u).x(u) + \lambda.t(u).x(u))\{x \in S\})$. This latter term minus $\lambda.T$ gives $\omega(\lambda)$. We call the dual problem (D) of the problem (P), the problem that consists of maximizing $\omega(\lambda)$. From (2) and (3), for each $\lambda \in R^+$ and for each $x \in S$ that verified the problem (P):

$$\omega(\lambda) \leq f(x) \tag{4}$$

In fact, $\omega(\lambda) \leq L(x, \lambda)$, so $\omega(\lambda) \leq f(x) + \lambda.g(x)$, and then $\omega(\lambda) \leq f(x)$ as $\lambda.g(x)$ is a negative term.

Let $\omega^* = max(\omega(\lambda))$ and $f^* = min(f(x))\{x \in S; g(x) \leq 0\}$. From (4), for each $\lambda \in R^+$, we have $\omega(\lambda) \leq f^*$, then we have:

$$\omega^* \leq f^* \tag{5}$$

ω^* is by definition the biggest minor of f from $\omega(\lambda)$.

2.3   Dual Problem Resolution

The resolution of the dual problem (D) consists of finding ω^* that verify the equation below:

$$\omega^* = max(\omega(\lambda)) = max(min(L(x, \lambda))\{x \in S\})\{\lambda \in R^+\} \tag{6}$$

The algorithm consists of constructing increasing sets $S_1 \subset S_2 \subset ... \subset S_k \subset ...$ S, where S is the set of all possible paths. We construct a family of functions $\omega_k(\lambda) = min(L(x, \lambda))\{x \in S_k\}$, and we calculate $\omega_k^* = max(\omega_k(\lambda))\{\lambda > 0\}$ such as $\omega_k^* = \omega_k(\lambda_k)$. For each $\lambda \in R^+$, and $k \geq 1$, as $S_{k-1} \subset S_k \subset S$, and by definition $\omega_k(\lambda) = min(L(x, \lambda))\{x \in S_k\}$, we have: $\omega(\lambda) \leq \omega_k(\lambda) \leq \omega_{k-1}(\lambda)$, in fact, the minimum of a set that is included in another set is bigger or equal than the minimum of the set that includes it, and therefore we have:

$$\omega^* \leq \omega_k^* \tag{7}$$

And as by definition ω^* is the max of $\omega(\lambda)$: for each $k \geq 1$ and $1 \leq j \leq k$ we have:

$$max(\omega(\lambda_j)) \leq \omega^* \tag{8}$$

From (7) and (8), for k $\geq$ 1 and 1 $\leq$ j $\leq$ k, we have the inequality:

$$max\left(\omega\left(\lambda_j\right)\right) \leq \omega^* \leq \omega_k^* \tag{9}$$

The algorithm starts by an initial set S_2, and at each iteration it constructs the next set until it reaches an iteration k for which $\omega_k^* = max(\omega(\lambda_j))$ 1 $\leq$ j $\leq$ k, then it stops and $\omega^* = \omega_k^*$ which represents the solution of the problem (D) and in the same time gives a minor to the problem (P).

2.4   Lagrangian Relaxation Algorithm

The steps of the algorithm are summarized in Fig. 4. In the first iteration, the algorithm begins with a set (S_2) composed of two paths (x_0: corresponding path of optimal path in term of time of execution (cost = 0), and x_1: corresponding path of optimal path in term of cost (time = 0)), those two paths are computed using Dijkstra's algorithm. ω_2^* is then calculated as follow:

$$\omega_2^* = max(\omega_2(\lambda)) = max(min(L(x, \lambda))\{x \in S_2\})\{\lambda \in R^+\} \tag{10}$$

with $L(x, \lambda)$ $\{x \in S_2\} = (L(x_0, \lambda),$ $L(x_1, \lambda))$ and $L(x_0, \lambda) = f(x_0) + \lambda.g(x_0),$ $L(x_0, \lambda) = \sum c(u).x_0(u) + \lambda.(\sum t(u).x_0(u) - T) = cost(x_0) + (time(x_0) - T).\lambda$ and $L(x_1, \lambda) = cost(x_1) + (time(x_1) - T).\lambda$. Then, ω_2^* is obtained at λ_2. We then calculate $\omega(\lambda_2)$ using

---

1-   Start with a set $S_2 = \{x_0, x_1\}$ such as:
    a. x_0 is the optimal path in sense of execution time (for example, all blocks are in hardware)
    b. x_1 is the optimal path in sense of cost (for example, all blocks are in software)
2- x_0 is realizable if time(x_0) - T < 0, if x_0 is not realizable, the algorithm does not converge and there is no solution to the problem (P).
3-   k = 2
4-   Calculate $\omega_k^* = max(\omega_k(\lambda))$
    $\omega_k^* = max\ (min(L(x, \lambda))\{x \in S_k\})\ \{\lambda \in R^+\}$
5-   Calculate $\omega(\lambda_k) = min\ \{\ x \in S\ \}\ L(x, \lambda_k)$
6- $\omega(\lambda_k)$ is obtained at x_k
    a.   if $\omega_k^* = max(\ \omega(\lambda_j)\ )$ $2 \leq j \leq k$, we stop and the solution is x_k
    b.   else :
        i. $S_{k+1} = S_k \cup \{x_k\}$,
        ii.   k = k+1
        iii. iterate from the step 4

Fig. 4.   LR algorithm

Dijkstra's algorithm, the corresponding optimal path is x_2. If $\omega(\lambda_2)$ is equal to ω_2^* the algorithm stops and the solution of the problem (P) is x_2, else the algorithm continues with the second iteration, in which a new set S_3 is constructed as: $S_3 = S_2 \cup \{x_2\}$, ω_3^* is then calculated in the same manner:

$$\omega_3^* = max(\omega_3(\lambda)) = max(min(L(x, \lambda))\{x \in S_3\})\{\lambda \in R^+\} \tag{11}$$

With $L(x, \lambda)\{x \in S_3\} = (L(x_0, \lambda), L(x_1, \lambda), L(x_2, \lambda))$, and $L(x_0, \lambda) = cost(x_0) + (time(x_0) - T).\lambda$ and $L(x_1, \lambda) = cost(x_1) + (time(x_1) - T).\lambda$ and $L(x_2, \lambda) = cost(x_2) + (time(x_2) - T).\lambda$. Then, ω_3^* is obtained at λ_3 (example in Fig. 5), $\omega(\lambda_3) = min \, L(x, \lambda_3) \, x \in S$ is then calculated, the corresponding optimal path is x_3, ω_3^* is then compared to $max(\omega(\lambda_2), \omega(\lambda_3))$. In this manner the algorithm iterate until it finds k for which $\omega_k^* = max(\omega(\lambda_j)) \, 2 \leq j \leq k$, this latter has x_k as corresponding optimal path, x_k is then the solution of the problem (P).

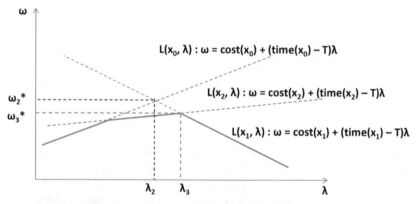

- Min($L(x_0, \lambda)$, $L(x_1, \lambda)$, $L(x_2, \lambda)$) is the concave function in bold
- Max of the concave function is $\omega_3{}^*$ obtained at λ_3

Fig. 5. Second iteration

3   Experiments

We implemented the algorithm in Java. To test the effectiveness of our approach, we made a series of tests, with an arbitrary number of blocks up to 1000. For each block, we randomly assign, a hardware cost (between 0 and 10), a software cost, a software to hardware cost and a hardware to software cost (values between: 0 and hardware cost), we also assign a randomly a software time execution (between 0 and 10), a hardware time execution, a hardware to software time execution and a software to hardware time execution (values between 0 and software time execution). We then assign an arbitrary value to the global time constraint that is greater than the minimum global time. We run the Genetic Algorithm (also implemented in Java) and the LR algorithm on each test, and compare the results of the two algorithms in term of global cost and global time. We also compare the execution time of the programs themselves.

Figure 6 shows the graph of the cost solution of the two algorithms LR and GA. The results show the LR and GA algorithms are almost identical when the number of blocks is small, but when the number of blocks increases, the LR algorithm gives greater results over the GA algorithm, and then it leads to a better optimization. It is noted that the algorithm converges practically for every test, this is particularly due to the fact that the graph used contains no feedback loops. The complexity of the algorithm is estimated to $O(n^3)$.

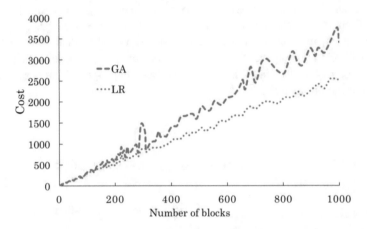

Fig. 6. GA vs. LR cost solution

The graph of Fig. 7 gives a comparison between the two algorithms in term of the execution time of the program itself. The results show that the execution time of LR algorithm is much better than GA algorithm especially when the number of blocks is high, which gives to the designer, the possibility of doing repeatedly multiple tests in a short time.

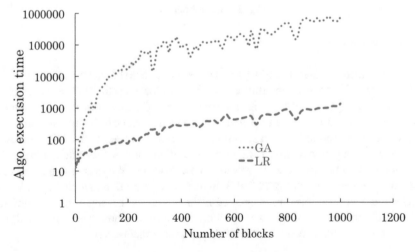

Fig. 7. GA vs. LR algorithm execution time

4   Conclusions

In this paper, we proposed a new heuristic approach to resolve the hardware software partitioning problem. The approach is based on the lagrangian relaxation method, this method gives an approximation to the exact solution in iterative way. The proposed algorithm is used to optimize the cost (hardware area) of an Embedded System (ES) while respecting a given execution time constraints. The results of different tests show that the LR algorithm leads to better optimization when comparing to GA algorithm, and especially when the number of the blocks increases. The approach can also be adapted to optimize the global time of execution under a given global cost constraint. It can also be used in case we have multiple hardware types (ASIC, FPGA, ...) and multiple software types (Multiprocessors, DSP, ASIP, ...). The DFG graph used in this article is straightforward, next work, is to study how to adapt this approach in case the DFG graph contains feedback loops, and also when the criteria involved in the partitioning problem are not limited to only the cost and the execution time. The global time execution calculation was based on the fact that the external systems are not taking into consideration; in a streaming case for example, the blocks are executed in parallel, and when a block is processing a data, its predecessor blocks are preparing the next data. In that case, the global time execution is related to the maximum execution time of the blocks, in the next work, we study the effect of this constraint on the partitioning problem optimization.

References

1. Schaumont, P.: A practical introduction to hardware/software codesign (2012)
2. Clausen, J.: Branch and bound algorithms-principles and examples. Department of Computer Science, University of Copenhagen, pp. 1–30 (1999)
3. Mann, Z.A., Orban, A., Arato, P.: Finding optimal hardware/software partitions. Formal Meth. Syst. Des. **31**(3), 241–263 (2007)
4. Niemann, R., Marwedel, P.: Hardware/software partitioning using integer programming. In: Proceedings of the 1996 European Conference on Design and Test, p. 473 (1996)
5. Knudsen, P.V., Madsen, J.: Pace: a dynamic programming algorithm for hardware/software partitioning. In: Proceedings of the 4th International Workshop on Hardware/Software Co-design, p. 85 (1996)
6. Eles, P., Peng, Z., Kuchcinski, K., Doboli, A.: Hardware/software partitioning with iterative improvement heuristics. In: Proceedings of the 9th International Symposium on System Synthesis, p. 71 (1996)
7. Banerjee, S., Dutt, N.: Very fast simulated annealing for hw-sw partitioning. Technical report, CECS-TR-04-17 (2004)
8. Zhao, X., Zhang, H., Jiang, Y., Song, S., Jiao, X., Gu, M.: An effective heuristic-based approach for partitioning. J. Appl. Math. **2013**, 1–8 (2013)
9. Saha, D., Mitra, R., Basu, A.: Hardware software partitioning using genetic algorithm. In: Proceedings of the Tenth International Conference on VLSI Design, pp. 155–160 (1997)
10. Purnaprajna, M., Reformat, M., Pedrycz, W.: Genetic algorithms for hardware-software partitioning and optimal resource allocation. J. Syst. Architect. **53**(7), 339–354 (2007)

11. Arato, P., Juhasz, S., Mann, Z.A., Orban, A., Papp, D.: Hardware-software partitioning in embedded system design. In: Proceedings of the 2003 IEEE International Symposium on Intelligent Signal Processing, pp. 197–202 (2003)
12. Chehida, K.B., Auguin, M.: Hw/sw partitioning approach for reconfigurable system design. In: Proceedings of the 2002 International Conference on Compilers, Architecture, and Synthesis for Embedded Systems, pp. 247–251 (2002)
13. Knerr, B., Holzer, M., Rupp, M.: Novel genome coding of genetic algorithms for the system partitioning problem. In: Proceedings of the 2007 International Symposium on Industrial Embedded Systems, pp. 134–141 (2007)
14. Li, S.G., Feng, F.J., Hu, H.J., Wang, C., Qi, D.: Hardware/software partitioning algorithm based on genetic algorithm. J. Comput. 9(6), 1309–1315 (2014)
15. Mudry, P.A., Zuerey, G., Tempesti, G.: A hybrid genetic algorithm for constrained hardware-software partitioning. In: Proceedings of the 2006 IEEE Design and Diagnostics of Electronic Circuits and systems, pp. 1–6 (2006)
16. Li, G., Feng, J., Wang, C., Wang, J.: Hardware/software partitioning algorithm based on the combination of genetic algorithm and tabu search. Eng. Rev. 34(2), 151–160 (2014)
17. Lin, G., Zhu, W., Ali, M.M.: A tabu search-based memetic algorithm for hardware/software partitioning. Math. Prob. Eng. 2014, 1–15 (2014)
18. Bhuvaneswari, M., Jagadeeswari, M.: Hardware/software partitioning for embedded systems. In: Application of Evolutionary Algorithms for Multi-objective Optimization in VLSI and Embedded Systems, pp. 21–36 (2015)
19. Lin, G.: An iterative greedy algorithm for hardware/software partitioning. In: Proceedings of 2013 Ninth International Conference on Natural Computation (ICNC), pp. 777–781 (2013)
20. Sim, J.E., Mitra, T., Wong, W.F.: Defining neighborhood relations for fast spatial-temporal partitioning of applications on reconfigurable architectures. In: Proceedings of 2008 International Conference on ICECE Technology, pp. 121–128 (2008)
21. Rini, D.P., Shamsuddin, S.M., Yuhaniz, S.S.: Particle swarm optimization: technique, system and challenges. Int. J. Comput. Appl. 14(1), 19–26 (2011)
22. Farmahini-Farahani, A., Kamal, M., Fakhraie, S.M., Safari, S.: HW/SW partitioning using discrete particle swarm. In: Proceedings of the 17th ACM Great Lakes symposium on VLSI, pp. 359–364 (2007)
23. Wu, J., Srikanthan, T., Lei, T.: Efficient heuristic algorithms for path-based hardware/software partitioning. Math. Comput. Model. 51(7), 974–984 (2010)
24. Fisher, M.L.: The lagrangian relaxation method for solving integer programming problems. Manage. Sci. 27(1), 1–18 (1981)
25. Czibula, O.G., Gu, H., Zinder, Y.: Lagrangian relaxation versus genetic algorithm based matheuristic for a large partitioning problem. Theor. Comput. Sci. (2017)

Multi-Agent System for Arabic Text Categorization

Mounir Gouiouez and Meryeme Hadni[(✉)]

USMBA, Fez, Morocco
mounir.gouiouez@gmail.com, meryemehadni@gmail.com

Abstract. Developing TC systems for Arabic documents is a challenging task due to the complex and rich nature of the Arabic language, and the way in which they are written according to its position in the sentence. Furthermore, Arabic is written from right to left, and its letters changing form according to their position in the word. There are various different methods for text categorization, including distance-based, decision tree-based methods, Bayesian naïf…etc. Furthermore, the large numbers of methods proposed are typically based on the classical Bag-of-Words model. In order to improve the accuracy of Arabic text categorization, therefore the accuracy of the results obtained, a new hybrid approach is proposed to improve the effectiveness of the automated techniques categorization. This paper presents the development of a concept and an associated architecture called the CAMATC (Cooperative Adaptive Multi-Agent System for Arabic Text Categorization), which is based on the combination of Multi-Agent Systems and the conceptual representation in the Arabic text categorization.

Keywords: Text categorization · Multi-Agent System · Graph-based · Named entities · BabelNet

1 Introduction

Text Classification (or Categorization) is a [1, 5] series of actions or steps taken in order to classify documents into a set of predefined categories. Using machine learning, the main objective of Text classification is to learn the automatically assignments of documents, according to a supervised learning approach.

TC techniques are used in many fields, paper archives, automated indexing of scientific articles…etc. The great majorities of these methods are designed to cover efficiency the documents written in the English language, and thus are not very applicable to documents written in the Arabic language, and the major difficulty, is the high dimensionality of the feature space (words or phrases) that occur in documents. Thus, the first issue that needs to be addressed in text categorization is to transform documents into a representation suitable in order to facilitate machine manipulation and retain much information as needed. The commonly used text representation is the Bag-Of-Words, which simply uses a set of words and the number of occurrences of the words to represent documents and categories. After counting the number of occurrences of a word w in a document, appropriate stemming algorithms [15] are applied to avoid needlessly large feature vectors.

© Springer International Publishing AG, part of Springer Nature 2018
M. Ben Ahmed and A. A. Boudhir (Eds.): SCAMS 2017, LNNS 37, pp. 161–174, 2018.
https://doi.org/10.1007/978-3-319-74500-8_15

To further reduce the number of measured terms, suitable methods can be used. Include the removal of stop words (non-informative terms) according to predefined corpus, and the construction of new features among different feature selection methods, such as χ^2 statistic, mutual information, term strength…etc. [2]. After selecting the terms, for each document a feature vector is generated, whose elements are the feature values of each term.

In Arabic language, the problem of Text Categorization (TC) is much more complicated than in other languages. Indeed, Arabic is a morphologically complex language that has large, agglutination and grammatical ambiguity, which can lead to uncertainty or inexactness of meaning. Hence, the current study sought to shed a light on these issues. This research proposes an automated system that can completely classify a given Arabic text. Existing work on TC has used many algorithms based on distance-based algorithms, Learning algorithms, and N-grams for searching text documents. However, No methods was performed to improve the documents written in the Arabic language.

In order to improve the accuracy of Arabic Text Categorization, therefore the accuracy of the results obtained, a new hybrid approach is proposed to improve the effectiveness of the automated techniques categorization. This paper presents the development of a concept and an associated architecture called CAMATC (Cooperative Adaptive Multi-Agent System for Arabic Text Categorization), which is based on the combination of Multi-Agent Systems and the conceptual representation in the Arabic Text Categorization.

2   The CAMATC Architecture

Multi-Agent System (MAS) has brought a new vision to study the complex situations with emphasizes the interactions of components of the systems. In literature, the MAS is one of the newest area of research in the artificial intelligence (AI), it has started in the early 90s with [17, 18], as an attempt to enrich the limits of classical AI [11]. The foundations of the MAS are interested in modeling phenomena with mental notions such as knowledge, intentions, choices, commitments [19] (Fig. 1).

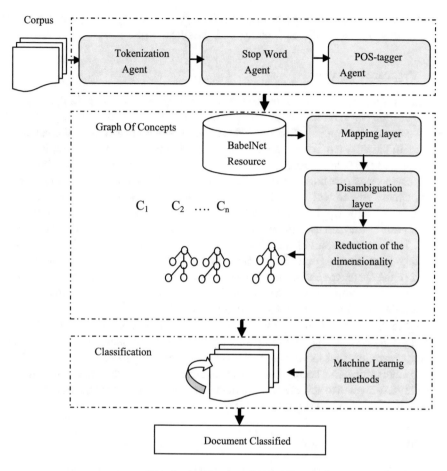

Fig. 1. Architecture CAMATC

There are various definitions of the concept agent [12–14] in the contemporary literature; however, the definition that we adopted, and which covers the characteristics of agents that we developed, is that proposed by Jennings, Sycara and Wooldridge [16]: For Jennings, Sycara and Wooldridge an agent is a computer system, located in an environment, which is autonomous and flexible to meet the objectives for which it was designed. As far as MAS is concerned, according to Ferber [18] is a system composed of the following: Environment, a set of objects in space; a set of agents who are active. Entities of the system, a set of relationships that binds objects together; a set of operations allowing agents to perceive, destroy, create, transform, and manipulate objects.

The CAMATC architecture, which stands Cooperative MultiAgent System for Arabic text categorization, is a generic MultiAgent architecture, aimed at preprocessing, mapping, disambiguation, reduction of the dimensionality and classification of the text documents according to the approach suggested. CAMATC agents can be cooperative, adaptive depending on their specific interaction in the architecture. CAMATC architecture encompasses four main levels (i.e., preprocessing phase, terms transformation, reduction of the dimensionality, and classification).

2.1 Preprocessing Phase

The first level relates to the preprocessing phase. It consists of generating a new text representation based on a set of previous steps. In this case, the first step is to represent the documents as segment of tokens, where each token corresponds to the value of a feature, in the this step Tokenization Agent (TA) divide text document into tokens by segmentation procedure based on punctuation and white spaces. After getting words from sentences, we use our own tool to produce all possible tokenizations for each word. Then, choose the correct one among multiple possible tokenizations for one word. nevertheless many words in documents repeat very frequently, they are essentially meaningless as they are used to join words together in a sentence like 'and', 'are', 'this' …etc. They are not useful in classification of documents. So they must be removed. In this stage Stop Word Agents (SWA) are aimed to filter and delete all the words which appear in the sentences and do not have any meaning or indications about the content of arabic list stopwords such as punctuations list (? ! …), pronouns list (هو ة الذي الثّ هما),…adverbs list (فوق تحت بّن)…etc.

In Arabic, the problem of POS-tagging is much more complicated than in other languages. Indeed, Arabic is a morphologically complex language that has numerous writing constraints such as vowels, agglutination and grammatical ambiguity, which can lead to ambiguities. In a sense, they can be considered as the core of the architecture. In fact, they are devoted to achieve to find suitable sentence by cooperating Hidden Markov Tagger and Based POS Tagger approaches.

2.2 Terms Transformation

In this section, we describe how the generic architecture has been customized to implement a system to perform text mapping and disambiguation.

Mapping Layer: At the mapping layer, agents play the role of wrappers, the terms is mapped into their corresponding concepts using predefined corpus. In particular, in the current implementation a set of agents wraps databases containing BabelNet [3] resource. This strategy replaces each term vector td by new entries for corpus concepts C appearing in the texts set. Thus, the vector td will be replaced by Cd where $Cd = (Cf(d, c1),.., Cf(d, cn))$. The concepts vector with $l = |C|$ and $Cf(d, c)$ denotes the frequency that a concept $c \in C$ appears in a text d. Furthermore, an agent wraps the adopted mapped that is a subset of the one proposed by BabelNet resource.

Disambiguation Layer: At the disambiguation layer, a population of agents manipulates the information belonging to the mapping level. The assignment of terms to concepts is ambiguous. Therefore, one entity may have several meanings and thus one entity may be mapped into several concepts. In this case, WSD allows us to find the most appropriate sense of the ambiguous entity. The main idea behind this work is to propose an efficient method for Arabic WSD. In this, to determine the most appropriate concept for an ambiguous entity in a sentence, we select the concepts that have a more semantic relationship with other concepts in the same local context (Algorithm 1).

Algorithm 1: A method for Arabic WSD.
W: Ambiguous entity.
S: Sentence containing w.
N: Number of the concepts of entity w.
LC={c1, c2, ..., cN} List of the concepts of W.
K: Number of concepts in the local context of W.
LW={c1, c2, ..., cK}List of the concepts of Local Context (± 2 terms).
BN: BabelNet ressource
SimWP(c_i, c_j): The Wu and Palmer similarity measure between two concepts c_i and c_j.
Begin
For each term W ϵ S do{
Map W into concepts using BN.
If W ϵ BN then LC={c1, c2, ..., cN}
End If
    FinPour
/* Calculate the score with the other concepts in the local context*/
S(C) $\leftarrow$ 0
For each concept c_i ϵ LC
{
For each concept $w_j \epsilon$ LW
S(c_i) $\leftarrow$ S(c_i) + Sim WP(c_i, w_j)
    }
/* Select the nearest concept*/
Cp(W)=Cp/maxi=1....N S(ci)=S(Cp)
End

The Reduction of the Dimensionality: Our corpus is very large, as it usually the case for text categorization application. The use of the concepts instead of the words reduces considerably the dimensionality of document to reduce further the size of the vectors, we use the CHI2 and CHIR methods for selecting only the most representative concepts.

Chi-Square: The Chi-Square statistics can be used to measure the degree of association between a term and a category [11]. Its application is based on the assumption that a term whose frequency strongly depends on the category in which it occurs will be more useful for discriminating it among other categories. For the purpose of dimensionality reduction, terms with small Chi-Square values are discarded. The Chi-Square multivariate is a supervised method allowing the selection of terms by taking into account not only their frequencies in each category but also the interaction of the terms between them and the interactions between the terms and the categories. The principal consists in extracting k better features characterizing best the category compared to the others, this for each category. An arithmetically simpler way of computing chi-square is the following:

$$X^2_{w,c} = \frac{n * \left(p(w,c) * p(\overline{w}, \overline{c}) - p(w,\overline{c}) * p(\overline{w},c) \right)^2}{p(w) * p(\overline{w}) * p(\overline{c}) * p(c)} \tag{1}$$

Where p(w, c) represents the probability that the documents in the category c contain the term w, p(w) represents the probability that the documents in the corpus contain the term w, and w(c) represents the probability that the documents in the corpus are in the category c, and so on. These probabilities are estimated by counting the occurrences of terms and categories in the corpus.

The feature selection method chi-square could be described as follows. For a corpus with m classes, the term-goodness of a term w is usually defined as either one of:

$$X^2_{max}(w) = \overset{max}{j} \left\{ X^2_{w,c_j} \right\} \tag{2}$$

$$X^2_{avg}(w) = \sum_{j=1}^{m} p(c_j) * X^2_{w,c_j} \tag{3}$$

Where $p(c_j)$: The probability of the documents to be in the category cj then, the terms whose term-goodness measure is lower than a certain threshold would be removed from the feature space. In other words, chi-square selects terms having strong dependency on categories.

CHIR: Our feature selection method CHIR uses $r\chi^2(w)$ to measure the term-goodness, and makes sure that the $r\chi^2$ statistic of each term represents only positive term-category dependency. The goal of this feature selection method is to find the terms that have strong positive dependency on certain categories in the corpus.

$$rX^2(w) = \sum_{j=1}^{m} p\left(R_{w,c_j}\right) X^2_{w,c_j} \text{ with } R_{w,c_j} > 1 \tag{4}$$

Where $p\left(R_{w,c_j}\right)$ is the weight of X^2_{w,c_j} in the corpus in terms of R_{w,c_j} and is defined as:

$$p(R_{w,c_j}) = \frac{R_{w,c_j}}{\sum_{j=1}^{m} R_{w,c_j}} \text{ with } R_{w,c_j} > 1 \tag{5}$$

Where $R_{w,c_j} = \dfrac{p(w,c)p(\overline{w},\overline{c}) - p(w,\overline{c})p(\overline{w},c)}{p(w)p(c)} + 1$

In other words, CHIR selects the terms which are relevant to categories and removes the irrelevant and redundant terms. The steps of CHIR to select q terms are as follows:

(1) For each distinct term in the corpus, calculate its $r\chi^2$ statistic.
(2) Sort the terms in descending order of their $r\chi^2$ statistics.
(3) Select the top q terms from the list.

2.3   Classification Phase

The classification phase consists in generating a weighted vector for all categories Graph, then using a machine learning methods to find the closest category.

Graph Construction

Here, we refer to a modeling approach to the Graph-of-Concepts (GoC) algorithm [12] to describe how a document can be represented by a graph. In general, the document $d \in D$ is represented by a graph $Gd = (V, E)$, where each node $v \in V$ corresponds to a entity $t \in T$ of the document d and the edges $e = (u,v)$ capture co-occurrence relations between entity u and v within a fixed-size sliding window of size w. The graph model needs several parameters to be specified during the construction phase.

Each document is represented by Graph of concepts weights that appeared in it (CF-IDF for concepts).

Weighting Concepts

The results of this step will be used to enrich the representing concepts graph of each document. When concepts are extracted from the document using BabelNet, selected concepts are weighted according to a variant TF.IDF noted CF.IDF:

The weight $W(C_d^i)$ of a concept C^i, in a document d is defined as the combined measure of its local centrality and its global centrality, formally:

$$W\left(C_d^i\right) = cc\left(C^i, d\right) * idc\left(C^i\right) \tag{6}$$

The local centrality of a concept C^i in a document d, noted $cc(C^i, d)$ based on its pertinence in the document, and its occurrence frequency. Formally:

$$cc\left(C^i, d\right) = \alpha * tf\left(C^i, d\right) + (1 - \alpha) \sum_{i \neq 1} Sim\left(C^i, C^1\right) \tag{7}$$

Where α is a weighting factor that balances the frequency in relation with the pertinence (this factor is determined by experimentation), $Sim(C^i, C^1)$ measures the semantic similarity between concepts C^i and C^1, $tf(C^i, d)$ is the occurrence frequency of the concepts C^i in the document d. $Sim\left(c^i, c^1\right)$ is calculated as follows:

$$Sim\left(c^i, c^1\right) = \frac{dist(c^i, c^1)}{|Arc(c^i)| + |Arc(c^1)|} * \frac{idc(c^i)}{idc(c^1)} \tag{8}$$

Where $dist(c^i, c^1)$: set of common concepts between c^i and c^1 and $|Arc(C)|$: set of concepts from root to C.

The global centrality of a concept is its discrimination in the collection. A concept which is central in too many documents is not discriminating. Considering that a concept C^i is central in a document d, if their centrality is superior to a fixed threshold s, the document centrality of the concept is defined as follows:

$$idc(c^i) = \frac{n}{N} \tag{9}$$

For text categorization we used the support vector machines, Naïve Bayes and its variants. These learning algorithms take as input feature vectors mentioned above. We consider the weight of each concept as a feature of the document.

Machine Learning Algorithms
After preprocessing and transformation the documents can be without difficulty represented in a form that can be used by a ML algorithm. Four algorithms are tested: Naïve Bayes, Multinomial Naïve Bayesian, Complement Naïve Bayesian and Support Vector Machines. To apply these algorithms the standard Bag of Concepts is used on the features that result from the previous step.

Support Vector Machine Classifier
Support Vector Machine (SVM) is a relatively new class of machine learning technique [14]. It is based on the principle of structural risk minimization, to construct a hyper plane or a set of hyper planes in a high dimensional space that separates the data into two sets or n sets with the maximum margin. A hyper plane with the maximum-margin has the distances from the hyper plane to points when the two sides are equal. Mathematically, SVMs use the sign function $f(x) = \text{sign}(wx + b)$, where w is a weighted vector in Rn. SVMs find the hyper plane $y = wx + b$ by separating the space Rn into two half spaces with the maximum-margin. Linear SVMs can be generalised for non-linear problems. To do so, the data is mapped into another space H and we perform the linear SVM algorithm over this new space. SVM has been successfully used on TC [11] and they showed better results than other machine learning techniques such as NB, decision trees, and KNN [7] with reference to accuracy.

Naïve Bayes Classifier
The Naïve Bayes (NB) classifier is a probabilistic model that uses the probabilities of terms and categories. The NB applied on the TC problem by the following Baye's theorem (Eq. 1).

$$p(c_i|d_j) = \frac{p(c_i) \cdot p(d_j|c_i)}{p(d_j)} \tag{10}$$

Where $p(c_i|d_j)$: is the probability of class given a document, or the probability that a given document D belongs to a given class $C.p(d_j)$: The probability of a document, we can notice that $p(d_j)$ is a Constance divider to every calculation, so we can ignore it. $p(c_i)$: The probability of a class (or category), we can compute it from the number of documents in the category divided by documents number in all categories. $p(d_j|c_i)$ represents the probability of document given class, and documents can be modelled as sets of words, thus the $p(d_j|c_i)$ can be written like:

$$p(d_j|c_i) = \prod p(word_i|c_i) \tag{11}$$

So

$$p(c_i|d_j) = p(c_i) \prod p(word_i|c_i) \tag{12}$$

Where $p(word_i|c_i)$: The probability that the i-th word of a given document occurs in a document from class C, and this can be computed as follows:

$$p(word_i|c_i) = (Tct + \lambda)/(Nc + \lambda V) \tag{13}$$

Where Tct: The number of times the word occurs in that category C. Nc: The number of words in category C. V: The size of the vocabulary table. λ: The positive constant, usually 1, or 0.5 to avoid zero probability.

Complement Naïve Bayesian Classifier

This classifier estimates the posterior probability as:

$P(c_i|d_j) = 1 - P(\bar{c}_i|d_j)$ where $\bar{c}_i$ indicates the complement of class. In this way, the probability $P(\bar{c}_i|d_j)$ can be easily estimated similarly as in the Naïve Bayes model:

$$P(\bar{c}_i|d_j) = \frac{p(d_j|\bar{c}_i)p(\bar{c}_i)}{p(d_j)} \tag{14}$$

Each $p(word_i|\bar{c}_i)$ is approximated by the frequency of the term $word_i$ in the complement $\bar{c}_i$. This approach is particularly suited when only few labeled examples are available for each category c_i.

Multinomial Model

The task of text classification can be approached from a Bayesian learning perspective, which assumes that the word distributions in documents are generated by a specific parametric model, and the parameters can be estimated from the training data. Equation 5 shows Multinominal Naive Bayes (MNB) model [6] which is one such parametric model commonly used in text classification:

$$p(c_i|d_j) = \frac{p(c_i) \prod p(word_i|c_i)^{f_i}}{p(d_j)} \tag{15}$$

Where f_i is the number of occurrences of a $word_i$ in a document, $p(word_i|c_i)$ is the conditional probability that a $word_i$ may happen in a document given the class value class, and n is the number of unique words appearing in the document.

The parameters in Eq. 5 can be estimated by a generative parameter learning approach, called frequency estimate (FE), which is simply the relative frequency in data

[2]. FE estimates the conditional probability $p(word_i|document)$ using the relative frequency of the $word_i$ in documents belonging to class.

$$p(word_i|c_i) = \frac{f_{ic}}{f_c} \tag{16}$$

where f_{ic} is the number of times that a $word_i$ appears in all documents with the class, and f_c is the total number of words in documents with class.

3   Evaluation and Discussion

3.1   Corpora Summary

We use various corpora to perform our experimentations, the corpora variations include small/large size corpus, with few and more categories. We used three corpora: CCA corpus, BBC-arabic corpus and EASC's corpus. The corpus of Contemporary Arabic (CCA Corpus) [15] was released from the University of Leeds by Latifa Al-Sulaiti and Eric Atwell. The corpus is classified to 5 categories (Autobiography 73, Health and Medicine 32, Science 70, Stories 58, Tourist and travel 60). The BBC Arabic corpus [15] is collected from BBC Arabic website bbcarabic.com, the corpus includes 4,763 text documents. Each text document belongs 1 of to 7 categories (Middle East News 2356, World News 1489, Business & Economy 296, Sports 219, International Press 49, Science & Technology 232, Art & Culture 122).

Figure 2 presents the district keywords and the number of text documents for each corpus.

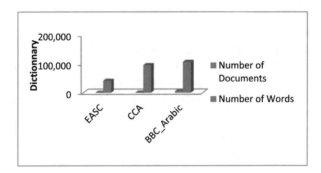

Fig. 2. Dictionary size for each corpus

The Essex Arabic Summaries Corpus (EASC) [11] is generated using http://www.mturk.com. The major feature of EASC is the fact that Names and extensions are formatted to be compatible with current evaluation systems. The data are available in two encoding formats UTF-8 and ISO-8859-6 (Arabic). Each text document belonging to 1 of 10 categories (Art and music 10, Education 07, Environnement 34, Finance 17, Health 17, Politics 21, Religion 08, Science and Technology 16, Sports 10, Tourism 14).

3.2 Experimental Configuration and Performance Measure

In this section, we present and analyze experimental results. Text Classification algorithms: SVM, NB, NBM, and CNB. We split each corpus to 2 parts (80% of the corpus for training and the remaining 20% for test). We could not run any classifier in batch mode because the corpora size is very large and did not fit to memory. All classifiers were run in incremental mode on 64-bit machine with 4 GB RAM.

The evaluation of the performance for classification model to classify documents into the correct category is conducted by using several mathematic rules such as recall, precision, and F-measure, which are defined as follows:

$$\text{Precision} = \frac{TP}{TP + FP} \text{ and Recall} = \frac{TP}{TP + FN} \tag{17}$$

where TP is the number of documents that are correctly assigned to the category, TN is the number of documents that are correctly assigned to the negative category, FP is the number of documents a system incorrectly assigned to the category, and FN are the number of documents that belonged to the category but are not assigned to the category. The success measure, namely, micro-F1 score, a well-known F1 measure, is selected for this study, which is calculated as follows:

$$\text{F1} - \text{measure} = \frac{2.\text{Precison}.\text{Recall}}{\text{precision} + \text{Recall}} \tag{18}$$

Experimental results investigate text representation and reduction dimensionality.

3.3 Dimensionality Reduction

After preprocessing phase, the vector representation is formed with concepts. In document, a large number of terms are irrelevant to the classification task and can be removed without affecting the classification accuracy. The mechanism that removes the irrelevant feature is called feature selection. Feature selection is the process of selecting the most representative subset that contains the most relevant terms for each category in the training set based on a few criteria.

Table 1. Effect of concepts selection criteria in categorization accuracy

Number of documents	DF	CHI	GSS	CHIR
2500	83,31	92,12	83,03	92,15
3000	83,35	93,16	83,49	92,53
3500	83,79	93,42	83,5	93,79
4000	84,01	93,71	83,84	94,08
4500	84,25	94,02	84,12	94,1
5000	84,42	94,39	84,73	94,48
Average	83,855	93,47	83,785	93,52

Table 1 displays the performance curves for SVM classifier after concepts selection using DF, CHI, GSS and CHIR thresholding. An observation merges from the categorization that DF and GSS thresholding have similar effects on the performance of the classifiers. The CHIR method gives a better result than other methods using SVM classifier and CCA corpus.

Feature selection aims to choose the most relevant words that distinguish between classes in the dataset. In our paper, we suggested DF, GSS, chi-2 testing (x2) and CHIR methods [11]. All these methods organize the features according to their importance to the category. The top ranking features from each category are then chosen and represented to the classification algorithm.

Table 1 shows the effect of concepts selection criteria in categorization accuracy using SVM method for Machine learning and CCA Corpus.

We can see from the results in Table 2 that MNB almost always performs worse than any of the other learning algorithms in the different corpora. This is consistent with previously published results [14].

Table 2. Comparaison of different machine learning using CHIR method

DataSet	EASC	CCA	BBC-arabic
SVM	92,03	94,52	89,45
NB	84,19	83,22	87,16
CNB	86,10	86,46	91,21
MNB	62,45	73,45	66,02

Our results (from Table 2) for the various classifiers have shown quite evident that the results for SVM combined by CHIR method for feature reduction are mostly better than NB and its variants.

3.4   Text Representation

There are two important modes in text mining such as: Bag of Words and Bag of Concepts representation. In our experiments, we decided to experiment on these two representations to compare these in our proposed Graph concepts representation (Fig. 3).

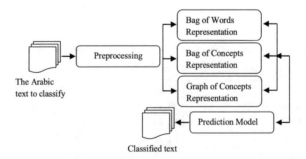

Fig. 3. Text representation

To evaluate the performance across categories, F-measure is averaged. The results of the approaches are shown in the following Table 3, and the best macro-averaged F-measure is bold. It is clear that the use Graph of concepts is a better way of representing Arabic texts, the performances as shown by the F-measure are better and that's true with the three classifiers. Of the three classifiers, the SVM classifier performs the best. That's what we expected.

Table 3. Text representation with different classifiers

Representation	NB	CNB	MNB	SVM
GOC	79%	80%	77%	83%
BOC	76%	73%	71%	82%
BOW	64%	78%	68%	70%

4   Conclusion

In this paper we have discussed the problem of text representation for Arabic text documents. We described main aspects of the systems available to process and analyze large Arabic documents in the field of Text Categorization. Then we introduced a novel approach using Multi-agent Text Categorization and BabelNet knowledge resource for Arabic text, which each document is represented by a graph that encodes relationships between the different named entities. This approach can be used in a case when the data is large (big data) and the hierarchical approach could not be reliable.

References

1. Sharef, B., Omar, N., Sharef, Z.: An automated Arabic text categorization based on the frequency ratio accumulation. Int. Arab. J. Inf. Technol. (IAJIT) **11**(2), 213–221 (2014)
2. Elberrichi1, Z., Rahmoun, A., Bentaalah, M.A.: Using WordNet for text categorization. Int. Arab. J. Inf. Technol. **5**(1) (2008)
3. Navigli, R., Ponzetto, S.P.: BabelNet: the automatic construction, evaluation and application of a wide-coverage multilingual semantic network. Artif. Intell. **193**, 217–250 (2012)
4. Hadni, M., Ouatik, S.A., Lachkar, A., Meknassi, M.: Hybrid Part-Of-Speech tagger for non-vocalized Arabic text. Int. J. Nat. Lang. Comput. (IJNLC) **2**(6) (2013)
5. Ciravegna F.: Flexible text classification for financial applications: the FACILE system. In: Proceedings of the 14th European Conference on Artificial Intelligence, Berlin, Germany, pp. 696–700 (2000)
6. He, J., Tan, A., Tan, C.: On machine learning methods for Chinese document categorization. Appl. Intell. **18**(3), 311–322 (2003)
7. Duwairi, R., Rania, A.-Z.: A hierarchical K-NN classifier for textual data. Int. Arab. J. Inf. Technol. **8**(3), 251–259 (2011)
8. Alsaleem, S.: Automated Arabic text categorization using SVM and NB. Int. Arab. J. e-Technol. **2**, 124–128 (2011)
9. Duwairi, R.: Arabic text categorization. Int. Arab. J. Inf. Technol. **4**, 125–132 (2007)
10. Abidi, K., Elberrichi, Z., Guissa, Y.T.: Arabic text categorization: a comparative study of different representation modes. J. Theor. Appl. Inf. Technol. **38**(1), 1–5 (2012)

11. Hadni, M., El Alaoui Ouatik, S., Lachkar, A.: Word sense disambiguation for Arabic text categorization. Int. Arab. J. Inf. Technol. (IAJIT) **13**(1A), 215–222 (2016)
12. Rousseau, F., Vazirgiannis, M.: Graph-of-word and TW-IDF: new approach to ad hoc IR. In: CIKM 2013: Proceedings of the 22nd ACM International Conference on Information and Knowledge Management, pp. 59–68 (2013)
13. Moore, J.L., Steinke, F., Tresp, V.: A novel metric for information retrieval in semantic networks. In: Proceedings of 3rd International Workshop on Inductive Reasoning and Machine Learning for the Semantic Web (IRMLeS 2011), Heraklion, Greece (2011)
14. Motaz, K.S.: The impact of text preprocessing and term weighting on Arabic text classification. Thésis, September 2010
15. Wooldridge, M., Jennings, N.R.: Intelligent agents: theory and practice. Knowl. Eng. Rev. **10**(2), 115–152 (1995)
16. Minsky, N.H.: The imposition of protocols over open distributed systems. IEEE Trans. Softw. Eng. **17**, 183–195 (1991)
17. Ferber, J.: Les Systèmes Multi-agents: Vers une Intelligence Collective. InterEditions, Paris (1995)
18. Shoham, Y.: Agent-oriented programming. AI **60**, 51–92 (1993)
19. Gouiouez, M., Rais, N., Idrissi, MA.: A new car-following model: as stochastic process using multi agent system. Int. J. Eng. Sci. Res. 2983–2989 (2013)

A Data Processing System to Monitor Emissions from Thermal Plants in Morocco

Mohamed Akram Zaytar[1](✉), Chaker El Amrani[1], Abderrahman El Kharrim[2], Mohamed Ben Ahmed[1], and Mohammed Bouhorma[1]

[1] Laboratory of Informatics, Systems and Télécommunications, Department of Computer Engineering, Abdelmalek Essaadi University, Route Ziaten, P.O. Box 416 Tangier, Morocco
MedAkramZaytar@gmail.com, celamarani@gmail.com,
med.benahmed@gmail.com, bouhorma@gmail.com
[2] Poly Disciplinary Faculty of Nador, University Mohammed the First,
P.O. Box 524 Oujda, Morocco
elkharrim@gmail.com

Abstract. This paper presents a data processing system comprised of multiple layers of computational processes that transform the raw binary meteorological data coming directly from two EUMETSAT Metop satellites to our servers, into a ready to visualise and interpret data stream in near real time using techniques varying from software automation, data preprocessing and general data analysis concepts. The proposed system handles the acquisition, decoding, cleaning, processing, and normalization of pollution data in our area of interest of Morocco.

Keywords: Data processing · Data aggregation · Data analysis · Data decoding
Data preprocessing · BUFR

1 Introduction

The impact of global air pollution on both the climate and the environment is a new focus in atmospheric science, over the last decade, air pollution's environmental threats significantly increased [1–3]. Generally speaking, climate change effects are indeed many and wide ranging [4]. There is no doubt that excessive levels of air pollution is causing significant damage to human and animal health as well as the wider environment. For these reasons, careful scientific research and monitoring of air pollutants is a necessity that must be exercised with a great deal of attention and precision.

At the present time, and as much as we rush to quickly infer and conclude from the climate or weather datasets we possess, most of the problems and difficulties we face hover around processing tasks, we spend most of our time preparing, cleaning, and transforming large volumes of datasets we receive. In our case, we will deal with data in a single format, called "BUFR", the data we process comes directly from the satellite's encoders and we receive the near real time data in bulks in 30 min intervals.

The main source of data the system processes is EUMETSAT, EUMETSAT is a global operational satellite agency at the heart of europe. The organization states that its purpose is to gather accurate and reliable satellite data on weather, climate and the

© Springer International Publishing AG, part of Springer Nature 2018
M. Ben Ahmed and A. A. Boudhir (Eds.): SCAMS 2017, LNNS 37, pp. 175–187, 2018.
https://doi.org/10.1007/978-3-319-74500-8_16

environment around the clock, and to deliver them to their member and cooperating states, international partners, and to users worldwide [5].

Specifically, The data the system processes comes directly from a type of satellites named Metop. Metop is a series of three polar orbiting meteorological satellites, two satellites are active, called Metop-A and Metop-B, both are in a lower polar orbit, at an altitude of approximately 817 km, they provide detailed observations of the global atmosphere, oceans and continents. The last satellite, Metop-C, is planned to be launched in 2018.

The system, in its turn, processes the data from its primitive raw BUFR format, which is a binary data format maintained by the world meteorological organization, to two exported comma separated files corresponding to Metop-A and Metop-B. The BUFR format is a somewhat controversial and hard to deal with data format that is table/message oriented and not supported by major data analysis tools, hence the difficulty to manipulate or aggregate its encoded values.

The software solution this paper presents is a system composed of two stacked layers. The first one decompresses and decode the BUFR data, and the second deals with preprocessing, cleaning and normalizing the dataset and finally combining the messages into one CSV file, 30 min at a time. This solution can be run as an automatic process in the server's scheduled jobs planner (such as cron).

The software solution proposed by this paper is a system that can be directly plugged into the end points of the near real time data stream, it allows for fast experimentation and visualization of already processed raw data points coming directly from the Metop-X satellites series, it will also result in a significant space and time reduction and an overall optimization of the internal research procedures since it focuses on interest areas and not the global scale.

2   Data Processing

2.1   The Dataset

The system primarily processes near time BUFR pollution data coming directly from two satellites orbiting in parallel to provide higher precision and accuracy, called Metop-A and Metop-B, the data in use comes directly from The IASI instrument, which is composed of a fourier transform spectrometer and an associated integrated Imaging subsystem (IIS). The fourier transform spectrometer provides infrared spectra with high resolution between 645 and 2760 cm^{-1}.

The main goal of IASI is to provide atmospheric emission spectra to derive temperature and humidity profiles with high vertical resolution and accuracy. Additionally, it is used for the determination of trace gases such as ozone, nitrous oxide, and carbon dioxide, as well as land and sea surface temperature and emissivity and cloud properties.

In terms of distance, IASI measures in the infrared part of the electromagnetic spectrum at a horizontal resolution of 12 km over a swath width of about 2200 km. with 14 orbits in a sun-synchronous mid-morning orbit (9:30 Local solar time equator crossing, descending node), global observations can be provided twice a day (every 12 h), the satellites take around 25 min to scan the area of interest (of Morocco), we get pollutant measurements of geographic points that are approximately 20 km apart from each other.

2.2 BUFR

BUFR, or the Binary Universal Form for the representation of meteorological data, is a binary data format, maintained by the World Meteorological Organization (WMO). BUFR is the result of a series of informal and formal "expert meetings" and periods of experimental usage by several meteorological data processing centers. BUFR was designed to be portable, compact, and universal. Any kind of data can be represented under BUFR, along with its specific spatial/temporal context and any other associated metadata, BUFR belongs to a category of table-driven code forms, where the meaning of data elements is determined by referring to a set of tables that are kept and maintained separately from the message itself, some [6] consider this a major weakness of the BUFR format, the fact that the correct tables are needed both to understand the meaning of the data and to parse the data, because when the tables are external, there is no foolproof to know when you have the correct tables (Fig. 1).

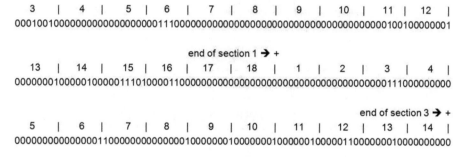

Fig. 1. BUFR representation example.

A typical BUFR message is composed of six sections, numbered zero through five.

- sections 0, 1 and 5 contain static metadata, mostly for message identification.
- section 2 is optional, if used, it may contain arbitrary data in any form wished for by the creator of the message (this is only advisable for local use).
- section 3 contains a sequence of the so-called "descriptors" that define the form and contents of the BUFR data product.
- section 4 is a bit stream containing the message's core data and metadata values as laid out by section 3.

2.3 The Problem

The inability of doing fast data experimentation, analysis and manipulation on BUFR data using the more popular data analysis libraries in either Python or R, prompts the following question: can we create a system that quickly transforms the binary data into a more suitable data format that is easy to experiment with? The problem at hand is around building a software solution capable of processing and presenting near real time BUFR data in a common file format that is independent of any external resources, as the server receives it.

2.4   From BUFR to CSV

The system directly receives the data from the server's data pipeline, which is composed of multiple compressed raw tar files with multiple BUFR files inside each one of the tar files, the system utilizes the received raw files to solve the problem of decoding, preprocessing and combining the data into CSV files showcasing the data points when the satellite scans the area of interest.

The following figure explains the procedure the system takes to preprocess and normalize the data (Fig. 2):

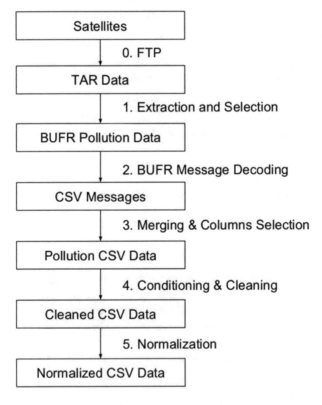

Fig. 2. Data pipeline graph.

In the first step, the system import the raw tar files through the FTP protocol, after extracting the compressed files the system gets multiple binary BUFR files name-coded following a strict naming convention in the following form "INSTRUMENT-ID

PRODUCT-TYPE PROCESSING-LEVEL SPACECRAFT-ID SENSING-START SENSING-END PROCESSING-MODE DISPOSITION-MODE PROCESSING-TIME", that corresponds to multiple important variables such as the instruments used, orbits, and time frames, the system filters the data based on orbits and timeframes to get BUFR pollution files in the area of interest of our choice using regular expressions on the names of the extracted files (under the pollution codename of "TRG"), what the system finally gets are multiple BUFR files corresponding to the area of interest that are ready to be decoded.

In the second step, the system uses a third party software solution called BUFREx-tract [7] to decode the BUFR files into bulks of exported text messages, each message containing a description of its columns and the values in each one in a "somewhat" CSV format.

In the third step, the system performs a fast selection/merge technique to combine all of the messages into two comma separated files corresponding to one scan, one file is for Metop-A and the second file is for Metop-B. Finally, we select the following columns of interest into the final CSV files (Table 1):

Table 1. The columns of the final CSV data file.

No.	Columns (every 12 h)	Unit
1	Year	Integer
2	Month	Integer
3	Day	Integer
4	Hour	Integer
5	Minute	Integer
6	Second	Integer
7	Latitude (high accuracy)	DEGREE
8	Longitude (high accuracy)	DEGREE
9	Integrated CH4 density	kg/m^2
10	Integrated CO2 density	kg/m^2
11	Integrated N2O density	kg/m^2

After exporting the necessary values into multiple structured CSV files, the system groups data points by exact geographical location and date and applies the mean function on the pollutant value to take the average of the possible duplicated measurements.

In the fourth step, the system deals with cleaning data points that are substantively unreasonable using logical conditions on the bounds of the values of CH4, CO2, and N2O.

As a General description of the process, Every half an hour, the system receives one compressed tar file through the server's end points, the system then automatically decompresses the file into BUFR BIN files, selects files corresponding to our area of interest, and decodes them using a third party software solution to the corresponding BUFR messages, finally it merges all messages into two CSV files corresponding to the satellites and cleans any out of bound values, this whole process results in a considerable reduction in data dimensionality and the space it occupies.

3 Results

3.1 Specifications

A C++/Python Implementation of this model was used to build and test the system using data analysis libraries (Numpy and Pandas) on a 8 GB RAM, 2 vCPUs setup.

3.2 Validation

For the numbers to be accurate, we ran the system continuously as a cron job for 15 days, we measured additional metrics manually to ensure the speed and overall reliability.

3.3 Results

The decompressing, decoding, merging, cleaning, and normalizing processes on the raw BUFR data took a relatively short amount of time and memory. Considering the volumes of data we originally processed at each analysis session, the process was fast and efficient, and our simple server setup was enough to transform the data in a matter of seconds. Here are the results of our experiments running the system on different volumes of Raw BUFR files (Fig. 3):

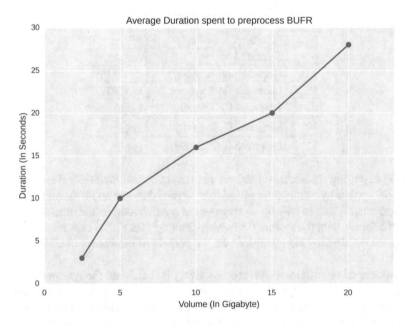

Fig. 3. The average duration spent to preprocess BUFR data.

These tests were conducted multiple times for each volume category, to ensure high accuracy, after analyzing the algorithm, we concluded that it runs on Linear time - O(n), making it strikingly fast for large volumes of Raw meteorological data. We can conclude that the system scales pretty well.

The system presents a data store that is efficient for highly structured, cleaned, space optimized files that are ready for practical scientific research and experimentation, some sample visualisations to follow (Figs. 4, 5 and 6):

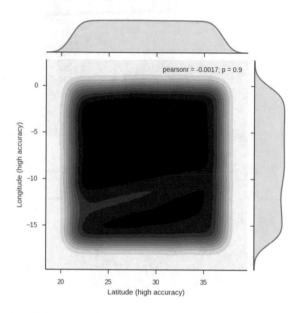

Fig. 4. Final processed geographical points distribution for Morocco.

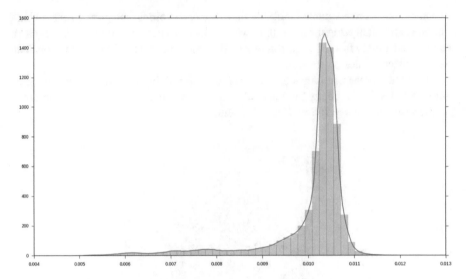

Fig. 5. CH4 distribution example in Morocco.

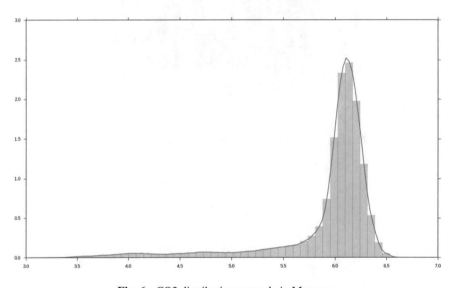

Fig. 6. CO2 distribution example in Morocco.

4   Case Study: Power Plants in Morocco

4.1   The Problem

To demonstrate some of the practical applications of fast processing and restructuring of data, we examine CO2 data points, after processing the data, we run fast visualisations to contrast any association between the thermal power plants and the density of CO_2 in the atmosphere, first we focus on the biggest power plants in Morocco:

- Jorf Lasfar Power Station (Fig. 7)

Fig. 7. A satellite image of Jorf Lasfar Power Station.

Latitude: 33.1047305
Longitude: −8.6376352
Fuel Type: Coal
Capacity: 1356 MW

- Tahaddart Thermal Power Station (Fig. 8)

Fig. 8. A satellite image of Tahaddart Thermal Power Station.

Latitude: 35.589
Longitude: −5.9868
Fuel Type: Natural Gas

Capacity: 384 MW
- Mohammedia Thermal Power Station (Fig. 9)

Fig. 9. A satellite image of Mohammedia Thermal Power Station.

Latitude: 33.682
Longitude: −7.4338
Fuel Type: Fuel Oil & Coal
Capacity: 600 MW

4.2 CO2 Data

The data used in this case study is a small alteration over two months worth of preprocessed CO2 Data, using this simple algorithm:

algorithm *average_geo_co2* **is**

    input: 2 Months of co2 density values of *Map Df*

    output: One Averaged set of data points of *Map Of*

    for each geographical point *p* **in** *Map* **do**

        ps <- getValuesOfPoint(*p*)

        Of[p] <- AVG(ps)

    return *Of*

The algorithm contrasts from the whole dataset without taking in each time interval series of data points, it averaged out the values of each geographic point in the interest map to produce a single map visualisation, creating two sets of points, one for the morning scans and the other is for the evening scan. We found a slight correlation in the night dataset (after a day of industrial activity).

4.3   Interpretation

The following visualisations have been produced (from 8 to 10 PM) (Figs. 10 and 11):

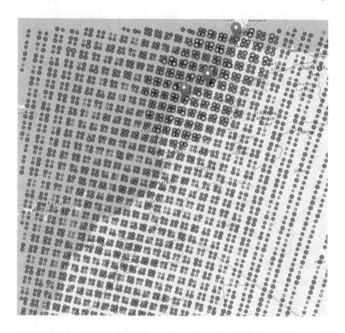

Fig. 10.   Visualised CO2 points in Morocco from Metop-A.

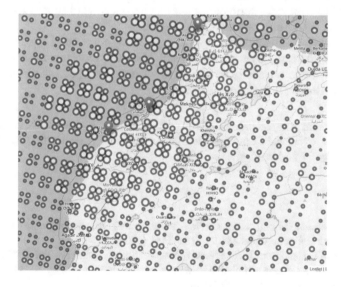

Fig. 11.   Zoomed in visualisation focusing on the power plants of interest from Metop-B.

Every Circle marker in these visualisations represent an averaged individual Geographic Data point, the radius of the circle represent the density percentage (relative to the minimum and maximum values of CO_2 values in the whole dataset). Generally, larger densities are concentrated in zones around the power plants of interest.

While we produced interesting visualisations that showcase a possible correlation between the industrial activities of three power plants in Morocco and the density of CO_2 in these zones, it is however not sufficient to conclude without considering other environmental variables, the density of CO_2 can be affected by multiple other environmental factors. However, what our system allows for is a fast hypothesis forming middleware that boosts our practical everyday data exploration and analysis tasks.

5   Conclusion

At the present time, the size and complexity of raw data is huge and continues to increase everyday. The use of data processing systems to store, process, and analyze data streams and create new data pipelines has changed how we discover and visualise big data in general. In this paper, we presented a software solution made of multiple stacked layers of subsystems that transform and process considerable volumes of raw pollution data in near real time, taking the data from the native file format to a structured, cleaned, normalized, and ready to be visualized and analyzed data store that is light and easy to experiment with. We look forward for our solution to further improve, empower, and accelerate the research process done on top of the EUMETSAT Data Stream interface.

Lastly, the paper presented a case study to demonstrate the possible relation between three power plant activities and the CO_2 densities in their respective location zones. In the goal of fast hypothesis formation and as an introduction to a more serious investigation into the whole phenomena of air pollution.

In the future, significant challenges and problems concerning Big Environmental Data must be addressed by the industry and academia, problems ranging from Data storage and systems scalability to new environmental data processing paradigms for pollution and the environment as a whole.

Acknowledgments. The authors are thankful to the Ministry of Higher Education and Scientific Research, and the National Centre for Scientific and Technical Research (CNRST) for funding this study, under the project: PPR/2015/7.

References

1. Pope III, C.A., et al.: Lung cancer, cardiopulmonary mortality, and long-term exposure to fine particulate air pollution. Jama **287**(9), 1132–1141 (2002)
2. Thomas, M.D., Hendricks, R.H.: Effects of air pollution on plants. Air Pollut. **239** (1961)
3. Dockery, D.W., et al.: An association between air pollution and mortality in six US cities. New Engl. J. Med. **329**(24), 1753–1759 (1993)
4. McMichael, A.J.: Climate Change and Human Health: Risks and Responses. World Health Organization, Geneva (2003)

5. EUMETSAT. http://www.eumetsat.int/website/home/AboutUs/index.html. Accessed 13 May 2017
6. Caron, J.L.: On the suitability of BUFR and GRIB for archiving data. AGU Fall Meeting Abstracts, vol. 1 (2011)
7. BUFR File Support Software. http://www.elnath.org.uk/. Accessed 13 May 2017

Detection and Classification of Malwares in Mobile Applications

Soussi Ilham[(✉)] and Abderrahim Ghadi

IT Department, Faculty of Science and Technology, University of Abdelmalek Essaadi (UAE),
90000 Tangier, Morocco
soussilham@gmail.com, ghadi05@gmail.com

Abstract. Android is the most widely used as mobile operation system, the thing that make it the target of many attackers, day by day the market of malicious applications develop and grow to reach every applications store. In this purpose many researches have been made to deal with threats through proposing different malware detection techniques. The main objective of this thesis is to detect malwares in android mobiles by applying reverse engineering techniques to unpack malicious apps code and classify them in order to detect whether the app is malicious or not. This paper aims to give a comprehensive account of proposed techniques to detect and classify malicious applications in android mobiles based on machine learning algorithms and a comparison between those techniques in order to conclude the most efficient one and their limitations with a view to ameliorate them.

Keywords: Machine learning · Android malicious · Detection · Classification
Reverse engineering

1 Introduction

Smartphone users number is increasing day by day specially smartphones based on android OS, on par with this evolution, applications stores witness a huge evolution in number of applications installed every day, which motivate attackers to develop malicious applications or clone known application for malicious purpose such as stealing user confidential and credential information and using it without his knowledge, moreover this malicious apps use an extra processing power of users device.

For this Reason, considerable attention has been paid by researchers to secure android devices that store wealth sensitive information of the users such as phone numbers, messages, credit card information, by proposing different techniques to detect this malicious applications, this lasts goes into 3 categories:

- Static analysis: analyze the application code after unpacking it in order to extract features that will be used in detecting whether the app is malicious or not without executing it, in general, this features are extracted from the manifest file or the java byte code.

© Springer International Publishing AG, part of Springer Nature 2018
M. Ben Ahmed and A. A. Boudhir (Eds.): SCAMS 2017, LNNS 37, pp. 188–199, 2018.
https://doi.org/10.1007/978-3-319-74500-8_17

- Dynamic analysis: is an analysis approach based on executing the application on real device or emulator to collect the features and capture the application behaviors in order to detect any malicious activities.
- Hybrid analysis: an approach that combine both approaches static and dynamic techniques.

The aim of this study is to propose a new approach to classify malwares in android devices into families after extracting the code by reverse engineering techniques in order to detect those malwares using machine learning algorithms, e.g. SVM (Support Vector Machine), naïve Bayes, etc.

From this point on, the remainder of the paper is organized as follows, a state of art section including previous works about classifying and detecting android malwares using machine learning algorithms and different methods of analysis, a discussion section presenting limitations of this works and a comparison between them, concluding by a conclusion section.

2 State of Art

In the last few years there has been a growing interest in android security, due to huge threats that harm android users and even companies by stealing and divulge their personal and confidential information, hence, several publications has appeared documenting this subject toward finding the most effective method to detect android malwares with the highest accuracy. Previous researches has demonstrated that machine learning algorithms are very efficient in classifying malwares, these algorithms are common used in most approaches in order to classify a sample to the appropriate family, the purpose here is detecting whether the application is malicious or benign. Those approaches can be classifier into three categories: static approach, dynamic approach, hybrid approach, we present in follow studies that have been made in each category:

2.1 Static Analysis

The static analysis is an approach used to identify behavior of applications without actually executing them through disassembling the compiled files of an app using reverse engineering tools.

Mohsin and Dongang [1] developed a static analysis framework that use reverse engineering to detect a malware, their experiment was tested on 1526 Google Play apps and 1259 Genome Malware apps. They used Androguard as a reverse engineering tool to disassembling a specific application by inputting it to the tool and decompiling its APK file, from these last they extract Dalvik byte code and manifest file that contain registered components. For each of these components the proposed detection system analyze this lasts and derives callback sequences upon the extracted event sequences. Hence a taint analysis is performed to detect malicious behavior from the permutations of callback sequence. Figure 1 shows the reverse-engineering life cycle model used for their reverse-engineering approach for detecting malwares, this model include state and transitions were previously omitted by the android system.

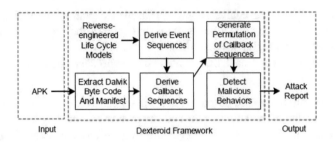

Fig. 1. Dexteroid framework [1]

Their study is based in first place on event sequence based analysis, they concentrate in their experiments on attacks that intend to omit states and transitions by invoking set of callbacks in a specific order.

The developed approach is an ensemble of algorithms, as first event sequence algorithm that derives all event sequences which were previously invoked on android system, it explore all events that can be trigged from static state like staticPostResumed state, those events allow to move an activity from the static state (staticPostResumed) and getting it back to it. Second step is to derive callback sequences from event sequences for both activity and life cycle model, for that, they use the life cycle callbacks which is defined in the activity code, for activity they extracted 26 event sequences and removed the duplication to evade from event sequences that produce same callbacks sequences, same thing for service life cycle model, by removing duplication and for 15 event sequences they extract 10 unique event sequences, subsequently, they analyzed all extracted callbacks sequences in purpose to detect any malicious behavior.

In order to detect any attack intended by the attackers, they analyzed any possible combination of callback ordering, by a proposed permutation sequence which is based on permutation unit as all subsequences that move an activity state and getting it back to its StaticPostResumed state, in addition to that, AUI-callbacks and miscellaneous are considered as permutations units also by virtue of being randomly invoked on StaticPostResumed state.

The strong point of this approach that they evade resource-exhaustive owing to generating permutation for all permutations units by using m-way permutation. Furthermore, they avoided aliasing of an object during analysis by reaching the object-sensitive level using entry instances consisting of a name field and an entry details type field, which will be used as a symbol table entry to make a shallow-copied object in order to reflect all changes made on that shallow-copied object to all aliases of same object. Otherwise, this approach has drawbacks also, including its limitations in detecting specific attacks such as information leakage and sending SMS to premium-rate numbers. Additionally, not all event sequences produced by event permutations is occurred in an applications which increase the false positive results.

This last approach is a static approach based on callbacks sequences to detect malicious activities on specific android application, in a different manner Mercaldo et al. [2] proposed a different static approach that is based on update attack.

Update attack is a type of attacks that targeted the update of an apparently an harmful application, the user install an application from an android application market that appears exhibit from any malicious behavior to make the anti-malware scan obtain a positive result at the first stage of the app life at victims device. Just as the user lunch update of that app a payload is downloaded on its device containing the malicious activities.

For Implementation, they used 2 datasets of malware samples: Drebin project, Genoma one to retrieved 4 families from these datasets that use update attack (Plankton, AnserverBot, BaseBridge, DroidKungFuUpdate).

This approach was the first that use model checking method for update attack. For the methodology as first step, they define a transform operator that translate the Java Bytecode into CCS process specifications in order to describe APK programs through CCS. As second step, recognizing the features that differently the update attack from all other malware families, they used for that an m-calculus logic formulae.

Strong points of this static approach based on malwares that use update attack technique is the higher percent of accuracy due to its possibility to divulge the part of the code that implement the payload download, in this way the mission of the anti-malwares to detect and dissect the malware become easier. In addition, this approach use the model checking method to identify the update attack wish allow them to get higher accuracy.

Simultaneously it has drawbacks also, as a static approach in this special type of attacks that targeted the update of the app is inefficient because the app itself is begin completely, the malware is only implemented in the payload of the application. Furthermore, it's difficult to detect this kind of attacks at all times by cause of unknowing when the payload will be installed or integrated in application, without forgetting the complication of distinguishing between illicit or and licit update.

As reported by Lihui Chen [3], machine learning algorithms are very useful and efficient in detecting malwares, this study analyze and compare various aspects and researches made in this field and they conclude the efficacy of such technique, it draws its strength from its flexibility and ability to adapt automatically to emerging of new features and evolution of new malwares that arise almost every day. As a result, they proposed a framework implemented in Python and Java code named it DroidOL based on machine leaning techniques. Authors analyze the application using static analysis to extract features that will be used by a classifier of online learning to detect the malicious activities.

As first step to detect a malware in this approach, they begin by invoking a static analysis using Soot as workbench on a set of applications to extract their ICFG subgraph which is the feature used in this study using WL kernel, just as building a feature vectors of all the apps, a PA classifier is performed to detect the malware. Online learning proved its effectiveness using variable features-set which make the model adaptive to new data and new features that emerge every day and that make a challenge in most models. Additionally, online learning benefits from the fast model update which is at least daily, this enhance the accuracy of classification by extracting the most recent features, moreover, the ability of extending the batch size occur in online learning algorithms is efficacy in learning maximum of a malware similars which increase the accuracy of classification and detection of malware, besides, their method is efficient in large-scale malware detection.

Yerima et al. [4] also used online learning as a technique in detecting malwares but differently from the last approach they used Bayesian classification in detecting and classifying malwares. The Bayesian classifier consist of 2 stages "learning and detecting", the first stage use a set of different malware samples and benign application in the wild in order to collect wanted features that will be reduced next by a feature reduction function according to its probability to occur in malware and benign apps.

The strong points in their choice of classifier is the fast performance in classification without computational overhead. In addition to ability to model both expert and learning easily in comparison to other algorithms.

For their experiment they used 2000 APKs, 1000 malware samples from 49 families and 1000 benign of different categories to wrap a vast variety of applications used in the world, the result obtained is very satisfying compared with other signature-based experiments.

2.2 Dynamic Analysis

Dynamic analysis is based on executing or emulating the suspicious application on secure and sandboxed environment and monitoring its activities and behaviors in order to detect whether it is malicious or benign app.

Previous researches indicate that dynamic based detectors are more efficient in detecting malwares which are otherwise difficult to detect using static approaches.

Pektas et al. [5] proposed a dynamic approach based on run-time behaviors to collect features and built features vector using a classifier based on machine learning algorithm. For their experiment they used cuckoo sandbox as an analysis tool that execute the app in a sandboxed environment to gather behavioral features from monitoring the run-time activities, and to test the effectiveness of their classifier their used 2000 malware samples belong to 18 different families from VirusShare, in addition to that a web anti-virus scanner Virustotal was used by authors to label the malware samples used in their experiment.

The strong points of this dynamic based approach are its Strength against obfuscation techniques since the analysis is made on the executable file and the manifest file that contain all the information necessarily to gather the features also it has been demonstrated by the authors that the run time behavior based approach is more efficient than signature-based approaches since the huge amount of malwares that appear almost every day.

Still there limitations for this approach as consumption of computing resources, for 2000 samples using 5 machines it took them 5 days thus to minimize the run-time of the classifier appending more resources are needed, moreover the classifier developed by the authors is feeble against some type of malwares.

In the Same context, Dash et al. [6] proposed a dynamic analysis approach based also on runtime behavior but differently from the previous mentioned work, this study used SVM as classifier, they named their detection system DroidScribe, a dynamic analysis system based on runtime behavior using support vector machine as a machine learning algorithm to classify malwares. DroidScribe derived runtime behaviors from system calls monitored during runtime analysis and use them to classify the malware into families. DroidScribe use CopperDroid as dynamic analysis framework to extract features during analysis including system calls and decoded binder communication and abstracted behavioral patterns, upon this extracted features the SVM based multi-class classifier using a training set of malware samples gathered from Genome Project and Drebin as datasets of malware samples to classify malwares into their families.

As advantages of such study are using conformal prediction that produce set of predictions instead of single one which increase the accuracy of classification and correct errors made by SVM classifier that may misclassified some samples because it has to make a choice between two classes that match in characteristics.

DroidScribe is robust in detecting malwares with sparse runtime behavior by improving the decision made by SVM classifier to predict set of best matches of malwares family instead of being forced to choice a single one which are indistinguishable during training by applying conformal prediction.

Otherwise, drawbacks of this approach is the limited information gained from analysis by cause of low-level events, moreover using CopperDroid as dynamic analysis framework make DroidScribe inherit its drawbacks as the ability of malwares that can bypass the screening test and the restrained resources during analysis. DroidScribe is vulnerable in detecting mimicry attacks and randomly added system calls and actions that change patters in system calls because the changes are not clearly visible.

Jang et al. [7] proposed a dynamic approach based on behavior similarity matching with profile, authors developed a hybrid anti-malware system named it Andro-profile utilize capabilities of both on and off device.

On device is the user application installed on his device, off device is a remote server that analyze and profile malicious behaviors of the suspicious app. Once the user made a request in the client application, the request will be sent to remote server and more precisely to repository component to search in its database if there an analysis result made by analyzer component that fulfill the user request, if not the repository component fetch the crawler component.

Thus the remote server contain three components:

- **Crawler** which is responsible for crawling the app in repositories (app markets). In case it couldn't crawl it, the client application on user device collect and send the app information to the server as hash digit of the apk file.
- **Repository component** responsible for testing if hash digit of the apk file crawled is similar to the suspicious apps hash digit, if positive then the app is duplicated and will be discarded, if not thus the repository component send the app to the analyze component.
- **Analyzer component** analyze the app and send the results to client application to exhibit it to user screen, and send it to repository component to add the analysis result in its database.

The procedure of andro-profile is described in Fig. 2:

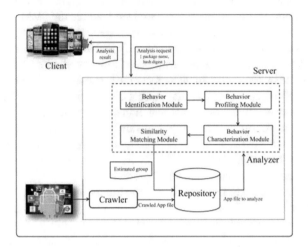

Fig. 2. Overall procedure of andro-profiler [7]

Strong points of this study, a short run-time as 55 s/MB to classify each malware, since andro-profiler is developed in lower level language it make it faster in classifying and profiling malwares.

The approach is efficient in detecting 0-day malwares and give higher accuracy in classifying malwares to their families using performance metrics as false positive and false negative results.

Andro-profiler has limitation also as failing in detecting premium rate SMS type of malwares, in addition to depending on emulator version used for analysis.

Moreover, it is limited in detecting malwares that are executed under specific conditions as (SDK version, cellular network; connection status, time, place), which make andro-profiler does not consider all malicious behaviors as privilege escalation and command and control (C and C) attack.

Malik and Khatter [8] proposed a dynamic approach based on system call pattern as a feature to detect and classify a malware into its family, to test the effectiveness of their approach, they developed a system that extract system call trace from application in order to analyze it and compare it to other benign apps.

For this experiment, 345 malicious samples from 10 different families including (FakeInstaller, Opfake, Plankton, DroidKungFu, BaseBridge, Iconosys, Kmin, Adrd and Gappusin) were used to extract their system call trace and they compare it to system call pattern of 300 benign apps, they observe that malwares execute more often some system calls than normal applications do.

Using this type of features to detect malwares still limited in number of malwares detected, a combination with other features is needed to improve the accuracy and effectiveness of this intrusion-detection system, moreover approaches that are pure based on system call doesn't get enough semantic content and information to perform the classification and detection procedure.

2.2.1   Emulator vs. Device in Dynamic Analysis

Dynamic analysis is analyzing a suspicious app by executing it on a sandboxed environment whether was the host device or an emulator, choice of this last influence the effectiveness of detection system. And to ensure best accuracy of detection, a most interested approach to this issue has been proposed by Alzayalte et al. [9], an approach that compare experiments of dynamic analysis applied on features that were extracted from on device and on emulator analysis.

As first step of their experiment is extracting features for supervised learning, they implemented a tool that automatically extract dynamic features for android mobile which is Dynalog, a dynamic analysis framework which invoke application on a sandboxed environment in a emulator to extract desired features.

The execution time of each app is 300 s on emulator same on device, therefore the behaviors are being logged in log several dynamic behaviors to later extract features including API calls and Intents as shown in Fig. 3.

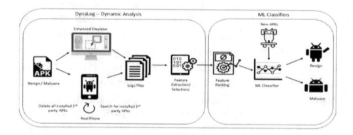

Fig. 3. Extracting features based on emulator and real phone using Dynalog [9]

Dataset that has been used is on total 2444, 1222 of them malware samples from genome project belong to 49 families, and other 1222 are benign apps from Intel Security. They get as result of this experiment, from 1222 malware samples 1205 were detected successfully on phone which give it 98.6% of accuracy, when 939 only on emulator with 76.84% accuracy. And from 1222 benign apps, 1097 successfully detected on phone with 90% accuracy versus 64.27% on emulator with 786 successful detected app. Table 1 shows the percentage of each experiment.

Table 1. Percentage of the successful analysis on android apps using Dynalog [9]

	Emulator	Phone
Malware samples	76.84%	98.6%
Benign samples	64.27%	90%
Total	70.5%	94.3%

Furthermore, authors used machine learning tools to analyze extracted features from both environments to compare the effectiveness of emulators and phones as detection analysis environments, by applying different machine learning algorithms (e.g. Support Vector Machine, Nave Bayes, Simple Logistic, Multilayer Perception, Partial decision

Trees, Random Forest, J48 decision). This experiment demonstrate that using real device for extracting features is most effective method for higher accuracy of detection.

Authors conclude their article by result that dynamic analysis based on real device environment is more efficient in detecting malwares due to its robustness against anti-emulation methods that detect emulators and virtual environment and then hide the malware malicious behaviors.

2.3 Hybrid Analysis

Considering disadvantages and advantages of both static and dynamic approaches and to solve this issue to get higher accuracy of detection of android malwares, many researchers have proposed various approaches of combination of both methods static and dynamic to benefits from both of their advantages and to avoid all drawbacks.

Zhang Yan [10] proposed an hybrid approach based on runtime system calls, this proposed approach consist of a dynamic analysis that is responsible for collecting system-calling data and on other hand a static analysis to process and analyze collected data in order to detect whether the app is malicious or benign by processing the extracted features into a robust computing server named detection server.

As a first procedure of this experiment is collecting data from known benign and malwares, using dynamic analysis on an executed app on device to extract its system calling data about individual and sequential system call with different depth, therefore a malicious pattern set and a normal pattern set is build influenced by system call pattern extracted from app runtime which will be used in comparison with apps patterns to detect the good or bad of this last.

On the other hand, for an unknown malwares, author collect apps runtime system calling data including individual and sequential system call with different depth and extract the patterns to compare them with normal and malicious patterns sets to detect whether the app is malicious or benign.

Next procedure of this experiment is processing of data based on static analysis, for this procedure author used an external detection server for better performance due to limitations of storage and performance of mobile devices. He transferred the collected data to the detection server, by using net link technology that write collected data into a local file including calling data and runtime process information, these lasts will be sent to the server in order for further analysis and processing using python script.

Using an external server for detecting made the approach very robust in detecting android malware because of the limitations in mobile analysis performance and storage capability, moreover the time sufficient of executing of app to perform the extraction of features is only 2 h which make the proposed approach very fast in detection. As result of their experiment using 100 malware samples and 104 benign samples, they get a high value of accuracy equals to 90%.

In the fact, the proposed approach doesn't have many drawbacks, we can only mention the ability to improve the accuracy by improving the malware and benign samples, and still suffer from the limitation of mobile devices in performance of gathering data during runtime of application.

Kapratwar [11] proposed also a hybrid approach but differently from previous research it is based on two features: static feature considered is permission extracted from the Manifest file, while dynamic analysis is based on system call frequency extracted at runtime of application. The main aim of this study is to test the effectiveness of each method (static and dynamic) individually and the effectiveness of their combination.

The first phase of this experiment was the reverse engineering of the app to gather the apk package that contain manifest file necessarily for extracting permissions features using apktool. Thus, author extract permissions features from android manifest file using a developed xml parser for each application and sent it to a feature vector generator in order to build a feature vector.

Therefore a feature selection phase is necessarily to reduce unused and redundant features using Information Gain as feature selection method.

Next phase of this experiment was the extraction of system call using dynamic analysis by applying an emulator integrated in Android Studio, the main objective of this phase is calculating the frequency of occurrence of system call in application which allow analyzer to detect malware, because malwares invoke some system call more often than normal apps.

This approach use machine learning algorithm to classify malwares into families for detection, and in order to choose the most efficient algorithm, author perform a test with algorithms of machine learning and they demonstrate that Random Forest classifier is the most effective machine learning algorithm to detect malwares.

As result of this experiment, author demonstrate that permissions data is more efficient in detecting android malwares than system call frequency, but with the combination of two of them, author gain more accuracy and fetch better results.

The proposed approach still need some improvement as expanding the datasets used in experiment, focusing on the static method and improving it.

3   Discussion and Conclusion

This paper is a modest contribution to the ongoing discussions about security of android mobile, the main aim of this study is to highlight different category of detection and classification based approaches in order to study different methods and compare them with each other to reveal the most effective one.

Different categories of detection approaches has been high-lighted as static, dynamic and hybrid based analysis, as mentioned in the first section, many researchers have demonstrate that static analysis grant higher accuracy because its informative and fast in detection and with less resource consumption, but it still limited in different ways as against obfuscation codes that make almost impossible to apply patterns match techniques, also static analysis is weak against dynamic code loading. Using callback sequences have gain higher accuracy but with high false positive results and it is limited to some kind of attacks, as for the update attack approach, static analysis is weak in such type of attacks because the monitored app is completely benign, only the payload integrated in it that has malicious code. Moreover researchers demonstrate the effectiveness of applying machine learning techniques in detecting and classifying malwares due to its scalability and flexibility.

For the second section, some researchers used runtime behavior as features for dynamically analyze an application during its runtime using machine learning algorithm, both approaches were robust in detecting obfuscation code but they suffer from consumption of computing resources and limited information gathered during analysis. Same as using system call feature that doesn't get enough information for detection which make the combination with other features a necessary matter.

Moreover using on device environment for detection system is more effective and give better accuracy results than emulator environments that suffer from limitation against malwares that use anti-emulator techniques to detect virtual environment and then change their malicious behavior. Thus, above researches demonstrated that dynamic based approaches are efficient against obfuscation code but consume computing resources and suffer from anti-emulation in case of using emulator environment.

And to avoid the drawbacks of this two methods (static and dynamic), researches proposed an hybrid approach that benefit from the advantages of static and dynamic analysis by combining static method in processing and analyzing for better performance and dynamic analysis for fast collect of data, in another way, using both methods based feature as permissions data from manifest file and system call frequency collected during app runtime give better result when authors combine them than experiments when they used each method individually.

4   Conclusion

From the outcome of our investigation it is possible to conclude that hybrid based approach is more efficient in analysis because it combine the advantages of static analysis as rich information gained and dynamic analysis as robustness against obfuscation code, moreover we can conclude also that machine learning techniques are very efficient in classifying and detecting android malwares.

As mentioned in this paper previous works used machine learning algorithm as SVM and Bayesian classification and achieved high accuracy results of detecting android malwares.

References

1. Junaid, M., Liu, D., Kung, D.: Dexteroid: detecting malicious behaviors in Android apps using reverse-engineered life cycle models. Comput. Secur. **59**, 92–117 (2016). https://doi.org/10.1016/j.cose.2016.01.008
2. Mercaldo, F., Nardone, V., Santone, A., Visaggio, C.A.: Download malware? No, thanks. How formal methods can block update attacks (2016)
3. Narayanan, A., Yang, L., Chen, L., Jinliang, L.: Adaptive and scalable android malware detection through online learning. ArXiv160607150 Cs (2016)

4. Yerima, S.Y., Sezer, S., McWilliams, G., Muttik, I.: A new android malware detection approach using Bayesian classification. In: 2013 IEEE 27th International Conference on Advanced Information Networking and Application (AINA)
5. Pektaş, A., Çavdar, M., Acarman, T.: Android malware classification by applying online machine learning, pp. 72–80 (2016)
6. Dash, S., Suarez-Tangil, G., Khan, S., et al.: DroidScribe: classifying android malware based on runtime behavior (2016)
7. Jang, J., Yun, J., Mohaisen, A., et al.: Detecting and classifying method based on similarity matching of Android malware behavior with profile. SpringerPlus **5** (2016). https://doi.org/10.1186/s40064-016-1861-x
8. Malik, S., Khatter, K.: System call analysis of Android malware families. Indian J. Sci. Technol. **9** (2016). https://doi.org/10.17485/ijst/2016/v9i21/90273
9. Alzaylaee, M.K., Yerima, S.Y., Sezer, S.: EMULATOR vs. REAL PHONE: Android malware detection using machine learning. ArXiv170310926 Cs, pp. 65–72. https://doi.org/10.1145/3041008.3041010
10. Tong, F., Yan, Z.: A hybrid approach of mobile malware detection in Android. J. Parallel Distrib. Comput. **103**, 22–31 (2017). https://doi.org/10.1016/j.jpdc.2016.10.012
11. Kapratwar, A.: Static and dynamic analysis for Android malware detection. Masters projects (2016)

Head Gesture Recognition Using Optical Flow Based Background Subtraction

Soukaina Chraa Mesbahi$^{(\boxtimes)}$, Mohamed Adnane Mahraz, Jamal Riffi, and Hamid Tairi

LIIAN, Department of Computer Science, Faculty of Science Dhar El Mahraz,
University Sidi Mohamed Ben Abdellah, BP 1796 Fez, Morocco
`mesbahisoukaina@yahoo.fr`, `adnane_1@yahoo.fr`,
`riffi.jamal@gmail.com`, `htairi@yahoo.fr`

Abstract. This paper presents a technique of real time head gesture recognition system. The primary objective is to implement system that can detect the movement of the head in different directions. The method comprises Gaussian mixture model GMM for background subtraction accompanied by optical flow algorithm, which contributed us the required information respecting head movement. An idea is given regarding the intensity variation between the frames of inputted video. This variation in intensity is used to determine the optical flow and the sum of the velocity vectors of the foreground image. Using the median filter to remove noise from an image, such noise reduction is a typical pre-processing step to improve the results of later processing. In our experiments, we tried to determine the movement of the head in different directions: left, right, up and down.

Keywords: Head gesture · GMM · Background subtraction · Optical flow

1 Introduction

Gestures are an important aspect of human interaction, they can originate from any bodily motion or state but commonly originate from face or hand. A gesture is a form of non-verbal communication in which visible bodily actions communicate particular message, either in place of speech or together and in parallel with words.

The use of gesture as a natural interface becomes a motivating force for modeling, analyzing and recognition of movement. Human machine intelligent interaction needs vision based motion estimation, which requires many interdisciplinary studies. Gesture recognition has long been researched with 2D vision, but with the advent of 3D sensor technology [1], its applications are now more diverse, spanning a variety of markets.

Gesture recognition is the mathematical interpretation of a human motion by a computing device. It provides a redundant form of communication between the user and the robot. There are a great number of devices that are tested to sense body position and orientation, facial expression and other aspects of human behavior or state that can be used to determine the communication between the human and the environment.

To backing gesture recognition, human body movement must be tracked and interpreted in order to recognize the meaningful gestures. Several methods have been developed for both hand gesture and body gesture recognition.

© Springer International Publishing AG, part of Springer Nature 2018
M. Ben Ahmed and A. A. Boudhir (Eds.): SCAMS 2017, LNNS 37, pp. 200–211, 2018.
https://doi.org/10.1007/978-3-319-74500-8_18

Other than the gesture complexities like variability and flexibility of structure of hand and head other challenges include the shape of gestures, real time application issues, presence of background noise and variations in illumination conditions.

In this paper of gesture recognition, we combine Gaussian Mixture Model (GMM) [2] for the background subtraction and Optical Flow method for the determination of head movement. The Mixture of Gaussians method is widely used for the background modeling since it was proposed by Friedman and Russell [3]. Stauffer [4] presented an adaptive background mixture model by a mixture of K Gaussian distributions. There are many optical flow methods, among which we tried the Gunnar Farneback optical flow algorithm [5].

Optical flow method can detect the moving object even when the camera moves, but it needs more time for its computational complexity, and it is very sensitive to the noise.

The essential purpose of the project is to devise a real-time head movement detection and gesture recognition system.

In this paper, we propose a gesture recognition system where motion detection takes by input from a simple camera and needful processing steps is done to recognize the gestures from a live video.

The rest of this paper is organized as follows, Sect. 2 describes the different materials and methods. Section 3 describes the proposed system. Section 4 describes results. Finally, Sect. 5 presents the conclusion.

2 Materials and Methods

2.1 Background Subtraction

Background subtraction is a widely used approach for detecting moving objects from static cameras. The justification of the approach is detection of moving objects from the difference between the current frame and a reference frame, often referred to as the "background image" or "background model". Background subtraction is done primarily if the image in question is a part of a video stream [2]. Background subtraction, also known as foreground detection, is a technique in the fields of image processing and computer vision wherein an image's foreground is extracted for further processing (object recognition etc...). For example, traffic monitoring counting vehicles, detecting and tracking vehicles. Generally, an image's regions of interest are objects (humans, cars, text etc.) in its foreground. After the stage of image pre-processing which may include image denoising, post processing like morphology etc. In all these cases, we must first extract the person or vehicles alone. Technically, extract the foreground in static background movement.

Generally based on a static background hypothesis, which is often not applicable in real environments. With indoor scenes, reflections or animated images on screens lead to background changes. Similarly, due to wind, rain or illumination changes brought by weather, static backgrounds methods have difficulties with outdoor scenes [6].

The simplest background model assumes that the relative intensity values for each pixel are modeled by a united-modal distribution. Stauffer and Grimson [4] proposed to model the value of each pixel as a mixture of Gaussians "MOG" (Mixture Of Gaussians) and use an approximation technique to update the model.

2.1.1   The Gaussian Mixture Model

Over time, different background objects are likely to appear at a same (i, j) pixel location. When this is due to a permanent change in the scene's geometry, all the models reviewed so far will, more or less promptly, adapt so as to reflect the value of the current background object. However, sometimes the changes in the background object are not permanent and appear at a rate faster than that of the background update.

In [7], Stauffer and Grimson raised the case for a multi-valued background model able to cope with multiple background objects. Actually, the model proposed in [7] can be more properly defined an image model as it provides a description of both foreground and background values.

In the context of a traffic surveillance system, Friedman and Russel [3] proposed to model each background pixel using a mixture of three Gaussians corresponding to road, vehicle and shadows. This model is initialized using an EM algorithm [8].

Then, the Gaussians are manually labeled in a heuristic manner as follows: the darkest component is labeled as shadow, in the remaining two components; the one with the largest variance is labeled as vehicle and the other one as road. This remains fixed for all the process giving lack of adaptation to changes over time. For the foreground detection, each pixel is compared with each Gaussian and is classified according to it corresponding Gaussian. The maintenance is made using an incremental EM algorithm for real time consideration.

Stauffer and Grimson [4] generalized this idea by modeling the recent history of the color features of each pixel $\{x_1, \ldots, x_t\}$ by a mixture of K Gaussians. We remind below the algorithm [9]. First, each pixel is characterized by its intensity in the RGB color space. Then, the probability of observing the current pixel value is considered given by the following formula in the multidimensional case:

$$P(X_t) = \sum_{i=1}^{k} w_{i,t} \cdot \eta(X_t, \mu_{i,t}, \sum_{i,t}) \tag{1}$$

Where the parameters are K is the number of distributions, $w_{i,t}$ is a weight associated to the ith Gaussian at time t with mean $\mu_{i,t}$ and standard deviation $\sum_{i,t}$.η is a Gaussian probability density function:

$$\eta(X_t, \mu, \Sigma) = \frac{1}{(2\Pi)^{n/2}|\Sigma|^{1/2}} e^{-\frac{1}{2}(x_t - \mu) \overset{-1}{\Sigma}(X_t - \mu)} \tag{2}$$

For computational reasons, Stauffer and Grimson [4] assumed that the RGB color components are independent and have the same variances. So, the covariance matrix is of the form:

$$\sum_{i,t} = \sigma_{(i,t)}^2 I \tag{3}$$

So, each pixel is characterized by a mixture of K Gaussians. Once the background model is defined, the different parameters of the mixture of Gaussians must be initialized. The parameters of the MOG's model are the number of Gaussians K, the weight $w_{i,t}$ associated to the ith Gaussian at time t, the mean $\mu_{i,t}$ and the covariance matrix $\sum_{i,t}$ [9].

Background Model Estimation. To extract the foreground, the background is modeled by selecting the clusters that effectively represent the background. The background changes less frequently in comparison to the foreground. Thus, the clusters with higher weights and lower variances represent it. The distributions are sorted by the descending values of ω/σ [10].

Then the first B distributions satisfying the following criteria are chosen to represent the background model, where

$$B = argmin_b(\sum_{i=1}^{b} w_i > T) \tag{4}$$

Here, T represents a threshold to determine the minimum amount of data that constitute the background.

Expectation-Maximization (EM). An iterative method to find maximum likelihood [8] or maximum a posteriori (MAP) estimates of parameters in statistical models, where the model depends on unobserved latent variables. The EM iteration alternates between performing an expectation (E) step, which creates a function for the expectation of the log-likelihood evaluated using the current estimate for the parameters, and maximization (M) step, which computes parameters maximizing the expected log-likelihood found on the E step. These parameter-estimates are then used to determine the distribution of the latent variables in the next step [8].

2.2   Optical Flow: Gunnar Farneback

The Gunnar Farneback method [11] is a two-frame motion estimation algorithm. Gunnar Farneback uses quadratic polynomials to approximate the motion between the frames. This can be done efficiently by using the polynomial expansion transform. In the case of Gunnar Farneback the point of interest is quadratic polynomials that produces the local signal model expressed in a local coordinate system such that:

$$f(x) \sim x^T A x + b^T x + c \tag{5}$$

Where A is a symmetric matrix, b a vector and c a scalar. The coefficients are estimated from a weighted least squares fit to the signal values in the neighborhood. The weighting has two components called certainty and applicability. These terms are the same as in normalized convolution [5, 12, 13], which polynomial expansion is based on.

– Displacement Estimation

Since the result of polynomial expansion is that each neighborhood is approximated by a polynomial, we start by analyzing what happens if a polynomial undergoes an ideal translation. Consider the exact quadratic polynomial [11]:

$$f_1(x) = x^T A_1 x + b_1^T x + c_1 \tag{6}$$

And construct a new signal f_2 by a global displacement by d

$$
\begin{aligned}
f_2(x) = f_1(x - d) &= (x - d)^T A_1 (x - d) + b_1^T (x - d) + c_1 \\
&= x^T A_1 x + (b_1 - 2A_1 d)^T x + d^T A_1 d - b_1^T d + c_1 \\
&= x^T A_2 x + b_2^T x + c_1.
\end{aligned} \tag{7}
$$

Equating the coefficients in the quadratic polynomials yields

$$
\begin{aligned}
A_2 &= A_1 \\
b_2 &= b_1 - 2A_1 d, \\
c_2 &= d^T A_1 d - b_1^T d + c_1.
\end{aligned} \tag{8}
$$

The key observation is that by Eq. (8) we can solve for the translation d, at least if A1 is non-singular,

$$
\begin{aligned}
2A_1 d &= -(b_2 - b_1), \\
d &= -\frac{1}{2} A_1^{-1} (b_2 - b_1).
\end{aligned} \tag{9}
$$

We note that this observation holds for any signal dimensionality (Fig. 1).

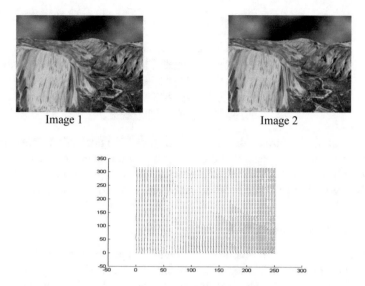

Image 1   Image 2

Fig. 1. Optical flow result, vectors showing the direction and intensity of flow of each pixel.

3 Proposed System

The webcam camera is used as an input video, frames are captured and used in further processing. The main information was gathered from the image, so the subtraction of the background for the entire process was performed. There are many approaches, which furnish helpful tool for image subtraction (Fig. 2). Among those Gaussian Mixture Model (GMM) [2] gave the best result (Table 1).

– Background subtraction MOG

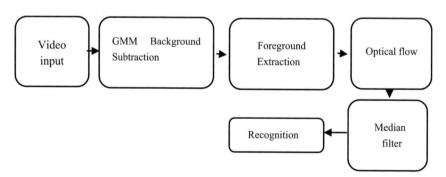

Fig. 2. Block diagram

Table 1. Different methods of background subtraction

Algorithm
Background subtraction
Background subtraction MOG
Background subtraction MOG2
Background subtraction GMG

It is a Gaussian Mixture-based Background/Foreground segmentation algorithm. It uses a method to model each background pixel by a mixture of K Gaussian distributions (K = 3 to 5). The weights of the mixture represent the time proportions that those colors stay in the scene. The probable background colors are the ones, which stay longer, and more static [14].

– Background subtraction MOG2

It is also a Gaussian Mixture-based background/ foreground segmentation algorithm. One important feature of this algorithm is that it selects the appropriate number of Gaussian distribution for each pixel. It provides better adaptability to varying scenes due illumination changes etc [15, 16].

– Background subtraction GMG

This algorithm combines statistical background image estimation and per-pixel Bayesian segmentation [17].

Figure 3 shows the results found after using the three different methods of background subtraction.

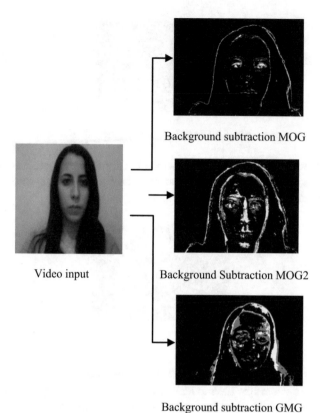

Background subtraction MOG

Video input

Background Subtraction MOG2

Background subtraction GMG

Fig. 3. Results after background subtraction

Now, we use a technique to extract foreground using the background subtraction MOG2 to follow the moving object with the real color intensities (Fig. 4).

Fig. 4. Image foreground extraction with real color intensities

In the next step, we applied the Gunnar Farneback algorithm to the foreground to determine the movement of pixels remaining in the foreground. Gunnar Farneback is a dense optical flow algorithm because it computes the optical flow for all pixels in the image (Fig. 5).

Fig. 5. The vectors of the optical flow

After that, we apply the median filter which is a digital filtering technique often used to remove noise from an image. Such noise reduction is a typical pre-processing step to improve the results of later processing (Fig. 6).

Fig. 6. Median filter

Many features are accompanied with the image. The information from the foreground is used to determine the movement of the head, and those features provide us the corresponding gesture. Here the optical flow vectors acted as the feature vector [18].

The last step is to compute Sx and Sy, which represent the summation of the velocities calculated by the optical flow along the x and y axis [18].

4   Results and Discussion

In this section, we show experimental results of the proposed method using background subtraction with Gaussian Mixture Model and optical flow.

In Table 2, Sx and Sy are the sums of the both optical flow vectors. Some summed values obtained shown below [18]:

Table 2. The head movement and and the corresponding values of Sx and Sy

Value of Sx	Value of Sy	Results	Recognition
0.0013	0.7410	Right	Right
−0.0400	−0.1455	Left	Left
0.3968	−0.0634	Down	Down
−0.0012	0.1238	Up	Up

According to the results found, we found that the detection of head movement can be generalized according to the signs of Sx and Sy. Table 3 presents the different possibilities signs of Sx and Sy.

Table 3. Results from Sx and Sy of the both optical flow vectors

Value of Sx	Value of Sy	Conclusion
+Ve	−Ve	DOWN
−Ve	−Ve	LEFT
+Ve	+Ve	RIGHT
−Ve	+Ve	UP

Figure 7 represents the recognition of the head using the proposed system.

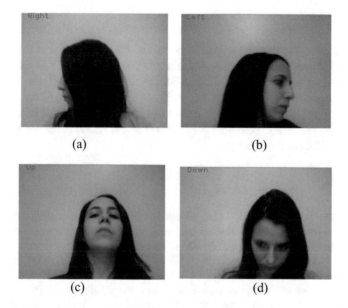

(a)   (b)

(c)   (d)

Fig. 7. Different position of the movement of the head, (a)- Head is moving towards left, (b)- Head is moving towards right, (c)- Head is moving towards up, (d)- Head is moving towards down.

The success rate and the comparison of this project with some previous work are given below [18] (Tables 4 and 5):

Table 4. Success rate

Experiments	Number of experiments	Number of correct recognition	Success rate
Left	10	6	60%
Right	10	8	80%
Up	10	10	100%
Down	10	7	70%

Table 5. Comparative study between our work and other approaches

Name of the technique used	Success rate	Comparative study
Optical flow	97.4%	Noise caused unintended head Motion
SVM	85.3%	High false positive rate
Our work	77.5%	- Less noise
		- Improved processing rate
		- Accurate time derivatives

The success rate has been calculated and found to be 77.5% with the use of threshold based classification. The time taken to process was obtained to be 6.05 frames per second.

5   Conclusions

In this paper, we proposed a system based on the GMM model and the optical flow method to recognize the movement of the head. We used the MOG2 approach as the main background subtraction algorithm, with median filtering to eliminate noise.

It can be deduced that, based on the result and the success rate found, the proposed system is very robust and allows the head to be detected with great efficiency. Other improvements are being studied based on other classifiers such as SVM, neural networks, and the ACP method.

References

1. Holte, M.B., Tran, C., Trivedi, M.M., Moeslund, T.B.: Human pose estimation and activity recognition from multi-view videos: comparative explorations of recent developments. IEEE J. Sel. Top. Signal Process. **6**(5), 538–552 (2012)
2. Mukherjee, S., Das, K.: An adaptive GMM approach to background subtraction for application in real time surveillance. Int. J. Res. Eng. Technol. **2**(1), 125–129 (2013)
3. Friedman, N., Russell, S.: Image segmentation in video sequences: a probabilistic approach. In: Proceedings Thirteenth Conference on Uncertainty in Artificial Intelligence (UAI 1997), pp. 175–181 (1997)
4. Stauffer, C., Grimson, W.: Adaptive background mixture models for real-time tracking (PDF). In: IEEE Computer Society Conference on Computer Vision and Pattern Recognition, vol. 2, pp. 246–252 (1999). https://doi.org/10.1109/cvpr.1999.784637
5. Farneback, G.: Polynomial expansion for orientation and motion estimation. Ph.D. thesis, Linkoping University, Sweden, SE-581 83 Linkoping, Sweden Dissertation No 790, ISBN 91-7373-475-6 (2002)
6. Piccardi, M.: Background subtraction techniques: a review (PDF). In: IEEE International Conference on Systems, Man and Cybernetics, vol. 4, pp. 3099–3104 (2004). https://doi.org/10.1109/icsmc.2004.1400815
7. Dempster, A., Laird, N., Rubin, D.: Maximum likelihood from incomplete data via the EM algorithm. J. Roy. Stat. Soc. Ser. B Methodol. **39**(1), 1–38 (1977)
8. Balakrishnan, S., Wainwright, M.J., Yu, B.: Statistical guarantees for the EM algorithm: from population to sample-based analysis. CoRR abs/1408.2156 (2014)

9. Bouwmans, T., El Baf, F., El Vachon, B.: Background modeling using mixture of Gaussians for foreground detection - a survey: (PDF). Recent Pat. Comput. Sci. **1**, 219–237 (2008)

10. Dibyendu, M., Wu, Q.M.J., Thanh, N.M.: Gaussian mixture model with advanced distance measure based on support weights and histogram of gradients for background suppression. IEEE Trans. Ind. Inf. **PP**(99), 1–11 (2014)

11. Farnebck, G.: Two-frame motion estimation based on polynomial expansion. In: Bigun, J., Gustavsson, T. (eds.) Proceedings of the 13th Scandinavian Conference on Image Analysis (SCIA 2003), pp. 363–370. Springer, Heidelberg (2003)

12. Knutsson, H., Westin, C.F.: Normalized and differential convolution: methods for interpolation and filtering of incomplete and uncertain data. In: Proceedings of IEEE Computer Society Conference on Computer Vision and Pattern Recognition, New York City, USA, 515–523. IEEE (1993)

13. Westin, C.F.: A tensor framework for multidimensional signal processing. Ph.D. thesis, Linkoping University, Sweden, SE-581 83 Linkoping, Sweden Dissertation No 348. ISBN 91-7871-421-4 (1994)

14. KadewTraKuPong, P., Bowden, R.: An improved adaptive background mixture model for realtime tracking with shadow detection. In: Proceedings of 2nd European Workshop on Advanced Video Based Surveillance Systems, AVBS01, September 2001

15. Zivkovic, Z.: Improved adaptive Gaussian mixture model for background subtraction. In: Proceedings of the 17th International Conference on Pattern Recognition, vol. 2, pp. 28–31 (2004)

16. Zivkovic, Z., van der Heijden, F.: Efficient adaptive density estimation per image pixel for the task of background subtraction. Pattern Recogn. Lett. **27**(7), 773–780 (2006)

17. Godbehere, A.B., Matsukawa, A., Goldberg, K.Y.: Visual tracking of human visitors under variable-lighting conditions for a responsive audio art installation. In: ACC, pp. 4305–4312 (2012)

18. Saikia, P., Das, K.: Hand gesture recognition using optical flow based classification with reinforcement of GMM based background subtraction. Int. J. Comput. Appl. 0975-8887, March 2013

An Intelligent Model for Enterprise Resource Planning Selection Based on BP Neural Network

Amine Elyacoubi[1]([⊠]), Hicham Attariuas[2], and Noura Aknin[2]

[1] Science Faculty, University Abdelmalek Essaidi,
26, Street Al Andaloussie Res Anatolia Office Building,
3rd floor N° 11, Tangier, Morocco
a.elycoubi@yahoo.fr
[2] Science Faculty, University Abdelmalek Essaidi,
Avenue de la Palestine Mhanech, Tétouan BP 2222, Morocco
Attariuas.hicham@gmail.com, Aknin_noura@yahoo.fr

Abstract. Enterprise resource planning (ERP) is the managing business system that allows an enterprise of any organization to utilize a collection of integrated applications to manage its business and automate many back office functions related to technology. The selection itself of a suitable ERP is one of the most important parts in the implementation. This paper attempts to use artificial neural networks to choose an ideal ERP for any enterprise. This paper constructs a three-level BP neural network to analyze the principle and model of a suitable ERP. By using the samples to train and inspect the BP neural network, we conclude that the application of BP neural networks is an effective method to forecast suitable ERP. Thus the purpose of this study is to requite mainly three factors among the many others that influence the choice of a suitable ERP. By using statistics in several investigation-filled samples, we can collect a database for many cases that can in return help us create a model that manages the choice of an ideal ERP for the company and reduces the costs of failure.

Keywords: Enterprise resource planning · ERP · ERP implementation
BP neural network

1 Introduction

The 21st century is regarded as the age in which informationization has become a massive and important concept in our professional activity. Additionally, computers have been used for decades as tools by companies that are involved in commerce in general, to achieve profitability and improve management. The information technologies such as material requirements planning (MRP) and manufacturing resource planning (MRPII) are built up from early versions of inventory control software.

Nowadays, Enterprise resource planning (ERP) has come to maturity by incorporating more functions such as logistics management, financial management, asset management and human resource management [1].

© Springer International Publishing AG, part of Springer Nature 2018
M. Ben Ahmed and A. A. Boudhir (Eds.): SCAMS 2017, LNNS 37, pp. 212–222, 2018.
https://doi.org/10.1007/978-3-319-74500-8_19

The selection of a suitable ERP decreases cost and increases interests for an enterprise. So, a rational supposition is that an ideal ERP is crucial in its implementation.

The artificial neural networks (ANNs) concept originates from biology. Its components are similar to and have the basic functions of neurons in an organism. The components are connected according to some pattern of connectivity, associated with different weights. The weight of a neural connection is updated by learning. ANNs posses the ability to identify nonlinear patterns by learning from the data. It can also imitate the knowledge-level activities of experts either physically or functionally. Therefore, ANNs can be used in business and banking applications for decision making, forecasting and analysis. In order to search the optimal weights for neural networks, a number of algorithms have been coined and developed. The back propagation (BP) training algorithms are probably the most popular ones. The structure of BP neural networks consists of an input layer, an output layer, as well as a hidden layer. The numbers of the input and output layer nodes are decided by task requirements. The optimal number of hidden layer nodes is determined by certain testing experiments. The BP training algorithms are well known for that. Yet they may have a slow convergence in practice, and the search for the global minimum point of cost function may be trapped at local during gradient descent [4].

The remainder of the paper is organized as follows; the first section is for the abstract and introduction. The second section is dedicated to the literature review. The third one proposes a model of BP neural network for forecasting appropriate ERP for the company. The fourth section describes the sample selection and data analysis. As for the last section, it provides a summary to this work and conclusions.

2   Literature Review

Enterprise Resource Planning (ERP) is a standard of a complete set of enterprise management system. It emphasizes integration of the flow of information relating to the major functions of the firm.

The choice of a suitable ERP is the most important and crucial step. Many companies fail in their implementation due to their poor selection methods, which allows us to note that, the opinion of an expert is crucial for the right implementation to take place. There are several modules for implementing an ERP in a company, but the main one is the choice of an ERP.

There are numerous phases in the ERP implementation process. One of the earliest and most critical phases is the ERP selection phase. If an organization selects an inadequate ERP to fit their needs, the project will most likely destine to fail. Research and practice have provided several cases of ERP project failures because of a faulty selection process. No matter what amendments the adopting company undertakes in the later phases, if there is no fit, there is no success [2].

Recently, scholars and researchers have proposed many tricks to provide better selection methods for a suitable ERP, for example, *Se Hun Lim and Kyungdoo* Nam have analyzed an Artificial Neural Network Modeling in forecasting successful Implementation of ERP systems [6] (See Fig. 2).

Today, the quality of decision-making is a prime factor for success in top management which allowed us to choose the ANN view for its reputation; big business and banking applications for decision-making have used ANN [4].

In a survey of business applications from 1992 to 1998, Vellido et al. found neural networks are matured to offer real practical benefits. Consequently, it can be used to assist in selecting potential suppliers (partners). Since the back propagation (partners) neural networks was proposed by *Rumelhart D E ECT,* a number of network models have been developed with the BP neural network as the one most favored by neural networks researchers [4].

BP Neural Network

BP neural network is a multi-layer neural network of error back- propagation (see Fig. 1). It consists of three layers: a layer of "input" that is connected to a layer of "hidden" units, which is connected to a layer of "output" units. The activity of the input units represents the raw information that is fed into the network. The activity of each hidden unit is determined by the activities of the input units and the weights on the connections between the input and the hidden units. The behavior of the output units depends on the activity of the hidden units and the weights between the hidden and output units.

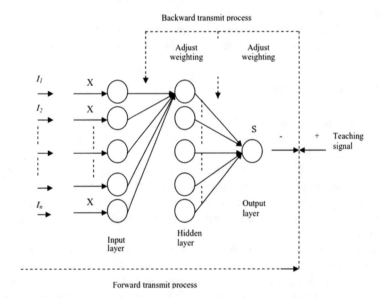

Fig. 1. Structure of three–layer BP neural network

The multi-layer perception is trained under supervision using the back propagation algorithm. By using a training algorithm to adapt the interconnection weights, BP neural network has the ability to implement a wide range of responses to the patterns in a given training set. The network functions in two stages during training: a forward pass and a backward pass. In the forward pass, the input vector is presented to the

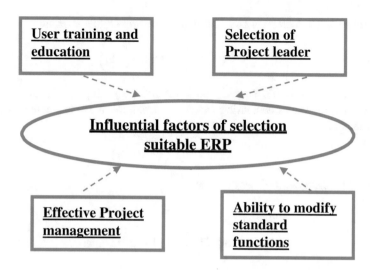

Fig. 2. Influential factors of selection ERP (Concluded by the other articles)

network, and the outputs of the units are propagated through each upper layer until the network output is generated. The difference (error) between the network output and the desired output is computed for each output unit. In the backward pass, starting from the network output, a function of the error is fed back through the network layers to the input stage. The interconnection weights are adjusted during the backward pass to minimize the error. The forward and backward passes are repeated until the network converges, that is, until a measure of the error is acceptably small.

The convergence of the BP neural network can be tested in several ways. The most practical test for network convergence is the error limit test, i.e., to test if the absolute difference between the desired response and the response for each output unit is below a small specified error limit. Alternatively, training could be terminated when the sum of the squares of the errors for all output units is below a specified limit.

The learning algorithm of BP neural network is shown as follows [19]:

1. Using random value (between 0 and 1) to initialize Wji and θj. Wji is the link weight from neuron i to neuron j, θj is neuron the threshold value of j (hidden layer and Output layer).
2. Using the deposing training sample collection {xP1} and the corresponding desired output collection {yP1}. The p, l delegate sample number and the difference Vector number separately;
3. Calculating each neuron output Oij:
   (1) Regarding the input layer neurons, the output is the same to its input,

$$Opi = xpi \qquad (1)$$

   xpi is the pth sample's ith value;

(2) Regarding hidden layer and output layer, neuron Output operation is as follows:

$$Oij = f\left(\sum wjiopi - \theta j\right) \tag{2}$$

Oij is the neuron i's output, also is the neuron j's Input, f (x) is a non-linear differentiable non-decreasing Function, generally, it is considered as sigmoid function,

$$f(x) = 1/(1 + e - x) \tag{3}$$

4. Calculating each neuron's error signal:

$$\text{Output level: } \delta ij = (yij - oij)\, oij\, (1 - oij) \tag{4}$$

$$\text{Input level: } \delta ij = oij\, (1 - oij)\Sigma\delta ijwij \tag{5}$$

5. Back-propagation, revises weight

$$Wj(t + 1) = wj(t) + \alpha\delta ijoij \tag{6}$$

α is the learning speed;
6. Calculating error

$$Ex = \{\Sigma\Sigma\}(opk - ypk)2/2 \tag{7}$$

If Ex is smaller than fitting error, finishes the network training; otherwise transfers it to 3, continues to train.

In my research I try to reconsider the factors that directly influence the proper choice of an ERP. First, I present the factors that directly influence the choice of an ERP with multiple articles in the same way. Then I present my 3 factors that I have reviewed from an investigation, I will speak in-depth about the investigation in chapter Analysis example.

3   Example Analysis

The empirical data for this research comes from the investigation in an online platform. Indeed, this survey consists of more than 48 well-chosen and well-defined questions. But our goal is to minimize the questions. The minimum possible is to have the most interesting factors that influences the choice of ERP, then use them as Layer BP neural network. I was able to have more than 80 filled with different consultants which gave me a database to base this research upon. This database consists of several replies from consultants in Morocco.

After the data collection, I found out that to have results in our research, the rematch of the moderating indicators that directly influences the final result is missed.

I used Software R statistical software that analyzes all the data and chooses the moderating indicators that influence greatly the final result.

Indeed I was able to work with the method logistic regression (Anova) which gave me 3 moderator indicators as I confirm this with another method of:

Step by step (see Fig. 3)

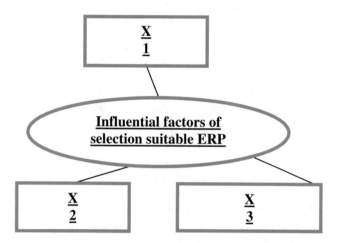

Fig. 3. Influential factors of selection ERP in this research (concluded by the software R)

Analysis of variance (Anova): is a collection of methods for comparing multiple means across different groups

ANOVA Model

If p is the number of factors, the ANOVA model is written as follows:

$$y_i = \beta_0 + \Sigma_{j=1...q}\beta_{k(i,j)}, j + \varepsilon_i$$

Where y_i is the value observed for the dependent variable for observation i, $k_{(i,j)}$ is the index of the category (or level) of factor j for observation i and εi is the error of the model.

The chart below shows data that could be analyzed using a 1-factor ANOVA. The factor has three categories. Data are orange points. The dashed green line is the grand mean and the short green lines are category averages. Note that we use arbitrarily the sum (ai) = 0 constraint, which means that β_0 corresponds to the grand mean.

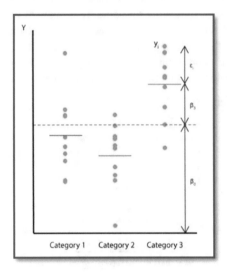

The hypotheses used in ANOVA are identical to those used in linear regression: the errors ε_i follow the same normal distribution $N(0, s)$ and are independent. It is recommended to check retrospectively that the underlying hypotheses have been correctly verified. The normality of the residues can be checked by analyzing certain charts or by using a normality test. The independence of the residues can be checked by analyzing certain charts or by using the Durbin Watson test.

Finally I had 3 moderator factors that I will use:

X1: Effective Project Management
X2: Level of implementation difficulty
X3: Percentage of coverage before adaptation

This paper constructs a three-layer BP neural network model which can forecast suitable ERP for any Enterprise. We forecast 80 samples of Expert opinion. The first 50 samples are used to train the neural network, and the last 30 samples are used to inspect the training effect. The result shows that BP neural network can be employed to the selection of suitable ERP for any enterprise. Compared to traditional methods, BP neural network has more advantages. The principle of BP neural network is simple. It has strong maneuverability, fast speed and satisfying fitting precision. The enterprises can employ it to avoid a great deal of calculation and human errors. At the same time, they don't need to waste much time and human resources on collecting the data or opinions of experts. So the BP neural network can help enterprises reduce costs greatly through the information which technologies of ERP can provide.

As seen in the three boxes: Current ERP is the choice of an expert, ERP estimated is our ERP that is concluded by our model of BP Neural network and the relative error (Table 1).

Table 1. Network training result

Question	Current ERP	ERP estimated	Relative error %
X1	5	4	10%
X2	6	5	10%
X3	12	10	20%
X4	4	3	10%
X5	5	4	10%
X6	4	3	10%
X7	6	4	20%
X8	6	5	10%
X9	7	6	10%
X10	5	4	10%
X11	4	3	10%
X12	5	4	10%
X13	6	4	20%
X14	7	6	10%
X15	3	4	10%
X16	3	4	10%
X17	6	4	20%
X18	5	4	10%
X19	4	3	10%
X20	7	5	20%
X21	4	3	10%
X22	3	3	0%
X23	5	4	10%
X24	4	3	10%
X25	6	7	10%
X26	3	4	10%
X27	5	5	0%
X28	5	7	20%
X29	5	4	20%
X30	3	4	20%

Note: ERP has been replaced by numbers for statistical software.

GRAPH OF CURVE RESULT

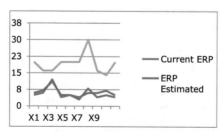

We can find that the simulated outputs are basically consistent with the desired outputs and all the relative errors are below 20%. We can also find that in the training samples or the inspection samples, the simulated rank of the samples are consistent with the actual ERP selection of the samples. It proves that the results for the forecasting are precise. We can adjust the weights and thresholds by training the network time after time to improve precision and reduce the relative error. By doing this, we can get simulated outputs which are quite accurate and close to the desired outputs (Table 2).

Table 2. Network inspecting result

Question	Current ERP	ERP estimated	Relative error %
X1	5	6	10%
X2	6	7	20%
X3	12	11	10%
X4	4	5	10%
X5	5	5	0%
X6	4	3	10%
X7	6	8	20%
X8	6	4	20%
X9	7	5	20%
X10	5	4	10%
X11	5	5	0%
X12	5	4	10%
X13	3	3	0%
X14	6	5	10%
X15	3	4	10%
X16	3	4	10%
X17	5	5	0%
X18	2	2	0%
X19	5	4	10%
X20	6	5	10%

GRAPH OF CURVE RESULT

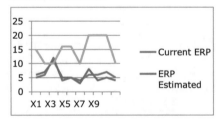

We can find that the simulated outputs are basically consistent with the desired outputs and the entire relative errors are below 20%.

In order for us to be successful, we need to use other means and other methods to compare the results and find the difference that compelled me to look for other methods in the same way. This allowed me to choose the linear regression as a linear approach for modeling the relationship between a scalar dependent variable y and one or more explanatory variables (or independent variables) denoted X. The case of one explanatory variable is called *simple linear regression*. For more than one explanatory variable, the process is called *multiple linear regressions* [1]. (This term is distinct from *multivariate linear regression*, where multiple correlated dependent variables are predicted, rather than a single scalar variable.)

4 Conclusion

In this research, we constructed a three-layer BP neural network model which can forecast and select the suitable ERP for any Enterprise. We forecast 80 samples of Expert opinions. The first 50 samples are used to train the neural network and the last 30 samples are used to inspect the training effect.

The result shows that BP neural network can be employed in the selection of suitable ERP for any enterprise.

The principle of BP neural network is simple. It has strong maneuverability, fast speed and satisfying fitting precision. The enterprises can employ it to avoid a great deal of calculation and human errors. At the same time, they don't need to spend much time and human resource on collecting the data. So the BP neural network can help enterprises reduce costs greatly. Besides this, the information technologies of ERP can provide data which are more comprehensive, accurate and timely for the application of neural network. It is an advisable approach for the enterprises which implement ERP to apply neural network. Therefore, we can conclude that the forecasting model based on three-layer BP neural network proposed in this paper is valuable in practice.

Acknowledgments. This research was supported by the University ABDELMALEK ESSAIDI Faculty of Science, Tetouan Morocco supervised by Mrs. Noura Aknin and Atariuass Hicham.

References

1. ERP Selection: The SMART Way MoutazHaddaraWesterdals - Oslo School of Arts, Communication and Technology, 0185 Oslo, Norway. Department of Computer Science, Electrical and Space Engineering, Luleå University of Technology (LTU), Porsön, SE-971 87 Luleå, Sweden
2. Zhang, P., Wang, S.: A study of neural networks in forecasting of logistics system. Wuhan University of Technology (2002)
3. Tavel, P.: Modeling and Simulation Design. AK Peters Ltd., Natick (2007)
4. Zhang, L., Wang, D., Chang, L.: A Model on Forecasting Safety Stock of ERP Based on BP Neural Network

5. Reuther, D., Chattopadhyay, G.: Critical Factors for Enterprise Resources Planning System Selection and Implementation Projects within Small to Medium Enterprises, Micreo Ltd. Queensland, Australia Mechanical, Manufacturing and Medical Engineering. Queensland University of Technology, Queensland, Australia
6. Lim, S.H., Nam, K.: Artificial Neural Network Modeling in Forecasting Successful Implementation of ERP Systems
7. Razmi, J., Sangari, M.S.: A hybrid multi-criteria decision making model for ERP system selection
8. Sun, H., Ni, W., Lam, R.: A step-by-step performance assessment and improvement method for ERP implementation: action case studies in Chinese companies
9. Velcu, O.: Strategic alignment of ERP implementation stages: an empirical investigation
10. Chen, G., Sai, Y., Zhang, J.: ERP Implementation Based on Risk Management Theory: Empirical Validation
11. Gupta, L., Wang, J., Charles, A., Kisatsky, P.: Prototype selection rules for neural network training. Pattern Recogn. **25–11**, 1401–1408 (1992)
12. Feng, W.: The Brief Introduction of Main Function Modules in ERP. Technol. Dev. Enterp. **5**, 61–62 (2006)
13. Longo, F., Mirabelli, G.: An advanced supply chain management tool based on modeling & simulation. Comput. Ind. Eng. https://doi.org/10.1016/j.cie.2007.09.008
14. Bansalk, K., Vadhavkar, S., Guptaa, A.: Neural networks based forecasting techniques for inventory control applications. Data Min. Knowl. Disc. **2**(1), 97–102 (1998)
15. Partovi, F.Y., Anandarajan, M.: Classifying inventory using an artificial neural network approach. Comput. Ind. Eng. **41**, 389–440 (2002)
16. He, Y., Li, F., Song, Z., Zhang, G.: Neural Networks Technology for Inventory Management
17. Zhang, P., Wang, S.: A study of neural networks in forecasting of logistics system. Wuhan University of Technology (2002)
18. Rumelhart, D.E., McClelland, J.D.: Parallel and Distributed Processing I, II. MIT Press, Cambridge (1986)

Discrete Wavelet Transform and Classifiers for Appliances Recognition

El Bouazzaoui Cherraqi[1(✉)] ⓘ, Nadia Oukrich[1] ⓘ,
Soufiane El Moumni[2] ⓘ, and Abdelilah Maach[1]

[1] Mohammadia School of Engineering, Mohammed V University in Rabat, Rabat, Morocco
elbouazzaouicherraqi@research.emi.ac.ma
[2] Laboratory of Information Processing, Hassan II-University of Casablanca,
Casablanca, Morocco

Abstract. Recognition of appliances' signatures is an important task in energy disaggregation applications. To save and manage energy, load signatures provided by appliances can be used to detect which appliance is used. In this study, we use a low frequency database to identify appliances based on discrete wavelet transform for features extraction and data dimensionality reduction. Further that, the accuracy of several classifiers is investigated. This paper aims to prove the effectiveness of DWT in load signatures recognition. Then, the best classifier for this studied task is selected.

Keywords: Classifiers · Discrete Wavelet Transform (DWT)
Appliances recognition

1 Introduction

Nowadays, electrical appliances dominate the energy consumption in residential sector. The most measurements used today are blind [1], which means that the consumption of individual units is ignored as it is described in Fig. 1. This disadvantage cannot give any indication about detailed consumption that down to the used appliances. To understand our energy consumption, appliances load monitoring [2, 3] becomes an essential element to reduce and save energy. Since its inception, Appliances Load Monitoring is based on load signatures to identify appliances, this signature can be defined as the unique electrical behavior of appliances when they turned on such as audio signals generated by human.

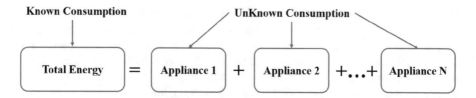

Fig. 1. Blind measurement of energy consumption

© Springer International Publishing AG, part of Springer Nature 2018
M. Ben Ahmed and A. A. Boudhir (Eds.): SCAMS 2017, LNNS 37, pp. 223–232, 2018.
https://doi.org/10.1007/978-3-319-74500-8_20

There are two approaches in Appliances load monitoring, the first one is Non-intrusive load monitoring [4] and the second one is intrusive load monitoring. For the first one, a unique sensor is used to collect information about energy consumption in household, this approach is proposed for the first time by Hart in 1992 [4], while, for the second approach, we use one sensor for each appliance or a group of appliances by dividing home into different zones of measurement. This method is accurate and simple. The only disadvantages are the high cost and the complexity of installation [2].

The challenge in appliance recognition is the large categories of appliances located at home, the collected databases are depended on the electrical line and transmission noise, and the preprocessing of these databases are widely different. In this paper, we use individual signatures to recognize appliances. In literatures, there are several public databases which are available for the scientific community, these datasets are divided into two groups which have low frequency (Freq ≤ 1 Hz): such as AMPds [5], ECO [6], GREEND [7], DRED [8], iAWE [9], REFIT [10], and Tracebase [11]. High frequency (Freq ≥ 50 Hz) such as WHITED [12] and COOLL [13]. According to the data acquisition frequency, the used load signatures in this work that belong to low frequency database named Tracebase, which is collected using a smart plug with a sampling frequency of 1 Hz. This appliances load signatures database contains individual signatures of 43 types of appliances from different residential devices and offices in Germany. Signatures are collected during 24 h in a separate file.

This work analyzes the accuracy of different classifiers to recognize appliances used in residential sector. The current implementation allowing to choose the most accurate classifier to solve the problem of appliances identification, two aspects are developed in the present paper: the features extracted using wavelet transform and the implementation of more than one classifier belonging to different families. In this study, the used classifiers are K Nearest Neighbors classifier (KNN), Support Vector Machine (SVM), Decision Tree classifier (DT), Random Forest classifier (RF), AdaBoost classifier (Adaboost), Gradient Boosting classifier (GB), GaussianNB (GNB), Linear Discriminant Analysis (LDA), Quadratic Discriminant Analysis (QDA).

The rest of this paper is organized as follows. Section 2 presents related works of appliances recognition. Section 3 describes, methodology of classification used in this paper. In Sect. 4 experimental results and classifiers performances are described. Finally, Sect. 5, we conclude this paper.

2   Related Work

In the residential sector there are several categories of appliances and this number increases accordingly to the growth of this sector [3]. According to electrical behavior of appliances, they are grouped into four groups as it described in [4] and developed in [2, 14]: appliances with two specific state ON/OFF such as toasters and lamps, appliances with a finite number of states, continuously variable appliances and permanent appliances, when the consumption maintained during a long time. The knowledge of these categories is useful to estimate the individual consumption of each appliance based on the change of event and the time between two successive states for the same device. Information down to the appliances level can reduce energy into 12% [15].

The common point between all proposed approaches in the appliance identification is the use of machine learning. Artificial neural network is used for appliances recognition based on low frequency sampling rate load signatures [24]. 10 features are used as inputs for ANN to perform the classification [15]. The heart of appliances recognition is the features extraction this stage is different from work to another according to the used method. In the paper of Reinhardt [11] 517 features are extracted from each device trace, only 15 relevant features are used in the classification and these final features are extracted using Weka toolkit to represent 33 categories of appliances. The classification is achieved using; random committee, Bayesian Network, J48, JRip, LogitBoost, Naive Bayes, Random Forest, and Random Tree the accuracy up to 95.5% obtained by Random Committee. In [16] the proposed appliances identification was built to recognize five classes of appliances, classification was performed using two classifier from two distinguished families, accuracy in this work up to 85%, This work was developed in [1] and the number of appliances categories was increased to 10 raw features that are analyzed using principal component analysis and the recognition of appliances achieved using SVM, 16 types of devices are investigated in this work given an accuracy up to 99.9%. Classification of ON/OFF appliances category was examined in [17], three classifiers namely Bayes Net, Random Forest and Hoeffding Tree are used to identify appliances through load signatures the proposed methods are implemented by using the WEKA software.

3   Methodology of Classification

Appliances recognition approach has been investigated through a several classifiers. In this work we use a real-world database with a large-scale modification. Table 1 shows details about the appliances used, we have chosen seven classes that are the most known in residential sector such as refrigerator, washing machine, laptop and so on. Our classification methodology is described in the Fig. 2.

Table 1. Number of instances prepared from Tracebase dataset for each appliance.

Appliance	Number of load signatures instances
Refrigerator	1026
Washing machine	56
Laptop-PC	482
Desktop-PC	1728
Monitor TFT	818
Router	455
Multimedia	111

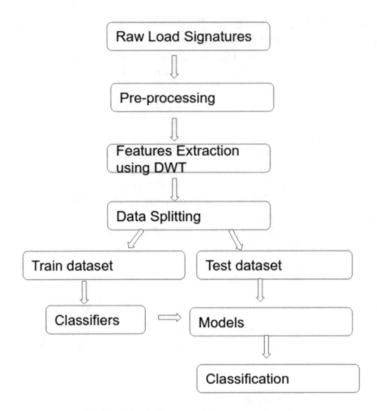

Fig. 2. Block diagram of the proposed study

3.1 Pre-processing

In the chosen database, we noticed that the collected data contains the raw time series of appliances which means the existence of duration when appliances are turned on and when they are inactive. In our application we are only interested on the tuned-on state, especially when the appliance generates its electrical behavior through a load signature. The tuned off periods of time are ignored here. Further to this, time series are subdivided into a several sub-sequences that describe each device.

3.2 Features Extraction Using DWT

Load signature is a sequence of values that varies during time, in this case values are real power. formally a load signature can be represented by $Ls = \{ls1, ls2, ls3,...,lsn\}$ where Ls is the whole load signature of appliance and ls is the recorded value of real power and n is total number of values. While the number of observation, high dimensionality and multivariate property makes the classification difficult. To reduce the size of data and denoised it Discrete Wavelet Transform is performed to extract approximate coefficients that represents the whole signal. Decomposition of signal is generated as a tree known as Mallat's decomposition tree that shown in Fig. 3, where Ls (n) is the load signature and the high and low filters are represented respectively by h and g, the first and second wavelet details are d1 and d2. The second level approximation is represented by a2 which represents the input of classifiers.

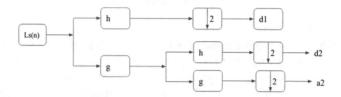

Fig. 3. Discrete wavelet transform with two levels

3.3 Splitting Database

The whole dataset that prepared for classification is splited randomly into two disjoint set, training and test. Different combinations are valuated to smooth results. The function used here to split database is implemented using the Sklearn library [9]. It allows a simple manipulation of database.

3.4 Classification

In our work, nine classifiers are implemented belonging to eight distinct families, namely, nearest-neighbors, support vector machines, decision trees, random forests, Boosting, Bayesian, discriminant analysis; Above, we briefly introduce the implemented classification algorithms.

K-Nearest Neighbors (KNN)
KNN is one of the most simple and fundamental classifier, based on the minimal distance between the training dataset and the testing dataset. This algorithm is widely used to solve the classification problems.

Support vector machine (SVM)
An SVM classifier is based on representation of outputs as points in space, every output category is separated by a clear gap. In test, new examples are conducted to one category based on which side of the gap they fall [18].

Decision Trees Classifier (DT)
DT is composed of nodes structured like a tree [19]. Nodes relate to direct edge from starting root node. Internal nodes are created with one incoming edge and producing two or more than two edges in this level. Values are compared to choose the right decision according to the feature. DT ending with terminal nodes.

Random Forest Classifier (RF)

Random forests use multiple of tree. Each tree depends on random vector values that tested independently with the same distribution for all trees in the forest. Each tree makes its own decision and the final decision is completed based on voting. Where the decision of a significant part of trees is the decision of overall outcome.

Adaboost Classifier (ADA)

An ADA classifier is a popular boosting technique [20]. This classifier helps to combine multiple classifiers that perform poorly into a single strong classifier. Beginning by fitting a classifier on the original dataset and then fits additional copies of the classifier on the same dataset but where the weights of incorrectly classified instances are adjusted such that subsequent classifiers focus more on difficult cases.

Gradient Boosting classifier (GB)

GB is one of the most powerful machine-learning techniques for building predictive models. GB approach is based on construction of new base-learners to be excellently linked with the negative gradient of the loss function that connected with the whole ensemble. The choice of the loss function may be arbitrary, with both of rich variety of loss functions derived distant and the possibility of applying one's own task-specific loss [21].

Gaussian NB (GNB)

GNB subdivided inputs into continuous variables and output as discrete variable variables. GNB is characterized by the conditional probability between features that given the label [22].

Supervised classification with conditional Gaussian networks increasing the structure complexity from naive Bayes [23].

Linear Discriminant Analysis (LDA) and Quadratic Discriminant Analysis (QDA)

LDA is widely used for dimensionality reduction and classification. This technique maximizes the ratio between classes. This approach is known as class dependent transformation. Also, it can be used to create independence between different classes [23], QDA has the same property as LDA but the observations in QDA are separated by a quadratic hyperplane.

4   Experimental Results

4.1   Implementation

Classifiers are implemented in python based on Sklearn modules [9]. This library contains a wide range of machine learning algorithms. Our implementation consists on bringing all classifiers in the same script and run them in the same python successively. The machine used here is a personal notebook with 2.4 GHz in processor and 4 GB of RAM, all this work has been set up under GNU/Linux environment.

4.2   Results

This study is based on preprocessing of Tracebase database to enhance classification. As it is described in the previous sections the main step in classification is feature extraction in order to simplify learning for classifiers. In our work features are extracted and dimensionality reduced using the DWT. The performance of classifiers is evaluated based on the reached accuracy by each classifier and the duration of providing results.

DT with accuracy 100% of and takes 1.07 s to map all database and create the model that used for the test, GB reaches 100% of precision but it takes a long time exactly 115.70 s compared with KNN which reaches 98.93% in only 2.27 s. Random Forest Classifier is the third best classifier, RF classifier is the faster classifier with 0.86 s in this application, but the accuracy is less than KNN, Gradient Boosting and DT. QDA reach less than 20% that showing the lowest accuracy among the nine classifiers performed in this work. Table 2 gives more details.

Table 2. Classifiers accuracies and training times

Classifier	Accuracy (%)	Time (S)
KNN	98.93	2.27
SVM	64.93	370.98
DT	100	1.07
RF	95.33	0.61
Adaboost	61.20	23.16
GBoosting	100	117.99
GaussianNB	66.53	0.91
LDA	69.06	50.62
QDA	19.06	27.33

To quantify accuracy of each classifier, penalization of false classification is performed by log loss function. Classifier with the maximum accuracy is minimizes log loss function. Mathematically this metric can be defined as negative log-likelihood of the desired labels given an observed probabilistic classifier's predictions. Figure 4 shows the variation of log loss function for all used classifiers. According to the negative log-likelihood GB is more accurate the only weakness of this algorithm is the running time compared to DT. The poor classification detected by Log loss function is obtained by the algorithm QDA.

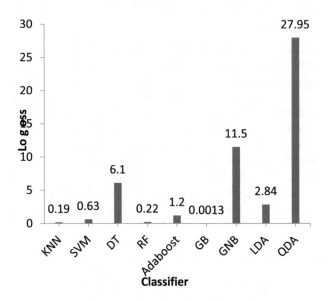

Fig. 4. Variations of Log loss function for the nine classifiers used

5   Conclusion

In this paper, load signatures recognition was presented using an automatic prepro-cessing based on Discrete Wavelet Transform. The aim of this work is to identify the most accurate classifier for appliances recognition using a low frequency sampling rate dataset. The methodology applied showed promising results proved by the well-known classifiers such as DT, KNN, LDA and so on. Training time and accuracy indicated that Decision Tree classifier was faster and more accurate for such prepared database. In the future we will analysis more load signatures dataset, and compare the powerful of classifiers in such complex problem. The improvement of appliances identification stage Non-intrusive load monitoring will help to save energy in the future.

References

1. Ridi, A., Gisler, C., Hennebert, J.: Automatic identification of electrical appliances using smart plugs. In: 2013 8th International Workshop on Systems, Signal Processing and Their Applications (WoSSPA), pp. 301–305 (2013)
2. Zoha, A., Gluhak, A., Imran, M.A., Rajasegarar, S.: Non-intrusive load monitoring approaches for disaggregated energy sensing: a survey. Sensors **12**(12), 16838–16866 (2012)
3. Ridi, A., Gisler, C., Hennebert, J.: A survey on intrusive load monitoring for appliance recognition. In: 2014 22nd International Conference on Pattern Recognition (ICPR), pp. 3702–3707 (2014)
4. Hart, G.W.: Nonintrusive appliance load monitoring. Proc. IEEE **80**(12), 1870–1891 (1992)

5. Makonin, S., Popowich, F., Bartram, L., Gill, B., Bajic, I.V.: AMPds: a public dataset for load disaggregation and eco-feedback research. In: 2013 IEEE Electrical Power & Energy Conference (EPEC), pp. 1–6 (2013)

6. Beckel, C., Kleiminger, W., Cicchetti, R., Staake, T., Santini, S.: The ECO data set and the performance of non-intrusive load monitoring algorithms. In: Proceedings of the 1st ACM Conference on Embedded Systems for Energy-Efficient Buildings, pp. 80–89 (2014)

7. Monacchi, A., Egarter, D., Elmenreich, W., D'Alessandro, S., Tonello, A.M.: GREEND: an energy consumption dataset of households in Italy and Austria. In: 2014 IEEE International Conference on Smart Grid Communications (SmartGridComm), pp. 511–516 (2014)

8. Uttama Nambi, A.S., Reyes Lua, A., Prasad, V.R.: Loced: location-aware energy disaggregation framework. In: Proceedings of the 2nd ACM International Conference on Embedded Systems for Energy-Efficient Built Environments, pp. 45–54 (2015)

9. Batra, N., Gulati, M., Singh, A., Srivastava, M.B.: It's different: insights into home energy consumption in India. In: Proceedings of the 5th ACM Workshop on Embedded Systems for Energy-Efficient Buildings, pp. 1–8 (2013)

10. Murray, D., Stankovic, L., Stankovic, V.: An electrical load measurements dataset of United Kingdom households from a two-year longitudinal study. Sci. Data **4**, 160122 (2017)

11. Reinhardt, A., et al.: On the accuracy of appliance identification based on distributed load metering data. In: Sustainable Internet and ICT for Sustainability (SustainIT 2012), pp. 1–9 (2012)

12. Kahl, M., Haq, A.U., Kriechbaumer, T., Jacobsen, H.-A.: Whited-a worldwide household and industry transient energy data set. In: 2016 Proceedings of the 3rd International Workshop on Non-Intrusive Load Monitoring (NILM) (2016)

13. Picon, T., et al.: COOLL: Controlled On/Off Loads Library, a Public Dataset of High-Sampled Electrical Signals for Appliance Identification, arXiv Preprint arXiv:161105803 (2016)

14. Baranski, M., Voss, J.: Nonintrusive appliance load monitoring based on an optical sensor. In: 2003 IEEE Bologna Power Tech Conference Proceedings, vol. 4, p. 8 (2003)

15. Paradiso, F., Paganelli, F., Luchetta, A., Giuli, D., Castrogiovanni, P.: ANN-based appliance recognition from low-frequency energy monitoring data. In: Proceedings of the 14th IEEE International Symposium on A World of Wireless, Mobile and Multimedia Networks (WoWMoM), pp. 1–6, June 2013

16. Zufferey, D., Gisler, C., Khaled, O.A., Hennebert, J.: Machine learning approaches for electric appliance classification. In: 2012 11th International Conference on Information Science, Signal Processing and their Applications (ISSPA), pp. 740–745 (2012)

17. Salihagić, E., Kevric, J., Doğru, N.: Classification of ON-OFF states of appliance consumption signatures. In: 2016 XI International Symposium on Telecommunications (BIHTEL), pp. 1–6 (2016)

18. Vapnik, V.: The support vector method of function estimation. In: Nonlinear Modeling, pp. 55–85. Springer (1998)

19. Nguyen, N.-T., Lee, H.-H.: Decision tree with optimal feature selection for bearing fault detection. J. Power Electron. **8**(1), 101–107 (2008)

20. Freund, Y., Schapire, R.E.: A desicion-theoretic generalization of on-line learning and an application to boosting. In: European Conference on Computational Learning Theory, pp. 23–37 (1995)

21. Natekin, A., Knoll, A.: Gradient boosting machines, a tutorial. Front. Neurorobotics **7**, 21 (2013)

22. Perez, A., Larranaga, P., Inza, I.: Supervised classification with conditional gaussian networks: increasing the structure complexity from naive bayes. Int. J. Approx. Reason. **43**(1), 1–25 (2006)

23. Balakrishnama, S., Ganapathiraju, A.: Linear discriminant analysis-a brief tutorial. Inst. Signal Inf. Process. **18** (1998)
24. Cherraqi, E.B., Maach, A.: Load signatures identification based on real power fluctuations. In: Noreddine, G., Kacprzyk, J. (eds.) International Conference on Information Technology and Communication Systems, ITCS 2017. AISC, vol. 640 (2018)

Cloud-Based Integrated Information System for Medical Offices

Sanou Landry[1], Anouar Dalli[1(✉)], and Seddik Bri[2]

[1] Ecole Nationale des Sciences Appliquées de Safi (ENSAS),
Université Cadi Ayyad, Marrakesh, Morocco
Anouar_dalli@yahoo.fr
[2] Groupe Matériaux et Instrumentations, Département Génie Electrique,
Ecole Supérieure de Technologie, Université Moulay Ismail, Meknes, Morocco
briseddik@gmail.com

Abstract. The abstract should summarize the contents of the paper in short terms, i.e. 150–250 words. During the last years, the global economic crisis has affected all domains, including the health sector. The latest developments and advancements in the computer science and information technology promise efficiency and reduced costs for management of medical offices. It provides easy access to critical information, thereby enabling management to make better decisions on time. The aim of the paper is to build an information management system for medical office environment based on an open Enterprise Resource Planning (ERP) software and cloud computing. This system enables improved patient care, patient safety, efficiency and reduced costs. It provides easy access to critical information, thereby enabling management to make better decisions on time. It is an efficient mean to analyze and evaluate in a realistic scenario the healthcare system performance in terms of reliability and efficiency.

Keywords: E-Health · Medical office · Cloud computing
Integration information system · Odoo

1 Introduction

A medical office is a medical facility where one or more doctors, with the help of some nurses provide treatment to patients. Medical offices provide consultations, routine cares and medicines prescriptions. Medical offices are the first place where patients go in case of issues, they maybe redirect to a hospital if their conditions require some more sophisticate treatments [1].

Medical offices can benefit a lot from an Integrated Information System. Such system will support the comprehensive information requirements for medical offices, including patient report management, appointment and financial management.

Computerization of medical records and documentations results in efficient data management and information dissemination [2]. Beside Integrated Systems, Cloud computing is a technology that allow enterprises to save money and boost their business. The cloud hosts the enterprise database and offer better scalability and security.

© Springer International Publishing AG, part of Springer Nature 2018
M. Ben Ahmed and A. A. Boudhir (Eds.): SCAMS 2017, LNNS 37, pp. 233–241, 2018.
https://doi.org/10.1007/978-3-319-74500-8_21

This paper presents a Cloud-Based Integrated Information System Solution that let multiple medical offices manage their activities in a single platform.

The paper is organized as follows. Section 2 present the approach use in this solution, Sect. 3 show the technical implementation, Sect. 4 present the benefits of the system.

2 Materials and Methods

2.1 Cloud Computing

In earlier times, there used to be a big worry about how and where to save the data but now the concept of cloud computing provides an efficient and economical way to deal with the same issue [3].

Cloud computing provides for different kinds of services. Through platform-as-a-service or PaaS, consumers can build and deploy their applications on the cloud provider's platform as and when needed. Through software as-a-service or SaaS, consumers use software services provided by the cloud providers. And finally through infrastructure-as-a-service or IaaS, consumers are provided with computing power and disk storage via virtual environments.

In a public cloud, an enterprise can offload its computing tasks to the external cloud provider whereas, in a private cloud, the computing services and resources remain within the perimeters of the organization's private network, so that it retains control of the computing tasks. A hybrid cloud is a combination of both private and public computing [4].

A hybrid model could be proposed to be used by organizations in the healthcare domain. The approach retains a private cloud for sensitive research activities but employs a public cloud for other services.

2.2 Odoo

The solution proposed is based on the Odoo server and this is for many reasons. Odoo is an exhaustive suite comprising of a wide array of business applications subsuming Project Management, CRM, Sales, Manufacturing, Warehouse Management, and Financial Management. Being open source. Odoo is free to download and easy to install. There is no licensing fee associated with Odoo and since, there is a vast community allied to this enterprise resource planning and management tool, it assures the continued support. Odoo is highly modular, thus it is easy to improve or develop additional features.

In term of security Odoo is open source, so the whole codebase is continuously under examination by Odoo users and contributors worldwide. Community bug reports are therefore one important source of feedback regarding security. Odoo is designed in a way that prevents introducing most common security vulnerabilities: SQL injections are prevented by the use of a higher-level API that does not require manual SQL queries; XSS attacks are prevented by the use of a high-level templating system that automatically escapes injected data; The framework prevents RPC access to private methods, making it harder to introduce exploitable vulnerabilities.

2.3   Architecture

Figure 1 shows a global view of the architecture. The cloud provider hold several instances of Odoo server where run all the services. There is a main server that handle all the registered Medical Offices. Additional servers are present to provide data redundancy. Those as use to maintain a high availability of the services and can also be used to balance the charge of the main server, resulting in better performance. Finally the database and users password will be encrypted and communication between Medical offices and the Cloud will be done through a firewall that insure that only authorizes users can access the system.

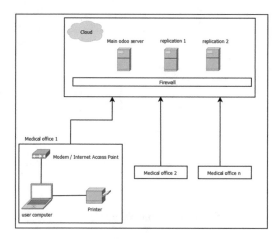

Fig. 1. Architecture

On the client side, medical offices just need some computers connected to internet and some printers to print reports or medicines prescriptions.

3   Implementation

Since the solution is cloud-based, all the technical part will be handle by the cloud-provider. Medical Offices will just have to subscribe to the solution and begin use it straight away.

The service provide will involves some system likes patient registration, patient appointment, patient billing, finance & accounts, medicine prescription, pharmacy management.

To access the service, client will need a computer per user, an access to internet and some printers. After the subscription, each user of the Medical Office will receive a Login and a Password that let them access the system at any time.

3.1 Patient First Visit and Consultation Process

Patients arrive at Medical Offices with or without a scheduled appointment. At this point they are received by an assistant or a nurse. If there is no scheduled appointment the assistant is responsible to register the appointment and direct the patient to the appropriate doctor (Fig. 2).

Fig. 2. Appointment management

If it is the first visit of patient, the assistant register the patient in the system. This registration consists of saving information like the patient name, telephone, address, email and may be a photo.

When the patient arrives to a doctor, the doctor already has access to the patient information and the motive of his visit. The doctor can then perform a consultation or prescribe some medicines.

The patient is then redirect to the cash register. After a consultation an invoice is automatically generated in the accounting system. The responsible to the cash register is only responsible to receive and confirm the patient payment.

3.2 Patient File Management

All Medical Offices get registered to the same service. Thus they share patient medical information. The patient file contains all information related to his previous consultation and even his ongoing treatment. The patient report management system is able to store file like patient's medical imaging. With that a patient can get to any Medical Office and ask to print him his complete medical report for a given period (Fig. 3).

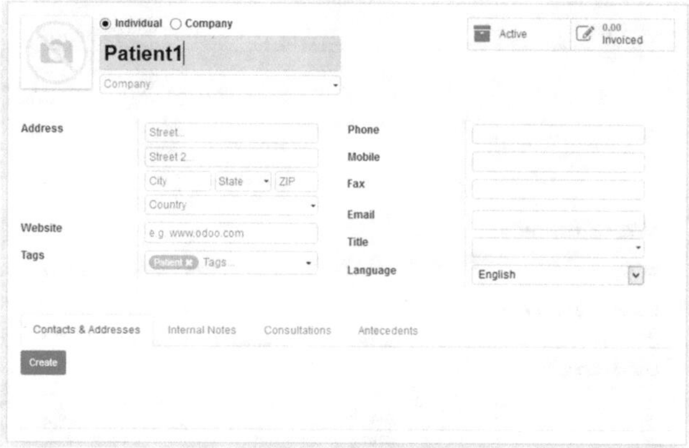

Fig. 3. Patient's information

3.3 Pharmacy Management

The pharmacy management or drug store management system make easy to follow medicines inventory and sales (Figs. 4 and 5).

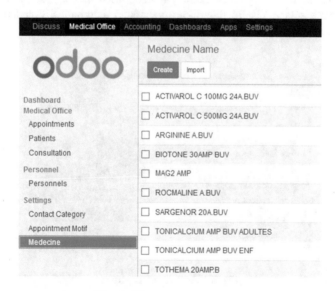

Fig. 4. Pharmacy management

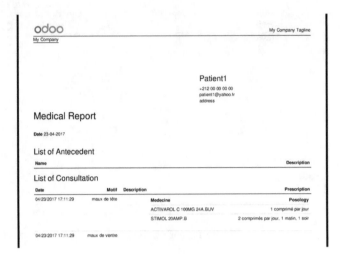

Fig. 5. Patient file printed

3.4 Finance Management

The accounting and finance system allow a global management of the Medical Office finances. It allows managing consultation payment, medicine payment, lease payment, human resource salary. Useful reports are generated automatically and the Medical Office can at any time get a clear vision of its finances (Fig. 6).

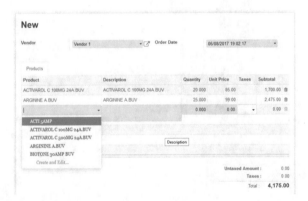

Fig. 6. Finance management

4 Experience in Other Countries

Many countries are already working to deploy electronic health record (EHR), or electronic medical record (EMR) systems. EHR systems are designed to store data accurately and to capture the state of a patient across time. It eliminates the need to track down a patient's previous paper medical records and assists in ensuring data is accurate and legible. It can reduce risk of data replication as there is only one modifiable file, which means the file is more likely up to date, and decreases risk of lost paperwork.

4.1   United States

In the United States, the Department of Veterans Affairs (VA) has the largest enterprise-wide health information system that includes an electronic medical record, known as the Veterans Health Information Systems and Technology Architecture (VistA).

4.2   France

By Act No. 2004-810 of 13 August 2004, the Government of France launched a project called Personal Medical File. In 2015 the project also known as DMP has been renamed Shared Medical File. The aim of the project is to provide medical professionals with the prior consent of the patient with medical information (medical history, laboratory analysis results, imaging, treatment in progress) from other health professionals (General practitioners, specialists, nurses or hospitals) defining a medical profile of each patient.

4.3   India

The Government of India, while unveiling of National Health Portal, has come out with guidelines for EHR standards in India. The document recommends set of standards to be followed by different healthcare service providers in India, so that medical data becomes portable and easily transferable. India is considering to set up a National eHealth Authority (NeHA) for standardization, storage and exchange of electronic health records of patients as part of the government's Digital India programme.

4.4   Austria

In December 2012 Austria introduced an Electronic Health Records Act (EHR-Act). These provisions are the legal foundation for a national EHR system based upon a substantial public interest according to Art 8(4) of the Data Protection Directive 95/46/EC. In compliance to the Data Protection Directive (DPD) national electronic health records could be based upon explicit consent, the necessity for healthcare purposes or substantial public interests. The Austria EHR-Act pursues an opt-out approach in order to harmonize the interests of public health and privacy in the best possible manner.

5 Discussion

5.1 Benefits

The proposed solution provides multiple benefits for Medical Offices and for Patients [5]. The first benefit for Medical Offices is that the solution is easy to access and does not cost much money. Since the solution is cloud based, Medical Offices will not have to buy expensive materials, some simple computers or tablets are enough and when they subscribe to the service they get access immediately to a bunch of features to manage all their activities. The second benefit for Medical Offices is that the solution will allow better communications. Doctors from different offices may communicate and help each other to provide better health care. Beside Medical Offices, the ones that will benefit the most of this solution is Patients. Since Patients records will track all their precedent consultations or treatments, any doctor can get an excellent view of a patient situation at any time and thus provide some better services [6, 7]. Another point is the decrease of information research time. Our solution come with an efficient search tool that let doctors find quickly the information they need in the database. Finally all the patients information gathered in the database may be useful for medical researches.

5.2 Security

A key point, of this solution is the security. The security is provided by the cloud provider by controlling who get access to the system via a firewall and by encrypting the database and the user password. In addition, Odoo is able to prevent the most common web attacks and medical office employees passwords are protected with industry-standard PBKDF2+SHA512 encryption (salted+stretched for thousands of rounds).

5.3 Improvements and Perspectives

Many things can be done to improve the proposed solution. One is to give more intelligence to the system and another one is to give patients access to their personal data.

By giving more intelligence to the system, the system will be able to generate diverse alerts and help doctors in their decisions. This is especially important during prescription, a clever system will be able prevent a doctor to prescribe a medicine that could triggered an allergy to a patient.

Involve patients in the management of their personal data is also an important point. A web interface that allow patients to consult their data will be very useful. In term of security patients could also use a smart card (like a bank card) to retain their most vital medical information but also to manage the access to their data. The card will be presented at each consultation and will allow doctors to unlock the patient medical file.

Finally we are aware that the management of finances on the cloud may be a sensitive subject. But this can be overcome simply by providing a software that let medical offices manage their finances locally.

6   Conclusion

Through this paper we have seen how an integrated Information System can improve Medical offices services. An Integrated Information System allows Medical Offices to manage easily their Patients appointments or their finances. By combining the Integrated Information System with the cloud we get a more enhance service that reduce cost and allow more collaboration for medical offices, all of this for a better care of Patients.

However implementing this management system is not only a technological challenge, but also a strategy decision one, regarding the following factors: government regulations, budget, available technologies, organizational culture, and these may affect the capacity to reach the desired goal.

Also, we suggest that building an e-Health on Cloud is a process that must evolve through at least four phases: determine the e-Health Cloud model, compare the offers of cloud providers, migrate the information to the data center, run and evaluate a pilot implementation.

References

1. Kobrinskii, B.A.: E-health and telemedicine: current state and future steps. E-Health Telecommun. Syst. Netw. **25**(3), 50–56 (2014)
2. Hasan, J.: Integrated Hospital Information System (HIS) special focus on BIRDEM hospital (600 beds). E-Health Telecommun. Syst. Netw. **60**(2), 29–35 (2013)
3. Singh, V.J., Singh, D.P., Bansal, K.L.: Proposed architecture: cloud based medical information retrival network. Int. J. Comput. Sci. Eng. Technol. **4**(5), 485–497 (2013)
4. Banica, L., Stefan, L.: Cloud-powered e-Health. Sci. Bull. Econ. Sci. **12**(1), 12–38 (2013)
5. Manfredi, S.: Performance evaluation of healthcare monitoring system over heterogeneous wireless networks. E-Health Telecommun. Syst. Netw. **7**(1), 27–36 (2012)
6. Marcu, R., Popescu, D., Danila, I.: Healthcare integration based on cloud computing. U.P.B. Sci. Bull. (Series C) **77**(2), 31–43 (2015)
7. Almuayqil, S., Atkins, A., Sharp, B.: Ranking of E-Health barriers faced by Saudi Arabian citizens, healthcare professionals and IT specialists in Saudi Arabia. Health **54**(8), 1004–1101 (2016)

A Game Theoretic Approach for Cloud Computing Security Assessment Using Moving Target Defense Mechanisms

Iman El Mir[1(✉)] ⓘ, Abdelkrim Haqiq[1], and Dong Seong Kim[2]

[1] Computer, Networks, Mobility and Modeling Laboratory,
Faculty of Sciences and Technology, Hassan 1st University, Settat, Morocco
iman.08.elmir@gmail.com, ahaqiq@gmail.com
[2] Department of Computer Science and Software Engineering,
University of Canterbury, Christchurch, New Zealand
dongseong.kim@canterbury.ac.nz

Abstract. Nowadays, the security of the Internet is always dressing severe challenges. To achieve network system's security with the complexity and the diversity of attack types is too difficult and costly. However, to make the network systems more resistant to attacks, Moving Target Defense (MTD) has been introduced to maximize the complexity and the uncertainty of various attacks. Indeed, these proactive defense techniques have been deployed as a game changer to prevent or delay attacks and limit their opportunity windows. This paper provides an important branch of MTD mechanisms named IP address hopping mechanism using detection systems for attack mitigation in the cloud environment. We have used a game theory approach to model the attack-defense interaction and analyze the effect of MTD on the payoff of both the attacker and the defender. Consequently, we have used Matlab simulator to validate the proposed model. Our approach demonstrated its feasibility in terms of attack mitigation in the cloud and proved the efficiency of IP address hopping mechanism to reduce the attacker's opportunity window.

Keywords: Moving target defense techniques · Game Theory
Nash Equilibrium · Bayesian game · Cloud computing security

1 Introduction

The attackers can profit from static configurations to easily exploit and discover the network vulnerabilities with the aim is to compromise the network and spread intrusions. But with the dynamic network reconfiguration especially the development of moving target defense techniques, we prevent the intruders from reaching their targets and achieve their objectives. The MTD allows an unpredictable change of the network configurations which make the attacker's effort more hard because he need frequently launch the reconnaissance phase to understand the feebleness of system otherwise he will attack the system based on a false report. Consequently, the attacking cost will be increased and the attack surface will be minimized.

© Springer International Publishing AG, part of Springer Nature 2018
M. Ben Ahmed and A. A. Boudhir (Eds.): SCAMS 2017, LNNS 37, pp. 242–254, 2018.
https://doi.org/10.1007/978-3-319-74500-8_22

The MTD techniques have been introduced so as to decrease the likelihood of attack being propagated successfully and its costs accumulated through the continued randomization the configuration of the network system such as IP addresses, network parameters, cryptographic keys and so one. There many challenges and requirements are related to the MTD deployment and implementation in which the aim is to optimize the configuration's randomization and evaluate the benefits and costs of MTDs techniques [1, 2]. In the last years, many research works have been made to deal with some of these challenges. In [3], Okhravi *et al.* have categorized the MTD techniques into five dominant domains which are networks, platforms, run-time environments, software, and data. They have discussed the benefits and drawbacks of the MTD's usage in these five domains. For example, Dynamic Run-time Environments domain to prevent attackers, exploit software vulnerabilities for compromising a machine. In the dynamic software domain, these techniques seek to modify the application while keeping the internal state deterministic relative just to the input. Hence in this manner, we guarantee the tolerance, the continuity of functionality while preventing the development and the launch of attacks. In [1], the authors have suggested a three-layer model in order to fill the gap. The first layer gathers the low-level context information of individual programs in a system. In the second layer, the interaction between different programs has been modeled while the third layer represents a user interface which outlines the effects of attacks and defenses and it is used to compare the effectiveness of MTD techniques. In [4], the authors presented a base for MTD theory definition which regroups the MTD System Theory and attacker Theory. They have talked about the three key problems of MTD systems, including the MTD problem in the term of how to tune the space configuration, to adapt to selection problem and the timing problem. The software-defined networking was used as a plate-form to implement the MTD approach. In [5], the authors have proposed MTD architecture based on OpenFlow so as to modify transparently the IP addresses according to a high level of unpredictability. Their objective is to protect the configuration's integrity and reduce functioning overhead. The MTD techniques can be presented as an effective countermeasure against attacks in order to well secure the cloud-hosted applications. Because both proactive and reactive measures are simultaneously applied. Based on SDN paradigms, MTD approach offers more benefits in term of enhancing security, service availability, maintaining network and computing resources [6]. Furthermore, Game Theory appeared as a suitable tool for attack defense modeling and increasing application for MTD approach implementation. The main contribution of this work is to develop a game-theoretic model based MTD techniques especially IP address hopping combined with detection systems to achieve more security in cloud computing environments. We have proposed to model our game as a Bayesian game. The defender is doubtful about the attacker's type, he cannot define if his adversary is experimented or not. His choice is based on his belief in the attacker's type. In Bayesian game [7], the payoffs are not common knowledge. The Bayesian game is defined as a game with incomplete information. Where Incomplete Information means that at least one player does not know someone else's payoffs.

The next Sect. 2, provides discussion about some Moving Target Defense solutions with special focus on attack-defense interaction. We introduce our system model using

Game Theory approach in Sect. 3. In Sect. 4, the game model was analyzed to discuss the pure and mixed Nash equilibrium possible. We have also implemented our model to carry out with some numerical results to validate the proposed game model. Finally, we conclude our paper in Sect. 5.

2   Related Works

Research works using MTD approach was gained momentum in securing the cloud-hosted applications. Among of these works are abstracted. In [8], the authors have introduced an MTD architecture which applies the proactive and reactive mechanisms of VM migration so as to promote the cloud-based application security and prevent the Denial of Service (DoS) attacks. The challenge of this work is the frequency optimization of migration and the reduction of attack risks. The challenge of this work is the frequency optimization of migration and the reduction of attack risks. The proposed solution implements the SDN controller based OpenFlow switches such that when an application is selected for migration to a new virtual machine, all users connected to this application will be redirected to this new target virtual machine. In order to mitigate the impact of DDoS attacks especially flooding attacks, the authors in [9] have proposed a framework based moving targets presented by dynamic and hidden proxy nodes. In [10], Jia *et al.* have presented cloud-enabled, shuffling based moving target defense techniques for DDoS attacks mitigation. Their approach aims to tune the victim servers into moving targets by changing the infected servers with the new replica servers in order to isolate and resist the DDoS attacks. In [11], they have suggested using the heterogeneous and dynamic attack surface for modeling the cloud-based services. They have discussed the conditions and the effectiveness of strategies of MTDs in defense against attack. In SDN environment, several efforts have been made to promote the cloud security using the MTD mechanisms. The SDN controller allows an efficient configuration, better performance in network services management. It consists of separation of the control from data planes. However, these features of SDN allow a good platform for Moving Target Defense solutions [12]. In [13], the authors discussed the advantages and drawbacks of some networked MTD techniques.

Basically, they suggested how to use SDN for implementing MTD migrations due to its flexibility and centralization of the control plane. However, their experimental results showed significantly the effectiveness of SDN so as to maximize the attack cost because the attacker must spend more time in configuration of the attack surface and traffic load. In [14], the live VM migration was implemented for attack prediction. Assuming that the virtual machines can be heterogeneous, the authors have suggested both the proactive and the reactive techniques for MTD so as to collect the heterogeneity for VM pool. Indeed, they are looking for optimization of cloud resources usage over migration. Self-Cleansing Intrusion Tolerance (SCIT) [2, 15, 16] enhances security through VM's rotating and re-imaging only the passive VMs by changing a single platform to its primitive state but doesn't guarantee the defense against the attackers which intrude the network using the same exploits. To evaluate the effectiveness of MTD techniques, the critical question is how to select the optimal strategy which can bring more benefits. In [17], the authors have used Markov game to study the optimal strategy which can be selected when MTD was applied. Their proposed model is based on non-cooperative game, and Markovian decision process to model the transition among multi-phases of MTD hopping. In [18], the authors considered the virtual coordinate systems to analyze the attack defense mechanisms. They have modeled their proposed framework using game theory to calculate the game equilibrium and discussed the advantages and drawbacks of both the defender and the attacker. IP address plays a vital role to establish the Internet communication. It represents the access point for the attacker for vulnerability exploitation and analysis and finally for attack propagation. In addition, port number also can be used as an important tool for many types of attacks. IP address hopping and port hopping mechanisms act as important branches of MTD techniques in order to protect the communications over the Internet [19]. Through the periodic change between addresses and devices, network address shuffling combines the IP protocols and port numbers to maximize the uncertainty and the effort of the attacker [20].

3   Game Model

3.1   Game Model Description

We present a game-theoretic approach for MTD considering two players the attacker and defender. By exploiting the existing vulnerabilities in the system, the attacker can launch a successful attack so as to compromise the target victim. The $V = \{v^1, v^2, \ldots, v^N\}$ is the set of the vulnerabilities existed in the system. We consider that the common knowledge to the defender and the attacker is the vulnerability information. Because the defender tries to minimize the number of the vulnerabilities which may be exploited by the attack in order to maximize the attack surface. In other words, the MTD remains powerful to reduce the attack surface and, the exploration surface. However, we can make different configurations at a different time. The $C = \{c^1, c^2, \ldots, c^M\}$ is the set of the configuration system. We can define the attack surface in function of the configuration system and the vulnerabilities existed in the system. $p(c^k)$ is the attack surface when the system has the c^k as the configuration of the network system. We suppose that attacker has N atomic attacks. $A = \{a^1, a^2, \ldots, a^N\}$ The attacker can successfully launch an attack a^k and produce a damage cost for the defender if he can understand the system configuration and well exploit its vulnerabilities. The cost incurred by the attacker depends on the configuration C defined by the defender and the attack A launched by the attacker. Where $cost(C) = \mu \times cost(A)$ with $\mu \in [0, 1]$. The purpose of the attacker is to compromise the system and generate an intrusion in the system. While the purpose of the defender is to change the system configuration and make the tasks of vulnerability detection and exploitation, very hard and difficult. In this way, the administrator can reduce the impact of the attacker and minimize the cost. For this reason, we can model this game as a two-person Bayesian game.

3.2 MTD Based IP Address Hopping

MTD techniques have been proposed, to prevent the cyber-attacks from destroying the computer network security by changing the network's state in some way (See Fig. 1). However, among the mechanisms of MTDs, IP address Hopping mechanism tends to modify periodically the IP addresses of connected machines in the network. In other words, we select all VMs affected with their IP addresses and after, we choose the new IP addresses existing in the list of available IP addresses for modifying the old by the new IP address; hence we make the attacker's effort too hard in the reconnaissance phase. The old IP address joined the list of available IP addresses for later re-used by another machine (See Algorithm 1).

3.3 Game Theory Model

We suggest that an attack vector, vector for compromise by the attacker is represented as a point of interaction with an asset, which can be accessed remotely, locally or with a physical access. The exploitation of these points of interaction provides for the attacker powerful entry points through potentially use of asset's value.

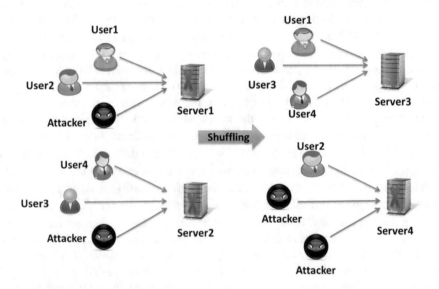

Fig. 1. MTD based IP address hopping for attack prevention

Algorithm 1 : IP address Hopping as a MTD' s mechanism

Input :
```
LA: List of affected VMs
LIP: List of available IP addresses


   While each VM ∈ LA do
        OIP : The old IP address
        NIP : The new chosen IP address
        OIP = V M.currentIP
        VM.currentIP  = NIP
        LIP = LIP + OIP

   EndWhile
```

Definition 1: System's Attack Surface *AS* is defined as the sum of all Available Points of Interaction (API) at a time *t*. $AS(t) = \sum_i API(i)(t)$. These *API* means attack vectors such as buffer overflows, networking protocols flaws, HTML emails.

Definition 2: Attack Surface Analysis is presented as a measure of all possible path combinations which an attacker may utilize to rob, damage, erase, or shift the available assets.

We assume that the two players are rational. We suppose that the players know well the system and they can predict more strategies to improve their expected payoffs. The attacker has two strategies, to attack (*A*) or Not attack (*NA*). The attacker can follow only one of the two strategies at a time. Its strategy is based on multiple stages such as network reconnaissance, to scan the critical vulnerabilities, to induce a buffer overflow. While the defender's strategies are summarized in the multi-stage process such as minimizing attack surface, reconfiguration of the network, IP address hopping. Then, the defender has also two strategies; To invest in security (*IS*) or Not to invest in security (*NIS*). The 2-tuple represents the actions concerned by each player of this game. For instance, the strategy profile (*IS, A*) means that the defender defends the system security using a specific strategy and the attacker launches an attack to violate the system security.

4   Game Model Analysis

Using the game described in the Tables 1 and 2; we define the possible Nash Equilibrium (NE).

Definition 4: The Nash Equilibrium (NE) represents the optimal outcome of the game because the player is not motivated to deviate from his selected strategy knowing that he considered the choice of his adversary. The players stay constant in their strategies. In other words, he will not acquire any more profit by playing other actions.

Table 1. Payoff matrix of the game for both players in case: attacker is experimented

Defender	Attacker	
	Attack (A)	Not Attack (NA)
To invest in security (IS)	$\delta\beta(2 \times TP - 1)G + (1 - \delta)\beta W - \delta Cost_{IS}$ $\delta\beta(1 - 2 \times TP)G + (1 - \delta)\beta W - \beta cost_A$	$-FP \times \delta Cost_{FD} - \delta Cost_{IS}$ 0
Not invest (NIS)	$-\beta W$; $\beta W - \beta cos\, t_A$	0 0

Table 2. Payoff matrix of the game for both players in case: attacker is not experimented.

Defender	Attacker
	Not Attack (NA)
To invest in security (IS)	$-\beta W$ 0
Not invest (NIS)	0 0

We assume that the game players are rational where each one among them plays his best strategy to act against his opponent and to maximize his payoff function.

A proactive MTD mechanism based IP address hopping shouldn't be triggered for all time, by cons, we can apply the IP address hopping method periodically basis on IDS alarm so as to control the network traffic. Whereas, for the rest of time, the provider stays listening to the IDS report. As well, the attacker need understand the system's configuration and analyze the potential vulnerabilities. Our game model is defined as a zero sum game (i.e. the losses of the defender represent benefits for the attacker and the losses incurred by the attacker are benefits for the defender). Our game model represents the interaction between the provider and the attacker using MTD techniques so as to secure our system network. The Game is modeled as repeated game according the MTD stages, i.e. at each stage, we have a specific configuration. At the next stage, the provider should change the IP address of VM concerned in order to perturb the attacker and to maximize his effort. Each one of the two players has two strategies. The provider applies a new reconfiguration *NRC* when the IDS is active during a time-driven IP address hopping period. It's denoted by *IS* (To Invest Security). The second strategy presents the no defense case where there is any reconfiguration and the IDS is inactive. It's denoted by *NIS* (No Invest Security). The pure strategy of the provider is as follows: $PS_{defender} = (IS, NIS)$. On the other side, in his first strategy (*A*), the attacker tries to understand the network configuration and to maximize his attack surface. For the second strategy, the attacker has not attack (*NA*). Therefore, the attacker's pure strategy is $PS_{attacker} = (A, NA)$. We denote the probability that the IP address hopping will be applied during a certain time as δ and the probability that the attacker will launch an attack as β. The used IDS is characterized by its detection rate

TP rate and its False positive *FP* where *TP*, *FP* $\in [0, 1]$. To analyze the game's utility. The payoffs for the two players are formulated in function of these parameters:

- *G*: The total benefit of the provider in terms of attack detection by IDS implementing and IP address hopping (IPH).
- *W*: The loss accumulated to the defender in the case the attack was successful.
- $Cost_{IS}$: The cost accumulated when IDS integrated with IPH as a MTD technique.
- $Cost_A$: The cost incurred of attacking.
- $Cost_{FD}$: The loss due to false detection.

We assume that $Cost_{IS}$, $Cost_A > 0$ and $G > Cost_{IS}$, $Cost_A$.

Considering the Table 1, for the profile strategy (*NI, A*), the defender's payoff is $-\beta W$ and the attacker's payoff is $\beta(W - Cost_A)$; it represents his benefit minus the cost due attacking. For the profile strategy (*IS, A*), the expected payoff of attack prevention is dependent on the value of *TP* rate, the period of time-driven IP address hopping and the duration of successful attack at the same time $\delta\beta$; $\delta\beta \times TP \times G - (1 - TP) \times G = \delta\beta(2 \times TP - 1)G$ Where $1 - FP$ represents the False Negative (FN) rate. The payoff of the attacker represents the loss of the defender, is equal to: $\delta\beta(1 - 2 \times FP)G + (1 - \delta)W$. Hence, the payoff of attacker is his benefits minus the cost of attacking. However, in the profile strategy (*IS, NA*), $-\delta FP \times Cost_{FD} - \delta Cost_{IS}$ is the cost of investment in security and losses due to false alarm. Finally, in the last profile strategy (*NIS, NA*) the payoff of both players is equal to zero. Any one of them has no action, the defender doesn't invest in security and also the attacker plays not attack action.

Nash Equilibrium analysis. We consider that the attacker has two types; he can play as experimented player or as not experimented attacker. We suppose that the defender takes in consideration the attacker's type. If the attacker is experimented, the probability is α. Otherwise; the probability is equal to $(1 - \alpha)$. Each one of the players tries to maximize its payoffs. We have modeled this attack-defense interaction as a Bayesian game for the reason that the defender seeks to maximize the likelihood of attack prevention and limit the attack surface while the attacker tries to minimize the attacking cost and the probability to be detected. Hence, in this subsection, we dwell on the Bayesian Nash Equilibrium of the proposed game to calculate the expected payoff of the players. The expected payoff of the provider when he plays his pure strategy (*IS*) depends on the pure strategies of the attacker (*A*) when its types is experimented and (*IA*) when he is not experimented. It's formulated as follows:

$$EP_{defender}(IS) = \alpha(\delta\beta(2TP - 1)G - (1 - \delta)\beta W - \delta Cost_{IS}) - (1 - \alpha)\beta W \quad (1)$$

And his expected payoff when he plays the (*NIS*) pure strategy is equal to:

$$EP_{defender}(NIS) = -\alpha\beta W \quad (2)$$

We have two cases:

First case: If $EP_{defender}(IS) > EP_{defender}(NIS)$, we see that when the defender plays (IS), the attacker should play his pure strategy (NA). Because, he doesn't need his best response. We conclude in this case, that there is not Bayesian Nash equilibrium.

Second case: If $EP_{defender}(IS) < EP_{defender}(NIS)$, the defender should play (NIS) as a best response. Hence, the profile strategy (NI by defender and A by the attacker if he is experimented or NA if he is not experimented) represents Bayesian Nash equilibrium. When the experimented attacker plays NA, the defender should play (NIS) as his dominant strategy. In the case that the defender chooses the (NIS) pure strategy, the experimented attacker needs to play A strategy. However, we conclude that the profile strategy (NIS by defender, NA for the two attacker's type (experimented and not experimented)) is not Bayesian Nash equilibrium. Considering the first case, we have concluded that there is not Bayesian Nash equilibrium but we can calculate the mixed Bayesian Nash equilibrium strategies. Let us define p the probability that the defender plays IS strategy and q the probability that the attacker plays A strategy. We formulated the expected payoff of the defender when he plays IS strategy as follows:

$$EP_{defender}(IS) = q\alpha(\delta\beta(2TP-1)G - (1-\delta)\beta W - \delta Cost_{IS})$$
$$- q(1-\alpha)(-FP \times \delta\cos t_{FD} - \delta cost_{IS}) - (1-\alpha)\beta W \qquad (3)$$

and his expected payoff when he adopts NIS strategy is equal to:

$$EP_{defender}(NIS) = -q\alpha\beta W \qquad (4)$$

The expected payoff of the attacker when he plays A strategy is

$$EP_{attacker}(A) = p(\delta\beta(1 - 2 \times TP)G + (1-\delta)\beta W - \beta cost_A) + (1-p)(\beta W - \beta Cost_A) \qquad (5)$$

and the expected payoff of the attacker applying his NA strategy is

$$EP_{attacker}(NA) = 0 \qquad (6)$$

By using the derivative rules, the Bayesian Nash equilibrium of this game are vectors (p, q).

$$\begin{cases} EP_{defender}(\text{IS}) = EP_{defender}(\text{NIS}) \\ EP_{attacker}(A) = EP_{attacker}(NA) \end{cases} \qquad (7)$$

The equilibrium strategy of the defender is equal to the probability that the defender triggers a defense mechanism to resist the attack with

$$p = \frac{W - cost_A}{\delta((2 \times TP - 1)G + W)} \tag{8}$$

While the equilibrium strategy of the experimented type of the attacker is equal to the probability that attacker launches an attack with probability:

$$q = \frac{\alpha\delta(FP \times cost_{FD} + cost_{IS}) + (1 - \alpha)\beta W}{\delta\beta[\alpha G(2TP - 1) + W] + \delta FP \times cost_{FD}} \tag{9}$$

It is found that the strategy which consists on p, q if attacker has experimented type, NA if the attacker has not experimented type represents a mixed strategy of the Bayesian Nash equilibrium.

In Fig. 2, we analyzed the effect of time-driven MTD (δ) and time-driven attack (β) on Expected payoff of defender. We remark that as δ increases the expected payoff of the defender increases and as β increases, the expected payoff of defender was reduced.

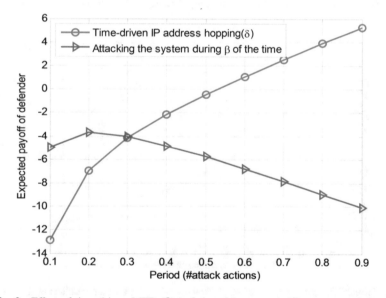

Fig. 2. Effect of time-driven MTD (δ) and time-driven attack (β) on expected payoff

In Figs. 3 and 4, we have evaluated the mixed Nash equilibrium strategies in function of time-driven $MTD(\delta)$ and detection rate (TP). We see that as δ and TP increase; the probability that the attacker plays (A) strategy decreases.

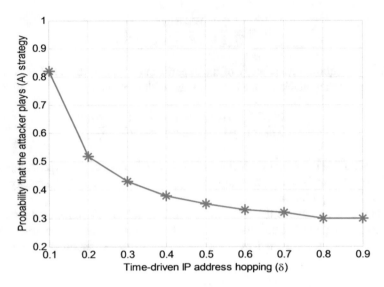

Fig. 3. Probability that the attacker plays strategy (A) vs. time-driven MTD (δ)

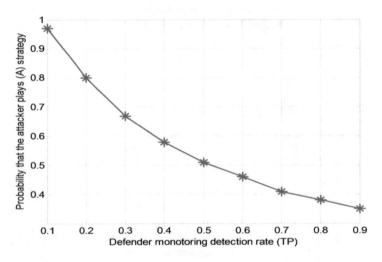

Fig. 4. Probability that the attacker plays strategy (A) vs. defender monitoring detection rate (TP)

5   Conclusion

In this paper, we have proposed to implement MTD mechanism at network level using random IP address changing. We suggested changing randomly the virtual machine IP address so as to make the attack surface unpredictable. We have modeled the interaction between administrator and attacker as two-person Bayesian game. We seek to change the virtual machine IP address for a certain time according the scanning report generated by IDS. We have calculated the expected payoff of the defender to evaluate and discuss the feasibility of the system defense proposed for attack attenuation and mitigation.

References

1. Xu, J., Guo, P., Zhao, M., Erbacher, R.F., Zhu, M., Liu, P.: Comparing different moving target defense techniques. In: Proceedings of the First ACM Workshop on Moving Target Defense, pp. 97–107. ACM (2014)
2. Carter, K.M., Riordan, J.F., Okhravi, H.: A game theoretic approach to strategy determination for dynamic platform defenses. In: Proceedings of the First ACM Workshop on Moving Target Defense, pp. 21–30. ACM (2014)
3. Okhravi, H., Hobson, T., Bigelow, D., Streilein, W.: Finding focus in the blur of moving-target techniques. IEEE Secur. Priv. **12**(2), 16–26 (2014)
4. Zhuang, R., DeLoach, S.A., Ou, X.: Towards a theory of moving target defense. In: Proceedings of the First ACM Workshop on Moving Target Defense, pp. 31–40. ACM (2014)
5. Jafarian, J.H., Al-Shaer, E., Duan, Q.: Openflow random host mutation: transparent moving target defense using software defined networking. In: Proceedings of the First Workshop on Hot Topics in Software Defined Networks, pp. 127–132. ACM (2012)
6. El Kafhali, S., Salah, K.: Stochastic modelling and analysis of cloud computing data center. In: Proceedings of 20th Conference on Innovations in Clouds, Internet and Networks, pp. 122–126. IEEE (2017)
7. Harsanyi, J.C.: Games with incomplete information played by Bayesian players, III: part I. The basic model. Manage. Sci. **50**(12-supp), 1804–1817 (2004)
8. Debroy, S., Calyam, P., Nguyen, M., Stage, A., Georgiev, V.: Frequency-minimal moving target defense using software-defined networking. In: Proceedings of the International Conference on Computing, Networking and Communications, pp. 1–6. IEEE (2016)
9. Wang, H., Jia, Q., Fleck, D., Powell, W., Li, F., Stavrou, A.: A moving target DDoS defense mechanism. Comput. Commun. **46**, 10–21 (2014)
10. Jia, Q., Wang, H., Fleck, D., Li, F., Stavrou, A., Powell, W.: Catch me if you can: a cloud-enabled DDoS defense. In: Proceedings of the 44th Annual IEEE/IFIP International Conference on Dependable Systems and Networks, pp. 264–275 (2014)
11. Peng, W., Li, F., Huang, C.-T., Zou, X.: A moving-target defense strategy for cloud-based services with heterogeneous and dynamic attack surfaces. In: Proceedings of the International Conference on Communications (ICC), pp. 804–809. IEEE (2014)
12. El Mir, I., Chowdhary, A., Huang, D., Pisharody, S., Kim, D.S., Haqiq, A.: Software defined stochastic model for moving target defense. In: Proceedings of the Third International Afro-European Conference for Industrial Advancement, pp. 188–197. Springer, Cham (2016)
13. Kampanakis, P., Perros, H., Beyene, T.: SDN-based solutions for moving target defense network protection. In: Proceedings of the 15th International Symposium on a World of Wireless, Mobile and Multimedia Networks (WoWMoM), pp. 1–6. IEEE (2014)
14. Zhuang, R., Zhang, S., Bardas, A., DeLoach, S.A., Ou, X., Singhal, A.: Investigating the application of moving target defenses to network security. In: Proceedings of the 6th International Symposium on Resilient Control Systems, pp. 162–169. IEEE (2013)
15. El Mir, I., Kim, D.S., Haqiq, A.: Cloud computing security modeling and analysis based on a self-cleansing intrusion tolerance technique. J. Inf. Assur. Secur. **11**(5), 273–282 (2016)

16. El Mir, I., Kim, D.S., Haqiq, A.: Security modeling and analysis of an intrusion tolerant cloud data center. In: Proceedings of the Third World Conference on Complex Systems (WCCS), pp. 1–6. IEEE (2015)
17. Lei, C., Ma, D.-H., Zhang, H.-Q.: Optimal strategy selection for moving target defense based on Markov game. IEEE Access **5**, 156–169 (2017)
18. Beckery, S., Seibert, J., Zage, D., Nita-Rotaru, C., Statey, R.: Applying game theory to analyze attack and defenses in virtual coordinate systems. In: IEEE/IFIP Proceedings of the 41st International Conference on Dependable Systems & Networks, pp. 133–144 (2011)
19. Cai, G., Wang, B., Wang, X., Yuan, Y., Li, S.: An introduction to network address shuffling. In: Proceedings of the 18th International Conference on Advanced Communication Technology, pp. 185–190. IEEE (2016)
20. Carroll, T.E., Crouse, M., Fulp, E.W., Berenhaut, K.S.: Analysis of network address shuffling as a moving target defense. In: Proceedings of the International Conference on Communications, pp. 701–706. IEEE (2014)

Ensuring Security in Cloud Computing Using Access Control: A Survey

Fatima Sifou[1]([⊠]), Ahmed Hammouch[1], and Ali Kartit[2]

[1] Laboratory for Research in Computer Sciences and Telecommunications
(LRIT), Faculty of Science, Mohammed V University, Rabat, Morocco
fatimasifou@gmail.com, hammouch_a@yohoo.fr
[2] Laboratory LTI, Department TRI, ENSA-El Jadida,
Chouaïb Doukkali University, Avenue Jabran Khalil Jabran,
BP 299, El Jadida, Morocco
alikartit@gmail.com

Abstract. Currently, cloud computing has widely been implemented in several organizations. This is due to the fact that this new model delivers cost-efficient, flexibility and scalability computing resources to the clients. This has led to a growing demand of cloud implementations in different sectors. Although cloud computing facilitates data management, it still faces various security problems. So, it is important to examine cloud security issues before moving to this model. In particular, we focus on existing access control mechanisms in cloud computing. Actually, there exist multiple schemes that ensure data protection, such as DAC (Discretionary Access Control), MAC (Mandatory Access Control), RBAC (Role Base Access Control), ORBAC (Organizational Role Based Access Control) and ABAC (Attributes Based Access Control). The main contribution of this paper is to select the appropriate access control model that meets cloud constraints. Based on these considerations, ABAC model is more suitable to cloud environments compared to other models. In this regard, we provide a comprehensive taxonomy of different ABAC schemes.

Keywords: Cloud computing · Access control · ABAC model
ABE schemes

1 Introduction

Cloud computing has been among the most important technologies used by companies because of its main features which prove advantageous for increasing flexibility, scalability, and reliability, thus, reducing the inherent computational cost and hardware. According to this paradigm, clients can efficiently share resources in terms of cost and regardless of the clients' location. Cloud can also offer to customer software, storage and networks. As indicated above cloud computing has many benefits but the most important one is that the clients don't have to buy resources or install the software which they need. However, the major challenge in cloud computing is data security and privacy. For example the owner of data may not have local control of data. There are further security issues of cloud computing including virtualization, database, networks

© Springer International Publishing AG, part of Springer Nature 2018
M. Ben Ahmed and A. A. Boudhir (Eds.): SCAMS 2017, LNNS 37, pp. 255–264, 2018.
https://doi.org/10.1007/978-3-319-74500-8_23

and access control. In this study we opt for access control as an essential component for ensuring data security. In fact, access control encompasses many procedures which define users who have permissions and authorizations to access specific cloud resources. The purpose of this method is to provide a verifiable system to protect data from unauthorized access and determine each client's access rights based upon pre-defined criteria. Consequently, access control is considered as a key component of cloud security. This paper is meant to discuss security issues for cloud computing and access control models as a solution [1]. This paper firstly provides an overview on cloud computing and security challenges akin to this model in Sect. 2. Section 3 dwells on access control and existing models in cloud computing. Section 4 explores the proposal model ABAC and their schemes. Finally, Sect. 5 concludes this paper and proposes future work.

2 An Overview on Cloud Computing

According to specialists cloud computing is one of the most discussed topic in the field of information technology owing to its flexibility, scalability, availability and cost efficient. It provides not only flexible and scalable IT, but it also enables consumers to share resources rather than having local servers. Furthermore, they can use these resources and pay for it as a service, thus taking advantage of the lower cost rate. Despite these advantages of cloud nothing is perfect. In fact, the real problem in this new trend is security including data intrusion, data loss and privacy. According to NIST view, cloud computing offers several service models, deployment models and some essential characteristics such as Self-services (On Demand), Broad Access, Rapid Elasticity, Resources Pooling, Scalability of Infrastructure and Measured Service etc. [1, 2].

2.1 Service Models

From NIST view, cloud computing offers three service models as described below.

Infrastructure as a Service (SaaS). Cloud providers only offer to clients the hardware like storage, networks, servers, and any other fundamental computing resources. In addition, they allow consumers to deploy and run their stuff including operating systems and applications. The consumer does not manage or control the infrastructure. Some examples of IaaS are: Amazon, GoGrid, 3 Tera, etc. [3].

Platform as a Service (PaaS). This model aims to enable customers to create their own applications in addition to infrastructure. The clients do not manage or control the underlying cloud infrastructure such as storage, servers, but have control of their applications. In this regard, various industry solutions are available: the Google App Engine, Force.com, Amazon Web Services Elastic Beanstalk, and the Microsoft Windows Azure platform.

Software as a Service (SaaS). The vendor allows to customer the capability for using provider's applications running on a cloud infrastructure. These applications are offered to customer as a service on demand. In this model, the clients only have to manage and maintain permissions and authorizations to their own data. So, the cloud providers have to manage or control the underlying cloud infrastructure, network, servers, operating systems, storage, or even individual application capabilities. SaaS is offered by many companies such as Google, Salesforce, Microsoft, Zoho etc.

2.2 Deployment Models

The cloud environment presents four deployment models as mentioned bellow.

Public Cloud. Cloud services are used by several organizations. Basically, resources are owned, managed and operated by cloud providers, this reduces IT costs. However, the sharing of resources among many public organizations causes a variety of security issues [4].

Private Cloud. In this model the cloud infrastructure is exclusively used by a single organization and its users. Actually, private cloud guarantees a high level of security though it is the most expensive model due to the fact that the owners manage themselves their resources because of its own management of resources.

Community Cloud. This kind of cloud is adopted by several organizations having similar interests and seeking to exploit the benefits of cloud computing. Moreover, it is managed and maintained by one or more of the organizations in the community. Since the security level of this model is often high, to opt for this option is more expensive than public cloud [5].

Hybrid Cloud. It is a composition of two or more cloud infrastructure models (private, public, or community). Furthermore, it aims to take advantage of the controlled environment in private clouds and rapid elasticity of public clouds. This model, however, suffers from its complexity [6].

2.3 Cloud Security Challenges

In spite of cloud advantages, security gaps deserve careful attention. Indeed, this new trend still faces many security challenges, especially on data theft, data loss and privacy. In this section we introduced some security issues.

Virtualization. It is an abstraction of hardware or Software that are able to run multiple applications concurrently. It also allows clients to create copy, share, migrate, and store data using virtual machines. However, this technology exposes systems to a variety of security risks which can be exploited by hackers due to the extra layer that need to secure [7].

Interoperability. This is the capability of many IT systems to work concurrently to exchange information and to use the information that has been exchanged. It also helps to increase client choice, competition and innovation. Furthermore, it allows different software components to cooperate, even if the interface and the programming language are different. Therefore, in order to allow users to choose the appropriate cloud platforms, organizations may resort to substituting cloud providers. Furthermore, there are some enterprises that demands different cloud platforms that match the various applications which meet their needs and services. This being the case, interoperability deems an essential component in cloud services [8, 9].

Data Storage. Cloud storage has made the burden of storing data, regardless of its size, an easy task. That is one major reason for the increasing demands of enterprises on cloud services. Not only can cloud provide huge spaces for data storage but it can also do the backup thereof. This goes in accordance with current trend of networking data storage with no need for the enterprises to set up the required architecture. These benefits that strengthen this framework should not be taken for granted. Actually, there are other factors that impinge on the effectiveness of the online data storage. Most importantly, security issues are almost the most debatable in this respect. As a case in point, users claim that cloud may cause data leakage or, frequently, temporary unavailability of the data stored particularly if the latter is dynamic. Consequently reluctant clients would not entrust the task of data storage to cloud services if there were no guarantee of permanent access to their information along with maintaining the integrity and privacy of the data [10].

Heterogeneity. Heterogeneity is a feature that seriously affects the Cloud system outcome. A common factor of heterogeneity is the use of various resources be they hardware or software for cloud environments. On the one hand, one should consider that one infrastructure is shared among different tenants, and this, subsequently, yields variable degrees of heterogeneity. On the other hand, the fact that cloud providers rely on in building services gives rise to problems inherent to the security requirements and the users' trust. Heterogeneity can also be imputed to the differential security performance of the various components in the cloud system [11]. Despite all these challenges, the concept of virtualization is deemed promising in achieving a great deal of homogeneity.

Access Control. In general, heterogeneity and diversity of services are the essential characteristic of cloud computing environments. Consequently, security measures must ensure fine-grained access control policies. More precisely, access control mechanisms need to be flexible enough to support cloud requirements, such as dynamic, context, or attribute. Normally, access control services are meant to integrate privacy-protection requirements, which are expressed via complex rules [11].

3   Access Control in Cloud Computing

Access control is a way to limit and control access to a system and its resources or information in order to decide which users are granted access. Therefore, it is considered as a key component of cloud security. For this reason, several access control models are explored below.

3.1   Mandatory Access Control (MAC)

MAC is actually advantageous in terms of the strong security it provides. It also offers centralized security policies. However, the complexity of its management and the increasing implementation cost remain some of its disadvantages [12, 13].

3.2   Discretionary Access Control (DAC)

DAC is known by the flexibility it exhibits vis-à-vis the access control model. In addition, it does not require a highly designed protection in the environment where it is used. Having said this, this model does have some flaws. First, DAC allows users to pass privileges among them. Second, It has been revealed that the DAC security system is vulnerable to Trojans in addition to the fact that Security policies can be changed by inserting malicious program [12].

3.3   Role Based Access Control (RBAC)

This model is feasible, for the security policy is very simple to manage. It could also be admired for its minimizing of the damage of information by intruders by means of assigning the roles based on the least privilege. Nevertheless, when a distributed and dynamic environment is involved, the implementation of RBAC proves difficult. Furthermore, it is impossible in this model to change access rights of the user without changing the roles of that user [12].

3.4   Organization Based-Access Control (OrBAC)

Access to cloud resources in this method is dependent mostly of the user's roles within an organization. Accordingly, various parameters have an impact on making parameters such as subject, action, object, view, and context. Among the arguments leveled in favor of this model is that it requires more scalability, and it easily manages changes in organization. Still, there is frequent criticism against this model. It shows a high degree of complexity and costs much in terms of computation [15–17].

3.5   Attribute Based Access Control (ABAC)

This model is based on the user's attributes such as date of birth and current location. So, its main function is to compare these attributes with the rules associated to each object to grant or deny access. This is in fact one advantage among others. ABAC is also more flexible in a dynamic and distributed environment in addition to its simplicity in implementation. The major downsides of this model, however, have to do with the long running time it requires as well as the difficulty to manage it [11].

4   Access Control Requirements

Various criteria are proposed in order to identify the weaknesses and strengths of existing access control mechanisms. These criteria are meant to help select the suitable model for cloud environment.

4.1   Reliability

Open and distributed systems are often subjected to security attacks. Therefore, reliability stands for the security features that make a system immune from potential attacks be they internal or external.

4.2   Flexibility

This feature allows a system to cope with any further changes and work in harmony with the different decisions akin to access control.

4.3   Dynamicity

Dynamicity is a key concept in cloud computing system. It reflects the ability of the system to proceed in an environment where data can be processed, exchanged or shared simply, securely and flexibly.

4.4   Easy in Administration

This requirement consists of making the access management less complex. This is due to the fact that handling access to resources in a distributed system such as cloud is very challenging in this regard.

4.5   Computational Costs

Actually, the incorporation of the aforementioned requirements while implementing access control mechanism induces increased running costs. Therefore, the computational costs should be taken into account in the design and choice of an access control model.

4.6   Fine-Grained Access

Since flexibility requirement in an access control system is highly demanded so that the user could acquire a set of permissions, security becomes an inherent component. It should meet the user's access needs along with the ability to stop unauthorized access and protect sensitive data [11, 14].

4.7   Security Policy Implementation

This requirement insures the integrity and safeguarding of information preventing, thereby, any potential loss of data. It is interestingly important to consider this requirement since the distributed system is large and open.

5   Proposed Model

The ABE model was proposed by Sahai and Waters in 2005. It aims to encrypt and decrypt data using user attributes and cipher-text attributes. This procedure is possible only if the set of attributes of the user key matches the attributes of the cipher-text. Most of the systems have applied this scheme to provide data confidentiality and integrity. There are basically three elements involved in this model namely authority, data owner, and data user. The authority is concerned with providing keys for data owners and users. As for the data owner, they encrypt data with a public key and a set of descriptive attributes. Thus, the role of a data user is to decrypt cipher-text with his private key, and then he can obtain the encrypted data, as shown in Fig. 1. However, the main issue in this method is that the owner of data has to use each time another public key for encrypting data. In this section, we discussed several ABE based access control schemes in order to choose the appropriate one to solve this problem [18].

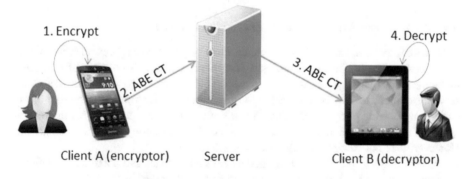

Fig. 1. Architecture of ABE scheme [18]

5.1   Key-Policy Attribute-Based Encryption Scheme

This scheme is based on the access policy. Its main function is to control and allows which user can decrypt cipher-text. Each user has a private key which is associated with access policy and the cipher-text contains a set of attributes. Therefore, if a set of attributes satisfies the access policy, the user will retrieve and decrypt the message. According to this method, users can have different access rights. So, different users can still manage to decrypt different pieces of data corresponding to the security policy [18].

5.2   Cipher Text-Policy Attribute-Based Encryption Scheme

In general, the CP-ABE scheme is designed to deal with situations that rely on expressive access policies to encrypt the target data. Technically, a user's private key is typically associated with an arbitrary number of attributes. So, clients need to specify an associated access structure over attributes in the encryption stage. Accordingly, users whose attributes pass through the cipher text's access structure have the ability to decrypt the secret message [19].

5.3   Attribute-Based Encryption Scheme with Non-monotonic Access Structure

This model is a new Attribute-Based Encryption scheme that includes explicit attributes for indicating the absence of attributes in encrypted data. In this approach, private keys are developed to support any access formula over attributes, especially non-monotone cases. Interestingly, this system can manage any access structure that uses Boolean formula such as AND, OR, NOT, and threshold operations. Consequently, this scheme is an appropriate concept to handle non-monotonic access structures by using secret sharing schemes. In this approach, a user can decrypt the secret data only if a given attribute does not exist among the attributes of the encrypted data. Basically, this access control model uses the broadcast revocation scheme that is based on bilinear groups to define attributes.

5.4   Hierarchical Attribute-Based Encryption Scheme

This scheme is composed of a hierarchical identity-based encryption scheme (HIBE) and a cipher text-policy attribute-based encryption scheme. It consists of five parts including: root master (RM), domains master (DM), data owner, data user, and cloud service storage. The RM is responsible for generating and distributing. The role of the data owner is encrypting and sharing data with users in cloud storage. This model can satisfy the property of fine grained access control, scalability and full delegation. However, it is not suitable for all cases but a few [20].

As discussed above, the existing models have advantages and potential limitations. In other words, each scheme is more appropriate for a specific situation. In this regards, Table 1 bellow presents the main benefits and drawbacks of access control models.

Table 1. A comparison between different attribute-based encryption schemes

Parameters	KP-ABE	CP-ABE	ABE-with non-monotonic	HABE
Fine-grained access control	Yes	Yes	Yes	Yes
Scalability	No	No	No	Yes
Computational costs	Yes	No	No	Yes
Collision resistant	Yes	Yes	Yes	Yes
User revocation	Yes	Yes	Yes	Yes

Although the majority of existing schemes are not able to satisfy the criteria of scalability, they can ensure the data confidentiality. In fact, they rely on the attributes in the user's private key to match the attributes in the cipher text. Fortunately, the HABE model is a promising scheme since it meets all the criteria.

6 Conclusion

Security in cloud computing is still the major obstacle that faces the implementation of this new concept. In this regards, we explored different available models in order to select the appropriate one. Accordingly, ABAC model is designed to meet cloud requirements, especially scalability and flexibility. Besides, this work provided a comprehensive evaluation of different attribute-based encryption schemes. Based on the results of that analysis, HABE scheme is the most suitable for cloud environment. In the future work, we intend to implement this proposed scheme in order to ensure data security. Moreover, we will extend the standard version of this cryptosystem to improve performance.

References

1. Prince, J.: Security issues and their solution in cloud computing. Int. J. Comput. Bus. (2012)
2. Vikas, M., Srishti, G., Jyoti, K.: Network security: security in cloud computing. Int. J. Eng. Comput. Sci. 3(1), 3643–3651 (2014)
3. Cloud computing security considerations, April 2011. Accessed Sept 2012
4. Ronald, L.K., Russel, D.V.: Cloud Security a Comprehensive Guide to Secure Cloud Computing (2011). https://23510310jarinfo.files.wordpress.com
5. All-of-Government Cloud Computing, Information Security and Privacy Considerations (2014). https://www.ict.govt.nz
6. Youssef, A.E.: Exploring cloud computing services and applications. J. Emerg. Trends Comput. Inf. Sci. 3(6) (2012)
7. Shyam, N.K., Amit, V.: A survey on secure cloud: security and privacy in cloud computing. Am. J. Syst. Softw. 4(1), 14–26 (2016). https://doi:10.12691/ajss-4-1-2
8. Rabi, P.P., Manas, R.P., Suresh, C.S.: Cloud computing: security issues and research challenges. Int. J. Comput. Sci. Inf. Technol. Secur. 1(2), 136–146 (2011)
9. Rohit, B., Sugata, S.: Survey on security issues in cloud computing and associated mitigation techniques. DBLP J. (2012)
10. Bhadauria, R., Chaki, R., Chaki, N., Sanyal, S.: Survey on security issues in cloud computing. ACTA TEHNICA CORVINIENSIS – Bulletin of Engineering, December 2014
11. Takabi, H., Joshi, J.B.D., Ahn, G.-J.: Security and privacy challenges in cloud computing environments. IEEE Secur. Priv. 8, 24–31 (2010)
12. Ed-Daibouni, M., Lebbat, A., Tallal, S., Medromi, H.: Toward a new extension of the access control model ABAC for cloud computing. Springer, Singapore (2016)
13. Aluva, R.: Enhancing cloud security through access control models: a survey. Int. J. Comput. Appl. (2015). https://doi.org/10.5120/19677-1400
14. Maw, H.A., Xiao, H., Christianson, B., Malco, J.A.: Survey of access control models in wireless sensor networks. J. Sensor Actuator Netw. 2014 3(2), 150–180 (2014). https://doi.org/10.3390/j3020150

15. Bokefode Jayant, D., Apte Sulabha, S.: Analysis of DAC MAC RBAC access control based models for security. Int. J. Comput. Appl. **104**, 0975–8887 (2014)
16. Majumder, A., Namasudra, S., Nath, S.: Taxonomy and classification of access control models for cloud environments. Springer, London (2014). https://doi.org/10.1007/978-1-4471-6452-4_2
17. Punithasurya, K., Jeba, P.S.: Analysis of different access control mechanism in cloud. Int. J. Appl. Inf. Syst. (IJAIS) **4**(2), 34–39 (2012)
18. Asim, M., Petkovic, M., Ignatengo, T.: Attribute-based encryption with encryption and decryption outsourcing. In: Australian Information Security Management Conference (2014)
19. Lee, C.-C., Chung, P.-S., Hwang, M.-S.: A survey on attribute-based encryption schemes of access control in cloud environment. Int. J. Netw. Secur. **15**(4), 231–240 (2013)
20. Tan, Y.-L., Goi, B.-M., Komiya, R., Tan, S.-Y.: A study of attribute-based encryption for body sensor networks. In: International Conference on Informatics Engineering and Information Science, pp. 238–247, November 2011

Fuzzy C-Means Based Hierarchical Routing Approach for Homogenous WSN

Aziz Mahboub[1], El Mokhtar En-Naimi[1(✉)] ![ORCID], Mounir Arioua[2], and Hamid Barkouk[1]

[1] LIST Laboratory, Department of Computer Sciences, FST of Tangier,
Abdelmalek Essaâdi University, Tangier, Morocco
amahboub@uae.ac.ma, ennaimi@gmail.com, barkouk@gmail.com
[2] Team of New Technology Trends, National School of Applied Sciences,
Abdelmalek Essaâdi University, Tetouan, Morocco
m.arioua@m.ieee.org

Abstract. The global challenge in wireless sensor networks is to extend the network's lifespan as long as possible. The sensor's battery has a limited life and unfeasible to be replaced, which eventually requires an energy efficient routing protocol. Clustering applied in routing has proven its ability to saving energy in sensor networks. The current paper proposes a new approach based on Fuzzy C-means and LEACH protocol to form the clusters and manage the transmission of data to the base station. A cluster estimation method was adopted as the basis for identifying fuzzy model. The proposed approach minimizes the energy consumption and prolongs the network lifetime of the sensor nodes.

Keywords: Wireless sensor network · Fuzzy C-means algorithm
Cluster estimation · Energy efficiency

1 Introduction

Wireless sensor networks are one of the main technologies dedicated for detection and monitoring of certain physical phenomena of the environment. For instance, detection and measurement of vibration, pressure, temperature and sound [1]. The researchers are more concerned about this technology. The WSN can be integrated into many applications domains, like street parking system, smart roads, and industrial monitoring [2]. Generally, WSN is formed by many self-organized sensor nodes which have limited energy, computational capabilities, and bandwidth [3–6], deployed in large numbers in a dedicated environments. In hostile environments, it's impractical to change the node's battery in operational mode. Each sensor node is supplied by an autonomous energy source (battery) and equipped with a micro-controller, memory, and a transceiver. However, the base station (Sink) collects data for processing and sending the data to the computer center. A WSN can have more than one sink [4].

The sensor network batteries have a fixed energy capacity, and this limitation has generated significant interest in using the important aspects of WSNs in order to increase battery life [5]. Optimization of the energy consumption is one of the noteworthy challenges in wireless sensor networks. For this reason, is mandatory to resolve the network

© Springer International Publishing AG, part of Springer Nature 2018
M. Ben Ahmed and A. A. Boudhir (Eds.): SCAMS 2017, LNNS 37, pp. 265–275, 2018.
https://doi.org/10.1007/978-3-319-74500-8_24

energy efficiency constraint in order to extend the system lifetime [4, 6]. Energy efficiency optimization is challenging due to the fact that network lifetime depends on several parameters, such as network architecture, routing protocols, channel characteristics, and radio consumption [5]. The energy required in communications is the important consuming part compared to processing process in wireless sensor networks. To address this limitation, a variety of techniques have been developed to reduce the communication energy consumption. Optimized routing techniques such as single-hop, multi-hop, or clustering have been widely adopted in WSN due to their energy effectiveness [7].

In clustering routing, the wireless sensor networks require to classify nodes into non-overlapping clusters in which a Cluster Head (CH) is set. The main objective of Cluster Head is aggregating and transmitting data to the sink via single hop or multi-hop which can be connected to a powerful computer via the internet or a satellite [3, 4, 6]. The Clustering scheme economizes energy in the way that it extends the lifetime of the network. As well it can be practical feature for a large sensor networks due to the fact that it's easier to manage the cluster heads than the entire network.

Several clustering routing protocols that use multi-paths have been proposed based on load balancing in the order of LEACH [4], SEP [8], MZ-LEACH [5], DEEC [4], KM-LEACH [9]. The clustering protocols can be classified into two types, the clustering algorithm with homogeneous schemes; where all the sensor nodes have the same energy. And heterogeneous clustering schemes, where all the sensor nodes are supplied with different amount of energy [8].

The low-energy adaptive clustering hierarchy (LEACH) protocol [9] uses a pure probabilistic model to select CHs and rotates the CHs periodically in order to balance energy consumption. However, in some cases, inefficient CHs can be selected. This due to the fact that LEACH depends only on a probabilistic model. Some cluster heads may be very close each other and can be located on the edge of WSNs. These inefficient cluster heads can have a negative impact on the network the energy efficiency.

This paper investigates the energy efficiency of clustering in routing protocol. We propose a substantial Fuzzy C-means clustering algorithm based on LEACH protocol which consists of using jointly fuzzy C-means clustering, LEACH, and subtractive clustering Method. The proposed approach takes into account the distribution uniformity of cluster heads in the network and the distances between nodes and cluster heads. The proposed hybrid algorithm is proved to be energy efficient and significantly prolongs the sensor network lifetime (Fig. 1).

The remainder of this paper is organized as follows. Section 2 discusses the outline of subtractive Clustering method. Section 3 defines the Fuzzy C-Means Clustering. The improved K-means algorithm is described in Sect. 4. Simulations and results are discussed in Sect. 5. The final section concludes the paper by setting the direction of future work.

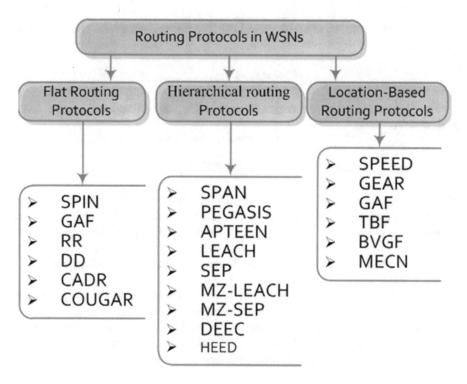

Fig. 1. Classification of routing protocols in WSNs.

2   Subtractive Clustering Method

In clustering approaches such as K-means, PAM, and Fuzzy C-means, the user has to fix the number of clusters in a dataset before to the application. This is considered as the main constraint of clustering algorithms, due to the fact that the user has to work out the optimal number of CHs through several tests to eventually define it. The choice of the adequate number of clusters is frequently ambiguous, with interpretations depending on the form and measure of the distribution of sensor nodes in the field and the desired clustering resolution of the user. Cluster analysis aims at identifying groups of similar objects and, therefore helps to discover distribution of patterns and interesting correlations in large data sets [11]. In 1995, Chiu developed an improved version of the mountain method [12], called the subtractive method, in which each data point is considered as a potential cluster centroid. We consider a collection of n sensor nodes $\{x1, x2...xn\}$ in sensor area. Every sensor node have a potential to become a cluster center and define a measure of the potential of sensor node x_i as [13].

$$M(x_i) = \sum_{j=1}^{n} e^{-\alpha\|x_i - x_j\|^2}$$ (1)

Where α is a positive constant and $\|x_i - x_j\|^2$ is the square of distance between x_i and x_j. Using this mountain function, cluster centroids are selected in a manner similar to that used in the original mountain method. Let M_1^* be the maximum value of the mountain function.

$$M_1^* = \max_i [M(x_i)] \tag{2}$$

Let x_i be the data point whose mountain value is M_1^*; this data point is selected as the first cluster centroid.

3   Fuzzy C-Means Clustering Algorithm

The fuzzy C-means clustering algorithm (FCM), developed by Dunn in 1974 and improved by Bezdek in 1987, has been widely studied and applied [14, 15]. FCM is an unsupervised clustering algorithm as k-means algorithm with the same objective of cluster division. Though, k-means is an algorithm based on hard set and FCM is an algorithm based on non-crisps method (all individuals are classified into two groups: 1 or 0) [16, 17].

This algorithm works by assigning affiliation to each sensor node corresponding to each cluster center on the basis of the distance between the cluster center and the sensor node. More the sensor node is close to the cluster center more is its membership towards the particular cluster center [18, 19]. The FCM algorithm is an iterative optimization algorithm that minimizes the following objective function.

$$f = \sum_{i=1}^{n} \sum_{j=1}^{C} u_{ij}^{m} \left\| x_i - CH_j \right\|^2 \tag{3}$$

Where n is the number of sensor nodes, c is the number of clusters, x_i is the ith sensor node, CH_j is the jth cluster center, u_{ij}^m is the degree of membership of the ith sensor node in the jth cluster, and m is a constant greater than 1 (typically m = 2). $\left\| x_i - CH_j \right\|^2$ represents the measure of the Euclidean distance between sensor node x_i and the cluster center CH_j.

The degree of membership u_{ij}^m and the cluster Head CH_j are defined as the following:

$$u_{ij} = \frac{1}{\sum_{k=1}^{c} \left(\frac{\left\| x_i - c_j \right\|}{\left\| x_i - c_k \right\|} \right)^{\frac{2}{m-1}}} \tag{4}$$

$$CH_j = \frac{\sum_{i=1}^{N} u_{ij}^m \cdot x_i}{\sum_{i=1}^{N} u_{ij}^m} \tag{5}$$

3.1 FCM Algorithm

```
Initialize membership u_{ij}
Find the fuzzy centroid CH_j for j = {1, 2, 3...c }
Update the fuzzy membership u_{ij}
Repeat steps 2, 3 until f(u_{ij}, CH_j) is no longer decreasing
```

4 Proposed Network Model

In this work, we proposed a new approach to deal with the constraint of energy consumption in WSNs. Here we present the outline of the proposed network model. The sensor nodes are deployed randomly in an area of 100 m^2, in order to incessantly monitor the environment. The data sensed by sensor nodes is forward to a base station located outside of the deployment area. Each sensor nodes can operate either in sensing mode to monitor the environment parameters and transmit it to the associated CH or in CH mode to gather data, compress it and forward to the base station. In addition, some assumptions are made as follows:

- The base station have illimited energy and computational power.
- The network is considered homogeneous and all sensor nodes have the same initial energy and they are static.
- Nodes have the capability of controlling the transmission power according to the distance of receiving nodes.
- Links are symmetric.

The proposed network model based on three algorithms firstly starts to use the subtractive clustering method to determine the adequate number of clusters. Then the FCM algorithm applied to form highly uniform clustering dispersion of nodes. The LEACH protocol is used in each cluster to construct the sub-clusters and to select the sub-clusters head and their members; which apply a pure probabilistic model to select CHs and rotates the CHs periodically in order to balance the energy consumption. Generally, the optimal number of CHs is estimated to be at 5% of the total number of nodes in each cluster [3–6].

Figure 2 presents a flowchart of the cluster formation process of the proposed approach, which combines the fuzzy C-means clustering, LEACH, and the subtractive clustering Method.

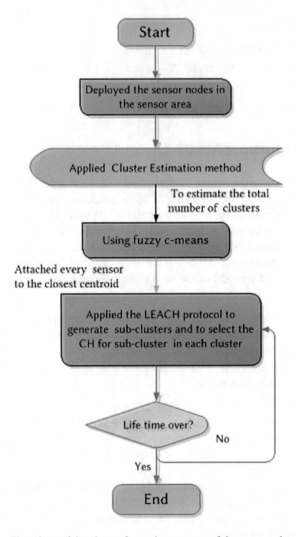

Fig. 2. Flowchart of the cluster formation process of the proposed approach

5   Simulations and Evaluation

All simulations were executed using MATLAB to analyze the total system energy and
the number of nodes alive. We started with a description of the significant metrics used
in the simulations. We provide the results of the used algorithm and then we show and
discuss all results with a comparative analysis. We consider 100 nodes randomly
deployed in an area of (100×100) m^2. The BS is located outside the area at the coordinate
of $(-50, 150)$ m. All nodes have a starting energy of 0.5 J and the base station is assumed
to have unlimited energy. Each node sends $L = 4000$ bit packets to the base station
through the sub-cluster head during each round (Fig. 3). The probability of being the

sub-cluster head is set to 0.05 (about 5% of nodes per round become sub-cluster heads). We use the first order model adopted by LEACH and SEP protocols [3–6]. This radio model uses both of the free space and multipath channels by taking into account the distance between the source and destination. So energy consumption for transmitting a packet of l bits in distance d is given by (6) and (7).

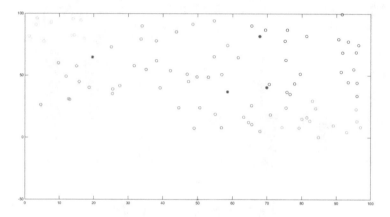

Fig. 3. Example of clustered network of $100 * 100 \, m^2$

The energy expended in the transmit electronics for free space propagation ETx-fs is described by:

$$E_{Tx-fs}(k, d) = K * \left(E_{elec} + \pounds_{fs} * d^2\right) \quad if \quad d < d_0 \tag{6}$$

And the energy expended in the transmit electronics for free multi-path propagation ETx-mp is given by:

$$E_{Tx-mp}(k, d) = K * \left(E_{elec} + \pounds_{amp} * d^4\right) \quad if \quad d \geq d_0 \tag{7}$$

Where: ETx is the electrical energy required to transmit a K-bit message over a distance d, Eelec corresponds to the energy per bit required in transmitting and receiving electronics to process the information. εfs and εamp are constants corresponding to the energy per bit required in the transmission amplifier to transmit an L-bit message over a distance d^2 and d^4 for free space and multi-path propagation modes, respectively. By equating formula (6) and (7), we determine the distance $d = d0$ when the propagation transition from the direct path to multi-path occurs:

$$d0 = \sqrt{\frac{\pounds_{fs}}{\pounds_{amp}}} \tag{8}$$

The comparison we carry out in this work between the proposed network model and LEACH protocol is based on some key performance parameters of wireless sensors network. Firstly, the network lifetime that describes the time interval between the start of the network operation and the death of the last sensor node. The second parameter is the number of dead nodes which is the total number of sensor nodes that have consumed all of their energy and are not able to do any kind of functionality (Table 1).

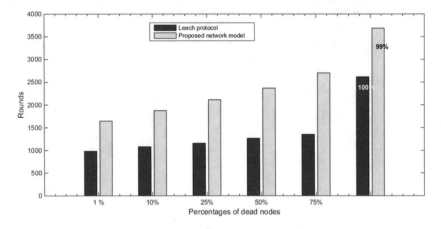

Fig. 4. Percentages of dead nodes

Table 1. Simulation parameters

Simulation area	100 m * 100 m
BS location	(50 m, −100 m)
Number of nodes	100
Transmission amplifier free space ε_{fs}	10 pJ/bit/m^2
Transmission amplifier, multipath ε_{mp}	0.0013 pJ/bit/m^4
Data aggregation energy	5 nJ/bit/msg
Transmission energy E_{Tx}	50 nJ
Receiving energy E_{RX}	50 nJ
Network dimension	100 m * 100 m
Signal wavelength λ	0.325 m
Antenna gain factor G_t, G_r	1
Message size (bits)	4000 bits
Initial node energy	0.5 J

The number of dead nodes versus transmission round of the proposed approach with LEACH protocol is presented in Fig. 5. The red curve indicates the proposed approach and the blue curve indicates LEACH protocol. The attained results have confirmed that the proposed network model provides the longest lifetime of the network compared to LEACH protocol. The obtained results show that the total energy of LEACH protocol and the proposed network scheme is completely consumed at 2617 rounds and after 4000 rounds, respectively. It's shown that that the proposed approach reduces the energy consumption by round, and extends the network lifetime by more than 80% compared to LEACH protocol.

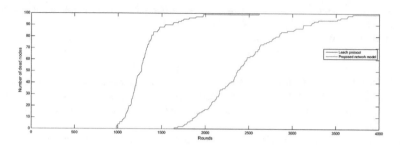

Fig. 5. Number of dead nodes versus transmission rounds

Figure 4 displays different percentages of dead nodes for the proposed scheme and LEACH protocol. The amber-coloured designates the proposed network model and the blue colored indicates the LEACH protocol. In each model, the following the following percentages 1%, 10%, 25%, 50% and 99% of dead nodes, were studied and analyzed. For LEACH protocol, 1%, 10%, 25%, 50%, 75% and 99% dead nodes occur at 984, 1084, 1159, 1267, 1354 and 2617 rounds respectively. The proposed model in round 1644 corresponds to 1% of dead nodes, 1873 to 10%, 2111 to 25%, 2369 to 50%, 2704 to 75% and 3690 rounds to 99% of dead nodes.

The obtained results have demonstrated that the proposed network model reduces the energy consumption by round, and extend the network lifespan. Therefore, due to the combination of the fuzzy C-means clustering, LEACH, and the subtractive clustering Method. The proposed approach provided the longest lifetime.

6   Conclusion

The main objective of this work is to propose a novel clustering routing approach based on three tools to deal with the problem of energy conservation of WSNs. Firstly; we using the subtractive clustering method to determine a correct number of clusters. The FCM algorithm applied to form highly uniform clustering of nodes and LEACH protocol applied for each cluster. The proposed approach minimizes the energy consumption, extends the network lifetime of the sensor nodes. The evolution and enhancement of the presented routing algorithm should be done in the future to support multipath and heterogeneous modes.

References

1. Güngör, V.Ç, Hancke, G.P.: Industrial Wireless Sensor Networks: Applications, Protocols, and Standards. CRC Press (83), pp. 1027–1040 (2013)
2. Heinzelman, W., Chandrakasan, A., Balakrisham, H.: Energy-efficient communication protocol for wireless microsensor networks. In: Proceeding of the 33rd Annual Hawaii International Conference on System Sciences, pp. 3005–3014 (2000)
3. Akyildiz, I.: A survey on sensor networks. IEEE Commun. Mag. **40**(8), 102–114 (2002)
4. Mahboub, A., Arioua, M., En-Naimi, E.M., Ezzazi, I.: Performance evolution of energy-efficient clustering algorithms in wireless sensor network. J. Theor. Appl. Inf. Technol. **83**(2), 1–8 (2016)
5. Mahboub, A., Arioua, M., En-Naimi, E.M., Ezzazi, I.: Multi-zonal approach for clustered wireless sensor networks. In: 2nd International Conference on Electrical and Information Technologies, ICEIT. IEEE (2016)
6. Zhu, P.: A new approach to sensor energy saving algorithm. TELKOMNIKA **11**(5), 2485–2489 (2013)
7. Velmurugan, T.: Performance based analysis between k-Means and fuzzy C-Means clustering algorithms for connection oriented telecommunication data. Appl. Soft Comput. **19**, 134–146 (2014)
8. Mahboub, A., Arioua, M., En-naimi, E.M., Ezzazi, I., El Oualkadi, A.: Multi-zonal approach clustering based on stable election protocol in heterogeneous wireless sensor networks. In: IEEE 4th Edition of the International Colloquium on Information Science and Technology CIST 2016 (2016)
9. Mahboub, A., Arioua, M., En-naimi, E.M.: Energy-efficient hybrid k-means algorithm for clustered wireless sensor networks. Int. J. Electr. Comput. Eng. (IJECE) **7**(4), 2054–2060 (2017)
10. Sharma, N., Verma, V.: Energy efficient LEACH protocol for wireless sensor network. Int. J. Inf. Netw. Secur. (IJINS) **2**(4), 333–338 (2013)
11. Kim, D.-W., Lee, K., Lee, D., Lee, K.H.: A kernel-based subtractive clustering method. Pattern Recogn. Lett. **26**(7), 879–891 (2005)
12. Yager, R.R., Filev, D.P.X.: Approximate clustering via the mountain method. IEEE Trans. Syst. Man Cybern. **24**(8), 1279–1284 (1994)
13. Biradar, R.V., Patil, V.C., Sawant, S.R., Mudholkar, R.R.: Classification and comparison of routing protocols in wireless sensor networks. Spec. Issue Ubiquit. Comput. Secur. Syst. **4**(2), 704–711 (2009)
14. Goyal, D., Tripathy, M.R.: Routing protocols in wireless sensor networks: a survey, pp. 474–480 (2012)
15. Kim, T., Bezdek, J.C., Hathaway, R.J.: Optimality tests for fixed points of the fuzzy c-means algorithm. Pattern Recogn. **21**(6), 651–663 (1988)
16. Kapitanova, K., Son, S.H., Kang, K.-D.: Using fuzzy logic for robust event detection in wireless sensor networks. Ad Hoc Netw. **10**(4), 709–722 (2012)
17. Lam, Q.-T., Horng, M.-F., Nguyen, T.-T., Lin, J.-N., Hsu, J.-P.: A high energy efficiency approach based on fuzzy clustering topology for long lifetime in wireless sensor networks. In: Nguyen, N., Trawiński, B., Katarzyniak, R., Jo, G.-S. (eds.) Advanced Methods for Computational Collective Intelligence, vol. 457, pp. 367–376. Springer, Heidelberg (2013)

18. Chattopadhyay, S., Pratihar, D.K., De Sarkar, S.C.: A comparative study of fuzzy c-means algorithm and entropy-based fuzzy clustering algorithms. Comput. Inform. **30**(4), 701–720 (2011)
19. Wang, H., Xu, Z., Pedrycz, W.: An overview on the roles of fuzzy set techniques in big data processing: Trends, challenges and opportunities. Knowl.-Based Syst. **118**, 15–30 (2017)

The Allocation of Submissions in Online Peer Assessment: What Can an Assessor Model Provide in This Context?

Mohamed-Amine Abrache[1]([⊠]) [iD], Aimad Qazdar[2],
Abdelkrim Bendou[1], and Chihab Cherkaoui[3]

[1] IRF-SIC Laboratory, FSA, Ibn Zohr University, Agadir, Morocco
mohamed-amine.abrache@edu.uiz.ac.ma
[2] GMES Laboratory, ENSA, Ibn Zohr University, Agadir, Morocco
[3] IRF-SIC Laboratory, ENCG, Ibn Zohr University, Agadir, Morocco

Abstract. In online learning environments, the individual characteristics of learners have necessarily an influence on the reliability and the credibility of peer assessment. This paper focuses on the stage of allocating students' submissions within online peer assessment process. Firstly, by providing an overview of the main applications of this process in online assessment tools and MOOCs. This overview considers mainly the methodologies of assigning submissions to learners for evaluation; and secondly, by proposing a model for the assessor based on the individual personal characteristics that shape her or his assessment profile. This profile plays a key role in the success of the peer assessment/feedback experience. We conclude this paper with a brief discussion of the potential that can provide the assessor model in the context of an approach that manages the allocation of submissions and considers the personal characteristics of the learners' community.

Keywords: Peer assessment · Peer review · Learner profile · Assessor model
MOOC · Online assessment tools · Allocation of submissions

1 Introduction

The concept of Massive Open Online Courses (MOOCs) represents an innovation in the field of distance education that uses the internet as a means of learning. It allows the participation of a large number of learners and creates a worldwide learning environment for students without being necessarily affiliated with a school or an institution [1].

Due to the massive nature of MOOCs, these learning environments are heterogeneous. They bring together a number of learners with varied experiences and previous knowledge [2]. Learners are also characterized by various behavior patterns which implies different learning styles [3].

This paper concerns the individual characteristics of the learners' community in the context of the teaching-learning-assessing/feedback process. It focuses on the assessment/feedback element of the process which represents one of the main challenges in the field of distance online learning, mainly in MOOCs [4].

© Springer International Publishing AG, part of Springer Nature 2018
M. Ben Ahmed and A. A. Boudhir (Eds.): SCAMS 2017, LNNS 37, pp. 276–287, 2018.
https://doi.org/10.1007/978-3-319-74500-8_25

The use of peer assessment in learning environments is intended to handle this evaluation challenge in a way that improves some learners' cognitive capacities as critical thinking and decision-making [5].

Peer assessment helps in engaging students into the learning process [6] and in reducing the drop-out rate through making MOOCs more interactive [7]. Indeed, student participation in peer assessment activities positively influences his\her completion rates [8].

Despite the reduction of assessment workload for the course staff thanks to peer assessment, the instructors are still concerned about the lack of the accuracy of learners' evaluation [9]. Providing a peer assessment process that ensures a high level of evaluation reliability and validity is the key element in supporting the credibility of its use.

The peer assessment process can be summarized as follows:

- Creation of the assessment by stating the assignment questions and determining the assessment criteria.
- Submission of the assignments by learners.
- Distribution of the submitted works among the participants for evaluation.
- Evaluation of each submission by a group of learners and the writing of feedback.
- Calculation of the final score.
- Delivering the score and feedback to each learner.

Since the optimization of the peer assessment process necessarily involves optimizing each of its steps; this paper provides an overview of the approaches applied in one of these steps, which is the allocation of submissions to assessors. At the same time, we are interested in the learner's assessment profile. We believe that the integration of a profile based on an assessor model into the repartition of assignments can be positively reflected in the accuracy and validity of the assessment because it considers the influence of learners' individual characteristics on the results of the evaluation.

The remainder of this paper is organized as follows. Section 2 presents an overview of the literature related to the assessment within the field of online learning environments. Section 3 outlines some existing peer assessment applications in online assessment tools and MOOCs. Section 4 represents the methodologies of allocation assessment in online peer assessment. In Sect. 5, we propose an assessor model that concerns the assessment profile of learners. Finally, the last section illustrates conclusions and perspectives for future works.

2 Background

Given the technologies used in the of online learning environments, assessment continues to be one of the flaws of many MOOCs, including those who have made significant investments in other aspects of online courses, such as the high-quality video production [10].

The assessment challenge is not only related to the massiveness of MOOCs but also to the modern pedagogical approaches, as problem-based learning and project-based learning [11]. The use of automatic correction techniques (e.g. multiple-choice and short answer questions, computer code, and vocabulary activities) could not ensure valid and sufficient assessment and feedback for such approaches.

An Automated Essay Scoring system (AES) was proposed as a plug-in within the Edx MOOC platform. The AES rely on natural language processing (NLP) algorithms to measure the writing quality of learners' essays [12]. However, the measure in the context of AEE (Automated Essay Evaluation) which integrates the AES, mainly concerns the syntax of essays and do not have the capacity to completely check the coherence of the content and the consistency of the statements linked to the semantics in different forms of writing [13].

According to Falchikov [14], effective assessment requires involving learners as active participants. Peer assessment is a strategy that stands on learners' participation and that can deal with different pedagogical approaches.

Topping defined the peer assessment as "an arrangement in which individuals consider the amount, level, value, worth, quality or success of the products or outcomes of learning of peers of similar status" [15].

Peer assessment represents an important element in the design of learning environments implementing a more participatory culture of learning [16]. Indeed, it contributes to the achievement of different evaluation goals as the initiation of "peer learning" processes by encouraging students to better interact with the learners' community.

In addition, the application peer assessment may follow various models according to different evaluation objectives as the "assessment as learning" objective that consists of providing the students with the ability to become self-directed learners who are not just engaged in thinking about the content of the course but also in the strategy of learning. Peer assessment can contribute to the achievement of the "assessment for learning" objective which represents the formative assessment that helps students to know their actual level and adjust their learning experience. A third evaluation objective in which peer assessment can take part is the "assessment of learning" which generally represents the summative type of evaluation [17]. Peer assessment exercise also promotes affective feedback and cognitive capacities [18].

3   Online Peer Assessment

An online peer assessment system is a web-based application that has been developed to manage and monitor peers grading and/or feedback process [19]. In this section, we give an overview of three online peer assessment tools, and then we show the features of the main implementations of peer assessment in MOOCs and essentially how they assign the submissions to participants.

3.1   CrowdGrader

CrowdGrader [20] is a peer grading and evaluation online tool that stands on a collaborative perspective of assessment to perform the evaluation process. Every participant is asked to evaluate four to six assignments, and then receives an overall crowd grade that represents a judgment not only on the quality of her or his work but also on the accuracy and the helpfulness of her or his feedback. The consideration of the quality of feedback in the calculation of the overall score motivates students to provide more accurate grades and more constructive reviews.

After submitting all the assignments by learners, the system uses an online algorithm that allocates new assignment to an assessor only after verifying the completion of the evaluation task that has been assigned to her or him before.

The accuracy grade gives an idea about the difference between the grade assigned by the reviewer and the global calculated grade assigned by the assessors of the same work. CrowdGrader gives the participants the opportunity to rate their received comments, which allows the calculation of a helpfulness grade for the feedback provided by each student.

3.2 Peerceptiv

Peerceptiv is a data driving peer assessment that supports students writing practices and aims to improve their critical thinking and stimulate their analytical reasoning. It was formerly called SWoRD (Scaffolded Writing and Rewriting in the Discipline) and has been developed to help students on providing effective and efficient quantitative feedback (numeric grades) and qualitative feedback (annotated comments) for the writing of their peers. After that the participants uploaded their work, the allocation of assignments occurs at random [21]. As for CrowdGrader, the system gives the possibility to rate the received comments of reviewers in order to estimate the helpfulness of their feedback, it measures inter-raters' reliability of the assessment that depends on the level of similarity between the assessor rating and the mean score assigned by the other reviewers of the same essay. Besides, teachers evaluate a number of submissions and then assigned them to the participants to be reevaluated. By doing so, the validity of the assessors is also measured through comparing their assessment to the rating provided by the teacher [22].

3.3 The Moodle Workshop

As a standard plug-in of the Moodle open-source LMS, Moodle Workshop module [23] for formative peer assessment is a collaborative grading system that allows students to assess each other's projects respecting different grading strategies specified by the course facilitators. It gives the instructors the possibility to edit the peer assessment settings to manage the assessment criteria, or to integrate some pre-evaluated samples for students to test their assessment ability, or to custom the feedback providing mechanisms, etc. The allocation of submissions in this tool is performed either manually or randomly and the final grade for a particular submission is assigned on the basis of the weighted average of the assessors' rating of this submission. The instructors may assign a higher weight for their assessment or replace the grade with a specific value, Moodle Workshop module gives grades for learners' assessment, by assigning higher grades to learners with the closed rating to the overall grade assigned to each assessment criteria [24].

3.4   Peer Assessment in MOOCs

As mentioned before, peer assessment is a solution that can address the problem of massiveness and the challenge of adopting some modern pedagogical approaches in MOOCs.

A common solution used in the context of MOOCs is the Calibrated Peer Review [CPR] [25], which represents a web tool used to ensure more validity in the context of peer assessment. The basic implementation of CPR relay on a calibration phase in which learners are asked to rate a number of assignments that have been previously prepared by the course staff as assessment training. Depending on their rating similarity to the staff provided rating, a Reviewer Competency Index (RCI) is assigned to each partici- pant to represent her or his evaluation competency. The learners with low RCI are invited to retake the exercise in order to improve their evaluation competency.

After the calibration phase, the system allocates each assignment randomly to a group of learners to be assessed. At the end of the evaluation exercise, an overall score is attributed to each assignment according to the weight that is attributed to its assessors. Indeed, the higher the assessor's RCI, the more weight is given to his rating. The CPR may also include a self-assessment phase in which the learners evaluate their own performance, and determine the concepts to improve.

Coursera is one of the main MOOC classes provider that has applied a Calibrated Peer Review [CPR] system with some differences. The principal difference is that the calibration occurs at the same time with the effective evaluation through taking a random sample of assignments submitted by the participants for an evaluation by the staff, then these assignments are randomly distributed to the assessors simultaneously with other assignments not included in the sample [31]. The participants evaluate in the same time all their allocated assignments without knowing which ones have been previously eval- uated.

As in Coursera, Edx learners are asked to assess some pre-evaluated assignments. Learners obtain the authorization to evaluate peers' submissions once they manage to assign similar marks compared with those provided by the instructors [26].

4   Allocation of Submissions

The optimization of peer assessment process aims to ensure more credibility of the assessment in conjunction with contributing to the motivation of learners to better engage in the course and in the writing of feedback. Optimizing the allocation of submissions in the context of peer assessment process can play an important role in achieving these goals.

The principle of submission priority [27] represents a mechanism that aims to balance the distribution of assignments and their number of evaluations. The allocation process characterizes each assignment by a priority level and a requested evaluations number. The principle consists of allocating the assignment with the higher priority, then decreasing it as well as the requested evaluations number of this assignment. Six hours later, if the assignment has been effectively evaluated, the system rewards the assessors who reviewed their assigned submissions by raising the priority of their assignments, so their work would be evaluated first. Otherwise, if the assignment has not been evaluated within six hours, its priority and requested evaluations number increase automatically so that it will be redistributed a second time.

In Coursera, the CPR is limited into calibrating the assessors and is not intended to exploit the allocation of the assignments in the optimization of the process. Besides, Edx's documentation encourages instructors to request for each submission a number of assessments beyond what they believe necessary; this has been suggested in order to overcome the situation where a participant did not assess her or his assigned submissions [26].

Based on our analyses of the two MOOCs's providers (Coursera and Edx) and the aforementioned online peer assessment tools, we noticed that the focus on the allocation of submissions is generally linked to the technical aspect of overcoming the problem of the non-evaluated submissions. While we believe that the introduction of the assessment profile and the evaluation competence of learners as parameters to be considered in the allocation of submissions may contribute effectively to the optimization of peer assessment process.

Some tools that belong to the field of the correction of evaluation results are generally based on the personal aspects of the assessors, as within the models of Piech [28] and Goldin [29] that stand on some unobservable estimated parameters such as the prejudices of the assessors to correct the rating. However, the criticism on these solutions lies in the fact that they do not interfere sufficiently with the allocation of submissions, because they are only applied at a level subsequent to what we are trying to optimize.

Furthermore, the use of a rating by a group of reviewers may reduce the influence of the 'weaker' assessors [30]. However, it cannot represent a radical solution for such problem. The random allocation of submissions implied the possibility that a particular assignment is evaluated by a group of learners made up entirely of novice assessors in terms of their evaluation competency. This situation, on the one hand, affects negatively the reliability of the evaluation, and, on the other, calls into question the credibility of the entire peer assessment process. Especially, as it is possible that another assignment may be evaluated by learners with more advanced capacities than the latter group, which implies that the evaluation conditions in this situation are not equal for each assignment and may have a negative impact on the motivation of learners not only for this form of assessment but for the course in general.

5   Assessor Model

In the context of MOOCS or other online learning tools, the creation of a learner assessment profile is an essential step in the direction of gathering information about the parameters that influence learners' assessment.

Indeed, investigating the personal individual characters that shape the learner assessment profiles allows the formation of an insight into the overall characteristics of the community that operates the assessment exercise, which can be used to optimize the peer assessment process. To this end, we propose an assessor model based on models and standards of learners modeling such as the "PAPI Learner" standard [31], or IMS

LIP (Information Management System - Learning Information Package) [32] or the model of Battou [33] proposed in the context of the ALS-CPL project. Battou defined the learner modeling as a process which covers the life cycle of the setting up of a learner model. This process includes the acquisition of knowledge about the learner, and then constructing, updating, maintaining and exploiting this model. In the context of the AeLF platform, a model similar to Battou's model has been used in order to create a framework that aims to supply an adaptive learning system (ALS) to e-learning platforms [34].

These previous findings allowed us to propose an assessor model that can be defined as a representation of the assessor which includes all the information and characteristics intrinsic to the evaluation process. In our context, this evaluation is a peer assessment within MOOC classes. The representation of information within the assessor model follows a qualitative approach in order to facilitate its acquisition. This model showed in Fig. 1 rely on different factors to characterize each assessor within the system:

- The ability to interact with the MOOC platform, and mainly with the assessment modules.
- The prior educational experience within the course domain.
- The prior assessment experiences.
- The effective assessment competence.

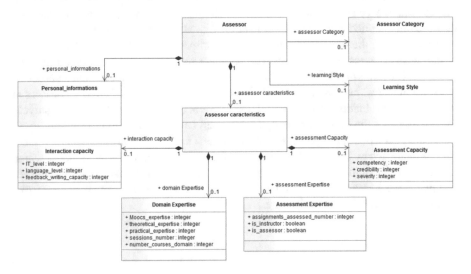

Fig. 1. The assessor model

On the other hand, there are other characteristics that may have a significant influence on learners' assessment capacity as their learning style.

5.1   Interaction Capabilities with the Platform

ICT competence and the language proficiency are two parameters that impact the inter-action of learners with the MOOC platform. Learners must have some basic competencies, including computing skills and the mastering of the language of the course [35]. To make appropriate judgments, the assessment experience depends more on the sufficient understanding of the subject and the peers' assignments. Language proficiency is also reflected in the feedback writing. The student resistance to peer feedback is rarer, compared to the relatively more frequent resistance to the formal peer assessment [36], which implicates that a good quality of feedback can be a means to ensure the acceptance of assessment results.

5.2   Expertise Within the Domain of the Course

Antecedent expertise in the domain of the course is a factor that influences not only the ability to achieve learning objectives but also the capacity to conduct a reliable assessment for peers' work. This domain expertise can be divided into three principal aspects: the previous experiences in MOOCs platform, the theoretical competence that underlies the course field, and the practical competence related to the application of academic knowledge in real-world situations. In addition, other factors may influence the domain's expertise such as the number of previous sessions of the course followed by the assessor, as well as the number of the completed courses related to the same educational specialty area.

5.3   Assessment Expertise

The developed experience through the assessment exercise is one of the main factors influencing the effectiveness of the evaluation [37]. Having an ability to interact with the platform or some antecedent knowledge of the domain is not always sufficient. The assessment competency influences the accurate identification of the mistakes and the gaps in peers' work, which is expressed in the relevance and the clarity of the feedback. This assessment expertise can be estimated according to the number of prior assessment experiences.

5.4   Assessment Capacity

Measuring learners' ability to assess is widely used in the field of the correction of peer assessment results. For illustration, we can mention two methods figuring in this context:

- The calculation of the credibility index (CI) [38] takes into consideration the assessors' precision, their consistency, and their transferability of accuracy among assignments. To measure the credibility index, each participant is asked to rate at least two pre-evaluated assignments. The assessment is applied by assigning three scores, which are her or his effective rating grade and the estimated highest and lowest possible grades for the same submission. The accuracy represents the gap between the instructor's and the learner's grades. The consistency is the stability of rating between the highest and lowest scores attributed to the same submission. Finally, the transferability measures the maintenance of the assessor's accuracy between different pre-evaluated assignments.

- The multi-faceted Rasch measurement model (FACETS) [39] proposed by Linacre (1989) consists of three facets: the student competencies, the difficulty of the field or the assignment, and the differences between the severity of assessors. This last facet implicates the need for a statistical adjustment because the severity of the assessors is a feature that should be included in the performance analysis of learners in order to improve the accuracy of their assessments.

The assessor model integrates three basic parameters that may represent globally the assessment capacity: the competency, the credibility, and the severity of the assessor.

5.5 Learning Style

The Felder-Silverman Learning Style Model (FSLSM) [40] has divided students' learning styles into four dimensions: active/reflective, sensory/intuitive, visual/verbal and sequential/global. Each learning style corresponds better to a specific teaching style, which is the same regarding the assessment criteria in the context of peer assessment. Indeed, some assessment criteria are better assessed by learners with specific learning styles more than others. According to this logic, Lan et al. [41] used the FSLSM to give a weight to learners' rating for each assessment criteria based on their learning style.

For example, in the context of a homework having as subjects: the development of a website, they gave a great weight to students with an active learning style for all the assessment criteria as the learners of this category tend to try things and prefer to learn by communicating and collaborating with other pairs, which is not the case for learners with a reflective style. For students with a visual learning style, they seem to be able to give more accurate assessment for the "Site Structure" criterion, as well as for the "Graphs" criterion that deals with the assessment of graphs on the website.

6 Conclusion and Perspectives

The assessor model proposed in this paper may play a role in the representation of the individual characteristics of learners related to assessment, so it may constitute a tool that takes a part in the reinforcement of the credibility and the accuracy of peer assessment.

We believe that the allocation of submissions is a critical phase in the context of the peer assessment process. The assessment profiles of learners created on the basis of the assessor model can be the source of relevant information for any solution introducing the learners' characteristics in the allocation of submissions.

A dynamic aspect of the model should be considered mainly in updating the assessment expertise and measuring the effective assessment competency of learners.

Our overview of the implemented allocation of submission methodologies led us to specify two basic objectives that should be targeted when assigning submissions to learners. The first is that every submission should be reviewed by a group of learners with different levels of assessment capacity in order to ensure the same condition of assessment for all submissions. This capacity may be measured through the data gathered based on the assessor model. The second objective is to provide a reasonable and fair number of assignments to be evaluated per assessor. This objective contributes to maintaining the motivation of learners to perform this exercise of assessment and to provide feedback.

It is possible to use the relation between assessment criteria and learning styles to create a database containing all the assessment criteria belonging to a particular domain with a specification of the evaluation weights for each criterion based on the learning styles of learners. In addition, the use of an evaluation that stands on various evaluation criteria may provide more validity and precision for the assessment.

Our current research work involves setting up a new algorithm for the allocation of submission based on the assessor model. We try to ensure a distribution with a reasonable evaluation charge, and equal evaluation conditions, which creates a contribution to the quest to improve the validity and the reliability of the assessment results and to the motivation of learners' engagement in learning activities.

References

1. Reis, R., Escudeiro, P.: The role of virtual worlds for enhancing student-student interaction in MOOCs. In: User-Centered Design Strategies for Massive Open Online Courses (MOOCs), pp. 208–221. IGI Global (2016)
2. Daradoumis, T., et al.: A review on massive e-learning (MOOC) design, delivery and assessment. In: 2013 Eighth International Conference on P2P, Parallel, Grid, Cloud and Internet Computing (3PGCIC). IEEE (2013)
3. Hmedna, B., et al.: Identifying and tracking learning styles in MOOCs: a neural networks approach. Int. J. Innov. Appl. Stud. **19**(2), 267 (2017)
4. Yousef, A.M.F., et al.: The impact of rubric-based peer assessment on feedback quality in blended MOOCs. In: International Conference on Computer Supported Education. Springer, Cham (2015)
5. Zhao, C., et al.: Exploring the role of assessment in developing learners' critical thinking in massive open online courses. In: European Conference on Massive Open Online Courses. Springer, Cham (2017)
6. Jiao, J., et al.: Improving learning in MOOCs through peer feedback: how is learning improved by providing and receiving feedback? In: Learning and Knowledge Analytics in Open Education, pp. 69–87. Springer, Cham (2017)
7. Khalil, H., Ebner, M.: MOOCs completion rates and possible methods to improve retention-a literature review. In: World Conference on Educational Multimedia, Hypermedia and Telecommunications (2014)
8. Ashenafi, M.M., Ronchetti, M., Riccardi, G.: Exploring the role of online peer-assessment as a tool of early intervention. In: SETE@ ICWL (2016)
9. Falchikov, N., Goldfinch, J.: Student peer assessment in higher education: a meta-analysis comparing peer and teacher marks. Rev. Educ. Res. **70**(3), 287–322 (2000)
10. Haber, J.: MOOCs, pp. 69–76. MIT Press, Cambridge (2014)
11. Shute, V.J., Rahimi, S.: Review of computer-based assessment for learning in elementary and secondary education. J. Comput. Assist. Learn. **33**(1), 1–19 (2017)

12. Balfour, S.P.: Assessing writing in MOOCs: automated essay scoring and calibrated peer review (TM). Res. Pract. Assess. **8**, 40–48 (2013)
13. Zupanc, K., Bosnić, Z.: Automated essay evaluation with semantic analysis. Knowl. Based Syst. **120**, 118–132 (2017)
14. Falchikov, N.: Involving students in assessment. Psychol. Learn. Teach. **3**(2), 102–108 (2004)
15. Topping, K.: Peer assessment between students in colleges and universities. Rev. Educ. Res. **68**(3), 249–276 (1998)
16. Kollar, I., Fischer, F.: Peer assessment as collaborative learning: a cognitive perspective. Learn. Instr. **20**(4), 344–348 (2010)
17. Gielen, S., et al.: Goals of peer assessment and their associated quality concepts. Stud. High. Educ. **36**(6), 719–735 (2011)
18. Lu, J., Law, N.: Online peer assessment: effects of cognitive and affective feedback. Instr. Sci. **40**(2), 257–275 (2012)
19. Babik, D., et al.: Probing the landscape: toward a systematic taxonomy of online peer assessment systems in education. In: EDM (Workshops) (2016)
20. de Alfaro, L., Shavlovsky, M.: CrowdGrader: a tool for crowdsourcing the evaluation of homework assignments. In: Proceedings of the 45th ACM Technical Symposium on Computer Science Education. ACM (2014)
21. Cho, K., Schunn, C.D.: Scaffolded writing and rewriting in the discipline: a web-based reciprocal peer review system. Comput. Educ. **48**(3), 409–426 (2007)
22. Schunn, C., Godley, A., DeMartino, S.: The reliability and validity of peer review of writing in high school AP English classes. J. Adolesc. Adult Lit. **60**(1), 13–23 (2016)
23. Rice, W.: Moodle E-Learning Course Development. Packt Publishing Ltd., Birmingham (2015)
24. Using Workshop – MoodleDocs. https://docs.moodle.org/29/en/Using_Workshop#Grade_for_assessment. Accessed 2017
25. Russell, J., et al.: Variability in students' evaluating processes in peer assessment with calibrated peer review. J. Comput. Assist. Learn. **33**(2), 178–190 (2017)
26. Edx: Open Response Assessments (2017). https://edx.readthedocs.io/projects/open-edx-building-and-running-a-course/en/named-release-birch/exercises_tools/open_response_assessments/index.html
27. Staubitz, T., et al.: Improving the peer assessment experience on MOOC platforms. In: Proceedings of the Third (2016) ACM Conference on Learning@ Scale. ACM (2016)
28. Piech, C., et al.: Tuned models of peer assessment in MOOCs. arXiv preprint arXiv:1307.2579 (2013)
29. Goldin, I.M., Ashley, K.D.: Peering inside peer review with Bayesian models. In: Artificial Intelligence in Education. Springer, Heidelberg (2011)
30. Van den Berg, I., Admiraal, W., Pilot, A.: Design principles and outcomes of peer assessment in higher education. Stud. Higher Educ. **31**(03), 341–356 (2006)
31. IEEE P1484.2/D7, 2000-11-28: Draft Standard for Learning Technology. Public and Private Information (PAPI) for Learners (PAPI Learner) (2002). http://ltsc.ieee.org/wg2/. Accessed 25 Oct 2002
32. Oubahssi, L., Grandbastien, M.: From learner information packages to student models: which continuum? In: International Conference on Intelligent Tutoring Systems. Springer, Heidelberg (2006)
33. Battou, A.: Approche granulaire des objets pédagogiques en vue de l'adaptabilité dans le cadre des Environnements Informatiques pour l'Apprentissage Humain (2012)
34. Qazdar, A., et al.: AeLF: mixing adaptive learning system with learning management system. Int. J. Comput. Appl. **119**(15), 1–8 (2015)

35. Fini, A.: The technological dimension of a massive open online course: the case of the CCK08 course tools. Int. Rev. Res. Open Distrib. Learn. **10**(5) (2009)
36. Brown, G.A., Bull, J., Pendlebury, M.: Assessing Student Learning in Higher Education. Routledge, New York (2013)
37. Jeffery, D., et al.: How to achieve accurate peer assessment for high value written assignments in a senior undergraduate course. Assess. Eval. Higher Educ. **41**(1), 127–140 (2016)
38. Xiong, Y., et al.: A proposed credibility index (CI) in peer assessment. In: Poster Presented at the Annual Meeting of the National Council on Measurement in Education, Philadelphia, PA (2014)
39. Engelhard, G.: Examining rater errors in the assessment of written composition with a Many-Faceted Rasch model. J. Educ. Meas. **31**(2), 93–112 (1994)
40. Felder, R.M., Silverman, L.K.: Learning and teaching styles in engineering education. Eng. Educ. **78**(7), 674–681 (1988)
41. Lan, C.H., Graf, S., Lai, K.R.: Enrichment of peer assessment with agent negotiation. IEEE Trans. Learn. Technol. **4**(1), 35–46 (2011)

Combined Mean Shift and Interactive Multiple Model for Visual Tracking by Fusing Multiple Cues

Younes Dhassi$^{(\boxtimes)}$ ⓘ and Abdellah Aarab

Laboratory of Electronics, Signals, Systems and Computers, Department of Physics,
Faculty of Sciences Dhar-Mahraz, Sidi Mohamed Ben Abdellah University, Fes, Morocco
dyounes2003@gmail.com

Abstract. To overcome the tracking issues caused by the complex environment namely illumination variation and background clutters, tracking algorithm was proposed based on multi-cues fusion to construct a robust appearance model, indeed the traditional Mean Shift (MS) estimate the state associated with each sub appearance model, and the interactive multiple model (IMM) adjusts the weights of different cues and then combine the sub appearance models to estimate the general state. The proposed method is tested on public videos that present different environment issues. Experiences and comparisons conducted show the robustness of our methods in challenging tracking conditions.

Keywords: Visual tracking · Mean shift · Interactive multiple models

1 Introduction

Object tracking is a common problem in the field of computer vision. The constant increase in the power of computers, the reduction in the cost of cameras and the increased need for video analysis have engendered a keen interest in object tracking algorithms. This type of treatment is today at the center of many applications multimedia in smart visual surveillance, human computer interaction, unmanned vehicles and telerobotics.

The tracking corresponds to the estimation of the location of the object in each of the images of a video sequence, the camera and/or the object being able to be simultaneously in motion. The localization process is based on the recognition of the object of interest from a set of visual characteristics such as color, shape, speed, etc. There are many challenging issues which make the development of a tracking method [1] very difficult, the major difficulty is the object appearance changing and the background confusion.

Most tracking algorithm are based on a single cue to represent the target, however one cue can't help building a robust appearance model, thus we propose in this paper to use multiple cues to build a robust and stable appearance model therefore dealing the problem of appearance variations, indeed we combine IMM [2] and Mean Shift [3] named IMM-MS based on multiple cues, three observation model are adopted, the Mean Shift stage estimate the state on the system for several sub appearance model and the IMM then dynamically adjusts the weights of different cues.

© Springer International Publishing AG, part of Springer Nature 2018
M. Ben Ahmed and A. A. Boudhir (Eds.): SCAMS 2017, LNNS 37, pp. 288–297, 2018.
https://doi.org/10.1007/978-3-319-74500-8_26

The remainder of the paper is organized as follows: A short overview of the related work is exposed in Sect. 2, the appearance modeling is presented in Sect. 3. In Sect. 4, the IMM-MS tracking algorithm is presented. Experimental results are shown in Sect. 5. Finally, we conclude the paper in Sect. 6.

2   Related Work

In last decade many visual tracking methods have been proposed, which can be categorized in two classes generative or discriminative. The generative methods represent the object by appearance model that can be obviously convoluted by a kernel, then the tracking process seeks the candidate whose observed appearance model is most similar to that of the template, among the popular generative methods one can quote, kernel-based object tracking [4], particle filter [5], interactive multiple model particle filter [6], robust online appearance models for visual tracking [7]. The discriminative methods try to separate the tracked object from the background by a binary classifier, the most widespread discriminative methods include online selection of discriminative tracking features [8], object tracking using incremental 2d-lda learning and Bayes inference [9], ensemble tracking [10]. Thereafter we will be limited to briefly presenting the most related methods to our own, especially the methods based on mean shift algorithm, for more details the reader can refer to a detailed review in [11]. In [12] multiple reference histograms obtained from multiple available prior views of the target are adopted as the appearance model, the authors propose an extension to the mean shift tracker, where the convex hull of these histograms is used as the target model. The authors in [13] represent the appearance model of the template and the candidate using background weighted histogram and color weighted histogram, the proposed method termed as adaptive pyramid mean shift uses pyramid analysis with adaptive levels and scales for better stability and robustness. In [14] color feature and motion information are combined into the mean shift tracking algorithm framework, the use of both features simultaneously increase the performance of the the MS algorithm, and, the Kalman filter is used to solve full occlusion problems. An integration strategy of of mean shift and SIFT feature tracking is adopted in [15], the proposed approach uses the measurements from mean shift and SIFT to pursue a maximum likelihood estimate employing the expectation–maximization algorithm. The methods above adopted single feature to model the appearance, to achieve robust and stable tracking, the use of a single cue is never enough. In [16] the others propose a multi-cue integration strategy, indeed the multiple cue integration technique based Mean-Shift framework assures robustness under various conditions. In our work we adopt multiple cues to model the appearance of the tracked object into mean-shift algorithm combined with IMM, IMM-MS dynamically adjusts the probabilities of models of different features.

3   Target Appearance Modeling

The target area is defined by a rectangular region, and then we use three features to describe the target's window, the RGB color space is the most common color mode, the components R, G and B are highly dependent and therefore the chromatic information is far from being adapted for direct use, wherefore we convert the target's window to HSV color space, which is considered close to the human conception. Three feature vectors are computed the color correlogram (CCOR) the edge orientation (EOH) and the local binary pattern histogram (LBPH).

3.1 Color Correllogram

The color correllogram [17] has been proposed to describe not only the color distribution of the pixels, but also the spatial correlation between the color peers. The correlogram specifies the probability of finding a pixel at a distance d of a pixel i which have the same color. Let I be the set of pixels of an image and $I_c(i)$ the set of pixels whose color is c. Then the color correllogram is defined by:

$$\gamma_i^d = \Pr_{p1 \in Ic(i), p2 \in I}[p2 \in Ic(i), |p1 - p2| = d] \tag{1}$$

In order to reduce computation time, H is quantized to 6 levels and S and V is quantized to 3 levels, thus the correlogram is created for each quantized channel, then the three correlogram are fused to built one correlogram vector.

3.2 Edge Orientation Histogram

The edge orientation [18] can strongly represent both shape and texture structure, indeed constitutes an informative mass of the image. We divide the target's windows into adjacent areas of small size called cells, and for each cell we compute the edge orientation histogram CHEO, finally the combination of CHEO Eq. 7 histograms forms the EOH vector.

The orientations are uniformly quantized into m bins (m = 6), $\theta_{x,y}$ denotes the orientation of the pixel (x,y) given by Eq. 2.

$$\theta_{x,y} = \frac{1}{2} arctan \left[\frac{2g_{xy}}{g_{xx} - g_{yy}} \right] \tag{2}$$

$$G_{x,y} = \left(\frac{1}{2} \left((g_{xx} + g_{yy}) + (g_{xx} + g_{yy})cos2\theta_{x,y} + 2g_{xy}sin2\theta_{x,y} \right) \right)^{1/2} \tag{3}$$

And the value of the rate of change at (x, y) in the direction of $\theta_{x,y}$ given by Eq. 3 Where

$$g_{xx} = \left| \frac{\partial H}{\partial x} \right|^2 + \left| \frac{\partial S}{\partial x} \right|^2 + \left| \frac{\partial V}{\partial x} \right|^2 \tag{4}$$

$$g_{xx} = \left| \frac{\partial H}{\partial y} \right|^2 + \left| \frac{\partial S}{\partial y} \right|^2 + \left| \frac{\partial V}{\partial y} \right|^2 \tag{5}$$

$$g_{xx} = \left|\frac{\partial H}{\partial x}\right|\left|\frac{\partial H}{\partial y}\right| + \left|\frac{\partial S}{\partial x}\right|\left|\frac{\partial S}{\partial y}\right| + \left|\frac{\partial V}{\partial x}\right|\left|\frac{\partial V}{\partial y}\right| \tag{6}$$

Then the edge orientation histogram is given as follow

$$CH_{EO}^{\theta} = \sum_x \sum_y G_{x,y} \delta_{\theta}\left(\theta_{x,y}\right) \tag{7}$$

Where δ_{θ} is the Kronecker delta function which associates the pixel (x, y) to the histogram bin θ.

3.3   Local Binary Pattern Histogram

The local binary pattern (LBP) [19] is widely used in classification and recognition applications, the Local Binary Patterns (LBP) have since been widely used for texture analysis of images. It's simple but effective texture operators that give a label to the pixels of an image according to their neighborhoods and is invariant to the monotonic changes of illumination.

We consider the channel V, the LBP code is calculated in a neighborhood of 8 pixels, in a radius 1, we then build the local binary pattern histogram (LBPH) Eq. 10.

$$LBP(x, y) = \sum_{n=0}^{7} 2^n s\left(I_{x,y}^v - I_n^v\right) \tag{8}$$

$$s(u) = \begin{cases} 1 & \text{si } u > O \\ 0 & \text{sinon} \end{cases} \tag{9}$$

Where $I_{x,y}^v$ and I_n^v are respectively are the value of the level v of the pixel in position x, y and that of its neighborhood. Then the local binary pattern histogram is given as follows

$$LBPH_m = \sum_x \sum_y LBP(x, y) \delta_m\left(I_{x,y}^v\right) \tag{10}$$

Where δ_m is the Kronecker delta function which associates the pixel (x, y) to the histogram bin m.

4   IMM-MS Tracking Algorithm

Originally, the Mean Shift procedure is an iterative search procedure of local maximum in a space, based on a gradient rise, and was subsequently adopted for real-time tracking of objects, indeed using the color density of the object. Tracking is performed from its initial position in the first frame, the object to be followed is modeled by a rectangular window, on which one calculates its distribution of color. The initial color distribution is referenced as a model, and is then compared to that of the candidate sites to determine the most likely position in the next image. The Mean Shift tracking system maximizes the appearance similarity iteratively by comparing the model histograms and that of the candidate object, the similarity between two histograms is defined in terms of Bhatta-charya coefficient. At each iteration, the Mean Shift vector is calculated such that the similarity between the histograms is increased. This process is repeated until convergence is achieved. Our work is based on the use of three sub observation model associated to three features generally denoted M_j {$j = 1,2,3$}, and which corresponds respectively to the color correllogram, the edge orientation histogram and the local binary pattern histogram, Therefore the IMM then dynamically adjusts the weights of different features, then we combine the Mean Shift algorithm with the IMM, the flowchart of the proposed framework is shown in Fig. 1.

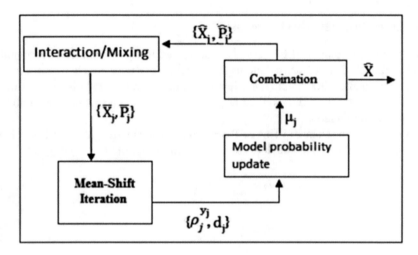

Fig. 1. The flowchart of the proposed framework.

We note that the IMM-MS has four stage, mixing/interaction stage, Mean Shift stage, mode probability update stage, and combination stage.

Step 1: interaction/mixing stage

The interacted mode state $\overline{X}_j(k-1/k-1)$ and its covariance $\overline{P}_j(k-1/k-1)$ are calculated by the mode probabilities $\{\mu_j(k-1)\}_{j=1}^{3}$ the mode transition probabilities p_{ij}, the mode state estimations $\hat{X}_j(k-1/k-1)$ and its covariance $\hat{P}_j(k-1/k-1)$. We have the following equations:

$$\overline{X}_j(k-1/k-1) = \sum_{i=1}^{3} \hat{X}_i(k-1/k-1)\mu_{i|j}(k-1) \tag{11}$$

$$\overline{P}_j(k-1/k-1) = \sum_{i=1}^{3} \left(\left(\hat{X}_i(k-1/k-1) - \overline{X}_j(k-1/k-1) \right) \left(\hat{X}_i(k-1/k-1) - \overline{X}_j(k-1/k-1) \right)^{T} + \hat{P}_i(k-1/k-1) \right) \mu_{i|j}(k-1) \tag{12}$$

$j = 1, 2, 3$ for the observation model

Where

$$\mu_{i|j}(k-1) = \frac{1}{c_j(k-1)} p_{ij} \mu_i(k-1) \tag{13}$$

is the mixing probability, and

$$c_j(k-1) = \sum_{i=1}^{M} p_{ij} \mu_i(k-1) \tag{14}$$

is the normalizing factor.

Step 2:   Mean Shift stage

We note $\left\{px_i\right\}_{i=1...nh}$ the set of coordinates of the pixels of the candidate centered on $\overline{X}_j$ denoted y_0 in the current frame.
The process of Mean Shift is described as follows:

i.   Compute the observation models associated with the state y_0
ii. Using the Bhattacharyya coefficient the distances between the template observation model and the target observation model are computed as follow:

$$\rho_j^{y_0} = \sum_{u=1}^{N_{bin}} \sqrt{M_j^{template} - M_j^{y_0}} \tag{15}$$

$$d_j = \sqrt{1 - \rho_j^{y_0}} \tag{16}$$

iii. Calculate weights $\left\{w_j^i\right\}_{i=1,..n_h}$ using

$$w_j^i = \sum_{u=1}^{N_{bin}} \delta\left(c(px_i) - u\right) \sqrt{\frac{M_j^{template}}{M_j^{y_0}}} \tag{17}$$

iv. From the Mean Shift vector, calculate the new state of the object

$$y_j = \frac{\sum_{i=1}^{n_h} x_i w_j^i g\left(\left\|\frac{y_0 - x_i}{h}\right\|^2\right)}{\sum_{i=1}^{n_h} w_j^i g\left(\left\|\frac{y_0 - x_i}{h}\right\|^2\right)} \tag{18}$$

Update $M_j^{y_j}$ then evaluate

$$\rho_j^{y_j} = \sum_{u=1}^{N_{bin}} \sqrt{M_j^{template} - M_j^{y_j}} \tag{19}$$

v. While $\rho_j^{y_j} < \rho_j^{y_0}$ do $y_j \leftarrow \dfrac{y_0 + y_j}{2}$

vi. If $\left\|y_j - y_0\right\| < \varepsilon$ stop, else $y_0 \leftarrow y_j$ and return to i step

vii.
$$\hat{X}_j(k/k) \leftarrow y_j \tag{20}$$

viii.
$$d_j = \sqrt{1 - \rho_j^{y_j}} \tag{21}$$

Step 3: Model probability update

The model probability of each observation model M_j^k is computed as

$$\mu_j(k) = \frac{\Lambda_j \sum_{i=1}^{M} p_{ij}\mu_i(k-1)}{\sum_{j=1}^{M} \Lambda_j \sum_{i=1}^{M} p_{ij}\mu_i(k-1)} \tag{22}$$

Where the likelihood function is calculated as

$$\Lambda_j = \frac{1}{\sqrt{2\pi}}\exp\left(-\frac{d_j^2}{2}\right) \tag{23}$$

Step 4: Combination

In this stage we combine the estimates associated with the three observation model by a weighted sum, and then the combined state estimate and its covariance are computed as

$$\hat{X}(k/k) = \sum_{j=1}^{M} \hat{X}_j(k/k)\mu_j(k), \tag{24}$$

$$\hat{P}\left(\frac{k}{k}\right) = \sum_{j=1}^{M} \mu_j(k) \\ \times \left\{ \hat{P}_j(k/k) + \left[\hat{X}_j(k/k) - \hat{X}(k/k)\right]\left[\hat{X}_j(k/k) - \hat{X}(k/k)\right]^{T} \right\} \tag{25}$$

5 Experimental Results

To test and evaluate the performance of our algorithm, we use two public test videos, they present challenging issues namely illumination changing and background clutters. We compare the obtained results by three methods, mean shift, particle filter, and our method, the two traditional methods of comparison are based on a single cue and uses the color histogram distribution to represent the appearance model. The first experiment, a man moves in an environment characterized by a variation of the illumination, some frame of tracking result are shown in Fig. 2, we can show that the proposed method tracks the Man efficiently despite the change of illumination in the area of the object while the other methods can't keep a close track which shows weakness and exposes a sensitivity to the challenging environment's condition.

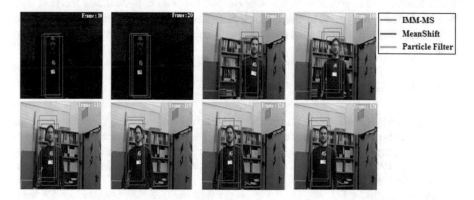

Fig. 2. Tracking results of the Soccer Man sequence

The second experiment CarDark, a car move in a challenging environment and background clutters, indeed the background near the target has the similar color or texture as the target, Fig. 3 exposes some frames of the experimental result of tracking, we can see that the environment issues makes tracking very difficult, indeed the comparison methods based on single cue perform a confused track throughout the video while the proposed algorithm track the target accurately and present better stability.

Fig. 3. Tracking results of the Soccer Man sequence

6 Conclusion

The IMM-MS is proposed in this paper, the Mean Shift algorithm is used under IMM framework, multiple cues are adopted namely color, edge and texture to represent the target's appearance, three appearance distribution are considered that is the color correllogram, the edge orientation histogram and Local binary pattern histogram. The Mean Shift stage estimate the state associated with each model and IMM is adopted to estimate the general state by combining the three appearance model. Using multiple cues to characterize the appearance of the target allows the tracker to be stable and robust to the illumination change and background clutters. The experimental results show that the proposed algorithm has superior tracking performance compared to classical tracking methods.

References

1. Li, X., Weiming, H., Shen, C., Zhang, Z., Dick, A., van den Hengel, A.: A survey of appearance models in visual object tracking. ACM Trans. Intell. Syst. Technol. **4**, 1–48 (2013)
2. Wu, S., Hong, L.: Hand tracking in a natural conversational environment by the interacting multiple model and probabilistic data association (IMM-PDA) algorithm. Pattern Recogn. **38**, 2143–2158 (2005)
3. Zhang, L., Zhang, D., Wu, C.: Robust mean-shift tracking with corrected background-weighted histogram. IET Comput. Vis. **6**, 62–69 (2012)
4. Comaniciu, D., Ramesh, V., Meer, P.: Kernel-based object tracking. IEEE Trans. Pattern Anal. Mach. Intell. **25**(5), 564–577 (2003)
5. Arulampalam, M., Maskell, S., Gordon, N., Clapp, T.: A tutorial on particle filters for online nonlinear/non-Gaussian Bayesian Tracking. IEEE Trans. Signal Process. **50**(2), 174–189 (2002)
6. Dou, J., Li, J.: Robust visual tracking based on interactive multiple model particle filter by integrating multiple cues. Neurocomputing **135**, 118–129 (2014)
7. Jepson, A.D., Fleet, D.J., El-Maraghi, T.F.: Robust online appearance models for visual tracking. In: CVPR (2001)
8. Collins, R.T., Liu, Y., Leordeanu, M.: Online selection of discriminative tracking features. IEEE Trans. Pattern Anal. Mach. Intell. **27**(10), 1631–1643 (2004)
9. Li, G., Liang, D., Huang, Q., Jiang, S., Gao, W.: Object tracking using incremental 2D-LDA learning and Bayes inference. In: ICIP (2008)
10. Avidan, S.: Ensemble tracking. In: CVPR (2005)
11. Wu, Y., Lim, J., Yang, M.-H.: Object tracking benchmark. IEEE Trans. Pattern Anal. Mach. Intell. **PP**(99) (2015)
12. Leichter, I., Lindenbaum, M., Rivlin, E.: Mean Shift tracking with multiple reference color histograms. Comput. Vis. Image Underst. **114**, 400–408 (2010)
13. Li, S.X., Chang, H.X., Zhu, C.F.: Adaptive pyramid mean shift for global real-time visual tracking. Image Vis. Comput. **28**(3), 424–437 (2010)
14. Mazinan, A.H., Amir-Latifi, A.: Applying mean shift, motion information and Kalman filtering approaches to object tracking. ISA Trans. **51**, 485–497 (2012)
15. Zhou, H., Yuan, Y., Shi, C.: Object tracking using SIFT features and mean shift. Comput. Vis. Image Underst. **113**, 345–352 (2009)
16. Liu, H., Ze, Y., Zha, H., Zou, Y., Zhang, L.: Robust human tracking based on multi-cue integration and mean-shift. Pattern Recogn. Lett. **30**, 827–837 (2009)
17. Soni, D., Mathai, K.J.: An efficient content based image retrieval system based on color space approach using color histogram and color correlogram. In: Fifth International Conference on Communication Systems and Network Technologies, pp. 488–492 (2015)

18. Vikhar, P., Karde, P.: Improved CBIR system using Edge Histogram Descriptor (EHD) and Support Vector Machine (SVM). In: International Conference on ICT in Business Industry & Government (ICTBIG), pp 1–5 (2016)
19. Kalakech, M., Porebski, A., Vandenbroucke, N., Hamad, D.: A new LBP histogram selection score for color texture classification. In: International Conference on Image Processing Theory, Tools and Applications (IPTA), pp. 242–247 (2015)

A New Task Scheduling Algorithm for Improving Tasks Execution Time in Cloud Computing

Naoufal Er-raji[✉], Faouzia Benabbou, and Ahmed Eddaoui

Laboratory of Modeling and Information Technology, Faculty of Sciences Ben M'sik,
University Hassan II, Casablanca, Morocco
na.erraji@gmail.com, faouziabenabbou@yahoo.fr,
ahmed_eddaoui@yahoo.fr

Abstract. The increasing demand in computing resources led companies as well as researchers to adopt new technologies and cloud computing is one of them. It consists of a collection of VMs (Virtual Machines) that are created under CPR (Cloud Provider Resources). Cloud Computing is still faces many challenges and task scheduling is one of them which have a very important role in determining the efficient tasks execution. So tasks should be scheduled efficiently such that the execution time can be reduced. Thus, to outperform this problem there is a need to implement a good task scheduling algorithm. In this paper, we address the problem of task scheduling through proposing a new task scheduling algorithm for efficient tasks execution. The proposed algorithm are tested using CloudSim simulator and the obtain result shows that the proposed algorithm gives better performance.

Keywords: Cloud computing · CloudSim · Mips · Task length
Task scheduling algorithm · Virtual Machine

1 Introduction

As many IT (Information Technology) giant definitions, the International Organization for Standardization defined the cloud computing as a paradigm for enabling network access to a scalable and elastic pool of shareable physical or virtual resources with self-service provisioning and administration on-demand. The cloud computing paradigm is composed of key characteristics, cloud computing roles and activities, cloud capabilities types, cloud service categories and cloud deployment models [1–4].

As depicted in the Fig. 1, in the cloud computing the users submit their tasks so as to be executed as soon as possible in the available shared pool of resource. In each cloud provider datacenter there is a component that control and manage the tasks scheduling in different resources (VMs).

© Springer International Publishing AG, part of Springer Nature 2018
M. Ben Ahmed and A. A. Boudhir (Eds.): SCAMS 2017, LNNS 37, pp. 298–304, 2018.
https://doi.org/10.1007/978-3-319-74500-8_27

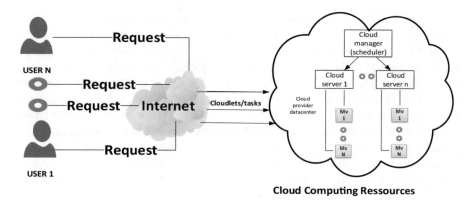

Fig. 1. Simple cloud architecture

Due to the increasing demand of cloud resources and the limited resources of CSP (Cloud Service Provider), this raises several challenges, and task scheduling is one of them. The task scheduling, is a process of choosing the best suitable available resources for execution the tasks or to allocate computer machines to tasks in such a manner that the execution time is minimized as possible [5, 6].

In this paper, we improve the tasks execution time by using the task length and the VM Mips (Million Instructions per Second) and the number of tasks per VM. The main objective of this paper is to minimize the tasks execution time so as to respect the QoS that are documented in the SLA.

The rest of the paper is organized as follows: Sect. 2 reviews the related works; Sect. 3 describes the proposed algorithm, Sect. 4 present experiments and results analysis. At last in Sect. 5 we make a conclusion and future works.

2   Related Work

Improving task scheduling in the cloud computing have been extensively studied. In this section, we present some works that take task scheduling in cloud environment as objective for their contributions.

In papers [7, 12] the proposed algorithms assign priority to different tasks according to task size such that one having highest size has the highest priority. Based on this priority the tasks are assigned to VMs. Saxena et al. [8] proposed an algorithm, which classify and group all tasks according to their deadline and cost constraints, and assign them to different priority queue (high, mid, low). Based on this priority queue the tasks are assigned to VMs. Sharma and Sharma [9] proposed a technique, which is based on credit system. Each task is assigned a unique credit based upon three parameters namely: Task Length Credit, Task priority credit, Task deadline credit. Based on these credits the tasks are assigned to VM. Lakraa et al. [10] proposed an algorithm, which assigns priority to different tasks according to the QoS (Quality of Service) of request task. High QoS task assigned with low QoS value and the low QoS task assigned with high QoS value. Hence, the task with lower QoS value is a high priority and the task with high

QoS value is a low priority. Based on these priorities the tasks are assigned to VMs. Kaur and Singh [11] proposed an algorithm, which assigns priority to different tasks according to specific attribute: User Level, Task urgency, Task Load and Time queuing up. Then, tasks are arranged in a sorted order by considering the calculated priority. Thus, the task with higher priority scheduled first. Based on this sort the tasks are assigned to VMs. Ijaz et al. [13] proposed a strategy, which assigns priority to different tasks according to the ALST (Absolute Latest Start Time) such that the one having minimal ALST among all the tasks has higher priority. Based on these priorities the tasks are assigned to VMs. Ghanbari and Othman [14] proposed an algorithm, which is based on the theory of a mathematical model named AHP (Analytical Hierarchy Process) for calculating the priority. It is a MCDM (Multi-Criteria Decision-Making) and MCDM (Multi-Attribute Decision-Making model). The proposed architecture is consisted of three levels namely: objective level attributes level and alternatives level. Based on these levels the tasks are assigned to VMs.

In the literature, there are many contributions. Which, make a comparison of approaches [3, 15–20].

3 Proposed Algorithm

In the literature the task is a single process or multiple processes which will be executed on a compute node presented by VM [21] in the cloud computing and, the task scheduling is a set of rules and parameters that maps the users' tasks on the multiple VMS so as to the completion time is minimized as possible [22]. In the other side the task scheduling algorithms are the responsible of distributing tasks submitted by users onto available resources (VMs) [23].

In this paper, we used the Quicksort as a sorting algorithm which quickly sort items by dividing a large array into two smaller arrays [24]. In the following the steps of our algorithm.

Algorithm: The Proposed Task Scheduling Algorithm

Receive the Tasks List
Receive the VMS List
Quicksort (Sort descending the VMs according to their Mips)
Quicksort (Sort descending the Tasks according to their Length)
NTPM (Nombre of Tasks Per Machine) = Tasks list size / VMs list size
Vmindex = 0
For all Tasks in the list of sorted Tasks do
For all VMS in the list of sorted VMS do
Schedule the current task to the VM with Vmindex as VM id;
If the current task index % NTPM = 0 and the current task index ! = 0 then
Vmindex = (Vmindex +1) % The VMs list size
End if
End for
End for

---

4   Experiments and Result Analysis

In order to verify our algorithm, we conducted experiment on Intel Core i5 Processor, Windows 7 platform and using CloudSim 3.0.3 simulator. The CloudSim is a framework developed by the GRIDS laboratory at the University of Melbourne. It provides basic classes for describing datacenter, VMs, users, computational resources, and policies for the management of different parts of the system [25]. The major benefit of simulation is that it allows cloud developers to test the performance of various scheduling policies in a controllable environment, before these policies are actually deployed in real cloud. This enables the developers to rectify their faults at an earlier stage.

We assume that CloudSim toolkit has been deployed in one datacenter having one host and thirty VMs. In the other side we have created 600 tasks that have to be executed in these VMs. The parameters setting of cloud simulator are shown in the following table (Table 1).

Table 1. Simulation configuration

Datacenter	Host	Virtual machine	Task
Arch: X86	Ram: 20480	Mips: 50–1000	FileSize: 300
OS: Linux	Storage: 1000000	Size: 1000	OutputSize: 300
VMM: Xen	BW: 1000000	Ram: 512	Length: 100–1000
Time-Zone: 10.0	VM_Scheduler: Space Shared	BW: 1000	
	PesNumber: 1	Deadline: 1–1000	
	VMM: Xen	Utilization Model: Full	
	Task Scheduler: Space-Shared		

In order to test the efficiency of our algorithm, FIFO (simplest scheduling algorithm, which works as the first task coming in the queue is executed first), Max-Min (simplest scheduling algorithm, which works as the task that has the maximum length will be assigned to the VM that has the minimum Mips), Max-Max (simplest scheduling algorithm which works as the task that has the maximum length will be assigned to the VM that has the maximum Mips) and the proposed algorithm are tested for their behavior by proposing different length for each task and different Mips for each VM. Based on these algorithms the tasks will be allocated to VMS. For this experiment we suppose that the tasks are coming in the same time (0.1).

The Experiment is conducted by varying the number of tasks like 100, 200, 300, 400, 500 and 600 respectively. The obtained results are shown in the following:

In order to simplify the comparison, we have based on the subtraction between different mentioned algorithms Total Execution Time and the proposed algorithm Total Execution Time, in the following the results obtained.

As shown in the Fig. 2, we found out that the proposed algorithm has minimum Total Execution Time along the tasks numbers. Also, in the Fig. 3 we found out that the difference between the Total Execution Times increases by adding more tasks. Thus, our algorithm is more efficient for the applications that have many tasks to be executed in the determinate deadline. Moreover, our algorithm can also improve the applications QOS by giving the application results in minimum time. Furthermore, this paper shows that CloudSim is very important to validate the research work in the cloud computing field.

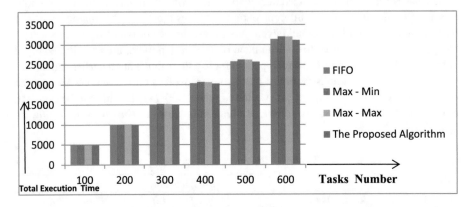

Fig. 2. Algorithms total execution time

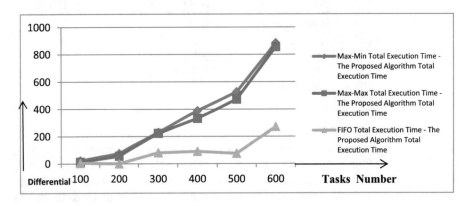

Fig. 3. Algorithms differential total execution time

5   Conclusion and Future Works

With the continuous growth of cloud computing, the task scheduling becomes a big challenge. So as to satisfy both the CSC as well as the CSP, efficient task scheduling algorithm is required. In this paper, we have addressed the problem of task scheduling by improving tasks execution time. The experimental results show that the proposed algorithm give minimum total tasks execution time. Finally, as future work, our research continues by enhancing this contribution as to add VM classification, task migration. Also, make in consideration the dependent tasks.

References

1. International Organization for Standardization "Information technology—Cloud computing —Overview and vocabulary" (2014)
2. Armbrust, M., Fox, A., Griffith, R., Joseph, A.D., Katz, R., Konwinski, A., Lee, G., Patterson, D., Rabkin, A., Stoica, I., Zaharia, M.: A view of cloud computing. Commun. ACM **53**(4), 50–58 (2010)
3. Singh, A., Kaur, I.: A survey on cloud computing and various scheduling algorithms. Int. J. Adv. Res. Comput. Sci. Manag. Stud. (IJARCSMS) **4**(2) (2016)
4. Definition of Cloud Computing. https://www.nist.gov/programs-projects/cloud-computing. Accessed 07 Mar 2017
5. Singh, R.M., Paul, S., Kumar, A.: Task scheduling in cloud computing: review. Int. J. Comput. Sci. Inf. Technol. (IJCSIT) **5**(6), 7940–7944 (2014)
6. Kaur, A., Kaur, U.: A survey for task scheduling in cloud computing. Int. J. Adv. Res. Comput. Sci. Softw. Eng. (IJARCSSE) **6**(5) (2016)
7. Lawanya Shri, M., Benjula Anbumalar, M.B., Santhi, K., Deepa, M.: Task scheduling based on efficient optimal algorithm in cloud computing environment. In: International Conference on Recent Research Development in Science, Engineering and Management (ICRRDSEM), May 2016
8. Saxena, D., Chauhan, R.K., Kait, R.: Dynamic fair priority optimization task scheduling algorithm in cloud computing: concepts and implementations. Int. J. Comput. Netw. Inf. Secur. (IJCNIS) **8**, 41 (2016)
9. Sharma, A., Sharma, S.: Credit based scheduling using deadline in cloud computing environment. Int. J. Innov. Res. Comput. Commun. Eng. (IJIRCCE) **4**(2) (2016)
10. Lakraa, A.V., Yadavb, D.K.: Multi-objective tasks scheduling algorithm for cloud computing throughput optimization. In: International Conference on Intelligent Computing, Communication & Convergence (ICCC) (2015)
11. Kaur, P., Singh, P.: Priority based scheduling algorithm with fast task completion rate in cloud. In: Advances in Computer Science and Information Technology (ACSIT), vol. 2, no. 10, April–June 2015
12. Agarwal, A., Jain, S.: Efficient optimal algorithm of task scheduling in cloud computing environment. Int. J. Comput. Trends Technol. (IJCTT) **9**(7), 344–349 (2014)
13. Ijaz, S., Munir, E.U., Anwar, W., Nasir, W.: Efficient scheduling strategy for task graphs in heterogeneous computing environment. Int. Arab J. Inf. Technol. (IAJIT) **10**(5), 486–492 (2013)
14. Ghanbari, S., Othman, M.: A priority based job scheduling algorithm in cloud computing. In: International Conference on Advances Science and Contemporary Engineering (ICASCE) (2012)
15. Sidhu, H.S.S.: Comparative analysis of scheduling algorithms of cloudsim in cloud computing. Int. J. Comput. Appl. **97**(16) (2014). (0975 – 8887)
16. Er-raji, N., Benabbou, F., Eddaoui, A.: Task scheduling algorithms in the cloud computing environment: survey and solutions. Int. J. Adv. Res. Comput. Sci. Softw. Eng. (IJARCSSE) **6**(1) (2016)
17. Jeyalakshmi, S., Sankarram, N.: Scheduling algorithms for cloud computing environments and research issues. J. Appl. Sci. Res. (JASR) **12**(3), 50–53 (2016)
18. Thaman, J., Singh, M.: Current perspective in task scheduling techniques in cloud computing: a review. Int. J. Found. Comput. Sci. Technol. (IJFCST) **6**(1), 65–85 (2016)
19. Singh, R., Agnihotri, E.M.: A review of cloud computing scheduling strategies. Int. J. Eng. Trends Appl. (IJETA) **3**(4) (2016)

20. Tilak, S., Patil, D.: A survey of various scheduling algorithms in cloud environment. Int. J. Eng. Inventions (IJEI) **1**(2), 36–39 (2012)
21. Patel, S., Bhoi, U.: Priority based job scheduling techniques in cloud computing: a systematic review. Int. J. Sci. Technol. Res. (IJSTR) **2**(11), 147–152 (2013)
22. Definition of Tasks. https://msdn.microsoft.com/en-us/library/bb525214(v=vs.85).aspx. Accessed viewed 07 Dec 2016
23. Akilandeswari, P., Srimathi, H.: Survey and analysis on task scheduling in cloud environment. Indian J. Sci. Technol. (INDJST) **9**(37) (2016)
24. Hoare, C.A.R.: Quicksort. Comput. J. **5**(1) (1962)
25. Buyya, R., Ranjan, R., Calheiros, R.N.: Modeling and Simulation of Scalable Cloud Computing Environments and the CloudSim Toolkit: Challenges and Opportunities (2009)

Big Data: Methods, Prospects, Techniques

Lamrani Kaoutar[✉], Abderrahim Ghadi, and Florent Kunalè Kudagba

List Laboratory, Faculty of Sciences and Techniques,
University Abdelmalek Essaadi, Tangier, Morocco
Kaoutar.lamrani1@gmail.com, ghadi05@gmail.com, kkunale@gmail.com

Abstract. Nowadays, Web content knows a rapid increase in syntactic data that makes their processing and storage difficult in classical systems. An alternative approach is to represent the Web in a more understandable form by the machines based on the initiative of the semantic web, on the new technologies and algorithms existing in parallelism, cloud computing, distributed systems and big data mining. These new intelligent techniques allow us to give new representations to the sources of the Web. Our research will develop around the semantic search of information on a set of massive, distributed, autonomous and heterogeneous Resource Description Framework (RDF) data. However, only a representation format of knowledge for their semantic access is not sufficient and we need strong response mechanisms to efficiently handle global and distributed queries on a set of RDF data marked by the dynamics and scalability of their content.

Keywords: Domain ontology · E-commerce · E-catalogues · Semantic web
Distributed queries · RDF · Cloud computing · Big data

1   Introduction

In our daily life massive amounts of data in e-commerce area are produced. Large volumes of commercial information in the web are recorded every day. More than the size of this big commercial data, the Variety, Velocity and Veracity characteristics are added to this data [1]. Whish mean that this wide variety of structured, semi-structured, and unstructured data needs a real-time processing, a new model of storage, efficient machine learning and data mining techniques.

The user of these e-commercial big data makes complex and multiple requests on his heterogeneous, distributed and autonomous sources to have information relevant to his search. This data is presented in the context of the semantic web described by the standard model RDF (Resource Description Framework) [2] presented by subject-predicate-object expression format and can be interpreted as a graph in which the subjects and objects are nodes and predicates are the edge of the graph.

In the web an architecture that plays the role of a virtual intermediate interface between the user request and between this big data is very important. This architecture which is going to be based on the cloud computing infrastructure illustrate to the user that he interrogate a heterogeneous and centralized data source by avoiding him to know the

© Springer International Publishing AG, part of Springer Nature 2018
M. Ben Ahmed and A. A. Boudhir (Eds.): SCAMS 2017, LNNS 37, pp. 305–312, 2018.
https://doi.org/10.1007/978-3-319-74500-8_28

relevant sources of his request, avoiding him to search the sources one by one with the different data forms and combining to him answers obtained from this different sources.

The big data processing of user request with the specific terms in this web intermediate architecture based cloud computing give an efficient quality of service in the search of information and a complete, coherent and optimal answer remains from several sources. Different methods and techniques of big data framework are presented in this paper to have an overview how to using this parallelism and distributed systems [3] to maintain and mapping RDF schema related to massive and distributed documents in commercial e-catalogue area.

The remainder of this paper is structured as follows. Section 2.1 broad an overview of big data mining methods. Section 2.2 discuss the big data processing methods. In Sect. 2.3 we discuss processing technologies. For the Sect. 2.4 an overview of Cloud computing infrastructure will be presented. Finally in Sect. 3 we present and discuss our thesis.

2   Related Works

2.1   Big Data Mining Methods

Distributed based data parallelism. At the current methodologies of data mining using a single personal computer to execute the data tasks become impossible with the massive amount of data produced in our time. For that it is necessary to use powerful computing environments to process this big data like parallelism and cloud computing framework. a distributed system are used to divide the problem into many tasks, each tasks is solved by using one or more processors that run concurrently in parallel and communicate with each other by message passing.

The main idea of data parallelism paradigm [4] is to divide a given large dataset D into in subsets (D1, D2 ... Dn) where each subset may or may not have duplicate data samples. Then, a specific data mining algorithm implemented in n local machines (or computer nodes) individually is performed over each subset. Finally, the n mining results are combined via one combination component to produce the final output. Differing from the baseline procedure how use the support vector machine (SVM) classification problem based dataset in a single machine. The distributed by data mining procedure borrows the divide-and-conquer principle where a given dataset is divided into n subsets for n computer nodes and the SVM algorithm is implemented in each computer node.

MapReduce based cloud computing. The MapReduce framework [4] is implemented in the cloud computing environment it is considered like a new generation of programming system for the parallel processing purpose. It is usually implemented in an open source software framework namely, apache Hadoop for distributed storage and processing over large scale datasets on computer clusters.

The MapReduce framework [4] is based on the following procedures. Firstly given a user-defined map function M, each map function turns each chunk of input data into a sequence of initial key-value pairs simultaneously in parallel on different local machines. Then, the map functions process these input data to produce a set of intermediate key-value pairs, which are collected by a master controller. Specifically, all intermediate values are grouped together, are associated with the same keys and are passed to the same machines. Next, the user-defined reduce function works on one key at a time, and combine all the values associated with that key to produce a possibly smaller set of values resulting in the final key-value pairs as the output.

2.2 Big Data Processing Methods

Hashing. In the hashing the search is made by index, the index search perform the entire search on the disk to retrieve a block. The hashing technique doesn't use an index structure to retrieve data from disk but he use a hash function to compute the location of the desired data on the disk [5]. Hash function 'h' is a mapping function that takes a value as an input and converts this value to a key (k). The value of k indicates where the data are placed. Hash files store the data in a bucket format who usually stores one disk block. Static and dynamic hashing are the two types of hashing. In static hashing, the hash function always computes the same address when a search key value is provided. The number of buckets remains the same for this type of hashing. Insertion, deletion, and search are all performed in static hashing. A problem arises when data quickly increase and buckets do not dynamically shrink. For that in dynamic hashing, the buckets are dynamically added and removed on demand also it performs querying, insertion, deletion, and update functions.

Indexing. Indexing approaches used to locate a data from large amount of complex dataset. In this context, various indexing procedures are used such as semantic indexing based approaches, file indexing, r-tree indexing, compact Steiner tree, and bitmap indexing [6]. The only problem with most of these indexing approaches is high retrieval cost. The development of efficient indexing techniques is a very popular research area at present and several new indexing schemes, such as VegaIndexer, sksOpen, CINTIA, IndexedFCP, and pLSM have been proposed for big data storage [6]. Although the new indexing schemes are helpful for big data storage, these schemas are in their infant stage.

Bloom filter. A bloom filter allows efficient dataset storage at the cost of the probability of a false positive based on membership queries [7, 8]. A bloom filter helps in performing a set membership tests and determining whether an element is a member of a particular set or not. False positives are possible, whereas false negatives are not. The bit vector is utilized as the data structure of bloom filters. Independent hash functions, including murmur, fnv series of hashes, and Jenkins hashes, are employed in bloom filters.

Parallel computing. Parallel computing utilizes several resources at a time to complete a task [9]. For big data, Hadoop provides the infrastructure for parallel computing in a distributed manner. Hadoop helps improve processing power by sharing the same data file among multiple servers. A complex problem is divided into multiple parts through parallel computing. Each part is then processed concurrently. The different forms of parallel computing include bit and instruction levels and task parallelism. Task parallelism helps achieve high performance for large-scale datasets. In parallel computing, multi-core and multiprocessor computers consist of multiple processing elements within a single machine. By contrast, clusters, MPPs, and grids use multiple computers to work on the same task.

2.3 Processing Technologies

Batch processing Framework. Apache Hadoop technologies is based in different field to process large amounts of data and performing the processing of data intensive applications [10] by using Map/Reduce programming model [11]. To process data at high speed Skytree Servec is utilized [12]. Tableau technologies also used to process large amount of datasets by using Tableau desktop, tableau public, and tableau server [13]. Karmasphere one of technologies performing business analysis to improve the parallel and distributed programs and scale up the capability of processing from a small to a large number of nodes [10]. Dryad technology based on data follow graph processing and consists of a cluster of computing nodes and a computer cluster to run programs in a distributed manner [14]. Pentaho is utilized to generate reports from a large volume of structured and unstructured data [15]. Last one is Talend Open Studio provides a graphical environment to conduct an analysis for big data applications.

Stream processing Framework. Storm, Splunk, S4, SAP Hana, SQLstream-s-Server and Apache Kafka technologies [16] are all used to process large amounts of data and managing this data in real time. For managing this large amounts of streaming data through in-memory analytics for decision-making apache kafka are used. A storm used to perform real time processing of massive amounts of data. Splunk utilized to capture indexes and correlates real-time data with the aim of generating reports, alerts, and visualizations from the repositories. SAP Hana technology provide real time analysis of business process. S4 is a general purpose and pluggable platform utilized to process unbounded data streams efficiently in a distributed, scalable and partially fault-tolerant system. SQLstream-s-server is also a platform to analyze a large volume of services and log files data in real-time.

2.4 Cloud Computing Infrastructure

Cloud computing is defined as a new technology enable the realization of a new computing model for the processing, storage technologies and for cheaper and powerful resources. The growing of this resource in the context of the semantic web which the data take the form of distributed RDF triple stores push this new computing model to use for her storing and querying a Hadoop/MapReduce paradigm in distributed file system.

Hadoop distributed file system (HDFS) [17] consider as the core component of Hadoop with not fully POSIX-compliant. It is used to store huge data sets from a large number of sources by abstraction of physical storage architecture in aims to give the user the illusion of querying a single and a centralized data source. HDFS allow two type of storage, the storage of file system metadata in main memory and application data on secondary storage. The former are stored on a namenode server and the latter on datanode server. All servers use TCP-based protocol to communicate with each other. HDFS known by its isolation property and carry out a chain of tasks to read a file. But it can serve all type of application it is limited only for batch processing such as MapReduce. An alternative of HDFS with the same design goal and architecture is Quantcast file system [18]. It consists of a single master node and thousands of datanodes. The metadata is stored in the memory of the master server and the application data stored on the disks of datanodes.

One of the ancestors of HDFS is the architecture of Ceph [19] considered one of the distributed file system has a cluster of metadata servers (MDS) which manages the namespace coherently And uses a dynamic subtree partitioning algorithm [20] in order to map and localizing the metadata on MDSs. Also GFS [21] is evolving into a distributed namespace implementation. It is a proprietary version of HDFS developed by Google, here architecture provide scalability and an efficient access to data by using a large cluster of commodity servers.

Lustre file system [22] is one of distributed file system enable the scalability for systems with an architecture that provide a height storage of clusters namespace on its roadmap for Lustre 2.2 release, composed of three component: a metadata serve which store metadata in a specific names and directories in the files, an object storage servers (OSS) and a file system client. IBM's General purpose file system (GPFS) [23] more than his character of distributed file system, it provide a set of feature of file management system include the virtualization storage, high availability, and automated storage management of a very large quantities of file data. Tahoe least-authority file system [24] provides a storage grid and designed to give secure and long term storage.

B-tree File system maintains metadata and application data, enable storage for small files, give an efficient writable snapshot, a compression and an online resize. The Blob-Seer file system [25] enables storage and accessing of a huge amount of binary data objects. It Known by his storing data as binary objects instead of actual data essentially by abstracting the data as a long bit sequences, which enables fine grain access and processing.

The Gfarm file system [26] is developed for Grid Data Farm architecture which supports high-performance distributed parallel computing for processing a group of files by a single program. XTreemFS [27] it is a distributed parallel file system developed mainly for grid computing environment. XtreemFS is an object-based file system. It consists of clients, object storage devices (OSDs), and metadata servers.

3   Proposed Work

The rapid development of technology produces a very large amount of data in multiple sources in e-commerce area accessible on the web. To manage this amount of data it is necessary to carry out complex and multiple queries through autonomous, heterogeneous and distributed data sources. To do this we propose an architecture based in cloud computing that represents the intermediary between the request of the user and between the large numbers of e-commercial data. This architecture gives to the user the possibility to request a homogeneous and centralized system in order to give the user a coherent response which can be combined from different sources (see Fig. 1).

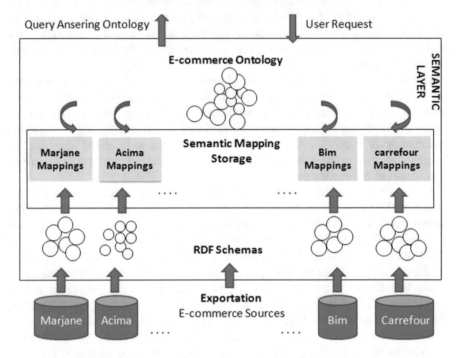

Fig. 1. Architecture of semantic mappings for distributed system.

In our thesis we proposed in first time to capitalize and describe knowledge of the e-commerce domain semantically based on the logic of description, the OWL-DL semantic language and the RDF-XML serialization. After we proposed a semi-automatic discovery strategy for semantic similarities between the global schemas presented by OWL-DL ontology and between the local schemas presented by the RDF schema. This semantic correspondence discovery based on a semantic technique that solve the problem of synonymous is a complements of terminology technique and structural technique, the first one calculates the similarity from the names associates to the concept of two ontologies, the second one the similarity between two entities is based on the taxonomic schema To solve the problem of homonyms. The semantic matches validated and detected by this strategy are stored in the cloud computing architecture to enabling user query accessing a centralized data.

This semantic knowledge presented previously including the semantic correspondence and the ontology are used to rewrite the global query send by the user to a form of a coherent and optimal combination of local queries that return partial responses. Query rewriting is an operation of reformulating queries to a mediator based on terms defined within domain ontology. The new query formulation is decomposed into as many relevant sources as our own integrative platform arranged to return computed partial responses. To search many relevant resources in sources we proposed to use processing data methods and a stream processing framework to have a real time response.

To resume, in our thesis the rewriting of requests in a logical and semantic framework is a delicate task for the data integration process which allows the user to send a global request of a set of terms that are presented in a global schema of our architecture for that a quasi-equivalent representation in the terms used in the local schemes is necessary. In general several sources of data can be used to respond to the same portion of the global query that is expressed as a conjunction of atomic constraints to be satisfied.

4 Conclusion

To benefit from the massive growth of semantic web, it is necessary to organize and manage data effectively to give an effective query response to the user request. For that various methods have been introduced and the fundamental purpose of these methods is to responding to the user queries in real time. in this paper we have present a different technique utilized for big data mining to describe a parallel computing environment and distributed system and how processing data to manage and analysis a various, velocity and voluminous data in this distributed and parallel systems, also the presence of these techniques lead to categorizes two kind of processing technology batch and stream technologies how support the cloud computing architecture.

References

1. Mayer-Schonberger, V., Cukier, K.: Big Data: A Revolution That Will Transform How We Live, Work, and Think. Eamon Dolan/Mariner Books (2014)
2. W3C. Rdf - semantic web standards. https://www.w3.org/TR/2014/NOTE-rdf11-primer-20140225/
3. Big data mining with parallel computing: a comparison of distributed and MapReduce methodologies. https://doi.org/10.1016/j.jss.2016.09.007
4. Tsai, C.-F., Lin, W.-C., Ke, S.-W.: Big data mining with parallel computing: a comparison of distributed and MapReduce methodologies. J. Syst. Softw. (2016). https://doi.org/10.1016/j.jss.2016.09.007
5. Odom, P.S., Massey, M.J.: Tiered hashing for data access. Google Patents (2003)
6. Gani, A., Siddiqa, A., Shamshirband, S., Hanum, F.: A survey on indexing techniques for big data: taxonomy and performance evaluation. Knowl. Inf. Syst. 46(2), 241–284 (2016)
7. Song, H., Dharmapurikar, S., Turner, J., Lockwood, J.: Fast hash table lookup using extended bloom filter: an aid to network processing. ACMSIGCOMM Comput. Commun. Rev. 35(4), 181–192 (2005)
8. Bloom, B.H.: Space/time trade-offs in hash coding with allowable errors. Commun. ACM 13(7), 422–426 (1970)
9. Richtárik, P., Takáč, M.: Parallel coordinate descent methods for big data optimization. arXiv preprint arXiv:1212.0873 (2012)
10. Shang, W., Jiang, Z.M., Hemmati, H., Adams, B., Hassan, A.E., Martin, P.: Assisting developers of big data analytics applications when deploying on hadoop clouds. In: Proceedings of the 2013 International Conference on Software Engineering, pp. 402–411. IEEE Press (2013)

11. Thusoo, A., Sarma, J.S., Jain, N., Shao, Z., Chakka, P., Anthony, S., et al.: Hive: a warehousing solution over a map-reduce framework. Proc. VLDB Endow. **2**(2), 1626–1629 (2009)
12. Han, J., Haihong, E., Le, G., Du, J.: Survey on NoSQL database. In: 2011 6th International Conference on Pervasive Computing and Applications (ICPCA), pp. 363–366. IEEE (2011)
13. Goranko, V., Kyrilov, A., Shkatov, D.: Tableau tool for testing satisfiability in LTL: implementation and experimental analysis. Electron. Notes Theor. Comput. Sci. **262**, 113–125 (2010)
14. Chen, C.P., Zhang, C.-Y.: Data-intensive applications, challenges, techniques and technologies: a survey on big data. Inf. Sci. **275**, 314–347 (2014)
15. Russom, P.: Big data analytics. In: TDWI Best Practices Report. Fourth Quarter (2011)
16. Big data: from beginning to future. http://doi.org/10.1016/j.ijinfomgt.2016.07.009
17. Shvachko, K., Kuang, H., Radia, S., Chansler, R.: The hadoop distributed file system. In: The 26th IEEE Symposium on Mass Storage System and Technologies (2010)
18. Ovsiannikov, M., Rus, S., Reeves, D., Sutter, P., Rao, S., Kelly, J.: The Quantcast file system. Proc. VLDB Endow. **6**(11), 1092–1101 (2013)
19. Weil, S.A., Brandt, S.A., Miller, E.L., Long, D.D.E., Maltzahn, C.: Ceph: a scalable, high performance distributed file system. In: Proceedings of the 7th Symposium on Operating Systems Design and Implementation (OSDI), pp. 307–320 (2006)
20. Weil, S.A., Pollack, K.T., Brandt, S.A., Miller, E.L.: Dynamic metadata management for petabyte-scale file systems. In: Proceedings of the 2004 ACM/IEEE Conference on Supercomputing, SC 2004, Washington, DC, USA, p. 4. IEEE Computer Society (2004)
21. Ghemawat, S., Gobioff, H., Leung, S.-T.: The Google file system. In: Peterson, L. (ed.) Proceedings of the Nineteenth ACM Symposium on Operating Systems Principles, October 2003, pp. 29–43. ACM, New York (2003)
22. Cluster File System Inc.: Lustre: a scalable, high-performance file system—White Paper. Cluster File Systems, Inc. (2002)
23. Fadden, S.: IBM general purpose file system—a White Paper (2012)
24. Wilcox-O'Hearn, Z., Warner, B.: Tahoe: the least-authority filesystem. In: Proceedings of the 4th ACM International Workshop on Storage Security and Survivability, StorageSS 2008, New York, NY, USA, pp. 21–26. Association for Computing Machinery (2008)
25. Nicolae, B., Antoniu, G., Bougé, L.: BlobSeer: how to enable efficient versioning for large object storage under heavy access concurrency. In: Proceedings of the 2009 EDBT/ICDT Workshops, New York, NY, USA, pp. 18–25. Association for Computing Machinery (2009)
26. Osamu, T., Hiraga, K., Soda, N.: Gfarm grid file system. New Gener. Comput. **28**(3), 257–275 (2010)
27. Hupfeld, F., Cortes, T., Kolbeck, B., Stender, J., Focht, E., Hess, M., Malo, J., Marti, J., Cesario, E.: The XtreemFS architecture—a case for object-based file systems in grids. Concurrency Comput. Pract. Experience **8**(17), 1–12 (2008)

CDT to Detect Co-residence in Cloud Computing

Hicham Boukhriss[1(✉)], Mustapha Hedabou[1], and Omar Boutkhoum[2]

[1] National School of Applied Sciences, University Cadi Ayyad, Safi, Morocco
boukhriss.hicham@gmail.com
[2] Faculty of the Sciences Semlalia, University Cadi Ayyad, Marrakech, Morocco

Abstract. Virtualization has been wide used for serving the ever growing computing demand, allowing cloud providers to instantiate multiple virtual machines (VMs) on a single set of physical resources. Customers use this shared resources without warned about the possibility of extracting or manipulating their sensitive data by an attacker who can co-resident his malicious VM with the target one. This paper presents the Co-residence Detection Technique (CDT), a method to analyze how attackers can co-resident with a target VM. Our method consists of three parts: (a) cartography cloud (b) co-residence test and (c) request for migration. We used Amazon EC2, GCE and Microsoft Azure as a case study to demonstrate that is possible to scan the local network and confirm co-residency with a target VM instance by using the network commands.

Keywords: Cloud computing · Co-residence · Migration · Detection
Virtualization · Cartography cloud

1 Introduction

Cloud computing has become the new buzz word through marketing and service offering from large groups. It is a technology enabling the localization of data and applications on dematerialized infrastructures accessible from the Internet.

The most important concept introduced by the cloud computing is the virtualization. This technology becomes rapidly the cornerstone for the security of critical computing systems reduces the total cost of ownership by using physical shared resources.

In the case of the network traffic, isolation depends completely on how the virtual environment is connected and the hypervisors conception must be flawless and without any bug. In most cases the virtual machine is connected to the host by means of a virtual switch, allowing the VM to use ARP spoofing [1, 2] to redirect packets coming and going to another virtual machine.

Recently, co-residence has attracted considerable attention from the researchers community in the field of cloud computing. Different methods are suggested and some of them are tested to make the malicious user sure if he is right co-residence, they are on the same physical host, with the victim machine. Depending on the fact that the attacker shares processor resources of his victim, he will easily retrieve private information such as CPU and cache usage and he may use it maliciously.

© Springer International Publishing AG, part of Springer Nature 2018
M. Ben Ahmed and A. A. Boudhir (Eds.): SCAMS 2017, LNNS 37, pp. 313–323, 2018.
https://doi.org/10.1007/978-3-319-74500-8_29

On the other hand, one must know that for every physical machine, all hosted virtual machines are managed by one of them called the manager Dom0. In fact, with a simple trace route, the attacker can verify that is managed or not by the same manager and therefore on the same host.

However, most previous works focus on how can we check if two instances are co-resident? [16], how to use the cache memory to detect co-residence? [17, 18], and if the latency network can lead to detect co-residence? [19].

In this paper, we developed a new method, called the CDT, to check the co-residence of two virtual machines. Based on the Result of our previous paper [3], which is illustrated as cartography, we propose a new detection algorithm that we assumed that it is applicable regardless of the cloud infrastructure. This algorithm consists of three parts chained and each of them increments the level of certainty of the co-residence.

The rest of this paper is organized as follows: First, we define and clarify the concepts of the most important key words in our research, in Sect. 2. The Sect. 3 reviews the related works on detection of co-residence. We focus in Sect. 4 on testing and demonstrating of our approach. Finally, we conclude.

2 Background

2.1 Detecting a Virtual Environment

Although the hypervisor is running on the same machine as the VM, it should be as transparent as possible. However, the hypervisor must store its code and information concerning the virtual machines [4] because these data can cause information leakage.

There are different reasons for wanting to detect the presence of a virtual machine. Recent malware have such detection systems to avoid ending up in a *honeypot* [5, 6]. An attacker can also search for a specific type of hypervisor to exploit a known vulnerability and take control of the physical machine.

2.2 Co-residence

Co-residence is being in the same physical machine with the target VM to extract its information in order to attack it. Attacks by co-residents follow the next two steps. First, the attacker determines a set of target VMs very precisely and then works on the co-location of their virtual machines with these targets on the same physical machines. Second, after co-residence is achieved, the attacker will construct different types of side channel attacks [7], to extract confidential information from the victim.

Note that this is different from [8–11], where the attacker does not have specific targets, and their goal is to obtain an unfair share of the cloud platforms capacity.

2.3 Dom 0

Most of the hypervisor administration facilities are accessible through a specially privileged virtual machine immediately instantiated after starting the hypervisor. This virtual machine, called "Domain 0" abstract Dom0, is a complete operating system (either Linux, BSD, Solaris …) whose core is specially modified to communicate with the underlying hypervisor [12]. Dom0 controls (start, stop, monitors) other non-privileged virtual machines (called "User domain" or domU) (Fig. 1).

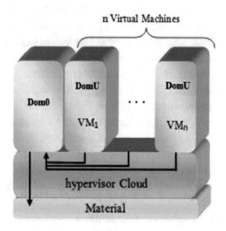

Fig. 1. Relationship between hypervisor, Hard and Soft ware

2.4   Migration

The migration of a virtual machine is the fact of moving this machine from one host to another [13]. It solves the problems of fault tolerance, continuity of services when a particular host must be replaced for maintenance tasks or upgrade hosted services.

There are two types of migration: (a) Cold Migration of a virtual machine is when we move all the specification _les and imaging systems from the origin host to a destination host machine. In a cold migration, the state of the machine is switched off during the transfer. (b) Live Migration is the process of transferring a virtual machine in real time while it is in operation from one host to another while maintaining the status of the machine during the process of transfer. This type of migration allows users to maintain services during the time of the process. The user of the services hosted on the machine has no awareness of any change in the host [14, 15].

3   Related Works

There are several publications covering the problem of showing the co-residence of two virtual machines: An overall summary of the different co-residences cases is given in [16]. In that paper, the authors mention 3 potential methods to check the co-residence of two instances. They confirm that the instances are considered co-resident if they match one of the three cases: (a) matching Dom0 IP address, (b) small packet round-trip times and (c) numerically close internal IP addresses. As result of their experiments, they notice that the round-trip time (RTT) required a "warm-up". There is always a delay between the first and the following reported sequence of RTT.

The idea of exploiting the cache memory to detect co-residence has been widely studied in different publications. Zhang et al. [17] bring a new defensive detection tool called HomeAlone. They proved that private data can be leaked across VM boundaries when the attacker exploits the reality that the cache unintentionally transmits information between VMs, in the same host, so one VM can knows information about the other by examining its cache footprint. In [18], by using the method of side channel attacks, which based on taking advantage of the responses of sharing resources, cunning users can readily hack private information from other co-resident VMs. VMs co-residency detection side channel attacks VCDS aims to get the location of the victim VM by analyzing the responses of the shared cache.

VCDS came to overcome the limitations of side channel attacks in term of the interferences introduced by the VMs and the noises introduced by the hardware features and software features.

To overcome the problem of determination of the co-residence when no cache is shared between VMs, Xu [19] suggested to test the network latency between two VMs. In fact, the round-trip times between co-resident VMs are necessarily shorter than those in the same LAN because the network communication between co-resident VMs usually does not go through the full TCP/IP stack. However, when it is difficult or even impossible, for one reason or another, to measure the network latency, the author proposes to check if the NIC (Network Interface Controller) is shared between VMs. In this case, it is possible to use iperf or ping.

In our proposed work, we tried to solve the problem of co-residence by concentrating on the network part which allows us to apply our method whatever the hardware or software infrastructure used.

4 Detecting Co-residence

This section first describes the steps of the proposed method. Secondly, demonstrates the test bed we build for our experiments.

4.1 Contribution

In cloud computing environments, it is possible to map the infrastructure and identify where a virtual machine resides. So, it is possible to instantiate new VMs until one is placed in co-residence with the target. After this investment the VM instantiated can extract confidential data from the legitimate VM [20].

Starting from this perspective, we have developed a simple and reliable method includes several sequenced steps to detect the co-residence of two VMs and which is explained in the flowchart of Fig. 2.

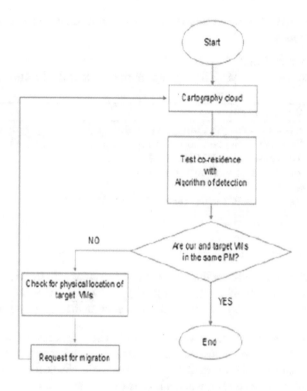

Fig. 2. Flowchart of the proposed method

Cartography cloud: It is based on the method of cartography, already mentioned in our article [3], we have determined the function fct_cart (,) (Algorithm 1) able to scan the network and give a clear idea on all the vertices of each virtual machine whose status is "up". The mapping function fct_cart (,) accepts two arguments and sends as a result the IP address of all active hosts from the local network. This function initiates a For loop which repeats a specific instruction several times equal to the network size (netmask) and a While Loop launches only if the variable stat is true (stat = up). Note that the variables used are: - i: integer and - stat: Boolean indicating the status of the scanned host which can be on (up) or off (down).

Algorithm 1 : Cartography Function

---
function $fct\_cart(\ ,\ )$
    for ($i = 1,\ i <= sizeof(netmask)\ ,\ i{+}{+}$) **do**
        while $stat = up$ **do**
            show the address IP of this VM in a vector $vVMi()$
        end while
    end for
end function

Co-residence test: To assume that two VMs are co-resident, they must verify two essential conditions:

- Sharing the same Dom0.
- The IP address of the VM victim belongs to the result of the function *fct_cart()*.

Algorithm 2 : Detection

    for for $i = 1$, $i <= sizeof$(vVMi()) , i++ **do**   //testing co-residence
        if the first hop = the last hop **and** the last hop = Dom0 IP **then**
            if IP adress of the VM victim $\in$ the result of fct_cart(,) **then**
                our VM and VMi are co-resident
            else
                Locate the PM using the method in the article [3]
            end if
        end if
    end for

If the result is true we admit that our VM is co-resident with the victim. Otherwise, we refer to our study explained in the article of [3] to locate the physical machine where the VM victim resides. Then we ask for the migration to the new physical machine.

Request for migration: There are several reasons why it is necessary to migrate operating systems, the most important being the workload balancing across physical servers. It is also necessary to migrate virtual machines when a physical host is defective or requires maintenance. And also, when a host is overloaded and it is no longer able to meet the demand, it is necessary to migrate the state of the virtual machine to a more powerful host, or less overloaded, who will be able to take over.

Algorithm 3 : Vm Migration-Based Placement

    for all the VMk on PMj **do**
        if $VMks$ migration takes place for VMi **then**
            for all the PMs except PMj **do**
                Find the best availabale PMs except PMj for VMk
                Select the best qualified victim VM running on PMj
            end for
        end if
        if migration constraint is satisfied **then**
            return true
        else
            return false
        end if
    end for

4.2   Implementation

We used a number of testbeds to evaluate our approach, as shown in Fig. 3. For the present experiment, we used three local Intel core i7 Dell machines, to launch VM instances, log instances information and run the co-residency detection test. Our general idea is to launch multiple VMs and detect the co-residence between all pairs of them.

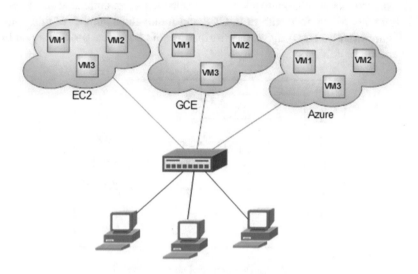

Fig. 3. Proposed solution scheme

To do so, we create three different accounts: one in Azure [21], another in Amazon EC2 [22], and the last one in Google GCE [23], to check the reliability of our solution whatever the cloud provider.

4.3   Evaluation

In this section, we aimed to evaluate the feasibility of our co-residency Detection Algorithm. As shown in Fig. 3, we realized the configuration already covered in the previous section, on the one hand, we used three physical servers' machines (DELL core i7) and on each of them we installed three VMs that will play the role of victims. On the other hand, we exploited three computers to play the role of attackers. To conclude, we had nine victims and three attackers are connected together by a switch Cisco. All experiments were conducted over 4 months between May to August 2016.

The experiment began when we launch a specific number of victims (we called it nV) and a different number of attackers (we called it nA) and each time we change the proportionally of those number in the condition that the victims must always be less or equal the attackers. We repeat each experiment several times to ensure the validity of the results.

At the beginning of our experience, we launched a command TCP SYN Traceroute to determine the IP address of the Dom0. In order for the attacker and the victim to be in co-residence, the IP address of the first jump must be the same as that of the victim.

Afterwards, one must check whether the IP address of the victim VM belongs to the result of the function fct_cart(), that is to say, we compare our IP address and each element of the vector in an incremental index.

In order to evaluate our approach, we used some tools like command-line interface (CLI) and putty to interface with EC2, GCE and Azure and a number of different test beds. Figure 4 (tables 1.2.3) shows the distributions of co-resident probability on EC2, Azure and GCE (Table 1).

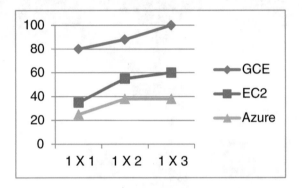

Fig. 4. Probability of co-residence

Table 1. Chance of Co-residence

Configuration	Chance		
	GCE	EC2	Azure
1 × 1	80%	35%	25%
1 × 2	88%	55%	38%
1 × 3	100%	60%	38%

The values shown in the tables include the average of tested Co-residence for several executions. At the beginning of each run, we started with the launching of all victims' VMs for a few minutes before the launching of the attackers and each time, we varied the number of victims instances and keep the number of attackers instances constant (Fig. 4).

As a result, we notice that the probability of detecting the co-residence increases as long as the number of the victim increases.

As our expectation, we notice an increase in the chances of co-residency with increasing number of attackers across all cloud providers. But for Azure the chance of co-residence remains stable even when the number of attackers is increased. This phenomenon can be due to a security strategy because Azure applies advanced analysis and performs in particular Machine Learning and behavior studies.

4.4   Our Algorithm in the Real Life

In the above section, we deployed our detection method on a small private cloud, where almost all the factors and the variables were under control. So, we decide to test our method in the real life to evaluate its effectiveness under harder circumstances and we included the time factor.

For this experiment we choose the server of an ecommerce store as a target machine to localize it and detect the co resident between our machine and this server in 24 h.

To do so, we divided the hours of the day into 6 equal intervals and we repeated the same experiment in each interval. The results obtained are shown in the table of the Table 2 (Fig. 5).

Table 2.   Results tables of the experiments

Interval (h)	Chance (%)
06–10	72
10–14	37
14–18	63
18–22	25
22–02	69
02–06	84

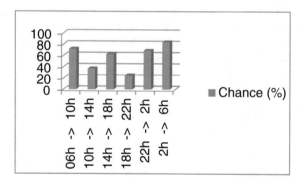

Fig. 5.   Chances of co-residence in 24 h

Discussion of the results: We notice that the periods of time when the chance of co residence was high are (06 h–10 h), (22 h–02 h) and (02 h–06 h). But it was more difficult to have a good precision in two particular periods (10 h–14 h) and (18 h–22 h) because this kind of websites has an important traffic, which affect our method, just before school or work time in the morning or at lunchtime, or right after school or work.

As *TCP SYN Traceroute* plays a critical role in our method, it has a significant impact on co-residence. This command sends out (TCP SYN) packets and waits for a response (ACK) from the server. When the web site traffic is high the server needs more time to send the response so the chances of getting the co residence decrease.

5   Conclusion

In this paper, we have shed light on detecting co-residence using the Co-residence Detection Technique which is based on algorithms. The strong point of this method is the attacker works without the cooperation of the victim VM because it does not need any sort of information's exchange with the target.

In order to demonstrate the feasibility of our detection method, we deploy it in multiple environments.

References

1. AbdelSalam, A.M., Elkilani, W.S., Amin, K.M.: An automated approach for preventing ARP spoofing attack using static ARP entries. (JACSA) Int. J. Adv. Comput. Sci. Appl. 5(1) (2004)
2. Srinath, D., Panimalar, S., JerrinSimla, A., Deepa, J.: Detection and prevention of ARP spoofing using centralized server. Int. J. Comput. Appl. 113(19) (2015). (0975 –8887)
3. Boukhriss, H., Mohammed, A.A., Hedabou, M.: New technique of localization a targeted virtual machine in a cloud platform. In: Proceedings 5th Workshop on Codes, Cryptography and Communication Systems, WCCCS 2014 (2014)
4. Upton, D.: Detection and subversion of virtual machines. University of Virginia, CS 851 - Virtual Machine (2006)
5. Zou, C.C., Cunningham, R.: Honeypot-aware advanced botnet construction and maintenance. In: International Conference on Dependable Systems and Networks DSN 2006 (2006)
6. Wang, P., Wu, L., Cunningham, R., Zou, C.C.: Honeypot detection in advanced botnet attacks. Int. J. Inf. Comput. Secur. 4(1), 30–51 (2010)
7. Del Pozo, S.M., Standaert, F.X., Kamel, D., Moradi, A.: Side-channel attacks from static power: when should we care? In: Proceedings of the 2015 Design, Automation & Test in Europe Conference & Exhibition, pp. 145–150 (2015)
8. Zhou, F.F., Goel, M., Desnoyers, P., Sundaram, R.: Scheduler vulnerabilities and coordinated attacks in cloud computing. In: Proceedings of Tenth IEEE International Symposium on Network Computing and Applications (NCA 2011), pp. 123–130 (2011)
9. Varadarajan, V., Kooburat, T., Farley, B., Ristenpart, T., Swift, M.: Resource-freeing attacks: improve your cloud performance (at Your Neighbors Expense). In: Proceedings of ACM Conference on Computer and Communications Security (CCS 2012), pp. 281–292 (2012)
10. Yang, Z., Fang, H., Wu, Y., Li, C., Zhao, B., Huang, H.H.: Understanding the effects of hypervisor I/O scheduling for virtual machine performance interference. In: Proceedings of Fourth IEEE International Conference on Cloud Computing Technology and Science (CloudCom 2012), pp. 34–41 (2012)
11. Bedi, H., Shiva, S.: Securing cloud infrastructure against co-resident DoS attacks using game theoretic defense mechanisms. In:Proceedings of International Conference on Advances in Computing, Communications and Informatics (ICACCI 2012), pp. 463–469 (2012)
12. Perez-Botero, D., Szefer, J., Lee, R.B.: Characterizing hypervisor vulnerabilities in cloud computing. In: Proceedings of the Workshop on Security in Cloud Computing (SCC). Princeton University, Princeton, NJ, USA (2013)
13. Zhao, M., Figueiredo, R.J.: Experimental study of virtual machine migration in support of reservation of cluster resources. In: Proceedings of the 2nd International Workshop on Virtualization Technology in Distributed Computing Article No. 5. Reno, Nevada (2007)

14. Clark, C., Fraser, K., Hand, S., Hansen, J.G., Jul, E., Limpach, C., Pratt, I., Warfield, A.: Live migration of virtual machines. In: Proceedings of the 2nd Conference on Symposium on Networked Systems Design & Implementation, vol. 2, pp. 273–286, (02–04 May 2005)
15. Ye, K., Jiang, X., Ma, R.: VC-migration: live migration of virtual clusters in the cloud grid computing (GRID). In: ACM/IEEE 13th International Conference (2012)
16. Ristenpart, T., Shacham, H., Savage, S.: Hey, you, get off of my cloud: exploring information leakage in third-party compute clouds. ACM (2009)
17. Zhang, Y., Oprea, A., Juels, A., Reiter, M.K.: HomeAlone: co-residency detection in the cloud via side-channel analysis. In: IEEE Symposium on Security and Privacy (SP) (2011)
18. Si, Y., Xiaolin, G., Jiancai, L., Xuejun, Z., Junfei, W.: Detecting VMs co-residency in the cloud: using cache-based side channels attacks. Elektronika Ir Elektrotechnika **19**(5), 73–78 (2013). ISSN 1392-1215
19. Xu, Y., Bailey, M., Kaustubh, Joshi, F.J.K., Hiltunen, M., Schlichting, R.: .An exploration of L2 cache covert channels in virtualized environments. In: Proceedings of the 3rd ACM Workshop on Cloud Computing Security Workshop, CCSW 2011, pp. 29–40 (2011)
20. Sevak, B.: Security against side channel attack in cloud computing. Int. J. Eng. Adv. Technol. (IJEAT). **2**(2), p. 183 (2012). ISSN: 2249 – 8958
21. Microsoft Azure: Cloud Computing Platform & Services. https://azure.microsoft.com/fr-fr/
22. Amazon Web Services (AWS) - Cloud Computing Services. https://aws.amazon.com/fr/
23. Compute Engine - IaaS – Google Cloud Platform. https://cloud.google.com/compute/

A Multi-layer Architecture for Services Management in IoT

Abderrahim Zannou[1]([✉]) ⓘ, Abdelhak Boulaalam[2], and El Habib Nfaoui[1]

[1] LIIAN Laboratory, University Sidi Mohamed Ben Abdellah, Fez, Morocco
{abderrahim.zannou,elhabib.nfaoui}@usmba.ac.ma
[2] LSI Laboratory, University Sidi Mohamed Ben Abdellah, Fez, Morocco
abdelhak.boulaalam@usmba.ac.ma

Abstract. Internet of things (IoT) is network of networks where, a massive number of objects/things are connected through wired/wireless communications and different infrastructures to provide value-added services. For this reason, controlling and managing the connected objects is a great challenge due to heterogeneity of objects, low power battery, limited memory and limited capacity of calculation. Also, missing standards for horizontal communication of objects is another interesting challenge. To overcome the problem of communication of objects and services management in IoT, recent studies have focused on applications in limited network where the communication is not standard and the dynamic service creation remains difficult. In this paper, we propose a global architecture for services management and communication which handles vertical and horizontal communications silos. It is composed of internal architecture that manages and creates dynamic services in internal network or in a device of IoT. Also objects in this architecture are classified in order to consider the specific domain or subdomain services.

Keywords: Internet of things · Dynamic service creation · Cloud computing
Semantic service · Horizontal and vertical communication · Ontologies

1 Introduction

The Internet of Things allows people and things to be connected anytime, anyplace, with anything and anyone, ideally using any path/network and any service [1]. The ultimate goal is to create "a better world for human beings", where objects around us know what we like, what we want, and what we need and act accordingly without explicit instructions [2].

Indeed, the IoT will open a full range of new possibilities, e.g., new business models and ecosystems. By utilizing IoT, we can facilitate the sharing of our existing devices, vehicles, building, etc., by embedding sensors and network connectivity that enable our things to collect and exchange data, and constitute the basic building blocks to progress towards unified information and communication technology platforms for a variety of applications [3–7].

© Springer International Publishing AG, part of Springer Nature 2018
M. Ben Ahmed and A. A. Boudhir (Eds.): SCAMS 2017, LNNS 37, pp. 324–334, 2018.
https://doi.org/10.1007/978-3-319-74500-8_30

While Internet of objects is composed of several technologies and variety of things such as mobile, phones, sensors, actuators Frequency IDentification (RFID) tags, etc. For resolving, heterogeneity of devices, there is much effort to create open platforms tried to resolve these problems but rest incomplete.

Thus, a massive amount of devices connected to the internet and the huge data associated result unorganized of objects communicated and the data realized by constraints objects is very difficult to stock. On other side, to execute service request must understand the detailed about request and treatment every-thing related with service request.

To overcome these limitations, we propose a global architecture for communication services for vertical and horizontal communication silos, and for management of services. We classified objects into four levels, the division is provided with capacity of object and kind of services. Our idea is to classify objects into four classes depending on their capacities and kind of services. In addition four layers are required for each object: Representation Layer, Semantic Service Layer, Constraints Layer, and Service Layer (e.g. a requested service will be sent directly to smart health care objects). As a result, the proposed architecture aim to realize organized communication of objects in internet of things and unified concept of management of services in IoT.

The remainder of the paper is organized as follows: Sect. 2 presents the IoT challengers. Section 3 present integration of Cloud Computing and internet of things. Following that, Sect. 4 gives some discussion of related work. Section 5 gives our vision of management of communication service in internet of things. Finally, we conclude this paper.

2   IoT Challengers

The rapid growth and the fast evolution of technologies open great challenges to the IoT applications. In fact, these challenges are essential for a correct application of the IoT in multiple domains. In what follow, we briefly describe each of these challenges.

Low power communication: Many IoT devices are small in size and do not have the continuous power source. Since IoT devices are typically untethered, they must survive on a battery or on power harvested from their environment. IoT devices typically require long lifetimes, further constraining power consumption.

It is important to consider resource constrictions, such as wakeup delays, power consumption, and limited battery and also packet size.

Interoperability: There is different platforms, systems and protocols in IoT, the challenge how to connect these different heterogeneous of devices, systems, platforms protocols etc. Interoperability should be handled by both the application developers and the device manufacturers in order to deliver the services regardless of the platform or hardware specification used by the customer.

Reliability: In IoT applications, the system should be correctly working, collecting data and communication protocols must be perfectly reliable. The object should be ready to make decision in emergency case.

Mobility: The IoT devices can move anywhere and anytime, their IP address and the architecture of network changing. In addition, mobility might result in a change of service provider.

Stability: New objects can be added in same network, the architecture of network change, the scalability of network is point specific of IoT. Need to design future network protocol and network architectures. The solution framework must support various aspects.

Availability: Service subscribers must have ability to access the object target in anytime and anywhere. Availability in IoT must perfectly work regardless of the technologies or protocols or service used in systems of IoT.

Management: Managing configuration, performances of objects and systems must execute periodic.

Security and privacy: Security and privacy considerations are not new in the context of information technology, the attributes of many IoT implementations present new and unique security challenges. Addressing these challenges and ensuring security in IoT products and services must be a fundamental priority. While these challengers depend of IoT devices, it is difficult to apply encryptions algorithms to objects due to limited energy and the capacity of processing. Many efforts of technical research challengers have been provided to resolve these challengers [10–18].

3   Integration of Cloud and IoT

Cloud computing and the IoT both serve to increase efficiency in our everyday tasks, and the two have a complimentary relationship. The IoT generates massive amounts of data, and cloud computing provides a pathway for that data to travel to its destination. Cloud Computing has the almost unlimited capacity of storage and processing power which is a more mature technology at least to a certain extent to solve the problem of most of the Internet of things [19–21].

With the rapid development of Internet, Cloud computing and Internet integration of medical monitoring and management platform is to provide new opportunities for the hospital, even in social fields [22–24].

The adoption of the Cloud computing concept enables new scenarios for smart services and applications based on the extension of Cloud through the—things [25]:

1. SaaS (Sensing as a Service), providing ubiquitous access to sensor data.
2. SAaaS (Sensing and Actuation as a Service), enabling automatic control logics implemented in the Cloud.
3. SEaaS (Sensor Event as a Service), dispatching messaging services triggered by sensor events.
4. SenaaS (Sensor as a Service), enabling ubiquitous management of remote sensors.
5. DBaaS (DataBase as a Service), enabling ubiquitous database management.
6. DaaS (Data as a Service), providing ubiquitous access to any data.

7. EaaS (Ethernet as a Service), providing ubiquitous connectivity to remote devices.
8. IPMaaS (Identity and Policy Management as a Service), enabling ubiquitous access to policy and identity management functionalities.
9. VSaaS (Video Surveillance as a Service), providing ubiquitous access to recorded video and implementing complex analyzes in the Cloud.

Cloud computing takes care of IoT devices and management of services, by providing a managed platform help to make critical decisions and strategies by connect devices to the cloud, analyze data from those devices in real time, and integrate data with organization applications and various web services. Other advantage of Cloud is incorporating virtualization, which is separating the hardware from the software. Due to multiples challengers of constraints object, the virtualization of configuration objects, service, and data is more important (Fig. 1).

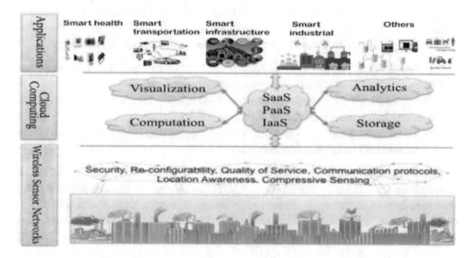

Fig. 1. Integration of cloud computing and IoT adopted from [9]

4   Related Work

Many proposals attempt to define an architectural model for IoT that are usually applicable to a specific application domain [26–31]. As far as we know, there is no suit able unified architecture till date that is appropriate for a global IoT infrastructure.

In [32], the work proposed architecture (SOCRADES) for an effective integration of the Internet of Things in enterprise services, The architecture implemented hides the heterogeneity of hardware, software, data formats and communication protocols that is present in today's embedded systems.

The specifications foster open and standardized communication via web services at all layers. The following layers can be distinguished: Application Interface, Service Management, Device Management, Security, Platform Abstraction and Devices.

The Service Oriented Architecture (SOA) based architecture for the IoT middleware proposed in [33]. It is quite similar to the scheme proposed in [32], which addresses the middleware issues with a complete and integrated architectural approach.

In [34] presents the Cloud architecture to accelerate service composition and rapid application development by inserting a special "Composition as a Service" layer for dynamic service composition (CM4SC). The CM4SC middleware encapsulates sets of fundamental services for executing the users' service requests and performing service composition. These services include process planning, service discovery, process generation, reasoning engine service, process execution, and monitoring, as detailed in [35].

In [36] proposed architecture (DIAT), the functionalities of IoT infrastructure are grouped into three layers Virtual Object Layer (VOL), Composite Virtual Object Layer (CVOL) and Service Layer (SL). The three layers are responsible for object virtualization, service composition and execution, and service creation and management respectively. In [37] proposed a unified semantic base for IoT that uses ontologies, resource, location, context & domain, policy and service ontologies. All these ontologies applied on architecture DIAT. In these two articles, there are three biggest limitations:

- It is very difficult to apply the architecture DIAT in constraints object and the workflow of management of service in constraints objects not defined.
- To study ontologies, we must have devices high capacity, thus in this case, we present our internal architecture for communication services solution.
- Horizontal communication service in two articles no defined, so we present global architecture for communication services.

Our internal architecture for communication services extend from architecture DIAT, in our vision we add a new layer named "Constraints Layer" to handles operations and situations depend of management service in constraints environment, and we have an intelligent layer depend with global architecture for communication services. Also, horizontal communication not defined, for this reason, a global architecture for communication services proposed to resolve this problem and in same time manages service IoT.

5   Proposed Architecture

5.1   Global Architecture for Communication Services

In global architecture for communication services, we define four levels of service, the categorization based the capacity of devices such as transportation, networking, collection data (sensor captured), processing, executing logic or physical (actuators), Store information and kind of service.

Level Service 1: The objects of this level are the small devices that have limited capacity and service(s). The devices of this level can directly communicate to other object service (independent service objects), like service; road safety, stop sign or indirectly through other devices or machines (dependent service objects), Like; fridge, human behavior that related with phone user, product for sale of smart companies etc.

The type of services in this low level categorized into two types:

- Private service where the service of device is available to an entity or set of entities.
- Public service where the service of device is available to all.

Level Service 2: Systems high capacity and the intelligent devices considering the object of level 2. They role is controlling communication and service management of object low level (objects of level 1). This level contains object that manage smart home, smart organization, smart company etc. Like servers, Clouds, smart phone etc. Smart systems and devices of this level contain special specifications considering IoT challenges (Fig. 2).

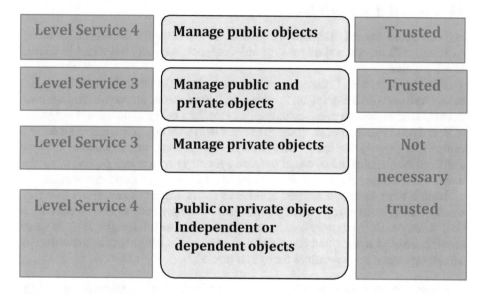

Fig. 2. A global architecture for communication services

Which contains generic platforms resolves IoT challenges and manages service in and out of different devices, in addition, to specific functions that processing service tasks such as service discovery, analytic, making decisions, virtualization etc.

Level Service 3: Systems and the intelligent machines trusted and have high capacity than the object of level 2. We devise these objects into two types:

Government devices: Designed to control devices that have public services like administration service, hospitals, driving laws etc.

Non-government devices: Designed to control for communication services between two organizations to resolve the horizontal for communication services.

Level Service 4: Systems and the intelligent machines government trusted and have high capacity than the object of level 3. The role of these systems to manage objects government of level 3. It is distribute roles and new services to execute in smart cities. Like smart country, smart space etc.

The main object of this separation of objects in IoT is to facilitate communication between devices that have IoT challengers by creating platforms that have high level of capacity to be root and an interface of vertical communication and horizontal communication. These platforms must contain also different functionalities of analytic, virtualization etc.

As result, the systems of smarty country contain and manage set of systems of smart city, in the same way these systems manage the low object. By this way, we construct a hierarchy of communication of object. The Internet of object communication will be organized.

5.2 Domains of Service in IoT

In life, there are several fields and in each field we find several areas or specializations. As result, each request and response of object applications can categorized in domain or subdomains.

Devised all services in domains, for that each service has domain service and each service also contains sub-services. In other side, these domains service we translate them to virtualization data service. Separate all services in a group of domains, the systems of level 2 and 3 to domains. Each server or cloud or system directed to manage a domain named root domain service, in this domain we find a set of subdomains, thus we distribute subdomains management to set of systems to realize task depend of subdomains, so we construct a tree of domains service managed by root domain service.

Each system manage a domain of services such as smart health, smart transport, smart farming, smart energy, smart planet, smart building, smart industry, smart education, smart weather. In smart health we separated also to subdomains such as smart hospital, smart pharmacy, and smart cabinet doctors. After that smart cabinet of doctors we can separate it into specialties for each it (Fig. 3).

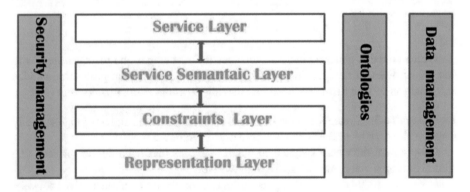

Fig. 3. Internal architecture for communication services

The objective of these subdivisions of domains is facilitate search in other way semantic service discovery of heterogeneous components and data integration, but also for the behavioral control and coordination of those components and finally co-creative value.

5.3   An Internal Architecture Communication Services

The final goal of management object in IoT is self-management of communication and services in IoT, in this section we design an internal architecture for communication services composed by four layers. These layers can exist in different devices. The main goal of this architecture is creating dynamic services in real time environments with associated global with architecture.

5.3.1   Service Layer

Service layer is responsible for allowing or disallowing of data, service user's applications and devices applications. By analyzing service request and data, service layer can create two types of services:

Static service where service request is clear, in this case the mission is sent directly to Semantic Service Layer.

Dynamic service where the layer creates self-service from data receipt, in this case the service sends all data needed to Semantic Service to create the dynamic service.

5.3.2   Semantic Service Layer

Semantic Service Layer is responsible to interpret meaning of service and data before service discovery and before service execution. When this layer receive a request service from Service Layer, it subdivides it a set of tasks, after that determine the domain of each task and their subdomains, also decide how these task can execute to reach final goal.

5.3.3   Constraint Layer

In IoT developing application services is still not an easy mission due to multiple challenges; these challengers are different from object to an another, from service to an another and depending on the situation as well. This layer subdivides all constraints into three categories:

Constraint object: defines constraints physical and programing of object service such as power energy, capacity RAM, capacity of storing, calculation, protocols of communications, system language etc.

Constraint service: defines constraints of services, the needs and resources must be contain in service provider to realize this service, such limited money credit, maximum of data storing etc.

Constraint link: In some cases, an object provider can't realize individually a service task; it searches of other resource to realize this task. But object intermediate service has also constraint must be respected to realize a service.

5.4   Representation Layer

The representation Layer (RL) is transformation digital of physical object, machine, device or any entities. Each object in IoT has specific role or composition of roles (tasks). RL Present set of functionalities, data, actions etc. This layer contain a module service well-structured and detailed such as type of service (public or private), domain and sub-domain of service, level of object, dependent or independent object etc.

RL can present service or set services e.g. For facilitate semantic search of service and data, the organization of data and service are important.

5.5   Data Management

Devices and entities database of internet of things can exist in three cases: internal database, external database or hybrid (between internal and external database). Also, data change in long time period time or in short time period. Furthermore, data exist in many forms xml, text, sql etc.

If object service is independent e.g. no managed by other devices, storing data pose a problem, in this case historical data will deleted, as result missing of benefits can take from this static data. As result Semantic Service Layer can't get historical data to realize a more specified service.

If object service is dependent e.g. managed by other device which storing data can realize by root object, if the root hasn't sufficient capacity of storing, the mission of storing affect to objects has high level (objects of level 3). For example person has object for blood pressure measurement, register data of measurement done in short duration, so after filling database of this object, the phone or other device manage storing historical data of this object, also if capacity of storage of root object full, historical data send to object of level 3 especially to server of smart health. In this case, a person's health status can be tracked.

For tracking a person's health status, the measurement data blood pressure not enough, thus need to carry other data whatever their value like time of each measurement, place of person during measurement and other health status of person. All that combined for create a semantic data of service. Furthermore, creating dynamic service in real time can be difficult when service task not clear, and require a set of complex tasks.

Finally, to realize a flexible semantic service, must organized a standard form, and carry all value dependent of this data.

6   Conclusion

In this paper, we designed and discussed a global architecture for IoT that facilitates and unifies communication services, handles horizontal and vertical communication silos and maps services. This architecture considers all elements such as heterogeneity domains of services in IoT and we believe that this architecture can be strongly user in multiple domain applications such as: Smart cities, PLM, Intelligent Manufacturing System, Intelligent Health care.

Four our future work, we will focus on the implementation of a prototype for the proposed architecture.

References

1. Guillemin, P., Friess, P.: Internet of things strategic research roadmap. The Cluster of European Research Projects, Technical report, September 2009. http://www.internet-of-thingsresearch.eu/pdf/IoT_Cluster_Strategic_Research_Agenda_2009.pdf. Accessed Aug 2011
2. Dohr, A., Modre-Opsrian, R., Drobics, M., Hayn, D., Schreier, G.: The Internet of things for ambient assisted living. In: 2010 Seventh International Conference on Information Technology: New Generations (ITNG), pp. 804–809 (2010)
3. Nasim, K., Mowla, N.I., Sharmin, N.: An Approach to IoT Data Management for an Intelligent Monitoring System on Refrigerator, September 2015
4. Kim, H., Lee, S., Shin, D.: Visual Information Priming in Internet of things: focusing on the interface of smart refrigerator, February 2017
5. Floarea, A.-D., Sgârciu, V.: Smart refrigerator: a next generation refrigerator connected to the IoT. In: Conference IEEE, 30 June–2 July 2016 (2016)
6. Pacheco, J., Satam, S., Hariri, S., Berkenbrock, H.: IoT security development framework for building trustworthy smart car services. In: Conference Paper, September 2016
7. Atif, Y., Dinga, J., Jeusfelda, M.A.: Internet of things approach to cloud-based smart car parking. Procedia Comput. Sci. **98**, 193–198 (2016)
8. Giusto, D., Iera, A., Morabito, G., Atzori, L. (eds.): The Internet of things. Springer (2010). ISBN: 978-1-4419-1673-0
9. Gubbi, J., Buyya, R., Marusic, S., Palaniswami, M.: Internet of things (IoT): a vision architectural elements and future directions. Future Gen. Comput. Syst. **29**(7), 1645–1660 (2013)
10. Borgia, E.: The Internet of things vision: key features, applications and open issues. Comput. Commun. **54**, 1–31 (2014). [CrossRef]
11. Jain, R.: Internet of things: challenges and issues. In: Proceedings of the 20th Annual Conference on Advanced Computing and Communications (ADCOM 2014), Bangalore, India, 19–22 September 2014 (2014)
12. Stankovic, J.A.: Research directions for the Internet of things. IEEE Int. Things J. **1**, 3–9 (2014). [CrossRef]
13. Mattern, F., Floerkemeier, C.: From the internet of computers to the Internet of things. In: From Active Data Management to Event-Based Systems and More, pp. 242–259. Springer, Berlin (2010)
14. Elkhodr, M., Shahrestani, S., Cheung, H.: The Internet of things: vision and challenges. In: Proceedings of the TENCON Spring Conference, Sydney, Australia, 17–19 April 2013, pp. 218–222 (2013)
15. Gubbi, J., Buyya, R., Marusic, S., Palaniswami, M.: Internet of things (IoT): a vision, architectural elements, and future directions. Future Gener. Comput. Syst. **29**, 1645–1660 (2013). [CrossRef]
16. Chen, S., Xu, H., Liu, D., Hu, B., Wang, H.: A vision of IoT: applications, challenges, and opportunities with China perspective. IEEE Internet Things J. **1**, 349–359 (2014). [CrossRef]
17. Muralidharan, S., Roy, A., Saxena, N.: An exhaustive review on Internet of things from Korea's perspective. Wirel. Pers. Commun. **90**, 1463–1486 (2016). [CrossRef]

18. Al-Fuqaha, A., Guizani, M., Mohammadi, M., Aledhari, M., Ayyash, M.: Internet of things: a survey on enabling technologies, protocols, and applications. IEEE Commun. Surv. Tutor. **17**, 2347–2376 (2015). [CrossRef]

19. Tao, F.: CCIoT-CMfg: cloud computing and Internet of things based cloud manufacturing service system, p. 1 (2014)

20. Chen, Y., Zhao, S., Zhai, Y.: Construction of intelligent logistics system by RFID of Internet of things based on cloud computing. J. Chem. Pharm. Res. **6**(7), 1676–1679 (2014)

21. Wang, H.Z.: Management of big data in the Internet of things in agriculture based on cloud computing. Appl. Mech. Mater. **548**, 1438–1444 (2014)

22. Soldatos, J.: Design principles for utility–driven services and cloud–based computing modelling for the Internet of things. Int. J. Web Grid Serv. **10**(2), 139–167 (2014)

23. Xie, F., Liang, C.Z.: Research of Internet of things based on cloud computing. Appl. Mech. Mater. **443**, 589–593 (2014)

24. Fang, S.: An integrated system for regional environmental monitoring and management based on Internet of things. IEEE Trans. Ind. Inf. **10**(2), 1596–1605 (2014)

25. Jadhav, R., Kulkarni, R., Perur, S., Kulkarni, G.L., Kunchur, P.: Prominence of Internet of things with cloud: a survey. Int. J. Emerg. Res. Manag. Technol. **6**, 40–43 (2017)

26. Kum, S.W., Kang, M., Park, J.-I.: IoT delegate: smart home framework for heterogeneous IoT service collaboration. KSII Trans. Int. Inf. Syst. **10**(8), 3958–3971 (2016)

27. Kim, H.-Y.: A design and implementation of a framework for games in IoT. J. Supercomputing **74**, 1–13 (2017)

28. Badave, P.M., Karthikeyan, B., Badave, S.M., Mahajan, S.B., Sanjeevikumar, P., Gill, G.S.: Health monitoring system of solar photovoltaic panel: an internet of things application. Electr. Eng. (2016)

29. Ferrández-Pastor, F.J., García-Chamizo, J.M., Nieto-Hidalgo, M., Mora-Martínez, J.: Developing ubiquitous sensor network platform using internet of things: application in precision agriculture. Sensors **16**(8), 1141 (2016)

30. Sijun, G., Zhang, Y., Zhou, X., Zheng, L.: Design of Internet of things application and service detecting system in agriculture, February 2015

31. Castellani, A., Bui, N., Casari, P., Rossi, M., Shelby, Z., Zorzi, M.: Architecture and protocols for the Internet of things: a case study. In: 2010 8th IEEE International Conference on IEEE Pervasive Computing and Communications Workshops (PERCOM Workshops), pp. 678–683 (2010)

32. Spiess, P., Karnouskos, S., Guinard, D., Savio, D., Baecker, O., Souza, L., Trifa, V.: SOA-based integration of the Internet of things in enterprise services. In: Proceedings of IEEE ICWS 2009, Los Angeles, CA, USA, July 2009

33. Atzoria, L., Ierab, A., Morabitoc, G.: The Internet of things: a survey (2010)

34. Zhou, J., Leppänen, T., Harjula, E., CloudThings: a common architecture for integrating the Internet of things with cloud computing. In: IEEE 17th International Conference on Computer Supported Cooperative Work in Design (2013)

35. Zhou, J., Athukorala, K., Gilman, E., Riekki, J., Ylianttila, M.: Cloud architecture for dynamic service composition. Int. J. Grid High Perform. Comput. **4**(2), 17–31 (2012)

36. Sarkar, C.: DIAT: a scalable distributed architecture for IoT, June 2015

37. Akshay Uttama Nambi, S.N., Chayan Sarkar, R., Prasad, V., Rahim, A.: A unified semantic knowledge base for IoT (2013)

Backpropagation Issues with Deep Feedforward Neural Networks

Anas El Korchi$^{(\boxtimes)}$ and Youssef Ghanou

High School of Technology, University Moulay Ismail, Meknes, Morocco
anaselkorchi@gmail.com

Abstract. Backpropagation is currently the most widely applied neural network architecture. However, for some cases this architecture is less efficient while dealing with deep neural networks [8, 9] as the learning process becomes slower and the sensitivity of the neural network increases. This paper presents an experimental study of different backpropagation architectures in term of deepness of the neural network with different learning rate and activation functions in order to determine the relation between those elements and their impact on the convergence of the backpropagation.

Keywords: Vanishing gradient descend problem · Neural networks
Deep learning

1 Introduction

Without a question, backpropagation is the most widely used neural network architecture. This popularity primarily revolves around the ability of the backpropagation network to learn complicated multidimensional mappings [1].

However, when dealing with a deep neural network [2] this algorithm runs into a problem known as vanishing gradients [3]. Informally, backpropagation computes the derivatives for all network weights using chain rule, and for deep layers the chain becomes too long and derivatives very hard to estimate reliably. So the algorithm becomes less efficient or almost impossible to converge.

The rest of this paper is organized as follows. Section 1 will describe the testing datasets Sinus, Ionosphere and Iris. In Sect. 2 will describe the structure of each neural network and will explain the experimental results obtained for each architecture. In Sect. 3 will conclude this paper.

2 Testing Datasets

The main testing dataset is the Sinus as it provides a clean and infinite testing examples automatically generated based on the mathematical function Sinus. Besides of the sinus dataset, MNIST [4], Ionosphere [5] and Iris [6] datasets were used during the experimental study with a vision to provide the necessary variety of data to study the behavior of the neural network with different problems. Table 1 shows a summary of the main

© Springer International Publishing AG, part of Springer Nature 2018
M. Ben Ahmed and A. A. Boudhir (Eds.): SCAMS 2017, LNNS 37, pp. 335–343, 2018.
https://doi.org/10.1007/978-3-319-74500-8_31

features of each dataset where the number of attributes refer to the number of columns in each dataset and the number of instances represents the number of rows in each dataset.

Table 1. Datasets used during the experimental study

No	Dataset	No of attribute	No of Instances
1	Sinus	3	1100
2	MNIST	784	60 000
3	Ionosphere	34	351
4	Iris	4	150

2.1 Sinus

The sinus dataset is a generated synthetic data based on the mathematical function Sinus, each row of this dataset does contains three randomly generated values between 0 and 1, those values are used as input and the sinus of the some of those three variables is our target variable.

$$Sin(X_1 + X_2 + X_3) = Y \tag{1}$$

This dataset has a training set of 1000 examples and a testing set of 100 examples. The Sinus was chosen as main experimental dataset as it is made from a real mathematical relation between its inputs and output which present a perfect testing/training environment for the neural network.

2.2 MNIST

The MNIST handwritten digits dataset consists of 28×28 black and white images, each containing a digit 0 to 9 (10-classes), each digit of the 60 000 training images is represented as a vector composed of 784 elements which refer to the value of each pixel, Pixel values are between 0 and 255. 0 means background (white), 255 means foreground (black).

2.3 Ionosphere

The Ionosphere is a dataset collected from a radar data. This radar data was collected by a system in Goose Bay, Labrador. The targets were free electrons in the ionosphere. "Good" radar returns are those showing evidence of some type of structure in the ionosphere. "Bad" returns are those that do not; their signals pass through the ionosphere. Received signals were processed using an autocorrelation function whose arguments are the time of a pulse and the pulse number. There were 17 pulse numbers for the Goose Bay system. Instances in this database are described by 2 attributes per pulse number, corresponding to the complex values returned by the function resulting from the complex electromagnetic signal.

2.4 Iris

The IRIS data sets consists of 3 different types of irises (Setosa, Versicolour, and Virginica) petal and sepal length, stored in a 150 × 4 numpy.ndarray.

The rows being the samples and the columns being: Sepal Length, Sepal Width, Petal Length and Petal Width.

3 Experimental Setup and Results

In the first part of this section we introduce the activation functions used during this experimental study.

In the second part we describe the structure of three different artificial neural networks and how their convergence varies in function of their deepness using Logistic Sigmoid as an activation function.

In the third part will present the possible refinement that could be applied to a deep neural network to avoid the vanishing gradient problem by using RELU as an activation function in the hidden layers of the neural network, also will describe the experimental results on the other data sets Ionosphere and Iris.

3.1 Activation Functions

Sigmoid
Sigmoid is a mathematical function having a characteristic S shaped curve, in our case it does refers to the logistic function represented as:

$$f(x) = \frac{1}{1 + e^{-x}} \tag{2}$$

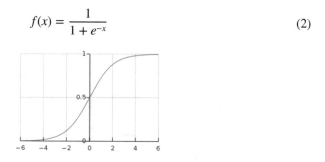

Fig. 1. Logistic sigmoid curve

RELU
The rectifier is a mathematic function defined as:

$$f(x) = \max(0, x) \tag{3}$$

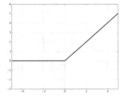

Fig. 2. The RELU curve

3.2 Experimental Study

In the experimental study we have used three different neural networks. Each neural network takes three inputs, each layer is composed of three neurons, the weights of each neuron are initialized with a random value generated in the range [−0.5, 0.5], the biases of each neuron are initialized with the value 1, the activation function used for each neuron is Logistic Sigmoid.

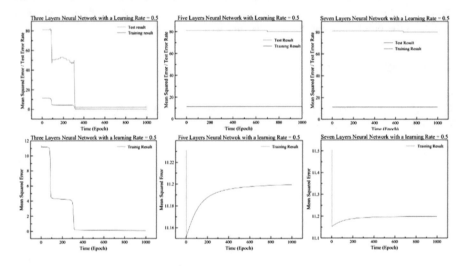

Fig. 3. Test/training result of different neural network architectures

Most remarkably the best neural network convergence found was for the three-layer neural network, the training for deep neural networks (Five/Seven Layers) was hard and slower, besides of the slowness of the learning process the sensitivity of a deep neural network was higher than a less deep neural network, both neural networks with five and seven layers did deviate after few iterations.

Those results can be explained as the backpropagation was suffering from the gradient descent problem when dealing with deep neural networks (more than one hidden layer) which is a difficulty found in training artificial neural networks with gradient-based learning methods and backpropagation. In such methods, each of the neural network's weights receives an update proportional to the gradient of the error function with respect to the current weight in each iteration of training. Traditional activation functions such as the logistic function have gradients in the range (0, 1), and backpropagation computes gradients by the chain rule. This has the effect of multiplying n of these small numbers to compute gradients of the "front" layers in an n-layer network, meaning that the gradient (error signal) decreases exponentially with n and the front layers receives a very low updates which make the training very slow.

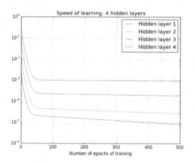

Fig. 4. Speed of the learning for each layer in a deep neural network [7]

As shown in the Fig. 2 the gradient tends to get smaller as we move backward through the hidden layers. This means that neurons in the earlier layers has a slower learning process than neurons in later layers.

To understand why the vanishing gradient problem occurs for deep neural networks, we can consider for example a sample deep neural network with 4 layers each one contains only one neuron (Fig. 5):

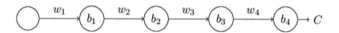

Fig. 5. Simple deep neural network

With w_1, w_2 ... are the weights and b_1, b_2 ... are the biases and C is the cost function of the neural network.

Based on the backpropagation algorithm the biases of each neuron will be updated using the formula:

$$\frac{\partial b_i}{\partial C} = f'(z_i) \prod_{j=i+1}^{n} w_j f_j'(z_j) \frac{\partial C}{\partial a_n} \tag{4}$$

where f is the activation function of each neuron,

$z_i = w_i a_{i-1} + b_i$ is the weighted input of the neuron I, n is the number of layers inside the neural network and $\frac{\delta C}{\delta a_n}$ is the cost function.

From this expression the bias of the first layer will be updated by:

$$\frac{\delta C}{\delta b_1} = w_2 f_2'(z_2) w_3 f_3'(z_3) f_4'(z4) w_4 \frac{\delta C}{\delta a_4} \tag{5}$$

Excepting the very last term, this expression is a product of terms of the form $w_i f_i'(z_i)$.

Knowing that the derivative of the sigmoidal function has a maximum value at the point $f'(0) = 1/4$ and all the weights of the neural network are initialized with a value in range of [-0.5, 0.5] we can deduce that $w_i f_i'(z_i) \leq 1/4$. And when we take a product of many such terms, the product will tend to exponentially decrease: the more terms which mean the more we would have a deep neural network, the smaller the product will be, which means that the weights of the first layers will be updated with a very small values and so the neural network will be hard to train.

However, the vanishing gradient problem depends on the choice of the activation function. Many common activation functions (e.g. sigmoid or tanh) 'squash' their input into a very small output range in a very non-linear fashion. For example, sigmoid maps the real number line onto a "small" range of [0, 1]. As a result, there are large regions of the input space which are mapped to an extremely small range. In these regions of the input space, even a large change in the input will produce a small change in the output - hence the gradient is small. This problem can be avoided by using an activation functions which don't have this property of 'squashing' the input space into a small region. A popular choice is Rectified Linear Unit which maps x to $max(0, x)$.

After replacement of the activation function by RELU in the hidden layers of each neural network in Fig. 1 with the same architecture (each layer contains three neurons the weights of each neuron are initialized with a random value generated in the range [-0.5, 0.5], the biases of each neuron are initialized with the value 1), the neural network was able to avoid the vanishing gradient problem and it did converge after a few iterations Figs. 3 and 4 (Tables 2, 4 and 5).

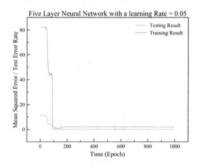

Fig. 6. Five Layers neural network with RELU in the hidden layers

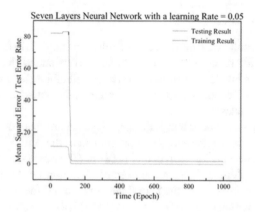

Fig. 7. Seven layers neural network with RELU in the hidden layers

Table 2. Testing result of Sinus dataset

Sinus dataset

	3L (Error rate)	5L (Error rate)	7L (Error rate)
Sigmoid	1%	82%	83%
RELU	1%	1%	2%

Table 3. Testing result of MNIST dataset

MNIST dataset

	3L (Error rate)	5L (Error rate)	7L (Error rate)
Sigmoid	3.2%	3.8%	10%
RELU	2%	1.8%	1.3%

Table 4. Testing result of Iris dataset

Iris dataset

	3L (Error rate)	5L (Error rate)	7L (Error rate)
Sigmoid	0.5%	36%	38%
RELU	0.6%	1%	1%

Table 5. Testing result of Ionosphere dataset

Ionosphere dataset

	3L (Error rate)	5L (Error rate)	7L (Error rate)
Sigmoid	2%	99%	99%
RELU	0.8%	1%	1%

3.3 Experimental Results

Our experimental results are summarized in Tables 1 through 3. In these tables each row denotes the activation function used in the hidden layers and each column denotes the number of the layers inside the neural network. For all tables, bold values indicate that the corresponding activation function's performance is statistically and significantly better than the other ones.

For all datasets we observe that for a three-layer neural network the change of the activation function doesn't rise a major improvement of the neural network in matter of its performance. However, for deep neural networks (5 and 7 layers) the change of the activation function (Replacement of the Sigmoid function by the *Relu*) does increase significantly the performance of the neural network, for all the dataset the backpropagation algorithm was able to converge by using the *RELU* activation function inside it's hidden neurons which demonstrate that the use of an activation function that doesn't squash it's input space into a small region can avoid successfully the vanishing gradient problem. In other side this results demonstrate also that activation function with a small input space range like *Sigmoid* or *Tanh* can't be used with deep neural networks (Figs. 6 and 7).

4   Conclusion

In this paper we described one of the main possible issues that the backpropagation algorithm could face when dealing with deep neural networks. We described the causes behind the vanishing gradient descend problem and the possible solutions that could be applied to avoid it.

From the results obtained we can conclude:

- Activation functions with a small input space like Sigmoid/Tanh can't be used with deep neural network at least inside it's hidden layers.
- Activation functions with a large input space like RELU are recommended to be used with deep neural network specially inside it's hidden layers as it helps to avoid the vanishing gradient descend problem.

References

1. Hecht-Nielsen, R.: Theory of the Backpropagation Network. Department of Electrical and Computer Engineering University of California at San Diego La Jolla (1992)
2. Schmidhuber, J.: Deep learning in neural networks: an overview. Neural Netw. **61**, 85–117 (2015). https://doi.org/10.1016/j.neunet.2014.09.003. arXiv:1404.7828
3. Hochreiter, S., Bengio, Y., Frasconi, P., Schmidhuber, J.: Gradient flow in recurrent nets: the difficulty of learning long-term dependencies. In: Kremer, S.C., Kolen, J.F. (eds.) A Field Guide to Dynamical Recurrent Neural Networks. IEEE Press (2001)
4. LeCun, Y., Cortes, C., Burges, C.J.C.: MNIST handwritten digit database, Yann LeCun, Corinna Cortes and Chris Burges. Accessed 17 Aug 2013
5. Sigillito, V.G., et al.: Classification of radar returns from the ionosphere using neural networks. Johns Hopkins APL Tech. Digest **10**(3), 262–266 (1989)

6. Fisher, R.A.: The use of multiple measurements in taxonomic problems. Ann. Eugenics Part II **7**, 179–188 (1936)
7. Nielsen, M.: Chapter 5: Why are deep neural networks hard to train. In: Neural Networks and Deep Learning (2017)
8. Ghanou, Y., Bencheikh, G.: Architecture optimization and training for the multilayer perceptron using ant system. IAENG Int. J. Comput. Sci. **43**(1), 20–26 (2016)
9. Ettaouil, M., Ghanou, Y.: Neural architectures optimization and genetic algorithms. WSEAS Trans. Comput. **8**(3), 526–537 (2009)

Contribution of Pedagogical Agents to Motivate Learners in Online Learning Environments: The Case of the PAOLE Agent

Abdelkrim Bendou$^{(\boxtimes)}$ ⓘ, Mohamed-Amine Abrache, and Chihab Cherkaoui

IRF-SIC Laboratory, FSA - Ibn Zohr University Agadir, Agadir, Morocco
k.bendou@uiz.ac.ma

Abstract. The work we present in this paper is a part of the general problem of drop-out in online learning environments (LMSs and MOOCs). It deals in particular with the motivation issue of learners to finish their courses. In this sense, we defend the idea of the interest of animated pedagogical agents to motivate learners and to adapt content, presentation and navigation to their profile. We therefore propose the first specifications of the PAOLE agent. The design of this agent is based on a new concept which we have called the Pedagogical Intervention. An intervention may be of different kinds, but it is more precisely used to overcome the current problem of abandonment of learners. It combines characteristics of intelligent agents like: autonomy, ability to perceive, to interact, to reason and to act; and some other characteristics of pedagogical agents as: observing, evaluating, adapting content, recommending, engaging, motivating, etc.

Keywords: Online learning · Environment · Adaptivity · Recommendation
Feedback · Pedagogical intervention · Pedagogical agent

1 Introduction

Over the last twenty years, several Online Learning Environments (OLE) have been developed. Among them we can mention: Adaptive Hypermedia Systems (AHS), Learning Management Systems (LMSs), Virtuel Learning Environments (VLE), Knowledge Management Systems (KMS) and very recently MOOCs [1, 2]. We can also highlight the different ways of learning in informal settings using social networks for learning, discussion forums, blogs, Wikis and access to free resources, etc.

In principle, such environments include course content delivery tools, synchronous and asynchronous activities, exercises and quiz modules, projects, games, workspaces for sharing resources. They also incorporate different types of multimedia learning resources: Videos, Webinars, Podcasts, Apps, etc.; but also various nomadic means of access to information: tablets, Smartphone, etc. [3]. These environments, both rich with open varied learning content as well as technologies for interacting and collaborating about this content; offers today new opportunities to learn to each connected resident of the planet [4].

© Springer International Publishing AG, part of Springer Nature 2018
M. Ben Ahmed and A. A. Boudhir (Eds.): SCAMS 2017, LNNS 37, pp. 344–356, 2018.
https://doi.org/10.1007/978-3-319-74500-8_32

In different types of online learning environments, the drop-out issue remains the most researched problem over the past ten years. In the particular case of MOOCs, for example, several studies show a large dropout rate estimated at 90% by most authors of the field [3, 9]. The causes of abandonment are related in particular to the lack of motivation and commitment of the learners. Other reasons may be due to occupation, lack of time, isolation, etc. We believe that scripting and non-adaptation of content is a secondary cause. This is also due, in our view to difficulties to keep in OLE a sufficient number of tutors, because of their massive dimension.

The current challenge in OLE is to keep learners motivated [4]. Some authors show that this is possible simply by adding some encouragement phrases above the statement of Mathematics exercises such as "Remember, the more you practice, the more you become intelligent, "or" This could be a difficult problem, but we know that you can get there (do it)", etc. Learner motivation could be improved by adding effective learning strategies. The multiplication of activities and strategies can lead to a large number of possibilities and new instructional interventions. Among these, other authors insist on increasing the number of exercises; adding videos, generating a feedback, asking learners to generate explanations before, during and after the learning process. Other flexibility factors can be obtained simply by increasing the learning time.

All of the points highlighted before lead us to think that a learner who remains isolated, without recommendations, and without pedagogical intervention, can only leads to failure. The advantage of integrating an animated pedagogical agent into MOOC courses seems interesting. In this sense, each learner has a personalized and "humanized" tutor, which makes the human-machine interaction more natural. Animated educational are supposed to play a playful role in guiding, reflecting and interacting with learners. Recent research shows that these characters can support the commitment and motivation of the learners [5]. Other studies point out that students interacting with animated pedagogical agents have been shown to demonstrate deeper learning and higher motivation [6].

In this paper we discuss the contribution of pedagogical agents, which are visible characters in learning environments designed to facilitate learning, to motivate learners and to engage them in an effective learning process. We present in particular an agent, able to propose alternatives and specific intervention strategies. This agent should be able to adapt learning pathways to a learner by proposing pedagogical interventions strategies, according to a number of indicators compiled by the learner model. Before going further into the specifications of the proposed agent, it seems important to go back on a particular OLE, which is the MOOC model. We have chosen this model of e-learning not only because it is new, but in contrast to distance learning platforms (LMSs), has generated new problems related to the notion of distance, which means that most of the learner leaves the course before the end. We note that some problems are almost identical in both the two types of environments, and can lead to common solutions.

In this paper, we first discuss the model of MOOCs with limitations and some improvements in part 2. In the third part, we present the advantages of integrating an animated educational agent into these e-learning environments. The last part will present our first specifications and developments of an animated pedagogical agent that takes into account our approach of improvement of the OLE in order to increase the motivation of learners.

2 The Drop-Out Problem

In this section, we propose to reconsider the concept of MOOC in order to characterize it while showing the limitations generated and the possible improvements. We are also interested in the reasons for the drop-out problem and the possible solutions based on adaptation strategies and the use of the pedagogical agents.

2.1 The Mooc Model

The MOOC model is a major educational innovation generating new ideas and challenges in online education such as massiveness, openness, accessibility, certification, peer assessment, nature and content programming, Etc.

In the MOOC model, we have gone from small groups of learners considered in traditional E-Learning contexts to a massive number reaching thousands of participants. Massiveness creates serious questions about how to manage large heterogeneous groups. The heterogeneity here refers to the nature of the enrolled students, without distinction of prerequisites, diplomas, age, language, etc. This implies in a way, a genuine 'democratization' of access to online resources. The certification produces new economic models [7]. Peer evaluation combined with heterogeneity creates problems of credibility. For example, it is unimaginable that the work of participants can be evaluated by children. Note also that the nature of the contents has also undergone changes using mainly videos programmed over shorter durations [8].

In the literature, we can distinguish two types of MOOCs: xMOOCs and cMOOCs. The xMOOCs take the traditional model of a transmissive approach by considering the teacher as an expert tutor and students as knowledge consumers. The cMOOCs are based on a connectivist approach, which views knowledge as being shared by the different participants, and learning as the process of generating those networks using online and social tools [5].

2.2 Limitation of Traditional Moocs

A detailed review of the literature shows that current MOOCs suffer from four main limitations, namely:

- **MOOCs Teach to a Certain Percentage of the learners**

  Hill [9] identified different types of MOOC participants: (1) No shows – register, but don't even login, (2) Observers – log in and read content, but do not engage, (3) Drop-Ins – want to achieve a specific goal, which once satisfied, ends the course for them, (4) Passive Participants – consume content, but don't do assignments, (5) Active Participants – fully intend to complete the course and all activities. In this classification, we can immediately see that only the fifth class is likely to complete a course. In addition, latest research demonstrates that among learners who complete courses; most of them seek tangible benefits such as getting a new job, starting a business, or completing prerequisites for an academic program.

- **Students Needs Assistance and Immediate Feedback**

Engaging students in the learning process is a big challenge for online learning environments. Designers and teachers must develop appropriate methods of engagement for online education. Research has shown that the higher the levels of interaction in a course, the more students develop positive attitudes towards courses. The feedback, encouragements and assistance are crucial in the success of a Learning Process. The instructor feedback reinforces the course material and encourages the students to become more engaged in the learning process. Thus as mentioned before, some authors show that this is possible simply by adding some encouragement sentences above the statement of Mathematics exercises such as "Remember, the more you practice, the more you become intelligent", or "this could be a difficult problem, but we know that you can do it easily", etc. At the same moment, appropriate assistance and help can avoid student disappointment, anxiety, and confusion and learning can be increased. Situations where learners require help and assistance are many; it may be such as help to solve problems, to manage their time, to retrieve best resources, etc.

- **Students Wants a great Adaptive instruction**

Even today, most of learning environments are still delivering the same educational content in the same way to learners with different profiles. Everyone nowadays knows that the learners are different, by their needs, expectations, interests, preferences, prerequisites, difficulties, facilities, performance, styles, etc. This difference is generally felt both by learners who are well advanced and at risk of being bored, but also by learners who have problems with their acquisition and are at risk of dropping out. In the context of MOOCs, learners are encouraged to read carefully the resources and participate in activities [10].

However, it is very difficult to track all activities and interactions in these tools because of the massiveness number of enrolled students. What is sought for learners is a tool which adapts its delivery and enhance motivation in different dimensions, ways and levels, namely the adaptation of content, presentation, navigation; but also through individual dimensions and/or collaborative, the fun and the massive one. The learner is also looking for tools that allow for a great openness (even massive and therefore very social), and that takes into account their different daily practices.

- **Students Needs Continuous Presence**

In online learning environments, learners are more likely to feel that "someone is there" when needed [11] and in general, they like to enjoy a strong interaction and feel that there is a "human presence" [12].

For Hersh [13], the more the exchanges that occur within an OLE have common features with those that occur in classrooms, the more students will feel connected and engaged in their learning tasks. McCluskey [14] found that the presence of teachers or tutors in an online course is an important factor influencing their success. Above all, he emphasized the presence that manifests itself through the various interventions such as: frequent feedback, clear communications, organizing and maintaining actions, providing students with clear goals, and strong direct instruction, etc.

2.3   The Causes of Drop-Out

As we have been able to show before, the main limitations related to the pedagogical model of MOOCS leads to dropout rates. This is a recurring problem that has resulted in a lot of recent research.

El Mhouti [15], presents a synthesis of the literature on the different reasons. He cites the main reasons as follows: no intention to complete, starting late, lack of time, course difficulty, lack of support, lack of digital or learning skills, bad experiences and expectations, peer reviewing, no adaptation is provided. Other authors point out other reasons for abandonment, such as: poor time management, lost rhythm, too difficult course, learning but not doing homework, poor course design [17].

Among the other reasons that have been highlighted in the literature: The difficulty of studying online courses after work, changes in job responsibilities, lack of support from family or employer, lack of feedback on teacher evaluations, lack of Interactions with other participants and teachers.

3   Pedagogical Agents to Avoid the Drop-Out Problem in OLE

3.1   Pedagogical Agents Vs Drop-Out

As we presented below, the quality of OLE environments depends essentially on their flexibility, the capacity to adapt the delivery and ability to provide feedback and recommendations to maintain students engaged on courses. In that direction, the opportunities of using pedagogical agents in OLE become more and more important. Pedagogical agents are mainly reactive, autonomous and proactive. They can improve interactions and support learning. Indeed, agents have other important characteristics such as ability to perceive, to communicate, reason and act in specialized fields [5].

They also have the ability to cooperate with other agents, which makes them effective in the context of these environments. Research in that domain is not new. Pedagogical agents have already made their proof to simulate collaborative and adaptive behaviors as they appear in some particular works. Different Intelligent Tutoring System (ITS) use agents as virtual entities emulating a human tutor adapting content to the learner's needs, profiles, preferences, rhythm, style [16, 18, 19]. Pedagogical agents are also used as learning aids and recommender agents to adapt content to user profiles [10]. Recent research indicates that agent's can learns from activities and the performance of a user or a group of users, and predict pedagogical decisions and interventions. All of this allows us to make the hypothesis that the agent-approach appears as an interesting technology and natural way to model adaptive feature in OLE.

3.2   Pedagogical Agents in Traditional Learning Environments

The Intelligent Tutoring System (ITS) is an historical precursor of the new learning environments, with promising results. The ITSs have previously relied on artificial intelligence techniques and had as main objective to simulate the trainer (or the interactions between the learner and the tutor). They constitute the first generation of learning environments which set up pedagogical agents. Pedagogical agents are agents whose function is educational or pedagogical and whose aim is to improve learning. A profound analysis of these pedagogical agents shows that they are very complex and efficient. We can underline for example, the agent STEVE [20], a personified agent working in a virtual training environment; or BAGHERA [21], which relies on a distributed multi-agent system where each agent can act as tutor, learner-assistant or as a teacher-assistant. Another famous intelligent agent is "AutoTutor" [22], which helps students learn new notions in Newtonian mechanics, computer science, or scientific reasoning through a natural language dialogue that it establishes with learners. Three types of educational agents have marked the history of ITSs: pedagogical agents, assistant agents and recommendation agents. Without going into the details of the differences between these agents, we will simply say that the main objective of agents is to play different and important roles in a learning environment such us being present and reactive in order to maintain the motivation of learners. Research suggests that pedagogical agents have the ability to play many roles in the multimedia learning environment, such as demonstrating, scaffolding, coaching, modeling and testing [23]. Animated educational agents are supposed to play a playful role in guiding, reflecting and interacting with learners. Recent research shows that these characters can support the commitment and motivation of the learners [5, 24, 25].

3.3 Pedagogical Agents in New Learning Environments (LMSs and MOOCs)

To our knowledge, little research has been carried out on the integration of pedagogical or recommendations agents in MOOCs and LMSs. As a well known agent in the context of LMS, we mention the agent ABITS (Agent Based Intelligent tutoring system) [26]. Suh and Lee [27] developed an extensible collaborative learning agent that was used to promote interaction among learners. Another example is the one proposed in the work of Lin [28] which have developed several agents communicating with the platform MOODLE using JADE. Other work are in the process of emergence in the context of MOOC learning environments, we so emphasize the work of [5, 10].

Additional research, although independent of LMS and MOOC platforms, claim that Pedagogical Agents foster Engagement, Motivation, and Responsibility [6, 29]. They suppose also that Pedagogical Agents are adaptable and versatile and can Address Learners'Sociocultural Needs. Similar research suggests integrating agents into MOOCs to adapt learning resources to the learner based on his preferences and learning style [15]. Research made in [30] proposes a Recommendation System for MOOCs based on the concept of generating predictions according to other learners' experiences.

Finally, although this research on the integration of agents in OLEs (LMS and MOOC) is rare, we find some attempts almost similar to the ITSs, but with the new vision and the new characteristics of the OLE. Interest is not the least, and the rest of this paper proposes the specifications of a pedagogical agent for these environments based on the concept of educational intervention and other characteristics of agents raised before.

4 The Paole Project

4.1 Paole Project

PAOLE (Pedagogical Agent for Online Learning Environments) is a project of the IRF-SIC Laboratory, university IBN ZOHR in Morocco, whose main goal is to examine what agent technologies can bring to the motivation of learners and to minimize the Drop–Out problem. It is a continuation of the various work carried out on personalization and adaptivity in OLE. In the first development of this Project, a roadmap has been drawn up to take into account a number of constraints, including the fact of taking account agent characteristics in general, but also the characteristics of pedagogical agents. It must also, as we will present next, address the problems highlighted at the beginning of this article on MOOCs.

4.2 Paole Specifications

A first review of the literature allowed us to distinguish some strategies of the pedagogical intervention of an agent, as we described above. First, we have mentioned four basic principles to be taken into account in MOOCs and OLE in general, namely: (1) the need to take into account the different profiles of learners, (2) continuous presence, (3) assistance and feedback, and (4) adaptation. Other general strategies can be emphasized, such as: rewarding the effort constantly for failures findings, add frequent and clear comments. We add to this the encouragement of collaboration in forums and social media. It is also necessary to reinforce the observation of traces of the learner and their analysis in order to determine a better intervention strategy. But it goes much better when the intervention is playful and in our case played by an animated agent, which sometimes can distract and consequently distress the learner in failure or in a demotivation state.

The work we present in this section is concerned with an attempt of specifications of an animated pedagogical agent, having the general capacities of the agents and the abilities that we have just described. From an educational point of view, he is classified in the category of a facilitator, but with an additional role of guide and companion. In this context, the pedagogical agent must interact with learners to minimize learner frustration, and enhance learning.

We propose a design based on the concept of pedagogical intervention which we describe in the following. The rest of the paper addresses an area, which to our knowledge is relatively unexplored in the field of OLE. This article draws inspiration from the general importance of the design of the intervention, situating it within the broader landscape of learning analysis, and then examines the specific issues of intervention design for The OLE. In the following, we refer to our agent by the abbreviation PAOLE (Pedagogical Agent for Online Learning Environments).

4.3 The Pedagogical Intervention

4.3.1 Structure of the Pedagogical Intervention

The pedagogical intervention structure constitutes the main component of PAOLE (see Fig. 1).

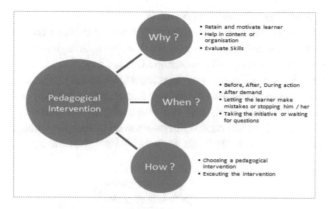

Fig. 1. Structure of the pedagogical intervention.

Although this notion of intervention is not very common in the literature, we have chose it to refer to the different actions of an agent related to the process of help, cooperation, collaboration, adaptation, observation, etc. We note that these actions are not of the same level of intervention and therefore the notion of intervention concerns different levels of learning. Lenoir [31] analyzes this notion in the context of traditional learning, and states that pedagogical intervention is a set of interactions between "the learner, learning objects and the teacher, in connection to the purposes underlying these reports". Following this research, the notion of intervention in PAOLE is the inter-action between the learner, the agent and the contents.

Our approach is to determine the importance of the intervention design, placing it in the largest landscape of educational agents. The pedagogical intervention design in this context is concerned with addressing questions such as: why intervene in learning and the teaching process, how should the agent intervene and why (Fig. 1).

To carry out its task, the pedagogical agent PAOLE will have to put in place an effective pedagogical strategy and, if possible, be able to change strategy according to the situation. The pedagogical agent here uses this model to select and adapt his role, his pedagogical strategy, and choose the activities he will put in place. As shown in the figure above, this model allows the answer to three main questions:

- **Why:** for this question, a pedagogical intervention occurs for several reasons, such as: motivating and retaining a learner, helping him to understand a concept or in order to solve a problem, etc.;
- **When:** the moment when the pedagogical intervention is carried out is important in learning. Intervention can take place before, during or after the learner's actions, i.e. the agent must choose the appropriate time either by taking the initiative or responding to the needs and questions of the learner;
- **How:** the answer to this question involves the selection of the most adapted intervention strategy to execute thereafter among a set of strategies.

4.3.2   Process of Educational Intervention

In this section, we present a set of processes that can be used by the agent PAOLE to design pedagogical interventions that support a productive learning. The different processes related to a pedagogical intervention that we have modeled are represented in the Fig. 2. We can distinguish six important processes, namely: (i) execute the session, (ii) observe learner, which can lead to analysis and update of the learner profile, (iii) the process of detecting a problem, (vi) the proposal and calculation of intervention strategies, (v) the choice of an adapted intervention and its implementation, and (vi) the operationalization of the strategy in an animated behavior of the agent.

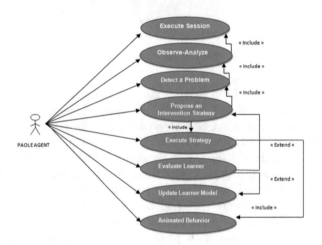

Fig. 2. The general use case of PAOLE

4.4   Modeling Elements of Paole

4.4.1   The Architecture of PAOLE

PAOLE is an autonomous Pedagogical Agent that supports human learning in online Learning Environments with the main objective of keeping them motivated.

From a computer architecture perspective, PAOLE consists of four components: (i) **a reasoning engine**, which monitors student's interactions and generates appropriate pedagogical interventions, (ii) **a behavior engine** responsible of generating behavior from primitive animations, sounds and speech elements, (iii) **a presentation manager**, which enables to present generate and present agents' animations, and (iv) a **communication module** for interactions with the other components of the learning environment as: the learner model and the domain model (see Fig. 3). PAOLE is currently under development. It is not created from scratch; we use the interface of the Microsoft agent, which was grafted for the first tests to the platform Moodle.

An animated agent action of PAOLE allows combining several elementary actions of the same character: messages, highlights of components, animations (show a component, applaud, greetings, etc.), and movements on the screen. Other high-level actions based in particular on intervention strategies are represented in the Fig. 4, in

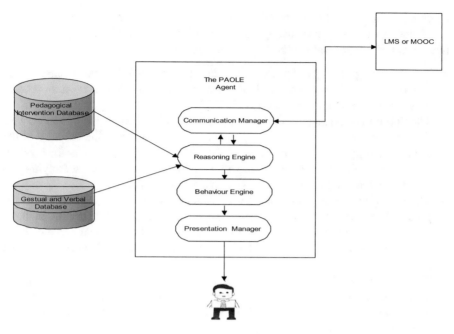

Fig. 3. The architecture of PAOLE.

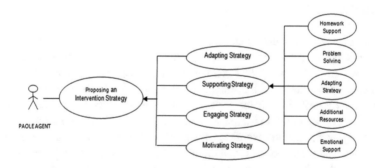

Fig. 4. A use case of PAOLE strategies

particular the strategies: adaptation, support, engagement and motivation that we present in the following.

4.4.2 Some Use Cases of PAOLE

Research suggests that pedagogical agents have the ability to play many roles in the multimedia learning environment, such as demonstrating, scaffolding, coaching, modeling and testing [23].

Research collect in this paper and developed by ourselves and others in the context of pedagogical agents provides further insight into the principles and processes we adopt as we described below. In this context, our agent can be characterized by four capabilities that guide its pedagogical intervention (Fig. 4). These capabilities are intervention strategies and can be summarized as follows:

- **The Adapting strategy:** by observing and analyzing the different behaviors and outcomes of the learner, the PAOLE agent can propose different types of adaptations: content, presentation, navigation, etc.;
- **The Supporting strategy:** a learner may at some point need help and support. Two situations are possible: either at the demand of the learner or at the initiative of the agent. The support that can be offered by the agent PAOLE may be at the level of: homework, solving a problem, providing additional resources, modifying the interface, or at another level such as the emotional support;
- **The Engaging strategy:** the engagement strategy involves engaging a learner in the proposed courses. This strategy will, for example, engage the learner in these courses by offering him varied and advanced content. Involvement can also be at the flexibility of the course agenda by proposing additional weeks to finish homework;
- **The Motivating strategy:** motivation can be expressed in questions addressed to the learner by the agent, by offering encouragement, providing feedback, encouraging collaboration, encouraging the learner to make other attempts in multiple choice questions, or providing memorable examples.

5 Conclusion

Following the various issues presented, and within the limits to be met by this paper, we hope to put a first draft of the current problems related to the issue of the Drop-Out Problem in Online Learning Environments. After examining the literature, we have given a characterization of the different solutions to keep learners motivated. We first examined the interest of animated pedagogical agents in order to carry out a continuous presence, to propose rapid assistance and feedback, adaptation to the learner model, etc. Other strategies for the intervention of educational agents have also been specified, such as: the encouragement of collaboration in forums and social media, the reinforcement of the observation of traces of the learner and their analysis in order to determine a better intervention strategy. We applied all these elements to give the first specifications of the agent PAOLE, emphasizing the strategies of adaptation, support, commitment and motivation. It is clear that several issues remain to be addressed to the expected system. Our work continues along these lines to try to finish a complete and a stable model of the PAOLE agent, which will be tested and validated in a wide spectrum.

For citations of references, we prefer the use of square brackets and consecutive numbers. Citations using labels or the author/year convention are also acceptable. The following bibliography provides a sample reference list with entries for journal articles [1], an LNCS chapter [2], a book [3], proceedings without editors [4], as well as a URL [5].

References

1. Wilen-Daugenti, T.: Technology and learning environments in higher education. Peter Lang, New York (2009)
2. Qazdar, A., Cherkaoui, C., Er-Raha, B., Mammass, D.: AeLF: Mixing Adaptive Learning System avec système de gestion de l'apprentissage. Int. J. Comput. Appl. **119**(15) (2015)
3. Cherkaoui, C., et al.: Un modèle d'adaptation dans les environnements d'apprentissage en ligne (LMS et MOOC). (SITA'2015), 10ème Conférence internationale sur les systèmes Intelligents. IEEE, 2015 (2015)
4. Bonk, C.J., Lee, M.M., Reeves, T.C., Reynolds, T.H. (eds.): MOOCs and Open Education. Routledge, New York (2015)
5. Daradoumis, T., Bassi, R., Xhafa, F., Caballé, S.: A review on massive e-learning (MOOC) design, delivery and assessment. In: P2P, Parallel, Grid, Cloud and Internet Computing Conference (3PGCIC). IEEE, October 2013
6. Baylor, A., Kim, Y.: Validating pedagogical agent roles: expert, motivator, and mentor. In: Lassner, D., McNaught, C. (eds.) Proceedings of EdMedia: World Conference on Educational Media and Technology 2003, pp. 463–466. AACE (2003)
7. Yuan, L., Powell, S., Cetis, J.: MOOCs and Open Education: Implications for Higher Education (2013)
8. Bakki, A., Oubahssi, L., Cherkaoui, C., George, S.: Motivation et engagement dans les MOOC: comment accroître la motivation de l'apprentissage en adaptant les scénarios pédagogiques? Dans Design for Teaching and Learning in a Networked World, pp. 556–559. Springer International Publishing (2015)
9. Hill, P.: The most thorough summary (to date) of MOOC completion rates, e-Literate, 26 February 2013
10. Zaïane, O.R.: Building a recommender agent for e-learning systems. In: Proceedings of International Conference on Computers in Education 2002, pp. 55–59. IEEE, December 2002
11. Galan, J.-P., Sabadie, W.: Evaluation du site Web: une approche par l'expérience de service, 17ème Congrès International de l'Association Française de Marketing, Deauville, pp. 1–26 (2001)
12. Gulz, A.: Benefits of virtual characters in computer based learning environments: claims and evidence. Int. J. Artif. Intell. Edu. **14**(3, 4), 313–334 (2004)
13. Hersh, D.E.: The Human Element (2016). https://www.insidehighered.com/news/2010/03/29/lms
14. McCluskey, F., Kupczynski, L., Ice, P., Wiesenmayer, R.: Student perceptions of the relationship between indicators of teaching presence and success in online courses. J. Interact. Online Learn. **9**, 23–43 (2010)
15. El Mhouti, A., Nasseh, A., Erradi, M.: Stimulate engagement and motivation in moocs using an ontologies based multi-agents system. Int. J. Intell. Syst. Appl. **8**(4) (2016)
16. Stoilescu, D.: Modalities of using learning objects for intelligent agents in learning. Int. J. Doctoral Stud. **4**, 49–64 (2009)
17. Nawrot, I., Doucet, A.: Building engagement for MOOC students: introducing support for time management on online learning platforms. In: Proceedings of the 23rd International Conference on World Wide Web, pp. 1077–1082. ACM (2014)
18. Frasson, C., Mengelle, T., Aimeur, E.: Using pedagogical agents in a multi-strategic intelligent tutoring system. In: Workshop on Pedagogical agents in AI-ED, vol. 97, pp. 40–47, August 1997

19. Johnson, W.L., Rickel, J.W., Lester, J.C.: Animated pedagogical agents: face-to-face interaction in interactive learning environments. Int. J. Artif. Intell. Edu. **11**(1), 47–78 (2000)
20. Rickel, J., Johnson, W.L.: Animated agents for procedural training in virtual reality: perception, cognition, and motor control. Appl. Artif. Intell. **13**(4–5), 343–382 (1999)
21. Pesty, S., Webber, C., Balacheff, N.: Baghera: une architecture multi-agents pour l'apprentissage humain. Agents Logiciels, Cooperation, Apprentissage et Activité Humaine ALCAA, pp. 204–214 (2001)
22. Graesser, A.C., Lu, S., Jackson, G.T., Mitchell, H.H., Ventura, M., Olney, A., Louwerse, M. M.: AutoTutor: a tutor with dialogue in natural language. Behav. Res. Methods **36**(2), 180–192 (2004)
23. Clarebout, G., Elen, J., Johnson, W.L., Shaw, E.: Animated pedagogical agents: an opportunity to be grasped? J. Edu. Multimedia Hypermedia **11**(3), 267–286 (2002)
24. Baylor, A.L., Chang, S.: Pedagogical agents as scaffolds: the role of feedback timing, number of agents, and adaptive feedback. In: International Conference of the Learning Sciences, Seattle, WA (2002)
25. Craig, S.D., Driscoll, D.M., Gholson, B.: Constructing knowledge from dialog in an intelligent tutoring system: interactive learning, vicarious learning, and pedagogical agents. J. Educ. Multimedia Hypermedia **13**(2), 163 (2004)
26. Capuano, N., Marsella, M., Salerno, S.: ABITS: an agent based Intelligent Tutoring System for distance learning. In: Proceedings of the International Workshop on Adaptive and Intelligent Web-Based Education Systems, ITS (2000)
27. Suh, H., Lee, S.: Collaborative learning agent for promoting group interaction. ETRI J. **28** (4), 461–474 (2006)
28. Lin, F.O.: Integrating JADE Agents into MOODLE (2010)
29. Veletsianos, G., Russell, G.S.: What do learners and pedagogical agents discuss when given opportunities for open-ended dialogue? J. Edu. Comput. Res. **48**(3), 381–401 (2013)
30. Garg, V., Tiwari, R., Gwalior, A.I.: Hybrid Massive Open Online Course (MOOC) recommendation system using machine learning. In: International Conference on Soft Computing Techniques in Engineering and Technology (2016). http://asctet.co.in/papers/OR0033.pdf
31. Lenoir, Y.: Relations entre interdisciplinarité et intégration des apprentissages dans l'enseignement des programmes d'études du primaire au Québec (Doctoral dissertation, Paris 7) (1991)

A Modular Multi-agent Architecture
for Smart Parking System

Khaoula Hassoune[✉] ⓘ, Wafaa Dachry, Fouad Moutaouakkil, and Hicham Medromi

(EAS-LRI) Systems Architecture Team, Hassan II University, ENSEM, Casablanca, Morocco
khaoula.hassoune@gmail.com

Abstract. Cities noticed that their drivers had real problems to find a parking space easily, the difficulty roots from not knowing where the parking spaces are available at the given time. In this paper we will design an automatic smart parking architecture using multi-agent and expert systems which are the main domains of artificial intelligence. AI is accomplished by studying how human brain thinks and how humans learn, decide, and work while trying to solve a problem, and then using the outcomes of this study as a basis of developing intelligent software and systems. Implementing this scalable and low cost car parking framework will provides a lot of services for the driver: driver guidance, automatic payment, parking lot retrieval, Gate management, security and low cost of implementation.

Keywords: Smart parking system · Multi-Agent Systems · Expert systems

1 Introduction

Researchers are recently turned to applying technologies for management of parking area by designing and implementation of a prototype system of smart parking that allows vehicle drivers to effectively find the free parking places, making a reservation and payment. In the future the demand for the intelligent parking service will increase because the rapid growth in the automotive industries. The automatic management of parking lots by accurate monitoring and providing service to the customers and administrators is provided by such emerging services. An effective solution to this service can be provided by many new technologies.

This paper describes a dynamic architecture for management of a smart parking system based on multi-agent and expert systems. One of the most characteristics is the use of intelligent agents as the main components which focus on distributing the majority of the system's functionalities into processes. The paper is organized as follows: after a brief introduction we will discuss in the second section a state of the art of smart parking systems, and then in the third section we present the concepts of multi-agents and expert systems. Finally, we present a modular architecture for new smart parking management. The last section concludes our work and draws some perspectives.

© Springer International Publishing AG, part of Springer Nature 2018
M. Ben Ahmed and A. A. Boudhir (Eds.): SCAMS 2017, LNNS 37, pp. 357–365, 2018.
https://doi.org/10.1007/978-3-319-74500-8_33

2 State of Art

Every day vehicle drivers have to find a vacant parking space especially during the rush hours. It is time-consuming and it is leading to more traffic, and air pollution.

The authors in [1–3] present the design and implementation of a smart parking system based on wireless sensor networks that allow vehicle drivers to find the free parking places. Also in [4] the author presents a wireless system for locating parking spots remotely via smartphone. This system automates the process of locating an available parking spot and paying for it.

Authors in [5] have proposed a scalable and low cost car parking framework (CFP) based on the integration of networked sensors and RFID technologies. These include driver guidance, automatic payment, parking lot retrieval, security and vandalism detection.

In others studies the authors have choose to design an automatic smart parking using internet of things which enables the user to find the nearest parking area and the available slot in that area [6–8].

From the previous state of art we remark that Researchers have promoted some services to the detriment of others. For that reason we will propose a new architecture that is based on multi-agent and expert systems. We should integrate the two different technologies together in order to achieve a system which is the most efficient, reliable, secure and inexpensive.

The proposed model is a modular multi-agent architecture where all processes are managed and controlled by different types of agents which are able to propose solutions, cooperate, on very dynamic environments and face real problems.

There are different kinds of agents in the architecture, each one with specific roles, capabilities and characteristics. This fact facilitates the flexibility of the architecture in incorporating new agents.

Our system will be divided in two architectures:

- Multi-agent architecture.
- Hardware architecture.

3 Overview of the Global Multi-agent Architecture

Many drivers had real problems to find a parking space easily especially during peak hours, the difficulty roots from not knowing where the parking spaces are available. Even if this is known, many vehicles may pursue a small number of parking spaces which in turn leads to traffic congestion. The traffic in parking space has been an area of concern in majority of cities. So, parking monitoring is an important solution.

From the previous state of the art it is noted that most authors have divided their architectures in different modules but no one of them have taken the preferences of the driver into consideration. For that reason we will propose a multi-agent system that is capable of providing parking services to the driver based on their personal preferences.

In this section, we give an overview of the multi-agent architecture which provides a high level model for smart parking management (Fig. 1).

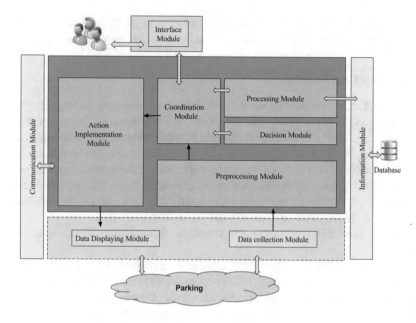

Fig. 1. Global smart parking architecture.

In order to understand deeply the common architecture, we describe below each layer of it.

- Decision Module: provide knowledge related to the regulations for agents. It is an expert system and its knowledge base mainly contains information related to the environmental performance and legislations.
- Communication Module: This layer ensures communications between all the layers of the architecture.
- Processing Module: This layer contains different agents, which can be implemented, responding to communication layer's alert and users requests.
- Preprocessing Module: Is responsible for the pre-processing of data captured from the environment (Sensors, Cameras, RFID).
- Coordination Module: It has the role of displaying information to the user in a suitable manner taking into account the constraints of his device. And it is responsible for the transmission of the user request to a specific agent of processing Module.
- Interface Module: It is responsible for capturing the user's query, as well as displaying the results.
- Action Implementation Module: Represents the set of behaviors necessary for the Parking control.
- Database: This part of the system includes all the data and tables used by all components of the platform, including static data, indicators related to agents, and appropriate decisions to the various scenarios of behavior to be submitted to Query Agent depending on the state of the collaboration between process agents.

- Information Module: manages interaction between the platform agent's and database of the system. It retrieves adequate data and sends it to the concerned agent.

3.1 Proposed Architecture

The traffic on roads and parking space has been an area of concern in majority of cities. So, parking monitoring is an important solution. To avoid these problems, recently many new researches have been developed that help in solving the parking problems to a great extent but no system had taken the preferences of the driver into considerations. For that reason our major objective is to design and implement architecture of a smart parking.

Implementing this scalable and low cost car parking framework will provides a lot of services for the driver: driver guidance, automatic payment, parking lot retrieval, Gate management, security and low cost of implementation.

Our work is based on the multi-agent and expert systems approaches because of their benefits.

The combination of these two approaches will encompasses cooperation, resolution of complex problems, modularity, efficiency, reliability, reusability and lies under the conjunctive use of knowledge as behavioral models of the experts.

Our proposed solution mainly focuses on analyzing user's queries to find a vacant slot based on their preferences.

Artificial Intelligence

Artificial Intelligence is a way of making a computer, a computer-controlled robot, or a software think intelligently, in the similar manner the intelligent humans think.

AI is accomplished by studying how human brain thinks and how humans learn, decide, and work while trying to solve a problem, and then using the outcomes of this study as a basis of developing intelligent software and systems.

The goals of AI are:

- To Create Expert Systems: The systems which exhibit intelligent behavior, learn, demonstrate, explain, and advice its users.
- To Implement Human Intelligence in Machines: Creating systems that understand, think, learn, and behave like humans.

An AI system is composed of an agent and its environment. The agents act in their environment. The environment may contain other agents.

Agent and Multi Agent Systems (MAS)

Agents are sophisticated computer programs that act autonomously on behalf of their users, across open and distributed environments, to solve a growing number of complex problems. Increasingly, however, applications require multiple agents that can work together. A multi-agent system is a set of software agents that interact to solve problems that are beyond the individual capacities or knowledge of each individual agent. We call a "platform" whatever allows those agents to interact not taking into consideration the shape that such a platform can take (centralized or not, embedded into the agents or not, …). This platform usually provides agents with a set of services depending on the system needs and is considered as a tool for the agents, it does not exhibit an autonomous or pro-active behavior.

Agents, according to MAS community have the following properties:

- Autonomy: An agent possesses individual goals, resources and competences; as such it operates without direct human or other intervention, and has some degree of control over its actions and its internal state. One of the foremost consequences of agent autonomy is agent adaptability as an agent has the control over its own state and so can regulate its own functioning without outside assistance or supervision.
- Sociability: An agent can interact with other agents, and possibly humans, via some kind of agent communication language. Through this means, an agent is able to provide and ask for services.
- Reactivity: An agent perceives and acts, to some degree, on its close environment; it can respond in a timely fashion to changes that occur around it.
- Pro-activeness: Although some agents, called reactive agents, will simply act in response to stimulations from their environment, an agent may be able to exhibit goal-directed behavior by taking the initiative.

Expert System

One of the largest areas of applications of artificial intelligence is expert systems (ESs), or knowledge based systems as they are sometimes known. [9] Provides us with the following definition: An expert system is a computer program that represents and reasons with knowledge of some specialist subject with a view to solving problems or giving advice.

To solve expert-level problems, expert systems will need efficient access to a substantial domain knowledge base, and a reasoning mechanism to apply the knowledge to the problems they are given. Usually they will also need to be able to explain, to the users who rely on them, how they have reached their decisions. They will generally build upon the ideas of knowledge representation, production rules, search, and so on, that we have already covered.

Contribution of MAS and Expert Systems

The Multi-agent approach is justified by:

- Adaptation to reality
- Cooperation,
- The resolution of complex problems,
- Integration of incomplete expertise,
- Efficiency,
- Reliability,
- Reuse.

Expert systems have a lot of attractive features:

- Increased availability,
- Reduced cost,
- Reduced danger,

- Permanence,
- Increased reliability,
- Explanation,
- Fast and complete response at all times,
- Intelligent Database,
- Multiple expertise.

3.2   Description of the Proposed Architecture

The proposed architecture is able to connect the parking database where the parking information is stored and also connects to the knowledge base related to the environment (the environmental knowledge base). The environmental knowledge base will be shared between the agents in all stages of parking process.

The purpose of the system is to manage the parking places in a way to reduce the traffic congestion and time looking for free places. Agents require a knowledge representation in which to analyze and find solutions for helping users to find parking places.

The proposed architecture uses a rule-based reasoning to examine the proper solution for drivers. One of the most popular techniques used in the artificial intelligence is Rule-based reasoning. The rule-based architecture [10] has two major components: Knowledge based that contains the general knowledge about the problem which is a set of the production rules identified as "IF…THEN…" and inference engine is a mechanism to process the rules.

In our work, the environmental regulations such parking Entrance, Parking Exit and Mobile application have been modeled in form of IF-THEN rules.

In our multi-agent system, we apply a rule-based reasoning technique on the environmental regulations in order to make the right decision. Figure 2 shows the IF-THEN rules of parking entrance.

Now we will give an overview of the IF-THEN rules of Mobile application.

- Case1: Registered driver (Category = Normal). This means that the driver can make reservation on mobile application.

Rule1:   IF driver is registered and category is normal
THEN check status of the payment

Rule2:   IF driver is registered and category is normal and the status of payment is valid
THEN Check availability of parking spots

Rule3:   IF driver is registered and category is normal and the status of payment is Invalid
THEN Recharge subscription card

Rule4:   IF driver is registered and category is normal and the status of payment is valid
and parking spots available
THEN the driver can make a reservation

Rule5:   IF driver is registered and category is normal and the status of payment is valid
and parking available spots and a place is reserved
THEN the driver can make guidance

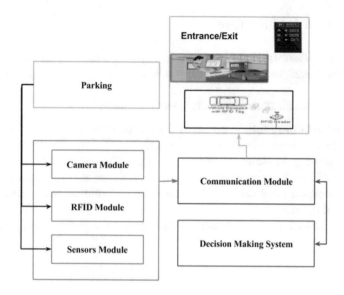

Fig. 2. Hardware architecture of smart parking system

- Case2: Registered driver (Category = VIP). The driver can't reserve on the application since a place has been allocated to him.

Rule1: IF driver is registered and category is VIP
        THEN check status of the payment

Rule2: IF driver is registered and category is VIP and the status of payment is valid
        THEN Check availability of parking spots

Rule3: IF driver is registered and category is VIP and the status of payment is valid
        THEN the driver can check his number place

Rule4: IF driver is registered and category is VIP and the status of payment is valid
        THEN the driver can check his number place

Rule5: IF driver is registered and category is VIP and the status of payment is valid
        THEN the driver can make guidance

- Case3: Simple user. The user must register and choose his/her category to have benefits.

Rule1: IF driver is new user
        THEN the user must register

Rule2: IF driver is new user and have chosen the category
        THEN make payment

4 Hardware Architecture

The physical (or Hardware) architecture of the system is made up of different systems interacting with each other to better carry out the different services requested by the driver.

Our Physical architecture is divided in two modules:

- **Data Collection Module:** It concerns the sensors, cameras and RFID Tag which capture the information from the environment.
- **Data Displaying Module:** acts directly on the LCD installed in the entrance of the parking, LED installed in every parking spots, and the gate management.
- **Communication Module:** This Module ensures communication between all the modules of physical architecture.
- **Decision making System:** is implemented to well lead the management of the parking.

5 Future Work

As a future work we will detail each module of our architecture, we will discuss the system architecture mainly agents characteristics and their behaviors. Also, we will describe the implementation of the proposed system including the interaction of agents and the connection between agents and knowledge bases.

Our objective is to validate the architecture that we propose in this paper by developing a distributed platform that provides a lot of services for the driver.

6 Conclusion

In this paper, we give an overview of different parking systems which was implemented by many researches to resolve the growing problem of traffic congestion, wasted time, wasting money, and help provide better public service, reduce car emissions and pollution. And we propose a multi-agent architecture which provides a high level model for smart parking management. For that reason we used different modern techniques such as Expert Systems and SMA. We have integrated the two different technologies together in order to achieve a system which is the most efficient, reliable, secure and inexpensive.

References

1. Yang, J., Portilla, J., Riesgo, T.: Smart parking service based on wireless sensor networks. In: IECON 2012-38th Annual Conference on IEEE Industrial Electronics Society, pp. 6029–6034. IEEE, October 2012
2. Yee, H.C., Rahayu, Y.: Monitoring parking space availability via ZigBee technology. Int. J. Future Comput. Commun. 3(6), 377 (2014)

3. Poojaa, A.: WSN based secure vehicle parking management and reservation system. In: National Conference on Research Advances in Communication, Computation, Electrical Science and Structures (NCRACCESS 2015) (2015)

4. Orrie, O., Silva, B., Hancke, G.P.: A wireless smart parking system. In: 41st Annual Conference of the IEEE Industrial Electronics Society, IECON 2015, pp. 004110–004114. IEEE, November 2015

5. Karbab, E., Djenouri, D., Boulkaboul, S., Bagula, A.: Car park management with networked wireless sensors and active RFID. In: 2015 IEEE International Conference on Electro/ Information Technology (EIT), pp. 373–378. IEEE, May 2015

6. Basavaraju, S.R.: Automatic smart parking system using Internet of Things (IoT). Int. J. Sci. Res. Publ. **5**(12), 629–632 (2015)

7. Suryady, Z., Sinniah, G.R., Haseeb, S., Siddique, M.T., Ezani, M.F.M.: Rapid development of smart parking system with cloud-based platforms. In: 2014 the 5th International Conference on Information and Communication Technology for the Muslim World (ICT4M), pp. 1–6. IEEE, November 2014

8. Gandhi, B.M.K., Rao, M.K.: A prototype for IoT based car parking management system for smart cities. Indian J. Sci. Technol. **9**(17), 1–6 (2016)

9. Jackson, P.: Introduction to Expert Systems, 3rd edn. Addison Wesley Longman, Harlow (1999)

10. Buchanan, B.G., Shortliffe, E.H. (eds.): Rule-Based Expert Systems. Addison-Wesley, Reading (1984)

Visual Vehicle Localization System for Smart Parking Application

Hicham Lahdili$^{(\boxtimes)}$ and Zine El Abidine Alaoui Ismaili

ENSIAS/Information, Communication and Embedded Systems (ICES) Team,
University Mohammed V, Rabat 10010, Morocco
dh.lahdili@gmail.com, z.alaoui@um5s.net.ma

Abstract. With a vision of proposing a fully automated parking management solution for smart parking, in which all the operations in the process of parking will be automated. And as a first step we will focus on vehicle localization inside parking based on image processing theory. Video based localization algorithms present an important interest in the field of intelligent video surveillance, the integration of such functionality in the surveillance system will revolt their classic roles. Navigation tool and other amazing systems can easily build based on such feature. This paper describes an implementation of a FPGA based real-time visual system for vehicle localization. Vehicle in the video frames are extracted after the application of the background subtraction method on the input image using a background reference image. The dynamic threshold used is computed by the Otsu method. Finally, the object mask resulting from the segmentation process is used to compute the relative distance to the camera based on the relation between the ratio of the size of a vehicle on the camera sensor and its size in real life which is a function of the camera focal length and distance between the vehicle and the camera. The experimental results show that the proposed system is sufficiently satisfying the real time constraint (under the 100 MHz frequency a 32 frames per second is achieved for the 1440 * 1080 resolution, and under 50 MHz frequency a 41 frames per second is achieved for the 640 * 480 resolution) with an accuracy error around the centimeter level.

Keywords: Smart parking · Vehicle localization · Image processing
Background subtraction · Otsu method · Real time · FPGA · HDL

1 Introduction

In nowadays, cities are constantly growing and with this swelling appear certain problems, one of these major challenges is the problem of parking management. In a vision of proposing a fully automated parking management solution, in which all the operations in the process of parking will be automated, comes our paper for presenting a visual vehicle localization system. The parking management system based on the vehicle localization information, and the available free place will generate commands with speed, direction and the angle of rotation information. The vehicle that will be equipped with a dedicated system designed for this purpose will execute the commands and move through the parking to its reserved place. As a result, the whole of the parking process will be done without human intervention.

© Springer International Publishing AG, part of Springer Nature 2018
M. Ben Ahmed and A. A. Boudhir (Eds.): SCAMS 2017, LNNS 37, pp. 366–378, 2018.
https://doi.org/10.1007/978-3-319-74500-8_34

The localization information generated by the visual localization system can be used beside of vehicle localization to build amazing tools for different purpose as guiding people during navigating in unfamiliar buildings like airport, museum, office building. And as our system is based on movement detection we can easily limit data storage for surveillance system to active area (The data issued from a camera will be stored only when moving objects are detected).

Despite the remarkable progress realized in image processing discipline, the complexity of the algorithms used (the increase in image resolutions, as well as the need to implement increasingly complex techniques) on the one hand, and the resources required (computational and memory) on the other hand, have made their implementation in embedded systems a challenging task. This complexity increases with the introduction of the constraints imposed by the standards according to the fields of application, ex: real time constraints, safety standards…

In this paper, we will present our proposal for a visual vehicle localization system (VLS), developed for static camera. Vehicles in the video frames are extracted after the application of the background subtraction method on the input image using a background reference image. The dynamic threshold used is computed by the Otsu method. Finally, the object mask resulting from the segmentation process was used to compute the relative distance to the camera based on the relation between the ratio of the size of the vehicle on the camera sensor and its size in real life which is a function of the camera focal length and distance between the vehicle and the camera.

The remainder of this work is organized as follows; Sect. 2 presents the related works, where the implemented algorithm and description of its parts is given in Sect. 3, in Sect. 4 the implementation board was presented in addition to the syntheses summary. Execution time characteristic of the implemented algorithm, as well as some experimental results are covered in Sect. 5. And finally Sect. 6 draws the conclusion.

2   Related Work

Positioning systems are more and more used in our daily life to perform more complex tasks such as navigation (everywhere in the globe and in all-weather conditions) or simply for pleasure like augmented reality games. The Global Positioning System (GPS) [8] which is the most widely used navigation system in the world receives signals from multiple satellites and employs a triangulation process to determine the physical localizations with an accuracy error around the meter level. The latter is in general acceptable for outdoor applications. Unfortunately, this system reaches its limitation within the buildings and closed environment because of the attenuation of electromagnetic waves. This limitation in addition to the low accuracy of traditional positioning system present the motivation to conduct various researches on different physical signal in a vision to build new positioning systems.

Wireless technologies such as ultrasonic [1, 2], infrared [3], radio-frequency based systems which may be radio-frequency identification [4], received signal strength of RF signals, Bluetooth and WIFI those systems are based on concepts like Time of arrival, Angle of arrival and Received signal strength indication to calculate the distance between the transmitters and the receiver [5–7]. The main disadvantage of the Wireless technologies that only object equipped with a receiver can be located.

Non-radio technologies like visual system. Different approaches are used in this category, visual marker based system [9]: markers are placed at specific locations, and when a device (mobile robot) identifies a marker, it can be localized thanks to the markers database. Map-based visual localization [10]: first a collection of successive image of an area (building, road, …) is used to build a dataset of images about this environment, then any device wants to be localized in this area will need just to take an image of its environment which will be compared to the dataset to find its current position. And finally real time visual localization system [17, 18]: the main idea behind those systems is based on the use of a camera network to localize generic objects such vehicle or people. Their major disadvantage is their low accuracy (0.37 m in the best case).

3   Proposed Algorithm

The main idea behind the proposed algorithm is that vehicle size in an image is a function of some camera characteristics (focal length and the sensor size) and vehicle's features such as weight, height and its distance to the camera. This means that for any vehicle with a known size in the field of view of a camera, we can estimate its distance to this one, if we can compute their dimensions in pixel in the image. And as we are only interested in moving objects, we will try firstly to find such objects, then extract their mask in the image, so their dimension, and finally compute the relative distance. The implemented system is described in the following diagram block (Fig. 1), which is composed of two main parts:

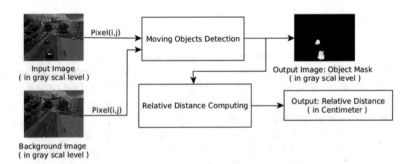

Fig. 1. Block diagram of the implemented VLS.

- Detection of vehicles in the scene.
- Calculation of the relative distance between the vehicle and the camera, then this distance will be used to compute the absolute distance.

The first step will be performed using the Background subtraction method, since our system will consist only of static cameras. The second step will be carried out by

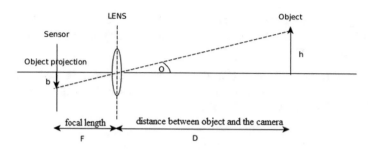

Fig. 2. Object projection on the camera sensor.

computing the size (in pixel) of the vehicle in motion and using some technical information about the vehicle and the camera we can deduce the distance between the vehicle and the camera.

3.1 Vehicle Detection

In the literature, several approaches are used for object detection, the most used are based on feature descriptor such the histogram of oriented gradient (HOG) [15, 16], in this approach the input image is converted to a feature vector, which simplifies the image by extracting useful information and throwing away extraneous information.

The computed feature vector is used by classification algorithms like Support Vector Machine and based on training data of positive (image with the object to be detected) and negative (image without the object to be detected) datasets. Objects are detected with a good accuracy. Unfortunately, object detectors can be painfully slow, especially when expensive features such as HOG need to be computed, it can really kill the performance. The background subtraction method presents another alternative, especially for indoor environments, where the lighting condition is approximately constant. And by taking into account their minimal implementation cost and its sufficient accuracy, as a result, we make the choice to go for background subtraction method.

Background Subtraction. The main idea of this method is based on the subtraction of the current image (in which the moving object is present) pixel-by-pixel from a reference background image as described in Eq. (1).

$$O(x, y, t) = |I(x, y, t) - B(x, y)| \tag{1}$$

Where:

$O(x,y,t)$ is the subtracted image.
$I(x, y, t)$ is the object image.
$B(x, y)$ is the background image.

This technique is designed for static camera, where the background is approximately the same in all frames, and by applying Eq. (1) background is removed from object image. As a result the histogram of the object image O(x, y, t) will be composed of two main pixel classes, background pixel class (near to the 0 gray scale level) and the object pixel class.

Otsu threshold. The second step in the background subtraction method is segmentation in order to get the foreground mask. The object image will be compared to a global threshold as presented in Eq. (2).

$$O(x, y, t) \leq T \tag{2}$$

Where:

O(x,y,t) is the subtracted image.
T is the threshold value.

If the pixel O(x, y, t) verifies Eq. (2), then it is considered as a foreground pixel, else it is a background pixel. T is a one global threshold, for all pixels in the image. And it needs to be a function of time, in other case the segmentation can easily be impacted by the environmental conditions change. The Otsu method will be used in order to meet this objective. The Otsu algorithm [13] is a popular dynamic thresholding method for image segmentation. Based on the idea that the image histogram can be divided into two classes, so it looks for a threshold that minimizes the variance for both classes. This way, each class will be as compact as possible. Only pixel value is taken into account for the Otsu algorithm the spatial relationship between pixels has no effect on the algorithm result, different regions with similar pixel value are treated as one region.

In Otsu method we exhaustively search for the threshold that minimizes the intra-class variance (the variance within the class), defined as a weighted sum of variances of the two classes:

$$\sigma w^2(t) = w(t) \cdot \sigma^2(t) + w'(t) \cdot \sigma'^2(t) \tag{3}$$

Where;

$\sigma w^2(t)$ is the intra-class variance
w and w' are the probabilities of the two classes separated by a threshold t
$\sigma^2(t)$ and $\sigma'^2(t)$ are variances of these two classes.

$$\sigma^2 = \sigma w^2(t) + \sigma b^2(t) \tag{4}$$

Where,

$$\sigma b^2(t) = w(t).w'(t).[\mu(t) - \mu'(t)]^2 \tag{5}$$

The Algorithm 1, which is based on the Eq. (5) step through all possible thresholds and keep the threshold value that maximizes the inter-class variance σb^2. The whole system computing the Background subtraction in addition to the Otsu method is described in the Fig. 3.

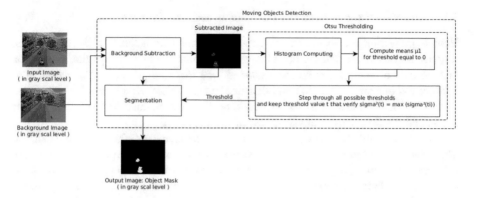

Fig. 3. Diagram block of the moving object detection module.

```
Algorithm 1: Otsu threshold
Input: grayscale_image
Output: Threshold
I = grayscale_image.weight * grayscale_image.height
σb² , σb_max, s, s' , w, w', μ,μ' ← 0
intensity ←255
histogram[255]
for j in 0 to intensity
      histogram[j] ← 0
for j in 0 to grayscale_image.weight
    for k in 0 to grayscale_image.height
          histogram[grayscale_image(j,k)] ← histo-
gram[grayscale_image(j,k)] +1
for j in 0 to intensity
       s ← s + histogram[j] * j
for j in 0 to intensity
       w ← w + histogram[j]
       if w = 0 then
            continue
       w'← I- w
       s' ← s' + histogram[j] * j
       μ ←  s'/w0
       μ' ← (s-s')/w'
       σb² ← w*w'*(μ-μ')²
       if  σb²>= σb_max then
            Threshold ← j
            σb_max ← σb²
Return: Threshold
```

3.2 Relative Distance Between the Vehicle and the Camera

Vehicle projection on camera sensor depend on three parameters vehicle dimensions, distance to the camera and the focal length as explained in Fig. 2. Equation 6 explains that the ratio of vehicle size on the sensor and the focal length is the same as the ratio between vehicle size in real life and distance to the vehicle. And in Eq. 7 vehicle size on the sensor is expressed as the vehicle size in pixels, divided by the image size in pixels and multiply by the physical size of the sensor.

The focal length and the sensor size are technical characteristic of the used camera. The vehicle size in pixels can be extracted from the segmented image (the output of the movement detection vehicle module). So we need just the vehicle size in real life to compute its relative distance, and as a result the absolute distance using the camera position information.

$$\tan(o) = \frac{b}{F} = \frac{h}{D} \rightarrow D = \frac{F.h}{b} \tag{6}$$

Where,
b   is the object size on the sensor.
H   is the object size in real life.
F   is the focal length (technical characteristic of the used camera).
D   is the distance between the object and the camera.

$$b = \frac{O.S}{I} \tag{7}$$

Where,
O   is the object size in pixels.
I   is the image size in pixels.
S   is the size of the sensor.

So, the whole equation can be rewritten as:

$$D = \frac{F.h.I}{O.S} \tag{8}$$

4 Implementation Using FPGA Board

Before going for the hardware implementation, a software implementation was already performed using a 650 MHz dual-core Cortex-A9 processor with an embedded Linux, but due to the limitation in the execution time 3 frame per second (fps), where a minimum of 30 fps is needed for real time video processing, we go for hardware acceleration, which presents an interesting choice for execution time improvement.

4.1 Board Description

With a vision to take the advantage of the parallelism feature provided by the field-programmable gate array (FPGA), we make the choice of implementing our system using the Artix-7 FPGA. The available fast RAM block space will be used to implement a dual - port RAM, which will be accelerate the treatment and give us more flexible architecture for our real-time image processing application (Fig. 5 and Table 1).

4.2 Simulation and Syntheses

The grayscale inputs images, foreground and background of the visual localization module are stored in two dual ports RAM, the process is started with computing the Otsu threshold for the subtracted image, then the segmentation is performed and the resulting object mask image is stored in a third dual port memory. The ISE Simulator (ISIM) was used to simulate the implemented system, the testbench (Fig. 4) reads the pixel's value of the foreground and background images (the ASCI PGM image format is used due to its simple manipulation) at each clock tick, and save the system result as an image in the same format. The inputs images Fig. 6(a) and (b) used in this example are part of the Background Models Challenge (BMC) [14], the resulting image from the background subtraction is presented in Fig. 6(c) and the object mask image is presented in Fig. 6(d). The simulation results of the presented example are presented in the Table 2.

Table 1. Implementation cost.

Logic utilization	Register	LUT	RAMB	BUFG
Used	837	448	41	1
%	1%	1%	29%	3%

Table 2. Simulation results.

Image resolution	640 * 480
Computed Otsu threshold	12
Computed object weight	49 pixels

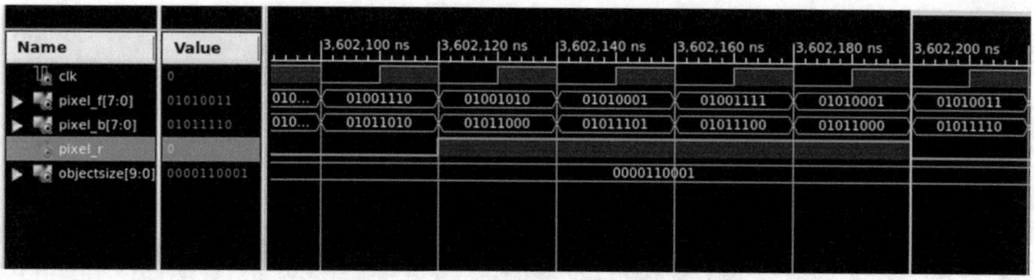

Fig. 4. System simulation using ISIM.

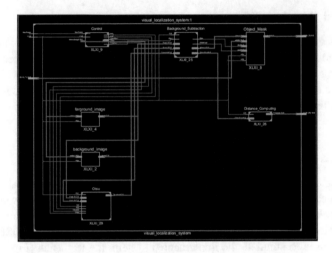

Fig. 5. RTL view.

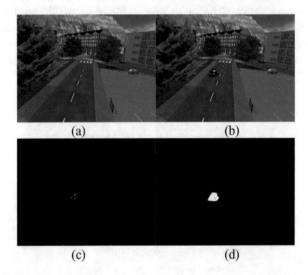

Fig. 6. (a) background image, (b) foreground image, (c) subtracted image, (d) object mask image.

5   Results and Evaluation

5.1   Temporal Analysis

The imaging device such as video cameras use the frames per second to explain the frequency (rate) at which an imaging device displays consecutive images called frames. The Phase Alternating Line (PAL) [12] and the National Television System Committee (NTSC) [11], which are the most used color encoding system for analogue television will be used as reference in our study. In the NTSC standard 30 frames are transmitted each second, each frame is made up of 525 individual scan lines. And in PAL 25 frames are transmitted each second, each frame is made up of 625 individual scan lines. Where a normal motion picture film is played back at 24 frames per second (fps). Thus we can consider that a real-time system is running at 30 fps.

This part, will evaluate the execution time for our system with different image resolution and frequency. The execution time characteristic of each block is presented in the Table 3. The execution time will be computed in function of the image resolution weight * height (W * H) using the operation number (clock ticks) unit Table 5. The detail of the Otsu threshold execution time is given in the Table 4.

Table 3. Block execution time in function of resolution and clock ticks.

Blocks	Background subtraction	Otsu	Segmentation	Relative distance
Operation number	W * H	W * H + 1792	W * H	W * H
Total	4 * W * H + 1792			

Table 4. Otsu block execution time in function of resolution and clock ticks.

Blocks	Histogram computing	Mean $\mu 0$ computing	Max σb^2 computing
Operation number	W * H + 256	256	1280
Total	W * H + 1792		

Table 5. Frames per second computing for different resolution in different clock frequency.

	640 * 480	1280 * 720
50 MHz	1 230 592 * 20 * $10^{\wedge -9}$ s = 0.024 s (>41 fps)	3 688 192 * 20 *$10^{\wedge -9}$ = 0.074 s (>13 fps)
75 MHz	1 230 592 * 13.3 * $10^{\wedge -9}$ s = 0.016 s (>62 fps)	3 688 192 * 13.3 *$10^{\wedge -9}$ = 0.049 s (>20 fps)
100 MHz	1 230592 * 10 * $10^{\wedge -9}$ s = 0.012 s (>83 fps)	3 688 192 * 10 *$10^{\wedge -9}$ = 0.036 s (>27 fps)

After some improvement of the implementation taking advantage of parallelism execution, the Background Subtraction and Histogram computing, Segmentation and Relative distance are grouped in the same block. Which means an optimization of 2W * H (Table 6). The new simulation values are given in Table 7. As a result, we notice that for 640 * 480 resolutions in the worst case frequency 50 MHz we achieve 83 fps more than the double of the needed fps. For 1280 * 720 resolutions a real time system can be built in frequency superior than the 56 MHz. And bigger resolution such 1440 * 1080 can be implemented under 100 MHz frequency.

Table 6. Block execution time in function of resolution and clock ticks.

Blocks	Background subtraction + Histogram computing	Otsu	Segmentation + Relative distance
Operation number	W * H + 256	1536	W * H
Total	2 * W * H + 1792		

Table 7. Frames per second computing for different resolution in different clock frequency.

	640 * 480	1280 * 720
50 MHz	0.012 s (>83 fps)	0.036 s (>27 fps)
75 MHz	0.008 s (>125 fps)	0.024 s (>41 fps)
100 MHz	0.006 s (>166 fps)	0.018 s (>55 fps)

5.2   Accuracy

In order to measure the real accuracy of our proposed system, we put a car in different distance from a fix camera Fig. 7 inside a small parking, and by using a fix camera the relative distance between the car and the camera is computed in real time from different distances. The real and computed distances are given in Table 8. Compared to similar work as [17, 18] as shown in Table 9, the maximal accuracy error of 0.1 M for the proposed system present an interesting improvement.

Fig. 7. Testing environment.

Table 8. Real and computed distance.

Real distance	Computed distance	Real distance	Computed distance
0.9 M	0.89 M	10 M	9.97 M
2 M	1.99 M	15 M	14.95 M
5 M	4.98 M	20 M	19.9 M

Table 9. Accuracy error compared to similar work.

Proposed system	System [18]	System [17]
0.1 M	0.34 M	0.8 M

6   Conclusion and Outlook

This paper presents an FPGA based Indoor real-time visual location system. The input image is segmented in order to extract the vehicle, this step is achieved using the background subtraction method, with a dynamic threshold computed using the Otsu method. The segmented image is then used to compute the relative distance to the camera. The proposed system is sufficiently satisfying the real time constraint 32 frames per second for the 1440 * 1080 resolution under 100 MHz frequency. In the next step, the presented algorithm will be improved and adapted in a vision of proposing a high accuracy visual navigation system for parking areas. This latter present one of the main parts in our approach of a fully automated parking solution, which will be presented in our future papers.

References

1. Ureña, J., Gualda, D., Hernández, Á., García, E., Villadangos, J.M., Pérez, M.C., García, J. C., García, J.J., Jiménez, A.: Ultrasonic local positioning system for mobile robot navigation: from low to high level processing. In: 2015 IEEE International Conference on Industrial Technology (ICIT), pp. 3440–3445, 17–19 March 2015. https://doi.org/10.1109/ICIT.2015. 7125610
2. Ijaz, F., Yang, H.K., Ahmad, A.W., Lee, C.: Indoor positioning: a review of indoor ultrasonic positioning systems. In: 2013 15th International Conference on Advanced Communications Technology (ICACT) (2013)
3. Huang, C.-H., Lee, L.-H., Ho, C.C.: Real-time RFID indoor positioning system based on Kalman-Filter drift removal and Heron-Bilateration location estimation. IEEE Trans. Instrum. Meas. **64**(3), 728–739 (2015). https://doi.org/10.1109/TIM.2014.2347691
4. Wang, C.-S., Chen, C.-L.: RFID-based and Kinect-based indoor positioning system. In: 2014 4th International Conference on Wireless Communications, Vehicular Technology, Information Theory and Aerospace & Electronic Systems (VITAE) (2014). https://doi.org/ 10.1109/vitae.2014.6934458
5. Xu, H., Ding, Y., Wang, R., Shen, W., Li, P.: A novel Radio Frequency Identification three-dimensional indoor positioning system based on trilateral positioning algorithm. J. Algorithms Comput. Technol. **10**(3), 158–168 (2016)

6. Wang, Y., Yang, X., Zhao, Y., Liu, Y., Cuthbert, L.: Bluetooth positioning using RSSI and triangulation methods. In: 2013 IEEE 10th Consumer Communications and Networking Conference (CCNC) (2013). https://doi.org/10.1109/CCNC.2013.6488558

7. Yang, C., Shao, H.: WiFi-based indoor positioning. IEEE Commun. Mag. **53**(3), 150–157 (2015). https://doi.org/10.1109/mcom.2015.7060497

8. Hofmann-Wellenhof, B., Lichtenegger, H., Collins, J.: Global Positioning System Theory and Practice, vol. 1. Springer, Vienna (2001)

9. La Delfa, G.C., Catania, V., Monteleone, S., De Paz, J.F., Bajo, J.: Computer vision based indoor navigation: a visual markers evaluation. In: Mohamed, A., Novais, P., Pereira, A., Villarrubia González, G., Fernández-Caballero, A. (eds.) Ambient Intelligence - Software and Applications. AISC, vol. 376, pp. 165–173. Springer, Cham (2015). https://doi.org/10.1007/978-3-319-19695-4_17

10. Wolcott, R.W., Eustice, R.M.: Visual localization within LIDAR maps for automated urban driving. In: IEEE/RSJ International Conference on Intelligent Robots and Systems (2014). https://doi.org/10.1109/iros.2014.6942558

11. https://en.wikipedia.org/wiki/NTSC

12. https://en.wikipedia.org/wiki/PAL

13. Vala, M.H.J., Baxi, A.: A review on Otsu image segmentation algorithm. Int. J. Adv. Res. Comput. Eng. Technol. (IJARCET) **2**(2), 387 (2013)

14. Background models challenge. http://bmc.iut-auvergne.com/

15. Dalal, N., Triggs, B.: Histograms of oriented gradients for human detection. In: 2005 IEEE Computer Society Conference on Computer Vision and Pattern Recognition (CVPR 2005). https://doi.org/10.1109/cvpr.2005.177

16. Malisiewicz, T., Gupta, A., Efros, A.A.: Ensemble of exemplar-SVMs for object detection and beyond. In: 2011 International Conference on Computer Vision (2011)

17. Einsiedler, J., Sawade, O., Schäufele, B., Witzke, M., Radusch, I.: Indoor micro navigation utilizing local infrastructure-based positioning. In: 2012 IEEE Intelligent Vehicles Symposium (2012). https://doi.org/10.1109/ivs.2012.6232262

18. Ibisch, A., Houben, S., Schlipsing, M., Kesten, R., Reimche, P., Schuller, F., Altinger, H.: Towards highly automated driving in a parking garage: general object localization and tracking using an environment-embedded camera system. In: 2014 IEEE Intelligent Vehicles Symposium (IV), 8–11 June 2014, Dearborn, Michigan, USA (2014). https://doi.org/10.1109/ivs.2014.6856567

RNA Secondary Structure an Overview

Abdelhakim El Fatmi[1,2(✉)], Arakil Chentoufi[1,2], M. Ali Bekri[1,2],
Said Benhlima[1,2], and Mohamed Sabbane[1,2]

[1] Faculty of Science, Moulay Ismail University, Meknes, Morocco
hakimelfatmi@gmail.com
[2] MACS Laboratory, Computer Science Department, Faculty of Science,
Moulay Ismail University, Zitoune, BP 11201 Meknes, Morocco

Abstract. It was believed that the single role of the Ribonucleic acid (RNA) is to carry the information necessitate to build a specific protein. Now it discovered that RNA has important and essential roles in many gene regulatory networks and many other cellular functions. Thus, the prediction of RNA structures becomes the subject of many studies in the last few years.

Determining the secondary structure of an RNA from its primary sequence is a challenging computational task. Various methods have been proposed to handle this problem. Initially, there are physical methods such as X-Ray, Crystallography, and Nuclear Magnetic Resonance. These methods are too costly, and they necessitate a lot of effort and so much time consuming. Therefore, the bioinformatics methods become highly needed.

In this paper, we will review the usually used approaches to predict RNA secondary structure counting the dynamic programming approach, the soft computing approach, the comparative approach, and the grammatical approach. Finally as perspective, we propose a method based on Genetic Algorithm principle and Greedy Randomized Adaptive Search Procedure (GRASP) method.

Keywords: Bioinformatics · Ribonucleic acid (RNA)
RNA secondary structure

1 Introduction

Ribonucleic acid (RNA) is a molecule consisting of a succession of nucleotides: Adenine (A), Cytosine (C), Guanine (G), and Uracil (U). In the search of the molecule structures, the last stage is the prediction of the three-dimensional shape of the molecule, a prediction with such precision with no information on the molecule except of its primary structure is largely out of reach for RNA molecules with a big size. The current physics methods such as X-Ray, Crystallography, and Nuclear Magnetic Resonance are too costly, they necessitate a lot of effort and so much time consuming [1]. That is why RNA secondary structure prediction has become crucial to the researchers and from the most important problematic in bioinformatics.

The secondary structure of an RNA sequence is formed when the RNA strand folds onto itself by forming hydrogen bonds between G-C, A-U, and G-U. So predicting RNA secondary structure return to predict the different hydrogen liaisons in an RNA molecule

© Springer International Publishing AG, part of Springer Nature 2018
M. Ben Ahmed and A. A. Boudhir (Eds.): SCAMS 2017, LNNS 37, pp. 379–388, 2018.
https://doi.org/10.1007/978-3-319-74500-8_35

using only the primary structure of the sequence. RNA secondary structure consists of many components, the stacked pairs or stems, the hairpin loops, the multi-branched loops, the internal loops, the bluge loops (Fig. 1), and a more complex topology called pseudoknot (Fig. 2). Pseudoknot is a special topology in RNA structures, this topology normally contains at least two stems such that the unpaired bases in a loop of a stem pair with bases outside the stem to form a new stem. Pseudoknots are more complex to predict compared to the other components, that is why many studies exclude its presence in the structure.

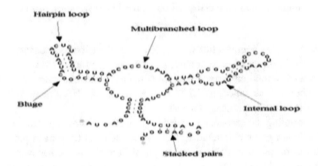

Fig. 1. Components of RNA secondary structure.

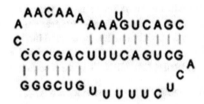

Fig. 2. A simple H-type pseudoknot.

2   Secondary Structure Prediction Approaches

RNA secondary structure prediction is a significant domain of research for many researchers. The prediction of the secondary structure represents an important step to solve many problematics related with the physical structure determination, such as the determination of the three dimensional structure and the interpretation of the biochemical abilities of the molecules.

In this section we will shed the light on the most popular approaches used to predict RNA secondary structure, we will start by Dynamic Programming approach (DPA), Soft Computing approach (SCA), Comparative approach (CA), and Grammatical approach (GA).

2.1 Dynamic Programming Approach (DPA)

DPA is based on the principle of dividing a complex problem into several subproblems, combining between this idea and the free energy minimization principle, various algorithms have been proposed to predict the RNA secondary structure. According to the free energy minimization principle, the secondary structure of an RNA sequence can be the most stable structure that has the lowest free energy.

The most basic DP algorithms provide simple secondary structure without including pseudoknots forms, as is augmented both time and space complexities. Among these algorithms, we can mention [2, 3].

Nussinov et al. [2] propose the first algorithm to predict RNA secondary structure based on the principle of the free energy minimization. In this algorithm the free energy is minimized when the number of base pairs is maximized. This algorithm takes $O(n^3)$ in time complexity. Later Zuker [3] proposes a famous algorithm to predict RNA secondary structure without pseudoknots called Mfold. This program predicts the secondary structure by minimizing the free energy according to the thermodynamic model proposed by Tinoco in [4].

There are other algorithms was developed specifically to predict pseudoknotted RNA secondary structure [5, 6].

In [5] Rivas and Eddy propose an algorithm able to generate secondary structure containing pseudoknot by the minimization of the free energy. This algorithm has a polynomial complexity $O(L^6)$ which in practice makes it difficult to use. Dirks and Pierce [6] develop an algorithm to predict RNA secondary structure but it restricted to only the simple type of pseudoknots which is H-type pseudoknot.

2.2 Soft Computing (SC)

Soft Computing (SC) is a group of methodologies that can work together or alone, to solve many real-life problems. SC methodologies exploit the tolerance of imprecision, uncertainty, approximate reasoning, and partial truth to give the optimal and the low solution cost [7].

Among the most used SC techniques to predict the RNA secondary structure there are Evolutionary Computation (i.e., Genetic algorithms), and Artificial Neural Networks (ANN).

Genetic algorithms (GAs). GAs are adaptive and powerful tools based on the principles of selection and evolution. GAs used to provide many solutions to a given problem [8].

GAs can be the preferable choice to handle various real-life problems when the search space is very large and complex, which make the conventional search methods useless.

To use GAs some operators should be defined such as selection, crossover, and mutation. And a fitness function to evaluate the quality of each solution. Predicting secondary structure using GAs is proposed in many works, among them, there are [9–12].

Based on the free energy minimization principle and the RNA folding pathways Van Batenburg et al. [9] present a GA for RNA secondary structure prediction. This algorithm starts by generating a list of possible stems, and then it creates the population by combining between the stems. Each element of the population is formulated by using the numbers 0 or 1. The crossover and the mutation are used as operators, and two types of fitness criterion have been used, the sum of stem length and the sum of stem stacking energies. Concerning the algorithm proposed by Wiese et al. [10], it generates all potential helices from the RNA sequence using a helix generation algorithm. Each helix is marked with a number ranging from 0 to n−1 where n is the number of all potential helices. The individuals of the population are presented by a combination of these numbers. The selection, the crossover, and the mutation are then applied to the solutions in an elitist model framework. Finally, the solutions with minimized free energy are accepted as possible solutions. Tong et al. [11] propose an algorithm called GAknot based on GA to predict RNA secondary structure including pseudoknots. GAknot starts by generating a set of helices, and then it creates each individual of the initial population by combining between these helices. GAknot uses three operators, crossover, replacement, and addition to provide an optimal solution. Finally, Shapiro and Navetta [12] develop another GA based on free energy minimization principle. The first step of this algorithm aims to generate a pool of stems, where each stem is presented as 4-tuple (start, stop, size, energy). Each individual of the population is created by selecting one stem and then adding stems to create a possible structure. The mutation and the crossover will be applied on each individual.

Artificial Neural Networks (ANNs). ANN is an information processing system consists of a high number of highly interconnected processing elements "neurons", functioning collectively to handle a particular problem. ANNs aim to treat artificial intelligence problems by acquiring knowledge through learning, also its have other employments such as the classification, the clustering and the prediction. ANN has been used to predict RNA secondary structure in some works such as [13, 14].

In [13], a Hopfield Neural Network (HNN) based parallel algorithm is presented for predicting RNA secondary structure. In this method the HNN is used to find the near-maximum independent set of an adjacent graph made of RNA base pairs, and then it computes the stable secondary structure of RNA. Koessler et al. [14] build a predictive model for RNA secondary structure using a graph-theoretic tree representation. They model the bonding of two RNA secondary structures to form a general secondary structure by a tree graph. This operation called merge. The resulting data from each merge operation is represented by a vector that will be used as input values for the neural network.

2.3   Comparative Approach (CA)

Since there is a close relationship between structure and function, it is therefore assumed that the sequences with same functions should have the same structures. The comparative approach research in the sequences homologous regions whose structure is retained. The CA is used when there is an alignment composed of sequences with the same function but of different species. The CA is considered more significant than the dynamic approach that uses the thermodynamic principle.

The first efficient algorithm using this approach was developed by Han and Kim [15]. This algorithm takes a set of aligned sequences, for which it performs the phylogenetic comparison, and searches for a certain number of most plausible common secondary structures. This algorithm consists of two main steps, the first step is to analyze the phylogenetic comparison, and the second step aims to select the optimal secondary structures.

Recently there are many other algorithms based on this approach, among them, we can mention [16–18].

DAFS [16] is used for aligning and folding RNA sequences. This algorithm calculates a pair-wise structural alignment. For a given unaligned two sequences DFAS uses two steps to predict the RNA secondary structure. The first step aims to compute two base-pairing probability matrices and an alignment-matching probability matrix. The second step serves to solve the integer programming (IP) problem of simultaneously aligning, and to fold the given sequences by dual decomposition to maximize the expected accuracy of the prediction. Turbofold [17] is an algorithm used to predict RNA secondary structure by estimates the base pairing probabilities. It takes as input a set of homologous RNA sequences, and then the base pairing probabilities for a sequence are estimated by combining intrinsic information, obtained from the sequence itself through the nearest neighbor thermodynamic model, with extrinsic information, obtained from the other sequences in the input set. Another algorithm based on CA to predict RNA secondary structure is RNA Sampler [18]. RNA Sampler is an iterative sampling algorithm that predicts RNA secondary structures in multiple unaligned sequences. It determinate the common structure between two sequences by probabilistically sampling aligned stems based on stem conservation calculated from intrasequence base pairing probabilities and intersequence base alignment probabilities.

2.4 Grammatical Approach (GRA)

In GRA, RNA secondary structure prediction is considered as a parsing problem [19]. The analysis of stochastic grammars is one of the most used techniques to analyze the information in bioinformatics. The principle of this technique return to the formal grammar that was developed as a model to analyze natural languages.

In the last few years, a significant care is given to context-free grammar (CFG) to predict the RNA structure, which took $O(n^3)$ time where n is the length of the input sequence. As extension of the CFG many methods have been developed to predict the RNA secondary structure with pseudoknots such as Scholastic Context Free Grammar (SCFG) [20–22], and Scholastic Multiple Context Free Grammar (SMCFG) [23]. SCFGs were used effectively to RNA secondary structure prediction in the early 90s, and it were used in combination with comparative methods in the late 90s [24]. Generally SCFG can be comparable to dynamic method on his generative power [19]. In the first works of predicting RNA secondary structure using GRA, Two sub-class of tree adjoining grammar (TAG) have been defined. The first is the simple linear tags (SL-TAG) and the second is the extended simple linear tags (ELS-TAG). SL-TAG cannot be used to predict RNA secondary structure with pseudoknots because of its limited representative power, while, ELS-TAG can define a pseudoknot structure grammatically

[19]. Among the most popular methods using SCFGs, there is the Pfold algorithm [25]. It is developed to create an evolutionary tree and a secondary structure from an aligned set of RNA sequences. There are other methods which predict the secondary structure from aligned RNA sequences like Turbofold [17], RNAalifold [26].

As improvement of Pfold algorithm, Sükösd et al. present PPfold algorithm [27]. It is a multithreaded version of Pfold, which is capable to predict the structure of large RNA alignment accurately on practical timescales. In [20] Garca describes a novel algorithm, based on pattern matching techniques, that uses a sequential approximation strategy to solve the original problem. This algorithm reduces the complexity to $O(n^2 \log(n))$, also it widens the maximum length of the sequence, as well as the capacity of analyzing several pseudoknots simultaneously.

Table 1 contains various methods developed to predict RNA secondary structure.

Table 1. Methods used in the RNA secondary structure prediction.

Method/Re	Year	Type	Input
[28]	2012	Comparative approach	K sequences
TurboKnot [29]	2012	Comparative approach	K sequences
IPknot [30]	2011	Soft Computing approach	Single RNA sequence
CyloFold [31]	2010	Soft Computing approach	Single RNA sequence
TT2NE [32]	2010	Soft Computing approach	Single RNA sequence
Tfold [33]	2010	Comparative approach	K sequences
[34]	2009	Soft Computing approach	Single RNA sequence
UNAFOLD [35]	2008	Dynamic approach	Single RNA sequence
PETFOLD [36]	2008	Comparative, Grammatical approaches	Multiple alignment
Pcluster [37]	2007	Comparative, Grammatical approaches	Multiple alignment
SimulFold [38]	2007	Comparative approach	K sequences
[39]	2007	Soft Computing approach	Single RNA sequence
[40]	2006	Soft Computing approach	Single RNA sequence
[41]	2006	Dynamic approach	Single RNA sequence
[42]	2006	Dynamic approach	Single RNA sequence
[43]	2004	Dynamic approach	Single RNA sequence
[44]	2004	Dynamic approach	Single RNA sequence
Vienna server [45]	2003	Dynamic approach	Single RNA sequence

3   Conclusion and Future Work

In this paper we have review the most used approaches to predict the RNA secondary structure, and also we have given several algorithms proposed to handle this problem. As future work we propose an idea of a new method for predicting RNA secondary structure including pseudoknots. The proposed method combines between two main techniques, genetic algorithm and GRASP method [46].

GA was already discussed in the previous section. Concerning the GRASP method, it was introduced by Feo and Resende [46]. GRASP combines the ad vantages of greedy heuristics, random search, and neighborhood methods.

This algorithm repeats a process consisting of two phases, the first is the construction phase and the second is the local search phase.

During the construction phase a list of feasible solutions is iteratively constructed. This list contains the best solutions, which are selected using a greedy function (F).

The local search phase comes in the wake of the construction phase, it aims to improve the solution obtained from the first phase by launching a local search to find the local optimum solution. Figure 3 represents a pseudo code of the proposed method.

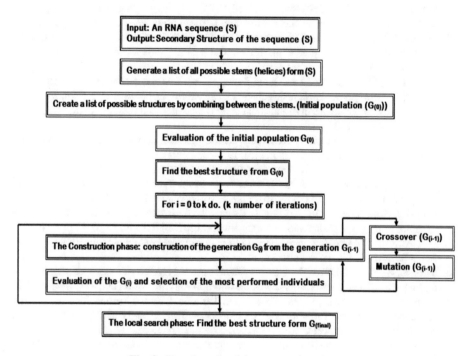

Fig. 3. Pseudo code of the proposed method.

References

1. Osman, M.N., Abdullah, R., AbdulRashid, N.: RNA secondary structure prediction using dynamic programming algorithm – a review and proposed work. In: 2010 International Symposium in Information Technology (ITSim), vol. 2, pp. 551–556. IEEE, June 2010
2. Nussinov, R., Pieczenik, G., Griggs, J.R., Kleitman, D.J.: Algorithms for loop matchings. SIAM J. Appl. Math. **35**(1), 68–82 (1978)
3. Zuker, M.: Mfold web server for nucleic acid folding and hybridization prediction. Nucleic Acids Res. **31**(13), 3406–3415 (2003)

4. Tinoco, I., Borer, P.N., Dengler, B., Levine, M.D., Uhlenbeck, O.C., Crothers, D.M., Gralla, J.: Improved estimation of secondary structure in ribonucleic acids. Nature **246**(150), 40–41 (1973)

5. Rivas, E., Eddy, S.R.: A dynamic programming algorithm for RNA structure prediction including pseudoknots. J. Mol. Biol. **285**(5), 2053–2068 (1999)

6. Dirks, R.M., Pierce, N.A.: A partition function algorithm for nucleic acid secondary structure including pseudoknots. J. Comput. Chem. **24**(13), 1664–1677 (2003)

7. Ray, S.S., Pal, S.K.: RNA secondary structure prediction using soft computing. IEEE/ACM Trans. Comput. Biol. Bioinf. **10**(1), 2–17 (2013)

8. Holland, J.H.: Outline for a logical theory of adaptive systems. J. ACM (JACM) **9**(3), 297–314 (1962)

9. Van Batenburg, F.H.D., Gultyaev, A.P., Pleij, C.W.: An APL-programmed genetic algorithm for the prediction of RNA secondary structure. J. Theor. Biol. **174**(3), 269–280 (1995)

10. Wiese, K.C., Deschenes, A.A., Hendriks, A.G.: RnaPredict—an evolutionary algorithm for RNA secondary structure prediction. IEEE/ACM Trans. Comput. Biol. Bioinform. (TCBB) **5**(1), 25–41 (2008)

11. Tong, K.K., Cheung, K.Y., Lee, K.H., Leung, K.S.: GAknot: RNA secondary structures prediction with pseudoknots using genetic algorithm. In: 2013 IEEE Symposium on Computational Intelligence in Bioinformatics and Computational Biology (CIBCB), pp. 136–142. IEEE, April 2013

12. Shapiro, B.A., Navetta, J.: A massively parallel genetic algorithm for RNA secondary structure prediction. J. Supercomput. **8**(3), 195–207 (1994)

13. Liu, Q., Ye, X., Zhang, Y.: A Hopfield neural network based algorithm for RNA secondary structure prediction. In: First International Multi-Symposiums on Computer and Computational Sciences, 2006, IMSCCS 2006, vol. 1, pp. 10–16. IEEE, June 2006

14. Koessler, D.R., Knisley, D.J., Knisley, J., Haynes, T.: A predictive model for secondary RNA structure using graph theory and a neural network. BMC Bioinform. **11**(6), S21 (2010)

15. Han, K., Kim, H.J.: Prediction of common folding structures of homologous RNAs. Nucleic Acids Res. **21**(5), 1251–1257 (1993)

16. Sato, K., Kato, Y., Akutsu, T., Asai, K., Sakakibara, Y.: DAFS: simultaneous aligning and folding of RNA sequences via dual decomposition. Bioinformatics **28**(24), 3218–3224 (2012)

17. Harmanci, A.O., Sharma, G., Mathews, D.H.: TurboFold: iterative probabilistic estimation of secondary structures for multiple RNA sequences. BMC Bioinform. **12**(1), 108 (2011)

18. Xu, X., Ji, Y., Stormo, G.D.: RNA Sampler: a new sampling based algorithm for common RNA secondary structure prediction and structural alignment. Bioinformatics **23**(15), 1883–1891 (2007)

19. Jiwan, A., Singh, S.: A review on RNA pseudoknot structure prediction techniques. In: 2012 International Conference on Computing, Electronics and Electrical Technologies (ICCEET), pp. 975–978. IEEE, March 2012

20. Garca, R.: Prediction of RNA pseudoknotted secondary structure using Stochastic Context Free Grammars (SCFG). CLEI Electron. J. **9**(2) (2006)

21. Sakakibara, Y., Brown, M., Hughey, R., Mian, I.S., Sjlander, K., Underwood, R.C., Haussler, D.: Stochastic context-free grammars for tRNA modeling. Nucleic Acids Res. **22**(23), 5112–5120 (1994)

22. Anderson, J.W., Tataru, P., Staines, J., Hein, J., Lyngs, R.: Evolving stochastic context-free grammars for RNA secondary structure prediction. BMC Bioinform. **13**(1), 78 (2012)

23. Kato, Y., Seki, H., Kasami, T.: RNA pseudoknotted structure prediction using stochastic multiple context-free grammar. IPSJ Digit. Courier **2**, 655–664 (2006)

24. Mizoguchi, N., Kato, Y., Seki, H.: A grammar-based approach to RNA pseudoknotted structure prediction for aligned sequences. In: 2011 IEEE 1st International Conference on Computational Advances in Bio and Medical Sciences (ICCABS), pp. 135–140. IEEE, February 2011

25. Knudsen, B., Hein, J.: Pfold: RNA secondary structure prediction using stochastic context-free grammars. Nucleic Acids Res. **31**(13), 3423–3428 (2003)

26. Hofacker, I.L., Fekete, M., Stadler, P.F.: Secondary structure prediction for aligned RNA sequences. J. Mol. Biol. **319**(5), 1059–1066 (2002)

27. Sükösd, Z., Knudsen, B., Værum, M., Kjems, J., Andersen, E.S.: Multithreaded comparative RNA secondary structure prediction using stochastic context-free grammars. BMC Bioinform. **12**(1), 103 (2011)

28. Doose, G., Metzler, D.: Bayesian sampling of evolutionarily conserved RNA secondary structures with pseudoknots. Bioinformatics **28**(17), 2242–2248 (2012)

29. Seetin, M.G., Mathews, D.H.: TurboKnot: rapid prediction of conserved RNA secondary structures including pseudoknots. Bioinformatics **28**(6), 792–798 (2012)

30. Sato, K., Kato, Y., Hamada, M., Akutsu, T., Asai, K.: IPknot: fast and accurate prediction of RNA secondary structures with pseudoknots using integer programming. Bioinformatics **27**(13), i85–i93 (2011)

31. Bindewald, E., Kluth, T., Shapiro, B.A.: CyloFold: secondary structure prediction including pseudoknots. Nucleic Acids Res. **38**(suppl 2), W368–W372 (2010)

32. Bon, M., Orland, H.: TT2NE: a novel algorithm to predict RNA secondary structures with pseudoknots. Nucleic Acids Res. **39**(14), e93 (2011)

33. Engelen, S., Tahi, F.: Tfold: efficient in silico prediction of non-coding RNA secondary structures. Nucleic Acids Res. **38**(7), 2453–2466 (2010)

34. Zou, Q., Zhao, T., Liu, Y., Guo, M.: Predicting RNA secondary structure based on the class information and Hopfield network. Comput. Biol. Med. **39**(3), 206–214 (2009)

35. Markham, N.R., Zuker, M.: UNAFold: software for nucleic acid folding and hybridization. In: Bioinformatics: Structure, Function and Applications, pp. 3–31 (2008)

36. Seemann, S.E., Gorodkin, J., Backofen, R.: Unifying evolutionary and thermodynamic information for RNA folding of multiple alignments. Nucleic Acids Res. **36**(20), 6355–6362 (2008)

37. Andersen, E.S., Lind-Thomsen, A., Knudsen, B., Kristensen, S.E., Havgaard, J.H., Torarinsson, E., Gorodkin, J.: Semiautomated improvement of RNA alignments. RNA **13**(11), 1850–1859 (2007)

38. Meyer, I.M., Mikls, I.: SimulFold: simultaneously inferring RNA structures including pseudoknots, alignments, and trees using a Bayesian MCMC framework. PLoS Comput. Biol. **3**(8), e149 (2007)

39. Zhang, T., Guo, M., Zou, Q.: RNA secondary structure prediction based on forest representation and genetic algorithm. In: Third International Conference on Natural Computation, 2007, ICNC 2007, vol. 4, pp. 370–374. IEEE, August 2007

40. Bindweed, E., Shapiro, B.A.: RNA secondary structure prediction from sequence alignments using a network of k-nearest neighbor classifiers. RNA **12**(3), 342–352 (2006)

41. Tan, G., Feng, S., Sun, N.: Locality and parallelism optimization for dynamic programming algorithm in bioinformatics. In: Proceedings of the 2006 ACM/IEEE Conference on Supercomputing, p. 78. ACM, November 2006

42. Namsrai, O.E., Jung, K.S., Kim, S., Ryu, K.H.: RNA secondary structure prediction with simple pseudo knots based on dynamic programming. In: International Conference on Intelligent Computing, pp. 303–311. Springer, Berlin, August 2006

43. Mathews, D.H., Disney, M.D., Childs, J.L., Schroeder, S.J., Zuker, M., Turner, D.H.: Incorporating chemical modification constraints into a dynamic programming algorithm for prediction of RNA secondary structure. Proc. Natl. Acad. Sci. U.S.A. **101**(19), 7287–7292 (2004)
44. Reeder, J., Giegerich, R.: Design, implementation and evaluation of a practical pseudoknot folding algorithm based on thermodynamics. BMC Bioinform. **5**(1), 104 (2004)
45. Hofacker, I.L.: Vienna RNA secondary structure server. Nucleic Acids Res. **31**(13), 3429–3431 (2003)
46. Feo, T.A., Resende, M.G.: Greedy randomized adaptive search procedures. J. Global Optim. **6**(2), 109–133 (1995)

A Game Theoretical Based Self-organization Dispatching Mechanism with IEEE802.16 Mesh Networks in Public Bicycle Station Scheduling

Jun Yu[✉] and Wei Zhang

China Design Group Co., Ltd., Nanjing 210009, China
yj_njut@163.com

Abstract. Recently, as the environmentally-friendly commuting, the public bicycle system is vigorously promoted by the government. The traditional dispatching manage is based on tree structure. The monitoring and dispatching manage center implements the bicycle scheduling by monitoring each public bicycle station's online available bicycle number. The borrowing and returning bicycle behavior is a dynamically random process. The communication between each public bicycle station engages the self-organization dispatch. This paper proposed a game theoretical based self-organization dispatching mechanism (GTSD). The main idea of GTSD is to put forward an analysis and resolve proposal in Station Scheduling with a analogous network ad-hoc thought. It periodically record station's online available bicycle number. Then GTSD directional send MSH-DSCH message to make path selection. GTSD sets a repeated game model to assess the advantage and disadvantage of path selection to optimize relaying strategy.

Keywords: Self-organization · Public bicycle station · IEEE802.16
Mesh networks · Game theoretical · Scheduling mechanism

1 Introduction

In recent time, the increase of private cars has brought more and more serious air pollution and traffic jams in our city, especially in the rush hour. It is necessary to promote the green travel at present. As an environmentally-friendly public transportation the public bicycle system draws attention to public concern. It is vigorously promoted by the government. Green travel as the theme to ride a bike in the form of more people concerned about the low carbon travel, and actively promote the choice of a healthy low-carbon way to travel. The monitoring and dispatching manage center is responsible for the public bicycle scheduling between each bicycle station, while the traditional dispatching manage is based on a analogous tree structure scheduling mechanism (Fig. 1). Each bicycle station establishes communication to the dispatching manage center without mutual communication between each bicycle station. The monitoring and dispatching manage center implements the bicycle scheduling by monitoring every public bicycle station's online available bicycle number. While, the online available

© Springer International Publishing AG, part of Springer Nature 2018
M. Ben Ahmed and A. A. Boudhir (Eds.): SCAMS 2017, LNNS 37, pp. 389–396, 2018.
https://doi.org/10.1007/978-3-319-74500-8_36

bicycle number is dynamic, cause the borrowing and returning bicycle behavior is a dynamically random process. The centralized manual scheduling is inefficient. So we proposed a game theoretical based self-organization dispatching mechanism (GTSD). It emphasizes the self-organization dispatch based on the distributed communication between each public bicycle station with application of the mobile Internet technique (Fig. 2). We improve the routing method in IEEE802.16 Mesh Networks [1–4] to enhancement the communication efficiency. We abstract the communication between each bicycle station to the mobile internet network routing project research.

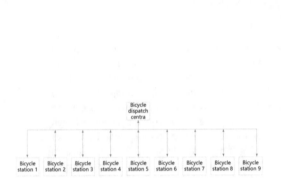

Fig. 1. Tree structure scheduling mechanism

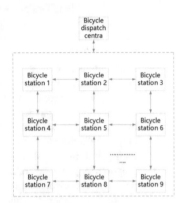

Fig. 2. Mesh self-organization dispatching

We abstract the communication between each bicycle station to the mobile internet network routing project research. As a emerging technology the WMN based on IEEE802.16 protocol has many researching space.

It is necessary to work out an appropriate routing protocol suiting for WMN. Relative representative thought is OLSR proposed by Yang etc. in document [5]. Document [6] figured out the low mobility network nodes bear low delay. Document [7] proposed to bring SINR being the routing metric. Document [8] worked out a routing protocol according to node's reputation. Document [9] mainly consider with path quality. This paper proposed a game theoretical based self-organization dispatching mechanism (GTSD). The main idea of GTSD is to put forward an analysis and resolve proposal in Station Scheduling with a analogous network ad-hoc thought. It periodically record station's online available bicycle number and forecast mainstream type next period.

The structure is: Sect. 2, introducing distributed coordinate scheduling in IEEE802.16 Mesh networks; Sect. 3, proposing A Game Theoretical Based Self-Organization Dispatching Mechanism (GTSD); making a conclusion in Sect. 4.

2 Distributed Coordinate Scheduling

2.1 The Frame Structure in Mesh Network

In TDMA mode, each time frame with fixed duration consists of a data sub-frame and a control sub-frame illustrated in Fig. 1. The data sub-frame with fixed number data min-slot up to 256 (hereafter slots) is an opportunity to transmit data. In the scheduling mod slots are regard as link bandwidth resources. Scheduling control sub-frame, network control sub-frame compose the control sub-frame. It negotiates bandwidth by broadcasting MSH-DSCH messages among neighbor nodes by a three handshakes procedure, during the scheduling control sub-frame. MSH-DSCH contains information elements(IEs): GrantIE(direction = 0), GrantIE(direction = 1), RequestIE (Fig. 3).

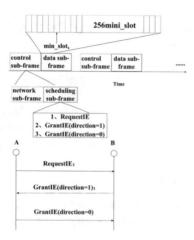

Fig. 3. Frame structure in Mesh

2.2 Coordinate Scheduling

In a collision-free style, nodes negotiate to allocate channel resources by distributed exchanging the scheduling message hop-by-hop. The requesting node A which wanting some slots to transmit data. Node A broadcasts MSH-DSCH with RequestIE to its neighbors (node B, node C, node D, node E) making a reservation of some slots, illustrated in Fig. 2. One of node A's neighbor (node B, node C, node D, node E) return MSH-DSCH with GrantIE(direction = 1), according to itself available free slots number. Finally, node A sends back GrantIE(direction = 0) to confirm the slots reservation. The next hop is randomly chosen in the distributed coordinate scheduling mode (Fig. 4).

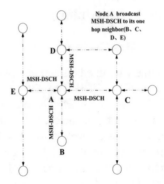

Fig. 4. The broadcast of MSH-DSCH

3   Game Theoretical Based Self-organization Dispatching Mechanism (GTSD)

In the Sect. 3, we proposed a Game Theoretical Based Self-Organization Dispatching Mechanism (GTSD). The nodes mentioned in Sect. 2 refer to the bicycle station in real scenario. We abstract the communication between each bicycle station to the mobile internet network routing project research. GTSD also improves the routing communication efficiency by excavating the rule in random routing process.

3.1   Self-organization Dispatching

Node's buffer queue runs in queuing model with single sever with rule being FCFS. We put forward such assumption on the businesses arriving characteristics:

(1) Non-aftereffect: business arriving with a mutual independence way in overlap time interval.
(2) Stability: for minimal Δt, during time interval $[t, t + \Delta t]$ no correlation between business's arriving and t value.
(3) Generalization: for minimal Δt, two or more businesses arriving probability is infinitesimal.

Also we assume the mean value of time scale span business arriving the queue to be a constant. The random process can be viewed as a Poisson process with λ density.

GRSD making statistics businesses every T, we can calculate some kind of business's arrival rate:

$$r(k) = (1 - d_r) \times \frac{Num}{T} + d_r \times r(k - 1) \tag{1}$$

Where, $d_r = e^{-\frac{T}{\mu}}$, k represents statistics time; μ is control parameter. Num means arrival numbers business, which represents the communication between each bicycle station.

GRSD making prediction on probability of business in next period:

$$P_{Num} = \frac{r(k)^{Num}}{Num!} e^{-r(k)}$$ (2)

GRSD bring in weighted smoothing disposal to preventing data bursting:

$$P_{Num(i)} = \zeta \times P_{Num(i-1)} + (1 - \zeta) \times P_{Num(i)}$$ (3)

Based on historical experience where ζ is set 0.3.

The source node should choose relative vacant node to be next hop. As the reference in document [4], it introduces node's vacant degree. GRSD proposes SC_i to reflect node n_i's vacant degree:

$$\text{Definition 1} \quad SC_i : SC_i = \frac{r(k)_i}{RcvR_i^2}$$ (4)

Where $RcvR_i$ represents business's receiving speed of node. The sampling period is T as same as the sampling value before, which is set to be 6 s in the simulation experiment; $r(k)_i$ represents business's sending speed of node. Receiving is the premise of sending, for this dominance, $RcvR_i$ has a higher order than $r(k)_i$ [4]. The node with a large value of SC_i will be more vacant and more reliability. It will get more transportation QoS guarantee when these nodes chosen to be next hop.

Every node broadcasts its identity $\{Identity | Identity = RT_{er}, nRt_{er}\}$, probability of mainstream type of arriving business $\{P_w | P_w = P_{RT}, P_{nRT}\}$, vacant degree SC_i to its neighbor. GTSD requires nodes to choose nRT_{ers} from its neighbor nodes $\{nRT_{er1}, nRT_{er2}, ..., nRT_{erq}\}$, then sends MSH-DCSH to node (from $\{nRT_{er1}, nRT_{er2}, ..., nRT_{erq}\}$) with a biggest SC_i. If two nodes accomplish the three handshakes procedure, the data will be transmitted during the slots they reserved in the three handshakes.

If there is only RT_{er} in node's neighbor nodes $\{RT_{er1}, RT_{er2}, ..., RT_{erq}\}$, then it sends MSH-DCSH to node (from $\{RT_{er1}, RT_{er2}, ..., RT_{erq}\}$) with lowest probability of mainstream type of arriving business to guarantee RT businesses' QoS.

Nodes make path election independently by exchanging scheduling information, the next hop selection strategy showed as Fig. 5.

3.2 Cooperation-Oriented Repeated Game Model

We draw on document [12]'s repeated game research model. We also abstract bicycle station to network node, applying mathematical statistics in routing mechanism. According to the basic cognition before, the three handshakes procedure will be promoted depending on the chosen node feeding back MSH-DSCH or not, when source node send MSH-DSCH to one neighbor node (hereafter chosen node). The wireless channel resource is constrained. If the chosen node respond to source node's request, choosing to feed back MSH-DSCH (GrantIE), then it would unavoidable consume slots resources, energy. The possible congestion and delay obviously degrades status of chosen node in "Mesh election". It would consume more time resources (slots) to relay

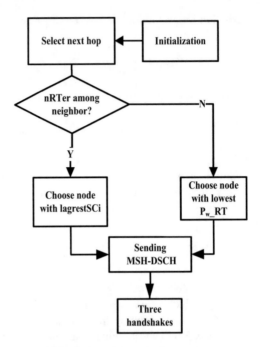

Fig. 5. Next hop selection strategy

other node's data, when it could have transmitted own data. We contract the behavior that the neighbor node evaluates the positive and negative to decide whether respond to source node's request, when receiving the RequestIE to a game action.

GTSD proposed a cooperation-oriented repeated game model $G(n, S, U)$.

(1) There existing $n(i = 1, 2, …, n)$ node for each single game.
(2) For every gamer node, there existing two strategies ($s_i = \{ZF, IG\}$); $S_i = s_i(s_1, s_2, …, s_n)$. Where the ZF meaning respondence probability s_{ZF}; the IG meaning denying probability $1 - s_{ZF}$.
(3) For each strategy, the benefit u_i, $U_i = u_i(u_1, u_2, …, u_n)$ for each game.

Bicycle station exchange own running conditions information and previous dispatch results. In the research model, gamer nodes knowing each others' profit function, strategy space. Each game runs in a asynchronously way. We define PR_i to value node's competitiveness:

$$\textbf{Definition 2 } PR_i : \ PR_i = \frac{S_{dn} + S_{db}}{fF_{dur}} \qquad (5)$$

Where fF_{dur} represents the mean maximum continuous time slot length. The S_{dn} represents answer slots numbers in fF_{dur}, value the probability to respond quest. The S_{db} means rejection slots numbers in fF_{dur}, value the probability to refuse request.

In the "Mesh election", node with higher PR_i would win out, so it could utilize more time source (slots) to transmit own data. GTST proposed a rule to increase PR_i aiming at encouraging more node respond for request actively:

$$u_{ci} = \lceil b_{PR} \times 256 \rceil \tag{6}$$

Where, u_c means node's benefit when respond to request thus choosing to cooperate. The b_{PRv} presents corresponding cooperation PR reward income. Slot sums to 256.

GTSD adjusts exp value (thus Transmission collision avoidance index) to punish refusing cooperation node, according to document [11]'s demonstration about the influence of exp on efficiency of three handshakes.

$$u_{rfi} = \alpha w \times |\exp^*| \tag{7}$$

Where, u_{rfi} presents benefit of refusing cooperation node. The $|\exp^*|$ means jitter of exp. The aw presents jitter exp value.

$$u_i = (u_{ci} - u_{wt}) \times s_{ZF} - u_{rfi} \times (1 - s_{ZF}) \tag{8}$$

Where u_i means node's exception profit function, u_{wt} presents node's communication QoS wastage when cooperation. We adopt document [15] to obtain the optimal solution of profit function's "Nash Equilibrium":

$$s_i(t+1) = s_i(t) + \theta_i \times s_i(t) \times \frac{\partial u_i(s)}{\partial s_i(t)} \tag{9}$$

Where $s_i(t)$ means presents node i's strategy for each game. The θ_i means speed's convergence parameter. For "Nash equilibrium", the adaptive dynamic distributed algorithm would save more resources than centralized algorithm of. It would automatic iterate node's game strategy $s_i(t)$ based on historical information.

$$U_i(s_1^*, s_2^*, \ldots, s_i^*) \tag{10}$$

GTSD will bring extra wastage, while in wireless mesh networks, the routing node tending to be static state equipments. These nodes do not be subject to power dissipation, so the extra wastage would be acceptable.

4   Conclusion

The announcement on that Mobile Internet-based bicycle station state and communication technologies are applied to the bicycle station dispatching information service is one of the important means to promote scheduling efficiency and will also greatly facilitate public travel. We proposed a repeated game research model GTSD and abstract bicycle station to network node with improving mathematical statistics in routing mechanism. GTSD also improves the routing communication efficiency by excavating the rule in random routing process.

References

1. Akyildiz, I.F., Wang, X., Wang, W.: Wireless mesh networks: a survey. Comput. Netw. **47**, 445–487 (2005)
2. He, J., Yang, K., Guild, K.: Application of IEEE802.16 mesh networks as the backhaul of multihop cellular networks. IEEE Commun. Mag. **45**(9), 82–91 (2007)
3. Akyild, I.F., Wang, X., Wang, W.: A survey on wireless mesh networks. Commun. Mag. **43**, 23–30 (2005)
4. Shen, C., Lu, Y.X.T.: The wireless Mesh network based on the synthetic criteria is protocol. Chin. J. Comput. **33**(12), 2300–2311 (2010)
5. Jacquet, P., Muhlethaler, P., Clausen, T., et al.: Optimized link state routing protocol for ad hoc networks. In: Proceedings of the IEEE International Multi Topic Conference, Pakistan, pp. 62–68 (2001)
6. Yang, Y., Wang, J.: Design guidelines for routing metrics in multihop wireless networks. In: Proceedings of the 27th IEEE Communication Society Conference on Computer Communications, Phoenix, pp. 2288–2296 (2008)
7. Elshaikh, M., Kamel, N., Awang, A.: High throughput routing algorithm metric for OLSR routing protocol in wireless mesh networks. In: Proceedings of the 5th International Colloquium on Signal Processing & Its Applications, Kuala Lumpurm, pp. 445–448 (2009)
8. Ding, Q., Jiang, M., Li, X., Zhou, X.: RePro: a reputation-based proactive routing protocol for the wireless mesh backbone. In: Proceedings of the 5th International Joint Conference on INC, IMS and IDC, Seoul, pp. 516–521 (2009)
9. De C Paschoalino, R., Madeira, E.R.M.: A scalable link quality routing protocol for multi-radio wireless mesh networks. In: Proceedings of the 16th International Conference on Communications and Networks, Honolulu, HI, pp. 1053–1058 (2007)
10. QoS technology introduction [EB/OL] (2008-05) (2013). http://www.h3c.com.cn/Products___Technology/Technology/QoS/Other_technology/Technology_recommend/200805/605881_30003_0.htm
11. Kang, X.J.: IEEE802.16 MeshNetwork node collaboration and distributed scheduling research. Department of Communication and Information System of Huazhong University of Science and Technology, Wuhan, Hubei Province (2007)
12. Feng, C.W.: WiMAX MeshResearch on network routing and scheduling algorithm. Comput. Eng. **37**(4), 93–95 (2011)
13. Cao, M., Ma, W., Zhang, Q., et al.: Modeling and performance analysis of the distributed scheduler in IEEE802.16 Mesh mode. In: Proceeding of the 6th ACM International Symposium on Mobile ad Hoc Networking and Computing, MobiHoc 2005, pp. 78–89. ACM Press, New York (2005)
14. You, L., Wang, Y.F.: Functional model and simulation of the IEEE802.16 Mesh mode based on NS2. Appl. Res. Comput. **8**(25), 2505–2508 (2008)
15. Cong, L.: Research on the allocation of wireless network resources based on game theory. Xidian University (2011)

Amazigh Speech Recognition System Based on CMUSphinx

Meryam Telmem[(✉)] and Youssef Ghanou[(✉)]

Team TIM, High School of Technology,
Moulay Ismail University, Meknes, Morocco
meryamtelmem@gmail.com, youssefghanou@yahoo.fr

Abstract. In this paper, we are proposing a new approach to build an Amazigh automated speech recognition system using Amazigh environment. This system is based on the open source CMU Sphinx-4, from the Carnegie Mellon University. CMU Sphinx is a large-vocabulary; speaker-independent, continuous speech recognition system based on discrete Hidden Markov Models (HMMs).

Keywords: Speech recognition · Amazigh language · HMMs
CMUSphinx-4 · Artificial intelligence

1 Introduction

Automatic speech recognition is a computer technique with the objective of transcribing a signal from speech into text. It is an area that is still emerging, and attracting the attention of the public as well as many researchers, and opens up the future. Towards a new man-machine generation. This importance is explained by the privileged position of speech as a vector of human information. The realization of a RAP system requires the contribution of several research domains: signal processing, mathematical models, algorithms, etc. [1–17].

A remarkable change in the state of the art makes the systems increasingly efficient with sufficient performance to be used in (many applications) many domains: Assistance to the autonomous life of people, Vocal control (in industry, medicine, aviation, toys, space), language learning and translation, indexing of large audiovisual databases, deep learning, etc. [1].

However, the speech signal is one of the most complex signals to characterize, which makes the task of a RAP system difficult. This complexity of the speech signal originates from the combination of several factors, the redundancy of the acoustic signal, the great inter and intra-speaker variability, the effects of the continuous speech coarticulation and the recording conditions. To overcome these difficulties, many mathematical methods and models have been developed, including dynamic comparison, neural networks [21], Vector Vector Machine SVM, stochastic Markov models and in particular The Hidden Markov Models HMM, which have become the perfect solution to the problems of automatic speech recognition.

© Springer International Publishing AG, part of Springer Nature 2018
M. Ben Ahmed and A. A. Boudhir (Eds.): SCAMS 2017, LNNS 37, pp. 397–410, 2018.
https://doi.org/10.1007/978-3-319-74500-8_37

Given the importance of ASR, several free software has been developed, Among the most famous: HTK [2] and CMU Sphinx [3], JULIUS, KALDI [4]. In this research work, we propose a novel approach to build an Amazigh automated speech recognition system, based on CMU Sphinx-4 which is based on Hidden Markov Models (HMMs). It is a flexible, modular and pluggable framework to help foster new innovations in the core research of HMM recognition systems. CMU Sphinx is used in this research work because of its high degree of flexibility and modularity [4].

The paper is organized as follows. Section 2 present the principle and the theory of speech recognition, Sect. 3 present in brief a description of the Amazigh language, Sect. 4 gives details about the steps to build the Amazigh Speech Recognition System, and the experimental results with The performance evaluation of the system is based on the Word error rate (WER). Finally, the conclusion is summarized in Sect. 5 with future work.

2 Related Works

This section presents some of the reported works available in the literature that are similar to the presented work. some of the works providing ASR system for others languages are (Kumar, K. (2012), Abushariah, M.A.A.M. (2012)).

El Ghazi et al. [7] have presented a system for automatic speech recognition on the Amazigh. used the hidden Markov model to model the phonetic units corresponding to words taken from the training base. The results obtained are very encouraging given the size of the training set and the number of people taken to the registration. To demonstrate the flexibility of the hidden Markov model we conducted a comparison of results obtained by the latter and dynamic programming.

Satori et al. [2–14] have developed of a speaker-independent continuous automatic Amazigh speech recognition system. The designed system is based on the Carnegie Mellon University Sphinx tools. In the training and testing phase an in house Amazigh_Alphadigits corpus was used. This corpus was collected in the framework of this work and consists of speech and their transcription of 60 Berber Moroccan speakers (30 males and 30 females) native of Tarifit Berber. The system obtained best performance of 92.89% when trained using 16 Gaussian Mixture models.

Kumar and Aggarwal [15], have built a connected-words speech recognition system for Hindi language. The system has been developed using hidden Markov model toolkit (HTK) that uses hidden Markov models (HMMs) for recognition.

Abushariah et al. [16] have proposed an efficient and effective framework for the design and development of a speaker-independent continuous automatic Arabic speech recognition system based on a phonetically rich and balanced speech corpus. The speech corpus contains a total of 415 sentences recorded by 40 (20 male and 20 female) Arabic native speakers from 11 different Arab countries representing the three major regions (Levant, Gulf, and Africa) in the Arab world. The proposed Arabic speech recognition system is based on the Carnegie Mellon University (CMU) Sphinx tools, and the Cambridge HTK tools were also used at some testing stages. The speech engine uses 3-emitting state Hidden Markov Models (HMM) for tri-phone based acoustic models. Based on experimental analysis of about 7 h of training speech data, the acoustic model is best using continuous observation's probability model of 16 Gaussian mixture distributions and the state distributions were tied to 500 senones. The language model contains both bi-grams and tri-grams. For similar speakers but different sentences, the system obtained a word recognition accuracy of 92.67% and 93.88% and a Word Error Rate (WER) of 11.27% and 10.07% with and without diacritical marks respectively. For different speakers with similar sentences, the system obtained a word recognition accuracy of 95.92% and 96.29% and a WER of 5.78% and 5.45% with and without diacritical marks respectively. Whereas different speakers and different sentences, the system obtained a word recognition accuracy of 89.08% and 90.23% and a WER of 15.59% and 14.44% with and without diacritical marks respectively.

Al-Qatab and Ainon [20] implemented an Arabic automatic speech recognition engine using HTK. The engine recognized both continuous speech as well as isolated words. The developed system used an Arabic dictionary built manually by the speech-sounds of 13 speakers and it used vocabulary of 33 words.

3 Amazigh Language

3.1 History

The Amazigh languages are a group of very closely related and similar languages and dialects spoken in Morocco, Algeria, Tunisia, Libya, and the Egyptian area of Siwa, as well as by large Amazigh communities in parts of Niger and Mali. In c, for example, Amazigh is divided into three regional varieties, with tariffs in the North, Tamazigh in Central and Southeast Morocco, and Tachelhite in the South-West and the High Atlas. Because of the problems of reliable language census, it is difficult to assess the exact number of speakers of Berber languages for each country [6] (Table 1).

Table 1. Number of Amazigh speakers by country

Pays	Appellation	Variétés linguistiques	Populations (millions)
Morocco	Amazigh	tachelhit, tamazight, tarifit, ghomara	15–20 (millions)
Algéria	Kabyle	kabyle, chaouia, tamazight, hassaniyya, tumzabt, taznatit	12–15 (millions)
Tunisia	Amazigh	chaouia, nafusi, sened, ghadamès	100 000
Libya	Tamacheq	nafusi, tamahaq, ghadamès, sawknah, awjilah	220 000 (env.)
Niger	Amazigh	tamajaq, tayart, touareg	720 000
Mauritanie	Zenaga	zenaga	200
Mali	Tamajeq-kida	tamajaq, tamasheq	440000 (env.)

This language have had a written tradition, on and off, for over 2000 years, although the tradition has been frequently disrupted by various invasions. It was first written in the Tifinagh alphabet, still used by the Tuareg, the oldest dated inscription is from about 200 BC. Later, between about 1000 AD and 1500 AD, it was written in the Arabic alphabet. Since the 20th century, it has often been written in the Latin alphabet, especially among the Kabylians.

3.2 Tifinagh

A modernized form of the Tifinagh alphabet was made official in Morocco in 2003, and a similar one is sparsely used Algeria. The Amazigh Latin alphabet is preferred by Moroccan Amazigh writers and is still predominant in Algeria (although unofficially). Mali and Niger recognized the Amazigh Latin alphabet and customized it to the Tuareg phonological system. Although, Tifinagh is still used in parts of Mali and Niger. Both Tifinagh and Latin scripts are increasingly being used in Morocco and parts of Algeria, while the Arabic script has been abandoned by Amazigh writers [7].

Only the IRCAM defined a precise order described by the expression below (a < b, means that a is sorted before b) (Table 2):

3.3 Phonetics

The graphic system of the standard amazighe proposed by the IRCAM comprises:

- 27 consonants of: labels (ⵇ, ⵀ, ⵛ), dental (+ , ∧ , ⴻ , E , ⵏ , ⵔ, ⵕ, ⵯ).
- the alveolar (⵿, ⵝ,   ⵝ,⵰) (ⵞ , ⵊ) ⵔ , ⵡ , ⵔ ˘ , ⵔ ˘ , ⵔ, ⵏ, ⵔ, ⵏ, ⵔ, ⵔ ⵔ.
- 2 semi-consonants: ⵢ and ⵓ.
- vowels: the full ones (ⵓ , ⵉ , ⵥ), neutral (ⵥ).

4   Automatic Speech Recognition

Given a speech signal, current automatic speech recognition systems are based on a statistical approach [8], a formalization proposal, a theory of information theory.

Fundamentally, the problem of speech recognition can be stated as follows. From acoustic observations X, the system looks for the sequence of words W * maximizing the following equation:

$$W^* = arg_w \max P(W|X) \tag{1}$$

After applying the Bayes theorem, this equation becomes:

$$W^* = arg_w \max \frac{P(X|W)P(W)}{P(X)} \tag{2}$$

P (X) is considered constant and removed from Eq. 2.

$$W^* = arg_w \max P(X|W)P(W) \tag{3}$$

Table 2. Official Table of the Tifinaghe alphabet as recommended by l'RCAM

Tifinaghe Ircam	Frensh correspondence	Enghlish correspondence	Value
ⴰ	YA	YA	a
ⴱ	YAB	YAB	b
ⴳ	YAG	YAG	g
ⴳⵯ	YAGW	YAGW	g^w
ⴷ	YAD	YAD	d
ⴹ	YADD	YADD	d
ⴻ	YEY	YEY	ə
ⴼ	YAF	YAF	f
ⴽ	YAK	YAK	k
ⴽⵯ	YAKW	YAKW	K^w
ⵀ	YAH	YAH	h
ⵃ	YAHH	YAHH	ħ
ⵄ	YAA	YAA	ʕ
ⵅ	YAKH	YAKH	x
ⵇ	YAQ	YAQ	q
ⵉ	YI	YI	i
ⵊ	YAJ	YAZH	ʒ
ⵍ	YAL	YAL	l
ⵎ	YAM	YAM	m
ⵏ	YAN	YAN	n
ⵓ	YOU	YOU	u
ⵔ	YAR	YAR	r
ⵕ	YARR	YARR	r
ⵖ	YAGH	YAGH	ɣ
ⵙ	YAS	YAS	s
ⵚ	YASS	YASS	s
ⵛ	YACH	YASH	ʃ
ⵜ	YAT	YAT	t
ⵟ	YATT	YATT	t
ⵡ	YAW	YAW	w
ⵢ	YAY	YAY	y
ⵣ	YAZ	YAZ	z
ⵥ	YAZZ	YAZZ	z

Where the term P (W) is estimated via the language model and P (X | W) corresponds to the probability given by the acoustic models. This type of approach makes it possible to integrate, in the same decision-making process, acoustic and linguistic information (Fig. 1).

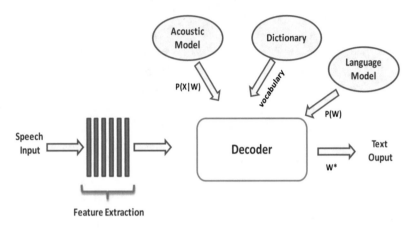

Fig. 1. Setups involved in ASR system.

4.1   Acoustic Analysis of the Signal of Speech

The relevant acoustic information of the speech signal is mainly in the bandwidth [50 Hz–8 kHz]. A signal parametrization system, also known as acoustic pre-processing, is required for signal shaping and calculation of coefficients. This step must be done carefully, as it contributes directly to the performance of the system.

The acoustic analysis is divided into three stages, an analog filtering, an analog/digital conversion and a calculation of coefficients.

4.2   Hidden Markov Model

The acoustic model is used to model the statistics of speech features for each speech unit such as a phone or a word. The Hidden Markov Model (HMM) is the *de facto* standard used in the state-of-the-art acoustic models. It is a powerful statistical method to model the observed data in a discrete-time series. An HMM is a structure formed by a group of states connected by transitions. Each transition is specified by its transition probability. The word *hidden* in HMMs is used to indicate that the assumed state sequence generating the output symbols is unknown. In speech recognition, state transitions are usually constrained to be from left to right or self repetition [9], called the left-to-right model as shown in Fig. 2.

Fig. 2. A left-to-right HMM model with three true states.

Each state of the HMM is usually represented by a Gaussian Mixture Model (GMM) to model the distribution of feature vectors for the given state. A GMM is a weighted sum of *M* component Gaussian densities and is described by Eq. (4).

$$P(x|\lambda) = \sum_{m=1}^{M} wi \, g\left(x|\mu i, \sum i\right) \tag{4}$$

5   CMU Sphinx

CMU Sphinx is a set of speech recognition development libraries and tools that can be linked in to speech-enable applications [10]. They have a number of packages for different tasks and applications:

- Pocketsphinx: Lightweight library of written recognition in C.
- Sphinxbase: support for libraries required by Pocketsphinx.
- Sphinx4: decoder for voice recognition search written in Java.
- CMUclmtk: Language model tools.
- Sphinxtrain: Acoustic model drive tool.
- Sphinx3: decoder for voice recognition search written in C.

5.1   Sphinx Train

SphinxTrain [10] Is the tool created by CMU for the development of acoustic models. It is a set of programs and documentation for realizing and constructing acoustic models for any language. SphinxTrain tool and which requires the installation of the libraries:

- ActivePerl: The tool to edit scripts for SphinxTrain and allows to work in a Unix-like.
- Microsoft Visual Studio: To compile sources In C to produce the Executables.

5.2   Architecture

The high level architecture for sphinx4 is relatively straightforward. As shown in the following Fig. 3, the architecture consists of the front end, the decoder, a knowledge base, and the application [4].

Front end: is responsible for gathering, annotating, and processing the input data. In addition, the front end extracts features from the input data to be read by the decoder.

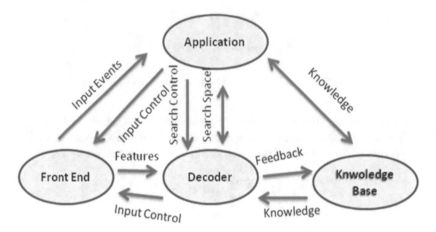

Fig. 3. Setups involved in ASR system.

The annotations provided by the front end include the beginning and ending of a data segment. Operations performed by the front end include preemphasis, noise cancellation, automatic gain control, end pointing, Fourier analysis, Mel spectrum filtering, cepstral extraction, etc.

Knowledge base: provides the information in the decoder needs to do its job. This information includes the acoustic model and the language model. The knowledge base can also receive feedback from the decoder, permitting the knowledge base to dynamically modify itself based upon successive search results. The modifications can include switching acoustic and/or language models as well as updating parameters such as mean and variance transformations for the acoustic models.

Decoder performs: the bulk of the work. It reads features from the front end, couples this with data from the knowledge base and feedback from the application, and performs a search to determine the most likely sequences of words that could be represented by a series of features. The term "search space" is used to describe the most likely sequences of words, and is dynamically updated by the decoder during the decoding process.

Application: may also receive events from the decoder while the decoder is working on a search. These events allow the application to monitor the decoding progress, but also allow the application to affect the decoding process before the decoding completes. Furthermore, the application can also update the knowledge base at any time.

6   Experiments and Results

This section describes our experience in creating and developing an ASR for the Amazigh language. The formation of the acoustic model is done by the SphinxTrain [4–8].

6.1 Corpus

Developing an ASR in a new language like the Amazigh language requires gathering a large amount of corpus viewing the statistical nature of models (HMMs) generalized in automatic speech recognition. So, it is a very tedious task if no corpus exists, Since we must then collect the necessary resources ourselves: speech signal, lexicon, textual corpus, etc. The corpus consists of the Alphabet (33 letters) Amazigh. 9 of Moroccan speakers, are invited to pronounce the letters ten times. The corpus comprises ten repetitions by each speaker of the same letter. Thus, the corpus consists of 2970 audio files (33 letters × 10 repetitions × 9 lecturers). The test database contains 330 audio files [4–11] (Table 3).

Table 3. Recording parameter used for the Preparation of the corpus.

Parameter	Value
Total number of audio files	2970
Corpus	33 letters Amazigh
Sampling	16 kHz, 16 bits
Wave format	Mono, wav

6.2 Dictionary

In a file extension dic, the correspondence it will be specified between the words of the file of transcription and the phonemes used in the file extension phone. The transcription using the Latin scriptes (Figs. 4 and 5).

```
YA      Y A
YAB     Y A B
YAG     Y A G
YAGW    Y A GW
YAD     Y A D
YADD    Y A DD
YEY     Y E Y
YEF     Y E F
YAK     Y A K
YAKW    Y A KW
YAH     Y A H
YAHH    Y A HH
YAA     YA A
YAKH    Y A KH
```

Fig. 4. Extract of the file tiftotal.dic in tiftotal application.

```
SIL A B Y G D J GW YA DD E
F K KW H HH KH Q I L M N OU
R RR GH S SS CH T TT W Z ZZ
```

Fig. 5. Extract of the tiftotal.phone writes by using the scriptes Latin

6.3 Language Model

Language model (language model or grammar Model) is a model that defines the use of words in An application. Each word in the model of Must be in the pronunciation dictionary

There are several types of models that describe language to recognize keyword lists, grammars and statistical language models, phonetic statistical language models. The choice from a language model depends on the application [12] (Fig. 6).

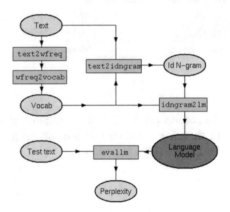

Fig. 6. Setups to building language model ARPA n-gram format with CMUCLMTK.

6.4 Acoustic Model

In the context of Markovian ASR, the acoustic model is generally an HMM, typically a three-state left-right HMM called Bakis, in a state associated with a phoneme. For example, the acoustic model for the word YAB which transcribes the alphabet ⊖ and which contains the phonemes Y A B can be represented by the HMM of the Fig. 2.

Once we have a corpus, we can move on to the stage of creating the acoustic model. It is a tiring and difficult step in view of the scarcity of relevant documentation. The steps of creating an acoustic model, even If CMU Sphinx has a relatively large community [13].

6.5   Results

A system for automatic recognition of As Sphinx 4 uses two dependent elements Of the language: The acoustic model and the language model. In our application we carried out the Modification of these two models as described in previously.

The sphinx 4 must be configured using a file Xml. Thus, the choice of algorithms, the extraction and Comparison of feature vectors and other Important aspects for the creation of a PCR system. Can be customized as required and Application.

The Word error rate [14] has become the standard measurement scheme to evaluate the performance of voice recognition systems [12]. We have an original text and a length recognition text of N words. From them, the words I have inserted. The word D has been deleted and the word S has been replaced. The error rate of Word is:

$$WER = (I + D + S)/N \qquad (5)$$

WER is generally measured by a percentage.

Number of States by HMM
In order to test the effect of the number of states change by HMM on the quality of the acoustic models the system was driven using the two configurations three or five states by HMM knowing that the Sphinx-4 system accepts only these 2 Configurations. Both models were then tested by the Word Error Rate, WER (Table 4).

Table 4. Number of states by HMM.

Number of states by HMM	WER
3	28.6
5	35.6

These results show that the best results were recorded for HMM = 3.

Number of Gaussian Probability Distributions In order to test the effect of change in the number of Gaussian probability distributions on the system performance, the latter was trained and tested for different Gaussian values ranging from 1 to 256 for the two cases of 3 and 5 states by HMM (Tables 5 and 6).

Table 5. Number of Gaussian probability distributions & HMM = 3.

Number of Gaussian	WER
1	35.9
2	12
4	28
8	28.6
16	33.7
32	39.5
64	53.8
128	81.5
256	89.4

Table 6. Number of Gaussian probability distributions & HMM = 5.

Number of Gaussian	WER
1	36.8
2	33.4
4	32.2
8	35.6
16	38.9
32	42.2
64	52.6
128	63.8
256	80.8

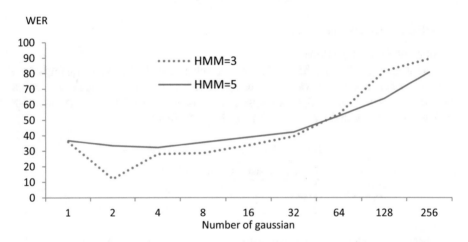

Fig. 7. Evolution of WER as a function of the number of Gaussian, in both cases 3 and 5 by HMM.

The system obtained best the best results when trained using 2 Gaussian Mixture models, and 3 number of states change by HMM (Fig. 7).

The presented work has been compared with the existing similar works. In paper El Ghazi et al. [7] have presented a system for automatic speech recognition on the Amazigh. used the hidden Markov model to model the phonetic units corresponding to words taken from the training base. The corpus consists of the database size used for this research work is 2000 words. The test database contains 330 audio files. However the system is giving good performance 90%, but the design is speaker specific and uses very smallvocabulary. In paper (Satori et al. [2–19]), they have proposed a spoken Arabic recognition system, where Arabic alphabets were investigated to form the ten Arabic digits (from zero to nine). The proposed system consists of two steps: - Mel Frequency Cepstral Coefficients (MFCC) features extraction; - Classification and recognition conducted by CMU Sphinx4 which is a speaker independent system based on hidden Markov model. The mean performance results reached, when realizing three tests, were between 83.33% and 96.67%. Although the system performance is good, the vocabulary size is small.

Our work joins the results of literature, System is highly efficient produces 88% of accuracy. Vieu the vocabulary size is relatively small.

7   Conclusion

In this paper, the system of automatic speech recognition system for Amazigh language was developed. This system is based on the open source CMU Sphinx-4, from the Carnegie Mellon University. The important components of ASR system are feature extraction, acoustic modeling, pronunciation and modeling using HMM. The database size for this research work is 2970 words and produces 88% of accuracy.

The originality of our work lies in the fact of approaching the Amazigh language it is considered among the first works treating this language using the Open Source CMU Sphinx.

However, we consider that this is a preliminary work which will enable us subsequently to fulfill the basic objective that is an independent speaker recognition system for the Amazigh language. In perspective, we can extend the application for more vocabulary size of the Amazigh language.

References

1. Lecouteux, B.: Reconnaissance automatique de la parole guidée par des transcriptions a priori. Doctoral thesis. Université d'Avignon (2008)
2. Satori, H., Harti, M., Chanfour, N.: Arabic speech recognition system based on CMUSphinx. IEEE. Program. I In: International Symposium on Computational Intelligence and Intelligent Informatics, ISCIII 2007 (2007). http://ieeexplore.ieee.org/document/4218391
3. Ghanou, Y., Bencheikh, G.: Architecture optimization and training for the multilayer perceptron using ant system. IAENG Int. J. Comput. Sci. **43**(1), 20–26 (2016)
4. Douib, W.: Reconnaissance automatique de la parole arabe par cmu sphinx 4. Doctoral thesis. Université Ferhat Abbas de Sétif 1 (2013)
5. https://en.wikipedia.org/wiki/Berber_languages
6. Amour, M., Bouhjar, A., Boukhris, F.: 2004 IRCAM: publication: "initiation à la langue Amazigh" (2004)
7. El Ghazi, A., Daoui, C., Idrissi, N.: Automatic speech recognition system concerning the moroccan dialecte (darija and tamazight). Int. J. Eng. Sci. Technol. (IJEST), **2012**, **4**(3), 966–975 (2012). ISSN 0975-5462
8. Ulucinar, B.: Master thesis report (2007)
9. Carnegie Mellon University. Sphinx-4. http://cmusphinx.sourceforge.net
10. Alotaibi, Y.A.: Investigating spoken Arabic digits in speech recognition setting. Inf. Comput. Sci. **173**, 115 (2005)
11. http://cmusphinx.sourceforge.net/wiki/tutoriallm
12. http://cmusphinx.sourceforge.net/wiki/tutorialam
13. http://cmusphinx.sourceforge.net/wiki/tutorialconcepts
14. Ettaouil, M., Ghanou, Y.: Neural architectures optimization and Genetic algorithms. Wseas Trans. Comput. **8**(3), 526–537 (2009)

15. Kumar, K., Aggarwal, R.K., Jain, A.: A Hindi speech recognition system for connected words using HTK. Int. J. Comput. Syst. Eng. 1(1), 25–32 (2012)
16. Abushariah, M.A.A.M., Ainon, R., Zainuddin, R., Elshafei, M., Khalifa, O.O.: Arabic speaker-independent continuous automatic speech recognition based on a phonetically rich and balanced speech corpus. Int. Arab J. Inf. Technol. (IAJIT) 9(1), 84–93 (2012)
17. Young, S.: The HTK hidden Markov model toolkit: design and philosophy. Doctoral thesis. Cambridge University Engineering Department, UK, Technical report. CUED/FINFENG/ TR152, September 1994
18. Ali, A., Zhang, Y., Cakrdinal, P., Dahak, N., Vogel, S., Glass, J.: A complete kaldi recipe for building Arabic speech recognition systems. In: 2014 IEEE Spoken Language Technology Workshop (SLT), pp. 525–529. IEEE, December 2014
19. Satori, H., ElHaoussi, F.: Investigation Amazigh speech recognition using CMU tools. Int. J. Speech Technol. 17(3), 235–243 (2014)
20. Al-Qatab, B.A.Q., Ainon, R.N.: Arabic speech recognition using Hidden Markov Model Toolkit (HTK). Paper presented at International Symposium in Information Technology (ITSim). Kuala Lumpur, 15–17 June (2010)
21. Ettaouil, M., Ghanou, Y., El Moutaouakil, K., Lazaar, M.: Image medical compression by a new architecture optimization model for the Kohonen networks. Int. J. Comput. Theory Eng. 3(2), 204–210 (2011)

An Improved Competency Meta-model
for Adaptive Learning Systems

El Asame Maryam[1(✉)], Wakrim Mohamed[1], and Battou Amal[2]

[1] Engineering Sciences Laboratory, Faculty of Science, Agadir, Morocco
mariam.assam@gmail.com, m.wakrim@uiz.ac.ma
[2] IRF-SIC Laboratory, Faculty of Science, Agadir, Morocco
a.battou@uiz.ac.ma

Abstract. The competency-based approach offers the opportunity for learners to adapt and personalize their learning activities and develop their capacities moreover, it can be applied in many fields and different virtual learning environments. However, researchers and organizations have shown a growing interest in the competencies modeling, which have been formalized for implementation in virtual learning environment and specifically in Adaptive Learning Environment (ALS). As a result, the first step to apply the competency-based approach in such environment is to modeling competency concept. In this paper, we present our proposal competency meta-model based on the HR-XML specification (Human Resources XML) and the IMS Reusable Definition of Competency or Educational Objective (IMS RDCEO), in order to operationalize it in an ALS.

Keywords: Competency-based approach · Competency meta-model
Adaptive learning system · IMS RDCEO · HR-XML

1 Introduction

Actually, different companies and organizations have developed their own specification metadata according to their specific needs, regardless if those standards are interoperable or not. Additionally, in the educational domain, several metadata specifications have been proposed to describe different aspects of competency elements thus to facilitate interoperability, reuse and exchange units of learning between virtual learning systems in general and particularly across ALS.

According to [1, 2] an adaptive hypermedia system (AHS) attempt to personalize the learning goals, activities, competencies and experience for each learner. It builds user model, domain model and interaction model to adapt its contents, navigation, and interface to the user need. Therefore, we believe that without a good representation of the knowledge and competency, an adaptive hypermedia system will be unable to adapt to its users, to personalize learning activities.

In this sense, we need to provide learners a tool to personalize their learning processes and to evaluate their competencies and learning outcomes. In other words, we do not just need a qualitative structural representation of knowledge in activities and resources,

© Springer International Publishing AG, part of Springer Nature 2018
M. Ben Ahmed and A. A. Boudhir (Eds.): SCAMS 2017, LNNS 37, pp. 411–419, 2018.
https://doi.org/10.1007/978-3-319-74500-8_38

but also a quantitative one providing a metric for evaluating competency gaps during the learning process [3].

In this paper, we suggest a competency metadata model that describes the main elements of the competencies. However, the motivation of this work is to solve the limitations of IMS-RDCEO [4] and HR-XML [5].

The paper is organized as follows: the first section provides the impact of the competency based approach on adaptive learning systems, representing some definitions of competency concept. The second section reviews the existing schemas for competency description especially IMS RDCEO and HR-XML. The third section presents our proposed competency meta-model. The fourth section provides a discussion on the limitations of IMS RDCEO and HR-XML standards. The last section concludes the paper and mentions the topics for further research.

2 Impact of the Competency Based-Approach on Adaptive Learning Systems

In today's adaptive learning systems, the competency is becoming a central concept to structure different learning objects, pedagogical activities and connect different performance level. In addition, most companies and organizations are aware of the importance of competency development for life-long learning [6]. Within this context, in this section, we clarify what is a competency? and what can the competency-based approach bring to adaptive learning system?

2.1 What is a Competency?

There are variety ways to define competency concept in the literature [8–11] and [12]. According to [13], competency as "a set of personal characteristics (skills, knowledge, attitudes, etc.) that a person acquires or needs to acquire, in order to perform an activity inside a certain context with a specific performance level".

Friesen and Anderson defined competency as the integrated various personal features to accomplish an activity or function [14]. In addition, Paquette defines competency as a statement of principle that determines a ternary relationship between a public target, knowledge and a skill [9]. According to Le Boterf competency is the mobilization of all individual and environmental resources to perform a task in a specific situation [15]. For Lasanier, a competency is a skill that integrates ability and knowledge which refers to various domains: cognitive, affective, psychomotor and social [16].

As indicated in our initial work [13], our definition is founded on the relation between skills, knowledge, performance level and context. It is defined as a quadrilateral (S, K, P, C) where "S" is a generic skill (described by an action verb) from a taxonomy of skills, "K" is the specific knowledge on which the actors can practice the skills, "P" is a combination of performance criteria values and "C" is the context in which the skill is applied.

2.2 What Can the Competency-Based Approach Bring to Adaptive Learning System?

Virtual learning systems in general and ALS in particular need at least a clear design on how the content should be organized, what are relations between learning concepts, and a new actors' roles (learners, teachers…) [17].

The competency-based approach is a way to organize learning activities and structure learning objects. It is also focused on learner's learning, motivation, assessment and content personalization. However, the Impact of the competency based-approach on adaptive learning system is:

- Competency-based learning enhances actor's roles (teachers, learners…) to achieve the goal of competence.
- The basic characteristic of competency-based education is the progression of competency mastery does not depend on time; the learner can skip courses or activities if it is capable of demonstrating them. That is to say, we can measure learning regardless of how much time is the learner taking.
- Competency-based learning is a way of structuring learning activities so that the individual learner can meet a predetermined set of competencies. It gives learners various ways of learning and supportive environment for learning to achieve their needs.
- Individualization of learning is the fundamental requirement of competency-based approach; each student has a specific learning rhythm. It also defines what learners should know and be able to do, it understands the learner needs.
- The competency-based approach allows to clearly define learning paths within the organization or across e-learning system and enables employees to be more proactive and Increases the potential for job satisfaction organization.

3 Existing Schemas for Competency Modelling

IMS RDCEO [4] and HR-XML [5] are recent standardization and specification initiatives in education and training to describe, refer and exchange common definitions of competencies and enable interoperability between learning systems [7, 13]. In this section we present a description of the IMS RDCEO and HR-XML specifications.

3.1 IMS RDCEO Specification

The IMS reusable definition of competency or educational objective (RDCEO) provides an information model for describing, referencing, and exchanging competencies definitions. Its main purpose is to enable interoperability between learning systems.

According to RDCEO documents, "the word competency is used in a very general sense that includes skills, knowledge, tasks, and learning outcomes".

In addition, this specification proposes five elements for describing competencies: competency identification, competency title, competency description, competency definition, and optional meta-data that must be conform to IEEE (IEEE Learning Object Metadata).

3.2 HR-XML Competency Standards

The HR-XML consortium aims to enforce e-commerce and inter-company exchange of human resources data within a variety of business contexts. In addition, the HR-XML consortium has provided one of their schemas special for competencies recommendation. The competencies schema achieves the rating, measuring, and comparing a competency.

Competency can be defined by [5] "A specific, identifiable, de-finable, and measurable knowledge, skill, ability and/or other deployment-related characteristic (e.g. attitude, behavior, physical ability) which a human resource may possess, and which is necessary for, or material to, the performance of an activity within a specific business context".

This standard provides nine categories to define competency information: name, description, required, competencyId, taxonomyId, competency evidence, competency weight, and user area.

4   The Proposed Competency Meta-model: Structure and Design

In this section, we propose our competency meta-model where competency based on these core dimensions, the first one is the personal characteristics namely skills and knowledge where the generic skills are applying to a specific domain and acting on knowledge. The second dimension is the competency level that is used to demonstrate the personal's performance. The third dimension is the context in which the individual's competency is applied.

In this sense, in our meta-model of learning:

- Each training aims to develop learners' general competencies.
- Each general competency is composed of four specific competencies where one of them corresponds to a level of performance that can be: beginner, intermediate, advanced or mastery.
- Learner competencies are defined relative to a training domain and each specific competency is developed by associating a composition of activities performed by a group of actors in a learning environment.
- Each competency is defined by skill selected from predefined skill taxonomy.
- Each level is defined of a set of prerequisites that can be university degrees or competencies needed that describe the knowledge and the skills needed. Furthermore, we check these competencies through a diagnostic test at the beginning of the training.

More specifically, the main elements of the general competency are as follows:

- Domain: this element specifies of the domain of training that is describing by identifier, title and is composed of many sub-domains.
- Sub-domain: this element specifies of the sub-domain of learner competency that is describing by identifier, title and description.
- Identifier: a unique label that identifies this general competency.
- Title: a single text label for the general competency.
- Definition: a structured definition of the general competency.
- Specific competency: This element specifies the specific competencies description of a sub domain.

In the Fig. 1, the different elements of general competency are shown.

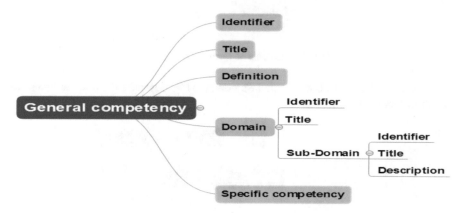

Fig. 1. Elements of general competency.

Regarding specific competency, their main elements are as follows:

- Identifier: a unique label that identifies this specific competency.
- Title: a single text label for the specific competency.
- Definition: a structured definition of the specific competency.
- Prerequisites: this element specifies the types of necessary prerequisites for access to a particular level of training. They can degrees that are required for a performance level or the necessary competencies (skills and Knowledge) to mastering a proficiency level.
- Description: this element specifies a complete description of the specific competency.
- Context: refers the environment in which the activities are carried.
- Performance level: refers to the performance level of the Competency that includes:
- Level: a text label that identifies different types of performance level.
- Scale: the qualitative or quantitative scales of the competency's level, each type includes a minimum value, maximum value, and threshold.

Figure 2, shows the upper elements of the specific competency.

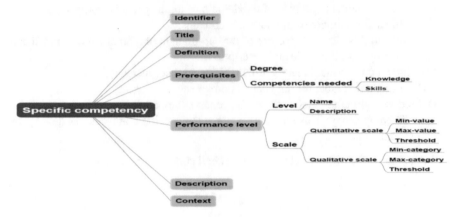

Fig. 2. Elements of specific competency.

To give more details about our meta-model, the Fig. 3 depicts the basic elements of the competency meta-model.

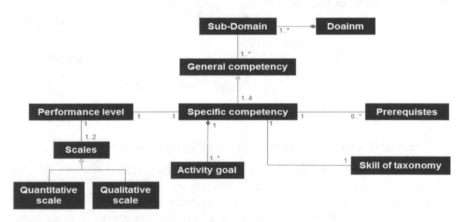

Fig. 3. Basic elements of the competency meta-model.

5   Discussion

First of all, it should be noted that the IMS-RDCEO specification facilitates establishing relations with other specifications that can guarantee a complete user modelling such as IMS-CP [18], IMS-LD [19], IMS-LIP [20] and IMS-QTI [21]. Moreover, the HR-XML standards offers a simple and flexible schema hat will be used within diversity contexts.

However, based on an analysis of the structure of IMS RDCEO and HR-XML specifications [22, 23] and [24], IMS-RDCEO and HR-XML standards do not address all facets of the competency concept. In addition, the description of competency elements is presented in narrative form. Thus, machines can face difficulties in searching and processing these elements. Furthermore, both specifications do not take into consideration proficiency level. Adds an important remark, both specifications do not take into account the context which the individual's competencies are acquired and applied.

According to [25, 26] "the IMS RDCEO needs a way to represent competency grading scale. This lack will affect the assessment of competencies". Moreover, it needs a way when assessing to represent the competency weighting factor of sub-competencies [23, 25].

To avoid these limitations, we have proposed a meta-model of competency that take into consideration, the examination of these specifications and the key dimensions of our competency definition that are (S: skills, K: knowledge, P: performance and C: context). In addition to facilitate the interoperability between the different systems, we reserve for each element a specific space. In addition, our proposal divides the general competency into specific competencies to help learners to evaluate competencies during the learning process.

We believe that our competency meta-model contains the basic elements of the competency-based approach, thus adapting learners' needs through e-learning systems.

6   Conclusion

In this paper, we have presented the impact of the competency-based approach on adaptive learning systems. Additionally, we have looked at the difficulties of existing competency standards namely IMS RDCEO and the HR-XML. However, within the context of building an adaptive e-learning system based on a hybrid pedagogical approach for improving learner's competencies, we proposed an improved competency meta-model based on IMS RDCEO and the HR-XML that is capable of accommodating the basic competency elements to adapt learner's needs, to personalize their learning activities and to demonstrate their performance levels.

Future work includes implementation of this competency meta-model and modelling other learning objects (activities and pedagogical resources).

References

1. Brusilovsky, P.: Adaptive navigation support in educational hypermedia: the role of student knowledge level and the case for meta-adaptation. Br. J. Educ. Technol. **34**(4), 487–497 (2003)
2. Cristea, A.I.: What can the semantic web do for adaptive educational hypermedia? Educ. Technol. Soc. **7**(4), 40–58 (2004)
3. Paquette, G.: An ontology and a software framework for competency modeling and management. Educ. Technol. Soc. **10**(3), 1–21 (2007)
4. IMS RDCEO: IMS Reusable Definition of Competency or Educational Objective (2002). http://www.imsglobal.org/competencies/index.html
5. HR-XML: HR-XML consortium competencies schema (measurable characteristics) (2006). www.hr-xml.org

6. Koper, R., Specht, M.: TenCompetence: lifelong competence development and learning. In: Sicilia, M.A. (ed.) Competencies in Organizational E-Learning: Concepts and Tools (2006)

7. Prins, F.J., Nadolski, R.J., Berlanga, A.J., Drachsler, H., Hummel, H.G.K., Koper, R.: Competence description for personal recommendations: the importance of identifying the complexity of learning and performance situations. Educ. Technol. Soc. 11(3), 141–152 (2008)

8. Guthrie, H.: Competence and competency-based training: what the literature says. National Centre for Vocational Education Research (2009)

9. Paquette, G.: Modélisation des connaissances et des compétences, pour concevoir et apprendre, pp. 187–188. Presses de l'Université du Québec, Sainte-Foy (2002)

10. Blum, W.: Quality teaching of mathematical modelling: what do we know, what can we do? In: The Proceedings of the 12th International Congress on Mathematical Education, pp. 73–96. Springer, Cham (2015)

11. Pijl-Zieber, E.M., Barton, S., Konkin, J., Awosoga, O., Caine, V.: Competence and competency-based nursing education: finding our way through the issues (2014)

12. Wang, N., Abel, M.H., Barthès, J.P., Negre, E.: Mining user competency from semantic trace. In: 2015 IEEE 19th International Conference on Computer Supported Cooperative Work in Design (CSCWD), pp. 48–53. IEEE, May 2015

13. El Asame, M., Wakrim, M.: Towards a competency model: a review of the literature and the competency standards. Educ. Inf. Technol., 1–12 (2017)

14. Friesen, N., Anderson, T.: Interaction for lifelong learning. Br. J. Educ. Technol. 35(6), 679–687 (2004)

15. Le Boterf, G.: Construire les compétences individuelles et collectives. Editions d'organisations, Paris (2000)

16. Boumane, A., Talbi, A., Tahon, C., Bouami, D.: Contribution to the modeling of competence. In: Modeling, Optimization and Simulation Systems: Challenges and Opportunities, MOSIM 2006, Rabat, Morocco (2006)

17. Abel, M.-H., Benayache, A., Lenne, D., Moulin, C., Barry, C., Chaput, B.: Ontology-based organizational memory for e-learning. Educ. Technol. Soc. 7(4), 98–111 (2004)

18. IMS Content Packaging Specification. v1.1.4 Final Specification (2004). http://www.imsglobal.org/content/packaging/index.html

19. IMS Learning Design. Version 1.0 Final Specification (2003). http://www.imsglobal.org/learningdesign/index.html

20. IMS Learner Information Package Accessibility for LIP. Version 1.0 Final Specification (2003). https://www.imsglobal.org/accessibility/acclipv1p0/imsacclip_bindv1p0.html

21. IMS Question and Test Interoperability. Version 1.2.1 Final Specification (2003). http://www.imsglobal.org/question/index.html

22. Sitthisak, O., Gilbert, L., Davis, H.C., Gobbi, M.: Adapting health care competencies to a formal competency model. In: The 7th IEEE International Conference on Advanced Learning Technologies, Niigata, Japan. IEEE Computer Society Press (2007)

23. Sampson, D., Fytros, D.: Competence models in technology-enhanced competence-based learning. In: Adelsberger, H.H., Kinshuk, Pawlowski, J.M., Sampson, D. (eds.) Handbook on Information Technologies for Education and Training, pp. 155–177. Springer-Verlag, Heidelberg (2008)

24. Sampson, D.G.: Competence-related metadata for educational resources that support lifelong competence development programmes. Educ. Technol. Soc. 12(4), 149–159 (2009)

25. Karampiperis, P., Sampson, D., Fytros, D.: Lifelong competence development: towards a common metadata model for competencies description – the case study of Europass Language passport. In: Proceedings of the 6th IEEE International Conference on Advanced Learning Technologies, 5th–7th July, Kerkrade, The Netherlands (2006)
26. García-Barriocanal, E., Sicilia, M.A., Sánchez-Alonso, S.: Computing with competencies: modelling organizational capacities. Expert Syst. Appl. **39**(2012), 12310–12318 (2012)

A Noise-Free Homomorphic Evaluation
of the AES Circuits to Optimize Secure Big
Data Storage in Cloud Computing

Ahmed EL-Yahyaoui$^{(\boxtimes)}$ ⓘ and Mohamed Dafir Ech-Chrif El Kettani

Information Security Research Team, CEDOC ST2I ENSIAS,
Mohammed V University, Rabat, Morocco
{ahmed_elyahyaoui, dafir.elkettani}@um5.ac.ma

Abstract. In this paper, we describe a homomorphic evaluation of the different AES circuits (AES-128, AES-192 and AES-256) using a noise-free fully homomorphic encryption scheme. This technique is supposed to be an efficient solution to optimize data storage in a context of outsourcing computations to a remote cloud computing as it is considered a powerful tool to minimize runtime in the client side. In this implementation, we will use a noise free quaternionique based fully homomorphic encryption scheme with different key sizes. Among the tools we are using in this work, a small laptop with characteristics: bi-cores Intel core i5 CPU running at 2.40 GHz, with 512 KB L2 cache and 4 GB of Random Access Memory. Our implementation takes about 18 min to evaluate an entire AES circuit using a key of 1024 bits for the fully homomorphic encryption scheme.

Keywords: AES · Evaluation · Fully homomorphic encryption
Optimization

1 Introduction

Fully homomorphic encryption (Fig. 1) was invented to allow computations over encrypted data. It has trivial applications to the security of big data and cloud computing. In fact, big data today is ubiquitous with the proliferation of Internet technologies in modern applications, social networking, airlines information, online shopping and so on. Broadly speaking, users can possess huge amount of data which is impossible to be processed locally given its cost and computing power. In this situation, users can enjoy unlimited computing power in the cloud to serve data processing and big data analytics. The security of cloud computing is not always trusted by the client, at least clients whose data privacy is substantial cannot trust any third part. For example, banking data and electronic health records are sensitive data that a simple information leakage can cause huge damage to owners.

Security is an integral requirement in the cloud computing which does have enormous attention from the research community. Among the fascinating available solutions, we find a type of encryption that allows performing computations over encrypted data. It is called fully homomorphic encryption and it was conjectured by

© Springer International Publishing AG, part of Springer Nature 2018
M. Ben Ahmed and A. A. Boudhir (Eds.): SCAMS 2017, LNNS 37, pp. 420–431, 2018.
https://doi.org/10.1007/978-3-319-74500-8_39

Rivest, Adleman and Dertozous in 1978 under the name of privacy homomorphism [1]. In 2009 Craig Gentry presented the first effective solution [2] to this conjecture in a breakthrough work of thesis. Gentry's contribution is a historical framework of constructing fully homomorphic encryption schemes that is based on a bootstrapping theorem to refresh ciphertexts and reduce noise growth after processing. Bootstrapping technique influences performance capacities and runtime of the homomorphic encryption scheme, thus it affects negatively its application to the cloud computing.

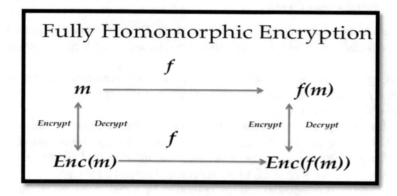

Fig. 1. Fully homomorphic encryption diagram

After Gentry's breakthrough [2], many improvements [3–6] are introduced to enhance the design of fully homomorphic encryption schemes and change for the better its implementation hopes. Concerning the design, some new techniques of noise refreshment are invented to avoid the cost of the bootstrapping method (bootstrap is called after each multiplication) and provide the homomorphic cryptosystem with intrinsic proficiencies. Modulus reduction [7], key switching [7] and flattening [8] are some of those new noise management techniques. Concerning the implementation, many practical contributions are done to reduce the runtime and minimize the resulting storage. Among these contributions we find: batching [9, 10] of a fully homomorphic encryption scheme to a scheme that supports encrypting and homomorphically processing a vector of plaintexts as a single ciphertext and evaluating homomorphically some specific symmetric encryption schemes (as AES circuit) [6] to simplify the client side encryption and minimize the server side data storage. Because a symmetric algorithm, like AES cryptosystem, is known to be very fast, secure and with low cost overhead. Nevertheless, these improvements are powerful, it remains insufficient to reach a practical fully homomorphic encryption scheme. Therefore, we can perform more optimizations at the implementation to get a faster and optimal fully homomorphic encryption scheme to be used in a context of big data and cloud computing security.

In this article, we will provide an efficient evaluation of the AES circuits based on a noise free fully homomorphic encryption scheme. Firstly, we will start by a mathematical background. Secondly, we will introduce a noise free fully homomorphic encryption scheme to be used in this work besides we will present different homo-morphic optimization techniques used in homomorphic encryption. Thirdly, we will detail the body section of our article about the homomorphic evaluation of the AES circuits. Finally, we will finish with a conclusion and perspectives.

2 Mathematical Background

2.1 Quaternionique Field $\mathbb{H}$

A quaternion is a number in his generalized sense. Quaternions encompass real and complex numbers in a number system where multiplication is no longer a commutative law.

The quaternions were introduced by the Irish mathematician William Rowan Hamilton in 1843. They now find applications in mathematics, physics, computer science and engineering.

Mathematically, the set of quaternions $\mathbb{H}$ is a non-commutative associative algebra on the field of real numbers $\mathbb{R}$ generated by three elements i, j and k satisfying rela-tions: $i^2 = j^2 = k^2 = i.j.k = -1$. Concretely, any quaternion q is written uniquely in the form: $q = a + bi + cj + dk$ where a, b, c and d are real numbers.

The operations of addition and multiplication by a real scalar are trivially done term to term, whereas the multiplication between two quaternions is termed by respecting the non-commutativity and the rules proper to i, j and k. For example, given $q = a + bi + cj + dk$ and $q' = a' + b'i + c'j + d'k$ we have $qq' = a_0 + b_0 i + c_0 j + d_0 k$ such that: $a_0 = aa' - (bb' + cc' + dd')$, $b_0 = ab' + a'b + cd' - c'd$, $c_0 = ac' - bd' + ca' + db'$ and $d_0 = ad' + bc' - cb' + a'd$.

The quaternion $\bar{q} = a - bi - cj - dk$ is the conjugate of q. $|q| = \sqrt{q\bar{q}} = \sqrt{a^2 + b^2 + c^2 + d^2}$ is the module of q. The real part of q is $Re(q) = \frac{q + \bar{q}}{2} = a$ and the imaginary part is $Im(q) = \frac{q - \bar{q}}{2} = bi + cj + dk$..

A quaternion q is invertible if and only if its modulus is non-zero, and we have $q^{-1} = \frac{1}{|q|^2} \bar{q}$.

2.2 Reduced Form of Quaternion

Quaternion can be represented in a more economical way, which considerably alle-viates the calculations and highlights interesting results. Indeed, it is easy to see that $\mathbb{H}$ is a $\mathbb{R}$-vectorial space of dimension 4, of which $(1, i, j, k)$ constitutes a direct orthonormal basis. We can thus separate the real component of the pure components, and we have for $q \in \mathbb{H}$, $q = (a, u)$ such that u is a vector of $\mathbb{R}^3$. So for $q = (a, u), q' = (a', v) \in \mathbb{H}$ and $\lambda \in \mathbb{R}$ we obtain:

1. $q + q' = (a + a', u + v)$ and $\lambda q = (\lambda a, \lambda u)$.
2. $qq' = (aa' - u.v, av + a'u + u \wedge v)$. Where $\wedge$ is the cross product of $\mathbb{R}^3$.
3. $\bar{q} = (a, -u)$ and $|q|^2 = a^2 + u^2$.

2.3 Ring of Lipschitz Integers

The set of quaternions defined as follows: $\mathbb{H}(\mathbb{Z}) = \{q = a + bi + cj + dk / a, b, c, d \in \mathbb{Z}\}$
Has a ring structure called the ring of Lipschitz integers. $\mathbb{H}(\mathbb{Z})$ is trivially non-commutative.

For r $n \in \mathbb{N}^*$, the set of quaternions: $\mathbb{H}(\mathbb{Z}/n\mathbb{Z}) = \{q = a + bi + cj + dk / a, b, c, d \in \mathbb{Z}/n\mathbb{Z}\}$
has the structure of a non-commutative ring.

A modular quaternion of Lipschitz $q \in \mathbb{H}(\mathbb{Z}/n\mathbb{Z})$ is invertible if and only if its module and the integer n are coprime numbers, i.e. $|q|^2 \wedge n = 1$.

2.4 Quaternionique Matrices $\mathbb{M}_2(\mathbb{H}(\mathbb{Z}/n\mathbb{Z}))$

The set of matrices $\mathbb{M}_2(\mathbb{H}(\mathbb{Z}/n\mathbb{Z}))$ describes the matrices with four inputs (two rows and two columns) which are quaternions of $\mathbb{H}(\mathbb{Z}/n\mathbb{Z})$. This set has a non-commutative ring structure.

There are two ways of multiplying the quaternion matrices: The Hamiltonian product, which respects the order of the factors, and the octonionique product, which does not respect it.

The Hamiltonian product is defined as for all matrices with coefficients in a ring (not necessarily commutative). For example:

$$U = \begin{pmatrix} u_{11} & u_{12} \\ u_{21} & u_{22} \end{pmatrix}, V = \begin{pmatrix} v_{11} & v_{12} \\ v_{21} & v_{22} \end{pmatrix} \Rightarrow UV = \begin{pmatrix} u_{11}v_{11} + u_{12}v_{21} & u_{11}v_{12} + u_{12}v_{22} \\ u_{21}v_{11} + u_{22}v_{21} & u_{21}v_{12} + u_{22}v_{22} \end{pmatrix}$$

The octonionique product does not respect the order of the factors: on the main diagonal, there is commutativity of the second products and on the second diagonal there is commutativity of the first products.

$$U = \begin{pmatrix} u_{11} & u_{12} \\ u_{21} & u_{22} \end{pmatrix}, V = \begin{pmatrix} v_{11} & v_{12} \\ v_{21} & v_{22} \end{pmatrix} \Rightarrow UV = \begin{pmatrix} u_{11}v_{11} + v_{21}u_{12} & v_{12}u_{11} + u_{12}v_{22} \\ v_{11}u_{21} + u_{22}v_{21} & u_{21}v_{12} + v_{22}u_{22} \end{pmatrix}$$

In our article we will adopt the Hamiltonian product as an operation of multiplication of the quaternionique matrices.

2.5 Schur Complement and Inversibility of Quaternionique Matrices

Let $\mathcal{R}$ be an arbitrary associative ring, a matrix $M \in \mathcal{R}^{n \times n}$ is supposed to be invertible if $\exists N \in \mathcal{R}^{n \times n}$ such that $MN = NM = I_n$ where N is necessarily unique.

The Schur complement method is a very powerful tool for calculating inverse of matrices in rings. Let $M \in \mathcal{R}^{n \times n}$ be a matrix per block satisfying: $M = \begin{pmatrix} A & B \\ C & D \end{pmatrix}$ such that $A \in \mathcal{R}^{k \times k}$.

Suppose that A is invertible, we have: $M = \begin{pmatrix} I_k & 0 \\ CA^{-1} & I_{n-k} \end{pmatrix} \begin{pmatrix} A & 0 \\ 0 & A_s \end{pmatrix} \begin{pmatrix} I_k & A^{-1}B \\ 0 & I_{n-k} \end{pmatrix}$

where $A_s = D - CA^{-1}B$ is the Schur complement of A in M.

The inversibility of A ensures that the matrix M is invertible if and only if A_s is invertible. The inverse of M is: $M^{-1} = \begin{pmatrix} I_k & -A^{-1}B \\ 0 & I_{n-k} \end{pmatrix} \begin{pmatrix} A^{-1} & 0 \\ 0 & A_s^{-1} \end{pmatrix}$ $\begin{pmatrix} I_k & 0 \\ -CA^{-1} & I_{n-k} \end{pmatrix} = \begin{pmatrix} A^{-1} + A^{-1}BA_s^{-1}CA^{-1} & -A^{-1}BA_s^{-1} \\ -A_s^{-1}CA^{-1} & A_s^{-1} \end{pmatrix}$. For a quaternionique

matrix $M = \begin{pmatrix} a & b \\ c & d \end{pmatrix} \in \mathcal{R}^{2\times2} = \mathbb{M}_2(\mathbb{H}(\mathbb{Z}/n\mathbb{Z}))$ where the quaternion a is invertible as well as its Schur complement $a_s = d - ca^{-1}b$ we have M is invertible and:

$$M^{-1} = \begin{pmatrix} a^{-1} + a^{-1}ba_s^{-1}ca^{-1} & -a^{-1}ba_s^{-1} \\ -a_s^{-1}ca^{-1} & a_s^{-1} \end{pmatrix}$$

Therefore, to randomly generate an invertible quaternionique matrix, it suffices to:

- Choose randomly three quaternions a, b and c for which a is invertible.
- Select randomly the fourth quaternion d such that the Schur complement $a_s = d - ca^{-1}b$ of a in M is invertible.

3   Homomorphic Encryption

In this work we will be based on the fully homomorphic encryption scheme described below. It is a noise-free cryptosystem, based on the Lipschitz quaternions and uses a homomorphic transform to encode bits into quaternions as it is described below:

3.1   Homomorphic Transform

Any bit $\sigma \in \mathbb{Z}/2\mathbb{Z} = \{0, 1\}$ can be encoded into a Lipschitz quaternion according to a homomorphic transform whose operations on the quaternions retain those on the bits. This transform can be given as follows:

$$bitToQuatern\colon \sigma \in \mathbb{Z}/2\mathbb{Z} \mapsto bitToQuatern(\sigma) = m + 2\ell i + pj + qk \in \mathbb{H}(\mathbb{Z})$$

such that m, ℓ, p, q are randomly chosen integers verifying the two conditions: $m \equiv \sigma[2]$ and $p \equiv q[2]$. The inverse transform that will be named $quaternToBit$ is given by: $quaternToBit(q) = Re(q)[2]$.

3.2   Key Generation

- Bob generates randomly two big prime numbers p and q.
- Then, he calculates $n = 2.p.q$.

- Bob generates randomly an invertible matrix: $K = \begin{pmatrix} a_{1,1} & a_{1,2} & a_{1,3} \\ a_{2,1} & a_{2,2} & a_{2,3} \\ a_{3,1} & a_{3,2} & a_{3,3} \end{pmatrix}$ $\in \mathbb{M}_3(\mathbb{H}(\mathbb{Z}/n\mathbb{Z}))$.
- Bob calculates the inverse of K, Which will be denoted K^{-1}.
- The secrete key is (K, K^{-1}).

3.3 Encryption

Lets $\sigma \in \mathbb{Z}/2\mathbb{Z} = \{0, 1\}$ be a clear text. To encrypt σ Bob proceed as follows:

- Using the transform $bitToQuatern$, Bob transforms σ into a quaternion: $m = bitToQuatern(\sigma) \in \mathbb{H}(\mathbb{Z}/n\mathbb{Z})$.
- Bob generates a matrix $M = \begin{pmatrix} m & r_3 & r_4 \\ 0 & r_1 & r_5 \\ 0 & 0 & r_2 \end{pmatrix} \in \mathbb{M}_3(\mathbb{H}(\mathbb{Z}/n\mathbb{Z}))$ such that $r_i \in$ $\mathbb{H}(\mathbb{Z}/n\mathbb{Z}) \forall i \in [\![1, 5]\!]$ are randomly generated with $|r_i| \equiv 0[2]$.
- The ciphertext σ of is $C = Enc(\sigma) = KMK^{-1} \in \mathbb{M}_3(\mathbb{H}(\mathbb{Z}/n\mathbb{Z}))$.

3.4 Decryption

Lets $C \in \mathbb{M}_3(\mathbb{H}(\mathbb{Z}/n\mathbb{Z}))$ be a ciphertext. To decrypt C Bob proceed as follows:

- He calculates $M = K^{-1}CK$ using his secrete key (K, K^{-1}).
- Then he takes the first input of the resulting matrix $m = (M)_{1,1}$.
- Finally, he recovers his clear message by calculating $\sigma = quaternToBit(m)$ using the $quaternToBit$ transform.

3.5 Addition and Multiplication

Let σ_1 and σ_2 be two clear texts and $C_1 = Enc(\sigma_1)$ and $C_2 = Enc(\sigma_2)$ be their ciphertexts respectively.

It is easy to verify, thanks to the $bitToQuatern$ transform, that:

(1) $C_{mult} = C_1 + C_2 = Enc(\sigma_1) + Enc(\sigma_2) = Enc(\sigma_1 \oplus \sigma_2)$. Such that $\oplus$ is the binary XOR.
(2) $C_{add} = C_1.C_2 = Enc(\sigma_1).Enc(\sigma_2) = Enc(\sigma_1 \otimes \sigma_2)$. Such that $\otimes$ is the binary AND.

4 Homomorphic Optimizations

Nowadays, we know many constructions of fully homomorphic encryption schemes that allow arbitrary constructions over encrypted cloud data. Since the first apparition of a fully homomorphic encryption scheme, a number of improvements have been carried out and different implementations have been achieved. In this part, we will take back to the literature and report the different optimization technics that are proposed to improve the efficiency of fully homomorphic encryption.

In a context of outsourcing computations to a remote cloud server (Fig. 1), a client that is characterized by its weak computation powers, low storage capacities and feeble processing sends its encrypted big data to an outsized server to take care of storage and processing. It is clear that we need two types of optimizations; we shall call it client-side and server-side optimizations.

Concerning the server-side improvement, a server should be able to easily evaluate functions on ciphertexts and store optimal ciphertexts in its databases. Thus, an effective optimization should take in consideration ciphertexts expansion and runtime of the homomorphic encryption scheme in server side. Among the solutions proposed to this paradigm is batching fully homomorphic encryption [9, 11] using Chinese Reminder Theorem (CRT), i.e. packing multiple plaintext messages into the slots of a single ciphertext. If this method allows executing many operations in just one operation, it has the drawbacks that it does not permit to reduce ciphertext expansion and it stretches the key size because the number of packed ciphertexts depends on the number of the key factors. A second method to improve server-side performances is to compress ciphertexts using a polynomial which coefficients are plaintext messages [12], this method cannot be used in client-side optimization, as the authors said, because there is no method to unpack a polynomial ciphertext. In terms of data traffic, this solution is optimal than the precedent because, here, the number of packed ciphertexts do not depend on the modulus factors as in batching method.

Concerning client-side optimization, a client should be able to easily encrypt and decrypt messages. Therefore, computations should be as fast as possible and the uploaded data to the cloud should be as short as possible. The first ideas for decreasing ciphertext expansion and improving runtime from client to cloud were proposed by Lauter et al. in [12]. The authors present trans-ciphering from a symmetric encryption algorithm to the fully homomorphic encryption scheme as a solution. Practically the client encrypt its data using a symmetric algorithm and encrypt the secrete key of this symmetric algorithm using the private key of his fully homomorphic encryption scheme and sends all ciphertexts to the cloud server. The cloud server store the encrypted symmetric secrete key and encrypted data in a large database. When the client ask the cloud to evaluate a function on his encrypted data, the cloud evaluate homomorphically the circuit of the symmetric algorithm using the public parameters of the fully homomorphic encryption scheme. This solution is efficient in client-side, because the encryption is very fast, and in storage, because the cloud will store ciphertexts issued from a symmetric algorithm, which sizes are equals to cleartexts. Many implementations of this proposition have been done after. For example in [6], Gentry, Halevi and Smart evaluated the AES circuit using the BGV encryption scheme [3], the homomorphic evaluation of an AES-128 circuit takes more than 17 min with bootstrapping.

Surely, we have changed for the better the situation and have minimized the cost of using a homomorphic scheme in practice. However, are these optimizations enough to get a practical fully homomorphic encryption scheme? Intuitively, the response is negative with no doubt (Fig. 2).

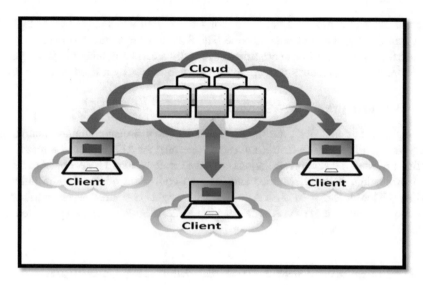

Fig. 2. Cloud computing context

In this work, we will focus on the third optimization solution and we will introduce an improvement to the trans-ciphering method. For that reason we use a noise-free fully homomorphic encryption scheme to evaluate homomorphically the AES circuits. We get ambitious result for different AES circuits (AES-128, AES-192 and AES-256) and different homomorphic key sizes.

5   Homomorphic Evaluation of the AES Circuits

Following we describe our homomorphic implementation of the AES circuits. Our implementation is JAVA programming language based.

5.1   An Overview of the AES Circuits

The AES circuits are three categories: AES-128, AES-192 and AES-256. Each AES circuit produces a cipher of its indexed size. AES round function operates on a 4 × 4 matrix of bytes. The key size used for an AES cipher specifies the number of repetitions of transformation rounds that convert the plaintext into the final ciphertext. The number of cycles of repetition are as follows:

- Ten (10) cycles of repetition for 128-bit keys.
- Twelve (12) cycles of repetition for 192-bit keys.
- Fourteen (14) cycles of repetition for 256-bit keys.

The basic operations that are performed during a round function are AddRoundKey, SubBytes, ShiftRows and MixColumns. The AddRoundKey is a combination of each byte of the current state with a block of the round key using the bitwise XOR.

The SubBytes operation is a non-linear substitution step where each byte is replaced with another according to a lookup table. The ShiftRows is a transposition step where the last three rows of the state are shifted cyclically a certain number of steps. Finally, the MixColumns operations pre-multiplies the state matrix by a fixed 4×4 matrix.

5.2   How to Represent an AES State

The ciphertext issued from our cryptosystem is a 3×3-quaternionique matrix that encrypts one binary bit (a matrix represents one bit); an AES cipher is a matrix that entries are bytes. To support operations we have created a new JAVA object to help in AES evaluation, it is ByteMatrice and it is the equivalent of the Byte primitive in JAVA. ByteMatrice is a JAVA class with eight attributes, each attribute is 3×3-quaternionique matrix. After evaluation each AES byte will be represented by a ByteMatrice element.

6   Homomorphic Evaluation of the Basic AES Operations

In this section, we describe the homomorphic evaluation of each AES operation.

6.1   AddRoundKey Operation

The AddRoundKey operation is a simple XOR operation between the current state and a block of the round key.

6.2   SubBytes Operation

This operation is the only non-linear transformation. It converts 8bit data to other 8 bit data. It uses operations over the field $GF(2^8)$ where elements are treated as bit strings with polynomial representation $G(X) = X^8 + X^4 + X^3 + X + 1$. Thus, every element of $GF(2^8)$ is a byte $b_7b_6b_5b_4b_3b_2b_1b_0$ considered as a polynomial:

$$b(X) = b_7X^7 + b_6X^6 + b_5X^5 + b_4X^4 + b_3X^3 + b_2X^2 + b_1X + b_0$$

Addition over $GF(2^8)$: $a_7a_6a_5a_4a_3a_2aa_0 + b_7b_6b_5b_4b_3b_2b_1b_0 = c_7c_6c_5c_4c_3c_2c_1c_0$ where $c(X) = a(X) + b(X)$ which also gives $c_i = a_i \oplus b_i$.

Multiplicat over $GF(2^8)$: $a_7a_6a_5a_4a_3a_2aa_0 \times b_7b_6b_5b_4b_3b_2b_1b_0 = c_7c_6c_5c_4c_3c_2c_1c_0$ where $c(X) = a(X) \times b(X) mod G(X)$.

The definition of SubBytes Transformation is the serial transformation of the following two steps:

1) Take the multiplicative inverse in Galois Field $GF(2^8)$, the element byte "00000000" = hex {00} is mapped to itself. Inverse Table is shown in the annex.
2) Then apply the affine transformation over $GF(2)$,. Binary "00000000" = Hex {00} will be transformed into "01100011" = {63}.

$$
\begin{bmatrix} b_7 \\ b_6 \\ b_5 \\ b_4 \\ b_3 \\ b_2 \\ b_1 \\ b_0 \end{bmatrix} = \begin{bmatrix} 1 & 1 & 1 & 1 & 1 & 1 & 0 & 0 \\ 0 & 1 & 1 & 1 & 1 & 1 & 0 & 0 \\ 0 & 0 & 1 & 1 & 1 & 1 & 1 & 0 \\ 0 & 0 & 0 & 1 & 1 & 1 & 1 & 1 \\ 1 & 0 & 0 & 0 & 1 & 1 & 1 & 1 \\ 1 & 1 & 0 & 0 & 0 & 1 & 1 & 1 \\ 1 & 1 & 1 & 0 & 0 & 0 & 1 & 1 \\ 1 & 1 & 1 & 1 & 0 & 0 & 0 & 1 \end{bmatrix} \times \begin{bmatrix} a_7 \\ a_6 \\ a_5 \\ a_4 \\ a_3 \\ a_2 \\ a_1 \\ a_0 \end{bmatrix} \oplus \begin{bmatrix} 0 \\ 1 \\ 1 \\ 0 \\ 0 \\ 0 \\ 1 \\ 1 \end{bmatrix}
$$

When evaluating homomorphically AES circuit we use the inverse of the SubBytes transformation as it is described below:

1) The inverse affine transformation is executed. The inverse transformation will be given in the following equation.

$$
\begin{bmatrix} a_7 \\ a_6 \\ a_5 \\ a_4 \\ a_3 \\ a_2 \\ a_1 \\ a_0 \end{bmatrix} = \begin{bmatrix} 0 & 1 & 0 & 1 & 0 & 0 & 1 & 0 \\ 0 & 0 & 1 & 0 & 1 & 0 & 0 & 1 \\ 1 & 0 & 0 & 1 & 0 & 1 & 0 & 0 \\ 0 & 1 & 0 & 0 & 1 & 0 & 1 & 0 \\ 0 & 0 & 1 & 0 & 0 & 1 & 0 & 1 \\ 1 & 0 & 0 & 1 & 0 & 0 & 1 & 0 \\ 0 & 1 & 0 & 0 & 1 & 0 & 0 & 1 \\ 1 & 0 & 1 & 0 & 0 & 1 & 0 & 0 \end{bmatrix} \times \begin{bmatrix} b_7 \\ b_6 \\ b_5 \\ b_4 \\ b_3 \\ b_2 \\ b_1 \\ b_0 \end{bmatrix} \oplus \begin{bmatrix} 0 \\ 0 \\ 0 \\ 0 \\ 0 \\ 1 \\ 0 \\ 1 \end{bmatrix}
$$

2) Then perform the same multiplicative inverse in Galois Field $GF(2^8)$, according to the Table 1.

6.3   ShiftRows Operation

The shifting of rows is a simple operation that only requires swapping of indices trivially handled in the code. This operation has no effect on the noise.

6.4   MixColumns Operation

The MixColumns operation consists in the application of a linear transformation to each column of the matrix state. Consider a column $c = (c_1, c_2, c_3, c_4)^t$ such that c_i is an element of $GF(2^8)$, each column c is transformed to another column d as following:

$$
\begin{bmatrix} d_0 \\ d_1 \\ d_2 \\ d_3 \end{bmatrix} = \begin{bmatrix} 02 & 03 & 01 & 01 \\ 01 & 02 & 03 & 01 \\ 01 & 01 & 02 & 03 \\ 03 & 01 & 01 & 01 \end{bmatrix} \times \begin{bmatrix} c_0 \\ c_1 \\ c_2 \\ c_3 \end{bmatrix}
$$

When evaluating homomorphically AES circuit we use the inverse of the Mix-Columns transformation to get c from d as it is described below:

$$
\begin{bmatrix} c_0 \\ c_1 \\ c_2 \\ c_3 \end{bmatrix} = \begin{bmatrix} 0E & 0B & 0D & 09 \\ 09 & 0E & 0B & 0D \\ 0D & 09 & 0E & 0B \\ 0B & 0D & 09 & 0E \end{bmatrix} \times \begin{bmatrix} d_0 \\ d_1 \\ d_2 \\ d_3 \end{bmatrix}
$$

6.5   Performances Results

The Table 1 shows the performances obtained after implementation. The implementation is done using a personal computer with characteristics: bi-cores Intel core i5 CPU running at 2.40 GHz, with 512 KB L2 cache and 4 GB of Random Access Memory. The present implementation is done under JAVA programming language using the IDE Eclipse platform.

Table 1. AES circuits evaluation performances

FHE param	AES 128	AES 192	AES 256
256 bits	3 min	4,36 min	4,71 min
512 bits	6 min	7 min	10 min
768 bits	12 min	14 min	16,45 min
1024 bits	18 min	21 min	24 min
2048 bits	57 min	1 h	1 h 18 min
4096 bits	2 h 33 min	2 h 51 min	3 h 38 min

For an AES-128 circuit and using an acceptable secure fully homomorphic encryption key of 1024 bits, we observe that we obtain good performances. The circuit is evaluated in about 18 min, which stays an ambitious runtime.

7   Conclusion and Perspectives

In this paper, we presented a homomorphic evaluation of the AES three circuits, AES-128, AES-192 and AES-256. We are employed a noise-free fully homomorphic encryption scheme based on matrix operations. From our best knowledge, this is the first such attempt for this category of homomorphic encryption schemes. The data are sent to the cloud server in an AES encrypted form, by using a fully homomorphic encryption scheme we decrypt homomorphicaly the encrypted mess and obtain an encrypted data under the homomorphic scheme.

To adapt homomorphic ciphertexts with AES inputs we have created a new JAVA primitive, ByteMatrice, it is the equivalent of the JAVA Byte primitive but with matrix elements. ByteMatrice allows as transforming an AES Byte input into a table of 8 matrices each matrix represent an encrypted bit.

We have obtained ambitious results for different security levels. For 1024 bits FHE key parameters an AES-128 circuit can be evaluated in less than 18 min on a small laptop. Performances can be improved by using a GPU implementation.

References

1. Rivest, R., Adleman, L., Dertouzos, M.: On data banks and privacy homorphisms. In: Foundataions of Secure Computataion, pp. 169–179. Academic Press (1978)
2. Gentry, C.: A fully homomorphic encryption scheme, September 2009. https://crypto.stanford.edu/craig/craig-thesis.pdf
3. Brakerski, Z., Gentry, C., Vaikantanathan, V.: Fully Homomorphique Encryption without Bootstrapping. http://eprint.iacr.org/2011/277
4. van Dijk, M., Gentry, C., Halevi, S., Vaikuntana, V.: Fully homomorphic encryption over the integers, Cryptology ePrint Archive, Report 2009/616 (2009)
5. Fan , J., Vercauteren, F.: Somewhat Practical Fully Homomorphic Encryption. http://eprint.iacr.org/2012/144
6. Gentry, C., Halevi, S., Smart, N.: Homomorphic Evaluation of the AES Circuit. https://eprint.iacr.org/2012/099.pdf
7. Brakerski, Z., Vaikantanathan, V.: Efficient Fully Homomorphic Encryption from (Standard) LWE. http://eprint.iacr.org/2011/344
8. Gentry, C., Sahai, A., Waters, B.: Homomorphic Encryption from Learning with Errors: Conceptually-Simpler, Asymptotically-Faster, Attribute-Based. http://eprint.iacr.org/2013/340
9. Cheon, J.H., Coron, J.-S., Kim, J., Lee, M.S., Lepoint, T., Tibouchi, M., Yun, A.: Batch Fully Homomorphic Encryption over the Integers. http://www.iacr.org/archive/eurocrypt2013/78810313/78810313.pdf
10. Nuida, K., Kurosawa, K.: (Batch) Fully Homomorphic Encryption over Integers for Non-Binary Message Spaces. https://eprint.iacr.org/2014/777.pdf
11. Gentry, C., Halevi, S., Smart, N.: Fully Homomorphic Encryption with Polylog Overhead. https://eprint.iacr.org/2011/566.pdf
12. Lauter, K., Naehrig, M., Vaikuntanathan, V.: Can Homomorphic Encryption be Practical? https://eprint.iacr.org/2011/405.pdf

Compact CPW-Fed Microstrip Octagonal Patch Antenna with Hilbert Fractal Slots for WLAN and WIMAX Applications

Mohamed Tarbouch[✉], Abdelkebir El amri, Hanae Terchoune, and Ouadiaa Barrou

RITM Laboratory, CED Engineering Sciences, Ecole Supérieure de Technologie,
Hassan II University of Casablanca, Casablanca, Morocco
mtarbouch@gmail.com

Abstract. In this paper, a Coplanar Wave Guide (CPW)-Fed microstrip octagonal patch antenna for WLAN and WIMAX Applications is proposed. The studied structure is suitable for 2.3/2.5/3.3/3.5/5/5.5 GHz WIMAX and for 3.6/2.4–2.5/4.9–5.9 GHz WLAN applications. The octagonal shape is obtained by cutting a small triangular part in the four angles of the rectangular microstrip patch antenna; in addition the using of CPW-Fed allows obtaining Ultra Wide Band (UWB) characteristics. The miniaturization in the antenna size for lower band is achieved by introducing the Hilbert fractal slots in the radiating element. The proposed antenna is designed on a single and a small substrate board of dimensions $46 \times 40 \times 1.6 \ mm^3$. Moreover the setup of Hilbert fractal slots allows obtaining lower resonant frequencies, more -10 dB bandwidths, more resonant frequencies and important gains. All the simulations were performed in CADFEKO, a Method of Moment (MoM) based solver.

Keywords: CADFEKO · CPW-Fed · Hilbert fractal · Miniaturization
WIMAX · WLAN

1 Introduction

WIMAX (Worldwide Interoperability for Microwave Access) is a wireless communications standard designed to provide a high speed data rates. Its capability to deliver high-speed Internet access and telephone services to subscribers enables new operators to compete in a number of different markets. In urban areas already covered by DSL (Digital Subscriber Line) and high-speed wireless Internet access, WIMAX allows new entrants in the telecommunication sector to compete with established fixed-line and wireless operators. The increased competition can result in cheaper broadband Internet access and telephony services for subscribers. In rural areas with limited access to DSL or cable Internet, WIMAX networks can offer cost-effective Internet access and may also encourage the UMTS/HSDPA (Universal Mobile Telecommunications System/ High Speed Downlink Packet Access) operators to extend their networks into these areas. WIMAX is designed to operate in the frequency range 2–66 GHz; however, most

© Springer International Publishing AG, part of Springer Nature 2018
M. Ben Ahmed and A. A. Boudhir (Eds.): SCAMS 2017, LNNS 37, pp. 432–444, 2018.
https://doi.org/10.1007/978-3-319-74500-8_40

interest is focusing on the 2–6 GHz range especially in the frequencies 2.3, 2.5, 3.3, 3.5,5 and 5.5 GHz [1–8].

WLAN (Wireless Local Area Networks) is a wireless computer network that links two or more devices using a wireless distribution method within a limited area such as a home, school, computer laboratory, or office building. This gives users the ability to move around within a local coverage area and still be connected to the network, and can provide a connection to the wider Internet. WLAN is designed to operate in the frequencies bands 2.4–2.5 GHz (802.11b/g/n), 3.6 GHz (802.11y), 4.9–5.9 GHz (802.11a/h/j/n/ac/y/p), 900 MHz (802.11ah) and 60 GHz (802.11ad) [1–11].

The miniaturization can affect radiation characteristics, bandwidth, gain, radiation efficiency and polarization purity. The miniaturization approaches are based on either geometric manipulation (the use of bend forms, meandered lines, PIFA shape, varying distance between feeder and short plate, using fractal geometries [12–15]) or material manipulation (Loading with a high-dielectric material, lumped elements, conductors, capacitors, short plate [16]), or the combination of two or more techniques [17]. Also several works [18–22] have appeared in the literature in which the size of the microstrip patch antenna has been reduced by introducing various types of slots in the microstrip patch antenna.

In this paper, the authors propose a compact CPW-Fed microstrip octagonal patch antenna with Hilbert fractal slots loading in the radiating element. In Sect. 2 we present the design of the CPW-fed octagonal patch antenna in three steps, in the first step, we describe the design procedure of the conventional rectangular patch antenna. In the second step we replace the Probe-fed by the CPW-fed to obtain the broadband characteristics; in the last step we cut of a small triangular shape in the four angles of the patch, those triangular shapes when properly established allow obtaining UWB characteristics with reasonable gain. Without the Hilbert fractal slots loading, the antenna covers the −10 dB bandwidth [2.76–6.30 GHz] only.

In the next section we present the literature review of the Hilbert fractal antennas and finally, we present the design of the CPW-fed octagonal patch antenna with Hilbert fractal slots. The setup of Hilbert fractal slots allows covering the lower bandwidth [2.3–2.5 GHz].

2 Design of CPW-Fed Octagonal Patch Antenna

2.1 Step 1: Design of Conventional Rectangular Patch Antenna

Based on the TLM (Transmission Line Model), the resonant frequency (fr) of the patch antenna (Fig. 1) is approximated by the Eq. (1):

$$f_r = \frac{c}{2W\sqrt{\dfrac{\varepsilon r + 1}{2}}} \tag{1}$$

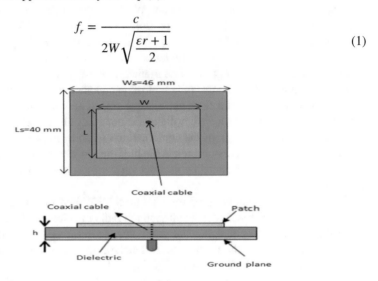

Fig. 1. The patch antenna

With:

c   = 3.108 ms−1, speed of the light
εr   = 4.4: the relative permittivity of the dielectric
W   : the width of the patch

For example, if fr = 3 GHz and εr = 4.4, the width of the patch is W = 30.43 mm. For the length of the patch we can follow the other formulas of designing the rectangular patch antenna that we can find in any good text book, the length of the patch is L = 23.43 mm.

Using the FEKO software to design a patch antenna with the following parameters: Ws = 46, Ls = 40, W = 30.43 mm, L = 23.43 mm, and h = 1.6 mm (Fig. 1). The S11 characteristic of the simulated structure is depicted in Fig. 2. From the results, it is evident that the antenna operates at the frequency 2.83 GHz whose S11 = −12.70 dB with −10 dB bandwidth of 70 MHz. We note that this value is different from the resonant frequency given by the TLM model, because the FEKO software is based on a full wave method: the Method of the Moment (MoM). The full wave methods like the MoM, the Finite Integral Technique (FIT), and Finite Element Method (FEM) are more accurate compared to the TLM method.

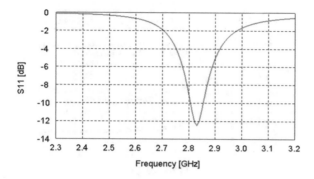

Fig. 2. Simulated S11 versus frequency graph of the patch antenna

In the next parametric studies of the antenna we will adopt the common antenna dimensions as Ws = 46, Ls = 40, W = 30.43 mm, L = 23.43 mm, and h = 1.6 mm.

2.2   Step 2: The Choice of Feeding Method of the Patch Antenna

There are many configurations that can be used to feed a patch antenna. In this section we compare the effect of three feeding methods: coaxial probe (Fig. 3a), Microstrip line with the ground plane in the reverse face of the patch (Fig. 3b), and Microstrip line with the ground plane in the same face of the patch (Fig. 3c).

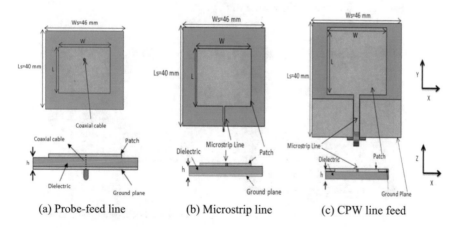

(a) Probe-feed line   (b) Microstrip line   (c) CPW line feed

Fig. 3. The different feeding methods of the patch antenna

The same microstrip patch antenna studied in step 1 using Probe-fed line will be fed with the two new modes (Microstrip line and CPW line). Figure 4 shows the difference between the S11 parameter versus frequencies for the three configurations. It's clear from the Fig. 4 that the CPW Feeding method allows having a wide bandwidth compared to two other feeding methods.

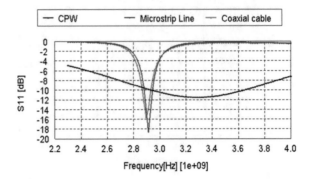

Fig. 4. Simulated S11 versus frequency of the patch antenna using the three methods of feeding

In the next steps of antennas design we will adopt the CPW feeding method.

2.3   Step 3: Design of the Octagonal Microstrip Patch Antenna

In this stage we cut a small triangular shape in the four angles of the microstrip patch antenna as shown in Fig. 5. With this way we obtain an octagonal patch antenna. To study the behavior of the obtained antenna regarding the dimension (Lcut and Wcut) of the triangular cut, we define a new parameter D as following: Lcut = L/D and Wcut = W/D. we set Ws = 46 mm, Ls = 40 mm, W = 30.43 mm, L = 23.43 mm, S = 0.6 mm, G = 0.3 mm, Wf = 3 mm and Yg = 14.5 mm.

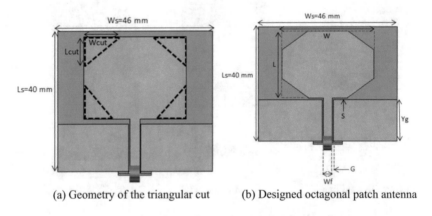

(a) Geometry of the triangular cut   (b) Designed octagonal patch antenna

Fig. 5. The design of the octagonal patch antenna

By varying the parameter D from 2.5 to 4.5, the S11 parameter of the octagonal patch antenna versus frequency is shown in Fig. 6. Table 1 summarizes the bandwidth obtained by varying the D parameter. Note that there is a close relation between the parameter D and the final size of the radiating element, as D decreases the total size of the octagonal patch decreases also.

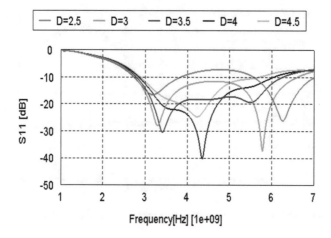

Fig. 6. Simulated S11 versus frequency by varying D

Table 1. Bandwidth versus the parameter D.

D	Bandwidth (GHz)
2.5	[2.69–3.88] & [5.52–7]
3	[2.72–6.65]
3.5	[2.76–6.3]
4	[2.80–6.11]
4.5	[2.84–5.74]

From Table 1 we note that cutting a triangular shape in the four angles of the microstrip patch antenna allows obtaining easily an UWB antenna. Also we can consider that D = 3.5 is the most adapted value in term of bandwidth, compactness and also adaptation in the whole operating frequency band. This value allows the antenna covering the −10 dB operating bandwidth of 3.54 GHz (2.76–6.3 GHz), that's means that the antenna is suitable for 3.3/3.5/5/5.5 GHz WIMAX and for 3.6/4.9–5.9 GHz WLAN applications

The analysis of the S11 parameter (Fig. 6) shows that, for the band of 1–7 GHz, the antenna have two resonant frequencies $fr1 = 3.41$ GHz and $fr2 = 5.52$ GHz. The total gain of this antenna varies between 2.1 dB and 5.6 dB in the −10 dB bandwidth (Fig. 7). Figure 8 shows the 3D Total gains for the two frequencies 3.41 and 5.52 GHz. We observe that the 3D-Total gain is Omnidirectional for the frequencies around 3.41 GHz and bidirectional for the frequencies around 5.52 GHz.

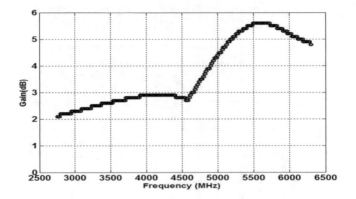

Fig. 7. The simulated Total gain versus the frequency of the CPW-Fed microstrip octagonal patch antenna

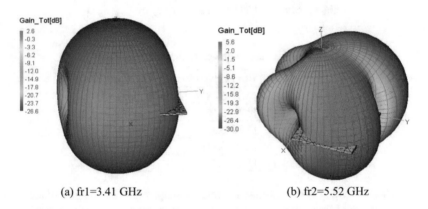

(a) fr1=3.41 GHz   (b) fr2=5.52 GHz

Fig. 8. Simulated 3D-Total gain for the frequencies 3.41 GHz and 5.52 GHz

3   Literature Review of Hilbert Fractal Antennas

3.1   The Generation of Hilbert Fractal Geometry

Fractals are geometric shapes, which cannot be defined using Euclidean geometry, are self-similar and repeating themselves on different scales like clouds, mountains, coasts, lightning, etc. [23, 24].

In the fractal structures, we use another dimension concept known as "the Hausdorff dimension" which defined by the Eq. (2) [13, 25].

$$h = \frac{\ln(n)}{\ln(R)} \tag{2}$$

Where the fractal is formed of "n" copies whose size has been reduced by a factor of "R".

The Hilbert curve is described for the first time by the German mathematician David Hilbert in 1891 [26]. The Hausdorff dimension is 2 and the method of construction is described in Fig. 9. D, d and b are the antenna width, fractal segment or spacing and antenna line width, respectively. For the first iteration, the antenna length and fractal segment are of the same dimensions. For the second and successive iterations, the fractal segment or spacing can be computed with proper design equations.

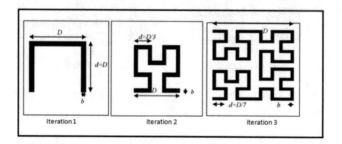

Fig. 9. The first three iterations of the HILBERT Curve Structure

3.2 The Use of the Hilbert Fractal on the Design of the Antennas

In 2003, GONZALEZ-ARBESÚ has studied the parameters and the behavior of 2D and 3D HILBERT curve wire antennas [27].

In 2006, MURAD has developed a modified patch antenna using the Hilbert curve. The proposed antenna is a miniaturized structure for the 2.45 GHz RFID applications [28].

In 2010, SANZ has studied the difference between the spiral wire antennas and the wire antennas based on the Hilbert geometries. He has showed that this second structure provides a higher bandwidth [29].

HUANG has set up a PIFA antenna based on a HILBERT structure, operational for the 2.4 GHz applications [30]. In 2012, Suganthi has studied some modified patch antennas based on the HILBERT structure. The proposed antennas are operational for medical applications [31].

4   Design of CPW-Fed Microstrip Octagonal Patch Antenna with Hilbert Fractal Slots

In this section we present the last step of Antenna design which consists of introducing the two first iterations of Hilbert fractal slot structure on the CPW-Fed octagonal patch antenna designed in Sect. 3. The final designed antennas are shown in Fig. 10.

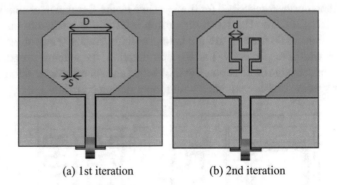

(a) 1st iteration   (b) 2nd iteration

Fig. 10. The CPW-Fed microstrip octagonal patch antennas with the two first iterations of the Hilbert fractal slots

We set Ws = 46 mm, Ls = 40 mm, W = 30.43 mm, L = 23.43 mm, Yg = 14.5 mm, Wf = 3 mm, G = 0.3 mm, S = 0.6 mm, S1 = 0.6 mm, D = 12.7 mm and d = 3.125 mm. Figure 11 shows a comparison of S11 parameter for the three designed structures: The simple CPW-Fed microstrip octagonal patch antenna, the microstrip octagonal patch antenna with first iteration of Hilbert slot and the microstrip patch antenna with the second iteration of Hilbert slot.

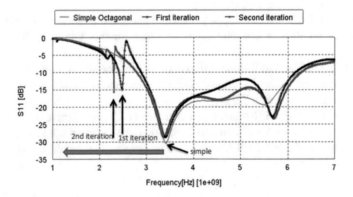

Fig. 11. The simulated S11 of the Octagonal patch antenna and for the two first iterations of Hilbert slotted antennas

We note that the setup of the first iteration of Hilbert fractal slot allows having 3 resonant frequencies: fr1 = 2.48 GHz, fr2 = 3.40 GHz and fr3 = 5.68 GHz

We note also that the first resonant frequency is lower than the first resonant frequency obtained with the simple octagonal patch antenna without the slot. Also, the two obtained −10 dB bandwidths are 150 MHz (2.38–2.53 GHz) and 3.33 GHz (2.86–6.19 GHz) (Fig. 11).

The setup of the Second iteration of Hilbert fractal slot allows having 3 resonant frequencies fr1 = 2.30 GHz, fr2 = 3.37 GHz and fr3 = 5.71 GHz, we note that the first resonant frequency becomes lower again. Also, the two −10 dB bandwidths are 70 MHz (2.27–2.34 GHz) and 3.55 GHz (2.77–6.32 GHz) (Fig. 11).

Figure 12 shows the behavior of the gain versus frequencies and versus iteration number. The total gain of the proposed antennas varies between 0.1 dB and 2.4 dB in the −10 dB lower operating bandwidth and varies between 1.7 dB and 5.8 dB in the −10 dB higher operating bandwidth.

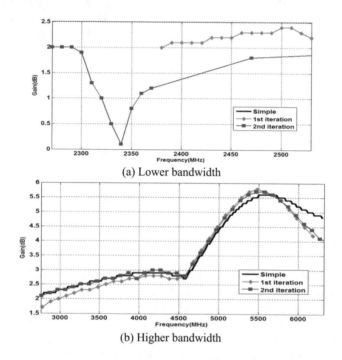

(a) Lower bandwidth

(b) Higher bandwidth

Fig. 12. The simulated Gain of the Octagonal patch antenna and for the two first iterations of Hilbert slotted antennas

Therefore the proposed antennas are suitable for 2.3/2.5/3.3/3.5/5/5.5 GHz WIMAX and for 3.6/2.4–2.5/4.9–5.9 GHz WLAN applications

Also, comparison between several different antennas for WLAN and WIMAX applications is illustrated in Table 2. From this table it's clear that our structures have a miniaturized design with very important gain compared to most other antennas. Moreover the proposed antennas have the wide bandwidth over all referenced antennas. This allows the designed antennas to cover all announced WLAN/WIMAX services.

Table 2. Comparison of the proposed antenna to others antennas for WLAN and WIMAX applications.

Ref.	Dimension (mm^2)	Size(cm^2)	Bandwidth(GHz)	Gain(dB)
[7]	50*42	21	2.4 to 3.4	2 to 2.6
			4.9 to 6.6	5 to 6
[32]	50*50	25	2.1 to 2.5	0 to 1.5
			3 to 4.2	2 to 2.2
			4.2 to 5.95	2.2 to 6.2
[4]	45*35	15.75	2.4 to 2.48	2.2 to 2.3
			5.7 to 5.9	2 to 2.3
			3.4 to 3.7	0.4 to 1
[2]	20*20	4	2.3 to 2.51	−1 to 0.7
			3.35 to 3.75	−1 to 1.4
			4.95 to 5.53	−0.8 to 1.8
[3]	40*40	16	3.28 to 3.51	0.6 to 2.2
			5.47 to 6.03	0.5 to 1
Our work	46*40	18.4	2.27 to 2.53	0.1 to 2.4
			2.76 to 6.32	1.7 to 5.8

5   Conclusion

In this paper, a CPW-Fed microstrip octagonal patch antennas for WLAN and WIMAX applications were proposed. The antennas have a small size of 18.4 cm2, which makes them suitable for use as internal antennas for embedded systems. By cutting a triangular shape in the four angles of the rectangular patch antenna, it was possible to implement UWB characteristics in the higher band. The setup of Hilbert fractal slots on the octagonal patch antenna allows generating new lower resonant frequencies.

The simulated results show that the studied antennas have an important gain and a good impedance matching. Thereby, this result makes the CPW-Fed microstrip octagonal patch antennas with Hilbert fractal slot an adequate candidates for some applications such as 2.3/2.5/3.3/3.5/5/5.5 GHz WIMAX and for 3.6/2.4–2.5/4.9–5.9 GHz WLAN. In the next work, fabrication and measurement should be done to confirm the simulated results.

References

1. Korowajczuk, L.: LTE, WIMAX, and WLAN Network Design, Optimization and Performance Analysis. Wiley, Chichester (2011)
2. Borhani, M., Rezaei, P., Valizade, A.: Design of a reconfigurable miniaturized microstrip antenna for switchable multiband systems. IEEE Antennas Wirel. Propag. Lett. **15**, 822–825 (2016)
3. Boukarkar, A., Lin, X.Q., Jiang, Y.: A dual-band frequency-tunable magnetic dipole antenna for WiMAX/WLAN applications. IEEE Antennas Wirel. Propag. Lett. **15**, 492–495 (2016)

4. Shi, X.W., Wu, T., Bai, H., Li, P.: Tri-band microstrip-fed monopole antenna with dualpolarisation characteristics for WLAN and WiMAX applications. Electron. Lett. **49**(25), 1597–1598 (2013)
5. Wen, J., Wenquan, C.: A novel UWB antenna with dual notched bands for WiMAX and WLAN applications. IEEE Antennas Wirel. Propag. Lett. **11**, 293–296 (2012)
6. Reha, A., El Amri, A., Benhmammouch, O., Oulad Said, A., El Ouadih, A., Bouchouirbat, M.: CPW-Fed H-tree fractal antenna for WLAN, WIMAX, RFID, C-band, HiperLAN, and UWB applications. Int. J. Microw. Wirel. Technol. 1–8 (2015)
7. Reha, A., El Amri, A., Benhmammouch, O., Oulad Said, A., El Ouadih, A., Bouchouirbat, M.: CPW-Fed slotted CANTOR set fractal antenna for WiMAX and WLAN applications. Int. J.Microw. Wirel. Technol. 1–7, May 2016
8. Basaran, S.C., Olgun, U., Sertel, K.: Multiband monopole antenna with complementary split-ring resonators for WLAN and WiMAX applications. Electron. Lett. **49**(10), 636–638 (2013)
9. Perahia, E., Stacey, R.: Next Generation Wireless LANs: 802.11n, 802.11ac, and Wi-Fi Direct, 2nd edn. Cambridge University Press, Cambridge (2013)
10. Gast, M.: 802.11 Wireless Networks: The Definitive Guide, 2nd edn. O'Reilly, Beijing (2005)
11. Bakariya, P.S., Dwari, S., Sarkar, M., Mandal, M.K.: Proximity-coupled microstrip antenna for bluetooth, WiMAX, and WLAN applications. IEEE Antennas Wirel. Propag. Lett. **14**, 755–758 (2015)
12. Chen, H.T., Wong, K.L., Chiou, T.W.: PIFA with a meandered and folded patch for the dual-band mobile phone application. IEEE Trans. Antennas Propag. **51**(9), 2468–2471 (2013)
13. Reha, A., El Amri, A., Benhmammouch, O., Oulad Said, A.: Fractal antennas : a novel miniaturization technique for wireless networks. Trans. Netw. Commun. **2**(5), 165–193 (2014)
14. Sun, S., Zhu, L.: Miniaturised patch hybrid couplers using asymmetrically loaded cross slots. IET Microw. Antennas Propag. **4**(9), 1427 (2010)
15. Chi, P.L., Waterhouse, R., Itoh, T.: Antenna miniaturization using slow wave enhancement factor from loaded transmission line models. IEEE Trans. Antennas Propag. **59**(1), 48–57 (2011)
16. Skrivervik, A.K., Zürcher, J.F., Staub, O., Mosig, J.R.: PCS antenna design: the challenge of miniaturization. IEEE Antenn Propag Mag. **43**(4), 12–27 (2001)
17. Reha, A., El Amri, A., Saih, M., Benhmammouch, O., Oulad Said, A.: The behavior of a CPW-Fed microstrip hexagonal patch antenna with H-Tree Fractal slots. Rev. Méditerranéenne Télécommunication **5**(2), 104–108 (2015)
18. Kakoyiannis, C.G., Constantinou, P.: A compact microstrip antenna with tapered peripheral slits for CubeSat RF payloads at 436 MHz: miniaturization techniques, design, and numerical results. In: Proceedings of the IEEE International Workshop on Satellite and Space Communications (IWSSC08), pp. 255–259 (2008)
19. Anguera, J., Boada, L., Puente, C., Borja, C., Soler, J.: Stacked H-shaped microstrip patch antenna. IEEE Trans. Antennas Propag. **52**(4), 983–993 (2004)
20. Bokhari, S.A., Zurcher, J.F., Mosig, J.R., Gardiol, F.E.: A small microstrip patch antenna with a convenient tuning option. IEEE Trans. Antennas Propag. **44**(11), 1521–1528 (1996)
21. Chatterjee, S., Ghosh, K., Paul, J., Chowdhury, S.K., Chanda, D., Sarkar, P.P.: Compact microstrip antenna for mobile communication. Microw. Opt. Technol. Lett **55**(5), 954–957 (2013)
22. Chen, W.S., Wu, C.K.: Wong, K.l.: Square-ring microstrip antenna with a crossstrip for compact circular polarization operation. IEEE Trans. Antennas Propag. **47**(10), 1566–1568 (1999)

23. Mandelbrot, B.B.: The Fractal Geometry of Nature. W.H. Freeman and Company, New York (1983)
24. Mandelbrot, B.B.: Les Objets Fractals. 4e éd., Flammarion (1995)
25. Falconer, K.J.: Fractal Geometry: Mathematical Foundations and Applications, 2nd edn. Wiley, Chichester (2003)
26. Hilbert, D.: Über die stetige Abbildung einer Linie auf ein Flächenstück. Math. Ann. **38**, 459–460 (1891)
27. Gonzalez-Arbesu, J.M., Blanch, S., Romeu, J.: THE Hilbert curve as a small self-resonant monopole from a practical point of vieW. Microwave Optical Technol. Lett. **39**(1), 45–49 (2003)
28. Murad, N.A., et al.: Hilbert curve fractal antenna for RFID application. In: Proceedings of International RF and Microwave Conference, 12–14 September 2006, Putrajaya, Malaysia, pp. 182–186 (2006)
29. Sanz, I., et al.: The Hilbert monopole revisited. In: Proceedings of the Fourth European Conference on Antennas and Propagation (EuCAP), pp. 1–4 (2010)
30. Huang, J.-T., Shiao, J.-H., Wu, J.-M.: A miniaturized Hilbert inverted-F antenna for wireless sensor network applications. IEEE Trans. Antennas Propag. **58**(9), 3100–3103 (2010)
31. Suganthi, S.: Study of compact Hilbert curve fractal antennas for implantable medical applications. Int. J. Emerg. Technol. Adv. Eng. **2**(10), 116–125 (2012)
32. Reha, A., El Amri, A.: Design, realization and measurements of CPW-Fed microstrip hexagonal patch antenna with H-tree fractal slots for WLAN and WIMAX applications. Int. J. Microw. Opt. Technol. **11**(4), 251–258 (2016)

MultiPrime Cloud-RSA Scheme to Promote Data Confidentiality in the Cloud Environment

Khalid El Makkaoui[1($\boxtimes$)], Abderrahim Beni-Hssane[2], and Abdellah Ezzati[1]

[1] LAVETE Laboratory, FST, Univ Hassan 1, B.P. 577, 26000 Settat, Morocco
kh.elmakkaoui@gmail.com, abdezzati@gmail.com
[2] LAROSERI Laboratory, Department of Computer Science, Sciences Faculty,
Chouaïb Doukkali University, El Jadida, Morocco
abenihssane@yahoo.fr

Abstract. Homomorphic encryption can be considered as an effective tool to overcome concerns over the confidentiality of sensitive data in the cloud. Since cloud environment is more threatened by attacks and since cloud consumers often use lightweight devices to access to cloud services, the homomorphic schemes must be promoted to work efficiently in terms of running time and security level. At EMENA-TSSL'16, we boosted the standard RSA cryptosystem at security level, Cloud-RSA. In this article, we suggest a fast variant of the Cloud-RSA for speeding up the Cloud-RSA algorithms. The fast variant is based on modifying the Cloud-RSA modulus structure and using the Chinese remainder theorem to decrypt.

Keywords: Cloud computing · Homomorphic encryption · Confidentiality
Speedup Cloud-RSA · Chinese remainder theorem (CRT)

1 Introduction

Recently, there has been an increasing adoption of cloud computing services; this is owing to benefit from the advantages offered by cloud providers such as: powerful computations, reducing costs, high services scalability and flexibility [1, 2]. However, concerns over the confidentiality of data are among the most important obstacles to widespread cloud services adoption [3–5].

Computing on encrypted data is an effective method to overcome these obstacles. Indeed, researchers have stressed a useful form of encryption, homomorphic encryption, which provides a third party with the ability to perform operations on encrypted data. This concept was firstly introduced by Rivest et al. in 1978 [6]. The encryption schemes which support the homomorphic properties are numerous e.g., [7–14]. The homomorphic encryption schemes can be separated into two categories: somewhat and full homomorphic encryption [15]. The encryption schemes of somewhat category support one homomorphic property (usually addition or multiplication) or more than one, but with a limited number of operations. The schemes of the second category support an arbitrary of operations.

© Springer International Publishing AG, part of Springer Nature 2018
M. Ben Ahmed and A. A. Boudhir (Eds.): SCAMS 2017, LNNS 37, pp. 445–452, 2018.
https://doi.org/10.1007/978-3-319-74500-8_41

Since encrypted data under homomorphic schemes is stored for a long time by the same encryption key in cloud servers and since cloud consumers often use lightweight devices to access to cloud computing services, these encryption schemes must be promoted to work efficiently in terms of running time and security level.

Thus, in [14], we boosted the standard RSA scheme [7], in terms of security level, by suggesting an enhanced version of it, called Cloud-RSA. The Cloud-RSA scheme provides the multiplicative homomorphism and resists more to confidentiality attacks.

In this paper, we take a fast variant of the RSA cryptosystem [16] as a basis to propose a new variant for speeding up the Cloud-RSA algorithms. Our new scheme uses a modulus formed from two or more distinct prime numbers, is based on modifying the structure of the Cloud-RSA exponents and using the Chinese remaindering to decrypt.

The paper is structured as follows. In the next section, we give the necessary requirements to prove the correctness of our new scheme. In Sect. 3, we review the Cloud-RSA scheme and present the algorithms of the proposed fast variant as well as its properties. Finally, we present our conclusions and future works.

2   Preliminaries

We introduce some theorems which are required to demonstrate the correctness of the proposed variant.

Theorem 1 (Chinese remainder theorem (CRT) [17]). *Let $a_0, a_1, \ldots, a_{k-1}$ be positive integers which are pairwise co-prime and let $b_0, b_1, \ldots, b_{k-1}$ be integers, then the following congruences:*

$$x \equiv b_i \,(mod\ a_i) \tag{1}$$

for $i = 0, 1, \ldots, k - 1$. Has a unique solution:

$$x \equiv b_0 M_0 M_0^{-1} + \cdots + b_{k-1} M_{k-1} M_{k-1}^{-1} (mod\ a)$$

where $a = a_0 \times \ldots \times a_{k-1} = a_i\, M_i$ and $M_i M_i^{-1} \equiv 1 \,(mod\ a_i)$.

Theorem 2 (Binomial theorem [18]). *Let $m \epsilon\, \mathbb{N}$ and $x, y \epsilon\, \mathbb{R}$, then*

$$(x + y)^m = \sum_{i=0}^{m} \binom{m}{i} x^{m-i} y^i \tag{2}$$

where $\binom{m}{i} = \dfrac{m!}{i!(m - i)!}$.

Theorem 3 (Fermat's Little Theorem [19]). *Let p be prime number and $a \epsilon\, \mathbb{Z}$ such that $gcd(a, p) = 1$, then*

$$a^{p-1} \equiv 1 (mod \, p) \tag{3}$$

3   Proposed Encryption Scheme

The aim of this work is to speed up efficiently the Cloud-RSA algorithms.

In practice of the Cloud-RSA scheme [14] one usually uses a small public exponent such as e $= 2^{16} + 1$ to preclude some exponent attacks. It follows that the Cloud-RSA encryption is much faster than the Cloud-RSA decryption.

In this section, we take a variant of the RSA scheme, namely MultiPrime RSA [16], as a basis to suggest a fast of variant the Cloud-RSA scheme. Let us refer to as Multi-Prime Cloud-RSA.

Key Generation

Input: Two large distinct primes p and q of the same bits long

- Compute $N = pq$ and $\phi(N) = (p-1)(q-1)$.
- Pick an integer e that is relatively prime to $\phi(N)$. Then determine the integer d satisfing $ed \equiv 1 (mod \, \phi(N))$.

Output: $(ek, \, pk)$
　　　The evaluation key is $ek = (N)$; the private key is $pk = (N, e, d)$.

Encryption

Input: Given pk, let $m_1, \ldots, m_k \in \mathbb{Z}_N$ be plaintexts.

- Compute the ciphertexts c_i as: $c_i \equiv m_i^e \, (mod \, N)$ where $1 \leq i \leq k$.

Output: $c_i = E \, (pk, \, m_i)$

Evaluation

Input: Given ek, let $c_1, \ldots, c_k \in \mathbb{Z}_N$ be ciphertexts.

- Compute $C \equiv c_1 \times \ldots \times c_k (mod \, N)$

Output: C as the result

Decryption

Input: Given pk and a ciphertext C

- Compute $C^d (mod \, N) \equiv m_1 \times \ldots \times m_k$

Output: $m_1 \times \ldots \times m_k = D \, (pk, \, C)$

Fig. 1. Cloud-RSA algorithms

In the following, we begin with a review of the Cloud-RSA encryption scheme. We then present the MultiPrime Cloud-RSA encryption scheme.

3.1   Review of the Cloud-RSA Scheme

The Cloud-RSA scheme [14] consists of key generation, encryption, evaluation and decryption algorithms, as depicted in Fig. 1.

3.2   MultiPrime Cloud-RSA Scheme

In this part, we describe the MultiPrime Cloud-RSA algorithms. We demonstrate the correctness of the algorithms. We then give its properties.

3.2.1   Algorithms

Now, we describe the key generation, encryption, evaluation and decryption algorithms of the MultiPrime Cloud-RSA scheme.

Key generation: The key generation algorithm works as follows:

Step 1: Generate k distinct primes $(p_1, p_2, \ldots, p_k)$ of the same bit-size where $k \geq 2$. Then compute $N = \prod_{i=1}^{k} p_i$ and $\phi(N) = \prod_{i=1}^{k} (p_i - 1)$.

Step 2: Choose an integer e that is relatively prime to $\phi(N)$. Then compute $d \equiv e^{-1}$ $(mod\ \phi(N))$.

Step 3: Compute

- $$d_i \equiv d\big(mod\ (p_i - 1)\big)$$

- $k_i = \dfrac{N}{p_i}\left(\dfrac{N}{p_i}\right)^{-1}$, where $1 \leq i \leq k$ such that $k_i \equiv 1\,(mod\ p_i)$.

The evaluation key is $ek = (N)$; the private key is $pk = \big(N, p_i, e, d_i, k_i\big)$. Where d_i's are the CRT-decryption exponents and k_i's are the CRT coefficients.

Encryption: It encrypts the data exactly as in the Cloud-RSA scheme. Given the private key pk, let $m \in \mathbb{Z}_N$ be a plaintext. The ciphertext c is computed as follows:

$$c = E(pk, m) \equiv m^e (mod\ N) \qquad (4)$$

Evaluation: Let $c_1, c_2 \ldots, c_k$ be encrypted data. Given the evaluation key, the cloud provider can compute

$$Eval\left(ek, c_1, c_2 \ldots, c_k\right) \equiv c_1 \times c_2 \times \ldots \times c_k (mod\,N)$$
$$\equiv m_1^e \times m_2^e \times \ldots \times m_k^e (mod\,N)$$
$$\equiv (m_1 \times m_2 \times \ldots \times m_k)^e (mod\,N)$$
$$\equiv E\left(pk, m_1 \times m_2 \times \ldots \times m_k\right)$$

where *Eval* is the function of evaluation.

Decryption: Given the private key *pk*. To decrypt the processed result $c = Eval\left(ek, c_1, c_2 \ldots, c_k\right)$, the cloud consumer uses the Chinese remainder theorem as follows:

Step 1: $c_i \equiv c\left(mod\,p_i\right)$

Step 2: $x_i \equiv c_i^{d_i}\left(mod\,p_i\right)$

Step 3: $m = m_1 \times m_2 \times \ldots \times m_k \equiv \sum_{i=1}^{k} x_i \times k_i (mod\,N)$.

3.2.2 Correctness Proof

In this part, we demonstrate the correctness of decryption algorithm. Let N be the product of k distinct primes and e, d be two integers satisfying $ed \equiv 1(mod\,\phi(N))$.

Suppose that $d_i \equiv d\left(mod\,p_i - 1\right)$ i.e., $\exists\, a_i \in \mathbb{N}$ such that:

$$d = d_i + a_i(p_i - 1)$$

where $1 \leq i \leq k$.

For any $c \in \mathbb{Z}_N$ is written as follows:

$$c = c_i + b_i p_i$$

where $1 \leq i \leq k$, $b_i \geq 0$ and $c_i \in \mathbb{Z}_{p_i}$.

Since p_i's are pairwise coprime, one can employ the CRT, Theorem 1, to decrypt c, one evaluates:

$$m_i \equiv c^d\left(mod\,p_i\right)$$

where $m_i \in \mathbb{Z}_{p_i}$ and $1 \leq i \leq k$.

Let us compute c^d using the *Theorem 2*:

$$c^d = (c_i + b_i p_i)^d$$
$$= \sum_{j=0}^{d} \binom{d}{j} c_i^{d-j} (b_i p_i)^j$$
$$= c_i^d + (b_i p_i)^d + \sum_{j=1}^{d-1} \binom{d}{j} c_i^{d-j} (b_i p_i)^j$$

Let us reduce m_i:

$$m_i \equiv c_i^d \left(mod \, p_i \right)$$
$$\equiv \left(c_i \right)^{d_i + a_i \left(p_i - 1 \right)} \left(mod \, p_i \right)$$
$$\equiv c_i^{d_i} \left(mod \, p_i \right), \text{ thanks to the } Theorem \, 3$$

Thus, we prove the correctness of the decryption algorithm.

3.2.3   Security Analysis

The security of the MultiPrime Cloud-RSA encryption scheme depends on the difficulty of factoring the modulus N formed of two or more distinct prime numbers as well as of finding the decryption exponents where the encryption exponent is also private. The fastest factorization algorithms (e.g., the Number Field Sieve [20] and the Elliptic Curve Method [21]) cannot take advantage of the modulus N structure. Even if we give the primes factorization of the modulus N. To expose the exponent d an adversary must try to find two exponents: e and d such that $ed = 1 (mod \, \phi(N))$ and that decrypted data are semantically correct.

To ensure a high security level, some recommendations to generate the RSA modulus must be taken into consideration. It has been recommended that the modulus N should have $1024 + 256x$ bits for $x \geq 0$ [22], the number of primes in N must be chosen carefully (e.g., for the modulus N of 1024-bit should not be formed of more than three primes) [23].

3.2.4   Performance Evaluation

The MultiPrime Cloud-RSA scheme has been implemented and compared with Cloud-RSA scheme. The performance evaluation, in terms of running time, has been measured using Python programming language.

Throughout the simulation, we have respected the recommendations for generating the key pair [22, 23]. We have set a modulus N of 1024 bits long formed of two distinct primes. We then have picked an encryption exponent $e = 2^{16} + 1$ that is relatively prime to $\phi(N)$.

The simulation results show that the MultiPrime Cloud-RSA scheme gets a decryption speedup by factor of 2.75 over the Cloud-RSA scheme.

4   Conclusion and Future Works

We have suggested a fast variant of the Cloud-RSA scheme, MultiPrime Cloud-RSA, to speed up its decryption algorithm. The MultiPrime Cloud-RSA uses a modulus N formed of two or more distinct primes and employs the Chinese remaindering to decrypt data. It provides the multiplicative homomorphism over the integers and its security relies on the difficulty of both factoring the modulus N and finding the decryption exponents where the encryption exponent e also is private. The simulation results show that the MultiPrime Cloud-RSA scheme offers a good performance, in terms of running time, in comparison with the Cloud-RSA scheme.

In our future works, we will try to speed up more the Cloud-RSA scheme with maintaining a high security level.

References

1. El Makkaoui, K., Ezzati, A., Beni-Hssane, A., Motamed, C.: Cloud security and privacy model for providing secure cloud services. In: 2016 2nd International Conference on Cloud Computing Technologies and Applications (CloudTech), pp. 81–86. IEEE (2016)
2. Xiang, C., Tang, C.: Efficient outsourcing schemes of modular exponentiations with checkability for untrusted cloud server. J. Ambient Intell. Humaniz. Comput. 6(1), 131–139 (2015)
3. Bennasar, H., Bendahmane, A., Essaaidi, M.: An overview of the state-of-the-art of cloud computing cyber-security. In: International Conference on Codes, Cryptology, and Information Security, pp. 56–67. Springer, Cham (2017)
4. Kiraz, M.S.: A comprehensive meta-analysis of cryptographic security mechanisms for cloud computing. J. Ambient Intell. Humaniz. Comput. 7(5), 731–760 (2016)
5. Alam, M., Emmanuel, N., Khan, T., Xiang, Y., Hassan, H.: Garbled role-based access control in the cloud. J. Ambient Intell. Humaniz. Comput. 8, 1–14 (2017)
6. Rivest, R., Adleman, L., Dertouzos, M.: On data banks and privacy homomorphisms. Found. Secur. Comput. 4, 169–180 (1978)
7. Rivest, R., Shamir, A., Adleman, L.: A method for obtaining digital signatures and public-key cryptosystems. Commun. ACM 21(2), 120–126 (1978)
8. ElGamal, T.: A public key cryptosystem and a signature scheme based on discrete logarithms. IEEE Trans. Inf. Theor. 31, 469–472 (1985)
9. El Makkaoui, K., Beni-Hssane, A., Ezzati, A.: Cloud-ElGamal: an efficient homomorphic encryption scheme. In: 2016 International Conference on Wireless Networks and Mobile Communications (WINCOM), pp. 63–66. IEEE (2016)
10. Paillier, P.: Public-key cryptosystems based on composite degree residuosity classes. In: EUROCRYPT 1999, pp. 223–238 (1999)
11. El Makkaoui, K., Ezzati, A., Beni-Hssane, A.: Securely adapt a Paillier encryption scheme to protect the data confidentiality in the cloud environment. In: Proceedings of the International Conference on Big Data and Advanced Wireless Technologies. ACM (2016)
12. Gentry, C.: A fully homomorphic encryption scheme. Doctoral Thesis. University of Stanford (2009). http://crypto.stanford.edu/craig
13. Chillotti, I., Gama, N., Georgieva, M., Izabachène, M.: Faster fully homomorphic encryption: bootstrapping in less than 0.1 seconds. In: ASIACRYPT 2016, vol. 10031, pp. 3–33 (2016)
14. El Makkaoui, K., Ezzati, A, Beni-Hssane, A.: Cloud-RSA: an enhanced homomorphic encryption scheme. In: Europe and MENA Cooperation Advances in Information and Communication Technologies, pp. 471–480. Springer, Cham (2017)
15. El-Yahyaoui, A., Elkettani, M.D.: Fully homomorphic encryption: state of art and comparison. Int. J. Comput. Sci. Inf. Secur. 14(4), 159–167 (2016)
16. Collins, T., Hopkins, D., Langford, S., Sabin, M.: Public Key Cryptographic Apparatus and Method. US Patent #5,848,159 (1997)
17. Wang, X., Xu, G., Wang, M., Meng, X.: Mathematical Foundations of Public Key Cryptography. CRC Press, Boca Raton (2015)

18. McGregor, C., Nimmo, J., Stothers, W.: Fundamentals of University Mathematics. Elsevier, Amsterdam (2010)
19. Shoup, V.: A Computational Introduction to Number Theory and Algebra. Cambridge University Press, Cambridge (2005)
20. Lenstra, A.K., Lenstra, H.W. (eds.): The Development of the Number Field Sieve, pp. 11–42. Springer, Heidelberg (1993)
21. Lenstra Jr., H.W.: Factoring integers with elliptic curves. Annal. Math. **126**, 649–673 (1987)
22. ANSI Standard X9.31-1998. Digital Signatures Using Reversible Public Key Cryptography for the Financial Services Industry (rDSA)
23. Wiener, M.: Cryptanalysis of short RSA secret exponents. IEEE Trans. Info. Theor. **36**(3), 553–558 (1990)

Relationship Between Smart City Drivers and Socially Cohesive Societies

Ayodeji E. Oke[(✉)] ⓘ, Clinton O. Aigbavboa, and Taniele K. Cane

University of Johannesburg, Johannesburg, South Africa
emayok@gmail.com

Abstract. The awareness, adoption and practice of smart city system has been on the increase among developers, regulators and other stakeholders of city planning and development. This study examines the relationship between drivers of smart cities and social inclusion of citizens in the societies with a view to enhancing urban citizens' identification and willingness to share skills and knowledge. Relevant literature materials relating to smart city concept and social cohesion were reviewed and variables obtained were further examined through the administration of questionnaires to experts and professionals concerned with the planning, development and regulation of cities. Majority of the respondents are willing to share knowledge and skills among themselves and prefers to be identified as an individual rather than by their race, gender or ethnicity. The study also assess the importance of the six smart city drivers. Therefore, smart city model, which is an urban design mechanism, has the capacity and capabilities to re-form and modify any environment provided the correct and rigid frameworks are built, adopted and properly followed.

Keywords: Smart city · Smart technology · Social cohesion
Sustainable construction

1 Introduction

The inception of the smart city model came about in the late 19th century, during a period in which smart growth movements, altered the manner in which: cities were designed, developed and coordinated [1]. The shift from the traditional methodology of urban planning to a more sustainable, innovative, collaborative as well as city specific methodology, has been identified as critical for the survival of the present and future cities [1, 2].

Due to the challenges stemming from the global trend in urbanization and globalization [3], the Smart City concept has received a vast amount of attention among researchers, municipal governments and large IT organisations, amongst others, these include IBM, Siemens [1] and Microsoft [4].

The rapid migration of the global population, into urban areas has revealed the need to adapt policy planning strategies, and to ensure that these urban areas are equipped with mechanisms to meet the needs of their current inhabitants without compromising the ability to meet the needs of the future inhabitants. Research conducted by the United Nations estimated 66% of the world's population will be living in urban areas by 2050;

© Springer International Publishing AG, part of Springer Nature 2018
M. Ben Ahmed and A. A. Boudhir (Eds.): SCAMS 2017, LNNS 37, pp. 453–462, 2018.
https://doi.org/10.1007/978-3-319-74500-8_42

furthermore 90% of this population shift is expected to take place in Asia and Africa [5] Through the transformation of cities to smart cities and the formation of smart cities based on the six fundamental components of the smart cities model developed by Giffinger et al. [6], cities will be empowered to integrate their different sectors, systems and sub-systems through the use of a network of ICTs [5]. This will provide city authorities with the tools to monitor every aspect of the city, including its utilities, which ultimately enabling them to communicate effectively with their citizens at the present moment and not hours or days later [7]. This research investigates the drivers for the development and transformation of cities into smart cities and their roles in enhancing the quality of life of the urban citizens.

2   Concepts of Socially Inclusive Cities

One of the biggest obstacles for nations to overcome in the 21st century is the increased migration trend of the global population into urban districts [3]. Conversely, the issue of globalization has led to the emergence of increased segregation among communities with diverse cultures, gender, age, race and unique stories of origin [8]. To this end, Urzua [9] defines globalization as a multi-dimensional process, which is comprised of a number of different elements, which are the development of a universal set of economic rules endorsed by the different countries administrative authorities, to make provision for these rules within their states' economy. These invariably improve their efficiency rates and surplus margins as well as the attractiveness of the country; the incorporation and establishment of innovative and technology savvy infrastructures; and the formation of a population which embraces the multiple cultures, races, genders and different levels of knowledge brought-forth by the different members of the society [3].

Moreover, the fine line between globalization and the formation of socially inclusive/cohesive cities is clearly seen. Social cohesion is defined by the UN Department of Economic and Social Affairs (UN DESA), as the formation of urban metropolitans, in which the inhabitants work collectively with the local and national authorities, to improve their own future prospective as well as the potential of the urban district they reside in, in a view to foster the implementation of equitable and co-operative strategies, therefore assisting the stakeholders in mitigating the prevalent divide within communities [10].

The study is focused on South Africa, which has been identified as a country that is rapidly expanding and improving [7]. The prevalent prejudices currently suffered by the country's population has been attributed to apartheid regime [11]. This is a time in which the white citizens of the country were seen as superior and the remaining ethnic groups were seen as inferior, thus they were restricted to what they could do, where they could go as well as where they could live [12]. Furthermore, the regime exposed these inferior members of society to an autocratic governance system. However, the end of this era in 1994 and the mobilization of the country's democratic government was welcomed by many, as it was a step in the right direction to the formation of an equitable society, that is, a society in which an individual's future opportunities were not influenced by their race and culture.

To this extent, there is a need for the development of mechanisms and policies that will assist in the formation of urban societies that support and are willing to share knowledge and skills with each other, and concurrently empowering fellow citizens to motivate and create a sense of belief in each other. In doing so, they stimulate the sense of belief and faith within themselves as individuals [13]. For this reason, the need to develop metropolises which are impartial, sustainable and thriving with opportunities for all is accentuated [10]. Furthermore, to highlight the importance of this phenomenon the organization of Economic Co-operation and Development (OECD) have established and mobilized a strategy to assist urban and national stakeholders, in addressing the global issue facing many societies from as early as the 19th century [14].

The smart city model is used to analyze and assess the progress local and national authorities are making in the emergence of socially cohesive societies on a global front. To this end, the model is comprised of three primary drivers, namely: (1) Social inclusion – the formation of a metropolitan which has the competence in creating a thorough equilibrium between its inhabitants and the financial, communal and administrative sectors of the region; (2) Social capital – the emergence of a morally transparent collaboration between local authorities and its citizens. In doing so they spur the perception of belonging; (3) Social mobility – which refers to the link between equitable prospective for all [10]. In support, the SADAC [12], stress the fundamental significance of the aspect of respect for oneself, their neighbors, communities and the country as a whole.

To this extent, a community, a society or a country, is not acknowledged as occupying a population in which the theory of social cohesion has been mastered, until such time, citizens are able to look at the individual next to them and not judge them by their; age, gender, race, cultural, deformities or economic status. Thus, a socially cohesive community is one in which citizens aspire to help empower those within their communities. Furthermore, a socially cohesive community is unified- in the sense that they acknowledge what they can/may achieve, if they are able to freely express their; minds, ideas, concerns and fears (SADAC 2012).

3   Smart Cities and Social Cohesion

The manner in which the developments in science, industry and technology have influenced the evolution of architecture in cities can be used to describe the history of urban districts. The link between technology and urban designs was first made during the 1850's. It was believed that the advances in technology would directly impact the future of society, economy and urban agglomerates [15].

The idea of a Garden city was developed by E. Howard in 1898; this idea has been identified by Hall and Tewdwr-Jones [16] and Angelidou [15], as one of the most influential city planning visions. Howard's idea was to create a hybrid city which maintained a constant equilibrium between meeting the needs of the city citizen and community, by combining the convenience advantages of the town with the environmental advantages of the country. The city was to be enclosed by open country land, and this land was then used to create an environment for urban activities such as public institutions [16].

The architectural designs of the industrial city; *'Une cite' 'industrielle'* developed by T. Garnier in 1904 emphasized the impact the advances in technology would yield on the future designs of cities. This shift in the design process of cities was further emphasized by, architect – A. Sant'Elia in the plans for project *'Citta' 'Nuova'* in 1913. These project plans envisioned a city which mimicked an efficient and fast paced machine, the inclusion of multi-storey structures as well as multi-levelled (lower ground, ground and aerial) transportation routes could be seen [15]. Following this vision, architect Le Corbusier in 1922, revealed plans for the development of a contemporary city; a city designed to cater for the needs of three million inhabitants. Equally important to this design, was the strategic positioning of multi-storey skyscrapers in the epicenter of the city, designed for both commercial and residential use.

Inspired by the advances in technology, architect T. Zenetos in the late 19th century introduced the Electronic Urbanism (EU) city model to the built environment. The designs of the EU model encompassed the use of atmospheric space to develop a wire web which was designed to function both structurally and electronically, thus enabling the transfer of information and communication among different spatial zones [17]. The wired city enabled city citizens; to live in their own electronically equipped bubble, thus allowing them to sink into their virtual home space a: any time and in any location. In addition, environmental sustainability was a key concern of Zenetos, this was illustrated by the way in which the structures he designed mirrored the spherical shape of the globe [17].

The real change in urban designs and development came about in the 1980's. During this time, city planners and urban design specialists; embraced the idea of city systems being connected through networks. They welcomed the advantages resulting from the boom in technology and the spread of ICT's into the daily lives of the urban citizen. In addition, the development of the World Wide Web (WWW) complimented their ideas of the future urban spaces being conceived as wired cities, intelligent cities, digital cities and virtual cities [15]. Furthermore, Aurigi [18] highlighted the critical role ICT's would play in the democratic government and the future administration of urban districts.

Studies conducted in the decade prior to the 20th century publicized, the effect the internet would have on the future of tangible cities. Emphasized in the research conducted by Crang and Graham [19], these theorists envisaged a city commissioned solely by the digital world – a world in which: trade relations, communication and information transferal would be diffused through a virtual web of networks. Although, the advances in digital technology and ICT; have significantly altered the manner in which cities around the globe function on a daily basis [20], the influential role they have performed in driving the development of innovative urban designs and plans for the future cities cannot be disregarded. Similarly, the manner in which they have revolutionized the functioning of city systems, ultimately improving the ability of the city to meet the ever growing needs of the urban population cannot be overemphasized [15]. Komninos [21] backed this view through the identification of the first intelligent city-Bletchley Park in the United Kingdom (UK) in 1939. The formation of the Bletchley Park city concentrated on the enabler technology formed for the development of knowledge and transmission of information in the urban region. Concurrently, it developed the individual and co-operative capabilities of the community, these included their ability to mitigate problems, improve manufacturing efficiencies and communication periods.

Furthermore, the origination of the creative city model did not take the blooming of the advances of technology into consideration. The model's objective was to implement initiatives which encouraged the development of civic alliances as well as non-civic partnerships. However, the inability of the creative city to formulate a comprehensive vision and development framework resulted in the emergence of a city model which was unable to cope with the rapid evolution of the globe and the subsequent challenges arising from the revolution [2].

The smart city model aims to guide local and national governments in the formulation and implementation of tactics, which will assist them in the mobilization of balanced strategies throughout the different city divisions. Moreover, these balanced strategies are unique to each city. Thus, on the one hand the strategies implemented should empower the city to overcome its developmental barriers and on the other had it should encourage the sustainable, innovative and equitable (trans)-formation of the city [21]. However, the successful mobilization of these balanced strategies throughout the metropolis is, directly linked to the cities smart technology capabilities. Equally important, is the amalgamation of the cities Smart infrastructure and Smart technology, as indicated in Fig. 1 [22].

Fig. 1. Smart technologies-integrating city systems *Source: City of Johannesburg [22]*

The link between technology, knowledge and innovation, encouraged the inclusion of knowledge management agendas in the global and local development policies of large international organizations such European Union (EU), United Nations (UN) and the World Bank [15]. For this reason, the knowledge economy has become a reality. The realism of this economic style and the improvements in technology over the recent decades have been identified as an essential facilitator in the establishment of the smart city model [15, 20]. In addition, smart city model was proposed as an amalgamation of the well-defined tactics employed in the intelligent city and the proposed objectives of the creative city [2].

4   Research Methodology

This study collected information from the relevant individuals through the use of close ended data collection instrument, which is a well-structured self-reporting questionnaire. The use of the self-reporting questionnaire as the data collection instrument suited the current study as it enabled objective and accurate collection of each individual respondent's opinion and particular knowledge on the smart city phenomenon, its characteristics as well as the holistic impact it has on the functioning and sustainable development of urban districts. However, to ensure the objectives of the study were met, 5-point Liker scales of reluctance (1 = extremely reluctant; 2 = reluctant; 3 = neutral; 4 = willing; and 5 = extremely willing) and agreement (1 = strongly disagree; 2 = disagree; 3 = neutral; 4 = agree; and 5-strongly agree) as well as 3-point (yes, no and unsure) Likert scales were used to quantify the opinions of the respondents. As a result, descriptive statistics of percentage, mean item score MIS), standard deviation (SD) and ranking system were used for the analyses of the findings.

The current study was conducted in the Gauteng province of South Africa. The province consists of three main metropolitan municipalities, namely: City of Johannesburg Metropolitan Municipality (COJ), City of Ekurhuleni Metropolitan Municipality (EKU) and City of Tshwane Metropolitan Municipality (COT). Furthermore, data was collected from town planners, project managers, contracts managers, architects, quantity surveyors, civil engineers, electrical engineers in telecommunications and information as well as data technicians currently practicing in the construction industry.

5   Findings and Discussion

5.1   Cities Promoting Social Cohesion

Table 1 presents the manner in which the sampled respondents, rate whether their city promotes social cohesion among their community. 25% said yes, 50% said No and 25% were Unsure. From the high number of disagreed citizens, it can immediately be deduced that the aspect of unified communities is far and few between within the province of Gauteng.

Table 1. Social cohesion among your community

City promoting social cohesion	%
Yes	25
No	50
Unsure	25

5.2   Citizens' Manner of Identification

Respondents were further asked to select the option they would prefer to identify themselves. The results in Table 2 revealed that 9.4% of the respondents prefers to be identified via their nationality, 3.1% identify themselves via their gender, 12.5% identify themselves through their race, while 75% wants to be identified just as an individuals. Previous study [13], present the following findings from the GCRO QOL Survey in 2015: 22% of the respondents identified themselves by their nationality, 20% by their race, 18% by their gender, 10% by their religion and 17% identified themselves as an individual. The results of the current study's findings and the findings reviewed in the literature do not reflect the same percentages; however, the current study's findings present a more favorable identification spectrum, as the manner in which citizens- identify themselves aggravates the issues of racism and inequalities in the study area.

Table 2.   Urban citizen's identification

Manner of identification	%
Nationality	9.4
Gender	3.1
Race	12.5
Individual	75.0

5.3   Willingness to Share Knowledge and Skills

Respondents were further asked to rate their willingness to share their knowledge and skills at their place of work as well as in their community.

The results in Table 3 revealed that 9.4% of the respondents were extremely reluctant in both their place of work and community, 6.3% of them were reluctant in their place of work and 18.8% in their community, 18.80% of them were neutral in their place of work and 34.40% in their community, 50.00% were willing in their place of work and 28.10% in their community, while 37.50% were extremely willing to do so in their place of work and 12.50% in their community. Therefore, 50% of the respondents are willing to share their knowledge and skills with their colleagues at work and 34.40% of them were neutral in sharing the skills and knowledge in their community. According to South African Department Arts and Culture [12] and South African Cities Network [23], the transferal of skills from one citizen to another, encourages the formation of a platform

Table 3.   Knowledge and skills sharing

Rating	Place of work %	In community %
Extremely reluctant	9.40	9.40
Reluctant	6.30	18.80
Neutral	18.80	34.40
Willing	50.00	28.10
Extremely willing	37.50	12.50

which enables members belonging to a group, to brainstorm, thereby increasing the knowledge capital of their fellow, work colleagues or community members.

5.4 Smart City Drivers and Social Cohesion

Based upon the computation of the MIS, the SD and the Ranking Index used (R), the findings revealing the extent to which the respondents agreed with the use of the six primary smart city drivers that can aid the process of forming socially cohesive societies. Table 4 presents, the manner in which individual smart city drivers can enhance the process of forming socially cohesive societies. Smart living is ranked first with an MIS of 4.09 and SD = 0.734, smart governance is ranked second with an MIS of 4.06 and SD = 0.948, smart economy is ranked third with an MIS of 4.03 and SD = 0.740, smart people is ranked fourth with an MIS of 3.97 and SD = 0.861, smart environment is ranked fifth with an MIS of 3.88 and SD = 0.707, and smart mobility is ranked sixth with the least MIS of 3.81 and SD = 0.821. Further, an average MIS of the primary drivers, reveals a group MIS of 3.97. The findings on the importance of smart living is supported by existing and similar studies [2, 24]. The studies stressed that the drivers aims to stimulate the desire of the city citizen to participate within their community, thus improving their social capital. Most importantly in order for countries to improve the levels of unification among their urban districts and citizens, there is a need to transform, modernize and make provision for increased basic service supplies as well as areas which are allocated for civic use.

Table 4. Smart city drivers and socially cohesive societies

Smart city drivers	MIS	SD	Rank
Smart living	4.09	0.734	1
Smart governance	4.06	0.948	2
Smart economy	4.03	0.740	3
Smart people	3.97	0.861	4
Smart environment	3.88	0.707	5
Smart mobility	3.81	0.821	6
Average	3.97		

6   Conclusion and Recomendations

Smart cities concept is useful in transforming the manner in which construction projects are designed and executed to include the principles of Green Designs, thus increasing the Sustainable Development of the industry. More so, careful thinking, investigating and brainstorming with the appropriate professionals in the urban district can enable the simultaneous development of a socially cohesive society. in addition, the rapid increases in urbanization, drastic changes in resource sustainability and securities as well as changes in the environment as a whole, require the need to be able to adapt as a city and as an individual when the force majeure prevail themselves.

The policies and literature materials that have been reviewed for the purpose of this study, specifically for the aspect of social cohesion do not correlate with the primary findings of the study. For this reason, it is recommended that-policies such as those developed by the South African Department Arts and Culture (SADAC), should be modified and during the modification process, workshops should be conducted with different communities across the country, thereby enhancing their abilities to obtain facts and opinions of the citizens.

Further to the findings, there is a need to investigate how smart cities will promote the mobilization of Public-Private-Partnerships (PPP) as well as improve the awareness levels of emerging procurement methods. Studies and workshops should also be conducted to improve the socially cohesive awareness levels of the citizens as this will aid the process of developing socially cohesive communities in the study area as well as nationally.

References

1. Harrison, C., Donnelly, I.A.: A Theory of Smart Cities. IBM, New York (2011)
2. Letaifa, S.B.: How to strategize smart cities: revealing the smart model. J. Bus. Res. **68**(14), 1415–1419 (2015)
3. Berrone, P., Ricart, J.E.: IESE: Cities in motion index 2016. IESE, University of Navarra Business School, New York, USA (2016)
4. City of Johannesburg 2016. City of Johannesburg. http://www.joburg.org.za/index.php?option=com_content&view=article&id=10522&catid=88&Itemid=226. Accessed 28 July 2016
5. Sha, R., Sonn, H.: The quest for smart cities: utilities management. IMIESA **40**(7), 24–29 (2015)
6. Giffinger, R.: Smart Cities- Ranking of European Medium Size Cities. Centre of Regional Science, Vienna (2007)
7. Das, D.K., Burger, E., Eromobor, S.: Indicative planning perspectives for development of Bloemfontein as a smart city in South Africa. Interim: Interdiscip. J. 11(1), 1–16 (2012)
8. City of Johannesburg Metropolitan Municipality 2012. Joburg 2040: Growth and development strategy (GDS 2040). Johannesburg, Gauteng: City of Johannesburg Metropolitan Municipality
9. Urzua, R.: International migration: Globalisation. UNESCO: International social science, 165421 (2000)
10. United Nations Department of Economics and Social Affairs 2012. United Nations Department of Economics and Social Affairs. http://www.un.org/en/development/desa/news/policy/perspectives-on-social-cohesion.html. Accessed 23 June 2016
11. Abrahams, C.: 20 years of social cohesion and nation-building in South Africa. J. South African Stud. 42(1), 95–96-107 (2016)
12. South African Department of Arts and Culture 2012. A national strategy for developing an inclusive and a cohesive South African society. Pretoria, Gauteng: Department: Arts and Culture-Republic of South Africa
13. Mosselson, A., Peberdy, S.: Gauteng city-region observatory quality of life survey 2015: 4. Headspace and happiness. Pretoria, South Africa: Gauteng City-Region Observatory (2016)

14. Bruhn, J.G.: Chapter 2: Concept of Social Cohesion. In: The Group Effect Social Cohesion and Health Outcomes, 1st edn., pp. 31–48. Springer Science and Business Media, LCC, United Kingdom, Europe (2009)

15. Angelidou, M.: Smart cities: A conjuncture of four forces. Cities, 4795-96-102 (2015)

16. Hall, P., Tewdwr-Jones, M.: Chapter 3: The seers: pioneer thinkers in urban planning, from 1880 to 1945. In: Urban and Regional Planning, 5th edn., p. 32. Routledge, New York (2011)

17. Kallipoliti, L.: Cloud colonies: Electronic urbanism and takes zenetos' city of the future in the 1960s, pp. 1678–1685 (2014)

18. Aurigi, A.: New technology, same dilemmas: Policy and design issues for the augmented city. J. Urban Technol. 13(3), 5–28 (2006)

19. Crang, M., Graham, S.: Sentient cities: Ambient intelligent and the politics of urban space. Inf. Commun. Soc. 10(6), 789–817 (2007)

20. Hollands, R.G.: Will the real smart city please stand up? City 12(3), 303–310 (2008)

21. Komninos, N.: Intelligent cities: variable geometries of spatial intelligence. Intell. Build. Int. 3(3), 172 (2011)

22. City of Johannesburg 2011. City of Johannesburg. http://www.joburg.org.za/index.php?option=com_content&view=article&id=7222:working-towards-a-smart-city&catid=88:news-update&Itemid=266. Accessed 28 July 2016

23. South African Cities Network 2016. State of south african cities report 2016. Johannesburg, Gauteng: South African Cities Network (SACN) (2016)

24. Alfano, A., Amitrano, C.C., Bifulco, F.: The 3rd Electronic Interdisciplinary Conference held in Italy: EIIC Interdisciplinary Conference (2014)

Appraisal of Smartization of Major Cities in South Africa

Ayodeji E. Oke[✉] ⑩, Clinton O. Aigbavboa, and Taniele K. Cane

University of Johannesburg, Johannesburg, South Africa
aoke@uj.ac.za

Abstract. The growing nature of cities through globalization and advancement in technology have necessitated for the transformation of infrastructures and other elements of the cities in the quest of meeting up with the sustainable development goals. This is known as smartization process, which implies that cities should be designed and modernized focusing on the social, environmental and financial benefits to the current and future citizens. Using the basis of smartization process, smart city initiates as well as cities in motion index (CIMI), this study examines the characteristics and global ranking of major cities in South Africa within a 3 years period. It could be observed that there has been a decline in average ranking of the selected cities by about 4.26% which is a concern not only for agencies of government responsible for city planning and development but also construction stakeholders that are involved in the design, development and management of cities and their infrastructures.

Keywords: City in motion · Information technology · Smart city
Sustainable construction

1 Introduction

The advances in modern technology and the rapid development of Information Technology (IT), digital technology (DT) and the Internet of Things (IOT) have formed the catalyst for the Smart City Model otherwise known as the Smartization process [1–3]. Letaifa [4] defines the smartization process as an urban design mechanism which modernizes and transforms infrastructures and urban districts into technological savvy hubs, with the ultimate vision of improving the standard of living, social and economic wellbeing of the city citizen.

The Sustainable Development of cities has become a growing concern, throughout the world. The rapid increase in urbanisation has posed a number of risks and concerns for the survival of urban metropolitans, thus the need to implement smarter, sustainable and resilient policies and models in city development; has altered the minds of urban planners, Architects and other professionals in the built environment (BE). The adaptability of these professionals, organisations and governments throughout the globe, has undoubtedly revolutionised urban planning and shed light on the advantages revealed; through the (trans)-formation of cities to Smart Cities. Smart Cities make use of the modern advances in technology to holistically integrate the city components and utilities.

© Springer International Publishing AG, part of Springer Nature 2018
M. Ben Ahmed and A. A. Boudhir (Eds.): SCAMS 2017, LNNS 37, pp. 463–469, 2018.
https://doi.org/10.1007/978-3-319-74500-8_43

Furthermore, the smart concept provides a stimulus for the development of an innovative environment, which drives the development of innovative city systems and multi-purpose devices such as the use of Smart Meters to monitor water and electricity consumption.

The idea of the simple yet complex urban model - the Smart City and the powerful tool it can form in not only driving the formation of sustainable and resilient cities, but the positive effects it has on the life of the city citizen is clearly accentuated in current literature. The (trans)-formation of cities to Smart Cities identifies the key role of the citizens as - the most important stakeholder of the city. For this reason, one of the major focuses of the process of "Smartization" is to improve the lives of the urban citizen through: improved service delivery, improved health care, individual security and the sustainable use of resources. Moreover, it aims to spur the formation of socially inclusive societies, which are identified as being: impartial, less poverty stricken and stimulate the perception of belonging among individuals.

The mix of elements comprising the fuel, propelling the engine of the smartization process are known as the smart city drivers [5]. They include smart living, smart governance, smart economy, smart people, smart environment and smart mobility. Using these drivers and based on the cities in motion index (CIMI), this study therefore examines the smartization of four major cities in South Africa, which are Johannesburg, Pretoria, Cape Town and Durban, with a view to understanding specific challenges to their stability and development, and proffer necessary solutions.

2   Smartization Process of Cities

To facilitate the ecosystem functionality of the smart city, the connection between the IOT, ICT, utilities, smart infrastructures, local authorities and fundamentally the key force heading the evolution - the citizen, must be established [2]. To this extent, the emergence of a well thought-out, strategic and rapid processing fiber and virtual network of ICT's are constructed throughout the inner-city of major cities [6].

The precise modus operandi of the smartization process as identified by Hollands [7] as well as Alfano et al. [8], concurrently contributes to the provision of a real-time control system for the citizens, commercial enterprises, emergency services as well as municipal governments among others [2, 6]. In addition, the wireless network enables real time communication between the different city utilities, institutions, commercial enterprises in both the private and public sector and the government [9], through the use of smart devices such as smart meters, digital sensors such as traffic sensors [1], as well as the use of surveillance cameras to improve public safety and decrease crime statistics [10, 11].

3   Influence of City Rankings on Smartization Process

Smart City initiatives have influenced the position and global ranking of major cities in the world. The purpose of city-rankings is to determine, assess and compare how cities throughout the globe fair against each other, based on economic growth; social development; and geographical features [5]. Moreover, cities throughout the globe are ranked on an annual bases on the Cities In Motion Index (CIMI). The index is facilitated by the IESE Business School at the University of Navarra, which aims to establish a system of measurement for the future sustainability of the world's major cities as well as the quality of life of the inhabitants. In addition to the above, the index is used to rank smart cities throughout the globe, therefore the mobilization of such indexes provides urban planners, city stakeholders and metropolitan governments with a realistic representation of the progress their urban districts are making in the (trans)-formation to sustainable goals and development.

Due to the imperative findings presented by the rankings; the authors of the index emphasis the ecosystem functionality of urban districts [9]. In doing so they stress that in order for urban administrators to address the developmental challenges faced by cities throughout the globe, the need to thoroughly understand the co-dependency of the different city networks and sub-networks as well as the trail of effect the different sectors have on one-another [2, 5, 7].

The findings of the first CIMI were published in 2013 based on the data collected in 2012; however for the purpose of this review only the findings from the 2013, 2014 and 2015 indexes were adopted as that of 2016 were not available at the time of this study. The CIMI rankings in 2013 had a global geographical coverage of 135 cities [12], in 2014 the geographical coverage increased to a total of 148 cities [13] while the largest increase in geographical coverage can be seen in the 2015 index with a total of 181 cities being analyzed [9]. Each of the cities were analyzed according to ten major characteristics and their specific indicators, as represented in the Table 1 [9].

To obtain each of the cities CIMI city ranking, a methodology known as DP2 was used. This specific methodology has been used by the IESE to enable each indicator comprising the ten characteristics, to be measured independently of each other by applying the same correction factor to each indicator. These correction factors are determined by using the complement of the coefficient of each determination for each of the indicators, thus providing a realistic value for each of the characteristics. Each of the characteristics are then multiplied by their relative weight factor and accumulated together, consequently providing each cities CIMI value. Furthermore, the CIMI value is also used to classify the cities into different groups according to their performance, these are outlined in the Berrone and Ricart's [9] report as; High (H) >90; Relatively High (RH) 90 > 60; Average (A) 60 > 45; and Low (L) <45.

Table 1. Characteristics and indicators for comparing cities on the CIMI

Characteristic	Indicators used to measure characteristic
1. Economy	Productivity level of the workforce, No. of Entrepreneurs, Gross Domestic Product (GDP) as well as the simplicity of starting a business.
2. Human Capital	Level of education, No. of universities and the Cultural attractiveness of the city i.e. No. of Museums, Art galleries.
3. Technology	No. of broadband users both wireless & fixed, No. of Facebook users per capita, Quality of the city council's web services, Innovative Index.
4. The Environment	Amount of CO_2 emissions released from burning of fossil fuels, Percentage of population with access to water supply, Pollution Index of city
5. International Outreach	Level competitiveness and attraction of foreign investors, No. of international tourists and No. of hotels per capita.
6. Social Cohesion	Attempts to determine the level of comradery between the different classes of the urban citizens through the measurement of the unemployment rate, ratio of women workers in public administration, Crime rate and Health index
7. Mobility & Transportation	No. of metro stations in the city, No. of public transport options available to commuters, No. of air routes in the city. Inefficiency Index (Time spent in traffic)
8. Governance	Range of government web services particularly at council level and the cities open data system
9. Urban Planning	Percentage population with access to sanitation facilities, No. of people per household, No. of Architecture firms designing for the city and No. of bicycle shops found in the city.
10. Public Management	Total tax rate paid by businesses, No. of embassies in the city, Reserves per capita in millions of current US Dollars

Source: Berrone and Ricart [9].

4 Findings

Table 2 represents the ranking of the four major cities in South Africa, which are Johannesburg, Pretoria, Cape Town and Durban on the CIMI. In 2013 the cities of Johannesburg and Pretoria for example, were ranked 118th with a CIMI of 22.16% and 117th with a CIMI of 22.17% respectively. Further, in 2014 the city of Johannesburg was ranked 130th with a CIMI of 43.12%, Pretoria on the hand was ranked 129th with a CIMI of 43.23%. Furthermore, in 2015 Johannesburg was ranked 140th with a CIMI of 51.49% and Pretoria was ranked 164th with a CIMI of 42.91%. Thus, the CIMI value for Johannesburg has increased by 29.33% over the 3 year period which is an 8.59% increase over Pretoria's increase of 20.74%.

Although, the increase in CIMI is positive, the decrease in global position of these cities on the rankings indicate that these urban societies have not mastered the balance between innovative and progressive development equally among the co-dependent

Table 2. Global ranking and the CIMI value of major South African cities

IESE-cities in motion index						
City	2015		2014		2013	
	Ranking	CIMI	Ranking	CIMI	Ranking	CIMI
Cape Town	120	56,92	117	49,11	119	21,95
Durban	159	44,45	124	44,96	104	29,33
Johannesburg	140	51,49	130	43,12	118	22,16
Pretoria	164	42,91	129	43,23	117	22,17

Source: Berrone and Ricart [9].

networks within the cities. As a result, Johannesburg is identified as a vulnerable unbalanced city; as it has a high growth rate in one of the characteristics but a low growth rate in the remaining characteristics. Similarly, Pretoria is classified as a vulnerable borderline, unbalanced and stagnant city, meaning across all ten characteristics there has been little or no growth over the 3 year period.

Upon determination of the CIMI values they can be further broken down and analyzed according to the ten characteristics they are comprised of. Therefore, one can obtain key information regarding the current status of the city and its development. On the one hand the characteristics which have increased in ranking over the 3 year period provide insight into which of the city systems have become the main focal point of the local and national government. On the other hand, the characteristics with a decline in ranking emphasis which city systems require the attention of the urban stakeholders for the future sustainability of the metropolis, thus assisting them to develop unique plans and strategies that will meet the present and future demands of their inhabitants as well as ensure their growth is holistic and dynamic [9]. Simultaneously, these will contribute to the key foundation pillars-drivers, which should be considered for the development of a resilient, socially cohesive and smart city.

To assist in the identification of the key foundation drivers for the (trans)-formation of the selected smart cities, Table 3 has been designed based on the findings of Berrone and Ricart [12], Berrone and Ricart [13] and Berrone and Ricart [9]. The table illustrate the global ranking of each characteristic for the cities. The rankings of Johannesburg and Pretoria have been analyzed to determine the characteristic's which had the most and least growth over the 3 year period.

Table 3. The CIMI global ranking of four major cities in South Africa

City Name	City Ranking			Economy			Human Capital			Social Cohesion			Environment			Public Management			Governance			Urban Planning			International Outreach			Technology			Mobility and Transportation		
	2013	2014	2015	2013	2014	2015	2013	2014	2015	2013	2014	2015	2013	2014	2015	2013	2014	2015	2013	2014	2015	2013	2014	2015	2013	2014	2015	2013	2014	2015	2013	2014	2015
Cape Town	119	117	120	99	110	145	113	121	89	131	142	144	44	88	116	44	68	37	24	35	41	55	83	129	85	113	100	76	124	109	100	137	162
Johannesburg	118	130	140	92	103	138	109	124	112	135	144	169	44	116	153	39	73	48	24	48	74	43	134	112	72	145	163	110	119	100	92	79	145
Durban	104	124	159	94	114	142	70	136	143	86	138	151	44	100	117	34	79	103	24	34	66	51	140	174	76	143	173	99	139	153	98	142	173
Pretoria	117	129	164	91	116	140	99	117	137	129	131	158	44	99	150	38	14	43	24	103	136	41	119	147	91	146	176	108	144	175	97	139	178

Global ranking for each of the ten characteristics for 2013, 2014 and 2015

Table 3 revealed that the variance between the 2015 and 2013 overall global ranking based on the calculated CIMI. Johannesburg's position decreased by 22 places which is equal to 12.2%. Pretoria's global position decreased by 47 places which is equal to 26%. However, over the 3 year period a total of 46 cities were added to the index which is equal to 25.4%. Therefore, the actual decrease in Johannesburg's global ranking is equal to 13.3% and Pretoria's global position had a slight increase of 0.55%.

The variance between the 2015 and 2013 ranking for each of the 10 characteristics were also calculated for the four cities. The variance of the 3 characteristics which revealed the highest growth (represented by blue on the table) and largest decline (represented by red on the table) on the rankings per characteristic are identified in ascending order as follows: In Johannesburg, the characteristics with the highest growth are: Technology with an increase of 10 positions; Public Management with an increase of 9 positions; and Human Capital with an increase of 3 positions. Similarly, in Pretoria the characteristics with the highest growth include: Human Capital with an increase of 38 positions; Social Cohesion with an increase of 29 positions; and Public Management with an increase of 5 positions.

Moreover, the characteristics with the largest decline in ranking are noted in Johannesburg as: Environment with a decline of 109 positions; International Outreach with a decline of 91 positions; and Urban Planning with a decline of 69 positions. In Pretoria the characteristics are outlined as: Governance with a decline of 112 positions; Environment and Urban Planning both have a decline of 106 positions; and 3 International Outreach with a decline of 91 positions. Therefore, the analysis of the decline in the above mentioned characteristics provide realistic guidelines for the urban planners and city stakeholders in determining the essential drivers for the transformation of the cities. For instance, to transform Johannesburg into a smart city, namely: Smart Environment; Smart Economy and Smart Mobility should be given necessary attention. Similarly, the Smart model drivers of: Smart Environment; Smart Governance; Smart Economy and Smart Mobility should be the main focus of Pretoria's transformation into a Smart City.

To demonstrate the fluctuations in the ranking of the major cities over the 3 year period; the mean of the four cities global ranking for each year were calculated and divided by the total number of cities analyzed in each annual ranking to obtain a percentage. Thus, global ranking for 2013, 2014 and 2015 amounted to 84.8%, 84.5% and 80.5% respectively. The variation between the years 2013 and 2015 revealed a decline of 4.26% in the cities' global ranking.

5   Conclusion

This study has been able to examine the link between the CIMI and the Smartization process as the CIMI characteristics and the indicators used to compute the characteristics provide realistic and city specific data. As a result, urban districts throughout the globe are assisted in the process of identifying and thoroughly defining, amongst others; the specific challenges hindering their future development and stability, issues concerning the functionality of the inner-city sectors, the general morale - productivity levels and knowledge capital of its inhabitants. The latter is identified as one of the critical components in the (trans)-formation of cities to Smart Cities, as failure to effectively understand these fundamental elements will result in a process of transformation that exposes undesired and disappointing results. This study has provided basic information and necessary guidelines for the urban planners, city stakeholders, developers, construction experts and other concerned stakeholders in determining the essential drivers for the transformation of the identified cities to smart and sustainable community for the benefit of the current and future citizens.

References

1. Harrison, C., Donnelly, I.A.: A Theory of Smart Cities. IBM, New York (2011)
2. Das, D.K.: Indicative planning perspectives for development of Bloemfontein as a smart city in South Africa. Interim Interdisc. J. 11(1), 1–16 (2012)
3. Sha, R., Sonn, H.: The quest for smart cities: utilities management. IMIESA 40(7), 24–29 (2015)
4. Letaifa, S.B.: How to strategize smart cities: revealing the smart model. J. Bus. Res. 68(14), 1415–1419 (2015)
5. Giffinger, R.: Smart cities – ranking of European medium size cities. Centre of Regional Science, Vienna (2007)
6. Komninos, N., Schaffers, H., Pallot, M.: eChallenges, e-2011 Conference by International Information Management Corporation (IIMC) Held in France (2011)
7. Hollands, R.G.: Will the real smart city please stand up? City 12(3), 303, 304–310 (2008)
8. Alfano, A., Amitrano, C.C., Bifulco, F.: Smart cities drivers and ICT: in search of relationships. In: EIIC Interdisciplinary Conference, EDIS-Publishing Institution of the University of Zilina (2014)
9. Berrone, P., Ricart, J.E.: IESE: Cities in Motion Index 2016. IESE: University of Navarra Business School, New York (2016)
10. Colldahl, C., Frey, S., Keleman, J.E.: Smart Cities: Strategic Sustainable Development for an Urban World. Blekinge Institute of Technology, Karlskrona (2013)
11. City of Johannesburg (2011). http://www.joburg.org.za/index.php?option=com_content&view=article&id=7222:working-towards-a-smart-city&catid=88:news-update&Itemid=266. Accessed 28 July 2016
12. Berrone, P., Ricart, J.E.: IESE Cities in Motion Index 2014. IESE Business School: University of Navarra, New York (2014)
13. Berrone, P., Ricart, J.E.: IESE Cities in Motion Index 2015. IESE Business School: University of Navarra, New York (2015)

Sentiment Analysis in Social Media with a Semantic Web Based Approach: Application to the French Presidential Elections 2017

Mohamed El Hamdouni[1(✉)], Hamza Hanafi[2], Adil Bouktib[2], Mohamed Bahra[2], and Abdelhadi Fennan[1]

[1] Faculty of Sciences and Technologies of Tangier, LIST Laboratory,
Abdelmalek Essaadi University, Tétouan, Morocco
mhd.elhamdouni@gmail.com
[2] Faculty of Sciences and Technologies of Tangier, Abdelmalek Essaadi University,
Tétouan, Morocco

Abstract. Sentiment Analysis is one of the recent fields that attracts the attention of many researchers to contribute in its improvement and to get the fruits of its applications. This paper presents a novel framework that aims to empower the sentiment analysis task by combining an unsupervised approach that relies on a lexicon-based strategy and domain ontologies to identify the sentiment of the textual content, and a visual sentiment ontology to analyze the emotions expressed by images, this hybrid method ensure the accuracy of results on data retrieved from social networks to get insights about the reaction of public towards a specific topic. The framework was put into test to analyze the data flowing in social networks during French elections 2017 to rank the candidates and detect the regions where they are leading and the results obtained were promising.

Keywords: Semantic web · Ontology · Sentiment analysis
Twitter and Facebook APIs

1 Introduction

Nowadays, the growth of social network platforms has gained a lot of attention from many organizations, companies and researchers because of the outburst of social data they made available, and partially accessible online, also for the reliability of emotions expressed spontaneously by people towards a topic without any social boundaries or psychological complexity, and moreover psychological researchers found that more the people are relaxed at their homes, the more their sentiment and expression are true [1].

In our paper, we seek to benefit from the tremendous development that artificial intelligence and semantic web knows, and also the tools they offer in order to extract the sentiments score of the contents posted by the actors of the social network and get insight about their opinions towards a specific topic, product, event etc.

A feature based sentiment analysis tool was proposed by [K. Vithiya Ruba and D. Venkatesan], which relies on building ontology for the interested domain, to help in getting more accuracy about the sentiment represented in the content. Another work was

© Springer International Publishing AG, part of Springer Nature 2018
M. Ben Ahmed and A. A. Boudhir (Eds.): SCAMS 2017, LNNS 37, pp. 470–482, 2018.
https://doi.org/10.1007/978-3-319-74500-8_44

done by [Cataldo Musto et al.], that present a domain-agnostic framework for intelligent processing of textual streams coming from social networks. And it was applied in two smart cities related scenarios; to monitor the recovering state of the social capital of L'Aquila's city after the dreadful earthquake of April 2009 and to build a hate map of the most at-risk areas of the Italian territory.

Our work was inspired from the both latter approaches, for the aim of implementing a framework that benefits from the advantages of using domain ontologies in the contextual processing of textual content. And we improved it by adding the analysis of visual content using a visual sentiment ontology tool. In order to explore the tendencies of the social network actors and get refined and rigorous results.

2 Social Network Analysis

Social Network Analysis (SNA) is one of the rising fields that benefits from many opportunities to develop and evolve due to the explosion of social data and the interactions between the different entities (people, organizations, businesses…), that are explicitly accessible online through social network platforms, SNA aims to recognize and take advantage of the key features of social networks in order to manage their life cycle and predict their evolution. And it offers efficient tools that allows researchers to make remarkable discoveries and valuable explanations of varied and complex real life situations in many domains such as economy, health, security etc. Which in many cases classical and statistical techniques fails to extract accurate and reasonable inference from it. This is made possible through the following steps:

Sampling data and the creation of a Graphical representation of the problem which helps to visualize and understand easily the key features of the network.

Selecting the appropriate metrics to get information about the actors of the network or about the structure of the whole network.

2.1 Algorithm Analysis

SNA algorithms are divided into two categories, the first one focus on the extraction of information about the actors of the network, whereas the second helps to understand the structure and the characteristics of the social network.

a. **Algorithms that extracts the most important actors in the social network:**

These algorithms highlights the most important actors and the strategic positions of the network, their main goal is the study of the centrality metric of the network. Hereafter, a brief description of various definitions of this metric:

Degree Centrality: The actor with the higher number of connections is considered to be the most important node in the graph.

Betweenness Centrality: This considers nodes that are more often on shortest path between other nodes as the most central.

Closeness Centrality: This considers as most central the nodes that have the smallest average length of the roads (sequence of relationships) linking an actor to others.

Egocentric Centrality: Centrality of a node on the subnetwork of its neighborhood.

b. **Algorithms that gives insights about the global structure of the network:**

Beside The centrality, there are more metrics that gives the necessary information to understand the social network, predict its evolutions to get more control over it, also gives the ability to recognize the distribution of individuals and activities, and measure its flexibility at communicating and distributing messages, and its mechanism of restructuring in case of failures.

Here we will present a brief description of the existing approaches adopted in the study of community detection algorithms to understand the distribution of actors in a social network. These algorithms are classified into two categories:

- **Hierarchical algorithms:** the main goal of these algorithms is the construction of a hierarchical tree of communities, called a dendrogram, namely a tree of denser and denser communities from top to bottom [2].
- **Heuristical algorithms:** they are based on heuristics related to the community structure of networks and to community characteristics [2].

2.2   Web Semantic Analysis

With the tremendous amount of digital content available on the web, their diverse origin and different formats, also the need for more precise and valuable results, great research and progress is ongoing in order to pass from the actual world web, which is concentrated on the interchange of documents, to the web of data referring by "Semantic Web" and defined by the World Wide Web Consortium (W3C) as follows:

"The Semantic Web provides a common framework that allows data to be shared and reused across application, enterprise, and community boundaries. It is a collaborative effort led by W3C. It is based on the Resource Description Framework (RDF)" [3].

Semantic web analysis is a new paradigm for semantic web that aims to offer new perspectives to understand of relations between contents, and give means to bind context to queries, in order to get precise data and eliminate the irrelevant results.

One of the pillars of Semantic Web is Ontologies, which defines the terms used to describe and represent an area of knowledge and define relations among them. And moreover, gives the systems the ability to recognize the contexts they are operating on and reasoning about those contexts.

3   Sentiment Analysis in Social Network

The goal of sentiment analysis called also Opinion Mining is to further enrich the comprehension of the content in natural language and to be able to extract subjective information from it, such as opinions and sentiments, in order to create organized and helpful knowledge to be used by either a decision support system or a decision maker. Alongside with the growth of online social applications a variety of algorithms were developed and improved to handle sentiment analysis tasks, which are considered as sentiment classification problems. The Fig. 1 shows a hierarchy of the state of the art algorithms used for opinion mining. Sentiment analysis can be classified into three approaches [4]:

- Machine learning approach which require the creation of a model by training the classifier with labeled examples. This approach can be divided into supervised and unsupervised learning methods. Where the first method make use of a large number of labeled training documents, and the second method is used when it is hard to find these labeled training documents.
- Lexicon-based approach use dictionaries of known and precompiled sentiment terms. This approach is divided into dictionary-based approach and corpus-based approach which use statistical or semantic methods to find sentiment polarity.
- The hybrid Approach combines both previous approaches, It employs the lexicon-based approach for sentiment scoring followed by training a classifier assign polarity to the entities in the newly find reviews. This approach is generally used since it achieves the best of both worlds, high accuracy from a powerful supervised learning algorithm and stability from lexicon based approach.

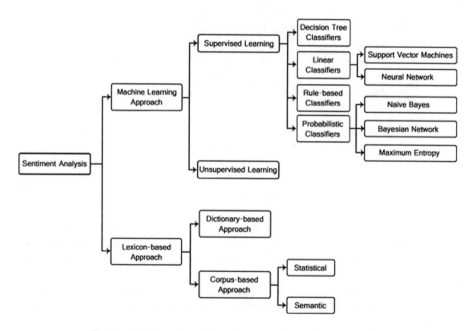

Fig. 1. The hierarchy of algorithms used in sentiment analysis.

The accuracy of this algorithms varies depending on the resources, domain and reliability of the used data, we can find that machine learning approach have a limitation associated with the model, which need to be trained with a large amount of data before start using it, this process consume time and the approach give low accuracy when the training data is not sufficient.

4   Our Framework

Our proposed framework is based on a set of modules that allow us to do sentiment analysis on a stream of social data coming from Twitter and Facebook APIs by implementing web semantic paradigms and domain ontologies to get more accurate results respecting the context of the analyzed content. The entry point of the framework is the definition of keywords to be extracted from social networks, then we proceed to analysis of this data by reasoning on our predefined ontologies, in order to produce valuable results for the end users in an analytic console.

The global architecture of our framework is described in Fig. 2. A description of each module is provided in the section below:

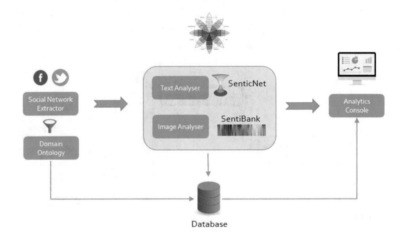

Fig. 2. Global architecture of our proposed framework.

Social network extractor: This is the starting point of our framework, it connects to Facebook and Twitter using official APIs to extract data based on the keywords predefined by the user, the retrieved content is stored in our database without any preprocessing.

Domain ontology: This module contains taxonomies of domains that we focus on in our study, these knowledge bases will contribute in extracting the refined sentiment scores.

Sentiment analysis: This is the core element of our framework, the main goal of this module is to recognize the sentiment stated or not in the extracted content, which is done by the following steps:

1. **Preprocessing:** This is an important step that helps to remove irrelevant and insignificant bits of content like urls, special characters, stop words and convert emoji's to their text correspondent.
2. **Text analyzer**: After preprocessing the content we move to the extraction of sentiments by adopting a lexicon based approach using SenticNet [5], a knowledge base

of 50,000 commonsense concepts made available online (https://pypi.python.org/pypi/senticnet) which we exploit in a python script to query the polarity value of the words or bigrams detected in the content. SenticNet is a rich knowledge base that leverages on data collected from multiple sources such as WordNet-Affect, Open Mind Common Sense and GECKA to associate polarities to concepts dynamically by relying on spreading activation, neural networks and the Hourglass Model [6] (Fig. 3).

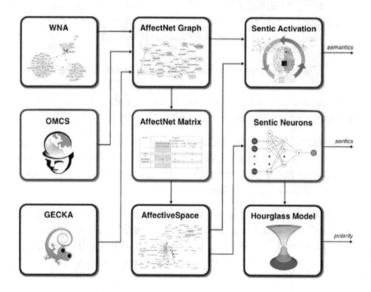

Fig. 3. SenticNet construction framework: this framework generates the semantics and sentics that form the SenticNet knowledge base.

SenticNet also covers more than 20,000 concepts in French which gives us the ability to process the content posted by French users through the social networks.

In our framework we query the sentiment score of each extracted word from the text and we calculate the mean value in order to get the global sentiment associated to the tweet or the Facebook comment, by using the equation below:

$$mean\,polarity = \frac{1}{count} \sum_{k=0}^{count} (word\,polarity)$$

The Fig. 4 shows an example of a tweet, after the preprocessing step, SenticNet calculate the polarities of the highlighted words and a mean value of the sentiment is associated to this tweet. And by applying the equation from above we get the following output:

$$Mean\,polarity = demain(0.021) + chemin(0.17) + victoire(0.76) = \frac{0.951}{3} = 0.317$$

Fig. 4. Example of a tweet mentioning Francois Fillion.

This result shows that a positive polarity is contained by this tweet, which tell that a positive sentiment was expressed by the user.

3. **Image analyzer:** If a media attachment is associated to the text, we use a visual sentiment ontology framework called SentiBank [7] which used an ontology of semantic concepts associated to emotions and structured as a bigrams of adjective-noun, and follow the schema below:

Our framework extract the media associated to the content and use SentiBank to extract the detected bigrams, the latter is passed to SenticNet to get their polarities, and calculate the mean polarity represented by the image which will be added to the mean polarity value of the text content of the proceeded tweet (Fig. 5).

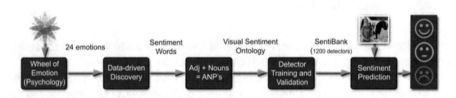

Fig. 5. Representation of SentiBank framework model.

By using the visual sentiment ontology on this example of attachment, SentiBank gives in output this set of bigrams:

busy_crowd
excited_student
anxious_crowd
angry_crowd
screaming_girls
drunk_girls

proud_parents
crowded_street
young_adul

Each of those results which are considered having a higher probability to be included in the image, will be splitted in order to pass the adjectives to the SenticNet to retrieve their polarities, and to calculate the sentiment represented by the image which is the mean polarity of calculated polarity (Fig. 6).

Fig. 6. An example of attachment associated to the tweet

Analytics console: This component gives the user a mean to visualize results in a variety of graphical representation such as a map and charts to help him make decisions, this was made by a java framework which is spring boot that offer us a simple way to handle connections to social networks via the spring social extension. The IHM was built on the stat of the art template engine which is Thymleaf. For the aim of giving ergonomic representation.

Database: This is the source of data from where the analytic console is fed, Our framework use NoSQL database engine to persist and retrieve content extracted by the official APIs, also the results of calculated sentiment of tweets and Facebook comments are stored in order to optimize the performance and reduce the response time of our framework, The main reasons that leads us to use a NoSQL database is that it allows us to store the extracted content into documents in form of JSON style documents which helps us to avoid separating the tweets or the comments into columns and to query them by different keywords, a perfect fit for that was MongoDB especially when our tweets have a geo-spatial coordinates property and MongoDB can manipulate them easily and supports geospatial indexing.

To benefit from the advantages of this engine we use Spring Data MongoDB project to interact with the multiple collections via a special helper class MongoTemplate that increases the performance and reduces the time consumed by common operations on them.

5   French Election Sentiment Analysis Platform

The elections that are taking part in France are raising a rich flow of user interaction full of emotions and opinions of different nature towards the candidates that are competing to get in the "Élysée palace". To put our framework into test we found that this event is a rich opportunity to evaluate our approach.

Those are the most powerful candidates who are leading the polls and generating an interesting amount of social data on Twitter and Facebook:

Emmanuel Macron.
François Fillon.
Marine Le Pen.
Benoît Hamon.
Jean-Luc Mélenchon.

To make the application of our framework on this event more valuable and to get precise knowledge about the reaction of the public toward it, a well-studied set of keywords was selected by using candidates' names and their syntaxes varieties, also the name of the political party they belong to.

In the first step we used the keywords that are related to the candidates to extract the relevant tweets by using Twitter Streaming API which allows us to get in real time all the activities mentioning a candidate. In case of Facebook we used the Graph Explorer API to get all comments of all posts in the wall of the candidates' page. We started this operation the 7th march 2017 until [current day] and we collected more than 15 million tweets and retweets, we stored all of it in a mongo database and we reached a disk space volume of over than 60 Gb.

The final output of our framework applied on the set of candidates mentioned above, is discussed in the next section and presented by the Fig. 7 below:

The analytics console of our platform contains four charts and a map. Hereafter, a brief interpretation of the results showed in each component:

Line chart: Our framework query our database to retrieve for each candidate, a set of tweets and Facebook comments posted in a period of time specified by the user. For each day of the latter, we calculate the percentage of positive polarities extracted from tweets and comments, the final results are presented in the line chart above. Which can help to get clear insight about the influence of a candidate on daily basis (Fig. 8).

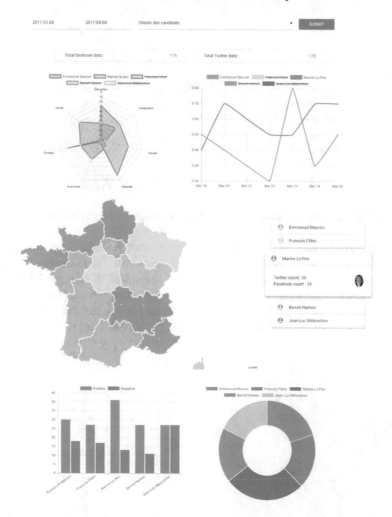

Fig. 7. Screenshot of the platform showing results for a query on 5 leading candidates

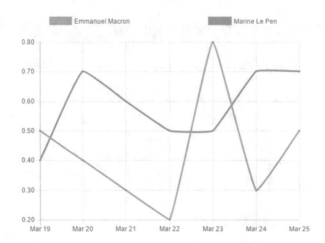

Fig. 8. Comparison between the evolution of Emmanuel Macron and Marine le Pen

Map: In order to show the regions where a candidate gained more positive sentiment results, our framework relies on geographical fields attached to the tweet, (Facebook comments were not represented in the map due to the lack of their geographical information), this process start by verifying the existence of coordinates fields, if they are not available, we move to get the place associated to the tweet, and if the both does not exist we pass to retrieve the location of the user who posted the tweet (Fig. 9).

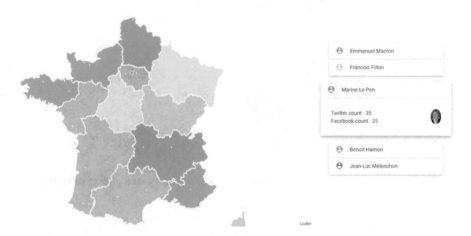

Fig. 9. Map represent the distribution of the winning candidates in France territories

Doughnut chart: The framework count tweets and Facebook comments that present positive sentiment toward each candidate and present in a doughnut results for each of the candidates to show the dominant one (Fig. 10).

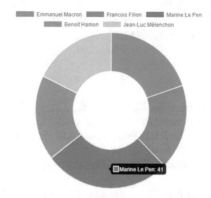

Fig. 10. Doughnutchart show the dominant candidates in France election

Bar chart: The process applied by the framework in order to present this type of chart, is by taking into consideration the number of positive and negative sentiment expressed by the users of twitter and Facebook applications (Fig. 11).

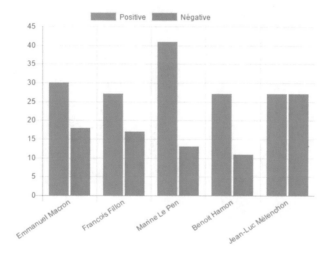

Fig. 11. Barchart display positive and negative sentiment for each candidates

6   Conclusion

Thanks to the combination of the web semantic paradigms and the lexicon-based approach in the sentiment analysis of the content available through the social network, we were successful in getting promising results that proves the efficiency of our framework.

In our future work we are going to study the use of an advanced equation to associate a more significant polarity value to the textual content and add a real time geo-graphical representation of data on the map. Also we will implement social network analysis algorithms to detect and explore communities that form the social network and demonstrate connections of their involved actors.

References

1. Abel, F., Geert-Jan Houben, C.H., Stronkman, R., Tao, K.: Twitcident: fighting fire with information from social web streams, pp. 305–308. ACM (2012)
2. Ereteo, G.: Analyse sémantique des réseaux sociaux. Autre [cs.OH]. Telecom ParisTech (2011)
3. W3C, http://www.w3.org/2001/sw/
4. Medhata, W., Hassanb, A., Korashyb, H.: Sentiment analysis algorithms and applications: JIS Juin (2015)
5. Orsucci, F., Paoloni, G., Fulcheri, M., Annunziato, M., Meloni, C.: Smart communities: social capital and psycho-social factors in SmartCities (2012)
6. Developing Smart Cities Services through Semantic Analysis of Social Streams. ACM (2015)
7. Borth, D., Ji, R., Chen, T., Breuel, T., Chang, S.-F.: Large-scale Visual Sentiment Ontology and Detectors Using Adjective Noun Pairs

A General Overview of 3D RNA Structure Prediction Approaches

Arakil Chentoufi$^{(\boxtimes)}$ ⓘ, Abdelhakim El Fatmi, Ali Bekri,
Said Benhlima, and Mohamed Sabbane

MACS Lab, Faculty of Science, Computer Science Department,
Moulay Ismail University, Meknes, Morocco
a.chentoufi@edu.umi.ac.ma

Abstract. The function of the RNA molecule depends on its three-dimensional (3D) structure. Therefore, understanding the role of the RNA molecule requires detailed knowledge of its 3D structure. Initially, this task was performed experimentally using X-ray crystallography and NMR spectroscopy, but this technique remains limited to small molecules and becomes more expensive for large molecules. For this reason, the number of RNA 3D structures in databases increases in a difficult way. In the other hand, the number of RNA sequences increases rapidly, due to the high development of the sequencing tools. In order to remedy this shortcoming, a number of computational methods have been developed based on different aspects, such as dynamic molecular fragments or pattern and the coarse-grained potentials to predict the RNA 3D structure. In this paper we give a general overview of these methods, their categories, their advantages and their drawbacks.

Keywords: RNA structure · RNA 3D structure prediction · Molecular dynamics
Knowledge-based energy function · Coarse-grained modeling

1 Introduction

In the cell, RNA molecules are involved in various biological processes. In the beginning, the well-known role of the RNA molecule, is to transport genetic information from the DNA molecule to the proteins. Recently, the scope of the RNA function beyond the central dogma has extended. In addition to the Messenger RNAs (mRNAs) that are responsible for the transmission of genetic information, various classes of RNA molecules have been discovered which are involved in a variety of functions, such as (tRNA molecules, RRNA) or enzymes (ribozymes) that play fundamental roles in translation machines. More recently, another type of RNAs, called non-coding RNAs with regulatory functions has been discovered, for instance small nuclear RNAs (snRNAs) which complex with proteins to form small nuclear ribonucleoproteins (snRNPs) involved in RNA splicing. RNA molecules have precise and complex 3D structures that allow them to perform their functions. However, understanding the exact mechanisms behind these functions is the main interest of the biological research community.

© Springer International Publishing AG, part of Springer Nature 2018
M. Ben Ahmed and A. A. Boudhir (Eds.): SCAMS 2017, LNNS 37, pp. 483–489, 2018.
https://doi.org/10.1007/978-3-319-74500-8_45

The list of RNAs discovered has been extended and the large number of them are still unknown to their 3D structure, experimental methods such as X-ray crystallography, nuclear magnetic resonance (NMR), X-rays (SAXS) and Cryo-electron microscopy (Cryo-EM), require a lot of time and money [1]. The RNA structure problem has a hierarchical nature, i.e., the secondary structure, which involves Watson-Crick canonical base pairs, is forms first. Then, the tertiary structure, which involves several other types of interaction [2]. In the literature, this problem can be divided into two subproblems: RNA secondary structure and RNA 3D structure. Both are approximately resolved and both have been addressed by computational methods.

Existing methods for RNA secondary structure prediction are sequence-based methods, they can be divided into two categories: Dynamic methods, which optimize a free-energy function, they perform well when the sequence alignments are available, such as Mfold [3] and RNAfold [4]. Stochastic methods such as CONTRAfold [5]. Most of these methods are limited to secondary structures without pseudo-knots, but some can predict pseudoknots, at a higher complexity cost. Here, we focus on the RNA 3D structure prediction. The current state of the art of the existing methods can be divided into three main approaches:

- Molecular dynamics.
- Fragments and motifs.
- Coarse-grained potential.

2   RNA 3D Prediction Tools

2.1   RNA 3D Prediction Tools Using Molecular Dynamics

iFoldRNA [6] a novel web-based methodology for RNA structure prediction with near atomic resolution accuracy and analysis of RNA folding thermodynamics. iFoldRNA predicts RNA tertiary conformation by using discrete molecular dynamics (DMD), it uses a coarse-grained representation of three beads-per-nucleotide (phosphate atom, the sugar and aromatic bases), it requires only an RNA sequence as input. This tool is powerful and fast for modeling small RNA molecules (<50 nt) at atomic resolution [2–5 Å root mean square deviation (RMSD) to its native structure]. For larger RNA molecules (>50 nt), it takes a long time to sample the conformational space. This method has been improved by incorporating some parameters including basic stacking, basic pairing and hydrophobic interactions obtained from experiments, to constrain the structures of larger RNA molecules [7]. iFoldRNA also provides additional information to the user, such as specific radius of gyration, heat, contact maps, simulation trajectories, RMSDs from native state, and fraction of native-like contacts. Availability: http://troll.med.unc.edu/ifoldrna.v2/index.php.

NAST [8] (The Nucleic Acid Simulation Tool) is a protocol for predicting the RNA 3D structure, which uses knowledge-based statistical potential and a coarse-grained model, i.e., each residue being represented by a single bead centered at its C'3. With this representation, NAST is capable of handling large molecule (>150 nt) such as the 158-residue P4-P6 or the 388-residue T. thermophila group I intron, with high velocity and with accuracy depending on the external restraints. It uses a molecular dynamics engine, to generate plausible 3D structures, and requires secondary constraints and tertiary contact information. NAST's energy function consists of four types of information: geometries from solved ribosome structures (distances, angles, and dihedrals be-tween C'3 atoms of two, three, and four sequential nucleotides, respectively), repulsive interactions between bases not farther than two positions away, ideal helical geometry for nucleotides participating in secondary structures, and long-range interactions between nucleotides participating in tertiary contacts. One advantage of NAST, is the ability to classify clusters of structurally similar decoys based on their compatibility with experimental data and its ability to construct missing loops.

Availability online: https://simtk.org/home/nast.

BARNACLE [9] (BAyesian network model of RNA using Circular distributions and maximum Likelihood Estimation) a probabilistic model that combines a dynamical Bayesian network (DBN) [10] with directional statistics, to predict the RNA tertiary structure. It is designed to overcome the problem of the discrete nature of the fragment assembly methods by using continuous space. For small molecules of RNA (<50 nt), BARNACLE can produce a reasonable RNA-like for 9 out of 10 test structures (<10 Å RMSD), it only requires the secondary structure information. For large RNA molecules and those with a complicated topology (junctions and long-distance contacts) are not yet predicted, due to too many degrees of freedom and the complexity of the probabilistic model.

Availability: http://sourceforge.net/projects/barnacle-rna/

2.2 RNA 3D Prediction Tools Using Fragments and Motifs

FARNA [11] (Fragment Assembly of RNA server) is an energy-based method for predicting the tertiary structure of RNA for a given sequence, with minimal free energy. It is inspired by the Rosetta methods of prediction of the tertiary protein structure [12]. To reduce conformational sampling space, this method constructs its 3D structure library to store the three-nucleotide fragments taken from the experimental structures and to build a plausible conformation, the short fragments are slipped together according to their purines and pyrimidine composition, and then a Monte Carlo simulation method is applied to assemble these fragments in reasonable native-like tertiary structure. The folding simulation process is guided by a knowledge-based energy function. Compared to other sampling strategies, FARNA is more efficient. However, FARNA can only predict the tertiary structure of small molecules of RNA (<40 nt), but for molecules of longer lengths or with complex topological structures, the challenge remains.

Availability: http://rosettaserver.graylab.jhu.edu

MC-Fold/MC-Sym [13, 14] is a pipeline program, which used to construct RNA 3D structures from the sequence, by using the coordinates and relations between the bases of known RNA structures. It is based on cyclic nucleotide motifs and uses the representation of all atoms. The procedure for constructing the RNA 3D structure begins by predicting the secondary structure by MC-Fold from the input sequence, with additional constraints, MC-Fold predict secondary structure more informative. Then, MC-Sym is applied to the MC-Fold output, to predict the RNA 3D structure using molecular dynamic simulations.

Availability: http://www.major.iric.ca/MC-Pipeline.

RNAComposer [15] is a fragment-based method for the fully automated prediction of RNA 3D structures from a user-defined secondary structure, which is divided into basic elements such as stems, loops (apical, internal, bulge, and n-way junctions), and single strands. These elements are translated into tertiary structure elements using the RNA FRABASE database [16]. To search for and select the appropriate item, certain criteria are used, such as secondary structure similarity, sequence similarity, purine and pyrimidine compatibility, source energy resolution and energy. The machine translation system also used NAB residues library [16] to repair the non-identical residues in alignment. RNAComposer merges the elements of the tertiary structure to assemble an initial RNA 3D structure. This initial model refines by using energy minimization in torsion angle space and the Cartesian coordinates using the CHARMM force field [17], to generate the final high quality RNA 3D model.

Availability: http://rnacomposer.ibch.poznan.pl.

2.3 Prediction by Coarse-Grained Potential

Several recent methods use a level of representation higher than the nucleotide, working directly on a set of nucleotides.

RNAJAG [18, 19] (RNA-Junction-As-Graph) is a sampling/data-mining module for predicting a tree graph of RNA junctions with global helical arrangements for a given secondary structure. This method involves two steps: Prediction of the junction topology using data-mining, these prediction results are used to construct the initial graph. Then, to sample the initial graph in 3D space, using knowledge-based statistical potentials derived from bending and torsion measures of internal loops as well as radii of gyration for known RNAs. Results show that RNAJAG advances folding structure prediction methods of large RNAs, and reproduces native-like folds of helical arrangements in most junctions (3- and 4- way junctions) tested. **RNAJAG module is available to academic users upon request.**

Lamiable et al. [20] have proposed a method for predicting the RNA 3D structure with a coarse grain representation. This method proceeds as follows: construction of a skeleton graph, which represents the coarse grain (helices and junctions between helices) of a given secondary structure without pseudo-knots. Then, the classification of the junctions found in 3D topological families, according to the study [21]. The third step consists of an initial integration without tertiary interaction by assembling the local forms obtained. Finally, the folding of the Initial Embedding.

GARN [22] (RNA sampling algorithm) is a coarse-grained method for sampling the 3D structure of RNA. The RNA molecule is represented by a graph similar to the representation used for the method proposed by Lamiable et al. This method used the game theory (regret minimization algorithms) and the Knowledge-based (KB) to reach the Nash equilibrium, which refers to the stable RNA 3D structure (Table 1).

Availability: http://garn.lri.fr/.

Table 1. Various methods for predicting RNA 3D structure

RNA 3D prediction tools		Advantages	Limitations	Input data
Molecular dynamics	iFoldRNA	Rapid conformational sampling ability	1- Small RNA molecules (<50 nt)	Sequence
			2- Errors increase as RNA Lengths	
	NAST	Performs well on molecules (<40 nt)	1- Few hundred nucleotides	Secondary structure
			2- Errors increase as RNA lengths	
			3- Accuracy depends on the external restraints	
	BARNACLE	1- Efficient sampling of RNA conformations in continuous space	1- Degrees of freedom	Secondary structure
		2- Captures several key features of RNA Structure	2- Long-range interactions	
Fragments and motifs	FARNA	Accurate high-resolution techniques	Errors increase as RNA lengths	Sequence
	MC-Fold/MC-Sym	Powerful for modeling small RNA	Computation time	Sequence & Secondary structure
	RNAComposer	Prediction of large RNA 3D structures of high quality	Accuracy depends on the RNA FRABASE dictionary	Sequence & Secondary Structure
Coarse-grained potential.	RNAJAG	1- Capturing full details of RNA's rich 3D architecture	1- Accuracy of both the junction family and coaxial stacking configurations	Secondary Structure
		2- Reduction of the RNA conformational space	2- Cannot account for protein-RNA interactions or solvent effects	
	Lamiable et al. [20]	Global shape of large molecules	1- Classification is limited.	Secondary Structure
			2- Parameters of cost functions junction	
	GARN	1- Build very large assemblies	The choice of the scoring functions	Secondary Structure
		2- Not require any fragment library to be available for SSEs		
		3- Fast and not dependent on templates or consecutive SSEs		

3   Conclusion

This article provides a general overview of existing methods used for prediction of RNA 3D structure. Several computational methods have been proposed to achieve RNA 3D structure from a secondary structure, a sequence or both. Which are provide reasonable accuracy for short RNA molecules. However, for long or complex RNA molecules, there are still many limitations as well as several challenges ahead. These methods can be divided into three main categories: molecular dynamics, fragments motifs and coarse-grained potential. For each category, we present briefly some recent methods by introducing their principle, strength, weakness and input data. This paper can serve as a basis for readers wishing to explore 3D RNA structure prediction.

References

1. Shapiro, B.A., Yingling, Y.G., Kasprzak, W., Bindewald, E.: Bridging the gap in RNA structure prediction. Curr. Opin. Struct. Biol. **17**(2), 157–165 (2007)
2. Tinoco, I., Bustamante, C.: How RNA folds. J. Mol. Biol. **293**(2), 271–281 (1999)
3. Zuker, M., Stiegler, P.: Optimal computer folding of large RNA sequences using thermodynamics and auxiliary information. Nucleic Acids Res. **9**(1), 133–148 (1981)
4. Hofacker, I.L., Fontana, W., Stadler, P.F., Bonhoeffer, L.S., Tacker, M., Schuster, P.: Fast folding and comparison of RNA secondary structures, Monatshefte fr Chemie/Chem. Mon. **125**(2), 167–188 (1994)
5. Do, C.B., Woods, D.A., Batzoglou, S.: Contrafold: RNA secondary structure prediction without physics-based models. Bioinformatics **22**(14), e90–e98 (2006)
6. Sharma, S., Ding, F., Dokholyan, N.V.: iFoldRNA: three-dimensional RNA structure prediction and folding. Bioinformatics **24**(17), 1951–1952 (2008)
7. Gherghe, C.M., Leonard, C.W., Ding, F., Dokholyan, N.V., Weeks, K.M.: Native like RNA tertiary structures using a sequence-encoded cleavage agent and refinement by discrete molecular dynamics. J. Am. Chem. Soc. **131**(7), 2541 (2009)
8. Jonikas, M.A., Radmer, R.J., Laederach, A., Das, R., Pearlman, S., Herschlag, D., Altman, R.B.: Coarse-grained modeling of large RNA molecules with knowledge-based potentials and structural filters. RNA **15**(2), 189–199 (2009)
9. Frellsen, J., Moltke, I., Thiim, M., Mardia, K.V., Ferkinghoff-Borg, J., Hamelryck, T.: A probabilistic model of RNA conformational space. PLoS Comput. Biol. **5**(6), e1000406 (2009)
10. Ghahramani, Z.: Learning dynamic bayesian networks. In: Adaptive processing of sequences and data structures, pp. 168–197. Springer, Heidelberg (1998)
11. Das, R., Baker, D.: Automated de novo prediction of native-like RNA tertiary structures. Proc. Nat. Acad. Sci. **104**(37), 14664–14669 (2007)
12. Yarov-Yarovoy, V., Schonbrun, J., Baker, D.: Multipass membrane protein structure prediction using Rosetta. Proteins Struct. Funct. Bioinf. **62**(4), 1010–1025 (2006)
13. Parisien, M., Major, F.: The MC-Fold and MC-Sym pipeline infers RNA structure from sequence data. Nature **452**(7183), 51–55 (2008)
14. Reinharz, V., Major, F., Waldisphl, J.: Towards 3D structure prediction of large RNA molecules: an integer programming framework to insert local 3D motifs in RNA secondary structure. Bioinformatics **28**(12), i207–i214 (2012)

15. Popenda, M., Szachniuk, M., Antczak, M., Purzycka, K.J., Lukasiak, P., Bartol, N., Blazewicz, J., Adamiak, R.W.: Automated 3D structure composition for large RNAs. Nucleic Acids Res. **40**(14), e112 (2012). p. gks339

16. Popenda, M., Szachniuk, M., Blazewicz, M., Wasik, S., Burke, E.K., Blazewicz, J., Adamiak, R.W.: RNA FRABASE 2.0: an advanced web-accessible database with the capacity to search the three-dimensional fragments within RNA structures. BMC Bioinf. **11**(1), 231 (2010)

17. Brooks, B.R., Brooks, C.L., MacKerell, A.D., Nilsson, L., Petrella, R.J., Roux, B., Won, Y., Archontis, G., Bartels, C., Boresch, S., et al.: CHARMM: the biomolecular simulation program. J. Comput. Chem. **30**(10), 1545–1614 (2009)

18. Laing, C., Jung, S., Kim, N., Elmetwaly, S., Zahran, M., Schlick, T.: Predicting helical topologies in RNA junctions as tree graphs. PLoS ONE **8**(8), e71947 (2013)

19. Kim, N., Laing, C., Elmetwaly, S., Jung, S., Curuksu, J., Schlick, T.: Graph-based sampling for approximating global helical topologies of RNA. Proc. Nat. Acad. Sci. **111**(11), 4079–4084 (2014)

20. Lamiable, A., Quessette, F., Vial, S., Barth, D., Denise, A.: An algorithmic game-theory approach for coarse-grain prediction of RNA 3D structure. IEEE/ACM Trans. Comput. Biol. Bioinf. (TCBB) **10**(1), 193–199 (2013)

21. Kamada, T., Kawai, S.: An algorithm for drawing general undirected graphs. Inf. Process. Lett. **31**(1), 7–15 (1989)

22. Boudard, M., Bernauer, J., Barth, D., Cohen, J., Denise, A.: GARN: sampling RNA 3D structure space with game theory and knowledge-based scoring strategies. PLoS ONE **10**(8), e0136444 (2015)

Proposition of a Parallel and Distributed Algorithm for the Dimensionality Reduction with Apache Spark

Abdelali Zbakh[1]([⊠])[iD], Zoubida Alaoui Mdaghri[1],
Mourad El Yadari[2], Abdelillah Benyoussef[1], and Abdellah El Kenz[1]

[1] Faculty of Sciences, University Mohammed V, Rabat, Morocco
zbakhabdou@gmail.com, zoubidaalaouimdaghri@gmail.com,
benyous.a@gmail.com, akenzele@yahoo.com
[2] University Moulay Ismail, Meknes, Morocco
mouradelyadari@gmail.com

Abstract. In recent years, the field of storage and data processing has known a radical evolution, because of the large mass of data generated every minute. As a result, traditional tools and algorithms have become incapable of following this exponential evolution and yielding results within a reasonable time. Among the solutions that can be adopted to solve this problem, is the use of distributed data storage and parallel processing. In our work we used the distributed platform Spark, and a massive data set called hyperspectral image. Indeed, a hyperspectral image processing, such as visualization and feature extraction, has to deal with the large dimensionality of the image. Several dimensions reduction techniques exist in the literature. In this paper, we proposed a distributed and parallel version of Principal Component Analysis (PCA).

Keywords: Distributed PCA · BIG DATA · Spark platform · Map-Reduce
Dimension reduction · Hyperspectral data

1 Introduction

The data collected today by the sensors increases rapidly and especially the hyperspectral data, which allow to give more physical information on the observed area.

The hyperspectral image is an image that represents the same scene following the hundreds of contiguous spectral bands in various wavelength ranges. The data of a hyperspectral image are organized in the form of a cube of three dimensions: Two dimensions denoted x and y represent the spatial dimensions and a spectral dimension denoted z (see Fig. 1) [1].

It will be noted that there are multispectral images composed of a dozen bands, while the hyperspectral image exceeds a hundred bands, which implies a significant requirement in terms of data processing and storage.

Unlike the classic color image, the hyperspectral image gives more physical information about each observed object of the scene. Thus the technique of

© Springer International Publishing AG, part of Springer Nature 2018
M. Ben Ahmed and A. A. Boudhir (Eds.): SCAMS 2017, LNNS 37, pp. 490–501, 2018.
https://doi.org/10.1007/978-3-319-74500-8_46

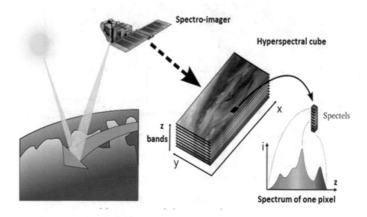

Fig. 1. Acquisition and decomposition of hyperspectral image

hyperspectral imaging is used in several fields, for example: geology, agriculture, town planning, forestry, in the military field.

To prepare hyperspectral image for visualization or further analysis such as classification, it is necessary to reduce the dimensions of the image to dimensions that can be analyzed by humans. Several dimensions reduction techniques exist. We find iterative versions and also parallel ones [2].

In this paper, we will propose a distributed and parallel version of the PCA dimension reduction algorithm that will be tested on the Spark platform using the MapReduce paradigm.

The rest of this paper is organized as follows: In Sect. 2, we will make an overview of the distributed parallel platforms most known in the field of BIG DATA processing. Thereafter, in Sect. 3, we will see the classic PCA dimension reduction technique and our proposed parallel distributed PCA. The tests of the proposed algorithm are in Sect. 4. Finally, we finish this paper with a conclusion and the future work.

2 Parallel and Distributed Platforms

In order to deal with BIG DATA such as our case hyperspectral images, we will use parallelized and distributed calculations in order to obtain results in a reasonable time.

Platforms that perform parallel distributed processing are multiplying in recent years. The two most recognized tools are Apache Hadoop and Apache Spark.

2.1 Apache Hadoop

Hadoop is among the most widely used, distributed platforms in the BIG DATA domain for storing data with his file manager named HDFS, and processing data with MapReduce on thousands of nodes [3]. (see Fig. 2).

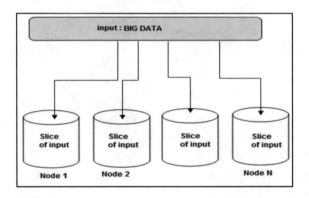

Fig. 2. HDFS abstraction of physical storage

2.2 Apache Spark

Apache Spark is an open source distributed platform for faster and sophisticated processing of BIG DATA, developed by AMPLab, from Berkeley University in 2009 and became an open source Apache project in 2010 [4].

Compared to other BIG DATA platforms, Apache Spark has many advantages:

- At storage levels: Spark allows to store the data in memory in the form of Resilient Distributed Data Set (RDD)
- At processing level: Spark extend Hadoop's Map-Reduce programming that works on disk to process RDDs in memory, allowing it to run programs in memory, up to 100 times faster than Hadoop MapReduce and in disk 10 Times faster
- In addition to the operations that exist in Hadoop (MapReduce), Apache Spark offers the possibility to work with SQL queries, streaming, graph processing, Learning machine… (see Fig. 3)

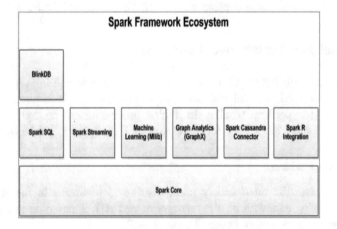

Fig. 3. Spark framework libraries

3   Dimensionality Reduction

Now to understand the information hidden in the hyperspectral image cube from human, or extract a useful part of the image, we often resort to visualization.

However, narrow human perception cannot visualize more than 3 hyperspectral bands. So before starting the visualization of our hyperspectral image, we must start by reducing the spectral bands to 3 without losing the quality of the information.

In the last few years, several techniques of dimensionality reduction have been made to reduce the hyperspectral data to a space of lower dimension, important examples include: ISOMAP, LLE, Laplacian eigenmap embedding, Hessian eigenmap embedding, conformal maps, diffusion maps and Principal Components Analysis (PCA) [5].

In the following, we will use PCA, the most popular technique in several domains: reduction of dimensionality, image processing, visualization of data and discovery of hidden models in the data [6].

3.1   Classic PCA Algorithm

Principal Component Analysis is a technique of reducing dimensions of a matrix of quantitative data. This method allows the dominant profiles to be extracted from the matrix [7]. The description of the classical PCA algorithm is as follows:

We assume that our hyperspectral image is a matrix of size (m = LxC, N) where L is the number of lines in the image, C is the number of columns and N is the number of bands with m ≫ N.

$$X \begin{bmatrix} X11 & \cdots & X1N \\ \vdots & \ddots & \vdots \\ Xm1 & \cdots & XmN \end{bmatrix}$$

Each line of the matrix X represents the pixel vector. For example the first pixel is represented by the vector: [X11, X12,..., X1N], with X1j is the value of the pixel 1 taken by the spectrum of number j.

Each column of the matrix X represents the values of all the pixels of the image taken by a spectrum. For example Xi1 = [X11, X21,..., Xm1] represents the data of the image taken by the spectrum 1.

To apply the PCA algorithm to the hyperspectral image X, the following steps are followed:

- Step 1: Calculate the reduced centered matrix of X denoted: XRC

$$XRC_{ij} = \frac{Xij - \overline{\overline{X_J}}}{\sigma j} \quad \text{for each } i = 1...m \text{ and for each } j = 1...N \quad (1)$$

$$\text{With } \overline{X_J} = \frac{1}{m} \sum_{i=1}^{m} Xij \quad \text{And} \quad \sigma j^2 = \frac{1}{m} \sum_{i=1}^{m} \left(Xij - \overline{X_J}\right)^2$$

In the formula 1, $\overline{X_J}$ denoted the average of column j and σj denoted the Standard deviation of column j.
- Step 2: Calculate the correlation matrix of size (N, N) denoted: Xcorr.

$$\text{Xcorr} = \frac{1}{m}\left(\text{XRC}^T.\text{XRC}\right) \tag{2}$$

In the formula 2, $\text{XRC}^T.\text{XRC}$ denoted the matrix product between the transpose of the matrix XRC and the matrix XRC
- Step 3: Calculate the eigenvalues and eigenvector of the Xcorr matrix denoted: $[\lambda, V]$
- Step 4: Sort the eigenvector in descending order of the eigenvalues and take the first k columns of V $(k < N)$
- Step 5: Project the matrix X on the vector V: U = X. V
- Step 6: use the new matrix U of size (m, k) for the visualization of the hyperspectral image

3.2   Distributed and Parallel PCA Algorithm

Related works:

There are currently two popular libraries that provide a parallel distributed implementation for the PCA algorithm: MLlib [8] on spark and the Mahout based on MapReduce [9]. In the Mllib library of Spark, we find an implementation for the parallel distributed PCA, but this implementation is done with the two languages: Sclala and Java. No implementation is made for the Python language.

In [6], Tarek et al. have shown that these two libraries do not allow a perfect analysis of a large mass of data and proposed a new PCA implementation, called sPCA. This proposed algorithm has a better scaling and accuracy than these competitors.

In [2], Zebin et al. proposed a new distributed parallel implementation for the PCA algorithm. The implementation is done using the Spark platform and the results obtained are compared with a serial implementation on Matlab and a parallel implementation on Hadoop. The comparison shows the efficiency of the proposed implementation in terms of precision and computation time.

In the following, we will propose a new implementaion for the parallel distributed PCA algorithm based on the Apache Spark platform using the Python programming language and which uses the distributed Mllib matrices.

Proposed implementation:

Since the hyperspectral image is a representation of the same scene with several spectral bands, we can decompose the hyperspectral image into several images, each image for a given spectrum (see Fig. 4).

The classic PCA algorithm requires intensive computation because of large hyperspectral image. We will present in this part a method of parallel distributed implementation of the algorithm using the Spark platform.

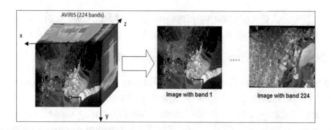

Fig. 4. Representation of the hyperspectral image by several images

First, we began by transforming the X matrix (see Fig. 5a) used to represent the hyperspectral image in the classic PCA to a vector of images denoted M, where each column of X is represented by an image in M (see Fig. 5b).

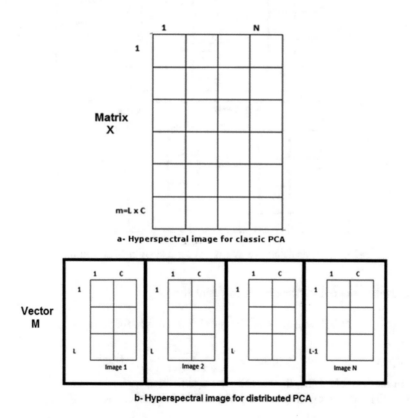

a- Hyperspectral image for classic PCA

b- Hyperspectral image for distributed PCA

Fig. 5. Representation of hyperspectral image for classic PCA and distributed PCA

Now, each image t of M denoted M_t is a matrix in our implementation (Represents an RDD in parallel distributed spark programming):

To make a parallel distributed implementation of PCA we will use the map reduce paradigm of spark. The proposed algorithm proceeds as follows

- Step 1: Calculate the reduced centered matrix of M:

As has been seen before, the matrix M contains several images and each image M_t is represented by a matrix (an RDD in Spark notation) of size (L, C) where L is the number of lines in image and C is the number of columns. Therefore, to calculate the reduced centered matrix of M denoted MCR, a parallel distributed computation is carried out of each image M_t (See graphical description of the algorithm in Fig. 6).

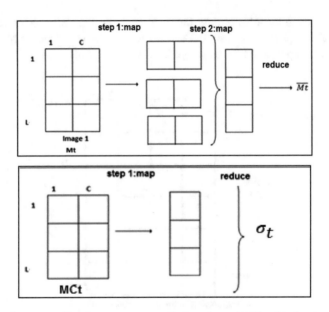

Fig. 6. Calculating the average of Mt and σt of MCt with Spark

- Calculate the reduced matrix of M denoted MC:

$$MC_t ij = M_t ij - \overline{M_t} \text{ for each } i = 1 \ldots L \text{ and for each } j = 1 \ldots C \qquad (3)$$

$$\text{with } \overline{M_t} = \sum_{i=1}^{L} \sum_{j=1}^{C} \left(\frac{1}{LxC} x M_t ij \right)$$

In the formula 3, $\overline{M_t}$ denoted the average of image M_t

- Calculate the reduced centered matrix of M denoted MCR:

$$MCR_t ij = \frac{MCtij}{\sigma_t} \text{ for each } i = 1 \ldots L \text{ and for each } j = 1 \ldots C \qquad (4)$$

$$\text{with } \sigma_t^2 = \frac{1}{LxC} \sum_{i=1}^{L} \sum_{j=1}^{C} (M_t ij - \overline{M_t})^2$$

$$\text{or } \sigma_t^2 = \frac{1}{LxC} \sum_{i=1}^{L} \sum_{j=1}^{C} (MC_t ij)^2$$

In the formula 4, σ_t denoted the standard deviation of image t

- Step 2: Calculate the MCR correlation matrix of size (N, N) denoted: Mcorr

According to step 1, the MCR is an images vector of size N. Each image represents a reduced centered matrix.

The next step is to calculate the correlation matrix of size (N, N), by making the matrix product, between the vector MCR^T and the vector MCR, using a distributed parallel computation MapReduce of Spark (See graphical description of the algorithm in Fig. 7).

$$\text{Mcorr} = \frac{1}{LxC} \left(MCR^T.MCR \right) \tag{5}$$

$$\text{Mcorr}_{t,k} = \frac{1}{LxC} \left(MCR_t.MCR_k \right) \tag{6}$$

$$\text{for each } t = 1 \ldots N \text{ and for each } k = 1 \ldots N$$

$$\text{with } MCR_t.MCR_k = \sum_{i=1}^{L} \sum_{j=1}^{C} MCR_t ij.MCR_k ij$$

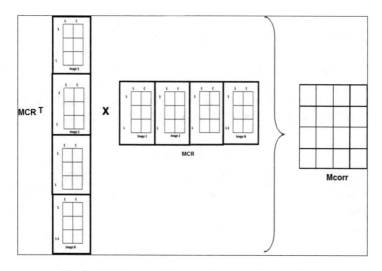

Fig. 7. Calculation of the correlation matrix with Spark

To calculate the value of each Mcorr$_{t,k}$ (Formula 6), the image MCR$_t$ is multiplied by the image MCR$_k$ pixel by pixel. Then we calculate the mean of result (See graphical description of the algorithm in Fig. 8).

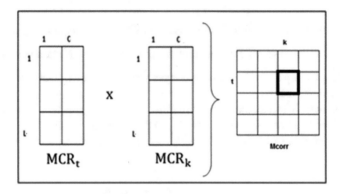

Fig. 8. Multiplication of two images

- Step 3: Calculate the eigenvalues and eigenvector of the Mcorr matrix: $[\lambda, V]$
- Step 4: Sort the eigenvector in descending order of the eigenvalues and take the first k columns of V $(k < N)$
- Step 5: Project the matrix X on the vector V: $U = X.V$
- Step 6: use the new matrix U of size (m, k) for the visualization of the hyperspectral image

4   Experimental and Computational Details

To test the validity of the proposed algorithm on hyperspectral images using the Apache Spark platform, we chose a set of open hyperspectral images of different sizes (see Table 1) and we tested our serial PCA algorithm, serial PCA of the Sklearn library of Python and parallel distributed PCA proposed on these datasets.

Table 1. Datasets

	Name	Spatial dimensions	Hyperspectral bands	Size
Dataset1	Moffett Field	500 × 500	3	5.3 MB
Dataset2	Moffett Field	500 × 500	10	17.5 MB
Dataset3	Moffett Field	1924 × 753	224	2.3 GB

The hyperspectral image used in our experiments for the classic PCA algorithm or the proposed PCA distributed algorithm is the Airborne Visible Infra-Red Imaging Spectrometer (AVIRIS) Moffett Field image with 224 spectral bands in the 2.5 nm to 400 nm, which was acquired on August 20, 1992 [10].

The three algorithms are implemented in Python 3 and are executed on several configurations (see Table 2) and the results of the experiments of PCA are given in Table 3.

Table 2. Configuration parameters

Classic PCA	Distributed and parallel PCA
CPU:Intel Core I5, 3.3 GHZ	Cluster Spark:
RAM: 4G	Master Node: 1
OS:Ubuntu 16.04 LTS	Slave Nodes: 4
	CPU of each node:Intel Core I5, 3.3 GHZ
	RAM of each node : 4G
	Network speed : 100 MB/s
	OS:Ubuntu 16.04 LTS

Table 3. The three most significant eigenvalues of PCA

	Our serial PCA	Sklearn serial PCA	Proposed distributed PCA
Dataset1	1.9371026343	1.9371026343	1.9371026343
	0.913755533084	0.913755533084	0.913755533084
	0.149141832615	0.149141832615	0.149141832615
Dataset2	8.12709201824	8.12709201825	8.12709201825
	0.880534232525	0.880534232525	0.880534232525
	0.797956006441	0.797956006442	0.797956006442
Dataset3	160.638762642	160.638762642	160.638762642
	28.0031335174	28.0031335174	28.0031335174
	14.5639033111	14.5639033111	14.5639033111

The visualization of the hyperspectral image after the application of the classic PCA algorithm or the proposed PCA distributed algorithm is given in Fig. 9.

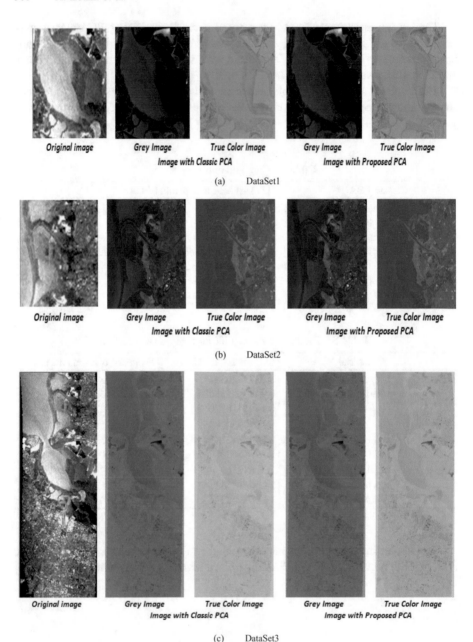

(a)   DataSet1

(b)   DataSet2

(c)   DataSet3

Fig. 9. Dataset Visualization,before and after application of classic PCA (our srial PCA or Sklearn serial PCA) and the proposed Distributed PCA

5   Conclusions

In this work, we proposed a parallel and distributed algorithm for dimensionality reduction called PCA. The algorithm is developed in Python 3 and tested on hyperspectral images using the Spark platform. The results coincide with the results of classic PCA and the visualization of the images after the application of our reduction algorithm confirms the validity of our algorithm.

In the next work we try to validate our algorithm based on the execution time of each method.

References

1. Mercier, L.: Système d'analyse et de visualisation d'images hyperspectrales appliqué aux sciences planétaires (2011)
2. Zebin, W., et al.: Parallel and distributed dimensionality reduction of hyperspectral data on cloud computing architectures. IEEE J. Sel. Top. Appl. Earth Observations Remote Sens. **9** (6), 2270–2278 (2016)
3. Apache Software Foundation. Official apache hadoop. http://hadoop.apache.org/. Accessed 10 July 2017
4. Apache Spark - Lightning-Fast Cluster Computing. http://spark.apache.org/. Accessed 10 July 2017
5. Van Der Maaten, L., Postma, E., Van den Herik, J.: Dimensionality reduction: a comparative. J. Mach. Learn. Res. **10**, 66–71 (2009)
6. Elgamal, T., et al.: sPCA: scalable principal component analysis for big data on distributed platforms. In: Proceedings of the 2015 ACM SIGMOD International Conference on Management of Data. ACM (2015)
7. Shlens, J.: A tutorial on principal component analysis. arXiv preprint arXiv:1404.1100 (2014)
8. MLlib machine learning library. https://spark.apache.org/mllib/. Accessed 10 July 2017
9. Mahout machine learning library. http://mahout.apache.org/. 10 July 2017
10. AVIRIS - Airborne Visible/Infrared Imaging Spectrometer - Data. http://aviris.jpl.nasa.gov/data/image_cube.html. 10 July 2017

Toward an Intelligent Traffic Management Based on Big Data for Smart City

Yassine Karouani[✉] ![ORCID] and Ziyati Elhoussaine[✉]

RITM Laboratory, Computer Science and Networks Team,
ENSEM-ESTC-UH2C, Casablanca, Morocco
ykarouani@gmail.com, ziyati@gmail.com

Abstract. It is anticipated that the Smart City research initiative will create new breakthroughs to revolutionize transportation system operations, infrastructure design, construction and management, as Big Data progresses. This latter will focus on the modeling, analysis and optimization of data-intensive intelligent transport systems, which will allow for more efficient system-wide operations. The focus is on the use of non-traditional data generated by smart city initiatives and emerging mobile applications, including data from social media, smart phones and more generally all connected objects. Research on this subject allows us to have a global view on the studies carried out in this field not on the infrastructure side but control and management of road traffic, based on the main objectives according to the users of the road. These objectives are the elaboration of a shortest path between a source and a destination, as well as the time required to traverse this path. We study different existing solutions such as solution employed by: Google, Japan (VICS, PCS) trying to find the advantage, the weak points and the common points to better bring out a new model which gathers the maximum advantages of these methods.

Keywords: Smart city · Big Data · Mongo DB · Traffic road · Traffic congestion
Vehicle routing

1 Introduction

Recently the entire world and in particular Morocco, everyone is looking for solution to improve traffic management and congestion. People for the sake of convenience use the automobile, and road traffic demand is on the increase. Traffic jam and an increase in traffic accidents have been the result, and the original objective of automobile transportation reaching one's destination with safety, comfort, confidence and speed has become more and more difficult to achieve. This problem is widely recognized. Nevertheless, traffic control is becoming more sophisticated thanks to new roads, road improvement, traffic signals and traffic control systems, and traffic safety education as administrative measure. The fact remains, however, that traffic continues to System increase, making road traffic more difficult and unpleasant. The objective of this document is to propose a solution and to highlight methods with the aim of improving the management of road traffic by respecting requirements for optimization and efficiency

© Springer International Publishing AG, part of Springer Nature 2018
M. Ben Ahmed and A. A. Boudhir (Eds.): SCAMS 2017, LNNS 37, pp. 502–514, 2018.
https://doi.org/10.1007/978-3-319-74500-8_47

in the functioning of social systems, including systems more sophisticated transportation, CO_2 emissions, and modernization of aging infrastructure. Today we have a lot of tools that can help us to collect information on urban activities such as people and traffic flows.

For example, smartphones, smart cards, and cameras, without forgetting the Internet. While information on these flows has been provided in the past only in the form of general public statistics. Currently the availability of these data extends to the sectors that use it; the amount of data received and collected is growing exponentially and the leaps in storage and computational power within the last decade underpin the unique selling proposition of Big Data of being able to provide better insights into various business processes or everyday life in a way that was not possible in the past.

The role of Big Data here is not just for massive data storage, but the most important is to analyze large datasets and discover the relationships between structured and unstructured datasets.

This article touches on two themes:

- Foundations and principles, discussions on some solutions for the management of existing road traffic.
- Applications, providing a variety of use cases of data analysis in the field of traffic management based on "Outsourcing", and GPS for data collection and large data processing side, analysis.

2   Related Works

In paper [1], authors implement a distributed system for solving the problem of garbage collection. The proposed solution tries to create a decentralized system for transportation and collection of garbage. An improved Ant Colony Optimization was implemented to create the best shortest routes that visit all collections points such as each collection point is visited once.

Authors in paper [2] review the applications of big data to support smart cities. It discusses and compares different definitions of the smart city and big data and explores the opportunities, challenges and benefits of incorporating big data applications for smart cities. In addition it attempts to identify the requirements that support the implementation of big data applications for smart city services. The review reveals that several opportunities are available for utilizing big data in smart cities; however, there are still many issues and challenges to be addressed to achieve better utilization of this technology.

Work [21] compare three evaluation methods that have been applied in their field tests in the past few years. The shortcoming and advantages of each method are presented. An improvement of Single Vehicle Test method is proposed to increase the reliability of reference value without jeopardizing the advantages of Single Vehicle Test. The comparison of old and new Single Vehicle Test method in their field tests shows the improvement can effectively increase the reliability of reference value, therefore increase the reliability of evaluation result.

A Smart and intelligent transportation is proposed in [22] using the graph technology, which analyzes the real time transport data using the parallel environment of Hadoop ecosystem. Based on the analysis, the system performs real time decisions related to any transport problem. The proposed system consists of various layers starts from capturing layer that collects the vehicular network data from the external vehicular network or road sensors. The whole system is implemented using apache Spark on the top of the Hadoop ecosystem in order to perform real time analysis and decision-making. For graph processing, Giraph, with the capability of parallel processing, is used, which dramatically increase the performance of the system.

3   Advanced Models and Architectures

The safety and comfort of motorists is a priority. That is why all stakeholders (state, local authorities, engineering offices, etc.) are mobilizing to optimize the use of infrastructures, that's why exist different solution and proposals such us:

3.1   Google Traffic

Early versions of Google Maps provided information to users about how long it would take to travel a particular road in heavy traffic conditions. Traffic information was based on historical traffic data and was not particularly accurate.

In this time Google offering traffic data based on information gathered anonymously from cellular phone users.

Google Traffic works by analyzing the data coming from Crowdsourced traffic data that's include a raw data of GPS determined locations transmitted by a large number of mobile phone users. By calculating the speed of users along a length of road, Google is able to generate a live traffic map (Fig. 1).

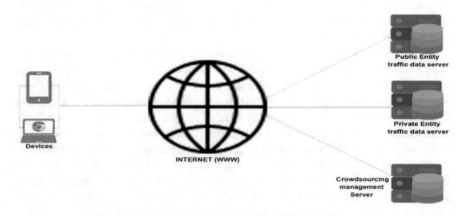

Fig. 1.   Collect traffic information for Google.

Google stated: "When we combine your speed with the speed of other phones on the road, across thousands of phones moving around a city at any given time, we can get a pretty good picture of live traffic conditions" [23].

Current and predicted traffic information is provided from incident data, traffic flow data, and media related to traffic received from multiple sources. The Crowdsourced data may be provided passively by applications on remote mobile devices or actively by users operating the remote mobile devices. An application on a mobile device may receive the multiple data types, aggregate and validate the data, and provides traffic information for a user. The traffic information may relate to the current position and route of the user or a future route. The present technology may also provide driving efficiency information such as fuel consumption data, carbon footprint data, and a driving rating for a user associated with a vehicle.

Figure 1 is a block diagram of an exemplary system for analyzing traffic data. System includes devices, network (internet), public entity traffic data server, private entity traffic data server, and crowd sourcing management server. Mobile device may communicate with network and be operated by a user in a moving vehicle. Mobile device may include circuitry, logic, software, and other components for determining a position, speed, and acceleration of the mobile device. In some embodiments, mobile device may include a global positioning system (GPS) mechanism for determining the location of the device. Based on the location data, the speed and acceleration of the device may be determined.

Network may communicate with devices, entity public and private data servers and crowdsourcing management server. Network may include a private network, public network, local area network, wide area network, the Internet, an intranet, and a combination of these networks.

Public entity traffic data server may include one or more servers, including one or more network servers, web servers, applications servers and database servers that provide traffic data by a public sector organization. The public sector organization may be, for example, the Department of Transportation or some other public entity. Public entity traffic data server may provide traffic data such as incident data for a planned or unplanned traffic incident, traffic speed and flow data, or traffic camera image and video data. A planned traffic incident may include data for a highway closure or construction work. An unplanned traffic incident may include data for a disabled vehicle or a car crash. The traffic speed and flow data may be determined from radar outposts, tollbooth data collection, or other data. Public sector traffic cams located on roadways may collect the traffic camera image and video.

Private entity traffic data server may include one or more servers, including one or more network servers, web servers, applications servers and database servers that provide traffic data such as incident data, traffic speed and flow data and traffic camera and image data. Examples of private entity traffic data servers are those provided by companies such as Inrix, Traffic Cast, Clear Channel, and Traffic.com.

Crowdsourcing management server may include one or more servers, including one or more network servers, web servers, applications servers and database servers, that receive crowd source data from devices, aggregate and organize the data, and provide traffic data to a device application on any of mobile devices. Data is received from a plurality of remote mobile device applications regarding current traffic information. The traffic information is aggregated to create a unified set of data and broadcast to mobile devices to which the data is relevant.

3.2   VICS: Vehicle Information and Communication System

The VICS [3–7] is a telecommunication system that transmits information such as congestion and regulation of traffic by detecting car movements with sensor installed on the road, this information is edited and processed by Vehicle Information and Communication System Center, and shown on the navigation screen by text or graphical form such as car-navigation system installed in individual cars (Fig. 2).

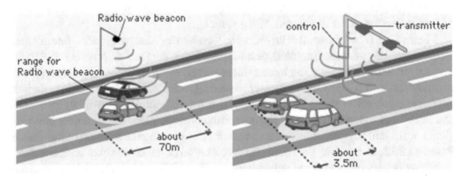

Fig. 2. Vehicle information and communication system.

First, a road traffic information collection system has already been built by the public sector. The information for the new information communications systems is already in existence and waiting for effective utilization.

Second, on board intelligence of vehicles is improved and popularized so that it can use the new information technologies. Already, approximately 400,000 vehicles have navigation systems installed, and this is a major factor in the successful building of a nation wide digital road map database.

Third, development of a mobile communications system, which provides information for such vehicle equipment, is near completion. Tests of the Advanced Mobile Traffic Information and Communication system (AMTICS), the Road and Automobile Communication System (RACS), and of FM multiplex broadcasting by the Telecommunications Technology Council indicate a promising future for these technologies.

3.3   PCS: Probe Car System

Acquisition of road traffic data is an important aspect of Road traffic Informatics systems. An innovative approach is utilizing the vehicles themselves as a source of real-time traffic data, functioning as roving traffic probes. This principle was probably first touched upon in the early 1970s in the course of a pilot with the Japanese Comprehensive Automobile Traffic Control System (CACS) [3–7], which was a government sponsored program aimed at the development of a route guidance and traffic information system. Besides inductive loops to transmit guidance information to CACS-equipped vehicles, also employment of the vehicles themselves to collect traffic data was propounded [8]. The traditional collection methods using roadside sensors are necessary but not sufficient because of their limited coverage and expensive costs of implementation and maintenance. To solve the limit of roadside sensors, the idea of collecting real time traffic data from in vehicle devices through mobile phones or GPS is quite popular known as probe car (floating car). Raw data of geolocalisation sent anonymously to a central processing center. After being collected and analyzed, useful information can be redistributed to the drivers on the road with real time (Fig. 3).

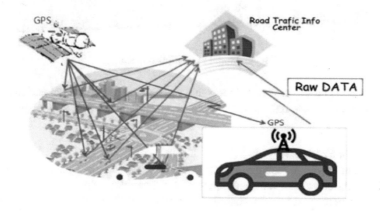

Fig. 3. Probe car system.

PCS [3–7] provides information on car movements and that on forecasting traffic congestion to individual drivers the same as VICS does. Different from VICS, this system collects traffic information from all cars, which are considered to be movable sensor units. Each car has a telecommunication unit and transmits several kinds of information such as position, velocity, and the status of the car to the central server.

4   An Advanced Model for Traffic Road Congestion in Smart City

Generally, the improvement of the roads is done by the installation of road equipment (signaling, radars, etc.) that can be expensive and long to set up [9].

Fortunately, the growth of intelligent transport systems (ITS) such as connected vehicles, driving assistants (GPS, radar warning devices, etc.) and smartphones has revolutionized traffic management.

4.1   System Model

The use of the Floating Car Data makes it possible to respond to the various issues of road safety and the reduction of congestion. In the Fig. 4 we explain our model for traffic congestion control in the smart city.

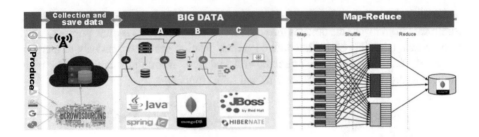

Fig. 4. Model for traffic congestion.

Our model is based on several layers that work in parallel to collect and manage all the data collected from outside and create several scenarios that will be used by all drivers.

Layer 1: Data Producers
This layer consists in producing the data that can be used in the system. It is the birth side of the ingredients of the final solution, for example: vehicles transmit the geo-location data generated by an installed GPS tag, pre-run data on the Web Services of Google, Bing, Tom-tom, Navcities…

Layer 2: Data Center
In this part it is a cloud where we store all the data received, and regenerated to a single format because they come in many different formats, that comes from sources installed and managed by us is encrypted and in hex format examples of resources used: Sensors on the roads, GPS devices on vehicles, applications on smartphones, and Data on the internet (Crowdsourcing) is come by the web services rest in general is in format JSON.

Layer 3: The Processing Center (Big DATA)
Big Data is the term that describes a set of non-structured and complex data, which come from our center data, it is used to store, manage, analyze and visualize information.

Big Data Analysis (BDA) is a way to innovate the marketing since it primarily generates a global revolution in the use of mass media and devices, given that treatment of data is going to create support processes which allow the integration of intelligent systems [10, 11], there upon BDA is implemented in order to deliver a better management of massive data in real time.

Nowadays, Big Data is enabling that great aspects of human life be studied as scientist and marketing area, given the volume of data generated daily and the analysis of its complexity [12]. Its challenge is in generated convert data in tools that allow to managers to solve problems acting rightly in making decision [13].

Our model depends the Big Data and these processing and analysis algorithms for improving the systems performance [14]. Some challenges and difficulties of Big Data's role in this model are shown below:

– **(A)** Data manipulation, storage availability, compatibility of systems, among others, attending to the autonomy and the adaptability of data for the development of applications [15]. Especially in a Smart Cites. Security and monitoring: To enhance user's reliability, sensible data should be protected and restricted to private access [16].

- **(B)** Technological development: It looks for reducing the heterogeneity between operative systems and to face the energetic barriers of devices [17].
- **(C)** Standardization: In order to avoid the wrong usage of critical information [18], the requirement is on the protection of sensible data and the use of it without authorization [16, 17] has block the manipulation of existent IoT's platforms. Privacy: People does not feel entrusted with the idea of sharing personal data with the world, they believe in technology but not in its managers [19, 20].

Layer 4: Map-Reduce

This layer for condensing large volumes of data into useful aggregated results. In this map-reduce operation, the map phase is applies to each input document (i.e. the documents in the collection that match the query condition). The map function emits key-value pairs. For those keys that have multiple values, and applies the reduce phase, which collects and condenses the aggregated data, then stores the results in a collection, the output of the reduce function may pass through a finalize function to further condense or process the results of the aggregation.

- **Map():** The Map worker parses key-value pairs out of the assigned split and passes each pair to the user-defined Map function. The intermediate key-value pairs produced by the Map function are buffered in memory. Periodically, the buffered pairs are written to local disk, partitioned into R regions for sharding purposes by the partitioning function that is given the key and the number of reducers R and returns the index of the desired reducer.
- **Shuffle:** the Map output to the Reduce processors: When ready, a reduce worker reads remotely the buffered data from the local disks of the map workers. When a reduce worker has read all intermediate data, it sorts the data by the intermediate keys so that all occurrences of the same key are grouped together. Typically many different keys map to the same reduce task.
- **Reduce():** The reduce worker iterates over the sorted intermediate data and for each unique intermediate key encountered, it passes the key and the corresponding set of intermediate values to the user's Reduce function.

Produce the final output: The final output is available in the R output files (one per reduce task).

4.2 Application

The block of code in Fig. 5 represent a function that helps us to display all the vehicles which find it at a distance given by parameter in the back office and the current position of the driver.

```
1  function predicte_state_off_all_viheculare_after_map_reduced() {
2      var results = [];
3      var all = db.device.find({}, {
4          id: 0,
5          Identifier: 1,
6          name: 1
7      });
8      var names = [];
9      var identifiers = [];
10     all.forEach(function(imei) {
11         identifiers.push("position_"+imei['identifier']);
12         names.push(imei['name']);
13     });
14     for (var i = 1; i < identifiers.length; i++) {
15         scol = db.getCollection(identifiers[i]);
16         scol.find().limit(1).sort({
17             date_gps: 1
18         }).forEach(function(data) {
19             results.push({
20                 name: names[i],
21                 position: data
22             });
23         });
24     }
25     return results;
26 };
```

Fig. 5. Magic function for state all vehicles in current time.

In input a request that's select all existence car in the system who can manipulate data represented by the variable "**all**", by looping on this last we obtain the list of the frames received from these vehicles (Fig. 6).

```
{
    "name" : "Moto1 11",
    "position" : {
        "_id" : ObjectId("589f45ec5d08412bf59b3d4f"),
        "_class" : "com.karouani.ensem.mongodb.models.Position",
        "rmuid" : "522613",
        "position_id" : 93617,
        "latitude" : 35.7614,
        "longitude" : -5.83877,
        "altitude" : 77,
        "speed" : 4,
        "direction" : 146,
        "raw_data" : "4D4347500075F907000004E8DE1F0400EB00200102000800F06B
F48E00B7EE01000000000000003992000000050F8164FF3A62B803151E00007F0000
00020A0E3A0B1204E00723",
        "date_creation" : "2016-04-18 12:58:01.0",
        "date_gps" : "2016-04-18 08:57:59.0",
        "date_cell" : "2016-04-18 11:58:01.0"
    }
}
```

Fig. 6. Result description as JSON format.

For check our solution we used this ingredient:

$$MacBook\ Pro\ I5 = CPU\ 2.4\ GHZ\ and\ 8G\ of\ RAM \tag{1}$$

$$Distance = 8\ km^2 \tag{2}$$

$$Number\ of\ Vehicle = 1200 \tag{3}$$

The result was given after 16 ms in the format Json in Fig. 6 which is composed of:

- **name:** The name represents the name of the vehicle tracker.
- **position:** the position object contains all the data required and received by the gps beacon installed in the vehicle, e.g. geolocation coordinates (latitude, longitude,

altitude), speed, date time to pick up and send the frame, reception frame And of course the direction where goes!

We make calculations to estimate the time consumed to generate the state of the traffic road in real time.

Suppose that Casablanca city contains 4 million vehicles [23] then mathematically speaking, if 1200 give the result in 16 ms then with 4 millions we get:

$$(4000000000 * 16)/1200 = 26666.67 \, \text{ms} \tag{4}$$

$$26666.67/(1000 * 60) = 0.889 \, \text{min} \tag{5}$$

In our solution for looking in the traffic real time you have just less than 1 min for got a result.

Figure 7 describes the front end in showing the result generated in maps using the Google services (Polylines, markers, and Heatmaps Layer), Jquery and Node JS.

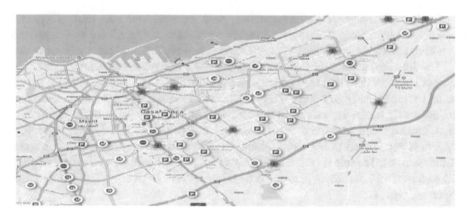

Fig. 7. Traffic result after map-reduce layer.

The colored lines representing traffic conditions on major highways refer to the speed at which one can travel on that road.

The dreaded red lines mean highway traffic is moving at less than 40 km/h and could indicate an accident or congestion on that route. Orange lines on the map mean traffic is moving faster, from 40 to 80 km/h, while green lines mean traffic is zipping along at 80 km/h or more. If you see gray lines, that means there's no traffic information available at the time and a red-black line refers to extremely slow or stopped traffic.

If you are looking at traffic on city streets, where the speed limits are much lower than on the highways, the colors take on more of a relative meaning. Red (or red-black) lines mean a lot of slow going and general congestion. Orange is a little better but still not the best for city travel, and green means traffic conditions are good.

Vehicle has 3 state: if speed is greater than 5 km/h and the ignition is on that's mine the vehicle is moving the sign is icon with the speed and a cap direction else ignition off and the speed is less than 5 km/h that's mine it's in parking with blue icon otherwise is stopped with the icon "STOP" color red example it's stop at a red traffic light. The info showing all others information related to the vehicle like date creating the data and Signal GPS…

5   Conclusion

Smart cities and urban planning leave a major impact on the development of the nations. It increases the decision power of the societies by making an intelligent and effective decision at the appropriate time. In this paper, we presented a solution for traffic management in a smart city; a four layer system is proposed to help drivers minimize travel time. We exploit the power of Big Data in the implementation of the system. In this model without the Big Data technology it was not possible to generate value in the information, since the platforms did not have speed for Processing Data, coverage to contain a large volume of information and the ability to categorize data according to their variety.

Taking advantage of the automobiles good freedom and free movability, and allowing each driver to freely select routes in a natural way, Our Model will contribute to traffic safety and smoothness. This will lead to more effective utilization of road assets.

Our solution combine the tree best solution in the world related to management traffic congestion is "VICS" this method collected information of all vehicles crossed a road using sensor installed, the PCS where we installed the device GPS into a vehicle to collect a data of the geolocalisation (Latitude, Longitude, speed, altitude, ignition… etc.), those method send the data to the central storage data (Layer 2) to save it into a system storage based in big data (Mongo DB) without missing that's part has some listner scheduled as role is collecting the data from the web using a web services REST (Crowdsourcing) in a major platform offer this solution like Google, Tom Tom, Waze, Open Signal… etc.

6   Perspectives

For the next works, we aim to enhance our model across the multiples stages mentioned above, and propose a short path to the front-end user by implementing some related algorithms, which take in consideration the complexity and near real-time processing, change map-reduce by apache spark, and java (Spring, hibernate…) by NODEJS

References

1. Elgarej, M., Mansouri, K., Youssfi, M.: An improved swarm optimization algorithm for vehicle path planning problem. In: 4th IEEE International Colloquium on Information Science and Technology (CiSt) (2016)
2. Al Nuaimi, E., Al Neyadi, H., Mohamed, N., Al-Jaroodi, J.: Applications of big data to smart cities. J. Internet Serv. Appl. **6**, 25 (2015)

3. Wang, S., Djahel, S., Zhang, Z., McManis, J.: Next road rerouting: a multiagent system for mitigating unexpected urban traffic congestion. IEEE Trans. Intell. Transp. Syst. **17**(10), 2888–2899 (2016)
4. Koyama, A., Inoue, D., Shoji, S.: An implementation of visualization system for vehicles and pedestrians. In: The 30th International Conference on Advanced Information Networking and Applications Workshops (WAINA) (2016)
5. Schmied, R., Moser, D., Waschl, H., del Re, L.: Scenario model predictive control for robust adaptive cruise control in multi-vehicle traffic situations. In: Intelligent Vehicles Symposium (IV). IEEE (2016)
6. Ma, D., Luo, X., Li, W., Jin, S., Guo, W., Wang, D.: Traffic demand estimation for lane groups at signal-controlled intersections using travel times from video-imaging detectors. IET Intell. Transp. Syst. **11**(4), 222–229 (2017)
7. El Hatri, C., Boumhidi, J.: Q-learning based intelligent multi-objective particle swarm optimization of light control for traffic urban congestion management. In: 4th IEEE International Colloquium on Information Science and Technology (CiSt) (2016)
8. Sánchez-Medina, J., Gálan-Moreno, M.J., Rubio-Royo, E.: Traffic signal optimization in "La Almozara" district in Saragossa under congestion conditions, using genetic algorithms, traffic microsimulation, and cluster computing. IEEE Trans. Intell. Transp. Syst. **11**(1), 132–141 (2010)
9. Ram, S., Wang, Y., Currim, F., Dong, F., Dantas, E., Sabóia, L.A.: SMARTBUS: a web application for smart urban mobility and transportation. In: 25th International Conference on World Wide Web Companion (2016)
10. Zhou, K., Fu, C., Yang, S.: Big Data driven smart energy management: from big data to big insights. Renew. Sustain. Energy Rev. **56**, 215–225 (2016)
11. Mohamed, N., Al-Jaroodi, J.: Real-time big data analytics: applications and challenges. In: 2014 International Conference on High Performance Computing and Simulation (HPCS), pp. 305–310 (2014)
12. Sharma, S.: Expanded cloud plumes hiding Big Data ecosystem. Future Gener. Comput. Syst. **59**, 63–92 (2016)
13. Xu, Z., Frankwick, G.L., Ramirez, E.: Effects of big data analytics and traditional marketing analytics on new product success: a knowledge fusion perspective. J. Bus. Res. **69**(5), 1562–1566 (2015). Glova, J., Sabol, T., Vajda, V.: Business models for the Internet of Things environment. Procedia Econ. Financ.
14. Kyriazis, D., Varvarigou, T.: Smart, autonomous and reliable Internet of Things. Procedia Comput. Sci. **21**, 442–448 (2013)
15. Henze, M., Hermerschmidt, L., Kerpen, D., Häußling, R., Rumpe, B., Wehrle, K.: A comprehensive approach to privacy in the cloud-based Internet of Things. Future Gener. Comput. Syst. **56**, 701–718 (2015)
16. Yan, Z., Zhang, P., Vasilakos, A.V.: A survey on trust management for Internet of Things. J. Netw. Comput. Appl. **42**, 120–134 (2014)
17. Botta, A., de Donato, W., Persico, V., Pescapé, A.: Integration of cloud computing and Internet of Things: a survey. Future Gener. Comput. Syst. **56**, 684–700 (2015)
18. Weber, R.H.: Internet of Things: privacy issues revisited. Comput. Law Secur. Rev. **31**(5), 618–627 (2015)
19. Lee, I., Lee, K.: The Internet of Things (IoT): applications, investments, and challenges for enterprises. Bus. Horiz. **58**(4), 431–440 (2015)
20. Liu, B., Li, L., Liu, K.: Study on the evaluation method of probe car system. In: IEEE Intelligent Vehicles Symposium (2010)

21. Rathore, M.M., Ahmad, A., Paul, A., Thikshaja, U.K.: Exploiting real-time big data to empower smart transportation using big graphs. In: 2016 IEEE Region 10 Symposium (TENSYMP), pp. 135–139 (2016)
22. Google Official Blog. https://googleblog.blogspot.com/2009/08/bright-side-of-sitting-in-traffic.html
23. La nouvelle Tribune du maroc. https://lnt.ma/parc-automobile-au-maroc-compte-3-437-948-unites-fin-2014/

Comparison of Fuzzy and Neural Networks Controller for MPPT of Photovoltaic Modules

Aouatif Ibnelouad[1]([✉]), Abdeljalil El Kari[1], Hassan Ayad[1],
and Mostafa Mjahed[2]

[1] Faculty of Sciences and Technologies, 112 Boulevard Abdelkrim Al Khattabi,
Marrakech, Morocco
a.ibnelouad@gmail.com
[2] Department of Mathematics and Systems, Royal School of Aeronautics,
Marrakech, Morocco

Abstract. The present paper proposes a comparison between two control methods for maximum power point tracking (MPPT) of a photovoltaic (PV) system under varying irradiation and temperature conditions: the fuzzy control method and the neural networks control method. The results of simulation obtained have been developed and analyzed by using Matlab/Simulink software for the both techniques have. The power transitions at varying irradiation and temperature conditions are observed and the power tracking time appreciated by the neural networks controller against the fuzzy logic controller has been evaluated.

Keywords: MPPT · Photovoltaic module · Neural networks controller
Fuzzy logic controller · Matlab/Simulink models

1 Introduction

Photovoltaic energy has nowadays an increased importance in electrical power applications, since it is considered as an essentially inexhaustible and broadly available energy resource [1]. In order to improve the efficiency of the photovoltaic generator (PV), in other words maximize the power delivered to the load connected to the terminals of the generator, several criteria for optimizing the efficiency of the photovoltaic system were applied [2, 3], and techniques were followed for good adaptation and high efficiency.

Among these techniques is the technique of Pursuit of the Power Point Maximal or "Maximum Power Point Tracker, MPPT" [4, 5], search technique optimal power points with intelligent methods Fuzzy logic and neural networks. The comparison between these techniques is set as a goal in the first steps, and then an improvement contribution is proposed.

The photovoltaic system consists of a photovoltaic panel with a power interface and a load. A simple DC/DC converter circuit (Boost) is used as interface between the PV panel and the charger.

© Springer International Publishing AG, part of Springer Nature 2018
M. Ben Ahmed and A. A. Boudhir (Eds.): SCAMS 2017, LNNS 37, pp. 515–527, 2018.
https://doi.org/10.1007/978-3-319-74500-8_48

The paper is structured as follows: Sect. 2 focuses on the model and characteristics of a PV module. Section 3 presents the proposed MPPT control strategies. Section 4 describes the detailed simulation results, followed by the conclusion in Sect. 5.

2 The Model and Characteristics of a PV Module

The PV solar module used in this study consists of polycrystalline silicon solar cells electrically. Its main electrical specifications are shown in Table 1.

Table 1. Technical data of the model manufacturer SUNPOWER SPR-305E (T = 25°, G = 1000 W/m^2)

Maximum power (W)	305
Open circuit voltage Voc (V)	64.2
Short-circuit current Isc (A)	5.96
Current at maximum power point Imp (A)	5.58
Voltage at maximum power point Vmp (V)	54.7

The mathematical models of the PV panel are defined below. Figure 1 shows the equivalent circuit of a solar panel. A solar panel is composed of several photovoltaic cells employing series or parallel or series–parallel external connections.

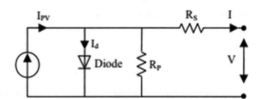

Fig. 1. Equivalent circuit of solar module

Equation (1) describes the I–V characteristic of a solar panel [6].

$$I = I_{pv} - Io\left[\exp\left(\frac{V + R_s I}{aV_t}\right) - 1\right] - \frac{V + R_s I}{R_P} \tag{1}$$

where, I_{pv} is the PV current; I_o is the saturated reverse current; "a" is a constant known as the diode ideality factor; R_s and R_p are the series and parallel equivalent resistances of the solar panel respectively; $V_t = \frac{N_s KT}{q}$ is the thermal voltage associated with the cells; N_s is the number of cells connected in series; q is the charge of the electron; K is the Boltzmann constant and T is the absolute temperature of the p–n junction.

I_{pv} has a linear relationship with light intensity and varies with temperature variations. I_o is dependent on temperature variations. Values of I_{pv} and Io are calculated from the following equations:

$$I_{pv} = (I_{pv,n} + K_1 \Delta T) \frac{G}{G_n} \qquad (2)$$

$$I_o = \frac{I_{sc,n} + k_1 \Delta T}{\exp(V_{OC,n} + K_v \Delta T)/a V_t - 1} \qquad (3)$$

In which $I_{pv,n}$, $V_{oc,n}$ and $I_{sc,n}$ are the PV current, open circuit voltage and short circuit current respectively under standard conditions (Tn = 25 °C and Gn = 1000 w/m²).

Kv is the ratio of the open circuit voltage to temperature; Ki is the coefficient of short-circuit current variation with temperature and $\Delta T = T - Tn$ is the deviation from standard temperature, G the light intensity [7].

For various values of the solar irradiance G, and cells' temperature TC, the I–V characteristics of the analyzed PV module are shown in Fig. 2.

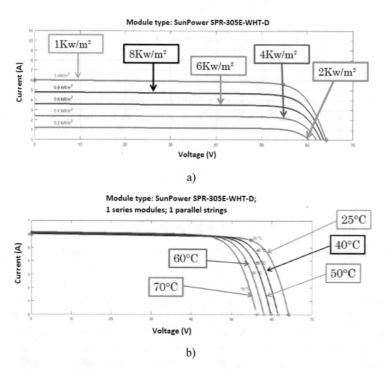

a)

b)

Fig. 2. The PV module PV characteristics for various values of G, and TC = 25 °C (a) and for G = 1000 W/m² and various values of T °C (b)

3 The MPPT Control Strategies

A MPPT controller is used for harvesting the maximum energy from the photovoltaic panel PV, and transporting that energy to the load, on condition that using an appropriate duty cycle to configure the DC/DC converter. A DC/DC converter ensures to transferring maximum energy from photovoltaic panel PV to load. A DC/DC converter is the interface that regulates the adaptation between the photovoltaic panel PV and the load to ensure our load closer to the MPP. In Fig. 3 is shown the block diagram of a PV module with MPPT controller.

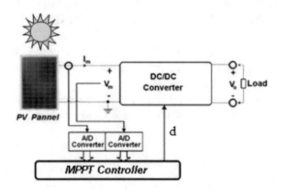

Fig. 3. The block diagram of a PV module with MPPT controller

The DC/DC converter typical application is to convert its input voltage V_{in} to a higher output voltage V_{out}, by varying the duty factor d, according to the equation as indicated below:

$$V_{out} = \frac{V_{in}}{(1 - d)} \tag{4}$$

3.1 The MPPT Controller with Fuzzy Logic

A Fuzzy Logic Control (FLC) is used to work as an MPPT controller that tracks the optimal operating point of a PV panel. Fuzzy Logic Control is one of the most commonly used technique in different engineering challenges of its multi-rule-based characteristics [8]. Fuzzy logic control has a simple and clear procedure because exact mathematical modeling and technical quantities of a system are not required for this controller [9]. The fuzzy controller consists of three blocks: the first block, fuzzification which numerical input variables are converted into linguistic variable based on a membership function. The second block is devoted to inference rules, while the last block is the defuzzification for returning to the real domain. This last operation uses the center of mass to determine the value of the output. Figure 4 shows the basic structure of the used MPPT Fuzzy controller [10]. For the MPPT controller with fuzzy logic, the inputs are taken as a change in power and voltage as well. There is a block for calculating the error (E) and the change of the error (CE) at sampling instants k [11]:

$$E(k) = \frac{P_{pv}(k) - P_{pv}(k-1)}{V_{pv}(k) - V_{pv}(k-1)} \qquad (5)$$

$$CE(k) = E(k) - E(k-1) \qquad (6)$$

Where, $P_{pv}(k)$ is the power delivered by PV module and $V_{pv}(k)$ is the terminal voltage of the module.

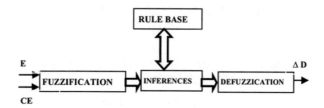

Fig. 4. Block diagram of the Fuzzy Logic controller

The Fuzzification

The resulting linguistic variables have been used for the MPPT fuzzy controller: PG (positive big), PP (Positive Small), ZE (Zero), NP (negative small), NG (large negative) for expressing the reel inputs and output variables. The Figs. 5 and 6 illustrate the membership functions of five fuzzy subassembly for the inputs variables and the output variable.

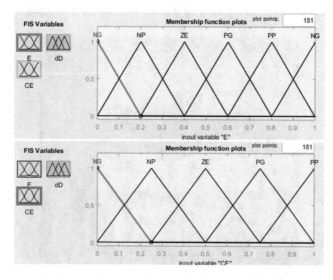

Fig. 5. Membership functions of the input variables E and CE

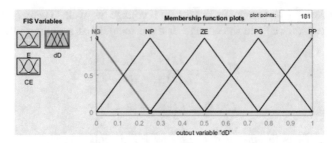

Fig. 6. Membership functions of output variables dD

Inference Rules

Table 2 shows the rules table of the fuzzy controller where all entries in the matrix are [11]:

Table 2. The fuzzy logic controller inference rule

E	dE				
	NG	NP	ZE	PG	PP
NG	PG	PG	PG	PG	PG
NP	PG	PP	PP	PP	ZE
ZE	PP	PP	ZE	NP	NP
PG	NP	NP	NP	NP	NP
PP	NG	NP	NP	NP	ZE
NG	PG	PG	PG	PG	PG

Defuzzication

The process of Defuzzification converts the inferred fuzzy control action into a numerical value at the output by making the combination of the outputs resulting from

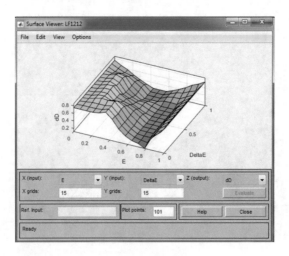

Fig. 7. The surface d = f (E, dE) of the MPPT controller output

each rule. In this paper the center of gravity defuzzifier, which is the most common one, is adopted. In the Fig. 7 is shown the surface output d = f (E, CE) of the MPPT controller.

3.2   The MPPT Controller with Neural Networks Controller

The new technique, which chooses the pursuit of the maximum power point, is the neural method. We will apply it to approximate the output, which is the voltage that corresponds to this power, as a function of illumination changes, and temperature, Is the tracking of the variation of the maximum power point. Where our system needs to evolve, quickly and efficiently.

Mathematical Modeling of the Biological Neuron
The mathematical model of an artificial neuron is illustrated in Fig. 8. A neuron consists essentially of an integrator that performs the weighted sum of its inputs. The result n of this sum is then transformed by a transfer function f, which produces the output D of the neuron. The R inputs of the neurons correspond to the vector $P = [p_1 p_2 \ldots \ldots p_R]^T$, while $W = [W_{1,1} W_{1,2} \ldots \ldots \ldots W_{1,R}]^T$, represents the vector of the weights of the neuron. The output n of the integrator is given by the following equation [12, 13]:

$$n = \sum_{j=1}^{R} w_{1,j} p_j - b$$
$$= w_{1,1} p_1 + w_{1,2} p_2 + \cdots + w_{1,R} p_R - b$$

(7)

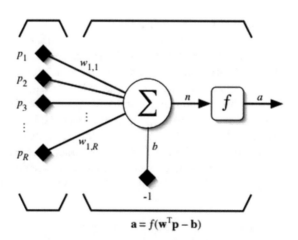

$$a = f(\mathbf{w}^T \mathbf{p} - \mathbf{b})$$

Fig. 8. The artificial neuron model

This can also be written in matrix form:

$$n = w^T p - b \tag{8}$$

$$a = f(n) = f\left(w^T p - b\right) \tag{9}$$

This output corresponds to a weighted sum of weights and inputs minus what is called the bias b of the neuron. The result n of the weighted sum is called the activation level of the neuron. The bias b is also called the activation threshold of the neuron. When the activation level reaches or exceeds the threshold b, then the argument of becomes positive (or zero). Otherwise, it is negative [12, 13].

There is an obvious analogy with biological neurons as shown in Table 3.

Table 3. Analogy between the biological neuron and the formal neuron

Biological neuron	Formal neuron
Synapes	Poids de connexions
Dendrites	Connexions d'autres neurones vers le neurone K
Axone	Connexions du neurone k vers d'autres neurones du réseau
Noyau	Fonction d'activation

Under Matlab/Simulink, the role of the neural network is to direct the controller to the region where the PPM is located. Thus, it is necessary to build the neural network, i.e. to prepare a learning base and to learn the network, and then implement this neural network in the control circuit.

Structure of an Artificial Neural Network

An array of artificial neurons is a set of neurons belonging to different layers connected together from the connections represented by weights. This network is responsible for receiving input signals and for providing an output signal as a function of these signals.

The structure of the connections between these different neurons determines the topology of the network. We distinguish several topologies: Feed forward-type multilayer neural network; recurrent network (looped network) and Cellular network.

The neural network used in our project is the feed forward type multilayer network.

Under MATLAB/SIMULINK, the Fig. 9 shows the structure of the neural network used in the control system. This network has an input layer containing two inputs, a hidden layer of ten neurons and an output layer containing a single neuron.

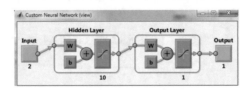

Fig. 9. Structure of the adopted neural networks

Learning

The purpose of learning is to estimate network parameters by minimizing an error function. Learning is supervised. The error function thus represents the distance that exists between the calculated response of the network and its desired response. The learning consists in applying to the network pairs of inputs and outputs (desired outputs), and then applying a learning algorithm to modify the various parameters of the network.

Under Matlab/Simulink, the prepared learning base is a Table 4, which indicates for each sunshine and temperature the voltage V_{mpp} corresponds to the maximum power point of the PS.

Table 4. Learning basis example

Nombre	Temperature (°C)	Irradiance (w/m²)	V_{max} (v)	I_{max} (A)	Puissance (Kw)
101	25	200	182.4	77.51	14.14
102	25	440	269.5	161.8	43.61
103	25	710	272.4	260.8	71.04
104	25	1000	273.7	366.8	100.4
105	26	340	266.7	125.1	33.36
106	26	760	271.9	279.1	75.89
107	26	900	272.1	330.8	90.02
108	26	1000	272.5	367.2	100.1
109	27	200	182.7	77.6	14.17
110	27	420	267.3	154.6	41.32
111	27	800	271	294.1	79.7
112	27	1000	271.8	367.2	99.79

At the end of the learning phase, we obtain the final neural network (Fig. 10) which give us a value very close to the exact value of the PPM. It admits as inputs the sunshine and temperature and as output, the voltage close to the PPM [14].

Fig. 10. Neural networks implemented under Simulink

4 Simulation Results

The simulation results acquired with Fuzzy and Neural Networks Controller, in checking the MPP of the analyzed PV module, for various values of solar irradiation G and cells' temperature T are given in Table 5. Therefore, this table confirm that the Neural Networks controller gives a quick response with stability around MPP than the Fuzzy logic controller.

Table 5. Simulation results of P_{max} checking comparison with fuzzy and neural networks controllers

G [w/m²]	T [°k]	T [°C]	P_{max} [w]	
			Neural networks controller	Fuzzy controller
1000	298	25	100,4	100
900	295	22	91,06	90,8
800	290	17	81,95	81,90
700	285	12	72,69	72,61
600	280	10	62,53	62,27

The Fig. 11 shows the evolution of the power produced by the PV module delivered MPP checking by the considered algorithms, under the solar irradiation $G = 1000$ W/m² and PV cells' temperature TC = 298 °K.

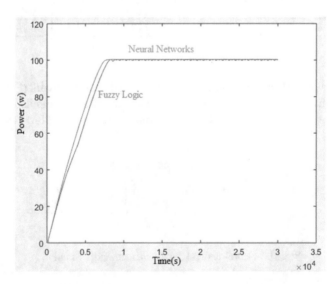

Fig. 11. The power evolution during MPP checking by the both algorithms at G = 1000 w/m² and T = 25 °C

The Fig. 12 shows the evolution of the power produced by the PV module delivered MPP checking by the considered algorithms, under the solar irradiation $G = 600$ W/m² and PV cells' temperature TC = 280 °K.

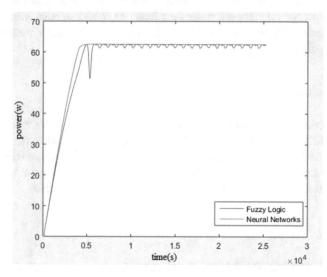

Fig. 12. The power evolution during MPP checking by the both algorithms at $G = 600$ w/m² and T = 10 °C

The Fig. 13 shows the evolution of the power produced by the PV module delivered MPP checking by the considered algorithms, for various values of solar irradiation G and cells' temperature T.

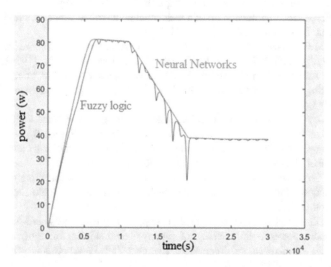

Fig. 13. The power evolution during MPP checking by the both fuzzy and neuron algorithms for various values of solar irradiation G and cells' temperature T

The simulations results illustrated in Figs. 11, 12 and 13, respectively for high conditions (G = 1000 w/m^2 and T = 25 °C), for low conditions (G = 600 w/m^2 and T = 10 °C) and for various values of solar irradiation G and cells' temperature T, confirm that the Neural Networks has good performance response such as rapidity, the time response of Neural Network faster than fuzzy logic and stability around the MPP, even if the shading phenomenon applied for PV modules that means the solar radiation decreases rapidly.

5   Conclusion

Two MPPT control strategies based on Fuzzy Logic and Neural Networks have been compared. It is found that the control of the DC/DC by the Neural Networks approach more well-founded than the other approach as fuzzy logic to extract the maximum power point, taking advantage of the adaptability and flexibility of the first, high efficiency of the second and fast response, power quality of the third.

References

1. Lalouni, S., Rekioua, D., Rekioua, T., Matagne, E.: Fuzzy logic control of stand-alone photovoltaic system with battery storage. Power Sources **193**, 899–907 (2009)
2. Glasner, I.: Advantage of boost vs. buck topology for maximum power point tracker in photovoltaic systems. TelAviv University, Faculty of Engineering, Department of Electrical Engineering, Israel, pp. 355–358. IEEE (1996)
3. Mujadi, E.: ANN based peak power tracking for PV supplied DC motors. Solar Energy **69** (4), 343–354 (2000)
4. Mujadi, E.: PV water pumping with a peck-power tracker using a simple six step square-wave inverter. IEEE Trans. Ind. Appl. **33**(3), 714–721 (1997)
5. Bose, B.K.: Microcomputer control of a residential photovoltaic power conditioning system. IEEE Trans. Ind. Appl. **IA-21**(5), 1182–1191 (1985)
6. Villalva, M.G., Gazoli, J.R., Filho, E.R.: Comprehensive approach to modeling and simulation of photovoltaic arrays. IEEE Trans. Power Electron. **24**(5), 1198–1208 (2009)
7. RezaReisi, A., Moradi, M.H., Jamasb, S.: Classification and comparison of maximum power point tracking techniques for photovoltaic system: a review. Renew. Sustain. Energy Rev. **19**, 433–443 (2013)
8. Ishaque, K., Abdullah, S.S., Ayob, S.M., Salam, Z.: Single input fuzzy logic controller for unmanned underwater vehicle. J. Intell. Robot. Syst. **59**, 87–100 (2010)
9. Chian-Song, C.: T-S fuzzy maximum power point tracking control of solar power generation systems. IEEE Trans. Energy Convers. **25**, 1123–1132 (2010)
10. Hatti, M.: Contrôleur flou pour la poursuite du point de puissance maximum d'un système photovoltaique, Huitieme Conference des Jeunes Chercheurs en Genie Electrique (JCGE 2008), Lyon (2008)
11. Aymen, J., Ons, Z., Craciunescu, A., et al.: Comparison of fuzzy and neuro-fuzzy controllers for maximum power point tracking of photovoltaic modules. In: Proceeding of IEEE International Conference on Renewable Energies and Power Quality (ICREPQ-2016) (2016). ISSN 2172-038

12. Tahar, R.: Application de l'intelligence artificielle au problème de la stabilité transitoire des réseaux électriques, Thèse magister (2005)
13. Parizeau, M.: Réseaux de neurones, Livre PDF (2004)
14. Harendi, A.: Modélisation et simulation d'un système photovoltaïque, UNIVERSITE KASDI MERBAH OUARGLA, Faculté des Sciences Appliquées, Département de Génie Electrique (2014)

A Ubiquitous Students Responses System for Connected Classrooms

Aimad Karkouch[1]([⊠]), Hajar Mousannif[2], and Hassan Al Moatassime[1]

[1] FSTG, Cadi Ayyad University, Marrakech, Morocco
aimad.karkouch@ced.uca.ac.ma
[2] FSSM, Cadi Ayyad University, Marrakech, Morocco

Abstract. Receiving feedbacks from students about their learning experience is a key part of any pedagogical approach. Students' feedbacks could be retrieved in a variety of ways using various Students Response Systems (SRS). A major drawback of existing SRS is their lack of seamless integration into learning environments. As such, they become a potential source of distraction for the learning process. We believe technology should blind seamlessly and provide support for pedagogy, thus, we propose a Ubiquitous Students Responses System (U-SRS) that is capable of continuously and seamlessly monitoring various students' learning performances features, making sense of them and providing insights for teachers, enabling them to adapt their pedagogical approach according to their students immediate needs. The proposed U-SRS takes advantages of machine learning and the Internet of Things paradigm to enable its services in connected classrooms. We present our solution's design and describe its architecture. We select a subset of relevant features, collected by connected objects, and used by machine learning algorithms to build learning performance predictive models. Finally, we highlight the advantages of U-SRS over exiting SRS solutions.

Keywords: Students Responses System · Students feedback
Internet of Things · Machine learning

1 Introduction

Receiving feedbacks from students about their learning experience is a key part of any pedagogical approach. In fact, these feedbacks carry information that allow measuring the effectiveness of the teaching process by comparing the desired outcome and the actual outcome (e.g. students' understanding levels) [1]. Moreover, knowing how well students comprehend the material they are taught, helps teachers and instructors effectively adapt their teaching behavior according to their students' specific needs as part of a Contingent Teaching approach [2]. For example, whenever students have difficulties understanding a concept or that they look confused, their teacher might proceed differently by giving more illustrative examples or by revisiting the ambiguous concepts from another perspective. This continuous shaping of the teaching process during a session supposes that teachers have the ability to successfully and accurately assess their students' comprehension state and their immediate needs (e.g. clarifying an

© Springer International Publishing AG, part of Springer Nature 2018
M. Ben Ahmed and A. A. Boudhir (Eds.): SCAMS 2017, LNNS 37, pp. 528–537, 2018.
https://doi.org/10.1007/978-3-319-74500-8_49

ambiguous concept). However, this is not a very common ability. Some teachers might have a wrong idea or even ignore their students' comprehension or knowledge about a subject they are attending [3, 4]. Many others found that some material they believed are obvious, often were not sufficiently understood by many students. A situation that went undetected because these students did not ask questions related to these areas [5]. While it is plausible that the ability to "see through" one's students' learning performances could be gained with experience [6], we cannot help but to wonder about unexperienced teachers and whether they are "blind" to their students inner states or not. Obviously, in one-on-one and small classes, it would be relatively easier for teachers to perceive students' states, and therefore adapt their pedagogy, because there is a close connection and more visibility. However, the bigger the class, the weaker the connection becomes and the teacher might lose touch with students that are not in the immediate vicinity which might eventually lead to the adoption of a one-fit-all pedagogical approach through lecturing [4].

Students' feedbacks could be retrieved in a variety of ways using various Students Response Systems (SRS, also known as Personal Response Systems). The most traditional and common form is by raising hands. While it is the most intuitive, it does suffer from many drawbacks. First, this feedback mechanism is inherently constrained to expressing one option (e.g. yes or no answers). When multiple options are available, many shows of hands are required which might cause losing time during sessions. Time can also be lost when counting and aggregating results especially in large classes with constrained visibility. Furthermore, this mechanism might not suit neither shy students [7] nor the ones concerned with the anonymity of their feedback [2]. We agree that raising hands gives an opportunity to some students to compare their answers, and therefore their understanding, with their pairs, however, we also believe that it may cause some others to feel less confident or be overwhelmed with the majority's answer. Thus, the student may let go of critical thinking and accept the most popular answer. An upgraded version of this mechanism consists of using colored cards. While it does offer a broader set of options to express in one single show, it still suffers from the other drawbacks we discussed (e.g. lack of anonymity, loss of time, etc.).

More advanced SRS rely on electronic components and various other platforms to convey students' feedback. Such SRS include clickers [3, 5], smartphones [8, 9] and social media [10, 11]. Clickers are handheld devices that look similar to a remote control with a single or multiple buttons which students can use to give feedback (usually answering a question posed during a teaching session). In attempt to overcome the costs of purchasing and maintaining clickers, smartphones were used, given their widespread, as SRS through clicker simulator mobile applications [8] or sending SMS [9]. Social media have grown in popularity especially among young people whom are usually attending a learning environment (e.g. school, college, etc.). They can be used as a far-reaching feedback conveying medium as they are widely and easily accessed (using smartphones for example), rich in content and most students are probably already familiar with.

These advanced SRS have a key advantage over the traditional raising hands approach: Giving the teacher instant insights extracted from the students' feedback. In the case of clickers, this is typically achieved by simply aggregating received answers and showing an overview of all feedbacks (e.g. percentage of each option). However, in the case of feedback conveyed in the form of SMS, Tweets or other textual forms such as Unit of Study Evaluations (USE, also known as Students Evaluations of Teaching or SET) [12], more complex data processing, such as Sentiment Analysis [10–13], is required to extract meaningful and actionable insights. Moreover, these advanced SRS are reported to effectively engage students [3, 5]. This could be justified by being suitable to a broader range of students including those seeking anonymity and those not able to overcome their shyness, however, it could also be just a novelty effect [2].

While having obvious advantages, these advanced SRS also suffer from various drawbacks. First, from a student perspective, these tools often require much more effort in order to provide a feedback (e.g. writing an SMS, Tweeting, filling a USE form, etc.). Some students even reject the idea of integrating SRS because of the efforts needed to engage with them [3]. Second, the communication mediums used, such as smartphones and social media, represent potential sources of distraction during teaching sessions (e.g. a ringing phone, notifications from social media, etc.). They can also impose more costs (e.g. SMS or Mobile Data – related carrier fees) and be limited as to the available space to express one's feedback (e.g. SMS and Tweets' limited number of allowed characters). Third, in order to take full advantage of this kind of SRS in place, teachers are required to allow enough time for feedback to be submitted after posing carefully designed and triggering questions capable of highlighting confusions in their students' understanding (e.g. "brain teasers" type questions) which makes the aforementioned SRS time consuming, usually resulting in less material covered [4]. Also, they are by design bound to a discrete measurement paradigm in a sense that, during an entire teaching session, they are only able to capture the students' understanding states on defined intervals or windows throughout the lesson, that is, only when students get the chance to provide a feedback following an instruction enquiry such as asking a question (e.g. "did you understand this concept?") to which the answer is given as a feedback.

The major drawback of existing technological-enabled SRS used in classrooms is that they do not blind in the learning environments. In fact, they can become intrusive to the learning process because of all the potential distractions resulting from engaging with them. To tackle this precise weakness, we propose the Ubiquitous Students Responses System (U-SRS), a feedback-conveying system that blinds seamlessly in the learning environment, that is effortless to engage with and which provides a continuous overview of students' states through monitoring, in a non-intrusive way, various facial, bodily and environmental features using connected smart devices. We share the same belief that such a seamless, continuous and effortless to operate solution is more effective in integrating and enhancing learning environment as teachers should keep focusing on the pedagogy rather than technology, and students on critical thinking rather than technological artifacts as suggested in [2].

The remainder of this article is as follows. Section 2 presents related works. In Sect. 3, we present the proposed U-SRS solution, its architecture and a subset of learning performances features it relies on. In Sect. 4, we compare our solution with existing SRS using various criteria. Finally, Sect. 5 concludes the article.

2   Related Work

Most existing technology-enabled SRS require a certain degree of effort both from teachers (e.g. preparing quality questions) and from students to provide textual feedback (e.g. writing an SMS or a Tweet) which are then processed using techniques, such as sentiment analysis of textual data, in order to provide a vision of students' understanding states [3, 8–11] which makes them potentially intrusive to the learning process. A better approach to monitoring learning performances which is both non-intrusive and continuous, i.e. providing feedback during the entire teaching session rather than on predefined intervals or windows, can be achieved through analyzing students' emotions by monitoring various facial and bodily features in addition to environmental parameters. In fact, some emotions (e.g. confusion and boredom) have been found to correlate with learning performances and outcomes in academic settings [6, 13–17]. Experiencing a broad set of emotions in learning environments have been justified with the importance of academic achievements in one's life and career which makes them an intense source of emotions [14]. Another explication states that affective states are almost always triggered by incoming new information. Thus, learning, as a process where new information are assimilated, almost always occurs within emotional episodes [17].

[16] investigated relevant emotions experienced by students while interacting with a virtual tutor in AutoTutor's (an Intelligent Tutoring System, ITS) tutorial sessions. The authors suggest that an emotion-aware learning environment could improve learning effectiveness and outcomes, so they conducted various studies and were able to identify a subset of frequently experienced emotions that could be tracked by monitoring facial, body posture and dialogue features in a non-intrusive fashion. The paper hinted at enhancing the existing AutoTutor system with pedagogy adjustment mechanisms based on the observed emotions, however, such capabilities were not shown to be implemented. In the same direction, [15] investigated how emotions experienced during AutoTutor sessions are correlated with learning outcomes, an aspect that was not investigated in [16]. A learning gain was computed for each participating student using their pre- and post-tests grades. Correlations between the most prominent detected emotions, reported by a human judge for each participant in a virtual tutoring session, and learning gains were then computed. The paper reports that learning gains are positively correlated with the confusion and flow emotions and negatively correlated with boredom. Also, due to a floor effect partially resulting from the low frequency of emotions reporting (30 s every 5 min), no significant correlation with the frustration and eureka emotions could be backed by the gathered data. The paper also investigated any potential correlation between the considered emotions and reported that only the pair boredom-flow registered a negative correlation. Finally, it was reported that confusion has a main effect on the learning, i.e. students that exhibited confusion had better gains than those who did not.

[6] proposes a theoretical model associating learning with affective states. Four quadrants are formed by the intersection of the emotion and learning axes; (i) constructive learning with positive emotions, (ii) constructive learning with negative emotions, (iii) Unlearning (e.g. discarding misconceptions) with negative emotions and

(iv) unlearning with positive emotions. Moving, typically in counter-clockwise direction as suggested by the authors and ideally starting from quadrant I or II, from one quadrant to another occurs as the learning process continues. The authors envision to build a Computerized Learning Companion with the ability to detect and act, i.e. adjust its pedagogical approach, upon learners' affective states, however, no details on its implementation were given.

All of these studies agree on the feasibility and relevance of emotions in learning environments. While we also adopt emotions as a medium of communicating students' feedbacks, we use a subset of relevant emotions. Moreover, environmental parameters have been reported to potentially have an impact on learning outcomes [16], however, as far as we are aware of, our solution is the first to actually incorporate environmental features as a measurable indicator of students' performances. More importantly, the three later aforementioned studies [6, 15, 16] envision building or enhancing existing virtual learning systems (e.g. AutoTutor) with capabilities to detect students emotions and act upon them by adjusting their pedagogical approach. We adopt a different approach as we are building a non-intrusive feedback system for actual classrooms. Further, in order to monitor students, we use smart devices which are capable of sensing their environment and communicating their readings, based on which, insights about students' learning performances are extracted and provided to teachers (instead of virtual tutors) in real-time.

3   *U-SRS*: A Ubiquitous Students Responses System for Connected Classrooms

Our U-SRS solution uses emotions, alongside other bodily, contextual and environmental parameters, as learning performance indicators in real classrooms instead of virtual learning environment. We empower teachers, instead of virtual tutors, with insights about their students' performances and comprehension states, so that they are able to proceed with adjustments in their pedagogy whenever needed.

3.1   System Overview

As depicted in Fig. 1, U-SRS uses smart devices embedded in ordinary classroom's tables which blind seamlessly in the learning environments allowing them to operate in a non-intrusive way avoiding any distractions to the learning process. These smart devices are equipped with vision, sensing and communication capabilities and they are trained to detect, at a much higher frequency than an ordinary human judge, each student's emotions, making sense of them and then reporting them to the teacher in real-time. The sensing capabilities are enabled by various modular sensors integrated with our core devices (i.e. connected objects) such as image, movement, temperature and luminosity sensors. The extracted insights are conveyed to teachers continuously and in real-time through various channels: Computers, smartphones and smart watches. These insights help teachers shape their pedagogic approach according to the perceived state of their students even those out of their direct line of sight.

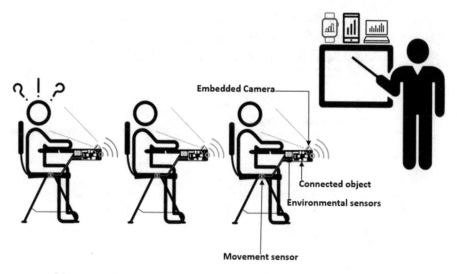

Fig. 1. An overview of the proposed ubiquitous Students Responses System

3.2 Pedagogic Uses

U-SRS's aim is to empower and enhance the adopted pedagogic approach by providing individual and classroom-wide insights. Among the pedagogic uses that could be enabled by this system, we state:

1. **Seamless learning feedback:** Teachers receive up-to-date data about their students' comprehension and learning performances, enabling them to timely address any confusions or ambiguous concepts.
2. **Seamless teaching feedback:** Teachers could improve their teaching methods and how they deliver knowledge by determining what went wrong in the session each time a drop in the perceived students' learning performances is reported.
3. **Classroom-wide awareness:** Instructors might decide to show the overall performances to students so that they could self-assess or compare to their peers' performances.

3.3 System Architecture

The various components of the U-SRS span three major layers as shown in Fig. 2:

1. In the **Physical Layer,** two major tasks are performed. First, the perception of the real-world phenomena (i.e. the learning process) which is carried by connected objects using sensor modules to continuously monitor students and other contextual parameters in the learning environment such as luminosity and temperature. These connected objects handle collecting, preprocessing and streaming preprocessed data to the server-side application. The second task is delivering learning performance insights to end-users (i.e. teachers) through various connected devices such as smart watches and smartphones. It is worth noting that the blue arrows in Fig. 2 refers to

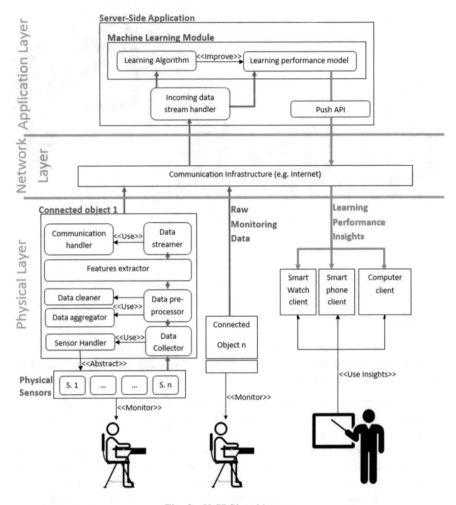

Fig. 2. U-SRS' architecture

the preprocessed performance data captured from the real world, while the green arrows refers to the learning performance insights provided by the U-SRS solution to teachers.

2. In the **Network Layer**, our solution use existing communication infrastructure such the Internet to convey both the collected data from the Physical Layer to the Application Layer and the performance insights in the opposite direction.

3. In the **Application Layer**, machine learning algorithms use the collected raw perceived data to train and build a predictive model for learning performances. Based on the features received from connected objects, insights about students' learning performances are extracted and sent back to teachers in order to adapt their pedagogy. The *Push API* module is used to deliver insights as soon as they become available.

3.4   Feature Engineering

There exist a plethora of parameters that could be collected in a learning environment, however, not all of them are significant for machine learning algorithms that aim to build a predictive model. In this section, we select a subset of these parameters that were showed in earlier studies to correlate with learning performances. The following Table 1 summarizes the chosen learning performances features subset.

Table 1. Learning performance indicators' chosen features

Category	Feature	Source	Supporting studies
Emotions	Confusion	Facial features (from embedded camera)	[6, 13–17]
	Flow		
	Eureka		
	Boredom		
	Frustration		
Body gestures	Movements/vibrations (i.e. fidgeting)	Movements sensor	[16]
Environment	Noise	Environmental sensors	[16]
	Luminosity		
	Temperature		
	Humidity		

4   *U-SRS* Comparison with Existing SRS

A cornerstone of our design of the U-SRS is its seamless integration in the learning environment. We believe technology should not become the focus of the learning process but rather support and empower the chosen pedagogy. Various advantages, over existing SRS solutions, stem from this key design principle, which we summarize in Table 2. The definition of the different used taxonomy criteria is as follow:

Table 2. U-SRS and existing SRS' comparison

	U-SRS	Raising hands	Clickers	Social media
Non-intrusive	√	√	×	×
Continuous	√	×	×	×
Feedback options	Many	Limited	Limited	Many
Real-time	√	×	√	√
Effortless	√	×	×	×
Anonymous	√	×	√	√
No-Triggers	√	×	×	×

1. **Non-Intrusive**: Qualifies a solution that seamlessly integrates the learning environment and is not a potential source of distraction (e.g. a ringing smartphone).
2. **Continuous**: Qualifies solutions capable of monitoring the learning performances during entire session with high frequencies of data collection rates and not just on discrete intervals (e.g. when a question is asked).
3. **Feedback options**: How many options are available to students when giving a feedback (e.g. choosing from a list of buttons or expressing using open text)?
4. **Real-Time**: Are students feedback conveyed in a real-time fashion or are delayed.
5. **Effortless**: Does the use of the SRS impose overheads on students (e.g. writing an SMS) or on teachers (e.g. designing brain-teasers-type question)?
6. **Anonymous**: Are the feedbacks publicly associated with students (Not suited for shy students)?
7. **No-Triggers**: Does the SRS need to trigger students (e.g. asking a question) in order to receive feedbacks?

It is worth noting that we did consider that the aspect of being a ubiquitous solution (i.e. no explicit interaction is required), may lead to not explicitly triggering mental processing in students' mind as opposed to clickers for examples. However, we believe teachers should be the one that prompt the students to actively think and to not stay dull by asking them direct, specific and more informed questions using the insights they receive from the U-SRS solution. What the proposed system does is pinpoint students that are not following up or are confused so that the teacher could help them.

5   Conclusion and Future Work

Students Responses Systems (SRS) are advanced feedbacks-conveying channels used to provide an overview of the comprehension states of students. They have a major role in helping teachers grasp a view of their audience's, i.e. students, performances. However, existing SRS fail to seamlessly integrate into learning environments given the overheads they impose. To tackle this problem, we propose a Ubiquitous Students Responses System (U-SRS) capable of blinding seamlessly into learning environments, continuously monitoring several learning performances indicators, making sense of them and finally providing insights to teachers about comprehension state of their students enabling them to adapt their pedagogy to their audience's needs in a timely manner. Our solution is built upon the Internet of Things paradigm to provide its pervasive services where the extracted insights are enabled using machine learning technologies. We highlighted the advantages of the U-SRS over existing SRS, its design and its architecture with all the constructing blocks. We also selected a subset of relevant features for our machine learning algorithms.

Our upcoming efforts will focus on experimenting with our solution in a real learning environment to assess its performances. Such experimentations have already been planned and preparations are mostly finished.

Acknowledgments. The work of A. Karkouch leading to these results has received funding from the Moroccan National Center for Scientific and Technical Research under the grant N° 18UCA2015.

References

1. Mory, E.H.: Feedback research revisited. Handb. Res. Educ. Commun. Technol. **2**, 745–784 (2004)
2. Draper, S.W., Brown, M.I.: Increasing interactivity in lectures using an electronic voting system. J. Comput. Assist. Learn. **20**(2), 81–94 (2004)
3. Gauci, S.A., Dantas, A.M., Williams, D.A., Kemm, R.E.: Promoting student-centered active learning in lectures with a personal response system. Adv. Physiol. Educ. **33**(1), 60–71 (2009)
4. Beatty, I.D.: Transforming student learning with classroom communication systems. Educ. Appl. Res. **2004**(3), 1–13 (2005)
5. FitzPatrick, K.A., Finn, K.E., Campisi, J.: Effect of personal response systems on student perception and academic performance in courses in a health sciences curriculum. Adv. Physiol. Educ. **35**(3), 280–289 (2011)
6. Kort, B., Reilly, R., Picard, R.W.: An affective model of interplay between emotions and learning: reengineering educational pedagogy-building a learning companion. In: Proceedings of the IEEE International Conference on Advanced Learning Technologies, ICALT 2001, pp. 43–46, February 2001
7. Denker, K.J.: Student response systems and facilitating the large lecture basic communication course: assessing engagement and learning. Commun. Teach. **27**(1), 50 (2013)
8. Teevan, J., Liebling, D., Paradiso, A., Suarez, C.G.J., von Veh, C., Gehring, D.: Displaying mobile feedback during a presentation. In: Proceedings of the 14th International Conference on Human-Computer Interaction with Mobile Devices and Services - MobileHCI 2012, p. 379 (2012)
9. Leong, C.K., Lee, Y.H., Mak, W.K.: Mining sentiments in SMS texts for teaching evaluation. Expert Syst. Appl. **39**(3), 2584–2589 (2012)
10. Altrabsheh, N., Cocea, M., Fallahkhair, S.: Predicting learning-related emotions from students' textual classroom feedback via Twitter. In: 8th International Conference on Educational Data Mining, pp. 436–439 (2015)
11. Altrabsheh, N., Gaber, M.M., Cocea, M.: SA-E: Sentiment analysis for education. Front. Artif. Intell. Appl. **255**, 353–362 (2013)
12. Calvo, R.A., Kim, S.M.: Sentiment analysis in student experiences of learning. In: Third International Conference on Educational Data Mining, pp. 111–120 (2010)
13. Feng, T., Qinghua, Z., Ruomeng, Z., Tonghao, C., Xinyan, J.: Can E-learner's emotion be recognized from interactive Chinese texts? In: Proceedings of the 13th International Conference on Computers Supported Cooperative Work in Design, CSCWD 2009, pp. 546–551 (2009)
14. Pekrun, R., Goetz, T., Titz, W., Perry, R.P.: Academic emotions in students' self-regulated learning and achievement: a program of qualitative and quantitative research. Adult Learn. **18**(2), 19 (2007)
15. Craig, S., Graesser, A., Sullins, J., Gholson, B.: Affect and learning: an exploratory look into the role of affect in learning with AutoTutor. J. Educ. Media **29**(3), 241–250 (2004)
16. D'Mello, S., Graesser, A., Picard, R.W.: Toward an affect-sensitive autotutor. IEEE Intell. Syst. **22**(4), 53–61 (2007)
17. Stein, N.L., Levine, L.J., Stein, N.L., Leventhal, B., Trabasso, T.: Making sense out of emotion: the representation and use of goal-structured knowledge. Psychol. Biol. approach. Emot. 45–73 (1990)

A Cyber-Physical Power Distribution Management System for Smart Buildings

Youssef Hamdaoui$^{(\boxtimes)}$ ⓘ and Abdelilah Maach$^{(\boxtimes)}$

Department of Computer Science, Mohammedia Engineering School (EMI),
Mohammed V University, Rabat, Morocco
`youssefhamdaoui@research.emi.ac.ma, maach@emi.ac.ma`

Abstract. Cyber-physical systems can collect information from different physical parameters through smart sensors and process this information in the cyber-world for best actuators control and energy efficiency. A smart building is such a good scope of application and presents a prototypical cyber-physical energy system because through sensing and computing, the system can analyze and make decisions based on contextual awareness information from the physical world and use this information in an intelligent way for smart distribution and smart balancing between power resources in outage mode. A new smart building architecture is proposed in this paper based on cyber physical manager for a dynamic selection of powered houses in an islanded building. The cyber physical manager control exchange with the main grid and also manage the power distribution into building based on analyzing of some parameters like real time consumption measurements collected by smart sensors/meters and batteries level charge. Prosumers production is also very important to be including in the building self powering in outage period. A simulation is done with results discussion for more optimization and to check the effectiveness and superiority of our autonomous power management system through on-site distributed generators within buildings to operate in islanding mode. Finally, This paper present a solution for new building generation that can have a self-supply based on distributed renewable energy and thereafter reduce costs for both government and costumers and create smart living space that becomes an important trend of future.

Keywords: Smart distribution system · Blackout · Emergency situation
Prosumers · Cyber physical system · Storage system · Classification
Smart building

1 Introduction

Modern buildings present the new vision of developed cities to resolve the fast growth power demand and also to integrate consumers as producer entity into the new smart grid. The integration of end costumer has a very important advantages because through renewable resource like solar panel, wind turbine and storage battery, costumers has the ability to produce energy for their own needing and also injecting extra power to neighbors as a service and to the main grid. Smart Grid (SG) is defined as new concept of existing electrical grid and present some advantages: always available, interactive, interconnected, Intelligent (capable of sensing system overloads, bi-directional communication and

© Springer International Publishing AG, part of Springer Nature 2018
M. Ben Ahmed and A. A. Boudhir (Eds.): SCAMS 2017, LNNS 37, pp. 538–550, 2018.
https://doi.org/10.1007/978-3-319-74500-8_50

rerouting power), Efficient (capable of meeting increased consumer demand without adding infrastructure), accommodating (accepting energy from any resources including solar and wind as easily and transparently and capable of integrating any ideas and technologies like energy storage technologies), motivating (enabling real-time communication between the consumer and utility so consumers can tailor their energy consumption based on individual preferences, like price and/or environmental concerns), resilient (increasingly resistant to attack and natural disasters as it becomes more decentralized) and green.

The result will be more efficient power systems capable of better managing the growing power consumption, providing fault resilience and seamlessly integrating Distributed Renewable Energy Resources (DRER) e.g. wind and solar [1, 2].

Both developed and developing countries will play a major role in the implementation of smart grid. In most countries, the energy is produced by power plants like nuclear, Fossil-fuel, Thermal, Hydroelectric, which requires regular investment in transmission and distribution system to maintain high degree of uninterrupted power system service quality and reliability for users, the reason why the integration of distributed energy resources impose itself. Key benefits of smart grid are uninterrupted power supply for all households [3], reduced transmission and distribution loss, high penetration of renewable energy sources, cyber secured electrical grid, large scale energy storage, and flexibility to consumers to interact with electricity market, market based electricity pricing and demand side management [4, 5].

IT technologies allow a fast communication (wired/wireless communication) to have real time measurements and to plan some real decisions depending on the context and situations like in outage case, When we detect an outage we have to intervene quickly to response the demands and ensure reliability with the same quality and service as main grid power, it's what we call the resiliency problem. Resiliency is required to improve grid reliability. The objectives is to respond to the growing demands, economic growth, security, quality of life of habitats and increased quality of life specially in buildings and generally in cities, thus to make them smarter. Cyber-physical systems can collects the awareness information from the various physical parameters through sensors and process this information in the cyber-world and control actuators, it is efficient to control, monitor and query on machinery, equipment, and personnel state. a modern building is such a good scope of cyber-physical systems application.

This paper looks smart buildings as a cyber-physical energy system with deeply coupled embedded sensing and networked information processing because through context-aware sensing and computation the (CPS) will be able to acquire contextual awareness information from the physical world and process this information in the cyber-world to make decision and ensure the optimized energy distribution by emergency. Integrating context awareness into the system increases the efficiency in terms of energy savings and was observed to be significant, around 30% and these systems will be able to anticipate needs and situations and react to the environment around them. In this paper, we propose an dynamic distribution concept based on some parameters collected by smart meters [6] and entities classification for smart buildings through an intelligent controller focusing on diverse requirement satisfaction, including response time, energy efficiency, real time sensor measurement, continuous supply, reliability and user's needs. The smart energy management system presented considers an smart distribution network that collect real-time data from a physical sensors and meters through a connected house controller that have a role of power switcher from on/off energy use in order to satisfy comfort conditions of houses and saving energy. The communication between building component is supposed done with a combination of wireless and wired network.

This paper provides an approach for smart buildings in islanding mode or when we have an outage. The rest of the paper is organized as follows: Sect. 2 present an different definition and concepts used by reviewing recent related works, Sect. 3 describes our proposal distribution building model, and Sect. 4 present the strategy of classification and our dynamic selection Algorithm with a simulation, Finally we conclude with discussion of results and decision generated for outage case studied.

2   Definition and Concepts

2.1   Intentional Islanding

An Outage or an Intentional Islanding is a condition in which a portion of the power grid, which contains both load and DG is isolated from the main utility caused by extreme weather or when the demand in the peak hours exceeds the supply [7, 8].

The major energy outage is mostly related to weather and more than 88% of costumers are affected by outages caused by weather as shown in Fig. 1. Maintenance problems, substation failures and transmission lines failures are low causes of the outages. These causes require further study to enhance the grid reliability.

The duration of interruptions might be from minutes to hours depending on the severity of the fault that occurred. The costs of an outage are considerable and not recoverable, called outage Costs. Outage detection is achieved by using islanding detection methods (IDM). Generally, IDMs are divided into local and remote methods communication based. Local methods are based on measurement of some parameters or variables on the MG side (frequency and voltage) [9] and include passive methods, active methods and hybrid method [10].

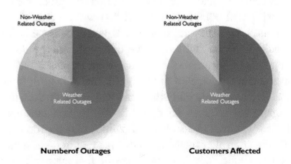

Fig. 1. Major weather-related power outages—those affecting $\geq 50,000$ customers—increased dramatically in the 2000s (Kenward and Raja 2014).

- Passive methods technique detects parameters change such as voltage and frequency and measure local, and do not affect the inverter output power quality.
- Active methods technique inject small disturbances (force voltage, frequency or network impedance change) into the micro grid and monitor the network response. Generally, the active techniques produce deterioration of the quality of power and cause instability problems.

Hybrid methods are an active and passive combination and used for complex systems and can improve multiple performance indices but Islanding detection time is prolonged.

2.2 Cyber-Physical System (CPS)

Smart With the requirement of low-carbon economy and energy crisis, many new information technologies have been applied to Cyber-Physical Energy Systems for saving energy and protecting environment, such as smart grid, electric vehicles and home automation. The objective is an energy efficiency and system reliable control. The CPS is composed of hardware and software and is an intelligent, programmable environment capable of performing metering, computations, numerical processing, running optimization subroutines, establishing bi-directional communication with the smart grid monitor and control centre and building controllers to make decisions based on the specified real time parameters.

The Cyber-Physical Energy Systems represent a new class of widely distributed and globally interconnected energy systems that integrate computation processes, communication processes and control processes [11]. For example when we are in sunny case we can detect rain with sensor and the CPS make a decision base on this context aware information to open all windows in building through controllers and actuators. Some important properties for an autonomic CPS are [12]:

- Self-adaptation.
- Self-organization.
- Self-optimization.
- Self-configuration.
- Self-protection.
- Self-healing.
- Self-description.
- Self-discovery.
- Self-energy-supplying.

The CPS Model consists to do an iterative 4 actions: Monitor, Analyze, Plan, and Execute. The objective of a CPS is to monitor and manage different controllers, sensors and actuators properties and collect information to analyze all context aware parameters, the analysis should be done in quasi real-time, Third step is to make decisions, schedules and executes the decided actions. When we are in outage case the CPS have o detect outage by one of IDMs and collect all information transmitted by different controllers connected to have the ability to lunch a islanded operation. Islanded operation will cover some loads during the outage period and depend on the distributed generation installed.

2.3 Smart Building

Optimizing energy efficiency in buildings is an integrated task that covers the whole lifecycle of the building. Smart building is new concept based evolution of renewable energy and of IT technologies. Building consumption account for 30–40% of total country energy consumption as mentioned in Indian study [13]. In the future, most of buildings will be self powered with an auto energy management system to control and manage distribution for energy efficiency and improve reliability when we have an outage and cut off main grid through a CPS [14].

Smart buildings are environments such as apartments, offices, museums, hospitals, schools, malls, university campuses, and outdoor areas that are enabled for the cooperation of objects (e.g., sensors, devices, appliances) and systems that have the capability to self-organize themselves. All connected smart components are controlled by a main controller. The main controller is responsible for the energy efficiency management and has to optimize the consumption into building and control different parameters for improving reliability and ensure a best quality of power. Figure 2 present an example of modern building that contain a PV installation, a building CPSM that monitor and manage the different state and some loads divided between houses using and EV that represent a mobile energy. The architecture is based on DER and can be powered also by utility grid to response extra loads in peaks case. Smart building can inject the surplus energy o the grid when the batteries are in full charge and all loads are covers by solar PV production.

A really smart building should increase the heating via his actuator on the heating, if the sensor on the solar collector indicates a high performance. A sensor like this produces much information; buildings with many sensors will generate a high amount of information. All these information have to be processed, complex event processing can do that job in this paper and we suppose a residential building that contains a combined DER (PV and Storage Batteries to power building).

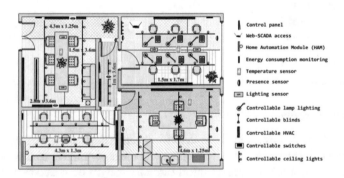

Fig. 2. House smart components

2.4 Smart Home with Smart Controllers and Sensors

Future Smart Homes (SH) can significantly contribute not only integrating energy generation system but also properly scheduling tasks to maximize the consumption of locally-generated energy and reduce the energy demand to the Grid during peak hours. SH contains some appliances and smart devices, which are centrally managed by a smart controller [15]. Controllers are a smart component which is responsible for transmitting information to the CPSM and executing decisions and centralize all information transmitted by the smart sensors, meters, devices and actuators inside house [16, 17] as shown in Fig. 2. Smart sensors and smart meters are used to collect data about the power consumption. Actuators are responsible of execution of decisions like close windows. Simple home become an interactive smart home with a storm of information and measurements, All information are used to control and optimize the energy consumed, secure house from dangers and create a easy live for humans.

As far as the wireless IoT is the main concern, many different wireless communication technologies and protocols can be used to connect smart devices and to allow monitoring of building energy the smart device such as Internet Protocol Version 6 (IPv6), over Low power Wireless Personal Area Networks (6LoWPAN), Zigbee, Bluetooth Low Energy (BLE), Z-Wave and Near Field Communication (NFC). They are short range standard network protocols, while SigFox and Cellular are Low Power Wide Area Network (LPWAN) [18].

2.5 Hybrid Distributed Energy Renewable

DER based distribution become a requirement for the industrial, commercial and residential customers because they require a high degree of reliability due to the increase of digital sophisticated systems but the problem is the complexity and costly of DER. Distributed generation (DG) can improve energy utilization efficiency and power supply reliability in outage mode [19]. When a fault happens in a distribution network, island operation and DG is require to continue supply demands.

Using a single renewable energy generation is not a best solution and a combination can be effective solution to unstable effects of electricity supply, for example PV systems have some problems like the efficiency of electric power and depend on weather conditions. Wind turbines are affected by system fault at t = 52,000 s that means power output of wind generation is reduced to zero and depends on weather conditions. A hybrid solar-wind-battery system can provide 100% of power supply for costumers, thus greatly decreasing the energy costs and increasing the continuity and reliability of power supply [20].

3 Distribution Scheme Model

3.1 Building Scheme Model

Figure 3 present our proposed building model. The building have an outside controller who is reasonable of switching between on main grid use when we are in the case of needing energy or off use when we have energy enough power to cover all loads. The DER are installed on the floor and we have a hybrid installation with inverter. The CPSM is responsible to monitor and control the production and the level of batteries. The main controller can make some decisions like charging batteries when not using all the electricity that the solar PV system is generating or when battery is

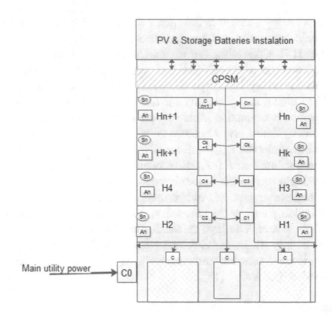

Fig. 3. Smart building components

fully charged and still have solar electricity, surplus energy will be exported to the grid or when the solar PV panels have a reduced or zero output and batteries are discharged then we can start use the main grid electricity. The house controller is connected to all sensors and actuators inside the house to monitor and Transmit information to the main controller (CPS). Electrical power sources are available on a scale (kW range) that makes them suitable for on-site generation in buildings.

3.2   Houses Classification

Houses are the component consume in the building micro grid and each one have his own weight because there is some costumer more important than others and need power with a high level of emergency like people with health problems. We choose in this paper a classification parameter to make a difference and an order for smart distribution. We define three category: (a) U_{C1}: high level of emergency, (b) U_{C2}: medium level emergency, (c) U_{C3}: low level emergency.

The classification is dynamic and depends on entity declaration or the CPS declaration for public entities like hospitals, Hotels or industries.

3.3   Dynamic Selection Method

Smart Our approach is to start to response emergencies demands first because it present a high priority, for example when we have a healthy problem and we use some critical machine that need a continuous electricity, after we response to houses with a medium priority like person who work from their houses and at the end we response to low priority demands.

After outage detection and after collecting all data needing for our islanded operation, selection algorithm has to define the islanded houses that can't be covered by our selection and the houses that we can be covered. The advantage in our method is that it is dynamic and not static because the outage context and the different status like solar production, batteries storage, classification, loads.... We suppose that all controllers have a switcher to have the ability of on/off building energy use. All house loads is divided between obligatory and facultative load. Obligatory load is the uninterruptable load that house and the optional is the interruptible load that can be covered by the main grid or by a self house production like electricity for washer machine.

The Distribution network contains different nodes as explain in the last section and each node has his weight into building. The selection construction model consist on the state of some parameters that we see more important and decisive in the selection layered tree, there is others parameters like house maturity, environment, weather context, etc....but in this paper we choose just 3 parameters:

- Costumer Category (U_C).
- Costumer emergency Demand (P).
- Costumer facultative demand (Q).
- Costumer production (C_p).

3.4 Selection Scheme Algorithm

The objective of our approach is to have a layered tree model with the most priority uninterrupted load first and the low priority interrupted load at the end. The selection is based on some parameters defined. Our approach give a value to distribution network and show smartness in covering loads based on aware inside building needs. The condition is that the total loads not exceed the capacity existing in the batteries and produced by solar PV. The capacity existing $P_{(total)}$ is the sum of different energy produced by DER (solar PV panels and energy stored in batteries). The concept of the smart distribution and the dynamic selection is to start satisfy the obligatory demand (P) and when we cover all P demands, we start satisfy the Q demands but we respect the selected entities order and if we have more capacity we verify entities not selected if we can cover it or not. The searching of entities is done with arbitrary searching because we have limited houses.

The dynamic approach consists to collect information in real time from different resources to know the existing power and the entities belonging building area, after updating data information the first thing is to identify the entities with high level of emergency and we start building our tree composed of selected entities. The idea is to obtain a one components vector (V[n]). The first components are houses with high priority followed by the medium and at the last the low emergency. The algorithm is described below in the Fig. 4.

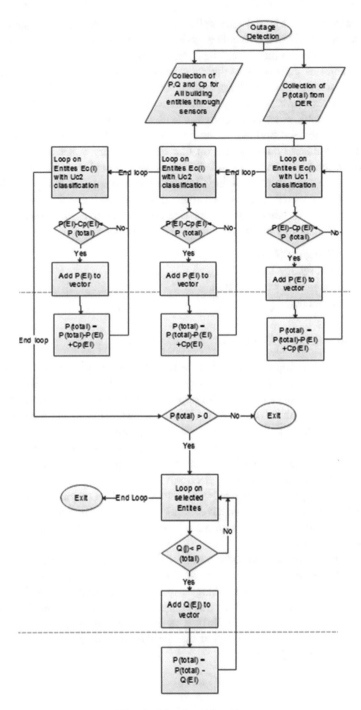

Fig. 4. Islanding Algorithm

The algorithm result is a vector with n elements presents houses powered during outage period.

4 Simulation and Analyses

4.1 Case of Study

Our Approach is implemented with Java language, Mysql Database, eclipse environment and based on house's consumption generated by Gridlab-d simulator.

As shown in Table 1, simulation will done for a building contains 30 house and some DGs (houses, classification are generated by a random function and loads

Table 1. Load characteristics

Entity	P(Kwh)	Q(Kwh)	Classification	C_p (Kwh)
E(1)	20	5	U_{C2}	0
E(2)	17	2	U_{C3}	0
E(3)	15	1	U_{C2}	5
E(4)	30	15	U_{C3}	0
E(5)	15	6	U_{C2}	0
E(6)	14	3	U_{C3}	7
E(7)	13	2	U_{C2}	0
E(8)	12	1	U_{C3}	0
E(9)	10	5	U_{C3}	0
E(10)	10	1	U_{C1}	0
E(11)	24	3	U_{C3}	0
E(12)	10	4	U_{C3}	0
E(13)	25	13	U_{C1}	8
E(14)	22	9	U_{C1}	9
E(15)	8	2	U_{C2}	0
E(16)	7	2	U_{C3}	10
E(17)	15	10	U_{C2}	0
E(18)	14	5	U_{C3}	0
E(19)	13	2	U_{C3}	0
E(20)	12	4	U_{C1}	0
E(21)	11	1	U_{C3}	0
E(22)	12	1	U_{C3}	0
E(23)	17	3	U_{C2}	15
E(24)	15	1	U_{C3}	0
E(25)	15	2	U_{C2}	0
E(26)	30	13	U_{C3}	10
E(27)	31	20	U_{C1}	15
E(28)	3	1	U_{C3}	0
E(29)	0	0	U_{C3}	0
E(30)	31	11	U_{C1}	0

generated by Gridlab-D simulator to have real cases). The test system adopts reactive local compensation, all power loss is ignored. We suppose for this simulation a total DGs active power/Kwh 450 kWh for one day power (PV/Storage Batteries).

4.2   Simulation Results and Discussion

According to the selection scheme algorithm proposed in the paper, the simulation steps and results are presented as follows.

The building contains 30 houses with different classification and contains some DERs, the algorithm search the entities with high priority and start looping on the query results, E_{10}, E_{13}, E_{14}, E_{20}, E_{27} and E_{30} are entities with classification U_{C1}, for each one we verify if his obligatory demand minus his self production $(P-C_p)$ is more than existing updated power P(total), if yes we add the entity to our vector, else we skip and loop for next entity. When we add an entity we update the $P(\text{total}) = P(\text{total}) - P + C_p$. After finish with entities with U_{C1}, we look for entities with U_{C2}, in our case we have E_1, E_3, E_5, E_7, E_{15}, E_{17}, E_{23} and E_{25} are entities with classification U_{C2}, for each one we verify if his obligatory demand minus his self production $(P-C_p)$ is more than existing updated power P(total), if yes we add the entity to our vector, else we skip and loop for next entity. When we add an entity we update the $P(\text{total}) = P(\text{total}) - P + C_p$. After finish with entities with U_{C2}, we look for entities with U_{C3}, in our case we have E_2, E_4, E_6, E_8, E_9, E_{11}, E_{12}, E_{16}, E_{18}, E_{19}, E_{21}, E_{22}, E_{24}, E_{26}, E_{28} and E_{29} are entities with classification U_{C3}, for each one we verify if his obligatory demand minus his self production $(P-C_p)$ is more than existing updated power P(total), if yes we add the entity to our vector, else we skip and loop for next entity. When we add an entity we update the $P(\text{total}) = P(\text{total}) - P + C_p$. After finish with last classification for obligatory demand, we verify the existing power depletion, in our case, after satisfy all selected entities p(total) = 450–392 = 58 kwh, then we look to power interrupted demand (Q) but we loop into our selected vector by order of selection, in our case we can satisfy E_{10}, E_{13}, E_{14}, E_{20}, E_{27} and E_{30} because the sum of their Q is 58 Kwh. The islanding algorithm exit in case of power depletion or in case of powering all entities. The result of our simulation is the vector $\mathbf{V} = [P(E_{10}), P(E_{13}), P(E_{14}), P(E_{20}), P(E_{27}), P(E_{30}),$ $P(E_1), P(E_2), P(E_3), P(E_4), P(E_5), P(E_6), P(E_7), P(E_8), P(E_9), P(E_{11}), P(E_{12}),$ $P(E_{15}), P(E_{16}), P(E_{17}), P(E_{18}), P(E_{19}), P(E_{21}), P(E_{22}), P(E_{23}), P(E_{24}), P(E_{25}),$ $P(E_{26}), P(E_{28}), Q(E_{29}), Q(E_{10}), Q(E_{13}), Q(E_{14}), Q(E_{20}), Q(E_{27}), Q(E_{30})]$.

5   Conclusion

In this work, we propose an efficient approach for power shortages problem that can be integrated in islanded building operation to improve access and provide consumer power throughout emergencies and grid outage. The proposed approach is based on a smart dynamic power balancing with dynamic parameters to control and perform the distributed energy in islanded building through a smart CPS and achieve a 100% self-powering. A next step of work is to make a model/prototype with Gridlab-D to simulate real scenarios and take more parameters in our islanding selection like weathers, building events, prosumers context, building aware to predict future loads and based on this predication information we will make a real smart decision to avoid outage problem and satisfy all demands and ensure a continuous power.

References

1. Zhabelova, G., Yang, C.-W., Patil, S., Pang, C., Yan, J.D., Shalyto, A., Vyatkin, V.: Cyber-physical components for heterogeneous modelling, validation and implementation of smart grid intelligence. In: 2014 12th IEEE International Conference on Industrial Informatics (INDIN), pp. 411–417. IEEE (2014)
2. Wang, Q., Zhang, C., Ding, Y., Xydis, G., Wang, J., Østergaard, J.: Review of real-time electricity markets for integrating Distributed Energy Resources and Demand Response. Appl. Energy **138**, 695–706 (2015)
3. Hamdaoui, Y., Maach, A.: A smart approach for intentional islanding based on dynamic selection algorithm in microgrid with distributed generation. In: 2017 International Conference on Big Data, Cloud and Application (BDCA), pp. 1–7. ACM (2017)
4. Guo, Y., Pan, M., Fang, Y.: Optimal power management of residential customers in the smart grid. IEEE Trans. Parallel Distrib. Syst. **23**, 1593–1606 (2012)
5. Taktak, E., Abdennadher, I., Rodriguez, I.B.: An adaptation approach for smart buildings. In: 2016 IEEE 18th International Conference on High Performance Computing and Communications, IEEE 14th International Conference on Smart City, IEEE 2nd International Conference on Data Science and Systems (HPCC/SmartCity/DSS), pp. 1107–1114 (2016)
6. Weiss, M., Helfenstein, A., Mattern, F., Staake, T.: Leveraging smart meter data to recognize home appliances. In: Proceedings of the 2012 IEEE International Conference on Pervasive Computing and Communications (PerCom), pp. 190–197. IEEE (2012)
7. Hamdaoui, Y., Maach, A.: A novel smart distribution system for an Islanded region. In: 2017 International Conference Advanced Information Technology, Services and Systems (AIT2S) (2017)
8. Hamdaoui, Y., Maach, A.: Dynamic balancing of powers in islanded microgrid using distributed energy resources and prosumers for efficient energy management. In: 2017 IEEE Smart Energy Grid Engineering (SEGE). IEEE (2017)
9. Guerrero, J.M., Vasquez, J.C., Matas, J., de Vicuna, L.G., Castilla, M.: Hierarchical control of droop-controlled AC and DC microgrids—a general approach toward standardization. IEEE Trans. Ind. Electron. **58**, 158–172 (2011)
10. Hamdaoui, Y., Maach, A.: Smart islanding in smart grids. In: 2016 IEEE Smart Energy Grid Engineering (SEGE), pp. 175–180. IEEE (2016)
11. Ge, Y., Dong, Y., Zhao, H.: A cyber-physical energy system architecture for electric vehicles charging application. In: 2012 12th International Conference on Quality Software (QSIC), pp. 246–250. IEEE (2012)
12. Gurgen, L., Gunalp, O., Benazzouz, Y., Gallissot, M.: Self-aware cyber-physical systems and applications in smart buildings and cities. In: Proceedings of the Conference on Design, Automation and Test in Europe, pp. 1149–1154. EDA Consortium (2013)
13. Nallamothu, B.K., Selvam, C., Srinivas, K., Prabhakaran, S.: Study on energy savings by using efficient utilities in buildings. In: Communication, Control and Intelligent Systems (CCIS), pp. 477–481. IEEE (2015)
14. Hamdaoui, Y., Maach, A.: Energy efficiency approach for smart building in islanding mode based on distributed energy resources. In: 2017 International Conference on Advanced Information Technology, Services and Systems (AIT2S) (2017)

15. Komninos, N., Philippou, E., Pitsillides, A.: Survey in smart grid and smart home security: issues, challenges and countermeasures. IEEE Commun. Surv. Tutor. **16**, 1933–1954 (2014)
16. Khanna, A.: Smart grid, smart controllers and home energy automation—creating the infrastructure for future. Smart Grid Renew. Energy. **03**, 165–174 (2012)
17. Oukrich, N., Elbouazzaoui, C., Maach, A., Bouchard, K., Elghanami, D.: Human activities recognition based on auto encoder pre-training and back-propogation algorithm. J. Theor. Appl. Inf. Technol. (2017) ISSN 1992-8645
18. Al-Sarawi, S., Anbar, M., Alieyan, K., Alzubaidi, M.: Internet of Things (IoT) communication protocols. In: 2017 8th International Conference on Information Technology (ICIT), pp. 685–690. IEEE (2017)
19. Hamdaoui, Y., Maach, A.: An intelligent islanding selection algorithm for optimizing the distribution network based on emergency classification. In: 2017 International Conference on Wireless Technologies, Embedded and Intelligent Systems (WITS), pp. 1–7. IEEE (2017)
20. Ma, T., Yang, H., Lu, L.: A feasibility study of a stand-alone hybrid solar–wind–battery system for a remote island. Appl. Energy **121**, 149–158 (2014)

Amazigh PoS Tagging Using Machine Learning Techniques

Amri Samir[1(✉)], Zenkouar Lahbib[1], and Outahajala Mohamed[2]

[1] LEC Laboratory, EMI School, University Med V, Rabat, Morocco
amri.samir@gmail.com
[2] CESIC Laboratory, IRCAM Institute, Rabat, Morocco

Abstract. Amazigh is a morphologically rich language, which presents a challenge for Part of Speech tagging. Part of Speech (POS) tagging is an important component for almost all Natural Language Processing (NLP) application areas.

Applying machine-learning techniques to the less computerized languages require development of appropriately tagged corpus. In this paper, we have developed POS taggers for Amazigh language, a less privileged language, using Conditional Random Field (CRF), Support Vector Machine (SVM) and TreeTagger system. We have manually annotated approximately 75000 tokens, collected from the written texts with a POS tagset of 28 tags defined for the Amazigh language. The POS taggers make use of the different contextual and orthographic word-level features. These features are language independent and applicable to other languages also. POS taggers have been trained, and tested with the same corpora. Evaluation results demonstrated the accuracies of 89.18%, 88.02% and 90.86% in the CRF, SVM and TreeTagger, respectively.

Keywords: Amazigh · Corpus · CRF · NLP · Machine learning
POS tagging

1 Introduction

Part of speech tagging is very significant pre-processing task for Natural language processing activities [1]. A Part of speech (POS) tagger has been developed in order to check off the words and punctuation in a textual matter having suitable POS labels of Amazigh text. POS tagging makes up a primal task for processing a natural language. It is built up using linguistic theory rule, random pattern and sometimes a combining both [1]. Our work shows the evolution of an easy and effective automatic tagger in support of inflectional and derivational morphologically rich language Amazigh. Amazigh language is morphologically rich with less linguistically peculiar patterns and rules and heavy annotated corpora and thus the development of POS tagger is a difficult task [2]. POS tagging is a phenomenon of allotting the words in a textual matter as matching to a picky component of speech. In general, POS tagging is as well denoted to as grammatical tagging of textual matter as representing to a specific component of speech because of both its definition and context.

A part-of-speech is a grammatical category, commonly including verbs, nouns, adjectives, adverbs, determiner, and so on. The Part of Speech tagger is an important

© Springer International Publishing AG, part of Springer Nature 2018
M. Ben Ahmed and A. A. Boudhir (Eds.): SCAMS 2017, LNNS 37, pp. 551–562, 2018.
https://doi.org/10.1007/978-3-319-74500-8_51

application of natural language processing. It is an important part of morphological analyzer. Part of speech tagging is the process of assigning a part of speech like noun, verb, preposition, pronoun, adverb, adjective or other lexical class marker to each word in a sentence. The major problems in the process of POS tagging are: ambiguous words and unknown words [3]. The first and foremost problem is with those words whose more than one tag can exist. These can an easy task for humans but not so for the automatic word taggers. In the process of tagging we can sometimes get such words that have different tag categories when they are used in different context. Thus it is a very tedious job. This phenomenon is known as lexical ambiguity. But while occupying the same part of speech many words can have multiple meanings. Ambiguous words are the major problem in the part of speech tagging. Many words can have tags which are more than one [4].

To overcome these problems, we propose the Amazigh Part-of-Speech tagging methods based on stochastic approaches.

The rest of this paper is organized as follows: In Sect. 2, we describe the related work in POS tagging, linguistic background is also presented in Sect. 3. Our data and material used are described in Sect. 4. Section 5 presents experimental results. Finally, Sect. 6 concludes the paper and describes the future works.

2   Related Work

Considerable amount of work has already been done in the field of POS tagging for English. Different approaches like the rule based approach, the stochastic approach and the transformation based learning approach along with modifications have been tried and implemented. However, if we look at the same scenario for Amazigh language, we find out that not much work has been done. The main reason for this is the unavailability of a considerable amount of annotated corpora, on which the tagging models could train to generate rules for the rule based and transformation based models and probability distributions for the stochastic models.

The previous works of POS tagging can be divided into three categories; rule based tagging [5], statistical tagging and hybrid tagging [6, 7]. A set of hand written rules are applied along with it the contextual information is used to assign POS tags to words in the rule based POS. The disadvantage of this system is that it doesn't work when the text is not known.

In the literature, many machine learning methods have been successfully applied for POS tagging, namely: the Hidden Markov Models (HMMs) [8], the transformation-based error driven system [9], the decision trees [10], the maximum entropy model [11], SVMs [12], CRFs [13]. Results produced by statistical taggers obtain about 95%–97% of correctly tagged words. There are also, hybrid methods that use both knowledge based and statistical resources.

There are different types of statistical tagging approaches discussed in this paper that, Conditional Random Field (CRF), Support Vector Machine (SVM) and TreeTagger. Along with this the studies done on the basis of comparisons and evaluation are also shown.

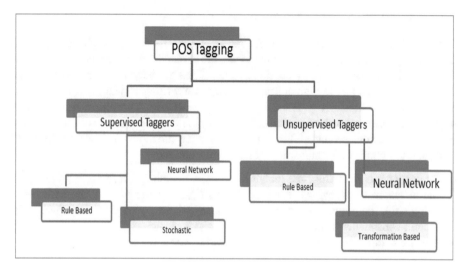

Fig. 1. POS classification

POS tagging works on different approaches. The different models of POS tagging are shown in Fig. 1.

3 Linguistic Background

3.1 Amazigh Language

Amazigh called also Berber belongs to the Hamito-Semitic "Afro-Asiatic" languages [14, 15]. Amazigh is spoken in Morocco, Algeria, Tunisia, Libya; it is also spoken by many other communities in parts of Niger and Mali. In Morocco Amazigh language uses different dialects in its standardization (Tachelhit, Tarifit and Tamazight). Amazigh has a complex morphology and the process of its standardization is performed via different dialects. Amazigh NLP presents many challenges for researchers. Its major features are:

- Amazigh has its own script: the Tifinagh, which is written from left to right. The transliteration into Latin alphabet is used in all the examples in this article.
- It does not contain uppercase.
- It presents for NLP ambiguities in grammar classes, named entities, meaning, etc. For example, grammatically the word «ⴰ ⵣ ⴰ» (tazla) can function as verb «ⴰ ⵓ ⴰ ⵣ ⴰ» meaning (over it) or as name (race), etc.
- As most languages whose research in NLP is new, Amazigh is not endowed with linguistic resources and NLP tools.

3.2   Amazigh Script

Like any language that passes through oral to written mode, Amazigh language has been in need of a graphic system. In Morocco, the choice is ultimately fell on Tifinaghe for technical, historical and symbolic reasons. Since the Royal declaration on 2003, Tifinaghe has become the official graphic system for writing Amazigh. Thus, IRCAM has developed an alphabet system called Tifinaghe-IRCAM. This alphabet is based on a graphic system towards phonological tendency. This system does not retain all the phonetic realizations produced, but only those that are functional. It is written from left to right and contains 33 graphemes which correspond to:

- 27 consonants including: the labials (ⵎ, ⴼ, ⵍ), dentals (ⵜ, ⴷ, ⴹ, ⴻ, ⵉ, ⵏ, ⵕ, ⵆ), the alveolars (ⵙ, ⵥ, ⵣ, ⵤ), the palatals (ⵛ, ⵊ), the velar (ⴽ, ⵅ), the labiovelars (ⴽ ⵯ, ⵅ ⵯ), the uvulars (ⵇ, ⵈ, ⵖ), the pharyngeals (ⵃ, ⵄ) and the laryngeal (ⵀ);
- 2 semi-consonants: ⵢ and ⵓ;
- 4 vowels: three full vowels ⵔ, ⵉ, ⵓ and neutral vowel (or schwa).⵰

3.3   Amazigh Morphological Properties

Language morphology is knowledge of the ways in which the language's words can have different surface representations. Hence, the Amazigh morphology is considered rich and complex in terms of its inflections involving infixation, prefixation and suffixation. The Amazigh morphology covers five main lexical categories, which are noun, verb, adverb, pronoun, and preposition [16].

We described below the famous syntactic categories of the Amazigh language:

- *Noun*
  Noun is a lexical unit, formed from a root and a pattern. It could occur in a simple form ('izm' the lion), compound form ('ighsdis' the rib), or derived one ('amkraz' the labourer). This unit varies in gender (masculine, feminine), number (singular, plural) and case (free case, construct case).
- *Verb*
  It has two forms: basic and derived forms. The basic form is composed of a root and a radical, while the derived one is based on the combination of a basic form and one of the following prefixes morphemes: 's'/'ss' indicating the factitive form, 'tt' marking the passive form, and 'm'/'mm' designating the reciprocal form. Whether basic or derived, the verb is conjugated in four aspects: aorist, imperfective, perfect, and negative perfect.
- *Particles*
  Particle is a function word that is not assignable to noun neither to verb. It contains pronouns, conjunctions, prepositions, aspectual, orientation and negative particles, adverbs, and subordinates. Generally, particles are uninflected words. However in Amazigh, some of these particles are flectional, such as the possessive and demonstrative pronouns.

4 Data and Material

Amazigh is a resource poor language. Applying statistical models to the POS tagging problem requires large amount of annotated corpus in order to achieve reasonable performance. But, annotated corpus for Amazigh is not available. So, we have used a manually annotated corpus of 75000 tokens with the 28 POS tags [17, 18].

In addition to this, we will focus on stochastic models which require large amount of hand labeled data in order to achieve reasonable performance. Though Hidden Markov Model (HMM) is one of the widely used techniques for POS tagging, it does not work well when small amount of labeled data are used to estimate the model parameters.

Incorporating diverse features in an HMM-based tagger is difficult and complicates the smoothing typically used in such taggers. In contrast a Conditional Random Field (CRF) based method [13] or a SVM based system [12] or decision tree based system can deal with diverse, overlapping features.

4.1 CRF Model

To use CRF for a sequential learning task like POS tagging, we should have a training set. Also, we would like to specify various features to encode various dependencies for the particular task at hand. The training set will be used to learn a model. The learned model, here, is a set of parameters (known as weights) which corresponds to the importance of the features used in this task. In this sense, the following are the steps to train our POS tagging model using the CRF:

- Create an instance of the feature generator class – FeatureGenImpl or a new feature generator that we might have created. Also, pass the required parameters like the modelGraph value and number of labels in the label set, to the constructor.
- Create an instance of the iitb.CRF class. This class is the interface for the training and inference routine of the package. We will need to pass the object of the feature generator just created as one of the parameters to the constructor of the iitb.CRF class.
- Read the instances of the training set in the objects of the class implementing the DataSequence interface. Also, create an object of the class implementing the DataIter interface to encapsulate these data sequences.
- Call the train routine of the feature generator class to train the dictionary. Pass the DataIter object to the train routine.
- Call the train routine of the CRF object. Again, we will have to pass the training set iterator object to this train routine.

4.2 SVM Model

One of the robust statistical learning models is Support Vector Machines (SVMs). SVM was first introduced by Vapnik [19]. SVM produces a binary classifier, the so-called optimal separating hyperplanes, through extremely non-linear mapping the input vectors into the high-dimensional feature space. SVM constructs linear model to estimate the decision function using non-linear class boundaries based on support vectors. If the data is linearly separated, SVM trains linear machines for an optimal hyperplane that separates the data without error and into the maximum distance between the hyperplane and the closest training points. The training points that are closest to the optimal separating hyperplane are called support vectors. All other training examples are irrelevant for determining the binary class boundaries. In general cases where the data is not linearly separated, SVM uses non-linear machines to find a hyperplane that minimize the number of errors for the training set.

SVMs are applied to text categorization, and are reported to have achieved high accuracy without falling into over-fitting even though with a large number of words taken as the features.

4.3 TreeTagger Model

TreeTagger is an HMM-based tagger, which differs from the traditional probabilistic taggers in the way it estimates the transition probabilities, it was developed by Helmut Schmid in the TC project at the Institute for Computational Linguistics of the University of Stuttgart [20].

TreeTagger has been successfully used to tag various languages including German, English, French, Italian, Dutch, Spanish, Bulgarian, Russian, Greek, Portuguese, Chinese, Swahili, Latin, Estonian and old French texts and is adaptable to other languages if a lexicon and a manually tagged training corpus are available.

In contrast to the N-gram taggers that typically use smoothing to deal with the sparsity of the training data, TreeTagger uses a binary decision tree, built recursively from a training set of trigrams, to obtain estimates of transition probabilities. The decision tree automatically determines the appropriate size of the context which is used to estimate the probabilities. Possible contexts are not only N-grams, but also other kinds of contexts such as e.g. (tag-1 = ADJ and tag-2 != ADJ and tag-2 != DET). The tagger is also capable of assigning tags to unknown words using a suffix lexicon organized in a tree for this purpose. At the tagging stage we have provided the tagger with lists of tokens to be added to the initial lexicon and choose the values to assign to tokens with zero frequencies. At the training stage we have provided the tagger with additional lexicon lists.

5 Experiment and Discussion

5.1 Test Environment

In Amazigh POS tagging, the ML tools of choice are data-mining-based tools that support the ML algorithms CRF++, YamCha, and TreeTagger. They all share the following features: a generic toolkit, language independence, absence of embedded linguistic resources, a requirement to be trained on a tagged corpus, the performance of sequence labeling classification using discriminative features, and a suitability for the pre-processing steps of NLP tasks.

- [1]**CRF++:** This is a free open source toolkit, written in C++, for learning CRF models in order to segment and annotate sequences of data. The toolkit is efficient in training and testing and can produce n-best outputs. It can be utilized in developing many NLP components for tasks such as text chunking and NER, and can handle large feature sets.
- [2]**Yamcha:** A commonly used free open source toolkit written in C++ for learning SVM models. This toolkit is generic, customizable, efficient, and has an open source text chunker. It has been utilized to develop NLP pre-processing tasks such as NER, POS tagging, base-NP chunking, text chunking, and partial chunking.
- [3]**TreeTagger:** A free POS tagger system presented in the Subsect. 4.3.

We have a total of 3 models as described in Sect. 3 under different stochastic tagging schemes. The same training text has been used to estimate the parameters for all the models. The model parameters for supervised SVM, CRF and TreeTagger models are estimated from the annotated text corpus. For semi-supervised learning, the SVM learned through supervised training is considered as the initial model. Further, a larger unlabeled training data has been used to re-estimate the model parameters of the semi-supervised HMM. The experiments were conducted with three different sizes (25 K, 50 K and 75 K words) of the training data to understand the relative performance of the models as we keep on increasing the size of the annotated data.

The Amazigh corpus was divided into ten approximately equal parts. From these ten different disjoint pairs of collected texts were created. In each pair there is a training set containing about 90% of running words from the corpus and a test set containing about 10% of running words from the corpus. A ten-fold cross-validation test was performed for the three remaining taggers.

All the models have been tested on a set of randomly drawn 400 sentences (5000 words) disjoint from the training corpus. It has been noted that 16% words in the open testing text are unknown with respect to the training set, which is also a little higher compared to the European languages [21].

The training data includes manually annotated 4830 sentences (approximately 75,000 words) for all supervised algorithms (CRF, SVM and TreeTagger). A fixed set of 1000 unlabeled sentences (approximately 20000 words) taken from Amazigh corpus are used to re-estimate the model parameter during semi-supervised learning.

5.2 Results and Discussion

When evaluating a supervised classifier, it is important to use fresh data that was not included in the training or dev-test set. Otherwise, our evaluation results may be unrealistically optimistic.

So, we define the tagging accuracy as the ratio of the correctly tagged words to the total number of words. Table 1 summarizes the final accuracies achieved by different learning methods with the varying size of the training data. Note that the baseline model

---

[1] http://crfpp.sourceforge.net/.

[2] http://chasen.org/ ∼ taku/software/yamcha/.

[3] http://www.cis.uni-muenchen.de/ ∼ schmid/tools/TreeTagger/.

(i.e., the tag probabilities depends only on the current word) has an accuracy of 78.8%. Results for ten-fold cross-validation testing for the three taggers are shown in Table 1. As can be seen from the table the TreeTagger gave best results, the CRF tagger came second and SVM gave the worst results of the three taggers. In that study the taggers were applied to and tested on the Amazigh corpus (75000 tokens) with 28 tags.

Table 1. Mean tagging accuracy for three POS taggers

Accuracy	CRF	SVM	TreeTagger
All words	89.18	88.02	90.86
Known words	91.24	91.06	92.94
Unknown words	68.20	58.17	76.30

Table 1 shows results for known words, unknown words and all words. Mean percentage of unknown words in the ten test sets was 7.04. TreeTagger shows overall best performance in tagging both known and unknown words. CRF seems to do better than SVM at tagging unknown words but does worse on known words than CRF. This is similar to what was seen in the experiment on Arabic text and indicates that the major difficulty in annotating Amazigh words stems from the difficulty in finding the correct tag for unknown words. Words belonging to the open word classes (nouns, adjectives and verbs) account for about 90% of unknown words in the test sets whereas words in these word classes account for just over 51% of all words in the test sets.

The similar works achieve an accuracy of 88.66% when using lexical features. The results of the tagger based on SVMs with lexical features gave an accuracy of 88.27% with a data set of 1438 sentences and applied to 15 tags [22].

We have chosen SVM, CRF and TreeTagger because they are open source, robust and efficient for POS tagging. By comparing our used techniques with others we have found that:

- Decision trees are automatically constructed tree-structured flowcharts that are used to assign labels to input values based on their features. Although they're easy to interpret, they are not very good at handling cases where feature values interact in determining the proper label.
- In naive Bayes classifiers, each feature independently contributes to the decision of which label should be used. This allows feature values to interact, but can be problematic when two or more features are highly correlated with one another.
- Maximum Entropy classifiers use a basic model that is similar to the model used by naive Bayes; however, they employ iterative optimization to find the set of feature weights that maximizes the probability of the training set.

Table 2. Tagging accuracy for POS taggers in last work

Accuracy	CRF	SVM	TreeTagger
All words	88.18	86.60	**89.26**

Comparing the actual obtained results with the results of our work reported in [18] (Table 2), we can confirm a significant improvement and these actual accuracies are very promising. This improvement is due essentially to two factors:

- The size of actual training corpus (75000 tokens instead of 60000 tokens).
- Most features used when training and generation of pos tagging model of the three techniques.

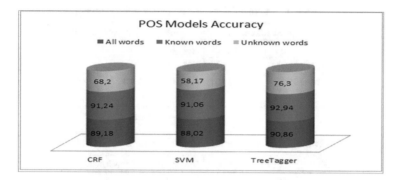

Fig. 2. Tagging accuracy for different techniques used

Figure 2 shows that the three taggers have different procedures for annotating unknown words and this is reflected in the difference in performance. Extensive analysis was performed of the errors made by the three different taggers. The analysis showed that the taggers make to a certain degree different types of errors. This can be used to combine the results of tagging with different taggers to improve tagging accuracy.

Moreover, to increase tagging accuracy it seems important to improve tagging of unknown words. This can be done in two ways, either by improving the methods that the taggers use for tagging unknown words or increasing the size of the lexicon used by the taggers.

5.3   Analysis of Error Types

Tables 3, 4 and 5 show the top four confusion classes for our three models. The most common types of errors are the confusion between proper noun and common noun and the confusion between adjective and common noun. These results from the fact that most of the proper nouns can be used as common nouns and most of the adjectives can be used as common nouns in Amazigh.

The current work looks at the existing tagsets of Amazigh being used for tagging corpora and analyzes them from two perspectives. First, the tagsets are analyzed to see

Table 3. Major misclassifications using TreeTagger

Tag	Total tokens	Error	Maximum misclassification
NP	500	102	NN (78)
ADJ	546	41	NN (28)
NN	2832	162	ADJ (122)
DET	190	13	PP (7)

Table 4. Major misclassifications using CRF model

Tag	Total tokens	Error	Maximum misclassification
NP	500	108	NN (82)
ADJ	546	51	NN (35)
NN	2832	182	ADJ (132)
DET	190	17	PP (9)

Table 5. Major misclassifications using SVM model

Tag	Total tokens	Error	Maximum misclassification
NP	500	116	NN (85)
ADJ	546	55	NN (40)
NN	2832	190	ADJ (141)
DET	190	19	PP (11)

their linguistic level differences. Second, they are compared based on their inter-tag confusion after training with three different POS taggers. These analyses are used to derive a more robust tagset. The results show that collapsing categories which are not syntactically motivated improves the tagging accuracy in general. Specifically, the different types of adverb are merged, because they may come in similar syntactic frames. Reduplicated categories are given the same category tag (instead of a special repetition tag). Units and dates are also not considered separately as the differences have been semantically motivated and they can be categorized with existing tags at syntactic level. Though, the measuring unit is currently treated as a noun, it could be collapsed as an adjective as well. The difference is sometimes lexical. NNP (Proper Noun) tag could also have been collapsed with NN (Common Noun), as Amazigh does not make clear between them at syntactic level. However, these two tags are kept separate due to their cross-linguistic importance.

Almost all the confusions are wrong assignment due to less number of instances in the training corpora, including errors due to long distance phenomena (Fig. 3).

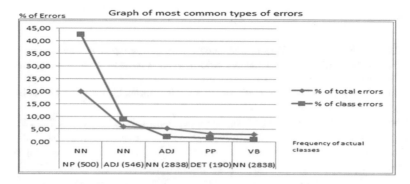

Fig. 3. Graph of most common error types

6 Conclusion and Perspectives

Natural Language is the medium for communication which is incorporated by every human being. One of the most important activities in processing natural languages is Part of Speech tagging. In POS Tagging we assign a Part of Speech tag to each word in a sentence and literature. POS tagging is one of the simplest, most constant and statistical model for many NLP application. POS Tagging is an initial stage of linguistics, text analysis like information retrieval, machine translator, text to speech synthesis, information extraction etc. Currently many tools are available to do this task of part of speech tagging. The POS taggers described here are very simple and efficient for Amazigh automatic tagging.

A stochastic approach includes frequency and probability or statistics. The problem with this approach is that it can come up with sequences of tags for sentences that are not acceptable according to the grammar rules of a language.

The performance of the current systems is good and the results achieved by these methods are excellent. We believe that future enhancements of this work would be to improve the tagging accuracy by increasing the size of tagged corpus and combining these single taggers. We plan also for the near future use other ML approaches and compare them to these approaches used in this word.

Finally, a hybrid solution for POS tagging in Amazigh can be proposed that can be used in other advanced NLP applications, which might use a combination of rule-based approach and the techniques mentioned earlier to achieve a significant gain in performance and performs with very good accuracy as English or other western languages in all domains.

References

1. Singh, J., Joshi, N., Mathur, I.: Development of Marathi part of speech tagger using statistical approach. In: International Conference on Advances in Computing, Communications and Informatics (2013)
2. Kumar, D., Singh Josan, G.: Part of speech tagger for morphologically rich Indian language: a survey. Int. J. Comput. Appl. **6**(5), 1–9 (2010)
3. Dhanalakshmi, V., Kumar, A., Shivapratap, G., Soman, K.P., Rajendran, S.: Tamil POS tagging using linear programming. Int. J. Recent Trends Eng. **1**(2), 166–169 (2009)
4. Kaur Sidhu, G., Kaur, N.: Role of machine translation and word sense disambiguation in natural language processing. IOSR J. Comput. Eng. (IOSR-JCE) **11**, 78–83 (2013)
5. Martin, J.H, Jurafsky, D.: Speech and Language Processing. International Edition (2010)
6. Van Guilder, L.: Automated Part of Speech Tagging: A Brief Overview, Handout for LING361. Georgetown University (1995)
7. Nakagawa, T., Uchimoto, K.: A hybrid approach to word segmentation and pos tagging. In: The 45th Annual Meeting of the ACL on Interactive Poster and Demonstration Sessions, pp. 217–220 (2007)
8. Charniak, E.: Statistical Language Learning. MIT Press, Cambridge (1993)
9. Brill, E.: Transformation-based error-driven learning and natural language processing: a case study in part-of-speech tagging. Comput. Linguist. **21**, 543–565 (1995)
10. Schmid, H.: Improvements in part-of-speech tagging with an application to German. In: Proceedings of the ACL SIGDAT-Workshop, pp. 13–26. Academic Publishers, Dordrecht (1999)
11. Ratnaparkhi, A.: A maximum entropy model for part-of-speech tagging. In: Proceedings of EMNLP, Philadelphia, USA (1996)
12. Kudo, T., Matsumoto, Y.: Use of support vector learning for chunk identification (2000)
13. Lafferty, J., McCallum, A., Pereira, F.: Conditional random fields: probabilistic models for segmenting and labeling sequence data. In: Proceedings of ICML 2001, pp. 282–289 (2001)
14. Chafiq, M.: [Forty four lessons in Amazigh]. éd. Arabo-africaines (1991)
15. Chaker, S.: Textes en linguistique berbère - introduction au domaine berbère, éditions du CNRS, pp. 232–242 (1984)
16. Boukhris, F., Boumalk, A., Elmoujahid, E., Souifi, H.: «La nouvelle grammaire de l'amazighe». IRCAM, Rabat (2008)
17. Amri, S., Zenkouar, L., Outahajala, M.: Amazigh part-of-speech tagging using Markov models and decision trees. IJCSIT J. **8**(5), 61–71 (2016)
18. Amri, S., Zenkouar, L., Outahajala, M.: Build a morphosyntaxically annotated amazigh corpus. In: Proceedings of the 2nd International Conference on Big Data, Cloud and Applications, Tetuan, Morocco (2017). https://doi.org/10.1145/3090354.3090362
19. Vapnik, V.N.: The Nature of Statistical Learning Theory. Springer, New York (1995)
20. Schmid, H.: Probabilistic part-of-speech tagging using decision trees. In: International Conference on New Methods in Language Processing, Manchester, UK, pp. 44–49 (1994)
21. Dermatas, E., George, K.: Automatic stochastic tagging of natural language texts. Comput. Linguist. **21**(2), 137–163 (1995)
22. Outahajala, M., Benajiba, Y., Rosso, P., Zenkouar, L.: POS tagging in amazigh using support vector machines and conditional random fields. In: Natural Language to Information Systems. LNCS, vol. 6716, pp. 238–241. Springer (2011). https://doi.org/10.1007/978-3-642-22327-3_28

Training and Evaluation of TreeTagger on Amazigh Corpus

Amri Samir[1(✉)], Zenkouar Lahbib[1], and Outahajala Mohamed[2]

[1] LEC Laboratory, EMI School, University Med V of Rabat, Rabat, Morocco
amri.samir@gmail.com
[2] CESIC Laboratory, IRCAM Institute, Rabat, Morocco

Abstract. Part of Speech (POS) tagging has high importance in the domain of Natural Language Processing (NLP). POS tagging determines grammatical category to any token, such as noun, verb, adjective, person, gender, etc. Some of the words are ambiguous in their categories and what tagging does is to clear of ambiguous word according to their context. Many taggers are designed with different approaches to reach high accuracy. In this paper we present a new tagging algorithm with a Machine Learning algorithm. This algorithm combines decision trees model and HMM model to tag Amazigh unknown words.

Part of Speech (POS) tagging is an essential part of text processing applications. A POS tagger assigns a tag to each word of its input text specifying its grammatical properties. One of the popular POS taggers is TreeTagger which was shown to have high accuracy in English and some other languages. It is always interesting to see how a method in one language performs on another language because it would give us insight into the difference and similarities of the languages. In case of statistical methods such as TreeTagger, this will have added practical advantages also. This paper presents creation of a POS tagged corpus and evaluation of TreeTagger on Amazigh text. The results of experiments on Amazigh text show that TreeTagger provides overall tagging accuracy of 93.15%, specifically, 93.78% on known words and 65.10% on unknown words.

Keywords: Amazigh · Corpus · TreeTagger · Machine learning
POS tagging

1 Introduction

Part-of-Speech (POS) tagging is an essential step to achieve the most natural language processing applications because it identifies the grammatical category of words belong text. Thus, POS taggers are an import and module for large public applications such as questions-answering systems, information extraction, information retrieval, machine translation… They can be used in many other applications such as text-to-speech or like a pre-processor for a parser; the parser can do it better but more expensive. In this paper, we decided to focus on POS tagging for the Amazigh language. Currently, TreeTagger (hence fore TT) is one of the most popular and most widely used tools thanks to its speed, its independent architecture of languages, and the quality of obtaining results. Therefore, we sought to develop a settings file TT for Amazigh.

© Springer International Publishing AG, part of Springer Nature 2018
M. Ben Ahmed and A. A. Boudhir (Eds.): SCAMS 2017, LNNS 37, pp. 563–573, 2018.
https://doi.org/10.1007/978-3-319-74500-8_52

Our work involves the construction of the dataset and the input pre-processing in order to run the two main modules: training program and tagger itself. For this reason, this work is the part of the still scarce set of tools and resources available for Amazigh automatic processing.

The rest of the paper is organized as follows. Section 2 puts the current article in context by overviewing related work. Section 3 describes the linguistic background of Amazigh language and presents the used Amazigh tagset and our training corpus. Experimental results are discussed in Sect. 4. Finally, we will report our conclusions and eventual future works in Sect. 5.

1.1 Tagging

The process of assigning a part-of-speech or lexical class marker to each word in a collection. There are many potential distinctions we can draw leading to potentially large tag sets.

To do POS tagging, we need to choose a standard set of tags to work with. We could pick very coarse tag sets as NN, VB, ADJ and ADV.

1.2 Problem

The major problems in the process of POS tagging are: Ambiguous words and unknown words. The first and foremost problem is with those words whose more than one tag can exist. This problem can be solved by emphasizing on context rather than single words. These can an easy task for humans, but not so for the automatic word taggers. In the process of tagging we can sometimes get such words that have different tag categories when they are used in different contexts. Thus, it is a very tedious job. This phenomenon is known as lexical ambiguity. But while occupying the same part of speech many words can have multiple meanings. Ambiguous words are the major problem in the part of speech tagging. Many words can have tags which are more than one. Some words can have different meaning in different context but they have same POS. In order to solve such problem single word is considered rather than the context.

2 Classification of POS Taggers and Motivation

2.1 Classification of POS tagger

A Part-Of-Speech Tagger (POS Tagger) is defined as a part of software which assigns parts of speech to every word of a language that it reads [1]. The approaches of POS tagging can be divided into three categories; rule based tagging, statistical tagging and hybrid tagging. A set of hand written rules are applied along with it the contextual information is used to assign POS tags to words in the rule based POS [2]. The disadvantage of this system is that it doesn't work when the text is not known. The problem being that it cannot predict the appropriate text. Thus, in order to achieve higher efficiency and accuracy in this system, exhaustive set of hand coded rules should be used. Frequency and probability are included in the statistical approach. The basic statistical approach works on the basis of the most frequently used tags for a specific word in the annotated training data and also this information is used to tag that word in the unannotated text. But the disadvantage of this system is that some sequences of tags can come up for sentences that are not correct according to the grammar rules of a certain language. Another approach is also there that is known as the hybrid approach. It may even perform better than statistical or rule based approaches. First of all the probabilistic features of the statistical method are used and then the set of hand coded language specific rules are applied in the hybrid approach.

In the area of POS tagging, many studies have been made. It reached excellent levels of performance through the use of discriminative models such as maximum entropy models [MaxEnt] [3–5], support vector machines [SVM] [6, 7] or Markov conditional fields [CRF] [8, 9].

Among stochastic models, bi-gram and tri-gram Hidden Markov Models (HMM) are quite popular. TNT [10] is a widely used stochastic trigram HMM tagger which uses a suffix analysis technique to estimate lexical probabilities for unknown tokens based on properties of the words in the training corpus which share the same suffix. The development of a stochastic tagger requires large amount of annotated text. Stochastic taggers with more than 95% word-level accuracy have been developed for English, German and other European languages, for which large labeled data are available.

Then decision trees have been used for POS tagging and parsing as in [11]. Decision tree induced from tagged corpora was used for part-of-speech disambiguation [12].

For Amazigh POS tagging, Outahajala et al. built a POS-tagger for Amazigh [13], as an under-resourced language. The data used to accomplish the work was manually collected and annotated. To help increase the performance of the tagger, they used machine learning techniques (SVM and CRF) and other resources or tools, such as dictionaries and word segmentation tools to process the text and extract features' sets consisting of lexical context and character n-grams. The corpus contained 20,000 tokens and was used to train their POS-tagger model. Therefore, there is a pressing necessity to develop an automatic Part-of-Speech tagger for Amazigh.

2.2   TreeTagger system

One common technique for predicting part-of-speech tags is decision tree learning [14, 15]. A decision tree learner is a machine learning algorithm that sequentially partitions the training data set based on attribute values, recursively partitioning the data until all attributes at a node can be classified with the same value [16].

In the same line, TreeTagger is a system for tagging text with part-of-speech and lemma information. It was developed in the TC project at University of Stuttgart. TreeTagger has been successfully used to tag German, English, French, Italian, Dutch, Spanish, Russian, etc. It is generally adaptable to other languages if a lexicon and a manually tagged training corpus are available.

The heart of TreeTagger, as its name suggests, is "estimation of transition probabilities with a binary decision tree" [15].

The initial stage of building the decision tree happens during the training phase. It will parse through the text and analyze trigrams, inserting each unigram into the tree.

Probabilities for which tag to use are determined for a given node of the tree based on the information obtained from the two previous nodes (trigram). Once the tree is created, its nodes are pruned. If the information gain of a particular node is below a defined threshold, its children nodes are removed. Figure 1 represents a simplified version of a decision tree using an example of Amazigh sentence.

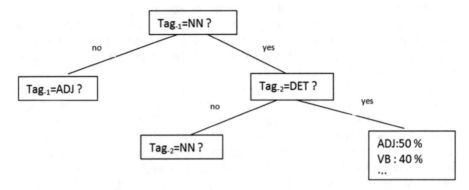

Fig. 1. A simplified decision tree from Amazigh text

2.3 Goals and Motivation

A lot of work has been done in part of speech tagging of several languages, such as English. While some work has been done on the part of speech tagging of Amazigh, the effort is still in its infancy. Very little work has been done previously with part of speech tagging of the Amazigh.

So, the primary goal of this work is to develop a reasonably good accuracy part-of-speech tagger for Amazigh. To address this broad objective, we identify the following goals:

- We wish to investigate different machine learning algorithms to develop a POS tagger for Amazigh.
- This work also includes the development of a reasonably good amount of annotated corpora for Amazigh, which will directly facilitate several NLP applications.
- Amazigh is a morphologically-rich language. We wish to use the morphological features of a word to enable us to develop a POS tagger with limited resource.
- Finally, we aim to explore the appropriateness of different machine learning techniques by a set of experiments and also a comparative study of the accuracies obtained by working with different POS tagging methods.

3   Linguistic Background

3.1   Amazigh Language

Amazigh called also Berber belongs to the Hamito-Semitic "Afro-Asiatic" languages [17]. Amazigh is spoken in Morocco, Algeria, Tunisia, Libya; it is also spoken by many other communities in parts of Niger and Mali. In Morocco Amazigh language uses different dialects in its standardization (Tachelhit, Tarifit and Tamazight). Amazigh has a complex morphology and the process of its standardization is performed via different dialects. Amazigh NLP presents many challenges for researchers. Its major features are:

- It has its own script: the Tifinagh, which is written from left to right. The translitiration into Latin alphabet is used in all the examples in this article.
- It does not contain uppercase.
- Like other natural language, Amazigh presents for NLP ambiguities in grammar classes, named entities, meaning, etc. For example, grammatically the word (tazla) can function as verb "ar tazla", meaning "over it" or as name "race", etc. At the semantic level, a word can have several meanings; for example, the word (axam) depending on the context can mean family or tent, etc.
- As most languages whose research in NLP is new, Amazigh is not endowed with linguistic resources and NLP tools.
- Amazigh, like most of the languages which have only recently started being investigated for NLP, still suffers from the scarcity of language processing tools and resources.

3.2   Amazigh Script

Like any language that passes through oral to written mode, Amazigh language has been in need of a graphic system. In Morocco, the choice ultimately fell on Tifinaghefor technical, historical and symbolic reasons. Since the Royal declaration on 2003, Tifinaghe has become the official graphic system for writing Amazigh. Thus, IRCAM has developed an alphabet system called Tifinaghe-IRCAM. This alphabet is based on a graphic system towards phonological tendency. This system does not retain all the phonetic realizations produced, but only those that are functional. It is written from left to right and contains 33 graphemes which correspond to:

- 27 consonants including: the labials (ⵝ, ⴱ, ⵛ), dentals (ⵜ, ⵏ, ⴻ, ⴻ, ⵉ, ⵎ, ⵇ, ⵀ), the alveolars (ⴰ, ⵌ, ⵂ, ⵥ), the palatals (ⵛ, ⵊ), the velar (ⵔ, ⵅ), the labiovelars (ⵔ ⵯ, ⵅ ⵯ), the uvulars (ⵒ, ⵅ, ⵖ), the pharyngeals (ⵄ, ⵃ) and the laryngeal (ⴹ);
- 2 semi-consonants: ⵢ and ⵓ;
- 4 vowels: three full vowels ⵄ, ⵣ, ⵣ and neutral vowel (or schwa) ⵣ.

3.3   Transliteration system

We have decided to use a specific writing system based on ASCII characters for technical reasons [6]. So, correspondences between the different writing systems and transliteration correspondences are shown in Table 1.

Table 1. Mapping from existing writing system and the chosen writing system

Tifinaghe Unicode		Transliteration		Chosen writing
Code	Character	Latin	Arabic	system
U+2D30	ⵀ	A	ا	A
U+2D31	ⴱ	B	ب	B
U+2D33	ⵅ	G	گ	G
U+2D33&U+2D6F	ⵅ ⵯ	Gw	گ	Gw
U+2D37	ⴷ	D	د	D
U+2D39	ⴹ	ḍ	ض	D
U+2D3B	ⴻ	E		E
U+2D3C	ⴼ	F	ف	F
U+2D3D	ⴽ	K	ک	K
U+2D3D&+2D6F	ⴽ ⵯ	Kw	+ گ	kw
U+2D40	ⵀ	H	ھ	H
U+2D43	ⴰ	ḥ	ح	H
U+2D44	ⵃ	E	ع	E
U+2D44	ⴳ	X	خ	X
U+2D45	ⵇ	Q	ق	Q
U+2D47	ⵉ	I	ي	I
U+2D47	ⵊ	J	ج	J
U+2D47	ⵍ	L	ل	L
U+2D47	ⵎ	M	م	M
U+2D47	ⵏ	N	ن	N
U+2D47	ⵓ	U	و	U
U+2D47	ⵔ	R	ر	R
U+2D47	ⵕ	ṛ	ر	R
U+2D47	ⵖ	Y	غ	G
U+2D47	ⵙ	S	س	S
U+2D47	ⵚ	ṣ	ص	S
U+2D47	ⵛ	C	ش	C
U+2D47	ⵜ	T	ت	T
U+2D47	ⵟ	ṭ	ط	T
U+2D47	ⵡ	W	ؤ	W
U+2D47	ⵢ	Y	ي	Y
U+2D47	ⵣ	Z	ز	Z

3.4   Tagsets and Corpus

AmazighTagset

A tagset is a collection of labels which represent word classes. A coarse-grained tag set might only distinguish main word classes such as adjectives or verbs, while more fine-grained tagsets also make distinctions within the broad word classes, e.g. distinguishing between verbs in past and future tense. This is an important step for a lexical labeling work to be based on the word classes of language and shall reflect all morphosyntactic relationships words of the Amazigh corpus (Table 2).

Table 2. Amazigh tag set

Tag	Attributes with the number of values
Noun	gender(3), number(3), state(2), derivation(2), Sub classification POS (4), number(3), gender(3), person(3)
Verb	gender(3), number(3), person(3), aspect(3), negation(2), form(2), derivation (2), voice(2)
Adjective	gender(3), number(3), state(2), derivation(2), POS subclassification (3)
Pronoun	gender(3), number(3), person(3), POS subclassification (7), deictic(3)
Determinant	gender(3), number (3), Sub classification POS (11), deictic(3)
Adverb	Sub classification POS (6)
Preposition	gender(3), number(3), person(3), number(3), gender(3)
Conjunction	POS subclassification(2)
Interjection	Interjection
Particle	POS subclassification (7)
Focalizer	Focalizer
Foreign word	POS subclassification (5), gender (3), number (3)
Punctuation	Type de la marque de ponctuation (16)

Corpus

A corpus is a collection of language data that are selected and organized according to explicit linguistic criteria to serve as a sample of jobs determined a language. Generally, a corpus contains up few millions of words and can be lemmatized and annotated with information about the parts of speech. Among the corpus, there is the British National Corpus [18] (100 million words) and the American National Corpus [19] (20 million words).

A balanced corpus would provide a wide selection of different types of texts and from various sources such as newspapers, books, encyclopedias or the web. For the Moroccan Amazigh language, it was difficult to find ready-made resources. We can just mention the manually annotated corpus of Outahajala et al. [20, 21]. This corpus contains 20 k words using a tagset described in Table 3 that is why we decided to build our own corpus. In order to have a vocabulary sufficiently large, we took texts from the web sites of some

Table 3. Constituents of Amazigh corpus

Source	%
Online newspapers and periodicals	22.7
Primary school textbooks	15
Texts from websites of organizations	10.4
Texts from government websites	8.6
Miscellany	16.5
Blog	15
Texts from website of IRCAM	12.8

Moroccan ministries, texts from IRCAM website[1] and from primary school textbooks…
etc. We have collected these different resources; after that, we have cleaned them and
convert them to text format especially UTF-8 Unicode. The Table 3 provides source
statistics of our corpus, which includes 6714 sentences (approximately 75,340 words).

4   Results and Discussion

4.1   Results

In the majority of the part of speech tagging approaches, the sample is often subdivided
into "training" and "test" sets. The training set is generally used for learning, i.e. fitting
the parameters of the tagger. The test set is for assessing the performance of the tagger.
In our experiments, we repeated the experiments five times and each time we used a
random sample of sentences, 20% from the gold corpus, as the test set and used the rest
of the files for the training. Table 4 shows the number of tokens and their percentages
in the training and test sets respectively.

Table 4. Number of tokens in training and test sets

Run	Training Tokens/Percent	Test Tokens/Percent	Total
1	60140/79.82	15200/20.18	75340
2	59540/79.02	15800/20.98	75340
3	60040/79.69	15300/20.31	75340
4	60480/80.27	14860/19.73	75340
5	59900/79.50	15440/20.50	75340
Avg.	60020/79.66	15320/20.34	

5-fold cross validation was used to calculate the accuracy of a tagger during all five
rounds of evaluation processes. We divided the gold standard data into five equal folds
using stratified sampling by the sentence length. Then we used with four folds of the

---

[1] www.ircam.ma.

data for training. In the training process, the tagger "learned" the language model from the training data set based on statistical possibility. The remaining one fold was for testing. During the testing process, the tagger assigned tags to the testing data set (a version of data without tags) with the learned language model. After this, we yielded a tagged version of the testing data set. We evaluated the tagged version of the testing data with its "gold standard" version, and get accuracy for one fold of the data set. After five times of training and testing, the whole data set was tested. We then calculated the average of five accuracies as the final accuracy of the tagger. The tagging accuracy, therefore describes how well the tagger performs.

Considering the tagging accuracy as the percentage of correctly assigned tags, we have evaluated the performance of the TreeTagger from two different aspects: (1) the overall accuracy (taking into account all tokens in the test corpus) and (2) the accuracy for known and unknown words, respectively. The latter is interesting since after training the tagger, it could be used for other text than the training text. It is interesting to know how it would cope with words that did not appear in its training. Tables 5, 6 and 7 depict the results of the experiments. For each run, Table 5 shows the percentage of seen words (words that exist in the training set), number of tokens in the test set, number of tokens correctly tagged and the percentage of accuracy for that run. Similarly, Table 6 shows the same for words that are new for the tagger. Table 7 shows the overall result for each run and its average in general:

1. The overall part-of-speech tagging accuracy is around 93.15%.
2. The accuracy for known tokens is significantly higher than that for unknown tokens (93.78% vs. 65.10%). It shows 28.68% points accuracy difference between the words seen before and those not seen before.

Table 5. Known tokens results

Run	Tokens	Correct	Accuracy
1	14870	13921	93.62
2	15300	14380	93.98
3	15100	14190	93.97
4	14562	13665	93.84
5	15120	14125	93.41
Avg.	15292	14339	93.78

Table 6. Unknown tokens results

Run	Tokens	Correct	Accuracy
1	330	220	66.67
2	500	312	62.28
3	200	134	67.00
4	298	203	68.12
5	320	204	63.73
Avg.	329.6	214.6	65.10

Table 7. Overall results

Run	Tokens	Correct	Accuracy
1	15200	14141	93.03
2	15800	14692	92.98
3	15300	14324	93.62
4	14860	13868	93.32
5	15440	14329	92.80
Avg.	15320	14270.8	93.15

4.2   Discussion

In our work, we used a data set of 75000 tokens and 28 tags, we have studied also the resulting tagged corpora and we concluded that Most of the errors could be categorized as follows:

- Errors in the case of the word are the highest. Those are partially due to the fact that some of the tags do not reflect the case of the word, and hence it is hard for the learner to conclude the reason of the next word being given its tag, examples of that are proper nouns, common noun and pronouns.
- Unknown proper nouns (of people and places) cannot be guessed. Only few rules may lead to realizing a proper noun. Having a large corpus would reduce this problem by inserting many names in the lexicon.
- A distinction between adverb and compounds is not easily guessed by our model.

Taking in consideration the tag set used we worked with, and the unavailability of a standard truth corpus, we think the results obtained here are very promising, and can be enhanced by many actions like: enlarging the training corpus, and enhancing the lexical analysis program.

5   Conclusion and Perspectives

Part-of-speech tagging now is a relatively mature field. Many techniques have been explored. Taggers originally intended as a pre-processing step for chunking and parsing, but today it issued for named entity recognition, in message extraction systems, for text-based information for speech recognition, for generating intonation in speech production systems and as a component in many other applications.

It has aided in many linguistic research on language usage. The Parts of Speech Tagging for Amazigh language using the statistical approach has been discussed. The system works fine with the Unicode data. The POS were able to assign tags to all the words in the test case. These also focus on the point that a statistical approach can also work well with highly morphologically and inflectionally rich languages like Amazigh.

As future work, there are some improvements to be considered to enhance the performance of the system. A corpus of different types of articles may be considered to train the model such as social media content, stories, and sport reports and so on in order to increase the accuracy of the system for these types. The variety and the size of the training data is an important factor to enhance the accuracy of the Amazigh POS tagger for assigning POS tags to unknown words correctly.

References

1. Manning, C., Schütze, H.: Foundations of Statistical Natural Language Processing. The MIT Press, Cambridge (1999)
2. Brill, E.: Transformation-based error-driven learning and natural language processing: A case study in part-of-speech tagging. Comput. Linguist. **21**(4), 543–565 (1995)
3. Ratnaparkhi, A.: A maximum entropy model for part-of-speech tagging. In: Proceedings of EMNLP, Philadelphia, USA (1996)
4. Toutanova, K., Manning, C.: Enriching the knowledge sources used in a maximum entropy part-of speech tagger. In: EMNLP/VLC 1999, pp. 63–71 (1999)
5. Toutanova, K., Dan, K., Manning, C., Yoram, S.: Feature-rich part-of-speech tagging with a cyclic dependency network. In: Proceedings of HLT-NAACL 2003, pp. 252–259 (2003)
6. Giménez, J., Màrquez, L.: SVMTool: a general POS tagger generator based on support vector machines. In: Proceedings of the 4th International Conference on Language Resources and Evaluation (LREC), Lisbon, Portugal, pp. 43–46 (2004)
7. Kudo, T., Matsumoto, Y.: Use of support vector learning for chunk identification (2000)
8. Lafferty, J., McCallum, A., Pereira, F.: Conditional random fields: probabilistic models for segmenting and labeling sequence data. In: Proceedings of ICML 2001, pp. 282–289 (2001)
9. Tsuruoka, Y., Tsujii, J., Ananiadou, S.: Fast full parsing by linear-chain conditional random fields. In: Proceedings of the 12th Conference of the European Chapter of the Association for Computational Linguistics (EACL 2009), pp. 790–798 (2009)
10. Brants, T.: TnT - a statistical part-of-speech tagger. In: ANLP 2000, Seattle, pp. 224–231 (2000)
11. Black, E., Jelinek, F., Lafferty, J., Mercer, R., Roukos, S.: Decision tree models applied to the labeling of text with parts-of-speech. In: Proceedings of the DARPA workshop on Speech and Natural Language, Harriman, New York (1992)
12. Màrquez, L., Rodríguez, H.: Part-of-speech tagging using decision trees. In: Nédellec, C., Rouveirol, C. (eds.) Proceedings of the 10th EuropeanConference on Machine Learning, ECML 1998. Lecture Notes in AI, Chemnitz, vol. 1398, pp. 25–36 (1998)
13. Outahajala, M., Benajiba, Y., Rosso, P., Zenkouar, L.: POS tagging in amazigh using support vector machines and conditional random fields. In: Natural Language to Information Systems. LNCS, vol. 6716, pp. 238–241. Springer, Heidelberg (2011). https://doi.org/10.1007/978-3-642-22327-3_28
14. Schmid, H.: Probabilistic part-of-speech tagging using decision trees. In: International Conference on New Methods in Language Processing, pp. 44–49 (1994)
15. Schmid, H.: Improvements in part-of-speech tagging with an application to German (1995)
16. Bishop, C.: Pattern Recognition and Machine Learning. Springer, New York (2006)
17. Chafiq, M.: [Forty four lessons in Amazigh]. éd. Arabo-africaines (1991)
18. Aston, G., Burnard, L.: The British National Corpus. Edinburgh University Press, 256 p. (1998)
19. Ide, N., Macleod, C., Grishman, R.: The american national corpus: a standardized resource of American English. In: Proceedings of Corpus Linguistics, vol. 3 (2001)
20. Outahajala, M., Rosso, P., Zenkouar, L.: Building an annotated corpus for Amazigh. In: Proceedings of 4th International Conference on Amazigh and ICT, Rabat, Morocco (2011)
21. Outahajala, M., Zenkouar, L., Rosso, P.: Construction d'un grand corpus annoté pour la langue Amazigh. La revue Etudes et Documents Berbères 33, pp.57–74 (2014)

Knowledge-Based Multicriteria Spatial Decision Support System (MC-SDSS) for Trends Assessment of Settlements Suitability

Waleed Lagrab[(⊠)] and Noura Aknin

Computer Science, Operational Research and Applied Statistics Laboratory,
Information Technology and Modeling Systems Research Unit, Abdelmalek
Essaâdi University, M'Hannech II, 93030 Tetouan, Morocco
waleed.lagrab@uae.ac.ma, aknin@ieee.org

Abstract. Spatial data mining is the discovery of interesting hidden patterns and characteristics that may be implicitly in spatial databases. This paper aims to produce a descriptive model for examining the suitability in settlements by applying various machine learning techniques to figure out the knowledge discovery in spatial databases (KDSD). The study illustrates the unique hallmark that characterizes the spatial data mining by conducting the data mining algorithms. Moreover, the study presents the importance of spatial data mining and discussed multiple data sets preprocessing, classification functions, clustering and outlier detection in directions supervised learning for extracting classification rules and assessing the local amenity based on rules reliability. The classification accuracy among the three methods of the classifier algorithms (Decision Tree, Rule-Based, and Bayesian) is also compared, thereby determining the most suitable classifier by experiments performance evaluation of the training and test set.

Keywords: SDSS · Data mining · Classification · Suitability analysis
Knowledge discovery · Educational facility

1 Introduction

Knowledge represents a significant explanatory role for humanity in all areas of life, whilst the human decision making requires a previous experience to make the right decision. Recently there have been many research activities on knowledge discovery in spatial database. Hence, among the data mining techniques developed in recent years, data mining methods are including generalization, characterization, clustering, classification, association, evolution, pattern matching, data visualization and meta-rule guided mining [1]. These methods could be adapted to analysis and improve the distribution of educational facility [2]. The selecting of the most appropriate site for a school is a significant consideration of the community. However, the locations, size, and shape of schools can essentially affect the educational quality and opportunities for students [3]. Educational facilities should meet the develop selection criteria carefully.

© Springer International Publishing AG, part of Springer Nature 2018
M. Ben Ahmed and A. A. Boudhir (Eds.): SCAMS 2017, LNNS 37, pp. 574–586, 2018.
https://doi.org/10.1007/978-3-319-74500-8_53

The criteria should be relied not only on the current requirements, but also the anticipated needs, in fact, this is not easy tasks. Thus, the construction proceedings should following multicriteria, which are listed in order of importance (i.e. Safety, location, environment, accessibility, availability, public service, and so on) [4, 5].

2   Related Work

The integration of data mining technics with GIS tools gives an extremely powerful to extract the knowledge from spatial data sets. Therefore, some of the extensive literature reviews regarding the suitability analysis based on SDSS. The spatial data mining has been published in many studies and surveys to shape a new method to incorporate and analysis the information spatially for example [6, 7]; these study highlighted the recent theoretical and applied research in spatial data mining and knowledge discovery. The existing methods for spatial data mining were also surveyed out their strengths and weaknesses which would be considered useful to future directions and suggestions for the spatial data mining [8].

Spatial data mining really is a promising area to determine many phenomena in the land cover, [9] presented spatial characteristics of underground pipeline network. A new study tried to present the land use of changing district with spatial data mining through the Geo-statistics method [10]. The suitability assessment has also been attracting a significant amount of research [11, 12]. In another trend, there are many types of research have already been performed in SDSS works (i.e., [13, 14]). Many attempts were conducted to develop the SDSS for instance, the development of a basic MCSDSS was illustrated by [15] that integrates spatial information techniques such as GIS and Multicriteria Analysis. Also [16] tried to develop spatial decision support system in Kenya to assist land use scientists, agricultural extension support personnel, and farmers to classify and characterize land quality. Moreover, new study demonstrated the validity of this methodological approach Southern Italy, the study applied on MC-SDSS to a study area of thirteen rural municipalities located in Apulia Region [17]. Previous studies based on knowledge discovery and the suitability performance evaluation. They attempted to develop a practical model or theoretical approach for suitability analysis to support the spatial decision making through using the GIS and data mining techniques. However, the finding varies among previous studies according to applied techniques and the data sets (or study area). In our study, we aim to develop a descriptive spatial data mining model using real world data sets of educational facilities that were gained from Mukalla city in Yemen.

3   Methodology

The classification is the major problem for information reclamation and employing data mining techniques to learn ranking functions that are viewed as a promising approach to information retrieval. To evaluate data mining classifiers, we have chosen a geographic database of Mukalla city as benchmark data sets. These data sets consist of all information about districts, blocks, educational facilities, and land use in the study area. Data sets connected with the semantic relationship, and contain kind of interesting attributes.

3.1   Data Description

The selected data set is represented spatial and non-spatial data about education services (Kindergarten, Elementary, and Secondary School), it has been distributed to many districts in the study area, and each district involves the area, population, land uses, and the relation between the district and neighborhood. Districts were divided into multiple blocks, and land uses have also been included inside blocks. Although, our data set contains a small amount of instance around 119 records. However, the increase in the number of sensitive attributes can probably improve the knowledge extraction from data set. Data mining requires error-free data, and miss value should also be handled to ensure the best data quality. The research has been adopted precisely to distinguish the authentic sources of evidence acquire, tools and methods for getting a more honest evaluation of this study. Therefore, a broader range of data acquired from a wide array of sources as (see Table 1).

Table 1. Data type and sources

Class	Type	#instant	#attribute	Data format	Data source	Date
School	KG	23	14	Geo-reference/Descriptive data	MoE[a], Survey field	2014
	ES	80				
	SS	23				
block	East	25	6	Descriptive data	Social affairs office,	2014
	Mid	66			GALSUP[b], MoS[c]	
	West	28				2015
Land use	Fuel	19	7	Geo-reference/Descriptive data	Oil office	2014
	Street	1			Open street map	2015
	Workshop	33			Industrial office	2014

[a]*Ministry of Education.*
[b]*General Authority for Land Survey & Urban Planning.*
[c]*Ministry of Statistics.*

3.2   Membership Function

Most important attributes in dataset are a nominal type. Mostly, these attributes give an indication meaningless. Therefore, it was converted to fuzzy membership value with a sensitive range in the classification process can be relied upon to produce rules. For example, to represent the distance between land use (fuel station, workshops, highway) and schools a fuzzy set might assign a degree of membership of 25 m is "V.Bad", 50 m is "Bad", 75 m is "Fair", 100 m as "Good", and more than 100 M as "Excellent". Multiple attributes in the data set have been converted into membership function such as (#schools, population, student capacity, building state, etc.). The value of these attributes was converted into the score (from nominal into numeric) based on schools, and transformed into membership function according to school's average in each block.

3.3 Feature Selection

The data mining prediction highly depends on attributes, whereas the data set contains a mixture of attributes, some of which are appropriate to making predictions. Feature Selection (Attribute Selection) is a mythology which supports the automated research for the superlative subset of attributes in the data set. Attribute has provided an ideal solution to the problem to be solved through feature subset evaluated. Many benefits can be obtained by applying the feature selection methods such as reduce the over fitting, enhance the classification accuracy, and reduce the training time. Waikato Environment for Knowledge Analysis (Weka) data mining platform involves many methods of attribute selection, for this purpose, we used two of the common methods of Attribute Evaluator: (1) WrapperSubsetEval; (2) ClassifierSubsetEval, the classifiers extract better classifications accuracy with few attributes, the Bestfirst search method has also been selected for 10-fold cross validation.

4 Clustering-Based Mixtures of Experts

Different methods of data collection have been used such as a formal document, observation, and survey. Then the data was analyzed through statistic and data mining programs such as (SPSS and Weka).

Mixture of experts (ME) relies on the feedback from various experts. A ME is one of the most common methods extensively in decision-making system (see Fig. 1). Whereas, the variance among experts is considered extremely significant for the success of the combination of methods conditions. Especially, in our case which needs to identify the situation of settlements in the study area in terms of the suitability of educational facilities. Thus, data acquired phase was followed by seeking professional experts on a specified subject help to specify significant attributes, although experts are capable of using their knowledge in the decision making, they are frequently incapable of formulating their knowledge explicitly in a form sufficiently systematic, correct, and complete to form a computer system. Many replies were obtained from experts to determine the amenity in settlements, the expert's opinions were categorized into three levels:

- Expert 1: #Schools, #Students, #Teachers, and #Classrooms.
- Expert 2: # Schools, Building State, Duration, Governmental Schools.
- Expert 3: District Area, Population Density, Extendibility, Land use (benefit, danger), Space Area, Accessibility, Distribution of other schools.

Those experts assisted in classifying districts into three classes (**Bad, Fair**, and **Good**).

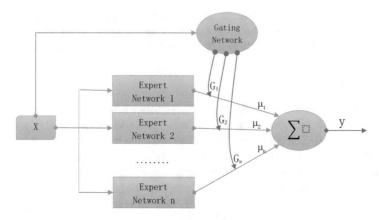

Fig. 1. Mixtures of expert's schema

5   Experiments Study and Discussions

The main aim of these experiments is to design and implement an automation model for Suitable Decision Support System (SuDSS). Several real-life data have been used in our experiments. These data are acquired from several resources in the study area as recently mentioned. Three methods of the most popular classification algorithms are conducted to produce a descriptive model. The experiments were carried out to build classification rule-set model and classify the city districts or blocks to three classes, these are (Good, Fair, Bad). Thus, the most important step is how to use the significant attribute to make sense of data, then how to build descriptive data mining model.

5.1   Classifier Performance Evaluation and Comparison

The classifier performance evolution considers a decisive phase for specifying the most suitable classifier or the greatest set of parameters for classifiers. The experiment has shown a good performance in the evolution of classifiers for the intrusion detection problem. However, the classifier misclassification rate is frequently due to lack of the misclassification of different classes might have more or less severe consequences.

The classification model can be further enhanced through a combination of relevant features and increase the observations, including new descriptors of spatial and non-spatial information. Many input variables may result in a model overfitting for most machine learning algorithms and the evolutionary data can be used to improve classifier performance.

From empirical comparisons, we gained many of the exceptional results. Interestingly, all classifiers providing better performance of accuracy level between 72.27 and 85.72, however, the decision tree classifier (DTCs) came in the first place. The classifier C4.5 is slightly more accurate than the Random Forest (RF) classifier, the C4.5 and RF predictors achieved the predictive strength of 85.72% and 84.03%, respectively. The Precision, Recall, and F-Measure of Bayesian, Rule-base classifier are also less than that of the DTCs as described in Table 2 and Fig. 2. As well as, The

JRip and decision table classifier realized best classification accuracy after the DTCs where the JRip attained 83.19% and 81.51 has been achieved by decision table classifier. Figure 2 presents the comparing classifier's parameters.

Table 2. Comparison of classification results

Classification	Classifier	CA (%)	TP	FP	Precision	Recall	F-Measure	ROC area
Bayesian	BayNet	72.27	0.72	0.15	0.714	0.723	0.713	0.885
	naïve Bayes	73.95	0.739	0.135	0.759	0.739	0.728	0.886
Decision tree	J48(C4.5)	85.72	0.857	0.085	0.866	0.857	0.854	0.908
	Randomforest	84.03	0.84	0.09	0.837	0.84	0.835	0.939
Rule–based	Decision table	81.51	0.815	0.113	0.819	0.815	0.794	0.885
	JRip	83.19	0.832	0.092	0.825	0.832	0.825	0.925

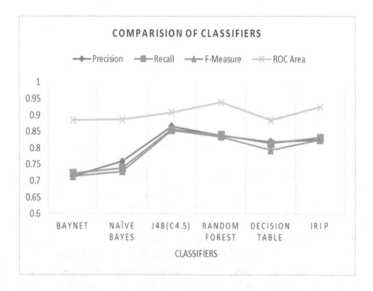

Fig. 2. Comparing classifier's parameters based on confusion matrix.

5.2 Data Generation and Comparing Classifier Results

The previous experiment on TrDS1 Figured out a good classification lead to supervised learning. In spite of a lack of enough observation. Consequently, we try to increase the number of observations by generating new instances randomly, so that the processing efficiency can be enhanced to better. Thus, the TrDS21 was magnified through increasing the instances number by "Weka- Resample" in pre-process data tool in "filters supervised" which contains filters for (discretization, normalization, resampling, attribute selection, transforming and combining attributes). Resample filter produces random subsamples of a data set using either sampling with replacement or without replacement. The TrDS1 must fit entirely in memory. A number of instances in the generated data set may be specified. Thus, the new generated data set (TrDS2) was duplicated to have 238 instances. Then, the new experiment was conducted on TrDS2 with the same classifiers using Weka's Experimenter Environment, and the experiment performs test and training set of the TrDS2 learning is presented in Table 3.

Table 3. The CA between original and generated data set.

Classification	Classifier	Original data set		Resample data set	
		TrDS1	TeDS1	TrDS2	TeDS2
Bayesian	BayNet	72.27	75.0	75.80	85.42
	naïve Bayes	73.95	75.0	79.54	81.25
Decision tree	J48(C4.5)	85.72	91.67	95.97	93.75
	Random forest	84.03	83.33	98.53	97.92
Rule-based	Decision table	81.51	83.33	85.92	87.5
	JRip	83.19	75.0	95.84	89.58

5.3   Experimental Results

The experiments were conducted using Weka data mining platform, it is an open source data mining tool for our experiment. WEKA is developed by the University of Waikato in New Zealand that implements data mining algorithms using the JAVA language, it is distributed under the GNU Public License. The eight commonly classifiers have been used to predict classification in training data sets, experimental results demonstrate the effectiveness and the efficiency of data mining classifiers to train spatial data sets.

We implemented the selected classifiers in data sets with different scenarios. Firstly, compared the CA and education time. The DT classifiers (C4.5 and RF) shown to be more efficient experimentally (see Table 2), experimental evidence exhibited that DT outperforms the six others classifiers algorithms in term of the classification accuracy, and it took to have the response time of fewer than two seconds. Secondly, the outlier elimination was conducted on the training set using InterquartileRange in Weka pre-pressing. Although the outlier and extreme value have been eliminated around 18 instances, the CA of most classifiers was not significantly affected.

Finally, we tried to magnify the number of observations of TrDS21 by Preprocess-Resample data tool to generate a new data set named TrDS2 with 238 labeled instances. The both data set TrDS1 and TrDS2 are split into a training set and a test set, the split percentage is 20% testing data set and 80% training data set. The experiment has been carried out on the TrDS2 by applying the same previous classifiers to evaluate the experimental performance of the generated data model with 10-fold cross validation; it is to be noted that the experimental results enhanced to better. Especially, the CA of decision trees classifiers. Whereas, the C4.5 achieved 95.97% in training set and 93.75% in the test set. In this experiment, it is also noted that the RF classifier has reached a 98.53% and 97.92, in training and test set, consequently. The JRip classifier has also achieved significant CA; experimental results are presented in Table 3; Figs. 3 and 4.

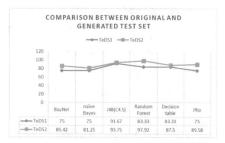

Fig. 3. Comparison between original and generated training set.

Fig. 4. Comparison between original and generated test set.

5.4   Rules Mining and Sequential Pattern

The main purpose of our experiment is to discover the meaningful pattern and rules. Generated classification rules mining that depending on the J48 classifier were the same attributes which Figured out as the most important attributes. Produced rules introduce the helpful features for decision making, and help to reduce the time and cost during the collection and data entry phase. However, the acquired attributes are frequently in large number. Thus, we cannot specify the either of them is important, as well-known plenty of attributes are often pretty good when they are carefully collected. Data mining algorithms have the ability to discover rules with most significant attributes. However, the discovery of rules process takes several steps, one of the most important is data clustering as mentioned earlier, this step depending on, the experts in the particular field, in our case need to determine the block's or district's suitability according to the existing educational facilities. Certainly, sometimes the expert's opinion is uncertainty and needs to improve by increases consult more experts. The mixture of expert's model is one of the best popular and interesting methods that has major potential to improve the performance in data mining. It works to divide the problem scope between experts as mentioned in Sect. 4.

J48 pruned tree – Rule set
  Std_Rate <= 0.0141
  | *Nbr_BlkRate* <= 0.979
  | | *ESF* <= 4.238: *Bad* (51.0)
  | | *ESF* > 4.238: *Fair* (2.0)
  | *Nbr_BlkRate* > 0.979
  | | *Nbr_Cap* <= 0.16: *Fair* (12.0/1.0)
  | | *Nbr_Cap* > 0.16: *Good* (3.0)
  Std_Rate > 0.0141
  | *Owner* <= 0.33: *Good* (9.0)
  | *Owner* > 0.33
  | | *Community* <= 0.5
  | | | *Nbr_Cap* <= 0.24: *Fair* (37.0/4.0)
  | | | *Nbr_Cap* > 0.24: *Good* (3.0)
  | | *Community* > 0.5: *Good* (2.0)

  Number of Leaves: 8
    Size of the tree: 15

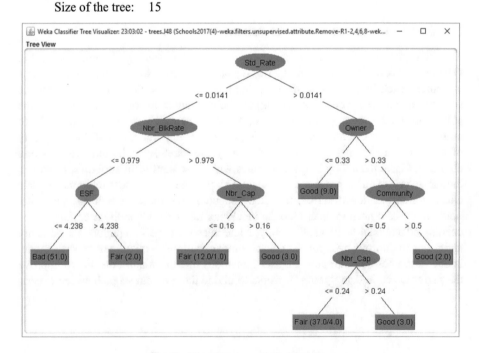

Fig. 5. Visualize pruned decision tree

JRip pruned rule set

$(Std\_Rate >= 0.0319)$ *and* $(Blk_SchRate >= 0.962) => $ *Cluster*
       $= Good$ $(14.0/1.0)$
$(Blk\_SchRate >= 1.735)$ *and* $(Std_Rate >= 0.0244) => $ *Cluster*
       $= Good$ $(3.0/0.0)$
$(Std_Rate >= 0.0138) => $ *Cluster* $= Fair$ $(36.0/3.0)$
$(Nbr\_BlkRate >= 1.028)$ *and* $(Nbr_Cap <= 0.15) => $ *Cluster*
       $= Fair$ $(10.0/1.0)$
$=> $ *Cluster* $= Bad$ $(56.0/6.0)$
   Number of Rules: 5
   Time is taken to build model: 0.05 seconds

5.5   Decision Tree Rules

Test mode: 10-fold cross-validation
Time is taken to build model: 0.09 s.

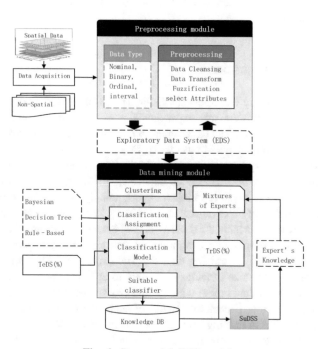

Fig. 6. Proposed SuDSS model

6   Proposed-Descriptive Model

The proposed model is a descriptive model of Suitability Decision Support System (SuDSS). The model aims at further quantifying the impact of the problems and solutions. By examining data mining algorithms in this way, the overall process flow is compared in Fig. 6.

Components of SuDSS model can be divided into the following phases:

- Data acquisition: The spatial data identifies the geographic location of features or boundaries and refers to as a line, point, and polygon. While non-spatial data or conventional data describes and defines the characteristics of spatial data.
- Preprocessing module: This module is responsible for the preparation and processing the database, and specifies the proper attribute for a data mining process. As well as, converts the information into reliable and sense data.
- Exploratory data system: Data visualization is a great manner helps to understand the nature of the data set through summarizing analyzing its key features and correlations.
- Data mining module: Data Mining is working to extract the knowledge by selecting the appropriate mining methods and building the descriptive model to carry on the mining.
- Knowledge discovery in database: The final step which figures out and discovers the useful KDSD rule and use the knowledge feedback to help in the spatial decision-making.

7   Results and Discussion

Three methods of learning classifiers (Decision Tree, Rule-Based, and Bayesian) were used to test the different data sets experimentally on the training and test set using Weka data mining software. The experiments were conducted using Weka data mining by applying six commonly classifiers (naïve Bayes, naïve Net, C4.5, randomForest, decision table, JRip). Experimental results exhibited the effectiveness and the efficiency of data mining classifiers to train spatial data sets. However, decision trees classifiers (C4.5 and RF) shown to be more efficient experimentally (see Table 2), we attempted to magnify the number of observations of TrDS21 by "Preprocess-Resample" data tool to generate a new data set named TrDS2 with 238 labeled instances. The both data set TrDS1 and TrDS2 are split into a training set and test set, split percentage is 20% testing data set and 80% for training data set. The experiment was carried out on the TrDS2 by applying the same previous classifiers to evaluate the experimental performance of the generated data model with 10-fold cross; we noted that the experimental results enhanced to better. Especially, the CA of decision trees classifiers. Whereas, the C4.5 achieved 95.97% in training set, and 93.75% in the test set. In this experiment, it is also noted that the RF classifier has reached a 98.53% and 97.92, in training and test set, consequently. The JRip classifier has also achieved significant CA; experimental results are presented in Table 3. Finally, the experiment generated classification rules mining that is relying on the J48 classifier where the selected attributes which have been figured out are the most important attributes (see Fig. 5).

8 Conclusions

The data acquired from the real world to describe a part of the earth's surface may include different kinds of information which translate natural phenomena and spatial data layers into computer-understandable entities. Therefore, many techniques and tools support the human in transforming the general data into useful knowledge. The spatial and nonspatial data have been collected about educational facilities and land use in Mukalla districts, and were stored in the major spatial geographic databases. In the study, we produced and examined a descriptive model to discover the suitability of settlements through the evaluation of the spatial suitability of educational services. Data mining algorithms have the ability to discover rules with most significant attributes, the produced rules introduce the helpful features for decision-making, thereby helping them to reduce the time and cost during the collection and data entry phase without decreasing the accuracy of classification. The SuDSS represents a descriptive model for discovering the suitability of settlements through the evaluation of the spatial suitability of educational services. The SuDSS can provide an opportunity to lead a study and fill the gap in this research field in the spatial decision by producing and performing the SDSS based on knowledge. It would be helpful for assessing, and improving the decision-making systems. Despite the comprehensiveness of this study. Nevertheless, this work has some significant limitations. First, the sample size was acquired is small, where the collection of the spatial data is not an easy issue. Therefore, data aggregation took a long time, it being available in multiple governmental sources. Second, the experiment deal with just three types of land use (Fuel station, workshop, high road), however, the educational facilities could be affected by another land use, especially in rural areas. Also, the performance evaluation of blocks does not take into account the block's elevation data (DEM). Finally, the assessment of multicriteria suitability relied dramatically on the human experts, and this assessment probably includes on sampling error, because no explicit criteria were adopted by the government to assist them to make a reliable evaluation. These limitations should be kept in mind when using this data. Thus, further researches are required in the future to overcome those limitations. As a future work, we would strive to work with local government and specialized experts in order to take on standard criteria for establishing the community services and land use in terms of the quality of building and equipment policy including human staff. Moreover, the process of acquiring spatial information and transfer it into reliable data is a hard to process, it requires lots of time and effort. We also plan to examine by experiments the possibility of merging the prediction module and SuDSS model as integrated Predictive-Descriptive Model (SuDSS-PDM).

References

1. Liao, S.-H., Chu, P.-H., Hsiao, P.-Y.: Data mining techniques and applications–a decade review from 2000 to 2011. Expert Syst. Appl. **39**, 11303–11311 (2012)
2. Lagrab, W., Aknin, N.: Analysis of educational services distribution-based geographic information system (GIS). Int. J. Sci. Technol. Res. **4**, 63–91 (2015)
3. Lagrab, W., Aknin, N.: A suitability analysis of elementary schools - based geographic information system (GIS). J. Theor. Appl. Inf. Technol. **95**, 731–742 (2017)
4. CA Dept. of Education: School Site Selection and Approval Guide - Facility Design (California Dept of Education) (2015). http://www.cde.ca.gov/ls/fa/sf/schoolsiteguide.asp
5. Moore, D.P.: Guide for Planning Educational Facilities. Planning Guide. Council of Educational Facility Planners International, Columbus (1991)
6. Kumar, A., Kakkar, A., Majumdar, R., Baghel, A.S.: Spatial data mining: recent trends and techniques. In: 2015 International Conference on Computer and Computational Sciences (ICCCS), pp. 39–43 (2015)
7. Mennis, J., Guo, D.: Spatial data mining and geographic knowledge discovery—an introduction. Comput. Environ. Urban Syst. **33**, 403–408 (2009)
8. Koperski, K., Adhikary, J., Han, J.: Spatial data mining: progress and challenges survey paper. In: Proceedings of ACM SIGMOD Workshop on Research Issues on Data Mining and Knowledge Discovery, Montreal, Canada, pp. 1–10. Citeseer (1996)
9. Shunzhi, Z., Wenxing, H., Qunyong, W., Maoqing, L.: Research on data mining model of GIS-based urban underground pipeline network. In: 2009 IEEE International Conference on Control and Automation, pp. 1515–1520 (2009)
10. Wang, Y., Chen, X.: Study on land use of changping district with spatial data mining method. In: Proceedings 2011 IEEE International Conference on Spatial Data Mining and Geographical Knowledge Services, pp. 218–222 (2011)
11. Kaundinya, D.P., Balachandra, P., Ravindranath, N.H., Ashok, V.: A GIS (geographical information system)-based spatial data mining approach for optimal location and capacity planning of distributed biomass power generation facilities: a case study of Tumkur district, India. Energy **52**, 77–88 (2013)
12. Ruiz, M.C., Romero, E., Pérez, M.A., Fernández, I.: Development and application of a multi-criteria spatial decision support system for planning sustainable industrial areas in Northern Spain. Autom. Constr. **22**, 320–333 (2012)
13. Ferretti, V., Montibeller, G.: Key challenges and meta-choices in designing and applying multi-criteria spatial decision support systems. Decis. Support Syst. **84**, 41–52 (2016)
14. Mason, S.O., Baltsavias, E.P., Bishop, I.: Spatial decision support systems for the management of informal settlements. Comput. Environ. Urban Syst. **21**, 189–208 (1997)
15. Bottero, M., Comino, E., Duriavig, M., Ferretti, V., Pomarico, S.: The application of a multicriteria spatial decision support system (MCSDSS) for the assessment of biodiversity conservation in the Province of Varese (Italy). Land Use Policy **30**, 730–738 (2013)
16. Ochola, W.O., Kerkides, P.: An integrated indicator-based spatial decision support system for land quality assessment in Kenya. Comput. Electron. Agric. **45**, 3–26 (2004)
17. Palmisano, G.O., Govindan, K., Boggia, A., Loisi, R.V., De Boni, A., Roma, R.: Local action groups and rural sustainable development. A spatial multiple criteria approach for efficient territorial planning. Land Use Policy **59**, 12–26 (2016)

Big Data Analytics: A Comparison of Tools and Applications

Imane El Alaoui[1,2(✉)] ⓘ, Youssef Gahi[3,4], Rochdi Messoussi[1],
Alexis Todoskoff[2], and Abdessamad Kobi[2]

[1] Laboratoire des Systèmes de Télécommunications et Ingénierie
de La Décision, University of Ibn Tofail, Kenitra, Morocco
{imane.el.alaoui,messoussi}@uit.ac.ma
[2] Laboratoire Angevin de Recherche en Ingénierie des Systèmes,
University of Angers, Angers, France
{alexis.todoskoff,abdessamad.kobi}@univ-angers.fr
[3] School of Electrical Engineering and Computer Science,
University of Ottawa, Ottawa, Canada
gahi.youssef@uit.ac.ma
[4] Ecore Nationale des Sciences Appliquées, University of Ibn Tofail,
Kenitra, Morocco

Abstract. With an ever-increasing amount of both data volume and variety, traditional data processing tools became unsuitable for the big data context. This has pushed toward the creation of specific processing tools that are well aligned with emerging needs. However, it is often hard to choose the adequate solution as the wide list of available tools are continuously changing. For this, we present in this paper both a literature review and a technical comparison of the most known analytics tools in order to help mapping it to different needs. Moreover, we underline how much important choosing the appropriate tool is acting for different kind of applications and especially for smart cities environment.

Keywords: Big data analytics tools · Big data tools' comparison
Smart cities

1 Introduction

With the rise of digital communication rate, the world became more connected. This has led to an explosion in term of the daily generated data, as in one second, google process 58,588 request, 2,561,661 emails are sent, 6,801 videos are viewed on YouTube and 2450 calls are made via Skype [1]. Thus, over than 2.5 quintillion of varied data are generated every day from different kind of sources [2], whereby it is categorized as follows:

Structured data: it is data that has a defined format. Structured data is often stored in a preformatted database or file that follows a specific template.
Semi-structured data: it is data that does not follow a predefined format but it is stored with an associated metadata.

© Springer International Publishing AG, part of Springer Nature 2018
M. Ben Ahmed and A. A. Boudhir (Eds.): SCAMS 2017, LNNS 37, pp. 587–601, 2018.
https://doi.org/10.1007/978-3-319-74500-8_54

Unstructured data: It is data that it is stored without any predefined format or metadata such as log and multimedia files.

However, big data is not only characterized by Volume and Variety. Organizations have also proposed additional characteristics in order to well describe the quality of a dataset. These characteristics are called 7V's, seven essential criteria, which were derived from the 3V basic definition introduced by Gartner and refers to; Volume, Variety, and Velocity [3]. The 7V criteria are defined as:

Volume: is the interesting amount of large data that could be a good opportunity to extract insights and make right decisions.

Variety: is the different possible forms of records. The more heterogeneous and varied the data is, the more effective could be the insight we extract from it.

Velocity: is the speed at which data is generated and how much fast data processing is. Generally, it is very important to extract value very quickly especially for real time systems.

Veracity: is the process of eliminating uncertain, imprecise, and inaccurate data. The generated data could be falsified and can lead to inaccurate decisions.

Visualization: is the fact of being able to visualize large and complex data in visible and understood manner.

Variability: is the fact of considering data whose meaning is constantly changing.

Value: is the possible meaningful value that could be extracted from a data set. The big data could not be important if it does not represent a potential insight.

Basically, these 7V make it possible to define an interesting big data. However, it is actually difficult to find a traditional database management tool that can handle a wide range of voluminous records, process data in real time, check data accuracy and support variability. Above all, a tool that gives the user a visibility of the data, through reporting tools, which allow making good decisions. To cover all these relevant needs, several tools and techniques have been proposed. The aim of this paper is to provide a literature review of this kind of tools and compare their characteristics. Moreover, we underline how much important choosing the appropriate tool is acting for different kind of applications.

The remainder of this paper is structured as follows. Section 2 presents the related work of big data analytics. Section 3 presents the ecosystem Hadoop which is the most commonly used platform for big data processing. Section 4 presents and compares the wildly known frameworks for big data processing. In Sect. 5, we present their implementations in various fields such as health, business and smart cities. In Sect. 6, we conclude the paper and provide some future work.

2 Related Work

The volume of the daily data is rapidly increasing, whereby it becomes difficult to process. To address this issue, several researchers have shown a keen interest to either evaluate the current state of big data and its analytics tools, or contribute in the setting up of some new big data processing platforms.

In [4], authors have focused on explaining the Hadoop concept as well as some Hadoop-based applications for different domains including healthcare, sport, market, business, network security, and education systems. In [5], Bajaber et al. have presented the era of big data 2.0 where several processing tools have been evolved to both enhance processing capabilities such as MapReduce 2.0 and propose some enterprise oriented solutions. In [6], a benchmark based on fault recovery ability as well as durability for modern distributed stream computing tools is presented. The paper provides a set of comparison results between Storm and Spark. Liu et al. [7] have overviewed open source Real-time/near real-time processing technologies while focusing on their architectures and platforms. In [8], authors have presented some real time big data analysis tools, and have categorized various studies based on both the used tools and the application type. They have focused on applications related to surveillance, environment, social media and health care. Researchers in [9] have discussed the big data analytics solutions. Further, they have exposed some research directions and open technique and platform-related issues for this area of research. In [10], Gong et al. have proposed the benchmarking and implementation of SMASH, a generic and highly scalable Cloud-based solution, to process large scale traffic data. Other researchers [11] have presented a resource allocation scheme for stream big data in a shared cloud. The purpose is to attain max-min fairness in the utilities of all the topologies' throughput.

It is important to underline that Hadoop is not the only attractive platform for big data processing, there is also Apache Spark. Whilst both tools are sometimes considered competitors, it is often admitted that they work even better when they are combined. Apache Spark has become increasingly suitable as top-level project for big data analysis, thus, many researches tend to focus on its enhancement. Gulzar et al. [12] have built a tool called BigDebug which provides interactive and real-time debug primitives for big data processing in Spark. In [13], researchers have presented an automatic check point algorithm to solve the spark long lineage problem with less impact on the global performance. In [14], NetSpark, an improved Spark framework is presented. This framework reduces the Spark task running time by combining network buffer management, RDMA technology (hardware-supported Remote Direct Memory Access) and optimization on data serialization. A strategy, called Multiple Phases Time Estimation (MPTE), has been presented in [15] to reduce the impact of straggler machines. Furthermore, scheduling of backup tasks has been enhanced by designing a new task scheduler.

Other studies have designed new frameworks based on spark in order to make big data analytics more powerful. In fact, Yan et al. [16] have created TR-Spark to face transient resources issues. This framework can run as a secondary background task on transient resources and allow more efficiency for spark-based applications. The design of this new framework is based on two principles: resource stability and data size reduction aware scheduling. The combination of these principles allows TR-Spark to adapt to the stability of the infrastructure. In order to better make business decisions, Park et al. [17] relies on spark to propose a goal-oriented big data analytics framework. This latter has been experimented on shipment decision.

Big data analytics tools could be used to extract insight in different domains and one of the most known applications is smart cities. However, as building smart cities encompasses various sectors, such as healthcare, education, transportation, safety, government and resource management, it requires a strong analysis of real time generated data from different sources. Hence, suitable tools should be employed to extract insights and then improve the day-to-day life. For this, there has been growing interest in considering data analysis while conceiving smart cities. Hashem al. [18] have emphasized the role of big data in this context and have proposed a business model that manages big data for smart cities. Furthermore, authors in [19] have reviewed the origin, the definitions, the issues and the applications domains of smart city. In [20], different definitions of the big data and smart city are discussed and compared. Further, the opportunities, the benefits and the challenges of integrating big data applications for smart cities are explored. Researchers in [21–23] have proposed different models' architectures and implementations for smart city development using Hadoop ecosystem. Gomes et al. [24] have also come up with a big data infrastructure model for a smart city project. These four previous models have been proposed in order to extract, store, process data in the context of smart cities. Whereas in [25], authors have enumerated the several barriers of smart city projects implementation.

It is worth noting that big data analytics frameworks are growing sharply and continuously changing. Hence, we provide in this contribution, an updated state of the art of the most known Hadoop-based big data analytics tools. Moreover, we present their implementations in different domain such as healthcare business and smart city.

In the next section, we present Hadoop ecosystem, the most used platform for big data processing.

3   Hadoop Ecosystem

Hadoop is the most commonly used framework for big data processing. It is an open source platform that was initially introduced in 2007 by Apache Software Foundation. Its principle is based on the distributed system [26] which consist in sharing the data and the processing between several interconnected machines. In fact, Hadoop distributes storage and computations across a set of grouped machines, called nodes, to form a set of interconnected machines called cluster. This latter is designed to scale up from one single node to thousands, whereby it is perceived as a single unit.

Hadoop is composed of two layers: a storage layer and a processing layer that we detail in sub-sections A and B respectively. Hadoop platform relies on a master/slave architecture, where the master consists of a NameNode and a JobTracker while the slaves consist of several TaskTrakers and DataNodes. There is also a secondary master that assure the high availability (HA) of the cluster. Here we highlight the role each component:

NameNode: reconstructs files from the blocks and manages the filesystem tree. It also manages makes all decisions regarding replication of blocks.
JobTracker: tracks jobs, manage the resources and restarts nodes in case of error.
TaskTrakers: consist of accepting the jobs and communicating the progress level to the JobTracker.
DataNodes: store data blocks and retrieve it on demand. They also communicate the blocks' list to the NameNode.

3.1   Storage Layer

HDFS [27] allows to store vast volumes of data on a large number of machines by providing a high-throughput access to the application data. It stores any type of data as HDFS files and split it into 64 MB blocks or more. Also, it provides a block replication system configured to 3 by default. HDFS relies on a uniform naming convention and a mapping scheme to keep tracking all files' location.

3.2   Processing Layer

Big data is not only about the original storage but also about the information that it contains. For this, Hadoop provides a processing layer. It allows resilient processing and distribution of large amounts of unstructured data in clusters where each node has its own storage. To distribute the data and collect the results, it employs a MapReduce mechanism. MapReduce processing engine was initially developed by google [28]. It is a parallel programming model designed to efficiently process vast amounts of data in the form of jobs [29]. Because cluster size does not affect the results of a processing task, tasks can be spread over an almost unlimited number of servers as follows: A MapReduce job splits the input data into independent blocks and consists of two phases: Map phase where data are organized into key/value pairs and are processed in a parallel manner. Then, the outputs of this phase are sorted and represent the reduce phase inputs. The Reduce phase combines all the intermediate results into one result. See Fig. 1. MapReduce and the Hadoop framework, therefore, simplify analytics tools development. However, although the high scalability and fault-tolerance, MapReduce has several limitations especially in the case of simultaneous execution and real-time processing [30]. For this, YARN (Yet-Another-Resource-Negotiator) has been integrated to support big data analysis evolution. YARN is placed on top of HDFS (Hadoop Distributed File System) to provide operating system capabilities for Big Data analytics applications [31]. This arrangement allows the simultaneous execution of multiple applications while providing better tracking of the data throughout its life cycle. It also allows mixing workloads in batch, interactive and in real time. YARN also maintains compatibility with MapReduce's Application Programming Interface (API). Its principle consists of dividing the JobTracker functionalities into two separate "daemons": RessourceManager, that arbitrates resources across all the applications (jobs in MapReduce), and ApplicationMaster that requests, launches and monitors the application. Thus, YARN makes MapReduce just one application like others that runs over it.

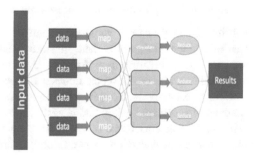

Fig. 1. MapReduce operating principle.

4   Open Source Big Data Analytics Tools

Hadoop relies on simple programming models and tools to ensure unstructured data processing and make it available on local machines. However, to convert the data into valuable information, it is necessary to have better analytical tools. These tools could be categorized into three categories:

4.1   Batch Processing Tools

They allow the data to be processed until it is exhausted at the input of the system. The processes are continuous and incremental, that means that the architecture will always take into account the new data without having to process the old ones again. To ensure consistency in the processing of these data, the results are visible and accessible only at the end of processing. The most known tools that allows processing under the batch mode are Map Reduce in its version Hadoop Apache Spark, Trident, and Flink.

4.2   Micro-batch Processing Tools

It is an intermediate technique between the batch and the stream processing. It divides streams into sequences of small chunks of data, which are then processed by batch. Thus, data is almost processed within real time. Spark and Storm with trident are the most known tools for micro-batch processing.

- **Spark** [32]: was initially developed at University of California, Berkeley [33]. It provides fault tolerance without replication and stores data in-memory. Further, Spark has libraries such as Spark Streaming, Spark SQL, Spark MLlib, SparkR that make big data analysis and processing simple. Moreover, Spark improves on processing speed issues by relying on in-memory computations as explained in [34, 35]. In fact, it is able to store up to 105 TB in only 23 min against 72 min with Hadoop native solutions.
- **Trident** [36]: a high-level abstraction layer for Storm, that is explained in next sub-section, can be used to accomplish state persistence. Trident also brings functionality similar to Spark, as it operates on mini-batches

4.3   Stream Processing Tools

Streaming tools can handle data streams, which is a continuous flow of incoming data records. The most known tools for this kind of data processing are:

- **Storm** [37]: is a real-time stream processor whose the stream data abstraction is called tuples which consist of an identifier and the data. In Storm, real-time computation consists of creating topologies and forming a directed acyclic graph. Its topology works as a data graph and is composed of inputs nodes and processing nodes which are called spouts and bolts respectively. To ensure the reprocessing of failed-tuples, Storm uses a mechanism of upstream backup and record acknowledgments.
- **Flink** [38]: is a distributed framework based on event stream processing. Flink also allows the use of micro-batches instead of pure stream. Jobs in Flink are compiled into

a directed graph of tasks, whereby stream data, called DataStream, are sequenced as partially ordered records. Flink is based on snapshot over distributed checkpoints that maintain the job status to provide fault tolerance. Apache Flink could be a good alternative to Hadoop MapReduce, Apache Spark and Apache Storm.

- **Samza** [39]: relies on YARN to decide about the resource distribution among applications. Samza is actually used to process the data received from Apache Kafka. This latter is a distributed publish-subscribe messaging system designed to collect and deliver high volumes of event data. On a late arrival, the framework will re-issue any relative result on the basis of the calculation window, since there is enough information at the local level to recalculate them.

There are many big data tools that help extracting insight, and it's important that the choice of using a particular tool can be defended. As Fault tolerance and latency are one of the very essential in big data processing, we provide a tools comparison in Table 1 based on these criteria as well as, auto-scale and other important characteristics.

Table 1. Comparison of Spark, Storm, Flink and Samza.

	MapReduce	Spark	Storm	Flink	Samza
Streaming model	Batch	Micro-batch	Naive stream/micro-batch	Naive stream/micro-batch	Naive stream
Latency	Low	Medium (depending on batch size)	Very low	Low	Low
Maturity	Low	High	High	Low	Medium
Auto-scaling	Yes	Yes	No	No	No
Fault tolerance mechanism	Data replication	RDD based check pointing	Record ACKs	Check pointing	Log-based
In-memory processing	No	Yes	Yes	Yes	Yes
Enterprise supports	No	Yes	No	No	No

Moreover, the Fig. 2 illustrates the research trends of these tools based on google statistics [40].

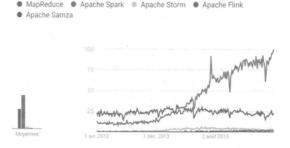

Fig. 2. Spark, Storm, Flink and Samza research trends.

In the next section, we present a set of applications that rely on this kind of analytics tools to extract valuable information for different domains. Further, we highlight the commonly used tools.

5   Applications and Their Commonly Used Tools

5.1   Application

Health. In health, the big data corresponds to all the socio-demographic and health data available from internal and public sources. Health data should be multiplied by 50 by 2020, notably through developments in genomics, connected medical equipment, the digitalization of patient records, and the use of mobile health applications and sensors Activity. The use of these data has many interests: identification of risk factors for disease, assistance in diagnosis, selection and monitoring of the effectiveness of treatments.

Making predictions of prognosis and diagnosis for patients: In [41], authors have developed a system that detect whether a patient is experiencing a cardiorespiratory spell. Researchers in [42] have proposed a system that exploits the health attributes in the users' tweets. Then, the system makes processing to predict his/her health status. In [43], a novel method based on k-Nearest Neighbors algorithm (KNN) to efficiently detect the outliers in large-scale healthcare data has been proposed. This method outperformed the KNN and Local Outlier Factor (LOF) in terms of accuracy and processing efficiency. Chen et al. [44] have used a spatial Durbin model and mortality data in 116 cities in China in order to measure health damage. Further, they estimate the annual death tolls and medical expenses caused by the air pollution.

Proposing modern model and system architecture to handle large amounts of information: Authors in [45] have proposed a model based on NoSQL to handle big data emanated from health information by relying on a cloud environment. The efficiency of the model was measured by comparing with a data relational database model, while considering query time, data preparation, flexibility, and extensibility parameters. Al Rasyid et al. [46] have built an application platform called EepisCure in order to load and visualize the data collected from raw sensors. These sensors form a Wireless Body Area Network which is a wireless network of wearable computing devices for monitoring the condition of a human body. Authors in [47] have come up with an architecture, called big health. The proposed system is based on big data and the health Internet of Things and allow extracting valuable health insight from the stored data. In [48] authors have focused connecting the air monitoring in household environment to a personal health reporting system in order to set up an alert system. Ta et al. [49] have proposed a generic architecture for healthcare analytics.

Reducing of treatment cost and improving population health: In [50], researchers have discussed the big data impact in the insurance sector and how it could transform rural healthcare. Kumar et al. [51] have analyzed patient's records, reports, symptoms, and feedbacks for e-health insurance in order to improve the quality of service. Hence, the new patient can see which hospital providing better treatment and also check its cost for a particular disease.

Business and marketing. Analyzing big data is becoming a key basis to improve quality in modern organizations. Many research studies have confirmed that relying on big data, for business and marketing domains, is a key success factor for a continuous development. Suguna et al. [52] have relied on Hadoop MapReduce to introduce a predictive prefetching system for E-commerce business activities.

In finance area, Dong et al. [53] have evaluated various models by focusing on a limited form of financial big data which is the intraday high-frequency data. This work is based on analyzing the high-frequency of Shanghai composite index to predict finance market. In [54], an approach for analyzing wind generation effect on the electricity price is proposed. The proposed approach has been applied to the Alberta market and has shown that the increased wind generation diminish the wholesale market.

In advertising area, Aivalis et al. [55] have developed an analytics application that evaluates the impact factor of two parameters (traffic and revenue) in order to provide target posting in social media. For the same topic, Deng et al. [56] have proposed a big data mobile marketing analytics framework to provide advertising and recommendation based on various criteria such as profiling, navigation history, localization and access behaviors.

In marketing area, a novel model has been proposed in [57] to predict the potential high-end luxury car buyers by using telecom big data mining. The contribution has also relied on social media to prove the efficiency of the model. Always while relying on social networks, Bollen et al. [58] have used twitter data to predict the evolution of Dow Jones Industrial Average markets by using twitter data. [59] has introduced a system that predicts the commerce websites consumers' behavior basing on their Facebook profiles. In [60], a big data approach that showed the correlation between sentiment analysis of social data and the stock performance is proposed.

Smart City. We reviewed several examples of big data applications which could be used to achieve a better quality of life for city residents in different domains. Therefore, these applications could be considered as guides to lead various smart city applications. For example, in order to improve the health of residents or to change the way that healthcare is delivered in a city, the previous applications in health sub-section could be a crucial element in smart city healthcare. In general, smart city involves the enhancement of several city components such as energy, transportation, government, public safety, as discussed in [20]. As a result, many governments have encouraged the development of smart cities around the world.

The following examples show how big data is greatly important to build a smart city in mainly 3 sub-domains:

The first one is the life enrichment that includes home, community, healthcare and education. To enhance this area, several works have been proposed. In [61], researchers have explored the enhancement of smart home services by integrating both big data analytics and social network services. Xu et al. [62] have built a model basing on Hadoop in order to predict user behavior in a traditional smart home system. Their purpose is to provide intelligent services that make easy the daily activities. For smart learning, Upudi et al. [63] have investigated the possibility of creating new framework for big data integration within smart learning environment. In [64], researchers have integrated social networking and big data analytics tools in order to provide a recommender system within smart education context. This system could provide for student information and contents such as areas of interest, friends and professors interested in the same field, videos and eBooks.

The second sub-domain is the public administration and services such as public safety supervision, smart traffic and smart tourism. In smart traffic, Raghothama et al. [65] have used GTFS data (General Transit Feed Specification) and data from other sources in order to understand factors influencing public transport delays in the cities. Authors in [66], have proposed a mechanism that helps mastering the vast volume of data gathered continuously from road sensors in order to set up a smart transportation system. In smart tourism context, Chua et al. [67] have introduced a novel approach based on geotagged social media data in order to characterize tourist flows. For smart safety, a novel approach is proposed in [68] and aim to minimize police time-to-arrival and the overall numbers of police on patrol by using real-time traffic analysis.

The third sub-domain consists of resource management including water, electricity and agriculture. To tackle these interesting topics, Yamini et al. [69] have proposed a system called Smart Home Energy Management System that monitors home devices energy consumption in a smart home. The proposed system aims at making energy usage more efficient. A framework based on Hadoop was designed in [70] to manage data in energy sector in order to reduce power wasting. In [71], authors used different kinds of sensors and come up with a smart drip irrigation system in order to make agriculture sustainable.

5.2   Commonly Used Tools

The following table classifies several researches by domains/sub-domains and by big data analytics tools used.

We noticed that Hadoop, MapReduce and spark are the most commonly used frameworks for big data processing in the academic researches. Although Strom, Flink and Samza real-time processing capabilities, have not been widely used in this area. We further noted that most of contributions did not underline why they favor some frameworks compared to others. For this, we present in Table 1, a comparison based on latency, fault tolerance and other characteristics of these frameworks. Further, a popularity comparison based on google research engine in Fig. 2 and their related work in Table 2 are explored

It is important to mention that big data processing tools are keeping changing, whereby new tools are presented every day. Therefore, the used tools in order to build processing-based environments should be well selected.

Table 2. Big data processing frameworks used in different fields.

		Hadoop	MapReduce	Spark	Storm	Flink	Samza	Other
Health	Making prediction	Yan et al. [43]		Nair et al. [42]				Thommandram et al. [41] Chen et al. [44]
	Proposing model	Al Rasyid et al. [46] Ma et al. [47] Ta et al. [49]		[47] [49]	[47]			Goli-Malekabadi et al. [45] Al Rasyid et al. [46]
	Reducing of treatment cost	Gupta et al. [50]	[50] Kumar et al. [51]	[50]				[50]
Smart city	Life enrichment	Xu et al. [62] Jagtap et al. [64]	[64]					[62]
	Public administration and services	Rathore et al. [66]		[66]				[66]
	Resource management	Vaidya et al. [70]	[70]					
Marketing		Suguna et al. [52] Deng et al. [56]	[52]	[56]				[56] Dong et al. [53] Zamani-Dehkordi et al. [54] Attigeri et al. [60]

6   Conclusions

In this work, we aimed to highlight big data era and its related characteristics. As big data analytics is the essential part of big data and can improve various services of our daily world, we reserved a big part of this contribution to present and compare the most commonly used tools and platforms for big data processing. Then, we presented a projection of big data processing tools in the context of various domains including health, business and smart city. We recall that the main purpose of this contribution is to provide an overview of various big data analytics tools and their applications.

As future contributions, we want to provide a functional comparison of these tools and a detailed overview of the whole big data value chain including collect, pre-processing, processing, visualization as well as machine learning platforms.

References

1. Internet Live Stats - Internet Usage & Social Media Statistics. http://www.internetlivestats.com/. Accessed 25 Mar 2017
2. Reinsel, D., Gantz, J.: Extracting Value from Chaos. IDC IVIEW, Sponsored by EMC (2011)
3. Laney, D.: 3D Data Management: Controlling Data Volume, Velocity, and Variety. META Group Inc., Stamford (2011)
4. Mohanty, S., Das, G., Suman, H., Maharana, P., Ratnakar, R.: A survey on working principle and application of Hadoop. Int. J. Adv. Innovative Res. **4**, 71–75 (2015)
5. Bajaber, F., Elshawi, R., Batarfi, O., Altalhi, A., Barnawi, A., Sakr, S.: Big data 2.0 processing systems: taxonomy and open challenges. J. Grid Comput. **14**(3), 379–405 (2016)

6. Lu, R., Wu, G., Xie, B., Hu, J.: Stream bench: towards benchmarking modern distributed stream computing frameworks. In: 2014 IEEE/ACM 7th International Conference on Utility and Cloud Computing, pp. 69–78 (2014)

7. Liu, X., Iftikhar, N., Xie, X.: Survey of real-time processing systems for big data. In: Proceedings of the 18th International Database Engineering & Applications Symposium, New York, NY, USA, pp. 356–361 (2014)

8. Yadranjiaghdam, B., Pool, N., Tabrizi, N.: A survey on real-time big data analytics: applications and tools. In: 2016 International Conference on Computational Science and Computational Intelligence (CSCI), pp. 404–409 (2016)

9. Tsai, C.-W., Lai, C.-F., Chao, H.-C., Vasilakos, A.V.: Big data analytics: a survey. J. Big Data 2(1), 21 (2015)

10. Gong, Y., Morandini, L., Sinnott, R.O.: The design and benchmarking of a cloud-based platform for processing and visualization of traffic data. In: 2017 IEEE International Conference on Big Data and Smart Computing (BigComp), pp. 13–20 (2017)

11. Jiang, Y., Huang, Z., Tsang, D.H.K.: Towards max-min fair resource allocation for stream big data analytics in shared clouds. IEEE Trans. Big Data PP(99), 1 (2017)

12. Gulzar, M.A., Interlandi, M., Condie, T., Kim, M.: BigDebug: interactive debugger for big data analytics in Apache Spark. In: Proceedings of the 2016 24th ACM SIGSOFT International Symposium on Foundations of Software Engineering, New York, USA, pp. 1033–1037 (2016)

13. Zhu, W., Chen, H., Hu, F.: ASC: improving spark driver performance with SPARK automatic checkpoint. In: 2016 18th International Conference on Advanced Communication Technology (ICACT), pp. 1–8 (2016)

14. Li, H., Chen, T., Xu, W.: Improving spark performance with zero-copy buffer management and RDMA. In: 2016 IEEE Conference on Computer Communications Workshops (INFOCOM WKSHPS), pp. 33–38 (2016)

15. Yang, H., Liu, X., Chen, S., Lei, Z., Du, H., Zhu, C.: Improving Spark performance with MPTE in heterogeneous environments. In: 2016 International Conference on Audio, Language and Image Processing (ICALIP), pp. 28–33 (2016)

16. Yan, Y., Gao, Y., Chen, Y., Guo, Z., Chen, B., Moscibroda, T.: TR-Spark: transient computing for big data analytics. In: Proceedings of the Seventh ACM Symposium on Cloud Computing, New York, USA, pp. 484–496 (2016)

17. Park, G., Park, S., Khan, L., Chung, L.: IRIS: a goal-oriented big data analytics framework on Spark for better business decisions. In: 2017 IEEE International Conference on Big Data and Smart Computing (BigComp), pp. 76–83 (2017)

18. Hashem, I.A.T., et al.: The role of big data in smart city. Int. J. Inf. Manag. 36(5), 748–758 (2016)

19. Yin, C., Xiong, Z., Chen, H., Wang, J., Cooper, D., David, B.: A literature survey on smart cities. Sci. China Inf. Sci. 58(10), 1–18 (2015)

20. Nuaimi, E.A., Neyadi, H.A., Mohamed, N., Al-Jaroodi, J.: Applications of big data to smart cities. J. Internet Serv. Appl. 6(1), 25 (2015)

21. Rathore, M.M., Ahmad, A., Paul, A.: IoT-based smart city development using big data analytical approach. In: 2016 IEEE International Conference on Automatica (ICA-ACCA), pp. 1–8 (2016)

22. Nathali Silva, B., Khan, M., Han, K.: Big data analytics embedded smart city architecture for performance enhancement through real-time data processing and decision-making. Wirel. Commun. Mob. Comput. 2017, e9429676 (2017)

23. Costa, C., Santos, M.Y.: BASIS: a big data architecture for smart cities. In: 2016 SAI Computing Conference (SAI), pp. 1247–1256 (2016)

24. Gomes, E., Dantas, M.A.R., de Macedo, D.D.J., Rolt, C.D., Brocardo, M.L., Foschini, L.: Towards an infrastructure to support big data for a smart city project. In: 2016 IEEE 25th International Conference on Enabling Technologies: Infrastructure for Collaborative Enterprises (WETICE), pp. 107–112 (2016)
25. Mosannenzadeh, F., Di Nucci, M.R., Vettorato, D.: Identifying and prioritizing barriers to implementation of smart energy city projects in Europe: an empirical approach. Energy Policy **105**, 191–201 (2017)
26. Coulouris, G., Dollimore, J., Kindberg, T., Blair, G.: Distributed Systems: Concepts and Design, 5th edn. Pearson, Boston (2011)
27. HDFS Architecture Guide. https://hadoop.apache.org/docs/r1.2.1/hdfs_design.html#Introduction. Accessed: 27 Mar 2017
28. Google Research Publication: MapReduce. https://research.google.com/archive/mapreduce.html. Accessed 21 Jan 2017
29. MapReduce Tutorial. https://hadoop.apache.org/docs/r1.2.1/mapred_tutorial.html. Accessed 27 Mar 2017
30. Lee, K.-H., Lee, Y.-J., Choi, H., Chung, Y.D., Moon, B.: Parallel data processing with mapreduce: a survey. SIGMOD Rec. **40**(4), 11–20 (2012)
31. Vavilapalli, V.K., et al.: Apache hadoop YARN: yet another resource negotiator. In: Proceedings of the 4th Annual Symposium on Cloud Computing, New York, USA, pp. 5:1–5:16 (2013)
32. Apache SparkTM - Lightning-Fast Cluster Computing. https://spark.apache.org/. Accessed 27 Mar 2017
33. Zaharia, M., Chowdhury, M., Franklin, M.J., Shenker, S., Stoica, I.: Spark: cluster computing with working sets. In: Proceedings of the 2nd USENIX Conference on Hot Topics in Cloud Computing, Berkeley, USA, p. 10 (2010)
34. Xin, R.: Spark officially sets a new record in large-scale sorting (2014). http://databricks.com/blog/2014/11/05/spark-officially-sets-a-new-record-in-large-scale-sorting.html. Accessed 27 Mar 2017
35. Sort Benchmark Home Page. http://sortbenchmark.org/. Accessed 27 Mar 2017
36. Trident Tutorial. http://storm.apache.org/releases/1.0.1/Trident-tutorial.html. Accessed 05 Apr 2017
37. Apache Storm: http://storm.apache.org/. Accessed 27 Mar 2017
38. Apache Flink: Scalable Stream and Batch Data Processing. https://flink.apache.org/. Accessed 27 Mar 2017
39. Samza: http://samza.apache.org/. Accessed 27 Mar 2017
40. Google Trends: Google Trends. https://g.co/trends/aes0h. Accessed 31 Mar 2017
41. Thommandram, A., Pugh, J.E., Eklund, J.M., McGregor, C., James, A.G.: Classifying neonatal spells using real-time temporal analysis of physiological data streams: algorithm development. In: 2013 IEEE Point-of-Care Healthcare Technologies (PHT), pp. 240–243 (2013)
42. Nair, L.R., Shetty, S.D., Shetty, S.D.: Applying Spark based machine learning model on streaming big data for health status prediction. Comput. Electr. Eng. (2017, in press)
43. Yan, K., You, X., Ji, X., Yin, G., Yang, F.: A hybrid outlier detection method for health care big data. In: 2016 IEEE International Conferences on Big Data and Cloud Computing (BDCloud), Social Computing and Networking (SocialCom), Sustainable Computing and Communications (SustainCom) (BDCloud-SocialCom-SustainCom), pp. 157–162 (2016)
44. Chen, X., Shao, S., Tian, Z., Xie, Z., Yin, P.: Impacts of air pollution and its spatial spillover effect on public health based on China's big data sample. J. Clean. Prod. **142**(Part 2), 915–925 (2017)

45. Goli-Malekabadi, Z., Sargolzaei-Javan, M., Akbari, M.K.: An effective model for store and retrieve big health data in cloud computing. Comput. Methods Programs Biomed. **132**, 75–82 (2016)

46. Al Rasyid, M.U.H., Yuwono, W., Muharom, S.A., Alasiry, A.H.: Building platform application big sensor data for e-health wireless body area network. In: 2016 International Electronics Symposium (IES), pp. 409–413 (2016)

47. Ma, Y., Wang, Y., Yang, J., Miao, Y., Li, W.: Big health application system based on health internet of things and big data. IEEE Access **PP**(99), 1 (2016)

48. Ho, K.F., Hirai, H.W., Kuo, Y.H., Meng, H.M., Tsoi, K.K.F.: Indoor air monitoring platform and personal health reporting system: big data analytics for public health research. In: 2015 IEEE International Congress on Big Data, pp. 309–312 (2015)

49. Ta, V.-D., Liu, C.-M., Nkabinde, G.W.: Big data stream computing in healthcare real-time analytics. In: 2016 IEEE International Conference on Cloud Computing and Big Data Analysis (ICCCBDA), pp. 37–42 (2016)

50. Gupta, S., Tripathi, P.: An emerging trend of big data analytics with health insurance in India. In: 2016 International Conference on Innovation and Challenges in Cyber Security (ICICCS-INBUSH), pp. 64–69 (2016)

51. Kumar, K.M., Tejasree, S., Swarnalatha, S.: Effective implementation of data segregation extraction using big data in E - health insurance as a service. In: 2016 3rd International Conference on Advanced Computing and Communication Systems (ICACCS), vol. 1, pp. 1–5 (2016)

52. Suguna, S., Vithya, M., Eunaicy, J.I.C.: Big data analysis in e-commerce system using HadoopMapReduce. In: 2016 International Conference on Inventive Computation Technologies (ICICT), vol. 2, pp. 1–6 (2016)

53. Dong, T., Yang, B., Tian, T.: Volatility analysis of Chinese stock market using high-frequency financial big data. In: 2015 IEEE International Conference on Smart City/SocialCom/SustainCom (SmartCity), pp. 769–774 (2015)

54. Zamani-Dehkordi, P., Rakai, L., Zareipour, H., Rosehart, W.: Big data analytics for modelling the impact of wind power generation on competitive electricity market prices. In: 2016 49th Hawaii International Conference on System Sciences (HICSS), pp. 2528–2535 (2016)

55. Aivalis, C.J., Gatziolis, K., Boucouvalas, A.C.: Evolving analytics for e-commerce applications: utilizing big data and social media extensions. In: 2016 International Conference on Telecommunications and Multimedia (TEMU), pp. 1–6 (2016)

56. Deng, L., Gao, J., Vuppalapati, C.: Building a big data analytics service framework for mobile advertising and marketing. In: 2015 IEEE First International Conference on Big Data Computing Service and Applications, pp. 256–266 (2015)

57. Zhang, H., Zhang, L., Cheng, X., Chen, W.: A novel precision marketing model based on telecom big data analysis for luxury cars. In: 2016 16th International Symposium on Communications and Information Technologies (ISCIT), pp. 307–311 (2016)

58. Bollen, J., Mao, H., Zeng, X.-J.: Twitter mood predicts the stock market. J. Comput. Sci. **2**(1), 1–8 (2011)

59. Zhang, Y., Pennacchiotti, M.: Predicting purchase behaviors from social media. In: Proceedings of the 22nd International Conference on World Wide Web, Rio de Janeiro, Brazil, pp. 1521–1532 (2013)

60. Attigeri, G.V., Pai, M.M.M., Pai, R.M., Nayak, A.: Stock market prediction: a big data approach. In: TENCON 2015 - 2015 IEEE Region 10 Conference, pp. 1–5 (2015)

61. Wich, M., Kramer, T.: Enrichment of smart home services by integrating social network services and big data analytics. In: 2016 49th Hawaii International Conference on System Sciences (HICSS), pp. 425–434 (2016)

62. Xu, G., Liu, M., Li, F., Zhang, F., Shen, W.: User behavior prediction model for smart home using parallelized neural network algorithm. In: 2016 IEEE 20th International Conference on Computer Supported Cooperative Work in Design (CSCWD), pp. 221–226 (2016)
63. Udupi, P.K., Malali, P., Noronha, H.: Big data integration for transition from e-learning to smart learning framework. In: 2016 3rd MEC International Conference on Big Data and Smart City (ICBDSC), pp. 1–4 (2016)
64. Jagtap, A., Bodkhe, B., Gaikwad, B., Kalyana, S.: Homogenizing social networking with smart education by means of machine learning and Hadoop: a case study. In: 2016 International Conference on Internet of Things and Applications (IOTA), pp. 85–90 (2016)
65. Raghothama, J., Shreenath, V.M., Meijer, S.: Analytics on public transport delays with spatial big data. In: Proceedings of the 5th ACM SIGSPATIAL International Workshop on Analytics for Big Geospatial Data, New York, USA, pp. 28–33 (2016)
66. Rathore, M.M., Ahmad, A., Paul, A., Jeon, G.: Efficient graph-oriented smart transportation using internet of things generated big data. In: 2015 11th International Conference on Signal-Image Technology Internet-Based Systems (SITIS), pp. 512–519 (2015)
67. Chua, A., Servillo, L., Marcheggiani, E., Moere, A.V.: Mapping cilento: using geotagged social media data to characterize tourist flows in southern Italy. Tour. Manag. **57**, 295–310 (2016)
68. Hochstetler, J., Hochstetler, L., Fu, S.: An optimal police patrol planning strategy for smart city safety. In: 2016 IEEE 18th International Conference on High Performance Computing and Communications; IEEE 14th International Conference on Smart City; IEEE 2nd International Conference on Data Science and Systems (HPCC/SmartCity/DSS), pp. 1256–1263 (2016)
69. Yamini, J., Babu, Y.R.: Design and implementation of smart home energy management system. In: 2016 International Conference on Communication and Electronics Systems (ICCES), pp. 1–4 (2016)
70. Vaidya, M., Deshpande, S.: Distributed data management in energy sector using Hadoop. In: 2015 IEEE Bombay Section Symposium (IBSS), pp. 1–6 (2015)
71. Kavianand, G., Nivas, V.M., Kiruthika, R., Lalitha, S.: Smart drip irrigation system for sustainable agriculture. In: 2016 IEEE Technological Innovations in ICT for Agriculture and Rural Development (TIAR), pp. 19–22 (2016)

Towards Smart Urban Freight Distribution Using Fleets of Modular Electric Vehicles

Dhekra Rezgui[1,2(✉)], Jouhaina Chaouachi Siala[2],
Wassila Aggoune-Mtalaa[3], and Hend Bouziri[4]

[1] Institut Supérieur de Gestion de Tunis, University of Tunis,
41, Rue de la Liberté, Bouchoucha, 2000 Le Bardo, Tunisia
dhekra.rezgui@live.fr
[2] IHEC, Carthage University-2016, Carthage, Tunisia
siala.jouhaina@gmail.com
[3] Luxembourg Institute of Science and Technology,
4362 Esch/Alzette G.D, Luxembourg
wassila.mtalaa@list.lu
[4] ESSECT, LARODEC Laboratory, University of Tunis, Tunis, Tunisia
hend.bouziri@gmail.com

Abstract. This work deals with the electric Modular Fleet Size and Mix Vehicle Routing Problem with Time Windows, which is an extension of the well-known Vehicle Routing Problem with Time Windows (VRPTW), where the fleet consists of electric modular vehicles (EMVs). An interesting feature of this work is that despite the fact that the vehicles have a limited range, they are allowed sometimes to recharge at certain customer locations in order to continue a tour. To tackle this problem, a comprehensive mathematical formulation is given in order to model the problem and the multiple constraints appeared due to the modularity, electric charging, time windows and capacity issues. Due to the NP-hardness of the problem, a memetic algorithm is designed for determining good quality solutions in reasonable computational times. Extensive computational experiments carried out on some benchmark instances show the effectiveness of both the problem formulation and the memetic algorithm.

Keywords: Urban logistic · Vehicle Routing Problem · Metaheuristics
Electric modular vehicles

1 Introduction

Transport contributes to harm the environment with pollution, noise and congestion. Concerning pollution, carbon dioxide (CO_2) has been the major contributor of the global warming effect on the Earth during the past decades. As for transport, the fastest growing emissions was always caused by emissions from the road sector, which increased by 68% since 1990 and reported approximately three quarters of transport emissions in 2013 [5]. For example, in 2013, transportation activities has caused approximately for nearly 23% of greenhouse gas (GHG) emissions in the European Union. Consequently, in E.U., policy makers implement measures to encourage improved vehicle efficiency in order to limit emissions from the transport sector. To

© Springer International Publishing AG, part of Springer Nature 2018
M. Ben Ahmed and A. A. Boudhir (Eds.): SCAMS 2017, LNNS 37, pp. 602–612, 2018.
https://doi.org/10.1007/978-3-319-74500-8_55

enhance the environment, many countries are adopting specific programs such as, for example financial incentives and prioritized access [12] in order to limit the access of the Internal combustion engine vehicles (ICEVs) to some urban areas and especially to promote Electric Vehicles (EVs) in their smart city planning Moreover, new concepts based on electrification of transportation in cities intend to sustain urban centres. In the last decades, several research have highlighted different aspects of green logistics integrating electric vehicles (EVs) into goods distribution. In recent years, the use of electric vehicles (EVs) in freight transportation gave birth to new variants of Vehicle Routing Problem (VRP). As its name suggests, the so-called (e-VRP) extends the VRP mainly to account for two constraining electric vehicle features: the long battery charging time and the short driving range. As a matter of fact, the limited battery EV capacity is a critical factor which makes vehicles detours to refuelling stations. Obviously, the required charging time depends on the size of the batteries and may vary from 30 min to several hours [17]. In practical situations, the interest of using EVs aims at minimizing respectively the fuel consumption and harmful emissions (e.g., CO2). Several projects address some kind of practices with EVs for goods distribution, most of these took place in Europe such as the one carried out by FedEx, General Electric, Coca-Cola, UPS, Hertz Staples, Frito-Lay and others [5].

In this paper, we are interested in a variant of the electric Fleet Size and Mix Vehicle Routing Problem which incorporates the possibility of recharging the electric modules at customer locations [1]. The available vehicles differ from the battery ones since the modules are autonomous in terms of consumption and electric charging. This is a completely new problem for which several constraints will be tackled. Concretely, this leads to a new class of VRP called eM-fleet size and Mix VRP with Time Windows integrating both the complexity of the original VRP and specific constraints induced by using electric and modular vehicles. Its objective is to minimize the acquisition cost, the total distance traveled and the recharging costs. We propose, as a resolution method, a hybrid approach that combines a Genetic Algorithm (GA) with a Local Search (LS) method. In other words, the resulting memetic algorithm is applied to this new type of problem. An extensive experimental study shows the effectiveness of including the modularity feature in electric vehicles for goods distribution.

This paper is organized as follows. In Sect. 2, we present the studied context of the eM- fleet size and Mix VRPs with Time Windows. Section 3 defines and models the problem. Section 4, discusses the adaptation of the memetic algorithm to address it. The computational results are presented and discussed in Sect. 5. Section 6 concludes the paper.

2 The Studied Context

The problem addressed in this paper aims to handle a set of customers within specific time windows by using a heterogeneous fleet of electric modular vehicles. The electric Modular fleet size and Mix VRPs with Time Windows as we name it can be viewed as a variant to the vehicle routing problem. Indeed, nowadays, researchers tend to pay close attention to the concept of VRP and there is a wide range of studies, which treat the vehicle routing problem in the supply chain. Our research combines two main concepts, which are the use of zero-emission vehicles in the context of the VRP in green city and the fleet size and mix VRP. In the following, we present some of the papers that are related to our work.

Braysy et al. [2] proposed an Effective Multi-restart Deterministic Annealing Meta-heuristic for the Fleet Size and Mix Vehicle Routing Problem with Time Windows (FSMVRPTW). Their model focused on the FSMVRPTW variant that was first defined by Liu and Shen [11] where each vehicle may have its own capacity and its own fixed cost. This problem was formulated as a mixed-integer linear program (MILP). The proposed solution approach was divided on three phases: the first phase focused on initial solutions generated by means of a savings-based heuristic combining diversification strategies with learning mechanisms. The second phase focused on reducing the number of vehicles with a new local search procedure and finally the last phase was generated to improve the partial solution obtained in the second phase by applying four local search operators. Instances were created based on the benchmark instances proposed by Liu and Shen [11], stemmed from the well-known VRPTW instances of Solomon [16]. The results showed that on the 168-benchmark instances of Liu and Shen [11], the proposed method outperforms best-known existing solutions.

Conrad and Figliozzi [4] have introduced a new variant of the VRP: the Recharging VRP (RVRP), where vehicles have a limited range and are allowed sometimes to recharge at certain customer locations in order to continue a tour. Two distance-constrained problems were proposed. The first was a capacitated problem (with relaxed customer time windows (CRVRP) while the second problem introduced hard customer time windows (CRVRP-TW) for which theoretical bounds were derived. In addition, a heuristic based on an iterative construction and improvement algorithm was employed to obtain the solutions of the RVRP. Results indicate that the average tour length is highly associated with the de-rived solution bounds.

Goncalves et al. [7] presented a model of VRP with pick-ups and deliveries which was applied to a real case of a Portuguese company that distributes batteries. Their problem was formulated as a mixed-integer linear program (MILP). Three scenarios were considered. The first one used a Vehicle Routing Problem with Pickups and Deliveries (VRPPD) for the classical fleet. In the second scenario, the fleet was divided into two types of vehicles, classical fleet and electric vehicles (EVs) without cargo capacity. The third scenario corresponded to the exclusive use of EVs with freight transportation capacity. Finally, after a comparative study of the three scenarios, the results showed that the first was the best alternative to the company. Indeed, the last two scenarios required an investment in the conversion into EVs, since their fixed costs have a big impact on the global structure of cost.

Erdogan and Miller-Hooks [6] proposed a Green VRP (GVRP) variant which determines vehicles routes and recharging of Alternative-Fuel powered Vehicles (AFVs) at alternative fuel stations (AFSs) simultaneously. This problem was formulated as a Mixed Integer Linear Program (MILP) which aims to minimize the travelled distance in a given day. The authors propose two construction heuristics, the Modified Clarke and Wright Savings heuristic and the Density-Based Clustering Algorithm. Numerical experiments were conducted using typical parameters. The results indicate that the feasibility of the problem depends on the number of customers and the number of AFSs. Moreover, as the number of the AFSs increases, more customers can be served and the total traveled distances decrease.

Bruglieri et al. [3] addressed the problem of serving a set of customers, within fixed time windows, by using Electric Vehicles (EVs) and considering their need to stop at the Recharging Stations (RSs) during the trips. The goal of the problem was to minimize the number of used EVs, the total Travel Time (TT), the total Recharging Time (RT) and the total Waiting Time (WT). As a solution method, the authors propose a Variable Neighborhood Search Branching (VNSB) combining the VNS approach with the Local Branching one. Numerical results on benchmark instances clearly show that the VNSB is suitable to detect good quality solution compared to a previous work, in which the battery level reached at each recharging station is always equal to the capacity.

Koc et al. [9] presented the Fleet Size and Mix Pollution Routing Problem (FSMPRP) which is an extension of the Pollution Routing problem (PRP) with an heterogeneous vehicle fleet. Their model deals with the minimization of a total cost function encompassing driver, vehicle, fuel and emissions costs. Their mathematical model was formulated as an integer linear program (ILP). As a solution method, they developed a powerful meta-heuristic approach which was well applied to large-size realistic benchmark instances. They proposed a hybrid evolutionary algorithm called the HEA ++ of Koc et al. [9].

Aggoune-Mtalaa et al. [1] have introduced a variant of the VRP, called eM-VRP which involves electric modular vehicles for goods distribution in urban environment. This problem has been modeled as a mixed integer linear program (MILP) in which the aim is to minimize transportation costs including both economical and environmental. An interesting feature of this work is that the resolution approach operates in two stages. In the first stage, they have developed a Module Routing Problem (MRP), which is considered as a reduction of the classical VRPTW, since the MRP is dedicated to assign customers to modules, considering time and capacity constraints. The latter was solved using an ad hoc genetic algorithm. Then in a second stage, a fusion of the routes is operated. To solve the resulting VRPTW problem, several meta-heuristic approaches were developed. The experimental study demonstrated the added value to use electric modular vehicles for city freight delivery.

Lin et al. [10] have introduced a general Electric Vehicle Routing Problem (EVRP), in which electric commercial vehicles with a limited range may recharge at a charging station during their daily delivery and pick-up tours. Their pro-posed EVRP takes into account the costs associated with not only the travel time but also the electricity consumption. This model considers the effect of vehicle load on energy (battery electricity) consumption. The problem was illustrated by a case study based on the real-world network setting in Texas.

More recently, Hiermann et al. [8] introduced the Electric Fleet Size and Mix Vehicle Routing Problem with Time Windows and Recharging Stations where vehicle types differ in their battery size, acquisition cost and transport capacity. Their model aims at minimizing the acquisition cost and the total distance traveled. Their problem is formulated as a mixed integer program. Some recharging policies were carried out. For instance, the recharge duration depends on the remaining charge level of the vehicle when arriving at a recharging station. To solve this problem, they proposed a hybrid heuristic, which combines an embedded local search with an Adaptive Large Neighborhood Search (ALNS). As a result, their experiments showed the competitiveness of the proposed approach comparing to the state of the art methods for solving the two E-VRPTW and FSMVRPTW problems. That is similar to the approach which we propose in this paper, excepted that the vehicles considered in the study are modular. This is what we present in the following sections.

3   Problem Definition and Formulation

3.1   Problem Definition

In this section, we address in details, the challenging issues associated with the use of new modular electric vehicles for urban freight distribution. The originality of the electric modular vehicles is that they are based on a modular and active frame system. This means that payload modules are designed in addition to a cab module (where the driver sits) in order to bring more space and flexibility. This enables the vehicle to drop o modules at respective delivery locations and pick them up later during a run. One of the interests for the vehicle to drop off a module is the possibility to overcome length or weight restrictions for delivery vehicles in certain areas. Releasing a module at a customer location can also help respecting delays for deliveries or permit to recharge the module battery if a charging terminal is available, allowing the vehicle to benefit later from this additional energy. Recharging the modules separately instead of the whole vehicle permits also to save time. An example of electric modular vehicles is shown in Fig. 1. For instance, a vehicle can have two or three modules in addition to the cabin module when it leaves the depot. It can serve a customer, let one of its modules there and continue a tour with the remaining modules. Another vehicle can then come to pick up the module later during a run. The fleet of resulting modular vehicles has then to be managed properly.

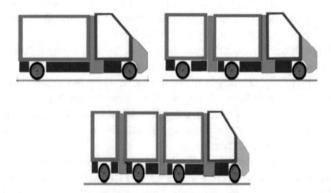

Fig. 1. Electric modular vehicles

3.2 A MILP Formulation of the eM-Fleet Size and Mix VRP with Time Windows

Our model is based on the use of electric vehicles. Therefore, some hypothesis have been set in order to define the problem. These assumptions are related basically to the recharging policy [13]. The resulting modelization of the eM-fleet size and Mix VRP with Time Windows has the objective to minimize the total traveling distance, the acquisition cost and the recharging cost.

Formally, the proposed formulation of the eM-fleet size and Mix VRP with Time Windows can be stated as follows: let $G = (N, A)$ be a complete directed graph where N denotes a set of customers. Further, let nodes 0 and $n + 1$ represent the beginning and the ending of the tour respectively, thus $N = \{0\}U\{1,,n\}U\{n + 1\}$. Then, let $A \subseteq N \times N$, denotes the arc set with $A = \{(i, j) \,|i, j \in N, \, i \neq j\}$. Each arc (i, j) is associated with a travel time t_{ij} and a distance d_{ij}. Each vehicle of type k, when it crosses an arc, consumes the amount $e^k \, dij$ of the remaining battery charge, where e^k denotes the energy consumption per distance unit travelled. A vehicle of type k has a capacity Q^k, a fixed cost c_f^k and a per unit-distance variable cost p^k. The cost of a vehicle of type k when it crossing the pair (i, j) is denoted by c_{ij}^k, which is obtained by multiplying the distance d_{ij} and the variable cost p^k. Four decisions variables are used: a binary variable x_{ij}^k indicates that a vehicle of type k travels from customer i to customer j. z_{ij}^{km} specifies if a vehicle of type k transports the module m from customer i to customer j is a binary decision variable, which indicates if customer i is served by module m. Last, r^k is the binary variable related to the recharging of vehicle type k at customer i.

$$\min \sum_{k \in V} cf^k \sum_{j \in N} x_{0j}^k + \sum_{k \in V} c_{ij}^k \sum_{i \in N0, j \in Nn+1, i \neq j} x_{ij}^k + \sum_{k \in V} c_r \sum_{i \in N} r_i^k \tag{1}$$

Subject to the following constraints

$$\sum_{k \in V} \sum_{j \in Nn+1, i \neq j} x_{ij}^k = 1 \qquad \forall \, i \in N \tag{2}$$

$$\sum_{k \in V} \sum_{j \in Nn+1, i \neq j} x_{ij}^k \leq 1 \qquad \forall \, i \in N \tag{3}$$

$$\sum_{I \in N0, i \neq j} x_{ij}^k - \sum_{j \in Nn+1, i \neq j} x_{ji}^k = 0 \qquad \forall \, j \in N \tag{4}$$

$$\lambda_p^m \leq \sum_{k \in V} \sum_{j \in N, p \neq j} Z_{pj}^{km} \qquad \forall \, m \in M, \forall \, p \in N \tag{5}$$

$$1 \leq \sum_{j \in N, p \neq j} Z_{pj}^{km} \leq 3 \qquad \forall \, k \in V, \forall \, i \in No, \forall \, j \in Nn+1 \tag{6}$$

$$a_j \leq \tau_j \leq b_j \qquad \forall j \in No, \forall \, j \in Nn+1 \tag{7}$$

$$\tau_i + \left(\max(s_i, h_i) + t_{ij}\right)x_{ij}^k - b_o\left(1 - x_{ij}^k\right) \leq \tau_j \, \forall k \in V, \forall i \in No, \forall j \in Nn+1, i \neq j \tag{8}$$

$$\tau_i + t_{ij}x_{ij}^k + w^k\left(E^k - y_i^k\right) + (b_o + w^k E^k)\left(1 - x_{ij}^k\right) \leq \tau_j$$

$$\forall k \in V, \forall i \in No, \forall j \in Nn+1, i \neq j$$

$$(9)$$

$$0 \leq u_j^k \leq u_i^k - q_i x_{ij}^k + Q^K\left(1 - x_{ij}^k\right) \quad \forall k \in V, \forall i \in No, \forall j \in Nn+1, i \neq j \quad (10)$$

$$0 \leq u_j^k \leq Q^k \quad \forall k \in V, \forall j \in Nn+1, i \neq j \quad (11)$$

$$\varepsilon \leq y_j^k \leq y_i^k - (e^k d_{ij})x_{ij}^k + E^k\left(1 - x_{ij}^k\right) \quad \forall k \in V, \forall i \in No, \forall j \in Nn+1, i \neq j \quad (12)$$

$$\varepsilon \leq y_j^k \leq E^k - (e^k d_{ij})x_{ij}^k \quad \forall k \in V, \forall i \in No, \forall j \in Nn+1, i \neq j \quad (13)$$

$$\varepsilon \leq y_j^k \leq \sum_{m \in M} l_i^m Z_{lj}^{km} \quad \forall k \in V, \forall m \in M, i \neq j \quad (14)$$

$$y_o^k = E^k \quad \forall k \in V \quad (15)$$

$$x_{ij}^k \in \{0.1\} \quad \forall k \in V, \forall i \in No, \forall j \in Nn+1 \quad (16)$$

The eM-fleet size and Mix VRP with Time Windows is a minimization problem as suggests formula (1). The objective function contains three terms: The first term is the sum of the costs of all the vehicles used, since x_{0j}^k indicates whether the vehicle k is used or not. Thus, if a vehicle leaves from the depot to any customer node, an acquisition cost f^k is added. The second term represents the total distance travelled by each electric vehicle in a given day. The third term is the total cost related to the recharging of the vehicle. Indeed, if the service time at a customer is finished and the charge level of the battery is below a given threshold, the vehicle will recharge until a threshold and a penalty g_r is added with the corresponding cost C_r.

Constraint (2) specifies that each customer is served exactly once and exactly by one vehicle, while constraint (3) ensures that each recharging station will have at most one successor node: a customer, a recharging station or the depot. Furthermore, constraint (4) stipulates that for each node, the number of arrivals must be equal to the number of departures. In addition, the series of inequalities. Constraint (5) ensures that if the module m serves customer p, then the module m should first leave customer p to serve customer j. In constraint (6) we enforce that each vehicle may carry at least one payload module to move forward the cabin module, and be composed of three modules at most at any given time. Besides, in constraint (7) each customer is visited within its specified time window, while constraints (8) and (9) guarantee the elimination of subtours. Moreover, (10) and (11) are capacity constraints. The constraint depicted in (10) ensures that the load of node j depends on the load of node i which takes into account the demand q_i, while constraint (11) ensures that the load u_j^k should not exceed the capacity Q^k of the vehicle.

The electric constraints are described by the inequalities (12)–(14). Constraint (12) ensures that the charge level of the vehicle at the node j depends on the energy consumed when it crosses the arc (i, j). Constraint (13) specifies that the remainder of the charge in node j is equal to the maximum load of the battery minus the charge consumed every long arc (i, j). Constraint (14) ensures that the level of electrical charge of the overall vehicle including the attached modules must not fall below a given threshold. Finally, in (15) we impose that the vehicle must be fully charged when it leaves the depot. Last, binary variables are defined in constraint (16).

4 Memetic Algorithm as a Solution Approach

In this work we intend first to assess to which extend an exact method such as the branch and bound technique used by the commercial solver CPLEX is able to address the problem. Then as far as we now that for large size problems, metaheuristics are suitable, we developed a typical memetic algorithm (MA) that combines the genetic algorithm with a local search, an approach that proved to be promising on routing and transportation problems such as the VRPTW with single or multiple depots. Different features need to be carefully considered in order to achieve good results, see [15]. Among these features, one could consider the initial solution representation, the fitness function and the crossover and mutation operators. As an initial solution, we used a simple representation in which a solution is represented with an array of n customer nodes which are served by the vehicles. Once individuals are created, they are ranked as per their fitness value calculated as the sum of the total distances traveled, acquisition cost and recharging cost. According to the results from the evaluation function, a new population is created after having applied operators such as the Tournament selection one and the Partial Mapping Crossover (PMX) because this latter has the advantage of providing best solutions in a short time frame as compared with others operators. Finally, the memetic method consists in enhancing the performance of the genetic algorithm (which yielded to preliminary encouraging results on this problem, see [14]), by introducing a local search as a mutation operator to intensify the child quality obtained with the crossover operator.

5 Preliminary Computational Results

As stated above, we first tried to implement our model into the commercial solver CPLEX to test it. Then in order to test our new approach, we applied the memetic algorithm to different classes of problems compared to the optimal solution found by CPLEX using the formulation depicted in Sect. 2. In this way, we have considered small instances of the so-called Solomon's benchmarks with different numbers of

Table 1. Comparison of results obtained with CPLEX and MA with 5 and 10 customers

Instance	CPLEX			MA		
	V	Obj	CPU	V	Obj	CPU
C101-5	2	342,42	0,30	2	343,07	0,23
C102-5	2	342,42	0,26	2	343,1	0,21
C201-5	2	1070,44	0,31	2	1071,71	0,15
C202-5	2	1070,44	0,29	2	1071,46	0,13
Rl0 1-5	2	309,48	0,31	2	311,57	0,14
R102-5	2	309,48	0,30	2	310,99	0,15
RC201-5	2	384,27	0,30	1	386,42	0,22
RC202-5	2	384,27	0,31	2	384,91	0,23
RC206-5	2	384,27	0,34	2	384,86	0,24
RC207-5	2	384,27	0,33	1	386,13	0,21
C101-10	2	351,80	13,91	3	352,81	0,45
C102-10	3	351,23	12,58	3	351,24	0,36
C201-10	2	1113,33	84,30	2	1116,32	0,51
C202-10	3	1110,89	101,42	2	1112,45	0,54
Rl0 1-10	2	352,69	7,23	2	353,12	0,41
R102-10	2	352,69	4,78	2	353,34	0,39
RC201-10	3	397,61	39,16	3	399,39	0,24
RC202-10	3	397,61	32,45	1	401,25	0,39
RC206-10	3	397,61	24,56	3	397,75	0,30
RC207-10	3	397,61	31,89	3	398,91	0,29

customers 5, 10, 15 or 20. Tables 1 and 2 present an overview of the results for both CPLEX and for the proposed meta-heuristic approach. The table columns refer respectively to: the instance name with the number of customers of the benchmark, the number of vehicles (V) and the value of the objective function (Obj) obtained with CPLEX and with our best solution denoted MA. We also provide the runtime in minutes in the columns denoted CPU.

Following these results, our memetic approach has appears to be competitive in terms of solutions quality in solving small eM-fleet size and Mix VRP with Time Windows instances to optimality with short computational times. The obtained results require in almost all cases, less or the same number of vehicles as compared to CPLEX. Concerning the traveled distance, our meta-heuristic is able to find near optimal solutions during the testing, but in a much faster way. To conclude, we can say that our memetic approach has proved to be efficient in solving the eM-fleet size and Mix VRP with Time Windows. We also observed that CPLEX is able to find optimal solutions for all the instances until 20 customers. For larger instances of the problem exact methods fail to obtain optimal solutions in a reasonable computation time. Our memetic algorithm.

Table 2. Comparison of results obtained with CPLEX and MA with 15 and 20 customers

Instance	CPLEX			MA		
	V	Obj	CPU	V	Obj	CPU
C101-15	2	378,12	110,11	2	379,80	11,56
C102-15	2	369,24	130,47	2	371,40	10,23
C201-15	2	1128,16	50,75	2	1128,23	11,08
C202-15	2	1128,16	53,48	2	1128,37	12,78
Rl0 1-15	3	399,35	62,05	3	402,07	11,14
R102-15	3	399,35	60,98	3	399,89	10,24
R201-15	3	659,35	79,25	3	660,18	12,61
RC201-15	2	422,74	14,47	2	423,11	9,92
RC202-15	3	417,44	13,77	3	418,88	8,11
RC206-15	2	422,74	10,91	2	422,86	6,76
RC207-15	3	417,74	13,64	3	419,19	9,87
C101-20	2	443,59	1123,58	2	407,1	19,14
C201-20	2	1146,22	669,25	2	1147,36	17,62
C202-20	2	1146,22	750,56	2	1147,77	18,21
R101-20	3	651,39	913,91	3	652,08	17,98
R102-20	3	537,77	848,78	3	538,25	17,55
R201-20	3	694,20	865,21	3	696,47	18,21
RC201-20	2	629,78	623,11	2	631,21	23,25
RC202-20	2	498,01	958,47	2	498,25	19,48
RC206-20	2	562,81	814,18	2	563,14	20,39
RC207-20	2	535,45	951,87	2	536,25	18,47

6   Conclusion and Future Work

In this paper, we intend to address a brand new problem of urban distribution involving modular electric vehicles. The first main innovation relies in the vehicle itself since it is under design. The originality of that vehicle is that it is based on a modular and active frame system. This means that payload modules are designed in addition to the cab module in order to bring more space and flexibility. This enables the vehicle to drop o modules at respective delivery locations and pick them up later during a run. Secondly, from the operational research side, new constraints linked with the modularity and the recharging aspects are considered. The aim is to combine the recent techniques, which have proven efficiency with those developed for this innovative electric modular vehicle. Technically, the problem has been modeled using a MILP program. Then it has been solved with the branch and bound technique used by the commercial solver CPLEX on several small size benchmark instances. The aim was to assess the limit of efficiency of an exact method on our formulation of the problem. A metaheuristic, namely the memetic approach has then been developed and tested on the same instances. It has been able to solve the problem optimally with less computational e orts, which validates the proposed approach. Further experiments will be achieved on greater size instances of the benchmark to prove the effectiveness of the metaheuristic.

References

1. Aggoune-Mtalaa, W., Habbas, Z., Ait Ouahmed, A., Khadraoui, D.: Solving new urban freight distribution problems involving modular electric vehicles. IET ITS **9**(6), 654–661 (2015)
2. Braysy, O., Dullaert, W., Hasle, G., Mester, D., Gendreau, M.: An effective multi-restart deterministic annealing metaheuristic for the fleet size and mix vehicle routing problem with time windows. Transp. Sci. **42**(3), 371–386 (2008)
3. Bruglieri, M., Pezzella, F., Pisacane, O., Suraci, S.: A variable neighborhood search branching for the electric vehicle routing problem with time windows. Elec. Notes Disc. Math. **47**, 221–228 (2015)
4. Conrad, R.G., Figliozzi, M.A.: The recharging vehicle routing problem. In: Proceedings Industrial Engineering Research Conference, Reno, CA (2011)
5. Electrification Coalition. State of the plug-in electric vehicle marke (2013). https://www.electrificationcoalition.org/StateEVMarket
6. Erdogan, S., Miller-Hooks, E.: A green vehicle routing problem. Transp. Res. Part E Logist. Transp. Rev. **48**(1), 100–114 (2012)
7. Goncalves, F., Cardoso, S.R., Relvas, S., Barbosa-Povoa, A.P.F.D.: Optimization of a distribution network using electric vehicles: a VRP problem. Technical report, CEG-IST, Technical University of Lisbon, Portugal (2011)
8. Hiermann, G., Puchinger, J., Hartl, R.F.: The electric fleet size and mix vehicle routing problem with time windows and recharging stations. Eur. J. Oper. Res. **252**(3), 995–1018 (2016)
9. Koc, C., Bektas, T., Jabali, O., Laporte, G.: The fleet size and mix pollution routing problem. CIRRELT-26 (2014)
10. Lin, J., Zhou, W., Wolfson, O.: Electric vehicle routing problem. In: International Conference on City Logistics, vol. 12, pp. 508–521 (2016)
11. Liu, F.H., Shen, S.Y.: A route-neighborhood-based metaheuristic for vehicle routing problem with time windows. Eur. J. Oper. Res. **118**, 485–504 (1999)
12. Pelletier, S., Jabali, O., Laporte, G.: Goods distribution with electric vehicles: review and research perspectives. CIRRELT-44 (2014)
13. Rezgui, D., Aggoune-Mtalaa, W., Bouziri, H.: Towards the electrification of urban freight delivery using modular vehicles. In: 10th IEEE SOLI Conference, vol. 6, pp. 154–159 (2015)
14. Rezgui, D., Aggoune-Mtalaa, W., Bouziri, H.: Improving the freight distribution in cities using electric modular vehicles. In: Most Innovative Research Award of the third International Conference on Green Supply Chain, London, United Kingdom (2016)
15. Rezgui, D., Chaouachi-Siala, J., Aggoune-Mtalaa, W., Bouziri, H.: Application of a memetic algorithm to the fleet size and mix vehicle routing problem with electric modular vehicles. In: GECCO 2017 Proceedings of the Genetic and Evolutionary Computation Conference, vol. 2, pp. 301–302 (2017)
16. Solomon, M.M.: Algorithms for the vehicle routing and scheduling problems with time window constraints. Oper. Res. **35**, 254–265 (1987)
17. U.S. Department of Energy. Plug-in electric vehicle handbook for fleet managers. Office of Energy Efficiency and Renewable Energy, National Renewable Energy Laboratory (NREL) (2012)

Entropic Method for 3D Point Cloud Simplification

Abdelaaziz Mahdaoui[1]([envelope]) [ORCID], A. Bouazi[2], A. Marhraoui Hsaini[2],
and E. H. Sbai[2]

[1] Department of Physics, Faculty of Science,
Moulay Ismail University, Meknes, Morocco
a.mahdaoui@edu.umi.ac.ma
[2] Ecole Supérieure de Technologie, Moulay Ismail University,
Meknes, Morocco

Abstract. To represent the surface of complex objects, the samples resulting from their digitization can contain a very large number of points. Simplification techniques analyse the relevance of the data. These simplification techniques provide models with fewer points than the original ones. Whereas reconstruction of a surface, with simplified point cloud, must be close to the original. In this article, we develop a method of simplification based on the concept of entropy, which is a mathematical function that intuitively corresponds to the amount of information this allows considering only relevant points.

Keywords: Simplification · Entropy · 3D point cloud · Density estimator
Mesh quality · Compactness

1 Introduction

It has been shown that three-dimensional (3D) scanning systems are highly developed. It is possible to obtain clouds containing millions of points. However, 3D scanners remain unable to determine the optimum points density to faithfully represent a surface. This leads to a significant redundancy of the data, which must be removed to limit computations that are necessary for the analysis and representation of the form.

The problem of simplifying point cloud can be formalized as follows: given a set of points X and a sampling surface S, find a sample of points X' with $|X'| \leq |X|$, Such that X' sampling a surface S' that is close to S. $|X'|$ is the cardinality of set X. This objective requires defining a measure of geometric error between the original and the simplified surface, for which the method will resort to estimate of global or local properties of the original surface. There are two main categories of algorithms to sample points: sub-sampling algorithms and resampling algorithms. The subsampling algorithms produce simplified sample points which are a subset of the original point cloud, while the resampling algorithms rely on estimating the properties of the sampled surface to compute new relevant points.

In the literature, the categories of simplification algorithms have been applied according to three main simplification schemes.

© Springer International Publishing AG, part of Springer Nature 2018
M. Ben Ahmed and A. A. Boudhir (Eds.): SCAMS 2017, LNNS 37, pp. 613–621, 2018.
https://doi.org/10.1007/978-3-319-74500-8_56

The first method is simplification by selection or calculation of points representing subsets of the initial sample. This method consists of decomposing the initial set into small regions, each of which is represented by a single point in the simplified sample [1–3]. The methods of this category are distinguished by the criteria defining the regions and their construction.

The second method is iterative simplification. The principle of iterative simplification is to remove points from the initial sample, incrementally per geometric or topologic criteria, locally measuring the redundancy of the data [4–10].

The third method is simplification by incremental sampling. Unlike iterative simplification, the simplified sample points can be constructed by incrementally enriching an initial subset of points, or sampling an implicit surface [7, 8, 11–18].

In this paper, we propose an iterative simplification technique based on the estimation of entropy. Thus, our method is based on the algorithm proposed by Wang et al. [19].

This paper is organized as the following: In Sect. 2, we recall the density function estimator and entropy definition. Afterwards, Sect. 3, present our 3D point cloud simplification algorithm based on the Shannon entropy [20]. While Sects. 4 and 5, lay out the results and the validation. Finally, we make conclusion.

2 Estimation of Density Function and Entropy Definition

There are several methods for density estimation, parametric and nonparametric methods. Nonparametric methods include the kernel density estimator, also known as the Parzen-Rosenblatt method [21, 22] and the K-NN method [23], each type has its advantages and disadvantages. For Parzen estimator, the bandwidth choice has strong impact on the quality of the estimated density [24]. In this paper, we will use a K-NN estimator to estimate the density function.

2.1   The Parzen-Rosenblatt Estimator

The kernel estimator was introduced in 1956 by Rosenblatt and then developed by Parzen in 1962 [21, 22]. Nonparametric estimation of the probability density of a distribution in dimension d can be defined by:

$$p(x) = \frac{1}{Nh^d} \sum_{i=1}^{N} K\left(\frac{X - x_i}{h}\right) \tag{1}$$

where h is the smoothing parameter ("Bandwidth") and K_d is the function of weight kernel. The kernel can be defined as a kernel product K_d:

$$K(X) = \prod_{j=1}^{j=d} K_j(x_j) \tag{2}$$

with $X = (x_1, \ldots, x_d) \in \mathbb{R}^d$ and the kernel $K(X)$ is a symmetric function satisfying the following conditions:

$K(X) \geq 0, K(X)$ must be continuous, $\int K(X)dX = 1$, $\int XK(X)dX = 0$ and $\int X^2 K(X)dX \neq 0$

The most frequently used kernel is a Gaussian Kernel:

$$K(X) = (2\pi)^{-d\backslash 2} \exp\left(-\frac{1}{2}X^T X\right) \tag{3}$$

2.2   The K Nearest Neighbors Estimator

The k nearest neighbors (K-NN) algorithm [23] is an attempt to estimate the non-parametric probability density function. The degree of estimation is controlled by a number k, which is the number of nearest neighbors, generally proportional to the sample size N. For each x where we want to estimate the density, we define distances between points of the sample and the x as following:

$$r_1(x) < \cdots < r_{k-1}(x) < r_k(x) < \cdots < r_N(x) \tag{4}$$

These distances are sorted in ascending order. The estimator with the method of nearest neighbor in dimension d can be defined as follows:

$$p(x) = \frac{k/N}{V_k(x)} = \frac{k/N}{C_d r_k(x)} \tag{5}$$

where $r_k(x)$ is the distance from x to the kth nearest point, $V_k(x)$ is the volume of a sphere of radius $r_k(x)$ and C_d is the volume of the unit sphere in d dimension.

The adjustment of the number k must be a function of the size N of the available sample in order to respect the constraints that ensure the convergence of the estimator. For a number N of observations, the number k can be calculated as follows:

$$k = k_0.\sqrt{Q} \tag{6}$$

By respecting these rules of adjustment, it's certain that the estimator converges when the number N increases indefinitely, whatever the value of k_0.

2.3   Entropy Definition

Claude Shannon introduced the concept of the entropy associated with a discrete random variable X as a basic concept in information theory [19, 20].

Let the distribution of probabilities $p = \{p_1, p_2, \ldots, p_N\}$ associated with the realizations of X. The Shannon entropy is calculated using the formula:

$$H(X) = -\sum_{i=1}^{N} p_i \log(p_i) \tag{7}$$

In the next section, we present our simplification approach based on the estimation of Shannon entropy.

3 Proposed Approach

We consider a point cloud $X = \{x_1, x_2, \ldots, x_N\}$ (see Fig. 1) at the input, our method begins with entropy estimation using all the points of X, denote $H(X)$. As previously indicated, the estimation of the density function is done by a K-NN estimator. Next, we estimate the entropy, denote $H(X - x_i)$, using $X - x_i$ with $i = 1, \ldots, N$. Then, we compute difference between $H(X)$ and $(X - x_i)$, denote $\Delta E_i = |H(X) - H(X - x_i)|$, $i = 1, \ldots, N$. Next, for $\Delta E_i \leq s$, s is the known threshold, the point x_i will be removed from point cloud X. If not the point x_i will be retained. At the end of the simplification algorithm, we obtain a simplified point cloud X' (see Fig. 2).

(a)   (b)   (c)   (d)

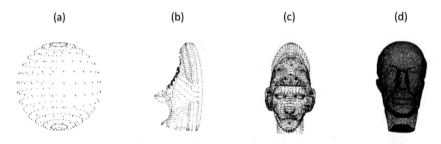

Fig. 1. Original point cloud, (a) sphere, (b) tennis shoe, (c) Atene, (d) Max Planck

(a)   (b)   (c)   (d)

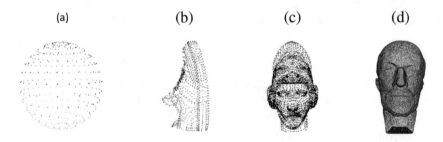

Fig. 2. Simplified point cloud (a) sphere, (b) tennis shoe, (c) Atene, (d) Max Planck

Simplification Algorithm

- Provide the dataset $X = \{x_1, x_2, \ldots, x_N\}$
- Provide the threshold s
- $X' = X$
- Calculate the entropy by using all data samples in X according to Eq. (7), denote this entropy $H(X)$.

- For i = 1 to N
  - Calculate the entropy $H(X - x_i)$, of point cloud X less the point x_i
  - Calculate $\Delta E_i = |H(X) - H(X - x_i)|$
  - If $\Delta E_i \leq s$ then $X' = X' - x_i$
  - End if
- End

4 Experiments and Results

We illustrate our simplification approach using three 3D models that represent real objects such as Max Planck (Fig. 1d), Atene (Fig. 1c) and tennis shoe (Fig. 1b). Also, we use a synthetic 3D model that represents a sphere (Fig. 1a). Figures 2(a–d) show simplification results on various sample points.

In the next section, we will validate the effectiveness of our proposed method. Actually, we have conducted a comparison between the original and the simplified point cloud. Accordingly, we will use a comparison between the original and simplified mesh.

5 Comparison Between Original and Simplified Mesh

In this section, we make a comparison between the original mesh and the one created from the simplified point cloud. To reconstruct the mesh, we use ball Pivoting method [25, 26] or Hsaini et al. method [27]. Then to measure the quality of the obtained meshes, we compute the quality of the triangles using the compactness formula proposed by Guéziec [28]:

$$c = \frac{4\sqrt{3}a}{l_1^2 + l_2^2 + l_3^2} \tag{8}$$

where l_i are the lengths of the edges of a triangle and a is the area of the triangle. We observe that this measure is equal to 1 for an equilateral triangle and 0 for a triangle whose vertices are collinear. According to [29], a triangle is of acceptable quality if $c \geq 0.6$.

In Figs. 3, 4, 5 and 6, we present the triangles compactness histogram of the two meshes. In each figure, the first line presents the reconstructed mesh from the original point cloud. The second line presents the simplified point cloud. Note that, the evaluation of the mesh quality is achieved by the compactness of the triangles.

Table 1 shows the percentage of triangles with a compactness greater than or equal to 0.6 for each 3D model. It is found that the percentage of the compactness is greater than 50% for most simplified point clouds of the 3D models. Thus, we observe that this value is superior for the simplified point cloud than it is for the original point cloud. The exception case is for the Atene model where it is observed this value is superior for the original point cloud.

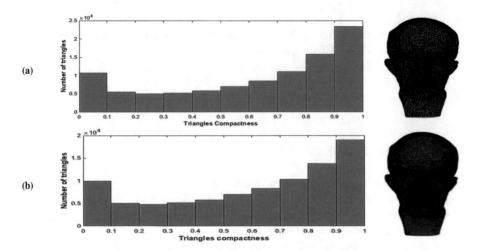

Fig. 3. Comparison of Max Planck mesh quality, (a) Original point cloud, (b) simplified point cloud

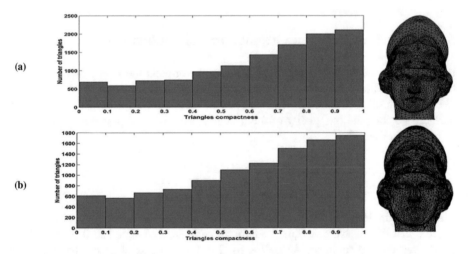

Fig. 4. Comparison of Atene mesh quality, (a) Original point cloud, (b) simplified point cloud

We have implemented our simplification method under MATLAB. The calculations are performed on a machine with an i3 CPU, 3.4 Ghz, and 2 GB of RAM.

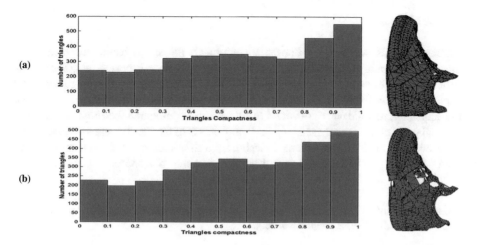

Fig. 5. Comparison of Tennis shoe mesh quality, (a) Original point cloud, (b) simplified point cloud

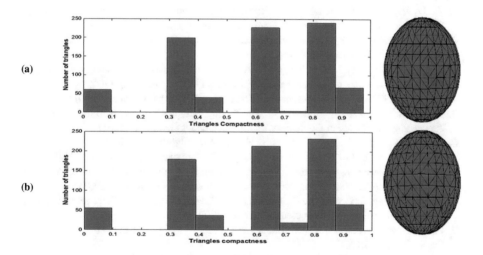

Fig. 6. Comparison of sphere mesh quality, (a) Original point cloud, (b) simplified point cloud

Table 1. Mesh compactness of different sample points

Object	Number of points		Compactness $\geq$ 0.6 (%)	
	Original	Simplified	Simplified	Original
Sphere	422	410	81.57%	64.27
Tennis shoe	1840	1692	50.45%	49.13
Aten	6942	6302	58.05%	59.89
Max Planck	49089	45303	57.58%	59.94

6 Conclusion

Throughout this work, we proposed an approach for simplification of dense and unstructured point cloud using the Shannon entropy with a k nearest neighbors estimator which allows the consideration of relevant data only. In the first stage, we have applied our simplification algorithm to different point cloud with different densities. Subsequently, to validate the obtained results, we have compared the simplified and the original point cloud. The measure of the compactness of the meshes shows that for each 3D model the quality of the obtained mesh from a simplified point cloud is greater than the one from an original cloud of points. It shows that our approach is efficient and effective.

Acknowledgments. The Max Planck, Atene and Tennis shoe models used in this paper are the courtesy of AIM@SHAPE shape repository.

References

1. Pauly, M., Gross, M., Kobbelt, L.: Efficient simplification of point-sampled surfaces. In: Proceedings of IEEE Visualization Conference (2002)
2. Wu, J., Kobbelt, L.P.: Optimized sub-sampling of point sets for surface splatting. In: Proceedings of Eurographics (2004)
3. Ohtake, Y., Belyaev, A.G., Seidel, H.-P.: An integrating approach to meshing scattered point data. In: Proceedings of Symposium on Solid and Physical Modeling (2005)
4. Linsen, L.: Point Cloud Representation. Universitat Karlsruhe, Germany (2001)
5. Dey, T.K., Giesen, J., Hudson, J.: Decimating samples for mesh simplification. In: Proceedings of Canadian Conference on Computational Geometry (2001)
6. Amenta, N., Choi, S., Dey, T., Leekha, N.; A simple algorithm for homeomorphic surface reconstruction. In: Proceedings of Symposium on Computational Geometry (2000)
7. Dey, T.K., Giesen, J., Hudson, J.: Sample shuffling for quality hierarchic surface meshing. In: Proceedings of 10th International Meshing Roundatble Conference (2001)
8. Alexa, M., Beh, J., Cohen-Or, D., Fleishman, S., Levin, D., Silva, C.: Point set surfaces. In: Proceedings of IEEE Visualization Conference (2001)
9. Garland, M., Heckbert, P.: Surface simplification using quadric error metrics. In: Proceedings of SIGGRAPH (1997)
10. Allègre, R., Chaine, R., Akkouche, S.: Convection-driven dynamic surface reconstruction. In: Proceedings of Shape Modeling International, IEEE Computer Society Press (2005)
11. Boissonnat, J.-D., Cazals, F.: Coarse-to-fine surface simplification with geometric guarantees. In: Proceedings of Eurographics (2001)
12. Boissonnat, J., Oudot, S.: Provably good surface sampling and approximation. In: SGP 2003 Proceedings of the 2003 Eurographics/ACM SIGGRAPH Symposium on Geometry Processing, Aachen, Germany (2003)
13. Boissonnat, J.-D., Oudot, S.: An effective condition for sampling surfaces with guarantees. In: SM 2004 Proceedings of the Ninth ACM Symposium on Solid Modeling and Applications, Genoa, Italy (2004)
14. Boissonnat, J.-D., Oudot, S.: Provably good sampling and meshing of surfaces. Graph. Models Solid Model. Theory Appl. **67**(5), 405–451 (2005)

15. Chew, L.P.: Guaranteed-quality mesh generation for curved surfaces. In: Proceedings of Symposium on Computational Geometry, San Diego, California, USA (1993)
16. Adamson, A., Alexa, M.: Approximating and intersecting surfaces from points. In: Proceedings of Symposium on Geometry Processing (2003)
17. Pauly, M., Gross, M.: Spectral processing of point-sampled geometry. In: SIGGRAPH 2001 Proceedings of the 28th Annual Conference on Computer Graphics and Interactive Techniques, New York, NY, USA (2001)
18. Witkin, A.P., Heckbert, P.S.: Using particles to sample and control implicit surfaces. In: SIGGRAPH 1994 Proceedings of the 21st Annual Conference on Computer Graphics and Interactive Techniques, New York, NY, USA (1994)
19. Wang, J., Li, X., Ni, J.: Probability density function estimation based on representative data samples. In: Communication Technology and Application (ICCTA 2011), IET International Conference, Beijing, China (2011)
20. Shannon, C.E., Weaver, W.: The Mathematical Theory of Communication. University of Illinois Press, Urbana (1949)
21. Parzen, E.: On estimation of a probability density function and modes. Ann. Math. Stat. **33** (3), 1065–1076 (1962)
22. Rosenblatt, M.: Remarks on some nonparametric estimates of a density function. Ann. Math. Stat. **27**(3), 832–837 (1956)
23. Silverman, B.W.: Density Estimation for Statistics and Data Analysis, Published in Monographs on Statistics and Applied Probability. Chapman and Hall press, London (1986)
24. Muller, H., Petersen, A.: Density Estimation Including Examples. Wiley StatsRef: Statistics Reference Online, pp. 1–12 (2016)
25. Bernardini, F., Mittleman, J., Rushmeier, H., Silva, C., Taubin, G.: The ball-pivoting algorithm for surface reconstruction. IEEE Trans. Vis. Comput. Graph. **5**(4), 349–359 (1999)
26. Mahdaoui, A., Bouazi, A., Hsaini, A.M., Sbai, E.: Comparative study of combinatorial 3D reconstruction algorithms. Int. J. Eng. Trends Technol. **48**(5), 247–251 (2017)
27. Marhraoui Hsaini, A., Bouazi, A., Mahdaoui, A., Sbai, E.H.: Reconstruction and adjustment of surfaces from a 3-D point cloud. Int. J. Comput. Trends Technol. (IJCTT) **37**(2), 105–109 (2016)
28. Guéziec, A.: Locally toleranced surface simplification. IEEE Trans. Vis. Comput. Graph. **5** (2), 168–189 (1999)
29. Randolph, E.B.: PLTMG: A Software Package for Solving Elliptic Partial Differential Equations, La Jolla, California 92093-0112 (2016)

Human Daily Activity Recognition Using Neural Networks and Ontology-Based Activity Representation

Nadia Oukrich$^{(\boxtimes)}$, El Bouazzaoui Cherraqi, and Abdelilah Maach

Mohammedia Engineering School, Ibn Sina, Agdal, B.P 765, Rabat, Morocco
nadiaoukrich@gmail.com, cherraqii@gmail.com

Abstract. In real-life people live together in the same place, recognize their activities is challenging than activities of one single resident, but essential to collect information about real life activities inside home, then ease the assisted living in the real environment. This paper presents a multilayer perceptron model and a supervised learning technique called backpropagation to train a neural network in order to recognize multi-users activities inside smart home, and select useful features according to minimum redundancy maximum relevance. The results show that different feature datasets and different number of neurons of hidden layer of neural network yield different activity recognition accuracy. The selection of suitable feature datasets increases the activity recognition accuracy and reduces the time of execution. Our experimental results show that we achieve an accuracy of 99% with the winner method and 96% with the threshold method, respectively, for recognizing multi-user activities.

Keywords: Multi-users · Activity recognition · Smart home
Multilayer perceptron · Back-propagation · Features selection
Mutual information

1 Introduction

Recognize human activities inside home can reduce costs of health and elderly care that exceed $7 trillion annually and rising [1], ensure comfort, homecare [2], safety, and reduces energy consumption. For these reasons, researchers and organizations focused in development of a real smart home project. A smart home is a normal house, but equipped with sensors and others technologies, which anticipates and responds to the needs and requirements of the elderly people, working to promote their luxury, convenience, security, and entertainment [3]. A key point in development of smart home is recognition of normal and daily routine activities of its residents. Human Activity Recognition (HAR) is a challenging and well-researched problem. In fact, a large number of research focuses in recognition of Activities of Daily Living [4] (ADLs) which means activities, performed in user daily routine, such as eating, cooking, sleeping, and toileting; there are various reasons why ADLs are the most covered in literature. Citing as examples, first those activities are general, real, and common between young and old people. Second, ADLs are the most use in standard tests of user

© Springer International Publishing AG, part of Springer Nature 2018
M. Ben Ahmed and A. A. Boudhir (Eds.): SCAMS 2017, LNNS 37, pp. 622–633, 2018.
https://doi.org/10.1007/978-3-319-74500-8_57

autonomy because disability with ADLs is the most common reason that older people live in nursing homes [5]. Finally, ADLs are the best suited for use as inputs to perform different home applications.

However in real-life, there are often multiple residents living in the same environment, and perform ADLs together or separately. Recognizing multi-user activities is more challenging than recognizing single-user activities. The main challenges are knows activities of every user at home and distinguish between two or more activities, which takes place at the same time. In this work, we use multilayer perceptron model, a type of Artificial Neural Network, the computation is performed using a set of simple units with weighted connections between them. Furthermore, Back Propagation (BP) algorithm [6, 7] to set the values of the weights. Otherwise, BP is a common method of training artificial neural networks in supervised learning method, which calculates the gradient of a loss function with respect to all the weights in the network. The gradient is fed to the optimization method, which in turn uses it to update the weights, in an attempt to minimize the loss function. Moreover, based on a ontological approach we propose a set of features adequate for multi-users, then we select the most relevant using a selection algorithm called, minimal-Redundancy-Maximal-Relevance criterion (mRMR) [8, 9] to obtain a set of subsets of features with high class-feature mutual information classification and the less feature-feature mutual information.

The rest of the paper is organized as follows. Section 2 reviews related work. Section 3 explain the proposed set of features based on an ontological approaches and introduce the optimization approach method for features selection. Then, it describes the designing of multilayer perceptron network using BP algorithm applied to recognize multi-users activities. Section 4 resume the test and results. Finally, Sect. 5 concludes.

2 Literature Review

2.1 Activity Recognition Approaches

Researchers classify activity recognition approaches into two categories. The first is based on the use of visual sensing facilities, example: camera, and exploit computer vision techniques to analyse visual observations for pattern recognition [10, 11]. Such solutions are challenged because of the potential for the violation of user privacy, the difficulty of extracting robust and informative features to infer high-level activities. The second category is based on the use of emerging and wearable sensor network technologies and using data mining and machine learning techniques to analyse sensors data and determine user's behaviour [12–19]. Sensors can be wearable [12] or fixed in doors or spatial place at home. Due to low cost and low power consumption, sensors based approach became a centre of interest at the last decade, researchers have commonly tested the machine learning algorithms such as knowledge-driven approach (KDA) [13], evolutionary ensembles model (EEM) [14], support vector machine (SVM) [15], Naïve Bayes (NB) classifier [16], hidden Markov model (HMM) [17], and conditional random fields (CRF) [18].

2.2   Neural Networks and Features Selection

Neural networks algorithm, used in this paper, were first published in 1960. In the years following, many new techniques have been developed in the field of neural networks, and the discipline is growing rapidly [19]. Neural network has proven successful in different fields [20–23] among them human activity recognition in smart home environments [23]. To have high accuracy in neural network using BP algorithm, researchers have studied indirect or direct means to select useful features, several artificial intelligence approaches were used to identify and select signal features which are the input to Neural Networks (NN) [24]. Features selection referred in this case represents distinctive information as inputs into the input layer. Moreover, feature extraction addresses the problem of finding the most compact and informative set of features, to improve the efficiency or data storage and processing [25]. Usually all features are not equally informative: some of them may be noisy [26] Usually, the best feature subset contains the least number of dimensions that contribute to higher recognition accuracy, therefore, it is necessary to remove the remaining and unimportant features to reduce the time execution and noise. In [27] submitted to 4th edition of international colloquium on Information science and technology in Tangier, we use back-propagation algorithm and a set of selected features in order to recognise activities of one single user. In this paper, we use an ontological approach to list more powerful features, select them based in mRMR method and use back propagation algorithm to recognise multi-users activities inside home.

2.3   Multi-users Activities

Multi-users activities can be classified into tree big categories:

(1)   Single user performs activities one by one
(2)   Multiple users perform the same activity together
(3)   Multiple users perform different activities independently [28].

To the best of our knowledge, there is less study on recognizing multi-user activities in a smart home environment. A series of research work has been done in the CASAS smart home project at WSU [29], in [30] a various kinds of sensors and a multiple users' preferences is done based on sensor readings. In [12] authors develop a multi-modal, wearable sensor platform to collect sensor data for multiple users. These works are still in a development phase because of the complexity of multi-users activities.

3   A Methodology for Activity Recognition

3.1   Activities of Daily Living

ADLs are defined as the routine activities that a person perform every day inside home without help. The ability of doing ADLs is crucial for health care to provide if required caring service. To validate our methodology for activity recognition, we relied on a smart home testbed located on Cairo in Egypt and is maintained as part of WSU Smart

Apartment ADL Multi-Resident Testbed [31] as presented in Fig. 2. The data was collected over three months while volunteer adult couple and a dog performed their normal work routines in the environment. In the smart home test bed activities to recognize are 13. These activities include both activities of the woman and the man at home. Below Table 1 describe tasks with the number of times the activity appears in the data.

Table 1. Present a description of tasks with the number of times the activity appears in the data.

Number	Name	Description	N° of time
Activity 1	Bed_to_toilet	One of the users move between bed and toilet at night	30
Activity 2	Breakfast	The two residents eat breakfast together	48
Activity 3	R1_sleep	Resident number 1 sleep	50
Activity 4	R1_wake	Resident number 1 wake	53
Activity 5	R1_work_in_office	Resident number 1 work in office at home	46
Activity 6	Dinner	The two residents eat dinner together	46
Activity 7	Laundry	One of the resident washes clothes	10
Activity 8	Leave_home	One of the residents leave home	69
Activity 9	Lunch	The two residents eat lunch together	37
Activity 10	Night_wandering	Residents wander around at night	67
Activity 11	R2_sleep	Resident number 2 sleep	52
Activity 12	R2_take_medicine	Resident number 2 take medicine	44
Activity 13	R2_wake	Resident number 2 wake up	52

3.2 Features Listing

To achieve a better representation of ADLs, we extract divers features for constructing the activity model. Features extraction refers to the process of extraction informations from raw data. Informations may include, time, location, duration, variance and average. An ontological approach is employed for features extraction, which is explicated in Fig. 1.

Based on this ontological approach, we have extracted 17 features, detailed below:

1.

$$S_i = \frac{1}{n_i} \sum_{k=1}^{n_i} S_{ik} \tag{1}$$

Si the means of Sensors ID of activity i, ni is the number of motion and door sensors noted in the dataset between the beginning and end of the activity, and Sik is the kth Sensor ID.

2. The logical value of the first Sensor ID triggered by the current activity;
3. The logical value of the second Sensor ID triggered by the current activity;
4. The logical value of the last Sensor ID triggered by the current activity;

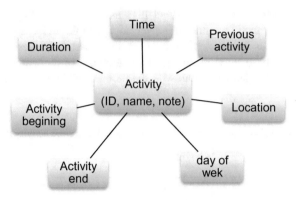

Fig. 1. Ontological representation of activity

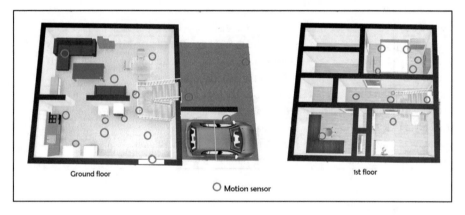

Fig. 2. Motion sensors location in the smart apartment testbed [31]

5. The logical value of before the last Sensor ID triggered by the current activity;
6. The name of the first sensor triggered by the activity;
7. The name of the last sensor triggered by the current activity;
8. The variance of all Sensor IDs triggered by the current activity:
9. The beginning time of the current activity;
10. The ending time of the current activity;
11. The duration of the current activity;
12. Day of week, which is converted into a value in the range of 0 to 6;
13. Previous activity, which represents the activity that occurred before the current activity;
14. Activity length, which is the number of instances between the beginig and the end of current activity;
15. The name of the dominant sensor durant the current activity;
16. Location of the dominant sensor;
17. Frequance of the dominant sensor.

3.3   Features Selection Method mRMR

Features selection, or variable subset selection is used for selection of a subset of features to construct models. Two important aspects of features selection are: minimum redundancy and maximum relevance.

Features values are uniformly distributed in different classes. If a feature is strongly differentially expressed for different classes, it should have large mutual information with classes. Thus, we use mutual information as a measure of relevance of features. For discrete variables, the mutual information I of two variables x and y is defined based on their joint probabilistic distribution $p(x,y)$ and the respective marginal probabilities $p(x)$ and $p(y)$:

$$I(x, y) = \sum_{i,j} p(x_i, y_i) log \frac{p(x_i, y_i)}{p(x_i)p(y_i)} \qquad (2)$$

For categorical variables, we use mutual information to measure the level of "similarity" between features. The idea of minimum redundancy is to select the feature such that they are mutually maximally dissimilar. Let $S = \{S1, S2, ...Ss\}$ denote the subset of features we are seeking. The minimum redundancy condition is:

$$min[W_1, W_s], \quad W_I = \frac{1}{s^2} \sum_{i,j \in S} I(f_i, f_j) \qquad (3)$$

Where $I(f_i, f_j)$ is the mutual information between feature f_i and f_j, s is the number of features in S and $W_I \in [W_1, W_n]$. To measure the level of discriminant powers of features when they are differentially expressed for different target classes $T = \{T1, T2, ... Tt\}$, we again use mutual information between targeted classes and the features. Thus the maximum relevance condition is to maximize the total relevance of all features in S:

$$max[V_1, V_s], \quad V_I = \frac{1}{s} \sum_{i \in S} I(f_i, T) \qquad (4)$$

Where $V_I \in [V_1, V_n]$. The mRMR features set is obtained by optimizing the conditions in Eqs. (3) and (4) simultaneously. Optimization of both conditions requires combining them into a single criterion function.

$$max\{VI - WI\}_{I=1}^{s} \qquad (5)$$

In this paper, in order to obtain better recognition accuracy, we have classified the features according to the mRMR score, in Table 3. Then we tested with data that contains the total features and after we eliminate one by one and we tested using back-propagation algorithm each time to compare results.

3.4   Model of Neural Network Using BP Algorithm

A Multilayer Perceptron medel is composed by three layers: The input layer, the hidden layer and the output layer, all linked by weighted connections. Back propagation algorithm attempts to associate a relationship between the input layer and the output layer, by computing the errors in the output layer and determine measures of hidden layer output errors, in way to adjust all weights connections (synaptic weights) of the network in iterative process that carried on until the errors decrease to a certain tolerance level. Initially, before adjust weights, they will be set randomly. Then the neuron learns from training subset and correct weights value, finally load to the testing mode. The Fig. 3 represent MP model which contains only one hidden layer. According to Kolmogorov's theorem, the network might be capable to have better performance using one hidden layer, just if the number of input neuron is n, and the inputs are normalized between 0 and 1, a network with only one hidden layer exactly map these inputs to the outputs.

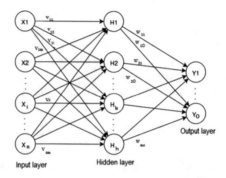

Fig. 3. Multilayer perceptron model

The active function f used in our work is a Sigmoid function:

$$f(x) = \frac{1}{1 + e^x} \tag{6}$$

The objective of BP approach is to minimize not only local error but also minimize the sum-squares-error function defined by:

$$E = \frac{1}{2}(D - O)^2 = \frac{1}{2}\sum_{j=1}^{m}(d_j - o_j)^2 \tag{7}$$

With o_j presnet the real vector and d_j present the target vector.

4   Experiments and Results

4.1   Parameters

The value of each feature is normalized as

$$X_n = \frac{X}{X_{max}} \qquad (8)$$

All 13 activities were performed in neural network using BP algorithm, output contains 13 neurons and only one node produces an output close to 1 when presents with a given activity and all others close to 0. In the test, the activity recognition accuracy is performed in two methods, first referring to a threshold to choose the result, in this paper we have chosen a threshold equal to 0.7, second called the winner which choose the maximum output vector value that correspond in recognition result. Table 2 describe other parameters of neural network using BP algorithm.

Table 2. Parameters of neural network using BP algorithm

Learning rate η	Number of iteration	Error threshold
0.1	2000	0.001

4.2 Determination of Feature Subset

Table 4 presents the comparison results of the two method of activity recognition accuracy of the different feature datasets and the performance measures of multilayer perceptron neural network using BP algorithm, the subsets are described based in mRMR classification in Table 3.

Subset 1: all features without selection;
Subset 2: all features are selected except $\{f_9\}$;
Subset 3: all features are selected except $\{f_9, f_{10}\}$;
Subset 4: all features are selected except $\{f_9, f_{10}, f_{11}\}$;
Subset 5: all features are selected except $\{f_9, f_{10}, f_{11}, f_{14}\}$;
Subset 6: all features are selected except $\{f_9, f_{10}, f_{11}, f_{14}, f_{13}\}$;
Subset 7: all features are selected except $\{f_9, f_{10}, f_{11}, f_{14}, f_{13}, f_{17}\}$.

Table 4 shows the comparison results of activity recognition accuracy performance of the seven different feature subsets. It can be seen that the activity recognition accuracy is lower for subset 1 a bit more in subset 2 ... etc. Subset 5 have relatively higher proportion of recognition accuracy. However, in the subset 6 and 7, the accuracy rate decreases. Obviously, if the number of features is quite low the recognition performance of neural network using BP algorithm it degrades.

Table 3. Features classification based in their mRMR score

9	f10	f11	f14	f13	f17	f6	f7	f15
3.59	3.57	3.32	2.89	2.18	2.15	1.93	1.90	1.60
f16	f4	f3	f1	f5	f2	f12	f8	
1.30	0.19	0.10	0.06	0.05	0.03	0.02	0.01	

Table 4. Comparison results of the two method of activity recognition accuracy of the different feature datasets

	Hidden neurons	Accuracy-threshold (%)	Accuracy-winner (%)
Subset1	17	88	92
	18	80	90
	19	82	90
	20	80	90
	21	82	92
	22	88	92
	23	90	90
	24	94	96
Subset 2	16	88	90
	17	96	98
	18	84	92
	19	84	92
	20	86	94
	21	80	88
	22	82	92
	23	84	90
Subset 3	15	80	96
	16	82	90
	17	90	96
	18	80	88
	19	86	94
	20	80	90
	21	74	86
	22	80	88
Subset 4	14	78	92
	15	82	92
	16	80	92
	17	94	96
	18	78	90
	19	82	92
	20	70	90
	21	76	90
Subset 5	13	74	82
	14	96	99
	15	74	86
	16	70	90
	17	94	99
	18	78	88
	19	74	94
	20	70	86

(*continued*)

Table 4. (*continued*)

	Hidden neurons	Accuracy-threshold (%)	Accuracy-winner (%)
Subset 6	12	84	94
	13	72	94
	14	72	90
	15	76	90
	16	74	88
	17	88	94
	18	74	86
	19	70	90
Subset 7	11	68	88
	12	70	86
	13	68	86
	14	78	96
	15	74	88
	16	66	86
	17	80	92

Selecting Subset 5 generates the best result and represents the relatively better recognition than others with fewer instances. Exactly, the recognition accuracy has been improved to 99% in the winner method and 96% in the threshold method for a number of neurons in the hidden layer equal to 14 with only 13 features. The variance of Sensor IDs with low mutual information score value means that redundant information degrades the recognition performance of neural network using BP algorithm. Therefore, the results indicate that the improper selection of Number of neurons increases the computational complexity and degrades the activity recognition accuracy.

In summary, considering the factors including total accuracy and training error convergence rate, the best features subset is set to Subset 5.

5   Conclusion

In this paper, we use backpropagation algorithm for training the network and mRMR technique to choose adequate features between proposed ones. We conclude that different features set and different number of neurons in hidden layer generate different multi-users activity recognition accuracy, the selection of unsuitable feature sets increases influence of noise and degrades the human activity recognition accuracy. To improve human activity recognition accuracy, an effective approach is to properly select the feature subsets. However, there is still a lot of room for development: future works will include experimenting on more efficient methods for activity recognition (e.g. deep learning), as well as construct activities models in order to recognise human activities in real time.

Acknowledgment. The data were prepared by WSU CASAS smart home project, which can be downloaded from WSU CASAS Datasets website [31].

References

1. IHS Smart Home Intelligence Service. https://technology.ihs.com/Services/526256/smart-home-intelligence-service. Accessed 14 Apr 2017
2. Moreno, L.V., Ruiz, M.L.M., Hernández, J.M., Duboy, M.Á.V., Lindén, M.: The role of smart homes in intelligent homecare and healthcare environments. Ambient Assisted Living and Enhanced Living Environments Principles, Technologies and Control, pp. 345–394 (2017)
3. Aldrich, F.K.: Smart homes: past, present and future. In: Inside the Smart Home, pp. 17–39 (2003)
4. Merrilees, J.: Activities of Daily Living, Reference Module in Neuroscience and Biobehavioral Psychology Encyclopedia of the Neurological Sciences, 2nd edn., pp. 47–48 (2014)
5. Fried, L.P., Guralnik, J.M.: Disability in older adults: evidence regarding significance, etiology, and risk. J. Am. Geriatr. Soc. **45**(1), 92–100 (1997)
6. Lavine, B.K.: Oklahoma State University, Stillwater: Instrumentation Metrics, Feed-Forward Neural Networks AZ, Elsevier books, USA (2009)
7. Rao, V.B.: C++ Neural Networks and Fuzzy Logic, chap. 7, M&T Books, IDG Books Worldwide, Pub. (1995)
8. Peng, H., Long, F., Ding, C.: Feature selection based on mutual information: criteria of max-dependency, max-relevance, and min-redundancy. IEEE Trans. Pattern Anal. Mach. Intell. **27**(8), 1226–1238 (2005)
9. Ding, C., Peng, H.: Minimum redundancy feature selection from microarray gene expression data. J. Bioinform. Comput. Biol. **3**(2), 185–205 (2005)
10. Fiore, L., Fehr, D., Bodor, R., Drenner, A., Somasundaram, G., Papanikolopoulos, N.: Multi-camera human activity monitoring. J. Intell. Robot. Syst. **52**(1), 5–43 (2008)
11. Moeslund, T.B., Hilton, A., Krüger, V.: A survey of advances in vision-based human motion capture and analysis. Comput. Vis. Image Underst. **104**(2), 90–126 (2006)
12. Wang, L., Gub, T., Taoa, X., Chenc, H., Lua, J.: Recognizing multi-user activities using wearable sensors in a smart home. Pervasive Mob. Comput. **7**, 287–298 (2011)
13. Chen, L., Nugent, C., Wang, H.: A knowledge-driven approach to activity recognition in smart homes. IEEE Trans. Knowl. Data Eng. **24**(6), 961–974 (2012)
14. Fahim, M., Fatima, I., Lee, S., Lee, Y.: EEM: evolutionary ensembles model for activity recognition in smart homes. Appl. Intell. **38**(1), 88–98 (2013)
15. Fahad, L.G., Rajarajan, M.: Integration of discriminative and generative models for activity recognition in smart homes. Appl. Soft Comput. **37**, 992–1001 (2015)
16. Borgesa, V., Jebersona, W.: Fortune at the bottom of the classifier pyramid: a novel approach to human activity recognition. In: International Conference on Information and Communication Technologies (ICICT 2014). Procedia Comput. Sci. **46**, 37–44 (2015)
17. Crandall, A., Cook, D.: Using a hidden Markov model for resident identification. In: Proceedings of the 6th International Conference on Intelligent Environments. Kuala Lumpur, Malaysia, pp. 74–79 (2010)
18. Zhan, K., Faux, S., Ramos, F.: Multi-scale conditional random fields for first-person activity recognition on elders and disabled patients. Pervasive Mob. Comput. **16**, 251–267 (2010)

19. Widrow, B., Lehr, M.A.: 30 years of adaptive neural networks: perceptron, madaline, and backpropagation. Proc. IEEE **78**(9), 1415–1442 (1990)
20. Mo, H., Wang, J., Niu, H.: Exponent back propagation neural network forecasting for financial cross-correlation relationship. Expert Syst. Appl. **53**, 106–116 (2016)
21. Velásco-Mejía, V.V.-B., Chávez-Ramírez, A.U., Torres-González, J., Reyes-Vidal, Y., Castañeda-Zaldivar, F.: Modeling and optimization of a pharmaceutical crystallization process by using neural networks and genetic algorithms. Powder Technol. **292**, 122–128 (2016)
22. Farhana, N.I.E., Abdul Majid, M.S., Paulraj, M.P., Ahmadhilmi, E., Fakhzan, M.N., Gibson, A.G.: A novel vibration based non-destructive testing for predicting glass fibre/matrix volume fraction in composites using a neural network model. Compos. Struct. **144**, 96–107 (2016)
23. Fang, H., He, L., Si, H., Liu, P., Xie, X.: Human activity recognition based on feature selection in smart home using back-propagation algorithm. ISA Trans. **53**, 1629–1638 (2014)
24. Kesharaju, M., Nagarajah, R.: Feature selection for neural network based defect classification of ceramic components using high frequency ultrasound, Ultrasonics (2015)
25. Isabelle, G., Steve, G., Massoud, N., Lotfi, Z.: Feature Extraction: Foundations and Applications. Springer, Berlin (2006)
26. Leray, P., Gallinari, P.: Feature selection with neural networks. Behaviormetrika **26**(1), 145–166 (1999)
27. Oukrich, N., Maach, A., Sabri, E., Mabrouk, E., Bouchard, K.: Activity recognition using back-propagation algorithm and minimum redundancy feature selection method. In: 4th IEEE International Colloquium on Information Science and Technology (CiSt), pp. 818–823. IEEE (2016)
28. Gu, T., Wu, Z., Wang, L., Tao, X., Lu, J.: Mining emerging patterns for recognizing activities of multiple users in pervasive computing. In: MobiQuitous 2009: The 6th Annual International Conference on Mobile and Ubiquitous Systems: Networking Services, pp. 1–10 (2009)
29. Rashidi, P., Youngblood, G., Cook, D., Das, S.: Inhabitant guidance of smart environments. In: Proceedings of the International Conference on Human–Computer Interaction, pp. 910–919 (2007)
30. Lin, Z., Fu, L.: Multi-user preference model and service provision in a smart home environment. In: Proceedings of IEEE International Conference on Automation Science and Engineering, CASE 2007, pp. 759–764 (2007)
31. Cook, D.: Learning setting-generalized activity mdoels for smart spaces. WSU CASAS smart home project. IEEE Intelligent Systems

Analysis of Energy Production and Consumption Prediction Approaches in Smart Grids

Atimad El Khaouat[(✉)] and Laila Benhlima

Mohammadia School of Engineers, Mohammed 5 University, Rabat, Morocco
elkhaouat.atimad@gmail.com, benhlima@emi.ac.ma

Abstract. The importance of energy prediction is to ensure Load balance, storage management, relevant integration of renewable resources... There are many scientific research efforts in this field based on different statistical methods and machine learning algorithms. In this paper we analyze four of prediction process in energy prediction in Smart Grids (SGs), especially energy consumption, production or load. This analysis is based on specific criteria and underlies advantages and limitations of each one.

Keywords: Smart Grids · Energy prediction · Consumption · Load forecasting
Analysis · Online/offline learning · Supervised/unsupervised learning
Algorithms · Prediction models

1 Introduction

Smart Grids use new information and communication technologies in order to optimize the electrical system, from the producer points to the consumer ones. Their main functions are to enable:

- Access in real-time to measurement through intelligent devices installed in SGs.
- High integration of renewable energies into the electricity network, based on information system allowing accurate and reliable prediction of production and consumption in the short, medium and long term.
- Vertical sharing of information, which makes the customer actor in the network.

The complexity of the SGs is that they include various components making them facing difficulties such as:

- Intermittent nature of renewable resources of energy,
- Planning of the energy production in order to ensure Demand Response,
- Consumer behavior which affects energy load because it varies according to different periods of the year, geographic area, climate changes...
- Distribution network characteristics.

All these difficulties increase the importance of prediction researches in Smart Grids. In fact, a number of studies are produced and concern power prediction in the short, medium and long-term [1], Demand Response prediction [2], load forecasting

© Springer International Publishing AG, part of Springer Nature 2018
M. Ben Ahmed and A. A. Boudhir (Eds.): SCAMS 2017, LNNS 37, pp. 634–642, 2018.
https://doi.org/10.1007/978-3-319-74500-8_58

[3], interruption prediction [4], and energy storage management [5]. These works are based on different data mining algorithms, statistical methods, and different techniques of data pre-processing.

Existing prediction solutions in Smart Grids are based either on statistical methods or on machine learning models incorporated with optimization functions or on hybrid methods that combine several models.

In this context, we propose an analysis of four newest prediction approaches in energy prediction precisely, in production, consumption and load forecasting.

The remainder of this paper is organized as following: the existing reviews in energy production, consumption and load forecasting in Smart Grids are conducted in Sect. 2. The evaluation of each the four selected prediction approach, subject of this study, is presented in Sect. 3, according to the defined criteria. A discussion section summarizes the pros and cons of each solution. Finally, the conclusion is drawn in Sect. 4.

2 Related Works

Prediction in Smart Grids interests increasingly researchers. These studies concern multiple domains such as anomaly prediction, storage management estimation, renewable energy forecasting [6] and so on. Number of research effort is done whose purpose is to evaluate energy prediction studies in order to classify them, compare them regarding accuracy, time computing, reliability...

Aman et al. [7] propose holistic measures for evaluating prediction models in Smart Grids based on a suite of performance measures to rationally compare models along the dimensions of scale independence, reliability, volatility and cost. This work includes both application independent and dependent measures such as Mean Absolute Percentage Error, coefficient of Variation of RMSE...

A cost-benefit analysis framework is formalized in [8] in order to define performance measures which are aligned with the main objectives of the end users, i.e., profit maximization in order to assist companies with selecting the classifier that maximize the profit.

Chelmis et al. [9] evaluated a wide range of popular methods for household electricity consumption forecasting in the absence of Demand Response based on actual data. This work is based on metrics such as: average deviance between predicted consumption, and actual consumption, percent error median and the mean absolute percent error.

Requirement, challenges and insights of prediction models for dynamic demand response are presented in [10]. This paper describes how dynamic demand response requires short term prediction to make real time adaptive decisions about curtailment, with comparing six short term electricity consumption prediction models.

Studies on wind power prediction techniques have been reviewed on [11] to summarize their own characteristics, compare them and evaluate them using parametric such as: prediction intervals, MAE, MAPE, NRMSE and NMAE.

The majority of existing reviewing studies are based on empirical experiments or analysis of known prediction algorithms in general and don't analyze the whole process. The particularity of our paper is that it analyzes the process of four of newest approaches developed in energy consumption and production forecasting including feature selection method, input data and so on. We then evaluate the techniques and characteristics used in prediction approach adopted in each analyzed study.

3   Analysis

Our analysis is done considering the following criteria:

- Used methods: For these criteria we analyze the used algorithms for prediction, if they are classical, new methods, hybrid ones or incorporating multiple algorithms.
- Approach type: Approaches used in the prediction process can be classified either supervised or unsupervised learning. Supervised learning requires a set of historical data for training model, or unsupervised learning [12]. They can be also classified as Offline/online learning [13]. Some approaches use offline learning, which consists of creating a static prediction model that, once established, will be applied subsequently to data that have the same characteristics as the training data used to construct the model. Online learning enables the model to be continuously adapted to the changing characteristics of the data.
- Selected features: this criterion aims to evaluate the nature of inputs data, their origin and feature selection algorithm if used.
- Complexity: the goal of this criterion is to determine the training time that takes the algorithms to generate prediction results.

In the next, we're presenting our review of the four selected prediction process.

3.1   Supervised Offline Model with Classical Algorithms

Yu et al. [14] use prediction models based on offline supervised machine learning, to guide the prediction of energy usage.

Selected Features. The relieff algorithm [15] was used to determine the importance of features, with assigning a weight to each feature to indicate its importance, then, the larger the weight, the higher the importance. As a result, the selected features as input data in this work are:

- Hourly historical energy usage.
- Weather information: Min T, Max T, Mean T, Mean W, Max W.

Used Methods. The three approaches used in this paper are classical methods, which are:

- Standard SVM;
- Least squares based SVM;
- Backward Propagation Neural Network (BPNN).

Complexity. Time complexity of the approaches is about $O(n^3)$ for SVM and $O(n^2)$ for Least squares SVM.

Concerning BPNN, it is difficult to determine its exact value of complexity because the paper [14] specifies only that the network consists of three layers without indicating the number of neurons in each layer and other useful information allowing us to estimate the complexity of the model.

According to experiment results provided by this work, among the three approaches presented in this paper SVM achieves the higher accuracy, and BPNN achieves the worst one, but concerning time overhead of those forecasting approaches, SVM record a higher time overhead compared to LS-SVM and BPNN.

3.2   Supervised Offline Model with Incorporated Algorithms

In this paper, Campos and da Silva [16] propose three supervised offline machine learning techniques to predict future load consumption, using two classical models which are Artificial Neural Networks (ANN) and Auto Regressive with Exogenous Inputs (ARX), then they present an incorporated model built from ANN and Genetic Algorithms (GA).

Selected Features. In this study, inputs are selected based on correlation between inputs and output. Then data are judged good as inputs if they have high correlation with system output which is the future load in this paper. Therefore, 11 kinds of inputs are used divided into three categories:

Past data:

- Load measure of day before at same time of load output.
- Load measure of last week at the same day and time of load output.
- Load measure of lasts 15 min.
- Load measure of lasts 30 min.
- Mean of the last 24 h of load measure.

Time:

- Hour of the load output.
- Minute of the load output.
- Week day of the load output.
- Month day of the load output.

Weather:

- Temperature.
- Relative humidity.

Data were filtered with a moving average filter, with window size of 3 samples to avoid training with wrong data, in order to eliminate noises from data.

Used Methods. In this work different models obtained through three different algorithms which are:

- Autoregressive with exogenous inputs (ARX): In this approach, the goal is to find a mathematical model that describes the relation between inputs and outputs through samples acquisition of the real process.

- Artificial Neural Network especially using multi-Layer Perceptron (MLP) with supervised training function Levenberg-Marquardt backpropagation and Gradient descent with momentum backpropagation as learning function.
- Neural Network and Genetic Algorithm: In this model GA is used as an optimization tool to improve the results of ANN, and that consist on selecting best data driving, among the 11 kinds presented thereafter.

Complexity. In this paper, we can't estimate complexity of used methods because of the absence of relevant parameters allowing us to determine it. But, according to experiment results done by authors of this paper, ANN working with GA is the longest because for each tested data driven there is a training phase.

Among the three approaches presented in the paper, the ANN incorporated with GA is the more accurate but has the disadvantage of a long computational time in training phase.

3.3   Supervised Online Model with Classical Algorithms

Aung et al. [17] present solution for peak load prediction (maximum use of electricity) problem of a future unit of time for a particular consumer entity (smart meter).

The solution follows an online learning approach and it is based on the support Vector Regression method.

Selected Features. In this paper features wasn't selected automatically by using one of feature selection algorithms, but empirically by encoding a feature vector with 32 attributes listed below:

- For attribute from 1 to 28: Peak load of previous 28 days (Pd-1 to Pd-28).
- Attribute 29: Average peak load of previous 7 days.
- Attribute 30: Average temperature of previous 7 days.
- Attribute 31: Forecasted average temperature of the day d.
- Attribute 32: Whether the day d is a holiday (weekend or public holiday).

Used Methods. To produce prediction model, the process adopted in this work is as follow: to predict the peak load of a day d (Pd), Pd is considered as a non-linear combination of the selected features. Then, the training process is divided on two steps:

> Step 1: is the initial training process, in this step, a least squares regressor is built by constructing for each "target" peak load value, in the historical record, a "feature vector" covering the attributes associated with the target. Then, the least-square SVR system is trained using a set of <feature vector, target> pairs for a large enough number of days.
>
> Step 2: A feature vector of the day d is constructed in the same manner as in the initial training step, using forecasted temperature of a day d and whether it is a holiday or not. Then, the feature vector of a day d is supplied to the regressor model built in the initial training step to generate the forecasted peak load value of this day.

So when the Pd is known for day d, the regressor model is updated, and that what make the model prediction online learning.

Complexity. Time complexity of this approach is about O (n^3) because it is based on Support Vector Regression.

The importance of this study is the utilization of an online prediction model, so it can ensure more accuracy and reliability with the development of big data concept in Smart Grids, where data is unstructured and heterogonous. But its principal limitation is the non utilization of an appropriate algorithm of feature selection.

3.4 Unsupervised Offline Model with Incorporated Algorithms

The work proposed by Mocanu et al. [18] aims at predicting energy consumption.

All the methods we have presented above are classified on supervised learning category, so historical data are required to produce energy prediction models. In this part we will analyze unsupervised offline method based on Reinforcement Learning, Transfer Learning and Deep Belief Networks.

Selected Features. Data are not needed to produce a prediction model but they are important to evaluate the result of the method, therefore, this study used a dataset provided by Baltimore Gas and Electric Company recorded over seven years (between January 2007 and January 2014).

Used Methods. In this paper reinforcement learning and transfer learning are used to predict energy. Deep Belief Network is also used to estimate continuous states in reinforcement learning.

- Reinforcement learning: This approach consists in learning how to map situations to actions in order to maximize a numerical reward signal. In this paper two reinforcement learning algorithms are used which are:
- Q-Learning: It can be used to find an optimal action-selection policy for any given finite Markov decision process [19].
- SARSA: State-Action-Reward-State-Action (SARSA) is a variation of Q-learning algorithm [20], which aims at using Q-learning as part of a Policy Iteration mechanism.
- Transfer learning: Transfer learning is the improvement of learning in a new task through the transfer of knowledge from a related task that has already been learned [21].
- Deep Belief Network: is a type of Deep Neural Network [22], it consists of two different types of neural networks: Belief Networks and Restricted Boltzmann Machines. In contrast to perceptron and backpropagation neural networks, DBN is unsupervised learning algorithm.

The process of prediction adopted in this work uses the different algorithms presented previously in this way:

- By including generalization of state space domain, energy prediction model for one building is first done. Then this model is transferred to other buildings using transfer learning.
- Usage of the two extended reinforcement learning algorithms to perform knowledge transfer between domains, which are State-Action-Reward-State-Action (SARSA)

algorithm and Q-learning algorithm by incorporating them with a Deep Belief Network, because the original form of these both methods cannot handle with continuous states space.

Complexity. This paper doesn't give information about complexity of the model, but we can estimate it by taking in consideration the fact that the model prediction is based on reinforcement learning algorithm, which have an exponential training time. However, according to [23], this complexity can be improved if there are no duplicate actions or if the domain prediction has some properties such as a linear upper action bound. In spite of this, in the best case the complexity is about $O(n^2)$. Thus, we can conclude that the model prediction used in this study is proportionally slow.

The importance of such a kind of models is to forecast energy consumption even if there is no historical data available. For example: in the cases of new or renovated buildings connected with the Smart Grids. But the high complexity of the model can be considered as a principal disadvantage of the approach.

3.5   Discussion

To summaries, analysis we carried out in this paper allow us to draw the following points:

- The most commonly used automatic learning methods for energy prediction are: SVM, SVR, and Neural Networks.
- Methods that ensure good accuracy are slower than others which have a good training time but a proportionally low precision.
- Feature selection is an important step in prediction process, moreover a good choice of feature selection algorithm contribute to enhance the accuracy of prediction model even if it can make it slower.
- Approaches based on supervised learning are more often used than those based on unsupervised learning. The importance of the latter is to be used in energy forecasting in the case where historical data are not available.
- Models based on online learning are recently applied in prediction in Smart Grids: they are more relevant (variety, velocity and unstructured nature of data generated).
- Models based on online learning are slower than those based on offline learning in run time ($O(n^3)$ just in the initial training step for the third approach).

4   Conclusion

The principal contribution of this paper is the evaluation of prediction process based on different criteria. The study focuses also on process produced on new papers with different characteristics that vary between supervised and unsupervised learning, online and offline learning or classical and hybrid methods. Our study highlights also data sets specifications, training time, and lists some of the advantages and disadvantages of each process. Then we have summarized our analysis with a discussion.

As a future work, we will work on proposing an online novel approach for energy production and user consumption to deal with velocity and variability of data in Smart Grids.

References

1. Kramer, O., Satzger, B., Lässig, J.: Power prediction in smart grids with evolutionary local kernel regression. Springer, Heidelberg (2010)
2. Cecati, C., Citro, C., Siano, P.: Combined operations of renewable energy systems and responsive demand in a smart grid. IEEE Trans. Sustain. Energy 2(4), 468–476 (2011)
3. Fan, S., Hyndman, R.J.: Short-term load forecasting based on a semi-parametric additive model. IEEE Trans. Power Syst. (2010)
4. Sarwat, A., Amini, M., Domijan Jr., A., Damnjanovic, A., Kaleem, F.: Weather-based interruption prediction in the smart grid utilizing chronological data. State Grid Electric Power Research Institute. https://doi.org/10.1007/s40565-015-0120-4
5. Narayanaswamy, B., Garg, V.K., Jayram, T.S.: Prediction based storage management in the smart grid. In: IEEE SmartGridComm 2012 Symposium - Support for Storage, Renewable Sources and MicroGrid. IBM Research, India Research Lab (2012)
6. Albadi, M.H., El-Saadany, E.F.: Overview of wind power intermittency impacts on power systems. Electr. Power Syst. Res. 80, 627–632 (2009)
7. Aman, S., Simmhan, Y., Prasanna, V.K.: Empirical comparison of prediction methods for electricity consumption forecasting. Trans. Knowl. Data Eng. (2014)
8. Verbraken, T., Verbeke, W., Baesens, B.: A novel profit maximizing metric for measuring classification performance of customer churn prediction models. IEEE Trans. Knowl. Data Eng. 25(5), 961–973 (2013)
9. Chelmis, C., Saeed, M.R., Frincu, M., Prasanna, V.K.: Curtailment Estimation Methods for Demand Response. Ming Hsieh Department of Electrical Engineering Department. University of Southern California
10. Aman, S., Frincuy, M., Chelmisy, C., Noory, M., Simmhanz, Y., Prasanna, V.K.: Prediction models for dynamic demand response: requirements, challenges, and insights, November 2015
11. Colak, I., Sagiroglu, S., Yesilbudak, M.: Data mining and wind power prediction: a literature review. Renew. Energy 46, 241–247 (2012)
12. Rossi, F.: Apprentissage Supervise. TELECOM ParisTech, Paris (2009)
13. Ben-David, S., Kushilevitz, E., Mansour, Y.: Online learning versus offline learning. Mach. Learn. 29, 45–63 (1997). Kluwer Academic Publishers. Manufactured in The Netherlands
14. Yu, W., An, D., Griffith, D., Yang, Q., Xu, G.: Towards statistical modeling and machine learning based energy usage forecasting in smart grid. In: RACS 2014 Proceedings of the 2014 ACM Research in Adaptive and Convergent Systems, vol. 15(1), March 2015
15. Durgabai, R.P.L.: Feature selection using Relieff algorithm. Int. J. Adv. Res. Comput. Commun. Eng. 3(10), 8215–8218 (2014)
16. Campos, B.P., da Silva, M.R.: Demand forecasting in residential distribution feeders in the context of smart grids
17. Aung, Z., Toukhy, M., Williams, J.R., Sanchez, A., Herrero, S.: Towards accurate electricity load forecasting in smart grids. In: The Fourth International Conference on Advances in Databases, Knowledge, and Data Applications, DBKDA (2012)

18. Mocanu, E., Nguyen, P.H., Kling, W.L., Gibescu, M.: Unsupervised energy prediction in a smart grid context using reinforcement cross-building transfer learning. Energy Build. **116**, 646–655 (2016)
19. Watkins, C.J.C.H., Dayan, P.: Q-learning. Mach. Learn. **8**, 279–292 (1992). Kluwer Academic Publishers, Boston. Manufactured in The Netherlands
20. Sutton, R.S., Barto, A.G.: Reinforcement Learning: An Introduction, pp. 141–144, 169–172
21. Torrey, L., Shavlik, J.: Transfer Learning. University of Wisconsin, Madison (2009)
22. Hebbo, H., Kim, J.W.: Classification with deep belief networks
23. Koenig, S., Simmons, R.G.: Complexity analysis of real-time reinforcement learning

Toward a Smart Tourism Recommender System: Applied to Tangier City

Zineb Aarab[1]([⊠]) [iD], Asmae Elghazi[1], Rajaa Saidi[1,2],
and Moulay Driss Rahmani[1]

[1] LRIT Associated Unit to CNRST (URAC 29), Faculty of Sciences,
Mohammed V University, BP 1014, Rabat, Morocco
aarab.zineb@gmail.com, as.elghazi@gmail.com,
r.saidi@insea.ac.ma, mrahmani@fsr.ac.ma
[2] SI2M Laboratory, INSEA, BP 6217, Rabat, Morocco

Abstract. Context-awareness has captivated a lot of attention, especially in the field of mobile and pervasive computing. Indeed, an important demand for real-world location data within the virtual world is growing. Yet, user's context is more than its location. Most information systems do not take into account the diversity of users' need and preferences because of its complexity to manage. Context-Aware Systems (CAS) are self-adaptive. Thus, automating the development of such systems and the interpretation of user's preferences is a real challenge. One way to achieve this challenge is by discovering knowledge for prediction regarding actions of users with the systems (their history and so on). In this article, we explore how context information can be exploit to create intelligent and adaptive recommender systems. It provides an overview of the multiform notion of context with a discussion of several context oriented approaches and systems, and illustrates the usage of such approaches in several application areas. In this paper, we present our vision of the notion of context through a context metamodel. We also present an effective tourist recommendation system that respect personal preferences and capture usage, personal, social and environmental contextual parameters. A case study applied has been conducted to Tangier, a Moroccan city where a new intelligent international city will be constructed for the next decade. The article concludes by discussing the challenges and future research directions for context-aware recommender systems.

Keywords: Context-Aware system · Mobile recommender system
Mobile tourism · Location-awareness · Adaptive system · Smart tourism
Smart tourism recommender systems

1 Introduction

In the last two decades, Context-Aware Systems have been an increasing and known research field inside scientific communities especially those related to recommendation, auto-adaptive, and information retrieval. The nucleus element of these emerging paradigms is the concept of the context. Indeed, various research papers have defined only context applications models or systems concerned worrying less about the metamodels that supports these models.

© Springer International Publishing AG, part of Springer Nature 2018
M. Ben Ahmed and A. A. Boudhir (Eds.): SCAMS 2017, LNNS 37, pp. 643–651, 2018.
https://doi.org/10.1007/978-3-319-74500-8_59

We can define a metamodel as the language and semantics to specify the particular model domains or applications [1]. Consequently, a context metamodel consists of organizing and presenting a more abstract view of context models in order to show their construction and manipulation. Metamodeling context is a major challenge for the identification of concepts involved in manipulating the context, their relationships, their formalization and presentation of their semantics [2]. Hence the goal of this paper is to present a generic context metamodel for CASs and the proposition of a tourism recommender system that is based on the proposed notion of the context.

One of the filed that can takes great profit of context-awareness research is the Tourism. Morocco is a fascinating country because it merges the African and Arab worlds, and their customs. Therefore, it attracts many tourists. Unfortunately, these sites don't receive the financial attention or support that they deserve. In order to highlight Moroccan tourism inserting tourist places in the numerical ecosystem of smart cities where economic, tourist, and logistic are considered together is required.

The paper is organized as follows: In Sect. 2 we will briefly present a state of the art about the notion of the context. Section 3 presents the description of the proposed context metamodel and the Smart Tourism Recommender Systems (STRSs) architecture. Section 4 presents the experimental results of the case study for two different tourism scenarios in the Tangier city and Sect. 5 makes the conclusion of our paper.

2 State of the Art: Context

The main motivation in the new computing paradigms (such as Ambient Intelligence, Pervasive Computing, Ubiquitous Computing, CASs, and so on) is decoupling users from computing devices [3]. In this regard, several context definitions were proposed in the literature. A discussion of the notion of context is presented in our previous work [4].

The word context, derived from the Latin con (with or together) and texere (to weave), describes a context not just as a profile, but as *an active process dealing with the way humans weave their experience within their whole environment, to give it meaning* [5]. A widely and uniquely accepted definition of this concept is not available in the literature. For example, [6] see the context as any information that can be used to characterize the situation of an entity. An entity could be a person, place, or object that is considered relevant to the interaction between a user and an application, including location, time, activities, and the preferences of each entity. In the last decade an important evolution of this notion comes out, the context was no longer considered as a state, but a part of a process in which users are involved [7].

Thus, defining the word 'context' still a big challenge and many researchers tried to find their own definition for what context actually includes. As proposed by [8] we can say that this term is usually adopted to indicate a set of attributes that characterize the capabilities of the access mechanism, users' preferences, information and services are delivered, these could include the access device (alike in the presence of strong heterogeneity of the devices) [8]; thus, sophisticated and general context models have been proposed. Additionally, the crucial standing of this notion has been the source of inspiration of several papers which gave multiple ramifications of its dimensions: spatial, spatial mobility, spatiotemporal, environment and personal dimension [9].

- The *spatial* dimension is the most generally examined one, which refers mainly the movement of people in space and having wireless access to information and services.
- The *spatial mobility* dimension is related to objects in the environment such as cars, small medical appliances, and so on. That may interact with other objects and users and subsequently influence the people using these devices.
- The *spatiotemporal* dimension is an extension of the spatial dimension, which describes just the side related to time and space, such as direction, speed and track.
- To improve the development of applications in the ambient-intelligence area, by collecting and analyzing ambient information, we must introduce the *environment dimension* which points out the ambient characteristics, such as temperature, light, humidity, noise and the orientation of movement of physical objects.
- Several studies [5, 9–12] focused on the user context, particularly user characteristics, which introduced *the personal-context dimensions*; it refers to physiological, mental, activity, and social context of the user.
- The *physiological* context: is commonly used in medical applications, the *physiological* profile may contain information such as pulse, blood, pressure, and weight, personal abilities and preferences too, which can be dynamically collected with body sensors.
- The *mental* context is useful in the case of medical analysis, it may include element such as mood, expertise, anger, and stress.
- The *activity* context; describes the activity of the user, it includes explicit goals of activity to be performed, activity breakdown structures, and role-playing information.
- The *social* context: is commonly used in social networks; it describes the social aspects of the context of the user. It may, for instance, contain information about friends, neighbors, coworkers and relatives.
- In addition, the term *social mobility* refers to the ways in which individuals can move across different social contexts, with the support of technology and services.

As we can see, the notion of context is not an easy concept to understand neither to surrounds. In one hand, to build context-aware information systems the consideration of the context is primordial. In the other hand, the context must be built and structured in an efficient way. Thus, context modeling is needed to understand and interpret dynamic context representations at a high-level abstraction in an unobtrusive way. The most important context modeling techniques are compared in [1], and listed as followings: key-value, markup schemes, graphical, object based, logic based, and ontology based modeling. Context metamodeling is a major challenge too, by identifying concepts involved in context manipulating, their relationships, formalization and semantics presentation. The most popular context metamodeling approaches are: metamodel centered ubiquitous web applications, UML Profile metamodels, metamodels supporting ontological models, high level abstraction and framework metamodel. Those approaches were discussed in details and compared in our previous work [1].

In the next section we will present the proposed generic context metamodel.

3 Toward a Smart Tourism Recommender System: Context Metamodel-Driven Definition

In this section, we present our definition of the concept of context through our context metamodel in the first subsection (Fig. 1), and then the architecture of our proposed Smart Tourism Recommender System in the second subsection (Fig. 2).

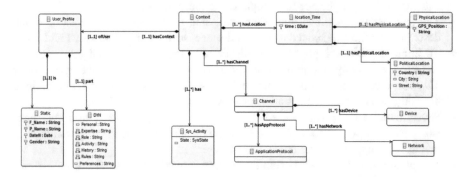

Fig. 1. Our proposed context metamodel.

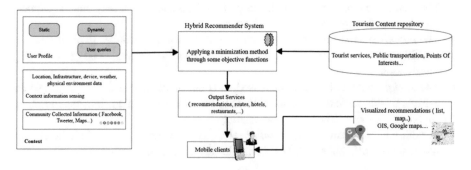

Fig. 2. The proposed tourism recommender system.

3.1 Toward a Context Metamodel-Driven Definition

We designed the conceptual basics of our framework first, starting with context metamodeling as our paramount concern. As you can see in Fig. 1, we define the context as an aggregation of fore different main groups of properties: the *channel* used by the user, a *location* and *time* description of the position of the user and the system, *System Activity* that describes the system states and a *user profile* which describes the user information in details (role, expertise, preferences and user activity…).

- A *channel* that defines the medium by which information of the context can be transmitted; it identifies the physical device and the connection used to access the application.

- The *System Activity* (Sys_Activity) that describes the system states weather is it stand-by or any other state (e.g. offering a service).
- A *location* and *time* description that identify the position of the user; where the user is located while interacting with the application.
- The *user profile* is, none other than, the information collected, structured, organized, and maintained about user-related data [13]. We classify the data concerning a user, as mentioned in the context metamodel in Fig. 1, into two different groups [13]:

(1) *Static data*: such as name address, age, gender, job, and preferences expressed explicitly by the user.
(2) *Dynamic data,* in which we can find:
(a) A *history of context action:* that the user has performed during past interactions with the application implemented on the platform.
(b) *User Preferences*: Expressed explicitly or formulated implicitly by means of association;
(c) *Rules* extracted from the above-mentioned history, or were set by a domain expert.
(d) *User Activity* that describes the movement of the user is he/she mobile or not (e.g. the activity of the patient, such as its pulse while he's/she's walking, in a health domain are very critical and must be monitored).

3.2   The Smart Tourism Recommender System Architecture

In this section, the Smart Tourism Recommender System Architecture is introduced and described (see Fig. 2). Three main components compose the system: the Context, the Tourism Content repository and the recommender system. While a "Tourism Recommender application" shows the output of the system. In the following paragraphs more details about the various modules will be given.

- The first module is the Context: which is composed of the user profile (the static and dynamic part and the explicit users queries), Context information sensing (i.e. the location, the characteristics of the device, the external physical environment e.g. weather conditions) and Community collected information (from Facebook, Tweeter, and other social networking application).
- The tourism content repository contains tourist services (like tourist sites), public transportation ….
- The recommender system that we adopt uses a hybrid approaches.

4   Case Study: Traveling to Tangier

In this section we present the experiment results of our Smart Tourism Recommender System starting from the premise:

The tourist just arrived to Tangier city and its current location is the train station, Fig. 3 show the nearest hotels in the area.

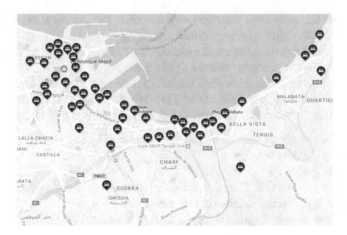

Fig. 3. The hotels surrounding the train station of Tangier city using The Booking application.

We have considered two different scenarios in this case study:

- Tourist 1: The first scenario is about an ordinary tourist aged between 40 and 65 years old where its profile does not show special requirements but we have chosen to give him/her the optimal hotel for two days (Fig. 4) according to the following criteria:

- The price.
- Air conditioner.

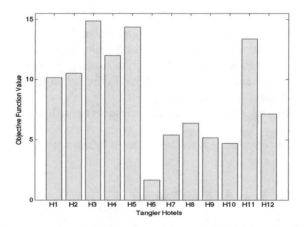

Fig. 4. The optimal choice among 12 hotels for the Tourist 1 using our STRS.

- TV.
- Swimming pool.
- Number of stars.
- Sports hall.
- The reviews.
- Other.

Figure 4 shows the results of 12 hotels with our approach. The hotel H3 which is Tarik Hotel is the best one for the Tourist 1. And Fig. 5 shows the resulting hotel for the Tourist 1 using Booking websites.

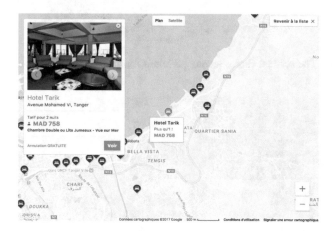

Fig. 5. The recommended hotel for the Tourist 1 using our STRS.

- Tourist 2: The second scenario is about a tourist student aged between 18 and 30, we have collected the following requirements from its profile:

- The price.
- The distance.
- Internet.
- The breakfast.
- Swimming pool.
- Air conditioner.
- The reviews.
- Other.

Our system have choosing, among 12 different hotels that satisfies the above requirements, the best chose for the Tourist 2 for two nights as it is shown respectively in Figs. 6 and 7.

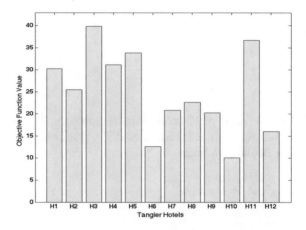

Fig. 6. The optimal choice among 12 hotels for the Tourist 2 using our STRS.

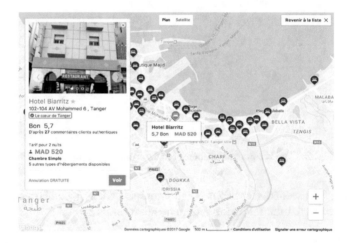

Fig. 7. The recommended hotel for the Tourist 2 using our STRS.

5   Conclusions and Future Work

This paper proposes the use of a context-aware approach for the most suitable services (hotel) for a user (the tourist in our case study) taken account of special requirements. The prime objective of this work is to find out an optimal or near optimal selection for recommendation advice to the other contemporary users. The system is based on our proposed context metamodel able to model a context. The proposed architecture has been implemented in a tourism recommender system according to the position identifying the user current context and its profile.

For a future work we want to explore more the user profile and the Tourism Content Repository to offer other services and to complete the implementation of our architecture.

References

1. Aarab, Z., Saidi, R., Rahmani, M.D.: Towards a framework for context-aware mobile information systems. In: 2014 Tenth International Conference on Signal-Image Technology and Internet-Based Systems (SITIS), pp. 694–701. IEEE (2014)
2. Ferscha, A., Vogl, S., Beer, W.: Context sensing, aggregation, representation and exploitation in wireless networks. Scalable Comput. Pract. Exp. 6(2), 71–81 (2001)
3. Adorni, M., Arcelli, F., Bandini, S., Baresi, L., Batini, C., Bianchi, A., Bianchini, D., Brioschi, M., Caforio, A., Cali, A.: Reference architecture and framework. In: Mobile Information Systems, pp. 25–46. Springer (2006)
4. Ali, R., Dalpiaz, F., Giorgini, P.: A goal-based framework for contextual requirements modeling and analysis. Requirements Eng. 15(4), 439–458 (2010)
5. Bolchini, C., Curino, C.A., Quintarelli, E., Schreiber, F.A., Tanca, L.: A data-oriented survey of context models. ACM Sigmod Rec. 36(4), 19–26 (2007)
6. Abowd, G., Dey, A., Brown, P., Davies, N., Smith, M., Steggles, P.: Towards a better understanding of context and context-awareness. In: Handheld and Ubiquitous Computing, pp. 304–307. Springer, Berlin, Heidelberg (1999)
7. Bolchini, D., Paolini, P.: Goal-driven requirements analysis for hypermedia-intensive Web applications. Requirements Eng. 9(2), 85–103 (2004)
8. Aarab, Z., Saidi, R., Rahmani, M.D.: Towards a framework for context-aware mobile information systems. In: 2014 Tenth International Conference on Signal-Image Technology and Internet-Based Systems (SITIS), pp. 694–701. IEEE, November 2014
9. Pernici, B.: Basic concepts. In: Mobile Information Systems, pp. 3–23. Springer (2006)
10. Strang, T., Linnhoff-Popien, C.: A context modeling survey. In: Workshop Proceedings, September 2004
11. Hong, J.Y., Suh, E.H., Kim, S.J.: Context-aware systems: a literature review and classification. Expert Syst. Appl. 36(4), 8509–8522 (2009)
12. Baldauf, M., Dustdar, S., Rosenberg, F.: A survey on context-aware systems. Int. J. Ad Hoc Ubiquit. Comput. 2(4), 263–277 (2007)
13. Corallo, A., Lorenzo, G., Solazzo, G., Arnone, D.: Knowledge-based tools for e-service profiling and mining. In: Mobile Information Systems, pp. 265–291 (2006)

Iris Recognition Algorithm Analysis
and Implementation

Siham Kichou[1(✉)], Abdessalam Ait Madi[1,2(✉)],
and Hassan Erguig[1(✉)]

[1] National School of Applied Sciences, Ibn Tofail University, Kenitra, Morocco
kichou.siham@gmail.com, aitmadi_abdessalam@yahoo.fr,
erguigh@yahoo.fr
[2] Faculty of Sciences and Technology,
Sidi Mohammed Ben Abdellah University, Fez, Morocco

Abstract. This paper focusses upon studying and implementing the iris recognition algorithm available on Open source and implemented by Masek and analysis of results using Chinese academy of sciences-institute of automation (CASIA) database.

Keywords: Iris · Recognition · Daugman algorithm · Segmentation
Iris code

1   Introduction

Individual authentication has always been an attractive purpose in embedded systems. Biometrics (or biometric authentication) refers to the identification of humans by their characteristics or traits. It employs physiological or behavioral characteristics to identify an individual. The physiological characteristics are iris, fingerprint, face and hand geometry. Voice, signature, gait and keystroke dynamics are considered as behavioral characteristics [2]. A biometric system can operate in two modes, Verification and Identification.

Verification systems answer the question "Is this the person X?" Under a verification system, an individual presents themselves as a specific person. The system checks their biometric against a biometric profile that already exists in the database linked to that person's file in order to find a match.

Verification systems are generally described as a 1-to-1 matching system because the system tries to match the biometric presented by the individual against a specific biometric already in the database.

Because verification systems only need to compare the presented biometric to a biometric reference stored in the system, they can generate results more quickly and are more accurate than identification systems, even when the size of the database increases.

Identification systems are different from verification systems because an identification system seeks to identify an unknown person, or unknown biometric. The system tries to answer the questions "Who is this person?" or "Who generated this biometric?" and must check the biometric presented against all others already in the database.

© Springer International Publishing AG, part of Springer Nature 2018
M. Ben Ahmed and A. A. Boudhir (Eds.): SCAMS 2017, LNNS 37, pp. 652–663, 2018.
https://doi.org/10.1007/978-3-319-74500-8_60

Identification systems are described as a 1-to-n matching system, where n is the total number of biometrics in the database.

This paper is organized in the following manner. The Sect. 2 presents an overview of Iris recognition. In Sect. 3 Daugman's approach is reviewed. The Sect. 4 presents the implementation and simulation of iris recognition. Finally, the Sect. 5 concludes the paper.

2 Iris Recognition

Among the numerous traits, iris recognition has attracted a lot of attention because it has many valuable features like higher speed, simplicity and accuracy compared to other biometric characters. Iris recognition depend on the unique patterns of the human iris to identify or verify the identity of an individual [3]. Iris is distinct for every person, even the twins have different iris patterns and it remains unchangeable for the whole of the life. Thus, this technology is now considered as providing positive identification of an individual without contact and at very high confidence levels. An iris is the colored area between dark paupil and bright sclera. Figure 1 represents the structure of Iris.

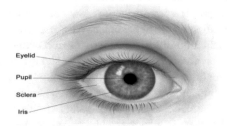

Fig. 1. Iris structure

2.1 Iris Recognition Main Steps

The main steps of an iris recognition system are:

- Segmentation: The inner and the outer boundaries of the iris are calculated.
- Normalization: Iris of different people may be captured in different size. The size may also vary for the same person because of the variation in illumination and other factors.
- Feature extraction: Iris provides abundant texture information. A feature vector is formed which consists of the ordered sequence of features extracted from the various representation of the iris images.
- Feature comparison: The feature vectors are classified through different techniques like Hamming Distance, weight vector and winner selection, dissimilarity function, etc.

Figure 2 describes the stages, which are used in Iris recognition.

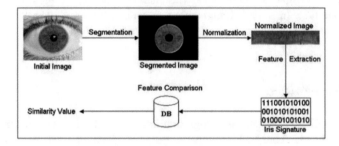

Fig. 2. Stages of iris recognition

The initial stage deals with iris segmentation. This process consists in to localize the iris inner (pupillary) and outer (scleria) borders. Eyelids and eyelashes that may cover the iris region are detected and removed. In order to compensate the varying size of the pupil it is common to translate the segmented iris part represented in the Cartesian coordinate system to a fixed length and dimensionless polar coordinate system. This is the normalization stage [4]. The iris image has low contrast and non-uniform illumination caused by the position of the light source. All these factors can be compensated by the image enhancement algorithms. Feature extraction uses texture analysis method to extract features from the normalized iris image. The significant features of the iris are extracted for precise identification purpose. After obtaining a feature set, commonly named biometric iris template, the final stage consists in the comparison between signatures, producing a numeric dissimilarity value. If this value is higher than a threshold, this means that there is a "non-match", meaning that each template belongs to different subjects. Otherwise, the system outputs a "match", meaning that both templates were extracted from the same person [5].

3   Daugman's Approach

Even if prototype systems had been proposed earlier, it was not until the early nineties that John Daugman implemented a working automated iris recognition system [1, 2]. The Daugman system is patented [5] and the rights are now owned by the company Iridian Technologies. Even though the Daugman system is the most effective and most well-known, many other systems have been established. The most prominent include the systems of Wildes et al. [4, 7], Boles and Boashash [8], Lim et al. [9], and Noh et al. [10].

The Daugman system has been tested under several studies, all reporting a zero failure rate. The Daugman system is claimed to be able to perfectly identify an individual, given millions of possibilities.

Compared with other biometric technologies, such as face, speech and finger recognition, iris recognition can easily be considered as the most consistent form of biometric technology [1].

3.1   Segmentation

The first processing step consists on locating the inner and outer boundaries of the iris. Integro-differential operator is used for locating the inner and outer boundaries of iris, as well as the upper and lower eyelids [6].

$$\max_{(r,x_0,y_0)} \left| G_\sigma(r) * \oint_{r,x_0,y_0} \frac{I(x, y)}{2\pi r} dS \right| \tag{1}$$

Where $I(x, y)$, r, $G(r)$ and S are the eye image, radius, Gaussian smoothing function and the counter of the circle respectively.

The operator computes the partial derivative of the average intensity of circle points, taking into account the increasing radius, r. After convolving the operator with Gaussian kernel, the maximum difference between inner and outer circle will define the center and radius of the iris boundary. For upper and lower eyelids detection, the path of outline integration is modified from circular to parabolic curve.

3.2   Normalization

In order to accomplish a size-invariant sampling of the authenticate iris pixel points, Daugman [7] applied a Rubber Sheet Model to map the sampled iris pixels from the Cartesian coordinates to the normalized Polar coordinates. It transforms the iris region from Cartesian coordinates (x, y) to Polar coordinates (r, θ) where r lies in the interval of $[0, 1]$ and θ is in the range of $[0, 2\pi]$. Figure 3 shows the rubber sheet model. The homogenous rubber sheet model accounts for pupil dilation, imaging distance and non-concentric pupil displacement. However, this algorithm does not compensate for the rotation variance.

$$I(x(r, \theta), y(r, \theta)) \rightarrow I(r, \theta) \tag{2}$$

Where $I(x,y)$ is the iris image, (x, y) are the Cartesian coordinates and (r, θ) are the polar coordinates, and

$$x(r, \theta) = (1 - r) xp(\theta) + r\, xs(\theta) \tag{3}$$

$$y(r, \theta) = (1 - r) yp(\theta) + r\, ys(\theta) \tag{4}$$

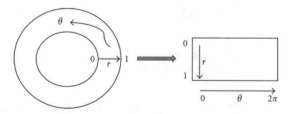

Fig. 3. Daugman's rubber sheet model

Where xp(θ) represents the abscissa of the point of the detected boundary of the pupil whose segment passing through this point and the center of the pupil makes an angle θ with a chosen direction. Similarly, yp(θ) represents the ordinate of this same point, then xs(θ) and ys(θ) represent the coordinates of the points obtained by the same principle but on the contour of the iris.

3.3   Feature Extraction

For Feature Extraction, Daugman [8] uses the Gabor 2D Filters. The Gabor filter could be seen as a Gaussian envelope multiplexed by a series of sinusoidal waves with different scales and rotations. It is used to extract the phase information.

As seen in the Fig. 4, the phase information is quantized into four quadrants in the complex plane. Each quadrant is represented by two bits phase information. Therefore, each pixel in the normalized image is demodulated into two bits code in the template. The phase information is extracted because it provides the significant information within the image. It does not depend on peripheral features, such as imaging contrast, illumination and camera gain.

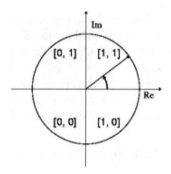

Fig. 4. The phase encoding principle four quadrants and two bits.

3.4   Matching

Daugman [9] employed the test of statistical independence to measure the similarity between two Iris templates. The test of statistical independence is implemented by the simple Boolean Exclusive-OR operator ⊗ applied to the 2048-bit phase vectors that encode any two iris patterns, masked ∩ by both of their corresponding mask bit vectors.

$$HD = \frac{\| (code\,A \otimes code\,B) \cap mask\,A \cap mask\,B \|}{\| mask\,A \cap mask\,B \|} \tag{5}$$

For Matching, Hamming distance was used which gives the number of positions at which the corresponding symbols are different. Hamming distance is defined as the fractional measure of dissimilarity between two binary templates.

$$HD = \frac{1}{N} \sum_{i=1}^{i=N} Ai \otimes Bi \qquad (6)$$

A value of zero represents a perfect match. The two templates that are completely independent gives a Hamming distance near to 0.5. A threshold is set to decide the two templates are from the same person or different persons.

4 Implementation of Iris Recognition Algorithm: Masek Open Source Code

The purpose will be to implement and test an open-source iris recognition system in order to verify the claimed performance of the technology. The development tool used will be MATLAB®. A rapid application development (RAD) method will be employed in order to produce results quickly. MATLAB® offers an excellent RAD environment, with its image processing toolbox, and high level programming methodology [10]. To test the system, a data set of eye images will be used as inputs: database of 756 greyscale eye images courtesy of The Chinese Academy of Sciences – Institute of Automation (CASIA) [14].

In Masek's [10] method, the automatic segmentation system is based on the Hough transform [10], and it is able to localize the circular iris and pupil regions. The extracted iris region is then normalized into a rectangular block with constant dimensions to account for imaging inconsistencies and finally, the phase data from 1D Log-Gabor [10] filters is extracted and quantized to four levels to encode the unique pattern of the iris into a bit-wise biometric template. The Hamming distance [10] is employed for classification of iris templates, and two templates are found to match if a test of statistical independence has failed. The input to the system is an eye image, and the output is an iris template, which will provide a mathematical representation of the iris region.

4.1 Segmentation

The Hough transform is used, which first involves Canny edge detection to generate edge map using Canny edge detection MATLAB function [11]. Eyelids detection is done using Hough transform [11, 12].

The Hough transform is a standard computer vision algorithm that can be used to determine the parameters of simple geometric objects, such as lines and circles, present in an image. The circular Hough transform can be employed to deduce the radius and center coordinates of the pupil and iris regions [10].

First, an edge map is generated by calculating the first derivatives of intensity values in an eye image and then thresholding the result. From the edge map, votes are cast in Hough space for the parameters of circles passing through each edge point. These parameters are the center coordinates xc and yc, and the radius r, which are able to define any circle according to the equation:

$$x_c^2 + y_c^2 - r^2 = 0 \qquad (7)$$

A maximum point in the Hough space will correspond to the radius and center coordinates of the circle best defined by the edge points. Wildes et al. and Kong and Zhang also make use of the parabolic Hough transform to detect the eyelids, approximating the upper and lower eyelids with parabolic arcs, which are represented as [10]:

$$\left(-(x - h_j)\sin\theta_j + (y - k_j)\cos\theta_j\right)^2 = a_j\left((x - h_j)\cos\theta_j + (y - k_j)\sin\theta_j\right) \qquad (8)$$

The range of radius values to search for was set manually, depending on the database used. For the CASIA database, values of the iris radius range from 90 to 150 pixels, while the pupil radius ranges from 28 to 75 pixels. In order to make the circle detection process more efficient and accurate, the Hough transform for the iris/sclera boundary was performed first, then the Hough transform for the iris/pupil boundary was performed within the iris region, instead of the whole eye region, since the pupil is always within the iris region. After this process was complete, six parameters are stored, the radius, and x and y center coordinates for both circles [10].

4.1.1 Open Source Matlab Functions

- createiristemplate: generates a biometric template from an iris eye image.
- segmentiris: performs automatic segmentation of the iris region from an eye image. Also isolates noise areas such as occluding eyelids and eyelashes.
- addcircle: circle generator for adding weights into a Hough accumulator array.
- adjgamma: for adjusting image gamma.
- Circlecoords: returns the pixel coordinates of a circle defined by the radius and x, y coordinates of its center.
- CANNY - Canny edge detection: function to perform Canny edge detection.
- Findcircle: returns the coordinates of a circle in an image using the Hough transform and Canny edge detection to create the edge map.
- Findline: returns the coordinates of a line in an image using the Hough transform and Canny edge detection to create the edge map.
- Houghcircle: takes an edge map image, and performs the Hough transform for finding circles in the image.
- HYSTHRESH - Hysteresis thresholding: Function performs hysteresis thresholding of an image.
- Linecoords: returns the x, y coordinates of positions along a line.
- NONMAXSUP: Function for performing non-maxima suppression on an image using an orientation image. It is assumed that the orientation image gives feature normal orientation angles in degrees (0–180).

4.1.2 Test Results

The automatic segmentation model proved to be successful. The CASIA database offers good segmentation, since those eye images had been taken specifically for iris recognition research and boundaries of iris pupil and sclera were clearly distinguished. For the CASIA database, the segmentation technique managed to correctly segment the iris region from 624 out of 756 eye images, which corresponds to a success rate of around 83% [10].

The issue is that images had slight intensity variances between the iris region and the pupil region as shown in Fig. 5. One problem confronted with the implementation was that it needed different parameters to be set for each database. These parameters were the radius of iris and pupil to search for, and threshold values for creating edge maps. However, for set-ups of iris recognition systems, these parameters would only need to be set once, since the camera, imaging distance, and lighting conditions would usually remain the same.

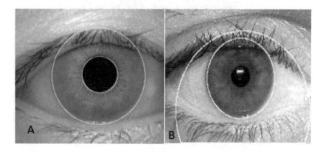

Fig. 5. A MATLAB successful result of segmented image for CASIA database, B. An example where segmentation fails for the CASIA database. Here there is little contrast between pupil and iris regions, so Canny edge detection fails to find the edges of the pupil

The eyelid detection system also proved quite successful, and managed to isolate most occluding eyelid regions. One problem was that it would sometimes isolate too much of the iris region, which could make the recognition process less accurate, since there is less iris information (Fig. 6). However, this is preferred over including too much of the iris region, if there is a high chance it would also include undetected eyelash and eyelid regions.

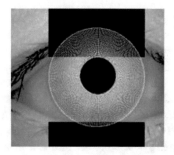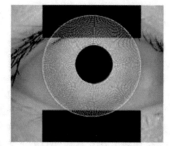

Fig. 6. Automatic segmentation of various images from 'CASIA' database. Black regions denote detected eyelid and eyelash regions.

The eyelash detection system implemented for the CASIA database also proved to be successful in isolating most of the eyelashes occurring within the iris region as shown in Fig. 7.

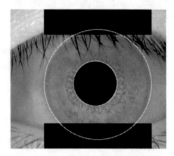

Fig. 7. The eyelash detection technique, eyelash regions are detected using thresholding and denoted as black.

A small issue was that areas where the eyelashes were light, such in the Fig. 8, were not detected. However, these undetected areas were minor when compared with the size of the iris region.

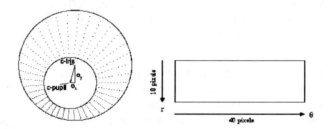

Fig. 8. Outline of the normalisation process with radial resolution of 10 pixels, and angular resolution of 40 pixels. Pupil displacement relative to the iris centre is exaggerated for illustration purposes.

4.2   Normalization

For normalization of iris region, a technique based on Daugman's rubber sheet model [13] is implemented. The center of the pupil is considered as the reference point and radial vectors pass through the iris region. A number of data points are selected along each radial line and this is defined as the radial resolution. The number of radial lines going around the iris region is defined as the angular resolution. A constant number of points are chosen along each radial line, so that a constant number of radial data points are taken, irrespective of how narrow or wide the radius is at a particular angle. The normalized pattern is created by backtracking to find the Cartesian coordinates of data points from the radial and angular positions in the normalized pattern.

4.2.1   Open Source Matlab Functions

- normalise iris - normalization of the iris region by unwrapping the circular region into a rectangular block of constant dimensions.

4.2.2   Test Results

The normalisation process proved to be successful and some results are shown in Fig. 9 below:

Fig. 9. Illustration of the normalisation process applied on an image of iris from CASIA database

4.3   Feature Encoding and Extraction

Feature encoding is implemented by convolving the normalized iris pattern with 1D Log-Gabor wavelets. The 2D normalized pattern is broken up into a number of 1D signals, and then these 1D signals are convolved with 1D Gabor wavelets. The output of filtering is then phase quantized to four levels using the Daugman method [1], with each filter producing two bits of data for each phasor. The output of phase quantization is chosen to be a grey code, so that when going from one quadrant to another, only 1 bit changes (Fig. 10).

4.3.1   Open Source Matlab Functions

- encode - generates a biometric template from the normalized iris region, also generates corresponding noise mask.
- gaborconvolve - function for convolving each row of an image with 1D log-Gabor filters.

4.4   Matching

For matching, the Hamming distance is chosen, since bit-wise comparison is required. The Hamming distance algorithm employed also incorporates noise masking, so that only significant bits are used in calculating the Hamming distance between two iris templates. When taking the Hamming distance, only those bits in the iris pattern that correspond to '0' bits in noise masks of both iris patterns is used in the calculation involved in the matching of pattern [10].

The example below illustrates the hamming distance for different templates (Table 1):

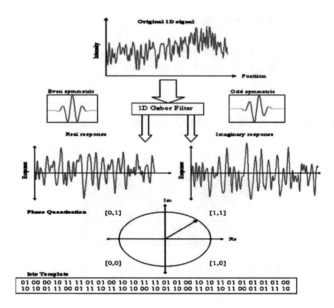

Fig. 10. An illustration of the feature encoding process.

Table 1. Examples of hamming distance

Iris template 1	Iris template 2	HD
10 00 11 00 10 01	10 00 11 00 10 01	0.83
00 11 00 10 00 10	00 11 00 10 00 10	0.00
00 11 00 10 01 10	01 10 00 11 00 10	0.33

4.4.1   Open Source Matlab Functions

gethamming distance - returns the Hamming Distance between two iris templates incorporates noise masks, so noise bits are not used for calculating the HD. shift bits - function to shift the bit-wise iris patterns in order to provide the best match. Each shift is by two bit values and left to right, since one-pixel value in the normalized iris pattern gives two bit values in the template.

5   Conclusion

This paper has presented an iris recognition system, which was tested using a universal open source database of greyscale eye images in order to verify the claimed performance of iris recognition technology.

A segmentation algorithm was presented, which localizes the iris region from an eye image and isolate eyelid and eyelash. In the implemented algorithm, automatic segmentation was achieved through the use of the circular Hough transform for localizing the iris and pupil regions, and the linear Hough transform for localizing occluding eyelids.

Then, the segmented iris region was normalized to eliminate dimensional irregularities between iris regions. This was realized by implementing a version of Daugman's rubber sheet model, where the iris is modelled as a flexible rubber sheet, which is unwrapped into a rectangular block with constant polar dimensions.

References

1. Chinese academy of sciences – institute of automation. Database of greyscale eye images. http://www.cbsr.ia.ac.cn/irisdatabase.htm
2. Kolodgy, C.: Authentication is something you have, something you know, and something you are when you add biometrics. idc, framingham, ma. Infoworld (2001)
3. Narote, S.P., Narote, A.S., Waghmare, L.M.: Iris based recognition system using wavelet transform. Int. J. comput. Sci. Netw. Secur. **9**, 101–102 (2009)
4. Proenca, H., Alexandre, L.A.: Iris recognition: measuring feature's quality for the feature selectio in unconstrained image capture environments. In: Proceedings of 2006 International Conference on Computational Intelligence for Homeland Security and Personal Safety, vol. 1, pp. 35–40 (2006)
5. Proenca, H., Alexandre, L.A.: Iris recognition: an analysis of the aliasing problem in the iris normalization stage. In: Proceedings of 2006 International Conference on Computational Intelligence and Security, vol. 2, pp. 1771–1774 (2006)
6. Daugman, J.: High confidence visual recognition of persons by a test of statistical independence. IEEE. Tans. Pattern Anal. Mach. Intell. **15**, 1148–1161 (1993)
7. Daugman, J.: How iris recognition works. IEEE Trans. Circuits. Syst. Video Technol. **14**(1), 21–30 (2004)
8. Daugman, J.: New methods in iris recognition. IEEE Trans. Syst. Man Cybern. **37**(5), 1167–1175 (2007)
9. Daugman, J.: The importance of being random: statistical principles of iris recognition. Pattern Recogn. **36**, 279–291 (2003)
10. Masek, L.: Recognition of human iris patterns for biometric identification. M.S. Thesis, University of Western Australia (2003)
11. Kovesi, P.: Matlab functions for computer vision and image analysis. http://www.cs.uwa.edu.au/~pk/research/matlabfns/index.html
12. Masek, L., Kovesi, P.: Matlab source code for a biometric identification system based on iris patterns. The School of Computer Science and software Engineering, The University of Western Australia (2003)
13. Daugman, J.: Biometric personal identification system based on iris analysis. US patent, patent number: 5,291,560 (1994)
14. Chinese academy of sciences – institute of automation. Database of 756 greyscale eye images. http://www.sinobiometrics.com version 1.0 (2003)

Optimization of Task Scheduling Algorithms for Cloud Computing: A Review

Gibet Tani Hicham$^{(\boxtimes)}$ and El Amrani Chaker

Laboratory of Informatics Systems and Telecommunications (LIST),
Department of Computer Engineering, Faculty of Sciences and Technologies,
Abdelmalek Essaadi University, Route Ziaten, B.P. 416, Tangier, Morocco
gibet.tani.hicham@gmail.com

Abstract. Cloud Computing is one of the most recognized computing paradigms in our time, where Information Technology (IT) resources and services are supplied on-demand over the Internet. Nowadays, almost everyone is using the cloud whether it is an email platform, an online storage or a complicated online datacenter for business high-scale deployments. Following the recent statistics and analytics, cloud computing business market is growing each year in term of demand and revenue. This business growth is pushing cloud providers to expand their infrastructures in order to manage the users' increasing level of requests. In the same context, this paper presents the importance of task scheduling algorithms in the optimization of the cloud providers' infrastructures exploitation, thus a significant cost reduction in term of new hardware investments and maintenance.

Keywords: Cloud Computing · Task scheduling · Scheduling algorithms
First Come First Served · Round Robin · Artificial neural networks

1 Introduction

In the past few years, there has been an ongoing investigation of Cloud Computing paradigm in the interest of enhancing the most substantial characteristics and aspects of this new model of computing. Cloud computing is a service-focused computing model that provide high quality and low-cost IT services in a pay-per-use manner in which guarantees are offered by the cloud service providers through customized Service Level Agreements (SLA) [1]. Actually, cloud users' needs only an internet connection in order to access one or many IT resources (Software, Platform, Storage, Hardware Servers…) as services [2]. With the increasing demand on cloud services, many cloud providers are struggling with new infrastructures deployments and management, in the aim of diversifying their cloud portfolio, and therefore promoting their profits. While the implementation of new IT infrastructures is expensive and time consuming, many studies are in progress in order to improve the existing infrastructures to host more and more services, thus reducing the cloud providers' expenses. Under the same topic, the current paper exhibits task scheduling as one of the most important characteristic of Cloud Computing model that needs more investigation in order to optimize the infrastructure utilization.

© Springer International Publishing AG, part of Springer Nature 2018
M. Ben Ahmed and A. A. Boudhir (Eds.): SCAMS 2017, LNNS 37, pp. 664–672, 2018.
https://doi.org/10.1007/978-3-319-74500-8_61

2   Cloud Computing Service Delivery Model

Cloud Computing proposes a new service delivery model for IT services, which can be abstracted to the following three basic layers [4, 5] (Fig. 1):

- Infrastructure as a Service (IaaS): This layer offers basic hardware access as a service. It covers storage services, computing resources, an entire server or a mix of all of these services in the form of a cloud datacenter.
- Platform as a Service (PaaS): This is the second layer that proposes IT platforms such as operating systems, middleware systems, database management systems…
- Software as a Service (SaaS): The last layer is a combination of applications that can be offered as a service to an end user: Email, document treatment, Agenda… This layer proposes also specific business applications such as: Customer Relationship Management, Invoicing, Accounting, Human Resources Management, Enterprise Resources planning and many more others…

Recently, many dedicated cloud computing service delivery models have been emerged based on the previous three models, such as:

- Database as a Service.
- Communication as a Service.
- Process as a Service.
- Security as a Service.

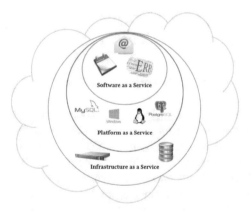

Fig. 1. Cloud Computing abstraction layers

Taking into account the specifications of this service delivery model, Cloud Computing architecture combines several modules communicating with each other over Application Programming Interfaces (API) [3] to better respond to each cloud service layer. Whereas the basic cloud modules are:

- Infrastructure: It consists of one or many clustered servers located on one or many datacenters (They can be geographically separated) connected over a private network. The infrastructure includes also the necessary networking components (Switches and routers) to interconnect the different hardware.

- Platform: It involves the essential operating systems to run the infrastructure. This module includes the required hypervisors for hardware virtualization.
- Cloud Services: It is the most important module and it comprises the different software that constitutes the cloud provider offer.

The functioning mechanism of cloud computing architecture consists of a front-end and a back-end. The front-end is what the cloud user sees, generally, a web based interface that can be accessed from any type of internet browsers over the public network however it can be a soft application developed to run on the user side terminal in order to access the required cloud services. On the other hand, the back-end is the three modules mentioned above, in addition to a cloud orchestration platform for the infrastructure provisioning and monitoring (Fig. 2).

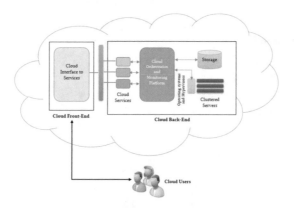

Fig. 2. Cloud Computing architecture

In the last past few years, the Cloud Computing market grew 28% to $110B in revenues in 2015. Whereas the public IaaS/PaaS services attained the highest growth rate of 51%, followed by private & hybrid cloud infrastructure services at 45% [6]. With this massive evolution in term of revenue comes an extensive development in term of new cloud infrastructure investments. While developing the infrastructure is the best way to generate more profits for cloud providers, it is becoming more difficult to manage and maintain, thus many research studies are ongoing to improve the existing cloud infrastructures in order to host more and more services and therefore reducing the costs of new infrastructures, which can be a major advantage for cloud providers.

3   Cloud Computing Characteristics

Cloud Computing paradigm is characterized with the following essential features [7]:

- On-demand self-service: it includes processing power, storage and virtual machines that will be allocated automatically to satisfy a cloud user request.

- Network Access: it is one of the mandatory characteristics of cloud computing in order to give access to the cloud services.
- Resource pooling: gathering the infrastructure resources needed to run a cloud service in a pool that will be assigned to a cloud user.
- Elasticity: This characteristic gives the cloud user the possibility to scale-up or scale-down cloud resources without having to worry about any new installation, updates or upgrades.

By studying these characteristics and features radiates the fact that the infrastructure is the most important aspect of cloud computing model and it will be discussed more on the following section.

4 Cloud Computing Infrastructure

The cloud computing infrastructure is a mandatory stage for creating a cloud service or offer. This stage incorporates two basic steps:

- Hardware implementation (Servers, Storage and Networking).
- Operating system (or hypervisor in case of virtualization) installation.

This two steps are required to create the essential resources for a cloud service execution:

- Compute resources.
- Memory.
- Storage.

This resources are managed by the operating system side and this is by controlling the access to each resource using specific scheduling algorithms. Despite the fact that these resources are used combined, the compute resources is the key element for cloud services, whereas it controls the celerity of execution and the number of services (tasks) that can be executed simultaneously.

To the extent of our knowledge, the computes resources scheduling algorithms (or task scheduling algorithms) used on the majority of the operating systems (or hypervisors) on the cloud, were developed on the past two decades (Like the First Come First Served algorithm FCFS) and lacks the intelligence required for such new computing model.

5 Task Scheduling Algorithms

The key purpose of scheduling algorithms is the appropriate allocation of task or a job to the appropriate resource [4]. While compute resources are the most significant mean for task execution, it is the first factor that should be addressed on a high level computing model such as cloud computing.

The most important Compute resources scheduling algorithms are the following:

- First Come First Served algorithm: Tasks get aligned in waiting queue and which come first get served. This algorithm is simple and fast [2, 8, 9].
- Round Robin algorithm: it was designed based on the distribution of Core Processing Unit (CPU) time among the scheduled tasks. All tasks get on a queue list whereas each task get a small unit of CPU time [2, 8, 9].
- Min–Min algorithm: In this algorithm, small tasks are executed first and large tasks are delayed [2, 8, 9].
- Max-Min algorithm: It consists of the same principal of Min-Min algorithm, this time larger tasks are executed at first [2, 8, 9].
- Priority scheduling algorithm: each task is assigned a priority index, and the task with the highest priority is allowed to run [2, 8, 9].

These algorithms are largely used in most cloud providers' infrastructures with some minor changes. While these algorithms still gives an outstanding performance, on most cases, they lack the required intelligence to adapt to cloud computing architecture and service delivery model. From this originates several research activities to add more intelligence to the existing algorithms and making them more effective in such innovative computing model and this will be discussed by reviewing the research literature on the next section.

6   Optimized Task Scheduling Algorithms for Cloud

In the past few years, several research activities have been conducted in order to optimize the cloud computing infrastructure and specifically speaking about task scheduling algorithms in the aim of enhancing the compute resources consumption. From these studies:

Hicham et al. [2] designed an algorithm based on the original Round Robin algorithm that calculates the time quantum (time portion that will be allocated to each task submitted for execution) from the average burst time of all tasks on a waiting list, hence, the algorithm produced an intermediate time quantum that will be recalculated dynamically from tasks that are ready for execution in the queue and this is a major enhancement that adapts the CPU resource allocation to the required jobs or tasks. The authors used the cloud computing simulation toolkit "CloudSIM" in order to carry out the necessary tests. The authors also proposed a second version of this Round Robin based algorithm [4] called Smarter Round Robin that injects a new layer to the first algorithm in order to adapt it to different situations that comes with the new delivery model of Cloud Computing.

Kumar et al. [10] proposed a genetic based algorithm that combines the principal of Min-Min and Max-Min scheduling algorithms in order to generate the initial population (tasks) that will be selected for execution and then this operation will be dynamically repeated to find the best solution (best response time). The researchers used "CloudSIM" simulation toolkit to examine their proposed algorithm.

Ghanbari et al. [11] introduced a priority based job scheduling algorithm which can be applied in cloud environments. The algorithm orders the jobs (tasks) submitted for execution based on priority on a list, then the list is compared to a matrix of available resources. Afterwards, the proposed algorithm select the job with the highest priority where the computing resources are available and executed, then the job is removed from the list and the same procedure is repeated for the rest of the jobs.

Tawfeek et al. [12] presented an Ant Colony Optimization scheduling algorithm for cloud computing. This algorithm uses a positive feedback mechanism and imitates the behavior of real ant colonies in nature to search for food and to connect to each other by pheromone laid on paths travelled. The algorithm Find the optimal resource allocation for tasks in a dynamic cloud system to minimize the make span of tasks on the entire system. The authors used CloudSIM simulation toolkit to test their solution and compared it to other scheduling algorithms such as First Come First Served and Round Robin.

Lawanya Shri et al. [13] proposed a scheduling algorithm named "GPA" (generalized priority algorithm). The algorithm start initially by prioritizing the tasks according to their size (highest size -> has highest rank). Afterwards, the algorithms execute the task with the highest rank and move down to the others.

7   Comparison of Methods and Discussions

According to Hicham et al. [4] (Fig. 3), the proposed Round Robin based algorithm showed a great benefit in regards to the average waiting time which represents the average time spent by each task on the waiting queue before getting to the CPU and that is a major improvement in comparison with the simple version of the Round Robin algorithm and also to different versions of the First Come first Served algorithm.

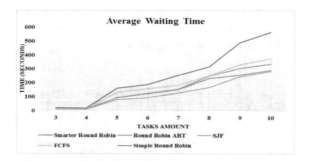

Fig. 3. Smarter Round Robin scheduling evaluation [4]

The authors in paper [10] demonstrated the performance of their proposed Genetic based algorithm according to the make-spans of scheduled tasks (The time elapsed from submission of the task until its execution). Figure 4 showed that the improved genetic algorithms proposed by Kumar et al. gave the best outcome.

In Fig. 5, Tawfeek et al. displays a comparison between their proposed Ant Colony based algorithm, FCFS and Round Robin algorithms in regards to the average

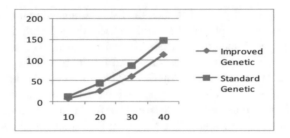

Fig. 4. Improved genetic scheduling evaluation [10]

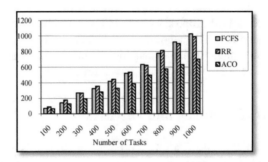

Fig. 5. Optimized Ant Colony scheduling evaluation [12]

make-span of scheduled tasks. Their proposed algorithms exhibits a better result than the other algorithms.

By looking at Fig. 6 we can see how the generalized priority scheduling algorithm (In yellow) proposed by Lawanya et al. [13] gave a notable result in comparison to the FCFS and Shorter Job First scheduling algorithms.

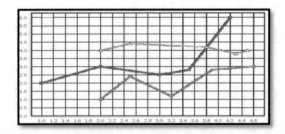

Fig. 6. Generalized priority scheduling evaluation [13]

By reviewing the outcomes mentioned above, we can sense that each one of the authors is trying to add more intelligence to the existent compute resources scheduling algorithms in order to improve its performance especially for cloud computing.

Accordingly, the optimized scheduling algorithms discussed above define the basic characteristics of a scheduling algorithm which can be deduced on: Tasks execution time (make-span), priority and foremost the tasks amount.

8   Conclusion

The cloud computing infrastructure is the most expensive component of this paradigm, due to the high costs of acquiring new hardware, maintenance, electricity, cooling systems and human resources. In this paper, the cloud computing infrastructure has been tackled, discussed and furthermore decomposed to numerous components in order to assess the significance of each part, thing that drove us to the most substantial one, which is compute resources scheduling algorithms (or task scheduling algorithms). Afterwards, several task scheduling algorithms has been analyzed based on literature in order to raise interest of cloud computing providers and investigators about this key factor of this revolutionary computing model.

References

1. Savitha, P., Geetha Reddy, J.: A review work on task scheduling in Cloud computing using genetic algorithm. Int. J. Sci. Technol. Res. 2(8), 241–245 (2013)
2. Hicham, G.T., Chaker, E.A.: Smarter Round Robin scheduling algorithm for Cloud computing and Big data. In: Proceeding of the 8th Edition of the International Colloquium on Scientific and Technological Strategic Intelligence (VSST 2016), Rabat, Morocco (2016)
3. Dinesh, B., Veera Mallu, B.: A review on Cloud computing. Int. J. Eng. Trends Technol. 4(4), 803–808 (2013)
4. Hicham, G.T., Chaker, E.A.: Cloud computing CPU allocation and scheduling algorithms using CloudSIM simulator. Int. J. Elec. Comput. Eng. 6(4), 1866 (2016)
5. Hu, F., Qiu, M., Li, J., Grant, T., Tylor, D., McCaleb, S., Butler, L., Hamner, R.: A review on Cloud computing: design challenges in architecture and security. J. Comput. Inf. Technol. 19, 22–55 (2011)
6. Columbus, L.: Roundup of Cloud computing Forecasts and Market Estimates. Forbes, March 2016
7. Sriram, I., Ali, K.H.: Research agenda in cloud technologies. In: 1st ACM Symposium on Cloud computing, SOCC 2010 (2010, Submitted)
8. Durga Lakshmi, R., Srinivasu, N.: A review and analysis of task scheduling algorithms in different Cloud computing environments. Int. J. Comput. Sci. Mob. Comput. 4(12), 235–241 (2015)
9. Santhosh, B., Harshitha, Prachi Kaneria, A., Manjaiah, D.H.: Comparative study of workflow scheduling algorithms in Cloud computing. Int. J. Innov. Res. Comput. Commun. Eng. 2(5), pp. 31–37 (2014)
10. Kumar, P., Verma, A.: Independent task scheduling in Cloud computing by improved genetic algorithm. Int. J. Adv. Res. Comput. Sci. Softw. Eng. 2(5), 137–142 (2012)
11. Ghanbari, S., Othmane, M.: A priority based job scheduling algorithm in cloud computing. Procedia Eng. 50, 778–785 (2012). Proceeding of the International Conference on Advances Science and Contemporary Engineering (ICASCE 2012)

12. Tawfeek, M., El-Sisi, A., Keshk, A., Torkey, F.: Cloud task scheduling based on Ant Colony Optimization. Int. Arab J. Inf. Technol. **12**(2), 64–69 (2015)
13. Lawanya Shri, M., Benjula Anbumalar, M.B., Santhi, K., Deep, M.: Task scheduling based on efficient optimal algorithm in cloud computing environment. Int. J. Adv. Technol. Eng. Sci. **4**(1) (2016)

Improving Online Search Process in the Big Data Environment Using Apache Spark

Karim Aoulad Abdelouarit$^{(\boxtimes)}$, Boubker Sbihi, and Noura Aknin

Information Technology and Modeling Systems Research Unit,
Computer Science, Operational Research and Applied Statistics Laboratory,
Abdelmalek Essaadi University, Tetuan, Morocco
Abdelouarit.karim@gmail.com, bsbihi@hotmail.com,
aknin_noura@yahoo.fr

Abstract. In this article, we study the use of the Apache Spark solution to improve the online search process from the Big Data flow as part of the refinement of our new Big-Learn solution for online search, used by a learner in an e-Learning environment. The purpose of this study is to evaluate the Spark system in terms of fast data processing, large-scale complex analysis, and also in terms of ease of use, execution and integration of this solution with other layers related to the data search process and especially with the Solr Framework. Apache Spark is considered better than Hadoop in terms of fast processing large data, and also in the real-time analysis. It is in this context that we propose to study the integration of the Spark solution in order to offer a technique that better processes the massive data and thus allows to improve and organize the results of the online search. Our solution is based on the combination of Spark technology for massive data processing, with the Solr search platform, in addition to the use of the Lucene engine for data indexing.

Keywords: Big Data · Online search · Spark · Hadoop · Solr

1 Introduction

The data explosion on the Internet and especially on the Web has made it difficult for traditional search engines to find interesting relationships between objects and to analyze and extract knowledge from the raw data generated by the Big Data phenomenon [10]. This article follows our previous work on the design and implementation of an architecture model for online search in the Big Data environment used by a learner in a context of learning or distance training, this solution is called Big-Learn [4]. As we have presented in our previous work, it is to design a complete system that relies on the collection of data corresponding to the user request of the online search, and then this data is processed by the MapReduce technique of the Hadoop tool and stored in its file system HDFS (Hadoop Distributed File System). Then, these data will be indexed by the Lucene engine so that they can be easily queried and returned to the user using the Solr Framework [2]. However, our solution architecture model as presented can be improved to cover a better process both in terms of data processing speed and in terms of complex and large scale analysis. The following paragraph describes

© Springer International Publishing AG, part of Springer Nature 2018
M. Ben Ahmed and A. A. Boudhir (Eds.): SCAMS 2017, LNNS 37, pp. 673–683, 2018.
https://doi.org/10.1007/978-3-319-74500-8_62

the use of Apache Spark in the Big Data environment by comparing the two systems: Hadoop and Spark specialized in the processing of massive and heterogeneous data; Sect. 3 presents the concepts and approaches of the Spark technique for processing voluminous data and its contribution to the online search process. Then we present in Sect. 4 the solution used, which is based on the Apache Spark tool integrated into our Big-Learn search system, which tends to the fast processing of raw data generated by the Big Data layer and to customization search results for better organization and presentation of information for the user of the online search. The last paragraph presents a general conclusion outlining a series of perspectives.

2   The Use of Apache Spark in the Big Data Environment

2.1   Spark and the Big Data Phenomenon

With the arrival of the Big Data phenomenon, existing data search algorithms are no longer able to function efficiently due to the enormous size of data to be processed [10]. Although search engines have been able to develop highly qualified skills, web pages generated by search engines are based on search terms that require sophisticated algorithms that can handle a large number of queries [2]. Hadoop is appeared at the origin of the Apache Nutch, which is an open source search engine. It was in February 2006, that part of Nutch became independent and created Hadoop. And since 2010, a new version of Hadoop began designing, called YARN (Yet Another Resource Negotiator). YARN changed the management of Hadoop's resources and supported a wide range of new applications with new functionalities [9].

In 2009, the AMP Lab group at the University of California, Berkeley, started the Apache Spark project to design a unified engine for distributed data processing. It went on to open source as an Apache project in 2010 [11]. Spark is a Big Data processing framework built to perform sophisticated analysis and designed for speed and ease of use, it exceeds Hadoop in the fast processing of large data, and also in the real-time analysis. It relies on a programming model similar to Hadoop's MapReduce but extends with an abstraction of data sharing called RDD (Resilient Distributed Datasets) [8]. Using this extension, Spark can capture a wide range of processing loads that previously needed separate engines including SQL, streaming, machine learning and graphics processing [11].

Apache Spark has several advantages over other big data technologies like Hadoop and Storm. First of all, Spark offers a unified and flexible framework to meet the needs of Big Data processing for different types of data (text, image, video, etc.). Then, the Spark system can run applications up to 100 times faster in memory on Hadoop clusters, 10 times faster on disk [11]. Apache Spark system is composed of Spark Core and a set of components in the form of libraries. The core consists of the distributed execution engine and the Java, Scala, and Python APIs offer a platform for distributed ETL application development. In addition, a set of libraries, built atop the core, allow diverse workloads for streaming, SQL, and machine learning. Figure 1 shows the ecosystem of Apache Spark with all their components and libraries.

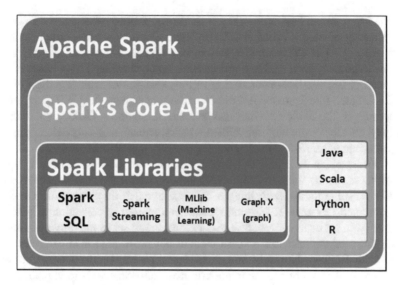

Fig. 1. The Apache Spark ecosystem.

As shown in this figure, Spark allows to quickly develop applications in Java, Scala or Python since it uses a unified API. In addition, it can be used interactively to query data from a Shell. Finally, in addition to the main API, Apache Spark contains additional libraries to work with Big Data and Machine Learning. These libraries include:

- Spark SQL: allows to execute SQL queries;
- Spark Streaming: allows real-time processing of flow data;
- Spark MLlib: is a Learning machine library that contains all classical learning algorithms and utilities;
- Spark Graph X: is designed for processing graphs.

Although Apache Spark is a new platform, it is constantly evolving with new additions and it has already been adopted as a real-time processing framework in many big companies, such as: Amazon, Yahoo, Samsung, Nokia, IBM, eBay, etc.

2.2 Spark Versus Hadoop

In this section, we present the main differences and similarities in the engines of both platforms Spark and Hadoop in order to explain which are the best scenarios for one platform or the other. Afterwards, we highlight the main differences according to several criteria implemented in both platforms.

Hadoop is positioned as a technology for processing large volumes of data. It is based on the MapReduce technique which exposes two functions called Map() and Reduce() which operate on pairs of (key, value). The Map() function stores the pairs of (key, value) and the Reduce() function reduces a set of pairs (key, value) that correspond to a single key in one (key, value) pair. Before reducing, the Framework run a grouping operation called shuffle [6]. Hadoop employ a resource management system called YARN (Yet Another Resource Negotiator) and uses the MapReduce technique to process the data. Each processing step is constituted by a phase of Map() and a phase of Reduce(). The output data of the execution of each step must be stored on the Distributed File System (HDFS) before the next step begins [7]. This approach is slow because of the replication and disk storage. In addition, Hadoop solutions generally rely on clusters, which are difficult to implement and administer. They also require the integration of several tools for the different Big Data usage cases (like Mahout for Machine Learning and Storm for flow processing). However, Spark is designed to cooperate with Hadoop, especially through the use of its HDFS storage system. To resolve these disadvantages, Spark introduces an abstraction called Resilient Distributed Datasets (RDD) that represents a read-only collection of partitioned objects across a set of machines. It allows to load a data set in memory once and read multiple times of the executor process without having to load it in each iteration as happens with Hadoop. Not only introduces Spark an in memory storage but also it provides a complete set of programming primitives allowing users to implement much more complex distributed applications than those implemented in the classic MapReduce. It allows to implement several distributed models like MapReduce, SQL like or even Pregel [10].

Spark enables the development of complex, multi-step data processing pipelines and data sharing in memory so that multiple jobs can work on the same dataset. Spark runs on HDFS infrastructures and offers additional features. Instead of seeing Spark as a replacement for Hadoop, it is more correct to say that it is an alternative to the MapReduce. Spark was not intended to replace Hadoop but to provide a complete and unified solution for Big Data processing. Table 1 outlines the major points of difference and improvement between the use of Hadoop and Spark.

As shown in this comparative table, the big difference between Hadoop and Spark is that Hadoop processes the data on the disk while Spark processes them in memory and tries to minimize disk usage. In this way, it achieves an increase in performance especially for applications that perform many iterative calculations on the same data.

Both Apache Spark and Hadoop MapReduce are popular frameworks today for cluster management and processing and analyzing big data. Both of them provides an abstraction level that directs the focus towards solving the real problem and lets the framework handle the low-level implementation details. The key difference is that Spark is a more general framework with a flexible pipeline of execution due to its DAG based execution engine where as MapReduce requires the problem to be reformed so it fits into the Map and Reduce paradigm [1].

Indeed, the choice between Apache Spark and MapReduce should not entirely land in the performance measured. Some thought should be given if the problem fits the MapReduce paradigm, if not then maybe Apache Spark can be the choice due to its flexibility with its execution engine and use of resilient distributed datasets.

Table 1. The differences between Hadoop and Spark.

	Hadoop	Spark
Data processing	Processes the data on the disk	Processes the data in memory and tries to minimize disk usage
Data analysis	For static data operations and applications that do not need direct results	Allows real-time data analysis and processing of multiple transactions
Data storage	HDFS	HDFS
Memory usage	Minimal, since it processes the data directly on the disk	Needs a lot of memory to cache the data
Infrastructure	Uses medium resources	Needs more resources and an expensive infrastructure to function properly
Performance	Less efficient, Performance is relatively poor when compared with Spark	In-memory performance
Security	Is more secure and has many security projects like Knox Gateway and Sentry	Can use Kerberos authentication, HDFS file permissions and encryption between nodes
Extensions	None, it needs additional platforms	Several modules for flow processing, SQL query, Machine Learning and graphic processing
Ease of use	Not easy to code and use	Provides a good API and is easy to code and use
Iterative processing	Every MapReduce job writes the data to the disk and the next iteration reads from the disk	Caches data in-memory
Fault tolerance	Its achieved by replicating the data in HDFS	Spark achieves fault tolerance by resilient distributed dataset (RDD) lineage
Runtime architecture	Every Mapper and Reducer runs in a separate JVM	Tasks are run in a preallocated executor JVM
Operations	Map() and Reduce()	Map(), Reduce(), Join(), Cogroup(), and many more
Execution model	Batch	Batch, Interactive, and Streaming

3   Towards a Fast and Real-Time Processing of Massive Data Using Spark

3.1   How Spark Processes the Massive Data

Apache Spark is another increasingly popular alternative to replacing MapReduce with a more efficient execution engine, but still uses Hadoop HDFS as a storage system for large datasets. Spark brings improvements to MapReduce through less expensive

Shuffle steps. Also, using memory storage and real-time processing, performance can be faster than other Big Data technologies. Spark maintains the intermediate results in memory rather than on disk, which is very useful especially when it is necessary to work several times on the same dataset. The runtime is designed to work both in memory and on disk. Operators run external operations when the data does not fit in memory, this allows larger datasets to be processed than the aggregated memory of a cluster. Spark tries to store as much memory as possible before switching to disk. Indeed, it allows to work with one part of the data in memory, another one on disk. It is necessary to review its data and use cases to assess its memory requirements, as Spark can offer significant performance benefits based on the work done in memory [11].

Spark was normally designed to work with static data through its Resilient Distributed Datasets (RDD). Spark uses batch program to deal with streams. This technique divides incoming data and processes small parts one at a time. The main advantage of this approach is that the structure chosen by Spark, called DStream, is a simple queue of RDDs. This technique allows users to switch between streaming and batch as both have the same API. Unlike Hadoop MapReduce, Spark has support for data re-utilization and iterations. Spark stores data in memory through iterations via explicit caching. However, Spark perform its executions as acyclic graph plans, which implies that it needs to schedule and run the same set of instructions in each iteration [5]. The following Fig. 2 present a comparative technique of Apache Spark and Hadoop's implementation of the MapReduce programming model. The performance analysis involves processing data to give an indication on how the fundamental differences between MapReduce and Spark differentiates the two framework in terms of performance when processing massive data.

As described in the figure, Spark performs operations in memory by copying the data from the distributed storage to RAM which is much faster. As a result, the Read/Write operation time is reduced. However, at the Hadoop-MapReduce the

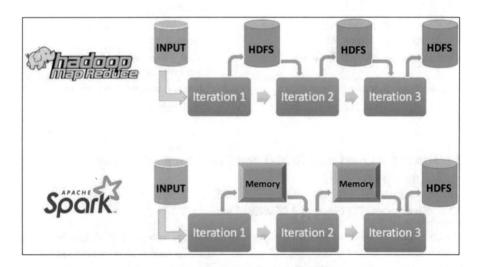

Fig. 2. Processing data by Spark and Hadoop.

process tends to be slow because each MapReduce operation stores the data on the disk. Consequently, multiple requests on the same dataset read data separately and create a lot of Read/Write operations on the disk. In addition to Spark's main APIs, the ecosystem contains additional libraries to work with Big Data and Learning machines. Table 2 shows all of these libraries with their descriptions.

Table 2. Spark framework libraries.

Description	
Spark Streaming	It can be used for real-time data flow processing. It is based on a "micro batch" processing mode and uses DStream real-time data, i.e. a series of Resilient Distributed Dataset (RDD)
Spark SQL	It allows to expose Spark datasets via JDBC API and to execute SQL queries using traditional BI and visualization tools. Spark SQL allows also to extract, transform and load data in different formats (JSON, Parquet, database) and expose them for ad-hoc queries
Spark MLLIB	It is a Machine Learning library that contains all the classical learning algorithms and utilities, such as classification, regression, clustering, collaborative filtering, dimension reduction, in addition to the underlying optimization primitives
Spark GraphX	It is the new API (in alpha version) for graph processing and parallelization of graphs. GraphX extends Spark's RDDs by introducing the Resilient Distributed Dataset Graph, an oriented multi-graph with properties attached to nodes and edges. To support these processes, GraphX exposes a set of basic operators, as well as an optimized variant of the Pregel API. In addition, GraphX includes an evergrowing collection of algorithms and builders to simplify graph analysis tasks

In addition to the libraries shown in Table 2, Spark allows other features that include:

- More Features other than Map() and Reduce() functions;
- The optimization of graphs of arbitrary operators;
- The lazy evaluation of queries, which helps to optimize the overall processing workflow;
- A concise and consistent APIs in Scala, Java and Python;
- An interactive Shell for Scala and Python (that is not yet available in Java).

Apache Spark is written in Scala language and runs on the JVM (Java Virtual Machine). The currently supported languages for application development are: Scala, Java, Python, Clojure and R.

3.2 Towards a Fast, Real-Time Processing Model for Online Search

With its data processing technology, Spark is able to cache data in memory using its resilient distributed data set (RDD) concept. These properties allow Spark to be faster for iterative algorithms because it avoids the use of the disk. A RDD is defined as a collection of records that can only be created from a stable storage or through transformation from other RDD's. The other features of an RDD is that it's a read-only after its creation and partitioned. In Apache Spark the programmer can access and create RDD's through the in-built language API either in Scala or Java. Indeed, the programmer can choose to either create the RDD from storage, a file or several files that is stored in any of the supported storage options like the HDFS. The other option is to create a RDD through transformations such as map and filter, their primary use is often to create new RDD's from already existing RDD's [1].

Therefore, in our architecture model for the online search process, the Spark system will be able to replace the conventional data collection and data mining layer with its distributed and real-time data processing technique and thus improve the operation of the massive data processing. Figure 3 shows the architecture model of the online search solution in the Big Data environment by integrating the Apache Spark system.

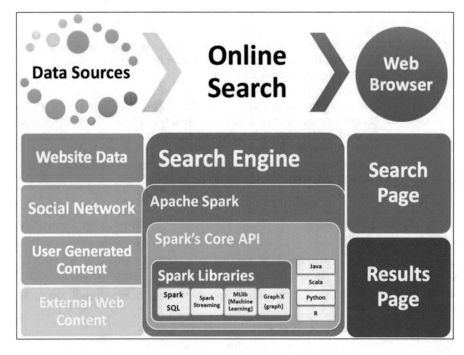

Fig. 3. Online search model using Spark in Big Data environment.

As shown in this figure, the user accesses the online search page via the Web browser to initiate his query. The search engine intercepts the user's request and starts searching the data on the Internet based on the entered keywords. In this context, the Spark system will intercept the collected data and process it at the same time it is stored in memory, without transmitting it to another engine as required by the previous Big Data processing systems like Hadoop's MapReduce technique. The collected data is referenced and indexed before presenting it to the user on the results page. The analyzed data includes all raw data from the Internet such as: websites, social networks, user-generated data and other external data sources than web data.

4   Using a Spark-Based System to Improve the Online Search Process

4.1   The Integration of Apache Spark into the Big-Learn System

The Big-Learn system for online search in Big Data environment, as already presented in our previous work [2] intercepts data from the Big Data layer via the Hadoop system which stores them in its HDFS storage system after their processing through its MapReduce technique. Thus, the data is loaded and transformed during the Map() phase, and then combined and saved during the Reduce() phase to write the Lucene index. The Lucene layer reads the data stored in HDFS, and stores it using Lucene Scheme, which in turn records the data as Lucene document in an index. Once all files are indexed to the Lucene layer, their queries are possible via the Solr layer. However, based on our assessment of both Spark and Hadoop solutions with respect to mass data processing, it is now essential to see Spark as a replacement for the Hadoop-MapReduce technique in the processing of raw data generated by the Big Data layer. Figure 4 shows the new architecture of the Big-Learn system after integrating Apache Spark system.

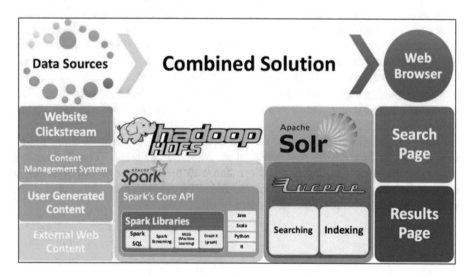

Fig. 4. Integrating Apache Spark in the Big-Learn system.

As shown in this figure of the new architecture, data coming from the Big Data will be intercepted and processed by the Apache Spark system this time and will always use the HDFS system for storing processed data. These data will be indexed at the Lucene layer so that they can be interrogated via the Solr interface.

4.2   Results and Discussion

Our study consists of the performance analysis between Apache Spark and MapReduce and how they can be applied for processing massive data and integrated with our Big-Learn Solution. Indeed, Both Spark and MapReduce are giants in the open-source big-data analytic world. At first glance we have the old and refined MapReduce framework that has many implementations in different languages, but on the other side we have a new but very promising framework Apache Spark that has one implementation but can be executed almost everywhere and can be seen as more flexible in that approach.

One of Apache Spark's key advantages is its ability and flexibility in working with all types and formats of unstructured data coming from different data sources such as text or CSV to well-structured data such as relational database. On the other hand, Apache Solr is a fast search platform built on top of Apache Lucene.

The combination of Apache Spark and Solr allows to easily explore a lot of data, and then provide the results quickly via a flexible search interface. Using this combination of open-source frameworks in a distributed environment is proven in this paper as a feasible approach to improve the online search process using large data sets.

5   Conclusion and Future Work

In this article, we have seen how the Apache Spark Framework, with its standard API, helps us in processing and analyzing data. We have seen also how Spark positions itself in relation to traditional MapReduce implementations like Apache Hadoop. Spark relies on the same file storage system as Hadoop, so it is possible to use Spark and Hadoop together in case applications have already been made with Hadoop. Spark can also combine other types of processing via its libraries such as Spark SQL, Spark Machine Learning and Spark Streaming. Thanks to its different modes of integration and adapters, Spark also allows interfacing with other technologies like Java, Scala, Python, etc.

As a perspective of this work, we will implement the scenario of using online search with this new architecture model of the Big-Learn solution. Then, in a second step, we will study the degree of integration and participation of our solution to improve learning and scientific research for students and which will take as case study students from Abdelmalek Essaadi University [3].

References

1. Andersson, L.: Natural Language Processing in a Distributed Environment: A comparative performance analysis of Apache Spark and Hadoop MapReduce (2016)
2. Abdelouarit, A.K., Sbihi, B., Aknin, N.: Towards an approach based on Hadoop to improve and organize online search results in big data environment. In: Proceedings of the International Conference on Communication, Management and Information Technology (ICCMIT 2016), pp. 543–550. CRC Press, July 2016

3. Abdelouarit, A.K., Sbihi, B., Aknin, N.: Solr, Lucene and Hadoop: Towards a Complete Solution to Improve Research in Big Data Environment (Case of The UAE) Mediterranean Congress Of Telecommunications (CMT 2016), pp. 12–13, May 2016

4. Abdelouarit, A.K., Sbihi, B., Aknin, N.: Big-learn: towards a tool based on big data to improve research in an e-learning environment. Int. J. Adv. Comput. Sci. Appl (IJACSA) **6** (10), 59–63 (2015)

5. García-Gil, D., Ramírez-Gallego, S., García, S., Herrera, F.: A comparison on scalability for batch big data processing on Apache Spark and Apache Flink. Big Data Anal. **2**(1), 1 (2017)

6. Kolosov, I., Gerasimov, S., Meshcheryakov, A.: Architecture of processing and analysis system for big astronomical data. arXiv preprint arXiv:1703.10979 (2017)

7. Kulkarni, A.P., Khandewal, M.: Survey on Hadoop and introduction to YARN. Int. J. Emerg. Technol. Adv. Eng. **4**(5), 82–87 (2014)

8. Lenka, R.K., Barik, R.K., Gupta, N., Ali, S.M., Rath, A., Dubey, H.: Comparative analysis of spatial Hadoop and geospark for geospatial big data analytics. arXiv preprint arXiv:1612. 07433 (2016)

9. Mavridis, I., Karatza, H.: Performance evaluation of cloud-based log file analysis with apache Hadoop and apache spark. J. Syst. Softw. **125**, 133–151 (2017)

10. Padillo, F., Luna, J.M., Ventura, S.: Exhaustive search algorithms to mine subgroups on big data using Apache Spark. Prog. Artifi. Intell. **6**(2), 1–14 (2017)

11. Zaharia, M., Xin, R.S., Wendell, P., Das, T., Armbrust, M., Dave, A., Ghodsi, A.: Apache spark: a unified engine for big data processing. Commun. ACM, **59**(11), 56–65 (2016)

The 2nd International Workshop on Smart Learning and Innovative Educations: SLIED'17

An E-learning Labs Platform for Moroccan Universities

Mohammed l'Bachir El Kbiach[1], Loubna Bounab[1],
Abderrahim Tahiri[1(✉)], Khalil El Hajjaji[1], Francisco Esquembre[2],
and Hassan Ezbakhe[1]

[1] Abdelmalek Essaädi University,
BP. 2121 M'Hannech II, 93030 Tetuan, Morocco
abderahim.tahiri@gmail.com
[2] Murcia University, Campus de Espinardo, 30100 Murcia, Spain

Abstract. Every year, several thousand students in Morocco participate in practical laboratory activities of science and technology in order to acquire new skills and verify equations already presented in the theoretical courses. In recent years, these practical activities have, unfortunately, been interrupted because of the problems of overstaffing of students in the faculties of science in Morocco. In order to remedy this situation and allow our students to stay in touch with the world of experience, we proposed the Experes project in the Erasmus+ framework to all Moroccan universities. This initiative is coordinated by the University of Murcia (Spain) and the Abdelmalek Essaâdi University (Morocco) with the involvement of all Moroccan universities, the ministry and European partners. In this paper we described the approach and the current results of the Experes project. This paper is presented according to six sections. The first gives an introduction to the work that has been done. The second section presents the Erasmus+ project Experes. The third part of the paper describes the methodology followed for the Experes project development. The fourth section is technical and in which we define the tools used and the output programmed applications of the project. The fifth part revolves around the testing phase of the applications produced by the Experes project. Finally, we conclude this article with a conclusion.

Keywords: E-learning · Labs platform · Experes Erasmus+ project

1 Introduction

Over the past decade, educational technologies have grown considerably, particularly in developing countries, thanks to the solutions they offer in the face of the many situations and constraints of modern educational systems: a sharp increase in the number of students, Human resources, low staffing rate, high capacity utilization rate, costly instrumentation, dilapidated and insufficient existing equipment, etc. These new technologies also offer the possibility of introducing new learning methodologies that are more flexible, adaptable to learners [1], exploiting the interactive features of innovative

© Springer International Publishing AG, part of Springer Nature 2018
M. Ben Ahmed and A. A. Boudhir (Eds.): SCAMS 2017, LNNS 37, pp. 687–696, 2018.
https://doi.org/10.1007/978-3-319-74500-8_63

techno-teaching devices, and allowing, among other things, collaborative work through more networks And the abundance of educational resources available on the net.

While the supply of e-learning sites in the field of theoretical education is increasing, especially when it comes to courses and exercises, the availability of scientific experiments and practical work remains rather limited [2], and Despite the developments in ICST. At the level of universities in the Maghreb, these e-TP e-tutorials are still in their embryonic stage, even if their implementation would enable learners to realize or follow experiments at a distance with less space constraints And on the other hand, to manage and optimize their learning.

In this context that the Erasmus+ project on ICTE (Information and Communication Technologies for Education) applied to scientific experiments - EXPERES has been proposed to the supervising Ministry and to all Moroccan universities. This project will allow the setting up of a virtual laboratory through a platform of practical work remotely [3]. These e-TPs will be proposed as part of pedagogical activities leading to a strengthening of science and technology education and to enhancing the quality of learners' knowledge in Moroccan universities.

To achieve this objective of learning practical instruction, the implementation of simulated practical exercises online will give students the opportunity to repeat the experiment as many times as they want, at any time and any place [4]. In addition to this flexibility, students will also have the advantage of communicating interactively with tutors who will be trained in parallel to answer their questions and the possible needs for formative evaluations on the educational platform to be developed and put online as part of this project. As a first step, we were interested in developing a virtual practical work platform dedicated solely to the physics and engineering sciences.

2   The Erasmus+ Project EXPERES

The Erasmus+ project on ICTE applied to scientific experiments - EXPERES was thus approved and supported by the Ministry of Education and all the Moroccan universities which were involved in the elaboration of the digital contents of dedicated e-labs in the sciences of physics and engineering. The coordination of this project was entrusted to the University of Murcia (Spain) in collaboration with Abdelmalek Essaadi University of Téouan (Morocco). Knowing that the Moroccan universities, members of this project, are: Ibn Toufail University of Kénitra, Ibn Zohr University of Agadir, Cadi Ayyad University of Marrakech, Sidi Mohammed Ben Adellah University of Fès, Hassan II University of Settat, Hassan I University of Casablanca, Sultan Moulay Sliman University of Beni Mellal, Moulay Ismail University of Meknes, Mohamed I Universityof Oujda, and Chouaib Doukkali University of EL Jadida. The other European partners of the project: University of Bologne, University of Vigo, KTH Institute of Stokholme, University of Léon and Erasmus Expertise Agency in France, have been selected for their experience in order to ensure quality training in e-learning and meet international standards. In the next section, we will describe the different elements on which was based the Experes methodological approach implementation.

3   Project Methodology

The methodological approach adopted to achieve the Experes expected results is based on a range of activities and actions of preparation, development and management throughout the project duration of three years. The implementation of this approach was based on 5 steps that are detailed in the following.

3.1   Study and Analysis

A careful analysis of the state of the premises with the Moroccan universities partners of the project was the first step. To this end, on the one hand, a summary census was carried out. It concerned both the level of scientific equipment and available human resources, but also, it listed the various constraints and specificities that could be linked to the implementation of e-Labs. On the other hand, an analysis of the technologies available and used within the universities has been carried out in order to better optimize the choice of the appropriate IT tools, hardware and software, which will be used in the e-Labs development phase.

An adequate survey targeting the e-TP experiments and meeting the specifications of the project was then initiated. In particular, it was necessary to define the type of structure capable of offering to the greatest number of students all the resources that would be available: digital documents, software tools, methodological guides, experimental protocols, etc.

A first assessment was then made on the basis of the answers provided by the partners. The analysis of these results made it possible to determine the practical content most appropriate for physics e-labs to integrate in the program of the license degree. The different activities of this stage constitute the first work package of the project. A total of 12 practical manipulations have been proposed. They are divided into four disciplinary fields, namely mechanics, electricity, optics and thermodynamics. These are mainly the same physics labs delivered in face-to-face semesters S1 to S4 in institutions of higher education in Morocco. These are precisely the following manipulations: Static and dynamic study of a spring, Simple pendulum, Conservation of mechanical energy, Measurement of resistances, Wheatstone Bridge, Cathodic oscilloscope, Prism, Lensometer, Diopter, Measurement of the gamma adiabatic coefficient of a gas, Calorimetry, and Thermal machines.

3.2   Conceptualization and Scripting

The second stage of this work involved the preparation of the conceptualization and scripting sheets for the 12 labs cited above. This work was distributed among the Moroccan universities with the support of the European universities partners of the project. Each Moroccan university has been in charge of the preparation of a TP in one of the four pre-defined themes. After several working meetings of the thematic teams and coordination between the partner universities of the project, several points were raised and defined more precisely, such as:

- Objectives definition;
- Labs contents (theory, simulation, audiovisuals, etc.);
- Prerequisites, target audience, keywords, etc.;
- Resources organization;
- Envisaged evaluation methods;
- Labs conduct.

Another issue raised and discussed by the various partners was the design of the overall system to be put in place at the end of the project. This concerns both the pedagogical aspect and the aspects related to the instrumentation and the tools and products needed to carry out these e-labs.

Design scenarios for these e-TPs have been developed. The evaluation of the relevance of the various actions and the acquisition of knowledge by the students should be considered in order to provide answers to any particular situation.

3.3 Project Team Qualification

The third stage of this project will be devoted to the qualification of teachers and technician personnel in the field of educational platforms management and e-labs programming. This type of training, conceived in collaboration with the European partners, was conceived in order to meet the expectations of the Moroccan universities. More specifically:

- Teacher training in the European partner institutions in order to acquire a good knowledge of the pedagogical methods used in e-learning, scripting, interactive multimedia developing, and tutoring and self-evaluation. This training will enable teachers to access information and skills that will enable them to effectively manage the e-TP that they will be responsible for;
- Training of media technicians and designers (basic techniques of multimedia and internet for e-learning and e-labs), in addition to training for administrators, animators and managers (e-learning device exploitation, organization modes, etc.).

3.4 Applications Labs Implementation

The fourth phase of the project consists of the implementation and implementation of online applications, with:

- Installation of equipment, software and organizational methods. It consists in the implementation and adaptation of a platform within the Moroccan universities allowing the assimilation of these methods of organization,
- Transfer and adaptation of existing distance learning courses to European partners and the development of new content, which will enable them to complement their offer.

Testing of e-TP by a sample of teachers to ensure proper design of the platform and evaluate the developed interface.

3.5   Project Activities Management

Finally, for this last step, a management, monitoring and evaluation strategy has also been defined. Indicators of progress of each activity carried out were introduced to measure over time the rate of achievement and the quality of each result, taking into account the assumptions and associated risks highlighted.

4   Labs Applications Development

4.1   Easy JavaScript Simulations Tool

Easy Java or JavaScript Simulations is a modeling and editing-creation tool that is specifically dedicated to the creation of discrete computer simulations to study and analyze a wide variety of phenomena ranging from the simplest to the most complex. These computer simulations as shown in the next figure are used to obtain numerical data from our models as a function of the progression in time and to show this data in a way that the human being can understand them.

EjsS has been designed to enable their users to work at a high conceptual level to focus most of their time on the scientific aspects of our simulation and to ask the computer to automatically perform all the necessary tasks but which are easily automated. Each tool, including EjsS, has a learning curve. Modeling is a science and an art. Easy Java/JavaScript Simulations is a tool that allows their users to express your knowledge of science by facilitating the necessary techniques of art, and by providing you with simple access to many examples of modeling created by other authors (Fig. 1).

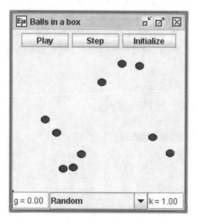

Fig. 1. An example of Ejs computer simulations

Easy Java Simulations is a modeling and authoring tool expressly devoted to this task. It has been designed to let its user work at a high conceptual level, using a set of simplified tools, and concentrating most of his/her time on the scientific aspects of our simulation, asking the computer to automatically perform all the other necessary but easily automated tasks.

Nevertheless, the final result, which is automatically generated by EJS from your description, can, in terms of efficiency and sophistication, be taken as the creation of a professional programmer.

In particular, EJS creates Java applications that are platform independent, or applets that can be visualized using any Web browser (and therefore distributed through the Internet), which read data across the net, and which can be controlled using scripts from within web pages.

There exist many programs that can be used to create Experes labs applications. What makes EJS different from most other products is that EJS is not designed to make life easier for professional programmers, but has been conceived by science teachers, for science teachers and students. That is, for people who are more interested in the content of the simulation, the simulated phenomenon itself, and much less in the technical aspects needed to build the simulation, and through this special feature, users of the Experes platform, Moroccan science teachers and students, without having advanced computer skills, will have the access to develop their own practical novel lab applications, so we guarantee the richness and durability of the Experes project services.

4.2   Applications Labs Experimentations

The main objective of the Experes project is the development of a digital interactive learning environment by creating a virtual laboratory offering License students in particular free access to practical activities without space and time constraints. 12 physics manipulations were then conceptualized, scripted and programmed by the 12 public Moroccan universities, supported by the supervisory Ministry and 6 European university institutions involved in the project.

The 12 manipulations of physics that covering the themes of optics, mechanics, electricity and thermodynamics, and they have been selected to perform the Experess labs applications are:

- Conservation of mechanical energy;
- Static and dynamic study of the spring;
- Single pendulum;
- Electrical resistances measurement;
- Wheatstone bridge;
- Cathodic oscilloscope;
- Prism;
- Diopter;
- Lensometer;
- Thermal machines;
- Gamma coefficient measurements;
- Calorimetry.

4.3 Applications Labs Package

All the labs applications have been programmed by EjsS modeling tool except the practical work of cathodic oscilloscope which has been programmed by HTML5 language. The results of the 12 applications labs are available at the universities web sites and among their results can be presented as follows:

- Conservation of mechanical energy

The programmed practical work on the conservation of mechanical energy is composed of two main environments: energy study of the vertical fall of a body environment: free fall and fall in a fluid, and energy study of an oscillating system environment: mass on spring (Fig. 2).

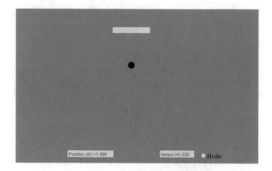

Fig. 2. Simulation of the vertical fall of a body experimentation

The student launches the manipulation by clicking the play button and observes the evolution of the kinetic energies EC, potential EP and total E = EC + EP (Fig. 3).

Fig. 3. Vertical fall of a body control panel interactive interface

After several manipulations, the student must select the correct answer (s) to the proposed questions in the application of this manipulation (Fig. 4).

- Cathodic oscilloscope

The objective of this application lab is to familiarize the students with the basic functions of the oscilloscope as in the reality, through measurements of continuous tension voltages, characteristics of alternating voltages and phase shift (Fig. 5).

As shown in the figure below, several activities are proposed in this lab framework which simulates the different characteristics and possibilities of the oscilloscope.

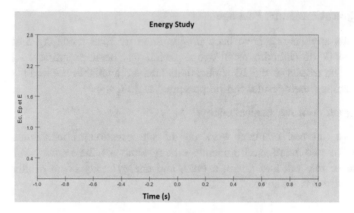

Fig. 4. Graphic presentation of student manipulations

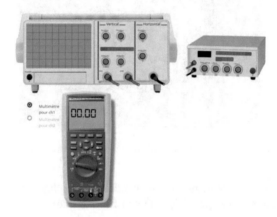

Fig. 5. Cathodic oscilloscope experimentation interactive interface

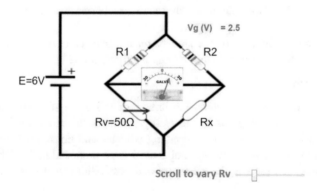

Fig. 6. Wheatstone bridge experimentation graphic

- Wheatstone bridge

The Wheatstone bridge lab is a manipulation used to measure unknown resistances. This application lab environment consists of four resistors (two known R1, R2, Rv variable and an Rx to be determined) and a galvanometer (Fig. 6).

It will, in the particular case of equilibrium, determine the value of the resistance Rx to be measured.

5 Publication and Evaluation

The phase of the testing of the labs applications is an important step for the final shaping of the labs and for the development of the Experes platform. Each Moroccan university participating in the Experes project has made the access available and public to the 12 developed applications labs for their students and teachers, also they have given them the opportunity to evaluate the applications labs environment pedagogically and technically.

At least, in each Moroccan Experes partner university, a group of 25 students, enrolled in undergraduate studies, under the supervision of their physics' teachers, have the opportunity to realize the different experimentations of the 12 labs applications. As approved by the Experes consortium, two questionnaires tests was adopted to evaluate the application lab experimentations, the first one for the teachers and the second one for the students.

The results of this project phase will be especially presented and discussed in a future work and will provide a detailed description of all the procedure of publication and evaluation of the Experes Labs.

6 Conclusion

In this work we presented the EXPERES project set up within the framework of Erasmus+ in Morocco. This project aims at the design and the on-line launching of virtual practical works of physics at the undergraduate university level. In our future work, we will present and discuss in detail the result of the survey carried out at the level of Moroccan universities on the situation of practical work of physique. We also plan to discuss the models of the conceptualization and script sheets as well as the simulation program that allowed the various activities to be carried out.

Acknowledgments. Our thanks to the European Commission for the Experes project financing. Our thanks are also devoted to all the Experes project participants and collaborators for their contribution and involvement.

References

1. Saoutarrih, M., Sedrat, N., Tahiri, A., Elkadiri, K.E.: Towards measuring learner's concentration in E-learning systems. (IJCT) Int. J. Comput. Tech. **2**(5), 27–29 (2015)
2. Nagel, C., Wolny, B.: E-learning in the Introductory Physics Lab. http://physics.univie.ac.at/eLearning/eLearnPhysik/dissemination/nagelwolnyproceedings.pdf
3. Kituyi, G., Tusubira, I.: A framework for the integration of e-learning in higher education institutions in developing countries. (IJEDICT) Int. J. Educ. Dev. Inf. Commun. Technol. **9**(2), 19–36 (2013)
4. Anđelković, T., Anđelković, D., Nikolić, Z.: The impact of E-learning in chemistry education. In: The Sixth International Conference on E-Learning (eLearning-2015), 24– 25 September 2015, Belgrade, Serbia (2015)
5. El Kabtane, H., Sadagal, M., El Adnani, M., Mourdi, Y.: An augmented reality approach to integrate practical activities in e-learning systems. Int. J. Adv. Comput. Sci. Appl. **7**(2), 107–117 (2016)
6. Washington, L.E., Sequera, J.L.C.: Model to implement virtual computing labs via cloud computing services. symmetry **9**(7), p. 117. https://doi.org/10.3390/sym9070117 (2017)
7. Deperlioglu, O., Kose, U., Yildirim, R.: Design and development of an e-learning environment for the course of electrical circuit analysis. Interdisc. J. E-Learn. Learn. Objects **8**(1), 51–63 (2012). ISSN: 2375-2084
8. Author, F., Author, S.: Title of a proceedings paper. In: Editor, F., Editor, S. (eds.) CONFERENCE 2016, LNCS, vol. 9999, pp. 1–13. Springer, Heidelberg (2016)

Using Decision Making AHP Method for the Choice of the Best Pedagogical Method for Developing Reading Skills for Young and Illiterate Public

Hasnââ Chaabi[1]([⊠]) [iD], Abdellah Azmani[1], and Amina Azmani[2]

[1] Laboratory of Informatics System and Telecommunication,
Faculty of Sciences and Techniques of Tangier, Tangier, Morocco
hasnaalchaabi@gmail.com, abdellah.azmani@gmail.com
[2] Laboratory of Methods Informatics and Enterprise,
Faculty of Sciences and Techniques of Tangier, Tangier, Morocco
amina.azmani@gmail.com

Abstract. The work presented in this paper is part of a global approach providing answers to solve the problems encountered by the Moroccan education system [2], especially in the primary school, as well as eradicating illiteracy. The article deals with the choice of the most suitable and efficient pedagogical reading skills method for young and illiterate public. The paper presents several pedagogical methods and focuses on their four most important (Phonics Method, Global Method, Mixed Method and Montessori Method). In order to choose the right reading teaching method, the paper proposes a study of a theoretical approach, based on the application of the decision making method AHP (Analytical Hierarchy Process) one of the MCDM/A methods (Multi-criteria decision-making and analysis methods).

The methodological context is based on an identification and representation of the most appropriate criteria and sub-criteria for the four pedagogical methods studied in order to rank them. Complementary research is in progress by the authors completes the work presented here, which focuses on other learning (calculation, comprehension, written and oral expression), including the integration of ICT (Information and communications technology) and multimedia.

An illustration of the application of this framework is also presented here.

Keywords: Education · Pedagogy · Primary school · Reading methods
Phonics Method · Global Method
Combination of phonics and whole language method · Mixed Method
Montessori Method · AHP · ICT · MCDM/A

1 Introduction

Education reduces poverty, increases employment opportunities and promotes economic prosperity. It offers everyone the chance to live a healthy life, builds democracy on a solid foundation, strengthens the autonomy of people and changes attitudes for a better and protected environment. On the other hand, the uneducated children of today

© Springer International Publishing AG, part of Springer Nature 2018
M. Ben Ahmed and A. A. Boudhir (Eds.): SCAMS 2017, LNNS 37, pp. 697–709, 2018.
https://doi.org/10.1007/978-3-319-74500-8_64

will be the illiterate adults of tomorrow. However, the World Monitoring Report on Education [22] shows that 81 per cent of pupils at the beginning of the primary cycle do not reach the minimum level in reading skills. Several international reports [21, 22] and rankings focus on the alarming situation of the Moroccan educational system and the worrying level of illiterate people in Morocco.

The work presented in this article is part of a global approach to help improve the Moroccan educational system (and even elsewhere) on the basis of scientific and pedagogical methods to provide concrete and effective solutions to improve learning skills in primary and for an illiterate population.

The article is structured around a section that presents the problem of primary education and recalls the pedagogical most used reading learning methods [13]. It then puts in application one of the MCDM/A methods of decision making, namely the AHP method.

This method is explained according to a search for classification of the 4 major methods used for learning to read (Syllabic, Global, Mixed and Montessori). This implementation begins by specifying the criteria and sub-criteria that relate to the learning of reading by a young or illiterate public. It continues with the treatment followed to arrive at an outcome that makes it possible to choose the most appropriate and efficient method of teaching. These analyzes are the result of some evaluations carried out at the primary level, nationally and internationally, in order to gauge the acquisition of basic skills in Reading and Mathematics. However, the application of the theoretical approach illustrated by this article will allow to frame and formulate a project of its experimentation in a school environment and also with the associations working in literacy.

2   Presentation of the Learning Context

2.1   Problematic

In many countries, mainly in Africa, the majority of pupils reach the end of primary school without achieving the most basic goals in reading, writing and arithmetic [6].

According to the report by the UNESCO [20] "… recent analyzes show that, in 21 out of 85 countries for which all the data are available, less than the half of the children learn only the basics. Seventeen are located in sub-Saharan Africa; the others are India, Mauritania, Morocco and Pakistan …".

This makes Morocco ranked among the 21 worst education systems in the world [20], and shows that our country is experiencing a major crisis in its primary learning system.

According to the Human Development Index (HDI) [5], Morocco is ranked in the bottom ranks [4]; 124th (out of 177) and the 11th among the 14 Arab countries.

2.2   Methods Used in Developing Reading Skills in the World

This article focuses on the development of the reading skills that is part of the process research of authors' work to make their contribution to the improvement of learning especially for the primary cycle.

The authors' investigations in the literature [1, 5, 10, 11, 13, 14], have revealed several pedagogical methods used in the development of the reading skills: The **Syllabic**, the **Whole language**, the **combination** of Phonics and Whole Language, the **Montessori**, the phonics, the Boscher, the natural, the Ideo-Visual, the interactive, the Borel-Maisonny phonetic and gestural, the phonomimic method, the Alphas method, the Jean-qui-rit method, the Bordesoules.

For the learning of reading, two opposite major theoretical currents stand out: the first one use a synthetic method which privileges an ascending approach going from the visual decoding to the production of meaning; the second is based on an analytical method which favors reasoning in a downward flow of information where the prior knowledge of the subject predominates [13].

The work presented here focuses on four main methods: syllabic and global methods (first current), mixed and Montessori methods (second current). The aforementioned methods constitute variants of the four.

3   Application of MCDM/A Methods for the Choice of the Appropriate Learning Method

The multi-criteria decision-making (MCDM) domain is divided into two sub-domains, Multi-Attribute Decision Making (MADM) and Multi-Objective Decision Making (MODM):

- MADM: selection of the best alternative in a finite set and predetermined set of alternatives [15];
- MODM: Selection of the best action in a continuous or discrete decision space [3]. The Multi-objective optimization is a branch of MODM [17].

Multi-criteria Methods: There are 3 families of approach to adopt. These approaches can be categorized in terms of aggregation, depending on how to aggregate judgments:

- **Complete aggregation:** a single criterion is distinguished on the basis of n starting criteria. The judgments are transitive (a > b, b > c therefore a > c).
- **Partial aggregation:** the output is an upgrade of the inputs. In this case there is incomparability between the criteria because they are from different origin.
- **Local aggregation:** the output is finding a better solution than the existing one.

The methodological framework for decision-making based on an MCDM method, in order to make a relevant choice among the four pedagogical methods studied, involves a representation and an identification of the criteria and sub-criteria. For this, we use here the Analytic Hierarchy Process method (AHP), which is part of the complete aggregation family.

3.1   General Presentation of the AHP Method

Created by Professor Thomas Saaty [18, 19], AHP is a multicriteria decision-making method that allows us to take into consideration several criteria in order to make the best decision.

When several criteria are considered, decision-making becomes critical. AHP helps in this process by hierarchically structuring the selection criteria. It is one of the strongest and the most used methods in the context of multi-criteria analysis.

This analysis makes it possible to formulate a conception, justification and transformation of preferences during the decision-making process.

The steps to be followed are:

- Development of an hierarchical structuring of the problem;
- Construction of a matrix of judgment;
- Determining a priority vector containing the weights of criteria;
- Study of the consistency of the matrix of judgment;
- Comparative study of alternatives to choose the best one.

Steps of an AHP analysis:
This section is dedicated to the study of the case on which the decision making method AHP has been applied for the choice of the pedagogical method most appropriate to the development of reading skills. Each pedagogical method uses more or less strongly a criterion than the other (to a certain degree).

The hierarchy of criteria reveals 3 major classes that describe how the child perceives the information presented to him, the ways in which he can reinforce his learning and the energy he must put into it:

- The «sensory organ» used by the child to acquire learning: vision (visual), hearing (auditory) and/or touch (Kinesthetic).
- The «acquisition mode» that allows him to consolidate his learning: repetition, intuition and/or understanding.
- The "energy" necessary to his learning: the flow (intensity or speed) and the effort (force) provided.

Figure 1 presents the hierarchy of criteria and sub-criteria [7] used to evaluate the pedagogical methods using the AHP method.

Sensory organ: this criterion has 3 sub-criteria:

- **Visual:** photographic memorization of the forms of words and/or sentences,
- **Auditory:** phonic correspondence following a logical progression of the letters' decryption,
- **Kinesthetic:** gestural mechanics that allows the child to feel what he reads to better remember.

Acquisition mode: this criterion includes 3 sub-criteria:

- **Repetition:** a training session during which the child repeats the same letters or repeats the same gestures,

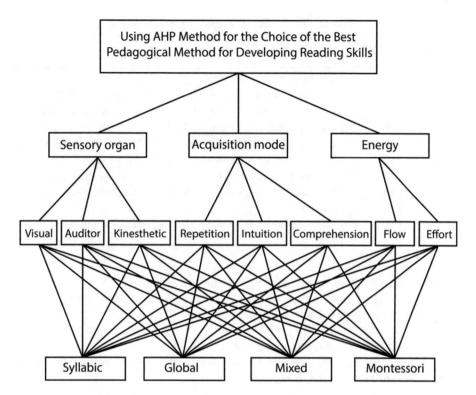

Fig. 1. Hierarchy of criteria and sub-criteria, using the AHP method, for evaluating a reading pedagogical method

- **Understanding:** the child has a mental representation attached to the word he is about to read,
- **Intuition:** in the context where there is emission of hypothesis or dependency of the context.

Energy: this criterion incorporates 2 sub-criteria:

- **Flow** (speed or **intensity**) of learning,
- **Mental and physical effort** provided by the child to assimilate the presented text.

Development of the Architecture of the AHP Method. Using the AHP method [8, 9] will allow us to determine, for each pedagogical method, the weight of use of the criteria and sub-criteria.

The step following the development of our architecture enables to construct the judgment matrix. For this we need to set the preference table that the Expert will use to define his preferences for each pair of criteria and sub-criteria.

These preferences expressed verbally or qualitatively will be numerically quantified according to the grid illustrated in Table 1. This is the fundamental scale of absolute value proposed by Professor Saaty [19].

Table 1. The fundamental scale of absolute value

Digital scale	Reciprocal	Linguistic scale
1	1	The 2 criteria are equivalent
2	1/2	The first criterion is almost equivalent to the second criterion
3	1/3	The first criterion is slightly important that the second criterion
4	1/4	The first criterion is moderately important than the second criterion
5	1/5	The 1st criterion is generally more important than the 2nd criterion
6	1/6	The first criterion is more important than the second criterion
7	1/7	The first criterion is much more important than the second criterion
8	1/8	The first criterion is extremely more important than the second criterion
9	1/9	The first criterion is infinitely more important than the second criterion

Construction of the Judgment Matrix. We present in this section the initial judgment matrices of the main criteria of the four selected pedagogical methods: the Syllabic, the global, the mixed and the Montessori.

All the data presented below are taken from the literature according to the results of experts of the different pedagogical methods [5, 10, 13].

The judgments are expressed according to wi/wj ratio which indicates the importance of the attribute 'i' with respect to 'j'.

$$
A = \begin{pmatrix} c_{11} & c_{12} & \cdots & \cdots & c_{1n} \\ c_{21} & c_{22} & \cdots & \cdots & \cdots \\ \cdots & \cdots & \ddots & \cdots & \cdots \\ \cdots & \cdots & \cdots & \ddots & \cdots \\ c_{n1} & \cdots & \cdots & \cdots & c_{nn} \end{pmatrix} = \left(c_{ij} \right)_{1 \le i,j \le n} \tag{1}
$$

The judgment matrix is a square matrix whose weights are w1, w2, ... The weights of the attributes are measured according to:

$$
c_{ij} = \frac{w_i}{w_j} \forall\, i,j = 1,\ldots,n \tag{2}
$$

Where: $c_{ij} = 1 \forall\, j = i$ and $c_{ij} = 1/c_{ji} \forall\, i,j$

The Tables 2, 3, 4 and 5 respectively present the judgment matrices including the calculation of the weights of the three criteria for the Syllabic, Global, Mixed and Montessori methods.

Let note that the syllabic method favors the 2nd criterion "Acquisition mode" with a weight of 0.723; the Global method privileges the 2nd and 3rd criterion with a weight of 0.428; and the Mixed and the Montessori methods put on the same pedestal the 1st and 3rd criterion with a weight equal to 0.428.

The following tables (Tables 2, 3, 4, and 5) represent the judgment matrices of the criteria according to preferences of each expert of the four pedagogical methods.[1]

Table 2. The 1st degree judgment matrix of criteria for the Syllabic method

Criterion/Criterion	Sensory organ	Acquisition mode	Energy	W
Sensory organ	1	1/5	3	0.193
Acquisition mode	5	1	7	0.723
Energy	1/3	1/7	1	0.083

Table 3. The 1st degree judgment matrix of criteria for the Global method

Criterion/Criterion	Sensory organ	Acquisition mode	Energy	W
Sensory organ	1	1/3	1/3	0.143
Acquisition mode	3	1	1	0.428
Energy	3	1	1	0.428

Table 4. The 1st degree judgment matrix of criteria for the mixed method

Criterion	Sensory organ	Acquisition mode	Energy	W
Sensory organ	1	3	1	0.428
Acquisition mode	1/3	1	1/3	0.143
Energy	1	3	1	0.428

Table 5. The 1st degree judgment matrix of criteria for the Montessori method

Criterion	Sensory organ	Acquisition mode	Energy	W
Sensory organ	1	1	4	0.458
Acquisition mode	1	1	3	0.416
Energy	¼	1/3	1	0.126

This matrix points out the fact that the 2nd criterion is infinitely more important than the 1st or 3rd criteria.

Determination of the Priority Vector. A normalized comparison matrix, such that the sum of the columns is equal to 1, is established in order to determine the relative weight of each criterion.

---

[1] The data used are derived from a literature investigation on the basis of coherence.

The weights of the attributes are measured with respect to each other according to the Eq. (3).

$$
w = \begin{pmatrix} w_1 \\ w_2 \\ \ddots \\ \ddots \\ w_n \end{pmatrix}
\tag{3}
$$

By using the above equations, the weights of the sub-criteria are calculated as presented in the following tables from Tables 6 7, 8, 9, 10, 11, 12, 13, 14, 15, 16 and 17.

Table 6. Weights of sub-criteria of the criterion Sensory organ for the syllabic method

Sensory organ			
Sub-criterion	Visual	Auditory	Kinesthetic
Weight of sub-criterion	0.267	0.669	0.064

Table 7. Weights of sub-criteria of the criterion Acquisition mode for the syllabic method

Acquisition mode			
Sub-criterion	Repetition	Intuition	Comprehension
Weight of sub-criterion	0.776	0.068	0.155

Table 8. Weights of sub-criteria of the criterion Energy for the syllabic method

Energy		
Sub-criterion	Flow	Effort
Weight of sub-criterion	0.5	0.5

Table 9. Weights of sub-criteria of the criterion Sensory organ for the Global method

Sensory organ			
Sub-criterion	Visual	Auditory	Kinesthetic
Weight of sub-criterion	0.723	0.193	0.083

Table 10. Weights of sub-criteria of the criterion Acquisition mode for the Global method

Acquisition mode			
Sub-criterion	Repetition	Intuition	Comprehension
Weight of sub-criterion	0.091	0.454	0.454

Table 11. Weights of sub-criteria of the criterion Energy for the Global method

Energy		
Sub-criterion	Flow	Effort
Weight of sub-criterion	0.5	0.5

Table 12. Weights of sub-criteria of the criterion Sensory organ for the Mixed method

Sensory organ			
Sub-criterion	Visual	Auditory	Kinesthetic
Weight of sub-criterion	0. 58	0. 349	0. 07

Table 13. Weights of sub-criteria of the criterion Acquisition mode for the Mixed method

Acquisition mode			
Sub-criterion	Repetition	Intuition	Comprehension
Weight of sub-criterion	0. 25	0. 25	0. 5

Table 14. Weights of sub-criteria of the criterion Energy for the Mixed method

Energy		
Sub-criterion	Flow	Effort
Weight of sub-criterion	0.334	0.666

Table 15. Weights of sub-criteria of the criterion Sensory organ for the Montessori method

Sensory organ			
Sub-criterion	Visual	Auditory	Kinesthetic
Weight of sub-criterion	0.346	0.11	0.544

Table 16. Weights of sub-criteria of the criterion Acquisition mode for the Montessori method

Acquisition mode			
Sub-criterion	Repetition	Intuition	Comprehension
Weight of sub-criterion	0.4	0.4	0.2

Table 17. Weights of sub-criteria of the criterion Energy for the Montessori method

Energy		
Sub-criterion	Flow	Effort
Weight of sub-criterion	0.5	0.5

Study of the Coherence. First, we have the Random Index (RI) which varies according to the number of criteria as the Table 18 is showing [16]:

Table 18. The Random Index (RI) [16]

n	1	2	3	4	5	6	7	8	9	10
RI	0	0	0.58	0.9	1.12	1.24	1.32	1.41	1.45	1.49

Next, the consistency indicator is calculated by the equation:

$$CI = \frac{\lambda \max - n}{n - 1} \tag{4}$$

Where:

$$\lambda_{max} = \sum_{i,j}^{n} \left(\frac{C_{ij}.W_j}{W_j} \right) \tag{5}$$

After the construction of judgment matrices and determination of priority vectors, the consistency of each matrix must be studied [18] by the calculation of the ratio CR, as follows:

To achieve this, a ratio is calculated to reflect the degree of consistency. A ratio more than 0.1 indicates a too high level of inconsistency.

$$CR = \frac{CI}{RI} \tag{6}$$

Table 19. Coherence Ratio for the four pedagogical methods

Method		CR
Syllabic	Criteria: Sensory organ, Acquisition mode, Energy	0.057
	Sub-criteria of criterion Sensory organ	0.025
	Sub-criteria of criterion Acquisition mode	0.071
	Sub-criteria of criterion Energy	0
Global	Criteria: Sensory organ, Acquisition mode, Energy	0
	Sub-criteria of criterion Sensory organ	0.057
	Sub-criteria of criterion Acquisition mode	0
	Sub-criteria of criterion Energy	0
Mixed	Criteria: Sensory organ, Acquisition mode, Energy	0
	Sub-criteria of criterion Sensory organ	0.028
	Sub-criteria of criterion Acquisition mode	0
	Sub-criteria of criterion Energy	0
Montessori	Criteria: Sensory organ, Acquisition mode, Energy	0.008
	Sub-criteria of criterion Sensory organ	0.046
	Sub-criteria of criterion Acquisition mode	0
	Sub-criteria of criterion Energy	0

In our case n = 3. By applying Eqs. (4), (5) and (6) we calculated the coherence ratio and we obtained, for the four methods a CR lower than 10% as shown in the Table 19. This means that the initial judgment matrix is coherent. Otherwise, a re-evaluation of the judgment matrix is necessary.

3.2   Classification of the Four Studied Pedagogical Methods

After checking the consistency, we have used the results of Tables (from Tables 6, 7, 8, 9, 10, 11, 12, 13, 14, 15, 16 and 17), which are shown graphically in Fig. 2.

The graph can be interpreted as follows:

The syllabic method takes more time and it is based on the auditory. It is very precise and it uses the deciphering principle.

The Global method is based on the development of intuition and it is interested in understanding the texts read. It requires more effort but it is fast and it introduces the learner to the world of reading by making him appreciate it.

The mixed method is based on the visual and auditory senses. It allows the development of the basic reading of the learner and motivates him to read because it promotes understanding.

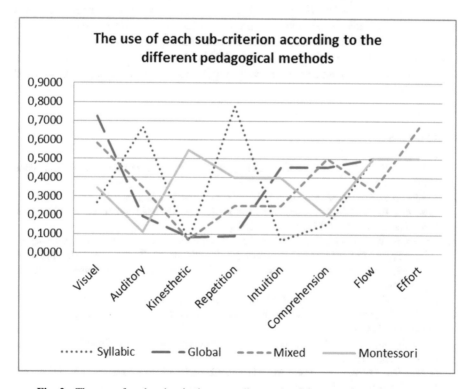

Fig. 2. The use of each sub-criterion according to the different pedagogical methods

The Montessori method is a multi-sensory reading method. It is based on the learner's desire to learn by combining the different channels and giving him autonomy.

4   Conclusion and Perspectives

The paper proposes to provide some answers to help improve primary education and takes into account the development of the reading skills by identifying the most appropriate pedagogical method to be used during the teaching process. Complementary research is in progress by the authors completes the work presented here, which focuses on other learning (calculation, comprehension, written and oral expression), including the integration of ICT and multimedia in a collaborative learning context. The article describes the application of the decision making method AHP, one of the MCDM/A methods to classify the four major methods used in the reading learning process, namely: Syllabic, Global, Mixed and Montessori.

The result, illustrated by the graph of Fig. 2, confirms the specificity of each of the 4 methods studied. Following the use of another level of AHP application on the basis of the aforementioned values (which will be the subject of a Next publication), the Global method is distinguished from the others.

In order to consolidate the approach taken by the authors, experimentation in a school and associative environment must be carried out. Therefore, the theoretical phase presented here is necessary to build the basis for experimentation in a real world. Discussions with the ministry and some NGOs (non-governmental organization), working in literacy projects, are underway for planning and conducting a tangible and costly evaluation.

References

1. N'Namdi, K.A.: Guide to teaching reading at the primary school level. UNESCO, Paris (2005)
2. Abadzi, H.: Instructional time loss and local-level governance. Prospects **37**, 3–16 (2007)
3. Aissanou, F.: Décisions multicritères dans les réseaux de télécommunication autonomes. Paris: Thèse numéro: 2012TELE0019 (2012)
4. Bougroum, M., Ibourk, A., Löwenthal, P.: La politique d'alphabétisation au Maroc: quel rôle pour le secteur associatif? Mondes en développement, No. 134, pp. 63–77, February 2006. https://doi.org/10.3917/med.134.0063,   https://www.cairn.info/revue-mondes-en-developpement-2006-2-page-63.htm
5. Bowman, F.L.: The Influence of Montessori-Based Literacy Instruction and Methods on Reading Achievement of Students in Grades 3, 4, 5, 6, and 7. Seton Hall University: Doctorat of Education (2013)
6. Brighelli, J.-P.: La Fabrique du crétin: la mort programmée de l'école. Jean-Claude Gawsewitch, Paris (2005)
7. Denzin, N., Lincoln, Y.: The SAGE Handbook of Qualitative Research. In: SAGE, éd. 4ème, Los Angeles (2011)

8. Diabagaté, A., Azmani, A., El Harzli, M.: The choice of the best proposal in tendering with AHP method: case of procurement of IT master plan's realization. Int. J. Inf. Technol. Comput. Sci. (IJITCS) **7**(12), 1–11 (2015)
9. El Haji, E., Azmani, A., El Harzli, M.: Using AHP method for educational and vocational guidance. Int. J. Inf. Technol. Comput. Sci. (IJITCS) **9**(12), 9–17 (2017)
10. Gombert, J.-É., Colé, P., Valdois, S., Goigoux, R., Mousty, P., Fayol, M.: Enseigner la lecture au cycle 2. Nathan Université (2000)
11. Grosselin, A.: De l'enseignement de la lecture par la méthode phonomimique. Revue pédagogique **2**(2), 517–530 (1881)
12. Hamouchi, A., Errougui, I., Boulaassass, B.: L'enseignement au Maroc, de l'approche par objectifs à l'approche par compétences: points de vue des enseignantes et enseignants. RADISMA, N°8 (2012)
13. Jamet, E.: Comment lisons-nous? Sci. Hum. (82), 20–25 (1998)
14. Janet, P., Dumas, G.: Journal de psychologie normale et pathologique. Presse universitaires de France-Paris (1926)
15. Lahby, M., Cherkaoui, L., Adib, A.: An intelligent network selection strategy based on MADM methods in heterogeneous networks. Int. J. Wirel. Mob. Netw. **4**(1), 83–96 (2012)
16. Lai, W., Han-lun, L., Qi, L., Jing-yi, C., Yi-jiao, C.: Study and implementation of fire sites planning based on GIS and AHP. Procedia Eng. **11**, 486–495 (2011). 1(1)
17. Sadjadi, S.J., Habibian, M., Khaledi, V.: A multi-objective decision making approach for solving quadratic multiple response surface problems. Math. Sci. **3**(32), 1595–1606 (2008)
18. Saaty, R.W.: The analytic hierarchy process—what it is and how it is used. Math. Model. **9** (3), 161–176 (1987)
19. Saaty, T.: Decision making with the analytic hierarchy process. Serv. Sci. **1**(1), 83–98 (2008)
20. UNESCO: Enseigner et apprendre: Atteindre la qualité pour tous. UNESCO, Paris, April 2013
21. UNESCO: Le rapport mondial de suivi sur l'EPT Education pour tous 2000–2015: Progrès et enjeux. France (2015)
22. UNESCO: Rapport mondial de suivi sur l'éducation. L'éducation pour les peuples et la planète: Créer des avenirs durables pour tous, Paris (2016)

Learners' Motivation Analysis
in Serious Games

Othman Bakkali Yedri(✉) ⓘ, Lotfi El Aachak, Amine Belahbib,
Hassan Zili, and Mohammed Bouhorma

Computer Science, Systems and Telecommunication Laboratory (LIST),
Faculty of Sciences and Technologies,
University Abdelmalek Essaadi, Tangier, Morocco
othmanbakkali@gmail.com, lotfil002@gmail.com,
xakiru@gmail.com, hassan.zili@gmail.com,
bouhorma@gmail.com

Abstract. Compared to traditional learning, serious games have a huge advantage in promoting learners motivation and positive feelings. Despite advantages and efforts invested by researchers to ensure the continuity of learning through serious games, many studies show that learners are able to abandon the experience in complete freedom, without achieving learning objectives. However, analyzing the motivational factors by maintaining a synergy between motivation and learning is the main key of success. In this paper, we will study in the first place similar works. Then we will present our motivational analysis approach based on a combination of several machine learning algorithms and learning analytics methods. Finally, a detailed discussion with the analysis of our obtained results will conclude the paper.

Keywords: Serious game · Learning outcomes · Game play · Experience
Adaptability · Game based learning · Service oriented architecture
Motivation · Data analysis · Expectation Maximization

1 Introduction

The term "serious games" has only risen to prominence in the last decade and also recently became accepted as an academic research topic [1]. However, the origin of the term can be traced back to Clark Abt who defines serious games as:

'We are concerned with serious games in the sense that these games have an explicit and careful thought-out educational purpose and are not intended to be played primarily for amusement [2]'.

Serious games as each interactive tools that let students practice, analyze their interactions, give feedback on learners actions, and help them to make progress [3]. Due to this variety of advantages, a growing number of professionals are looking for learning interactive tools to improve motivation in educational solutions [4]. However, it is almost universally accepted that there is a positive correlation between motivation and learning [5]. Indeed, motivation has a huge role in learning, and thus it is important that learning environment provides the learners with an appropriate level of challenge,

© Springer International Publishing AG, part of Springer Nature 2018
M. Ben Ahmed and A. A. Boudhir (Eds.): SCAMS 2017, LNNS 37, pp. 710–723, 2018.
https://doi.org/10.1007/978-3-319-74500-8_65

always balancing on limits of learners competences and skills [6]. Learners motivation have a significant role in an accomplishment learning process and a successful game for educational purpose [7].

Despite of their effectiveness in the term of learning [1, 8], professionals style suffering from students learning left due to a number of reasons which can be related to motivation, capacity and level of learning which differs from a learner to another. In this paper, we study the existing motivational models and their correlation with serious games. Then, we propose an approach to measure the learner motivation through the game and help the professionals to attract more learners' attention. Afterwards, we present an implementation of the proposal algorithm in a waste sorting serious game, which aims to teach kids how to recycle different waste. Finally, we conclude by an evaluation and discussion of how emotional state model proved successful along with an outlook on future research.

2 Theoretical Background

2.1 Related Works

At the beginning of century, a growing number of reformers are looking to computer and video games to improve motivation in educational settings [7, 9–12]. For example, Dewey, American philosopher, noted that interest and pleasure were powerful motivational forces, with regard to the specific problem of learning, asserted that people would be totally absorbed in the learning material only if they were sufficiently interested in learn [13]. Thus, serious games may enhance learning, not because of something inherent to video games, but because well-designed games utilize mechanics that increase engagement and motivation in their learners [14]. However, it is important to distinguish between engagement and motivation, although they are similar constructs, as a participant could be motivated to play a game, but if the game no longer offers adequate challenge, they may not be engaged by the game, potentially reducing future motivation [15].

Engagement is a psychological state experienced during activity that has both affective and cognitive components [16, 17]. In games, engagement comprises concepts of enjoyment, immersion, flow, and presence [18, 19]. Game mechanics that are thought to contribute to engagement include viscerally pleasing stimuli, interactivity/choice, clear goals/mechanics, feedback, novelty/exploration, and adaptive difficulty [11, 12].

The term motivation in psychology is a global concept for a variety of processes and effects whose common core is the realization that an organism selects a particular behavior because of expected consequences, and then implements it with some measure of energy, along a particular path [21]. Motivation is the set of forces/energies that drive an individual to act and/or allow him to control and regulate his behaviors.

Lack of intrinsic motivation = > less engagement

To summarize, serious games have always been considered to be intrinsically motivated, several mechanics employed in this purpose; the challenge, curiosity, control, interaction and simulation. The motivated learners are more engaged and enhanced skills performance; however learners with a negative emotional state appear a boredom behavior on their interaction with technology and decrease learning benefits [22].

2.2 Motivational Methodologies

In our study of motivation field, we find several works focused on the learner motivation in serious game in order to enhance the level of learner motivation related to their interaction with the game.

First, Clark in 2005 has stressed that the motivation in learning is defined as a pyramid in layers such as environmental factors, psychological factors, motivated behavior, knowledge and learning strategy, and performance. Then, James Keller, educational psychologist, ARCS KELLER model at Florida State University devised a motivational model based on a synthesis of existing research on psychological motivation [23]. His ARCS is an acronym that represents these four classes: Attention, Relevance, Confidence/Challenge, and Satisfaction/Success. Finally, Malone [24] goes much further because it provides a conceptual framework for explaining the four conditions that must be met for a video educational game is intrinsically motivating. For him, all the ingredients for good video games are a challenge, curiosity, control and fantasy.

To conclude, Malone model shows the main interest to summarize in a coherent and compact disparate elements belonging to various motivational theories. This method was also used by Keller [23] as part of another motivational model applied also to edutainment software. Although this meta-categories Keller (ARCS care for (A), relevance (R), confidence (C), and satisfaction (S)) are different from those that offer Malone, both designs are based on many common references in terms of motivational theories. However, unlike Malone, Keller goal is what he calls the motivational design, that is to say it provides more of a method of design and development.

2.3 Perception of Flow Condition

Flow theory [25] is based on a symbiotic experience between challenges and the skills that need to be implemented to address them. However, flow experience arises when skills are not overtaken or underused, when the challenge is optimal. When the individual plunges into the flow, the involvement in the activity is such that he forgets time, fatigue and everything around him except the activity itself. In this state, the individual operates to the maximum of his abilities and for the flow experience. The activity is performed for itself (as defined in the context of intrinsic motivation), even if the goal is not yet attaint (Fig. 1).

The calculation of state of flow after each interaction to adapt the game based on the mental state of a learner in order to achieve an optimal state of immersion to attract the learner to complete the experience.

Four symptoms of a task producing a flow state:

The loss of the ego: the concentration is only on the task to be accomplished, the primary physiological needs such as eating, drinking or sleeping are no longer perceived.

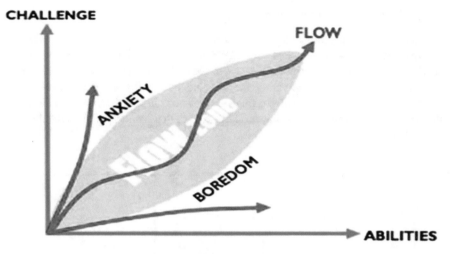

Fig. 1. Flow state [26]

Optimal concentration: the person is almost hypnotized by the task.

The alteration of the perception of time: the individual is no longer aware of the time that elapses while he performs his task.

Instant feedback: the person observes the difficulties and at the same time seeks solutions. Feedback is the reaction of learner about the result of the game.

The literature helps us to apply some concept as presented in the next section, in order to evaluate the impact of adaptation of the learning process through serious games.

3 Case of Study

The main objective of this current paper is to analyze the emotional state of learning across serious games according to Difficulty, learning objectives and learner competence during the game. The description of the proposed serious game and the establishment of the hall system will be described in this section.

Waste sorting serious game [27] is to teach kids how to recycle different waste. The learner should sort different waste into different trash, according to their types e.g. "paper, plastic, metal, glass, and organic, etc." The sorting is done by catching different objects generated randomly and dropping them in the appropriate container according to their types, this mechanism will be done by using either a mouse or an input device called leap motion.

The waste sorting serious game will be equipped by the timer, and the assessment system that evaluates the learners according to their performances; if they make a good choice the reward will be the gain of some points, although however in opposite case the punishment will be the loss of some points. With the assessment system, the timer, and the interactivity based on hand movement the proposed serious game will be more challenging and attractive especially for kids, it will allow them to live a beneficial and unforgettable experience.

3.1   The Proposed Architecture

The proposed architecture Fig. 2 is based on service oriented architecture, where the deployed serious games invokes several services e.g. "Motivational service, analysis service, etc." by the JavaScript API, each service has a specific task in order to improve the learning process by motivating the learning during the game sequence, all the learners behavior during the game sequence are saved into the database. In addition, all the services are developed by using Axis framework that facilitates the establishment of the web service by coding java classes.

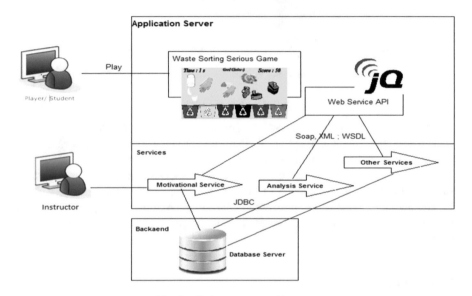

Fig. 2. Serious game architecture

This architecture can be reused, in other realization by invoking those services, independently of logic or technology used.

Motivational service has specific inputs (Difficulty Double; Competence Double; X Double; Y Double) and does a specific task, in order, to motivate the learner during the game.

> Input: Difficulty; Competence; X; Y
> Processing: Y= 1/2X-2 <= Flow <=Y= 1/2X+2
> Output: 0; 1; 2
> The result is resumed in the emotional state of the learner:
> 0 : The learner is in the flow state
> 1 : The learner is in the state of anxiety
> 2 : The learner is in the state of boredom

Analysis service has as a purpose to classify collected data into a cluster by using Expectation-Maximization [28] algorithm and the data will be analyzed with a decision tree by using C4.5 algorithm. Data will be saved into the database, as shown in the Table 1, these data will be operated also by the tools of learning analytics and educational data mining, to have a global view on the progression of all the learners.

Table 1. Database attributes

Knowledge	
✓	Id_session
✓	Name
✓	Age
✓	Sexe
✓	Average of response time
✓	Number of wrong answers
✓	Score
✓	Question
✓	Learning objectives
✓	Difficulty
✓	Learner Competence

3.2   Decision Tree

The current paper focused on the implementation of interest service as proof of concept of the global service. It is based on decision tree one of the learning machine algorithms

The proposed decision tree uses the C4.5 algorithm as learning algorithm, it is an algorithm used to generate a decision tree developed by Ross Quinlan often referred to statistical classifier [29]. It will be feed by the motivational output; in the next section, the result of such service will be discussed.

The decision tree is formalism for expressing such mappings and consists of nodes linked to several sub-trees and leafs or decision nodes labeled with a class which means the decision.

The C4.5 [29] is the learning algorithm that will generate the three according to the data saved in the database; the proposed algorithm is an extension of ID3 algorithm; it builds decision trees from a set of training data in the same way as ID3 by using the concept of information entropy (1) and The splitting criterion is the normalized information gain (2), the algorithm of C4.5 for building the decision tree is described below:

$$H(S) = -\sum_{x \in X} p(x) log_2 p(x) \qquad (1)$$

Where:

- S: The current (data) set for which entropy is being calculated.
- X: Set of classes in S.

- p (x): The proportion of the number of elements in class x to the number of elements in set S.

$$IG(A, S) = H(S) - \sum_{t \in T} p(t)H(t) \tag{2}$$

- H(S) Entropy of set S.
- T The subsets created from splitting set S by attribute A such that $S = U\ t \in T$.
- P (t) the proportion of the number of elements in to the number of elements in set S.
      H (t) Entropy of subset t.

Below the algorithm of C4.5:

- Check for base cases
- For each attribute a:
- Find the normalized information gain ratio from splitting on a.
- Let a_best be the attribute with the highest normalized information gain.
- Create a decision node that splits on a_best.
- Recur on the sub lists obtained by splitting on a_best, and add those nodes as children of node.

The Fig. 3 presents an example of the decision tree generated dynamically from database. The attributes "Inputs" that feed the decision tree are: difficulty, competence, state, id_session. By cons the classes "Outputs" are: Flow, Boredom and Anxiety.

The resulting decision tree provides the learner emotional state Flow, Anxiety and Boredom as the potential interest attributes of study.

Following the current analysis, the motivational services provides the employment of decision tree to help the definition of the learner emotional state based on data set of 96 students, but it requires more data set of play to propose good decisions.

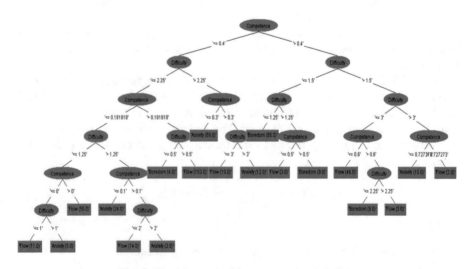

Fig. 3. Decision tree of learner emotional feeling

The establishment of such service by using several technologies/frameworks based on web service, during the development process has allowed a big flexibility to improve the approach and use other motivational factors as to progress of development. As known the service will be reused regardless to the technologies or the language used to develop such video games, therefore this solution will be generated to many other realizations by invoking the proposed services.

3.3   Clustering

The clustering is the task of grouping a set of elements in such a way that elements in the same group called a cluster. A Cluster is a collection of objects which are similar between them and are dissimilar to the objects belonging to other clusters.

Among the clustering algorithms there is Expectation Maximization (EM) algorithm; it's an iterative method for finding maximum likelihood or maximum a posteriori estimates of parameters in statistical models [20].

The EM iteration alternates between performing an expectation (E) step, which creates a function for the expectation of the log-likelihood evaluated using the current estimate for the parameters, and a maximization (M) step, which computes parameters maximizing the expected log likelihood found on the E step. These parameter-estimates are then used to determine the distribution of the latent variables in the next E step. The EM iteration consists of two steps expectation (E) step and maximization (M) steps.

The Expectation (E) Step: Each object assign to clusters with the center that is closest to the object. Assignment of object should be belonging to closest cluster.

The Maximization (M) step: For given cluster assignment, for each cluster algorithm adjusts the center so that, the sum of the distance from object and new center is minimized.

The obtained results will be interpreted and discussed in the next section, in order to evaluate the impact of the adaptation of the learning process through serious games.

4   Result and Discussion

The aim of this part is to analyze and predict the emotional state of the learner in an experiment in order to ensure optimal emotional state. However, we have seen the flow which is defined as an optimal psychological state that can be felt in various domains, particularly during an experiment and manifested during the perception of a balance between his personal skills and the difficulty of the task.

The game development consisted of two major parts:

Part 1: Difficulty generated arbitrary from database
Part 2: Difficulty is recognized by group of students according to the results of the first experience (competence, difficulty)

The current application has been experimented by data set of 96 young students. There are between 14 and 21 years old. They play the game in classroom supervised by their teachers. We find below the distribution of the data generated by learners which helps us to generate the decision tree.

The data obtained after several uses of the proposed game by several learners, as presented below in the Table 2, have been clustered in three categories by using Expectation-Maximization [28] algorithm, an algorithm often used in educational data mining to cluster learner's performances, the criteria used to cluster the learners are "id_session, Competence, right answers, wrong answers, and difficulty", Fig. 4.

Table 2. Database attributes

Level	Nbr student	Correct answers	Wrong answers	Nbr abandons
Non adapted game	96	376	564	18
Adapted game	96	792	168	3

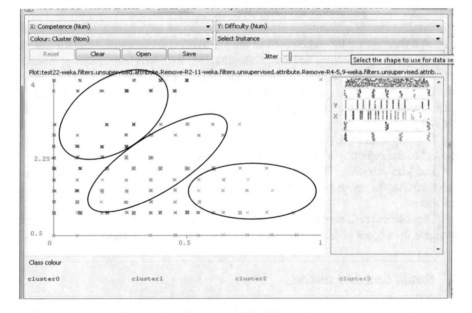

Fig. 4. Learners emotional state

As presented in Fig. 4, there are three categories of learners, clustered according to their emotional state "Flow, Boredom and Anxiety". 54% of them are in the Flow state; 26% are bored, by cons 20% are anxious, according to the given results the majority of the learners are satisfied but the rest have the ability to quit the game according to their feelings as presented in Fig. 5.

Another way to prove the effectiveness of the proposed serious game, and its impact on the learners, is comparing the adapted and the non-adapted game. This comparison is based on finding the difference of learning degree of each step "Before" the group that uses the non-adapted version of game and "After" the group that uses the adapted version. The Table 2 details this comparison.

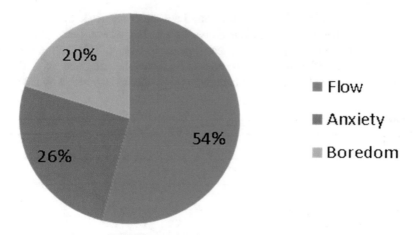

Fig. 5. Pie chart of emotional state before game adaptation

After this study we observed the importance of the adaptation of the game according to the level of competence of each student to keep them in an optimal emotional state during the experiment.

In order to prove the effectiveness of adapting serious games difficulty according to the learner performance, we take a sample of a learner results before and after adapting the game according to skill and difficulty level.

This graph shows that the learner may not finish the game because the questions are easy for him. As a result, the learner will have a better chance to leave the game. However, students who feel excited about playing a video game have a higher tendency to experience flow may not have helped them to learn from the game. After the game regulation difficulty, as presented in Fig. 6, according to the skill of the learner, we obtained the results as shown in the figure below.

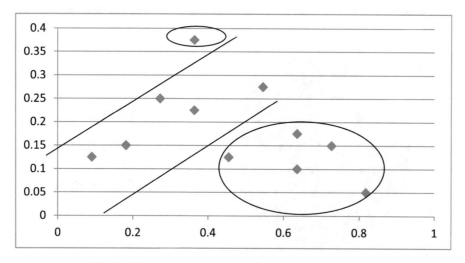

Fig. 6. Distrubtion of learner result before game adaptation

As present in the Fig. 7 there is harmony between competence and question difficulty. The combination of these element, results a feeling of well-being that the mere fact of being able to feel it justifies a great expense of energy. This feeling creates an order in our state of consciousness and strengthens the structure of self. Self-development occurs only when the interaction is experienced as positive by the person.

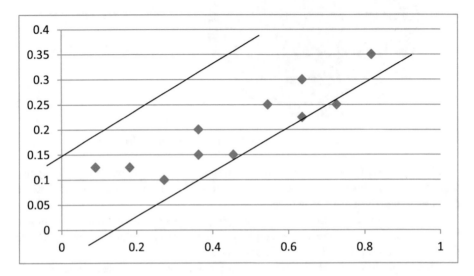

Fig. 7. Distrubtion of learner result after game adaptation

The pie chart Fig. 8 presents the distribution of learner interest according to parameters presented above after adapting the game difficulty according to learner's competence, which they have been designed: F, A, B. We notice that the Anxiety takes 15%, Boredom takes 11%, however the Flow state 74% of the distributions. The

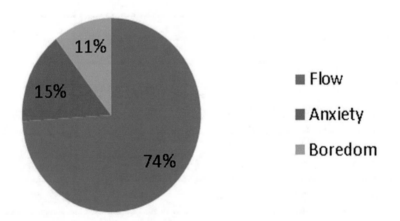

Fig. 8. Pie chart of emotional state after game adaptation

obtained results show that the adapted game is more beneficial for the learners than the non-adapted one.

The attractive field of motivation systems opens the opportunities not only for educational researchers but also for software engineers. The motivational system allows learners to be more immersed across the educational tools especially serious games. As presented below in the decision tree the players or learners are even in the flow emotional state if there is a harmony between competencies and difficulty.

The analysis results based on this study are:

Performance on a knowledge test
Perception of the state of flow during the gambling experience:
The main contribution of this research: A better understanding of Flow in an educational context; Detection of Students at Risk; A new theoretical model linking difficulty and competence; Improvement of the game according to the results of the experiments

The adaptation based on the learners' competence has proved its success, according to the obtained results of both learning analytics or the comparison of the tow version of the proposed game. Results show that students have more facilities to use the game and to progress through out problems after adapting the solution thanks to clustering result.

5 Conclusion

The objective of this paper was to analyze and predict the emotional state of a group based on the flow concept. In the first part, we have seen the motivational methodologies and the flow concept that was initiated by the research of Csikszentmihalyi which defines it as an optimal psychological state that can be felt in various fields and which manifests itself during the perception of a balance between personal competence and the demand of the task. Then, we have seen the case of study with a brief presentation of algorithms. Finally, we concluded with a discussion of obtaining results.

Education using games turns learning more enjoyable by maintaining a high interest. Assessments methodologies are likely to disrupt the interest. Our tendencies are embedding assessments of motivation behavior in serious game. Thus there would be implicit in the flow experience in a game.

The implementation of such service will allow the improvement of the learning process by attract the learner interest, and motivate him to continue learning with the game progression. Thanks to the obtained result the proposed solution will be more interesting with a large scale of data saved on knowledge base. Among the perspectives related to this work is the establishment of a system that will adapt serious games according to the prediction given by the proposed system, in order to increase learners' motivation, and keep them playing for more educational benefits.

References

1. Vermeulen, H., Gain, J.: Experimental methodology for evaluating learning in serious games
2. Abt, C.C.: Serious Games. University Press of America, Lanham (1987)
3. Heeren, B., Jeuring, J.: Feedback services for stepwise exercises. Sci. Comput. Program. **88**, 110–129 (2014)
4. Parsons, J., Taylor, L.: Improving student engagement. Curr. Issues Educ. **14**(1), 1–32 (2011)
5. Bixler, B.: Motivation and its relationship to the design of educational games. NMC Clevel., Ohio, vol. 10, no. 07 (2006). Accessed
6. Kickmeier-Rust, M.D., Albert, D.: Educationally adaptative: balancing serious games. Int. J. Comput. Sci. Sport **11**(1), 15 (2012)
7. Syufagi, M.A., Hariadi, M., Purnomo, M.H.: A motivation behavior classification based on multi objective optimization using learning vector quantization for serious games. Int. J. Comput. Appl. **57**(14), 23–30 (2012)
8. Mortara, M., Catalano, C.E., Fiucci, G., Derntl, M.: Evaluating the effectiveness of serious games for cultural awareness: the icura user study. In: De Gloria, A. (ed.) GALA 2013. LNCS, vol. 8605, pp. 276–289. Springer, Cham (2014). https://doi.org/10.1007/978-3-319-12157-4_22
9. Fenouillet, F., Kaplan, J., Yennek, N.: Serious games et motivation. In: 4eme Conference francophone sur les Environnements Informatiques pour l'Apprentissage Humain (EIAH 2009), vol. Actes de lAtelier Jeux Serieux: conception et usages (2009)
10. Fryer, L.K., Bovee, H.N.: Supporting students' motivation for e-learning: teachers matter on and offline. Internet High. Educ. **30**, 21–29 (2016)
11. Harandi, S.R.: Effects of e-learning on students' motivation. Procedia Soc. Behav. Sci. **181**, 423–430 (2015)
12. Leiker, A.M., Bruzi, A.T., Miller, M.W., Nelson, M., Wegman, R., Lohse, K.R.: The effects of autonomous difficulty selection on engagement, motivation, and learning in a motion-controlled video game task. Hum. Mov. Sci. **49**, 326–335 (2016)
13. Dewey, J.: Interest and Effort in Education. Riverside Press, Cambridge (1913)
14. Hunicke, R., LeBlanc, M., Zubek, R.: MDA: a formal approach to game design and game research. In: Proceedings of the AAAI Workshop on Challenges in Game AI, vol. 4 (2004)
15. Ryan, R.M., Deci, E.L.: Intrinsic and extrinsic motivations: classic definitions and new directions. Contemp. Educ. Psychol. **25**(1), 54–67 (2000)
16. Leiker, A.M., Miller, M., Brewer, L., Nelson, M., Siow, M., Lohse, K.: The relationship between engagement and neurophysiological measures of attention in motion-controlled video games: a randomized controlled trial. JMIR Serious Games **4**(1), e4 (2016)
17. O'Brien, H.L., Toms, E.G.: What is user engagement? a conceptual framework for defining user engagement with technology. J. Am. Soc. Inf. Sci. Technol. **59**(6), 938–955 (2008)
18. Boyle, E.A., Connolly, T.M., Hainey, T., Boyle, J.M.: Engagement in digital entertainment games: a systematic review. Comput. Hum. Behav. **28**(3), 771–780 (2012)
19. Zimmerli, L., Jacky, M., Lünenburger, L., Riener, R., Bolliger, M.: Increasing patient engagement during virtual reality-based motor rehabilitation. Arch. Phys. Med. Rehabil. **94** (9), 1737–1746 (2013)
20. Lohse, K., Shirzad, N., Verster, A., Hodges, N., Van der Loos, H.F.M.: Video games and rehabilitation: using design principles to enhance engagement in physical therapy. J. Neurol. Phys. Ther. **37**(4), 166–175 (2013)
21. Heckhausen, J., Baltes, P.B.: Perceived controllability of expected psychological change across adulthood and old age. J. Gerontol. **46**(4), P165–P173 (1991)

22. Derbali, L., Frasson, C.: assessment of learners' motivation during interactions with serious games: a study of some motivational strategies in food-force. Adv. Hum.-Comput. Interact. **2012**, 1–15 (2012)
23. Keller, J., Suzuki, K.: Learner motivation and E-learning design: a multinationally validated process. J. Educ. Media **29**(3), 229–239 (2004)
24. Malone, T., Lepper, M.R.: making learning fun: a taxonomy of intrinsic motivations for learning. In Snow, R., Farr, M.J. (eds.) Aptitude, Learning, and Instruction, Conative and Affective Process Analyses, vol. 3, Hillsdale, NJ (1987). http://ocw.metu.edu.tr/mod/resource/view.php?id=1311. Accessed 24 Sep 2016, CEIT506-EN
25. Csikszentmihalyi, M., LeFevre, J.: Optimal experience in work and leisure. J. Pers. Soc. Psychol. **56**(5), 815–822 (1989)
26. FlowFig1.png (298 × 246). http://edutechwiki.unige.ch/fmediawiki/images/4/46/FlowFig1.png. Accessed 10 Apr 2017
27. Lotfi, E., Amine, B., Mohammed, B.: Players performances analysis based on educational data mining case of study: interactive waste sorting serious game. Int. J. Comput. Appl. **108** (11), 13–18 (2014)
28. Dempster, A.P., Laird, N.M., Rubin, D.B.: Maximum likelihood from incomplete data via the EM algorithm. J. R. Stat. Soc. Ser. B Methodol. **39**(1), 1–38 (1977)
29. Salzberg, S.L.: C4. 5: Programs for machine learning by j. ross quinlan. morgan kaufmann publishers, inc., 1993. Mach. Learn. **16**(3), 235–240 (1994)

Scoring Candidates in the Adaptive Test

Tarik Hajji[1(✉)] ⓘ, Zakaria Itahriouan[1], Mohammed Ouazzani Jamil[1], and El Miloud Jaara[2]

[1] Private University of Fez (UPF), Fez, Morocco
{hajji,itahriouan,ouazzani}@upf.ac.ma
[2] Mohammed First University Oujda (UMP), Oujda, Morocco
emjaara@yahoo.fr

Abstract. In this paper, we present a methodology able to make an automatic and rapid evaluation of the candidates in adaptive tests. The objective is to evaluate candidate level related to specific skills using minimum number of asked questions. Our proposed approach is based on quadratic algorithm in terms of compatibility. We begin this work with an introduction to the adaptive tests where we present its limitations. The second part is about main algorithms used in this work. In the third part, we present our approach of relevant selecting items. The last part is about results and its analyses.

Keywords: Item response theory · Computing adaptive testing
Quadratic method

1 Introduction

1.1 Adaptive Tests

Adaptive tests have shown multiple advantages in learning domain, firstly, from tester side because he can have back the candidates' answers to improve its item bank. Secondly, the candidate benefits from optimized test according to his scope and his level. However, Adaptive tests has some limitations manifested in methodological constraints [1] (Constraint Programming, dynamic approache and numerical calculation) or theoretical constraints that are particularly delicate (e.g. difficulty to guarantee the absence of item differential functioning).

Current approaches of adaptive tests have a tendency to favor the information maximization (fisher's information, likelihood estimation [2]) provided after each answer according to previous responses. An iterative procedure for selecting items, respecting the area of coverage to evaluate, test validity and fairness of the scores awarded to the candidates. To overcome some of these constraints, we have designed a quadratic algorithm [3, 4] based on some models of item response theory, to improve the calculations and optimize resources.

1.2 Item Response Theory and Quadratic Algorithm

Item Response Theory (IRT) currently acts a central role in the analysis and study of adaptive testing. IRT models are effectively used to estimate the ability of a candidate

© Springer International Publishing AG, part of Springer Nature 2018
M. Ben Ahmed and A. A. Boudhir (Eds.): SCAMS 2017, LNNS 37, pp. 724–731, 2018.
https://doi.org/10.1007/978-3-319-74500-8_66

and to determine estimation parameters of item difficult. These lasts are obtained based on other parameters related to the item and candidate [5]. The IRT can be used in several domains: education, psychometrics, medicine…

Applicable logistical models of IRT to dichotomous items (where possible results are whether false (X = 0) or true (X = 1) [6]) consider that the probability that a candidate gives a good answer depends both on his skill and characteristics of the item. Thus, this probability increases with the skill:

$$P(x = 1|\theta) = c + (1 - c)\frac{1}{1 + e^{d*\alpha(b-\theta)}} \tag{1}$$

Where:

- c >= 0 is the index of guessing the item,
- θ is the skill level of the candidate,
- b is the index of difficulty.
- α the discrimination index (item is more discriminating than the other). It is proportional to the slope of the characteristic curve of the item when the skill level of the candidate is equal to the difficulty level of the item (θ = b);
- It is established that d = −1.7.

One advantage of the IRT is ability to position individuals and items on the same continuum. So if the item is too difficult, it means that the difference (θ - b) is large, and therefore a smaller probability to have a good answer. If this difference is negative, it means that the candidate has a high probability of responding correctly [7]. In general, an adaptive test is composed of four phases:

- The initial phase, in which the first or the items are selected and administered to respondent;
- The test phase, which is to select the next item to get the answer and to update the response pattern and the estimator of skill level.
- The stop phase, which interrupts the adaptive process when the stop criteria selected is satisfied;
- The final phase, which provides the final estimator skill level based on administered items.

There is two main parts in a CAT system: the estimation of the skill and the selection of the next item. The first gives an estimation of the ability of a candidate knowing his answers to the previous items, while the second allows you to select the next item, from the bank of items, based on the skill level.

We have to mention that the constitution of the bank of items is a central element [8]. An items bank is a large collection of available items, including questions already asked. In adaptive testing, we suppose that these items already calibrated by estimating their parameters in advance.

There are currently two main strategies for items selection based on Item Response Theory, which are generally used: the strategy of maximum information and Bayesian strategy. In our algorithm, after estimating the competence of the candidate consider ing its responses to all previous items, the function of information is been evaluated for different items at this level of competence, and we choose as next item the one who gives the following maximum information among the items included in the next step. Whenever the subject gives response to this new item, we reestimate θ and repeat the procedure. At the end of the test, the latest estimation of θ is the score obtained by the subject.

2   Quadratic Algorithm for Selecting the Next Item

In this section, we present a description [9] of a first version of a quadratic algorithm for estimating the skill of each candidate. We assume that we have only one parameter: the difficulty of the item b (Fig. 1).

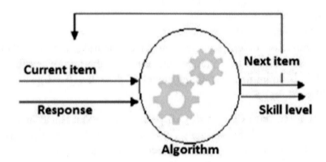

Fig. 1. Principal of algorithm

2.1   Initial Phase

Bank of items: contains 17 items with multiple choice question of 1 to 5, sorted in increasing order of difficulty (between 0 and 1) [10]:

- The first row contains labels item,
- The second row contains the identifier of the item,
- The third row contains the difficulty level of the item,
- The last row contains the correct answer number (Table 1).

The present item shave been extracted from 'Evapmib' database. A mathematics assessment base, online, developed by APMEP (association that represents mathematics teachers from kindergarten to university) in close collaboration with the National Educational Research Institute and Institute of research on mathematics Education [11]. This database is composed of evaluation questions from large or medium scale studies, each one is been provided with a set of descriptors: ID card, results recorded at various reversals, pedagogical and didactic analysis. The candidates are 50 students, with a level of skill already held (Table 2).

Table 1. Items bank

Item ID	Difficulty	Response
17	0,05	4
16	0,15	2
18	0,28	1
15	0,38	3
7	0,49	4
3	0,58	1
19	0,63	1
5	0,64	5
13	0,67	4
1	0,71	3
4	0,75	4
2	0,84	3
12	0,86	5
6	0,89	1
11	0,92	5
14	0,93	3
9	0,97	5

Table 2. Example of student ability

CandidatID	1	2	3	...	50
Ability	0.9	0.55	0.47		0.08

2.2 Test Procedure

The candidate receives the first item from the database of the items. According to his response, the first part calculates his (estimated) level of skill (Fig. 2).

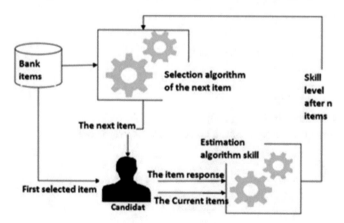

Fig. 2. Test procedure

The second part allows giving the next item, which corresponds to the skill level estimated by the first part. The individual responds to the second item, and so on, until there will be no item to administrate or there is a threshold value reached [7].

2.3 Selecting Items Part

The algorithm that we use in this work is based on a quadratic method for the research of the next question according to the last difficulty level reached:

As we see in the diagram above, once the candidate gives his first response to the first item, we try to give an item with a higher weight (difficulty) than last asked. If the answer is correct, we offer him another item with greater weight, otherwise, a less difficulty level item is proposed. The operation continue until there will be no question to ask which means we reached the max level of the candidate (Fig. 3).

According to this algorithm, candidates will not necessarily receive the same number of items. We present as follow an overview of the algorithm:

Table questions [or [Question, Difficulty, Answer]] ← input
Step 1: Sort the table according to the difficulty (indexed from 1 to n)

Step 2: Initialize the variables of the algorithm
- The increment= 0
- A positive temporary counter k = 1;
- A negative temporary storyteller g = 1;
- An algorithm stop test = true;
- Skill level Theta

Step 3: Find the item with max information (difficulty)
do {
    If response_of_question_number (increment) = true:
    Then
      increment = increment + 2(++k) ;
      test = true ;
      g = 0 ;
    Eles
      increment = increment – 2(++g) ;
      test = true ;
      k=0 ;
    End of if;
  } While(k !=0 && g !=0)
Returns the the level of the last correct answer ➔ output

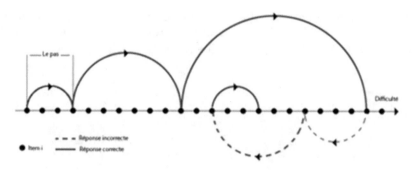

Fig. 3. Selecting items part

3   Results

After all the candidates have completed their tests, we retrieve the results as follows:

- The first line contains the correct answers for each item on line 3.
- The second line contains the levels of difficulty of each item.
- The first column identifies the candidate.
- The Second contains the estimated difficulty level before the test.
- The 17 next columns represent the candidate's answer for each item.
- The last column represents the max difficulty level achieved by the candidate (Fig. 4).

Réponse	4	2	···	1	3	4	1	1	5	4	···	5	3	5	5	2	
	0,05	0,1	···	28	0,38	0,49	0,58	0,63	0,64	0,6	···	,92	0,93	0,97	0,98	0,99	
ID	**Habilité**	i17	i1	···	i8	i15	i7	i3	i19	i5	i1!	···i11	i14	i9	i8	i10	**Habilité**
1	0,9	4	-	···	1	3	-	-	1	5	4	··· 5	2				0,92
2	0,55	4	-	···	1	3	4	2				···					0,49
3	0,47	4	-	···	.	3	4	2				···					0,49
4	0,33	4	-	···	1	1						···					0,28
5	0,54	4	2	···	.	3	4	2				···					0,49
6	0,26	4	-	···	1	1						···					0,28
7	0,62	4	2	···	.	-	4	1	1	4		···					0,63
8	0,62	4	2	···	1	3	4	1	1	3		···					0,63
9	0,63	4	2	···	.	-	4	-	1	2		···					0,63
10	0,64	4	2	···	.	3	.	1	.	5	3	···					0,64
11	0,87	4	-	···	1	3	-	1	1	5	-	···					0,86
12	0,84	4	-	···	.	3	4	-	-	-	4	···					0,84
13	0,81	4	2	···	.	3	-	1	-	5	-	···					0,75
14	0,58	4	-	···	1	-	4	2									0,49

Fig. 4. Extract from the bank of knowledge

Comparing the estimated skill level (based candidates) and those returned by the algorithm, there is a slight difference between the two results, as we see in Fig. 5(the blue is the estimated skill and orange is the value after the test).

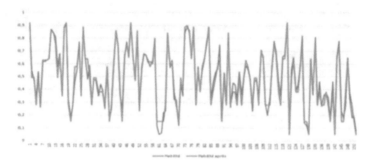

Fig. 5. Estimated skill and calculated skill

4   Conclusion and Perspectives

In this article, we have realized an algorithm of items selection to ask a candidate in an adaptive test. The proposed algorithm is newly based on quadratic approach considering only one parameter (the difficulty of each item). In order to have a better result, we will develop a new module to administer three bases of items with the same algorithm and calculate the standard deviation for each candidate. The smaller this value is, the more our algorithm is reliable. Theoretically, the algorithm is capable of handling a large number of items.

As perspective to this work, we consider developing another research where the objective is to make a faster and optimal algorithm. The next algorithm will be based on the following principles:

- Modeling the items base: Increasing the number of items in the used databases, instead of working with a single parameter. We will add other criteria related to the item in addition to the difficulty (discrimination, index guessing …).
- Improve the algorithm to administer any kind of questions.
- Using neural networks [12] for the selection of the next administered item, maximization of the information after a response to the current item and the estimation of the skill.
- Add a module to avoid administering easy items several times.

References

1. Rudner, L.M.: Online computer adaptive testing (CAT). http://echo.Edres.org:8080/scripts/cat/catdemo.htm
2. Johnson, M.S.: Marginal Maximum Likelihood estimation of item response models in

3. Jong, M.G.D., Steenkamp, J.E.M., Fox, J.P., Baumgartner, H.: Using item response theory to measure extreme response style in marketing research. J. Mark. Res. **45**(1), 104–115 (2006)
4. Popov, D.I.: Adaptive testing algorithm based on fuzzy logic. Int. J. Adv. Stud. **3**(4), 23–27 (2013)
5. Young, M.: The Technical Writer's Handbook. University Science, Mill Valley (1989)
6. Rizopoulos, D.: Item Response Theory in R using Package LTM. Department of Statistics and Mathematics, WU WirtschaftsuniversitÄat Wien (2010)
7. Kousholt, K., Hamre, B.: The students should be as clever as they can – the Danish case – similarities in the Danish School Reform and computer adaptive testing (CAT). Oxford Studies in Comparative Education (2015)
8. Baker, F.B., Kim, S.-H.: Item Response Theory: Parameter Estimation Techniques, 2nd edn. (2004)
9. Popov, D.I., Demidov, D.G., Denisov, D.A.: Adaptive method of knowledge evaluation during employee's validation (2014)
10. Photis NOBELIS. http://nobelis.eu/photis/Estimat/Proprietes/information_fisher.html
11. EVAPMIB. http://evapmib.apmep.fr/siteEvapmib/
12. Devouche, E.: Les banques d'items. Construction d'une banque pour le Test de Connais-sance du Français (2003)

Learning Management System
and the Underlying Learning Theories

Mohammed Ouadoud[1(✉)] ⓘ, Amel Nejjari[2(✉)],
Mohamed Yassin Chkouri[2(✉)], and Kamal Eddine El-Kadiri[1(✉)]

[1] LIROSA Lab, Faculty of Sciences, UAE, Tetouan, Morocco
mohammed.ouadoud@gmail.com
[2] SIGL Lab, National School of Applied Sciences, UAE, Tetouan, Morocco

Abstract. The problems of most Learning Management Systems (LMS) are first of all of a pedagogical nature and then of a technical one. Studying these problems, which are interrelated, provides a useful conceptual reference that enables us to design a new model for a more relevant solution. In this paper, a conceptual model of an LMS is presented, based on the hybridization between four learning theories, namely the traditional pedagogy, the behaviorism, the cognitivist, and the social constructivism. We will present at first each of these learning theories by discussing both their advantages and limits. Then, together the main principles of these learning theories and the technical functionalities of the proposed LMS that result from the hybridization of these principles are outlined, fit the needs of their final users, in particular learners.

Keywords: Learning management system · Learning theories
Conceptual model · The modeling of LMS

1 Introduction

During the last decade, e-learning platforms have evolved considerably. However, a number of comparative studies [7, 13, 17, 22, 23, 26, 29, 30, 33] have shown that their life cycle continue to change at a fast pace. Therefore, we have conducted a comparative and analytic study on free e-learning platforms based on our own approach of evaluating the e-learning platforms quality [1, 2, 11, 36]. Our main objective was to provide a useful tool that can help educational institutions to make the right and best choice among the available e-learning platforms. Different approaches of evaluating the e-learning platforms quality have been already proposed [24, 25], but no one of them has been adopted here because they focus only on technical aspects and neglect other important aspects such as security, maintainability, portability, compatibility, performance efficiency and usability.

In light of our studies and the previous ones we think that most of e-learning platforms including the LMS were initially developed a decade ago, based on a classical training model. The teacher is considered as the one who holds the knowledge and transmits it, according to different modalities, to future learners in order to foster their learning. They are mostly TMS[1], that is to say tools at the teacher's service to create and manage courses rather than at the service of learners and the learning process. Therefore, we have decided to work on a new conceptual model that combines between learning theories in order to promote both the teaching and learning processes.

The proposed LMS which results from our new conceptual model and which we plan to implement at a later stage, will be based on collaborative learning. Both teachers and learners are able to create, organize and propose different types of activities (forum, wiki, blog…) as they like. Furthermore, they are able to access and manage their interactions via these activities where and when they want according to their needs and objectives in terms of learning.

Our LMS will be built on the idea that we should give the same possibilities of action to both teachers and learners by distributing their control on the platform.

It is evident that the use of any tool in the field of education must be justified according to its pedagogical support and its capacity to address the real needs of its final users, particularly learners. However, it should be noted that although seen as an effective solution for overcoming space-time restrictions, the platforms might be an obstacle for the learning process to the extent that the pedagogical principles are neglected during their design. Thus, when designing our LMS we have tried to answer at first the following questions:

- How an LMS should be modeled to fit better the requirements of standards and norms of e-learning programs?
- To what degree of specificity could the LMS respond as an innovative technical system?
- To what degree of specificity the learning theories could promote the online learning?

These and many other questions were investigated within our work. Our objective was to test and to check if our hybridization is worthy and useful for the design, development and diffusion of e-learning systems, particularly the LMS.

The present study attempts to bring some light into the questions above by exposing at first the four learning theories that were judged the most important and relevant to our modeling, namely the traditional pedagogy, the behaviorism, the cognitivist, and the social constructivism. Then, these learning theories which have inspired for a long time the design of computer applications are combined and put into perspective with several emergent pedagogical functionalities to build an original modeling for our new LMS.

2 LMS and Online Learning

2.1 Definition

An LMS (Learning Management System) or e-learning platform is a software including a range of services that assist teachers with the management of their courses. It offers

---

[1] **TMS:** Teaching Management System

many services allowing the management of content, particularly by creating, importing and exporting learning objects. The set of the available tools in the LMS represent all these services that help in managing the teaching process and the interaction between users. More precisely these services are linked to the following variety of functionalities:

- The management of pedagogical content (creating, importing and exporting learning objects),
- The creation of individual's personal paths in the training modules,
- The availability of sharing tools,
- The distribution of communication tools,
- The student registration and the management of their files (training tracking and results),
- The distribution of online courses and many other pedagogical resources.

Figure 1 illustrates the general principle of the operation of an e-learning platform LMS by presenting the key features associated with the main actors: learners, teachers, tutors, coordinators, and administrators. The learner can consult and/or download the resources made at his/her disposal by the teacher, he/she can create his/her learning activities while following his/her progress in training. The teacher, who is responsible of one or more modules, can create and manage the educational content he/she wishes to broadcast via the platform. He/she can also build tools for monitoring learners' activities. The tutor accompanies and monitors each learner by providing the tools of communication and collaboration. Concerning the coordinator, he/she ensures the

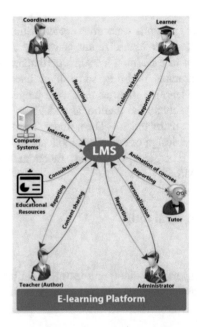

Fig. 1. The general architecture of an LMS

management of the overall system. Finally, the administrator is responsible for the customization of the platform having the rights of the administration deriving from it (system installation, maintenance, access management…).

2.2 Benefits of LMS

The LMS on which we increasingly rely as a means of learning have a considerable potential in the construction of knowledge and competence development. Thanks to the different services offered by these e-learning platforms, individuals can access and use interactively the multiple sources of information available to them everywhere, at all times. They can also compose customized training programs and thus develop their abilities to the highest level of their potential according to their needs [12].

Based on the work of De Vries [35], the main pedagogical functions that may be assigned to the LMS as computer applications for learning are:

- Presenting information,
- Providing exercises,
- Really teaching,
- Providing a space of exploration,
- Providing a space of exchange between educational actors (learners, teachers, tutors…

These different pedagogical functions, that correspond to one or many learning theories, allow the learner to acquire individual and collective knowledge according to the type of interaction that takes place between him/her and the sources of information made at his/her disposal. In practice, each individual has a set of tasks to deal with such as:

- Consulting and reading the pedagogical resources,
- Realizing the interactive exercises,
- Exploring the learning environment,
- Solving the problem situations,
- Discussing via synchronous and asynchronous tools of communication.

3 The Main Theoretical Currents

Education sciences draw their theoretical foundations, among others, in psychology, sociology, philosophy and cognitive science. This diversity of theoretical fields at the base of the different approaches to teaching and learning can sometimes be confusing insofar as some authors may find themselves inside of more than one theoretical current. Currently, a majority of educational theorists agree to group teaching and learning models according to four currents: traditional pedagogy, the behaviorist, the cognitivist and the social constructivism.

This paper describes the four previously mentioned currents, in a synthetic way that identify the main characteristics and technological adapted tools. In addition, it contains examples that illustrate the underlying key concepts and make the link with the LMS e-learning platform. Addressed in a historical perspective, this document intends to nourish the reflection of teachers who want to situate their educational practices inside a conceptual framework and who want to be able to appreciate the complexity and impact of their pedagogical actions.

Table 1 shows a schematic summary of the four main currents by linking them to the act designs of teaching and learning that correspond to them.

Table 1. Schematic presentation of the main theoretical currents

Traditional pedagogy	Behaviorist	Cognitivist	Social constructivist
Teaching is …			
Present information in a structured, hierarchical, and inductive way	Stimulate, create and reinforce appropriate observable behaviors	Present information in a structured, hierarchical, and deductive way	Organize learning situations conducive to dialogue with a view to provoke and resolve sociocognitive conflicts
Learning is …			
Following the course: unfolding the course and the tutor.	Associate, by conditioning, a reward to a specific response	Treat and store new information in an organized way	Co-construct his/her knowledge by comparing the representations with those of others
Appropriate teaching methods			
Learning by course, exercises and assessments	Assisted self-study program	Formal presentation, solving closed problems	Projects, discussions, exercises and work based learning

4   LMS and the Underlying Learning Theories

Although their considerable potential in the construction of knowledge and competence development, the LMS can generate a real pedagogical success only if, their use relies on solid and proven learning theories [12].

In the next part, we will evoke the transposition of the use of four learning theories in the design and development of LMS. For that purpose, we will do the correspondence between the tools available in LMS and the learning theories to which they refer. As a latter part will show, the hybridization of these learning theories that we have judged more important and relevant to our modeling work can only be a source of enrichment to improve the quality of online learning.

4.1   LMS and the Traditional Pedagogy

The conception of learning as supported by traditional pedagogy is essentially relying on a direct and systematic mode of transmission. Indeed, we put forward the authoritarian role of the teacher who must deliver fixed and unchanging knowledge, evaluate and involve learners by following the different stages of a pre-established scenario. From this perspective, learners are only passive recipients of information who respond ideally to external factors provided by their teacher in advance in a particular environment. In this way, they develop their knowledge. Among the main ideas that are associated with the traditional pedagogy [15], we mention:

- Lecture-based teaching: this idea generally refers to the teaching-centered pedagogy in which the teacher is the main provider of subject content to learners. The acquisition of knowledge is assessed through various operations of reproduction such as recitation, examination and practical exercises. Only the teacher has overall authority over learners who must follow his/her instructions and show goodwill to construct their knowledge in a more effective manner.
- The idea of transmission and reception: we consider that the teacher delivers knowledge in ways that are clear, concise and transparent and the learner receives it without any difficulty of memorization, understanding and reproduction. Trial and error learning seems not having its place in that perspective. The learner must listen, deploy efforts to study well and recite in accordance with the teacher expectations. The pedagogical relation is ideal from the teacher's point of view, the learner's spirit and the object of transmission.
- Individualism: the learner is a part of a group but still works individually. No exchange between learners is allowed. Obviously, this implies absence of debate, dialogue and communication. Everything is centralized around and by the teacher. No cognitive and social dimension exists in the learning process.
- The sanction: the role of teacher is to identify errors. Learners are classified in order to generate the spirit of competition between them. Those who fail to learn are those who commit one or many errors. Making errors is not considered as a necessary step for learning but it is seen as being the fault of the learner who had not shown goodwill to learn.

These main ideas from the traditional pedagogy have had an impact on the design and development of LMS, which focus on learning by reception. Indeed, this kind of software integrates different spaces in order to allow teachers organizing, structuring, exposing their knowledge and particularly assessing the learning progress. Thus, a central place is given to teachers, who do have the necessary tools to deliver knowledge and engage learners in the proposed learning activities of reproduction, consultation and execution.

In general, we think that the contribution of the traditional pedagogy is so valuable to the extent that it allows teachers to facilitate and assist learners by making at their disposal well-structured information based on over-prescriptive scenarios. However, traditional pedagogy has its limits that rely very much on the fact that learners who are considered as the main actors in the teaching learning process are widely neglected. Their needs are not taken into consideration and they are only seen as passive receivers of knowledge.

4.2   LMS and the Behaviorism

The behaviorism is a learning theory concerned with the study of human observable behavior without recourse to inner mental states [8]. It is built on the assumption that the brain is only a black box that no one can access.

The term "behaviorism" appeared at the beginning of 20th century in parallel with works of the American psychologist John Watson. This latter is considered as the pioneer of the behaviorism. He proposed making the general psychology a scientific discipline by using experimental laboratory methods to set exploitable results that can then statistically evaluated [9]. Works of the physiologist Ivan Pavlov on conditioning of animals influenced Watson. This leads him to admit that all behavior operate on a principle of "stimulus-response" or what is called "classical conditioning".

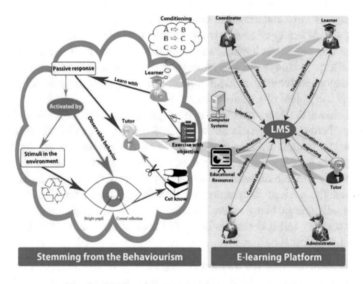

Fig. 2. LMS and underlying behaviorism model

For the advocates of the behaviorism (such as Pavlov and Skinner), the learning process is perceived in a very simplistic way as an external change in human behavior which results from a specific instrumental conditioning. This means that the confrontation of any individual with a discriminative stimulus inevitably leads usually to the emergence of constructed responses over time. To teach a certain skill, the behaviorist approach proposes to break it down into sub-objectives, which have to be simultaneously assimilated and mastered. In this perspective, the teacher should be able to present information to learners under restricted stimulus associated with reinforcement. Obviously, the learners' role here is to respond to these stimuli by adopting the expected behavior. The teacher also proposes progressive practical exercises that allow checking the acquisition of knowledge while giving positive and/or negative feedback based on the responses provided. The type of the pedagogical scenario that prevails in that case is the one, which highlights learning by reception-exercises-test (Fig. 2).

The mark of behaviorism can be found in the LMS which display systematic exercises allowing learning by repetition (trial and error) and in which the principles of conditioning are integrated.

4.3 LMS and the Cognitivist

Cognitivist is born at the same time as the Artificial Intelligence, in 1956. Miller and Bruner propose it in reaction to Behaviorism. It focuses on the ways of thinking and solving problems. Learning cannot be limited to a conditioned recording, but should rather be considered as requiring complex processing of the received information. Memory has its own structure, which involves the organization of information and the use of strategies to manage this organization [10].

Indeed, the initial questionings of Behaviorists designers goes back to the publication by Miller in 1956 of an article entitled "The magic number 7, more or less 2" [34] in which the physiological limits of human memory were highlighted. According to this author, the capacity of human memory is limited to seven isolated elements. Obviously, this is not compatible with the behaviorist design, which sees memory as a virgin receptacle in which knowledge accumulates.

Cognitive psychology considers that there are three broad categories of knowledge: declarative, procedural, and conditional knowledge. It invites the teacher to develop different strategies to facilitate the integration of each of them because they are represented differently in memory; the declarative knowledge gives an answer to the WHAT, the procedural knowledge to the HOW and the conditional knowledge to the WHEN and to the WHY [27].

There are different categories of cognitive strategies that contain several types of strategies. Furthermore, cognitive and metacognitive strategies can be the subject of a systematic teaching. In addition, the authors [21, 20] insist for that the teaching of these strategies be carried out in the learning context, in the program course. The teaching of these strategies will be effective if these strategies are integrated in the ordinary curricula, and presented to learners as a necessary means to the achievement of the learning objectives. However, quality education is not limited about telling learners what to do; it consists also about showing how to learn. Tardif [19] presents a learning model based on the importance of the gradual and effective appropriation of cognitive and metacognitive strategies. This model has aim to stimulate cognitive and emotional engagement, to show the learner how to treat the information in an adequate way and to bring the learner to make transfers.

For an LMS based on the cognitivist approach, the learner is an active information-processing system, similar to a computer: it perceives information that comes from the outside world, recognizes them, stores them in memory, then recovers them from his/her memory when he/she needs it to understand his/ her environment or resolve problems [32]. Teacher is the manager of learning, he guides, animates, directs, advises, explains, regulates, and remedies. Knowledge become an external reality that the learner must integrate into his/her mental patterns and reuse rather than acquire observable behaviors [31]. In addition, the favored teaching method leaves room to multiple learning pathways in order to take into account the different individual variables that can may influence the way in which learners' process information. The cognitivist teacher will be invited to use ICTs that promote high interactivity with learners, such as simulators, experiments and intelligent tutorials. However, the cognitivist model has an important limit, related to the fact that a well-structured material is not sufficient to ensure a learning. The motivation of the learners is a determining factor because it provides the required energy to perform learnings (Fig. 3).

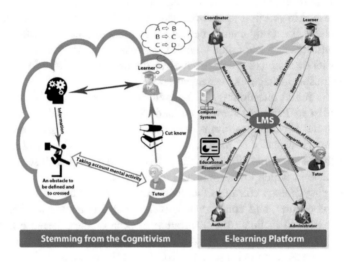

Fig. 3. LMS and underlying Cognitivist model

4.4   LMS and the Social Constructivism

The social constructivism is the fruit of the development of learning theories under the influence of some researchers, particularly Lev Vygotski in 1934 [27, 28], who wanted to depart from the behaviorism by integrating other factors that are able to positively influencing the knowledge acquisition. Thus, new ideas emerged in connection with the possible interaction of individuals with the environment.

The social constructivism outlines learning by construction in a community of learners. In this light, learners are expected to interact with the available human resources (teachers, tutors, other learners…) in the proposed learning environment. In this way, the learners' psychological functions increase through socio-cognitive conflicts that occur between them. These conflicts lead to the development of the zone of proximal development[2] [14] and thus facilitate the acquisition of knowledge.

Learning is seen as the process of acquisition of knowledge through the exchange between teachers and learners or between learners. These latter learn not only through the transmission of knowledge by their teacher but also through interactions [5]. According to this model, learning is a matter of the development of the zone of

---

[2] "The distance between actual development level as determined by independent problem solving and the level of potential development as determined through problem solving under adult guidance or in collaboration with more capable peers (Vygotsky, 1978, p. 86)".

proximal development: this zone includes the tasks that learners can achieve under the guidance of an adult; they are not very tough or so easy. The development of this zone is a sign that the learners' level of potential development increases efficiently [6].

The teacher's role is to define precisely this zone in order to design suitable exercises for learners. Furthermore, designing collaborative tasks, which involve discussions and exchange (socio-cognitive conflicts) between learners is so important in this model. Errors are considered as a point of support for the construction of new knowledge.

Based on the social constructivism approach (Fig. 4), the design of LMS were oriented towards integrating online communication and collaboration tools. In practice, a wide range of platforms, particularly the social constructivist ones, propose a set of tools, which allow sharing, exchanging and interacting in synchronous and asynchronous mode such as blogs, wikis, forums...

In summary, the ideas of social constructivist authors have highlighted the social nature of learning. Other authors have taken one-step further by emphasizing the distribution of intelligence between individuals and the environment. Furthermore, considering that learning occur in a social context is no longer enough to ensure deep learning. Indeed, working in groups can affect negatively the quality of learning if these following conditions are not taken into consideration: Learning styles, the way groups are formed, interaction modality, and the characteristics of tasks.

In addition, connectivism can considered as a branch of the social constructivism. It is not necessarily a learning theory, but rather a pragmatic concept of participatory teaching and learning [16], which is relying on assumptions of Latours Actor-Network-Theory [18]. If viewed as a theory by itself, it would also overlap with the social constructivist paradigm in terms of the importance of interaction in social structures.

George Siemens and Stephen Downes who developed the connectivism, they are based on the principles of connection, online networking and thus interactions between

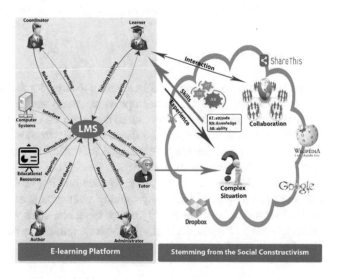

Fig. 4. LMS and underlying social constructivism model

objects of the world (material or symbolic). They stated that connectivism is based on the use of a network composed of nodes and connections as a central metaphor of learning [4]. In this metaphor, a node can be information, data, feelings, pictures or simulations.

The connectivism benefits [3] as a new learning theory reside on the importance given to the principle of connection which leads to the understanding of the learning process by describing how and why connections are formed in different levels: neural, cognitive/conceptual and social.

5 Synthesis

A great deal of research is focusing in one way or another on the platforms engineering for distance education, including LMS. For example Chekour [27], presented a synthesis of the main learning theories in the digital era, namely: the behaviorism, the cognitivism, the constructivism and the social constructivism. El-Mhouti [28] highlighted the ICT use in the service of active pedagogies, based on the social constructivist approach, the principles that structure the instructional design approaches, and the assessment of the social constructivist activities.

These works, among many others, emphasized the contribution of learning theories in the design and development of learning systems. The direct application of each of these theories allows particularly providing supporting methods to the design and development of LMS.

Based on these various research works, which seemed to us incomplete, we propose a modeling portrait of a new LMS platform [37]. This latter is anthropocentric and relies on a learning conception that is located at the intersection of the most used learning theories. Indeed, the idea is to orient the design work research towards a great and optimal compatibility between the services offered by e-learning platforms and the needs of all users, particularly learners, for better optimization of online learning.

6 Conclusions and Perspectives

In this paper, we propose the hybridization of four learning theories for the modeling of a new LMS platform. Our first motivation was to provide a more learner-centric LMS while opening it up dynamically to the teacher. Indeed the proposed LMS offers a range of customizable web services that fits users' needs. In this way, freedom of choice is left with regard to teaching and learning concerning the creation, adaptation, and personalization of various components of the LMS.

The modeling of the LMS is still taking place. We are looking at the implementation of its first prototype as part of university training with groups of teachers and learners. This will allow us to check the validity of our modeling work with the target audience and therefore take relevant decisions for better exploitation and wide dissemination of the LMS in the future.

References

1. Ouadoud, M., Chkouri, M.Y., Nejjari, A., El-Kadiri, K.E.: Studying and analyzing the evaluation dimensions of e-learning platforms relying on a software engineering approach. Int. J. Emerg. Technol. Learn. (iJET) **11**(1), 11–20 (2016). https://doi.org/10.3991/ijet.v11i01.4924

2. Ouadoud, M., Chkouri, M.Y., Nejjari, A., El-Kadiri, K.E.: Studying and comparing the free e-learning platforms. In: 2016 4th IEEE International Colloquium on Information Science and Technology (CiSt), pp. 581–586 (2016). http://dx.doi.org/10.1109/CIST.2016.7804953

3. Siemens, G.: Connectivism: a learning theory for the digital age. International journal of instructional technology and distance learning **2**(1), 3–10 (2005)

4. Duplàa, E., Talaat, N.: Connectivisme et formation en ligne. Distances et savoirs **9**(4), 541–564 (2012)

5. Doise, W., Mugny, G.: Le développement social de l'intelligence, vol. 1. Interéditions, Paris (1981)

6. Vygotsky, L.S.: Mind in Society: The Development of Higher Psychological Processes, Revised edn. Harvard University Press, Cambridge (1978)

7. The OVAREP (Observatoire des ressources multimédias), 2000 Étude comparative technique et pédagogique des plate-formes pour la formation ouverte et à distance

8. Good, T.L., Brophy, J.E.: Educational Psychology: A Realistic Approach, 4e éd. Longman, New York (1995)

9. Watson, J.: Le béhaviorisme. Editions Cepi, Paris (1972)

10. Crozat, S.: Éléments pour la conception industrialisée des supports pédagogiques numériques (Doctoral dissertation, Université de Technologie de Compiègne) (2002)

11. Ouadoud, M., Chkouri, M.Y., Nejjari, A., El-Kadiri, K.E.: Exploring a recommendation system of free e-learning platforms: functional architecture of the system. Int. J. Emerg. Technol. Learn. (iJET) **12**(02), 219–226 (2017). https://dx.doi.org/10.3991/ijet.v12i02.6381

12. G. Paquette, L'Ingénierie Pédagogique: Pour Construire l'Apprentissage en Réseau. PUQ, 2002

13. Dambreville, S.D.: Evaluer un dispositif de formation à distance Principes et retour d'expérience. Revue des Interactions Humaines Médiatisées **9**(2), 25–52 (2008)

14. Wake, J.D.: Evaluating the organising of a collaborative telelearning scenario from an instructor perspective œ an activity theoretical approach. Ph.D. dissertation in computer science, Department of Information Science, University of Bergen, December 2001. Thèse de doctorat

15. Morandi, F.: Introduction à la pédagogie. Armand Colin (2006)

16. Haug, S., Wedekind, J.: cMOOC – ein alternatives Lehr-/Lernszenarium? In: Schulmeister, R. (Hrsg.) MOOCs – Massive Open Online Courses. Offene Bildung Oder Geschäftsmodell? Waxmann, Münster (2013)

17. Dogbe-Semanou, D.A.K., Durand, A., Leproust, M., Vanderstichel, H.: Etude comparative de plates-formes de formation à distance. le cadre du Projet@ 2L, Octobre 2007

18. Anderson, T., Dron, J.: Three Generations of Distance Education Pedagogy (2011)

19. Tardif, J.: Pour un enseignement stratégique. L'apport de la psychologie cognitive. Logiques, Montréal, Québec (1992)

20. Weinstein, C., Mayer, R.: The teaching of learning strategies. In: Wittrock, M.C. (dir.) Handbook of Research on Teaching, 3e éd. Macmillan, New York (1986)

21. Pressley, M., Harris, K.: What we really know about strategy instruction. Educ. Leadership **48**(1), 31–35 (1990)

22. Dimet, B.: Etude comparative technique et pédagogique des plates-formes pour la formation ouverte et à distance, 15 January 2006

23. Menasri, S.: Etude comparative de plateformes d'enseignement en ligne (e-learning) utilisées dans un contexte universitaire, June 2004

24. Lablidi, A., Abourrich, A., Talbi, M.: Démarche préconisée pour évaluer une plate-forme, Revue de l'EPI (Enseignement Public et Informatique), December 2009

25. Aska, Le Préau & Klr.fr. Choisir une solution de téléformation: 2000 study: l'offre de plates-formes et de portails de téléformation/[Study conducted by Anne Bouthry, Patrick Chevalier, Serge Ravet, and al.] (2000)

26. Kaddouri, M., Bouamri, A.: Usage de plateformes d'enseignement à distance dans l'enseignement supérieur marocain: avantages pédagogiques et difficultés d'appropriation, Questions Vives. Recherches en éducation, vol. 7, no. 14, pp. 107–118, December 2010. http://questionsvives.revues.org/642

27. Chekour, M., Laafou, M., Janati-Idrissi, R.: « L'évolution des théories de l'apprentissage à l'ère du numérique » , Revue de l'EPI (Enseignement Public et Informatique), February 2015

28. El Mhouti, A., Nasseh, A., Erradi, M.: « Les TIC au service d'un enseignement-apprentissage socioconstructiviste » , Revue de l'EPI (Enseignement Public et Informatique), janv 2013

29. El Mawas, N., Oubahssi, L., Laforcade, P.: Étude comparative de plateformes de formation à distance. GRAPHIT-D2.5&2.2, 10 July 2014

30. Galloy, S., Haas, C., Lodewijck, M.: Analyse d'environnements de formation à distance. CIFoP – Le Centre Interuniversitaire de Formation Permanente, Mai-2002

31. Bibeau, R.: Les technologies de l'information et de la communication peuvent contribuer à améliorer les résultats scolaires des élèves. Revue de l'EPI, (94) (2007)

32. Bibeau, R.: École informatisée clés en main. Projet franco-québécois de recherche-action. Revue de l'EPI (Enseignement Public et Informatique), no. 82, pp. 137–147, June 1996

33. Graf, S., List, B.: An evaluation of open source e-learning platforms stressing adaptation issues. In: Fifth IEEE International Conference on Advanced Learning Technologies (ICALT 2005), pp. 163–165 (2005). http://dx.doi.org/10.1109/ICALT.2005.54

34. Kozanitis, A.: Les principaux courants théoriques de l'enseignement et de l'apprentissage: un point de vue historique, Bureau d'appui pédagogique - École polytechnique de Montréal, pp. 1–14, September 2005

35. Marquet, P.: Lorsque le développement des TIC et l'évolution des théories de l'apprentissage se croisent, Savoirs, no. 9, pp. 105–121, September 2005. http://dx.doi.org/10.3917/savo.009.0105

36. Ouadoud, M., Chkouri, M.Y., Nejjari, A.: LeaderTICE: a platforms recommendation system based on a comparative and evaluative study of free e-learning platforms. Int. J. Online Eng. (iJOE) **14**(1), 132–161 (2018). https://doi.org/10.3991/ijoe.v14i01.7865

37. Ouadoud, M., Nejjari, A., Chkouri, M.Y., Kadiri, K.E.E.: Educational modeling of a learning management system. In: 2017 International Conference on Electrical and Information Technologies (ICEIT), pp. 1–6 (2017). http://dx.doi.org/10.1109/EITech.2017.8255247

The 1st International Workshop on
Smart Healthcare: S-Health'17

Knowledge Acquisition for an Expert System for Diabetic

Ibrahim Mohamed Ahmed Ali[✉]

Information Technology Department, Faculty of Computer and Information Technology, Karray University, Khartoum, Sudan
ibrahim1630@gmail.com

Abstract. Diabetes is a serious health problem today. Most of the people are unaware that they are in risk of or may even have type-2 diabetes. Type-2 diabetes is becoming more common due to risk factors like older age, obesity, lack of exercise, family history of diabetes, heart diseases. Along with good lifestyle and healthy diet, reduces the risk of development of type 2 diabetes for treatment of elder people, proper care of diet, exercise and medication as well is more important. The research in developing intelligence knowledge base systems in diabetic domain is important for both health industry and diabetes patients. Recently expert systems technology provides an efficient tools for diagnosing diabetes and hence providing a sufficient treatment. The main challenge in building such systems is the knowledge acquisition and development of the knowledge base of these systems. Our research was motivated by the need of such an efficient tool. This paper presents the knowledge acquisition process for developing the knowledge base of diabetic type-2 diet.

Keywords: Expert systems · Knowledge representation · Diabetic diet
Type 2 diabetes · Rule-base

CCS Concepts: Artificial intelligence · Knowledge engineering

1 Introduction

Diabetes is one of the major risky diseases for health care in our lives. If people were aware of the factors of diabetes and know how much risks they are of getting diabetes, diabetes may be prevented early [1]. Type 2 diabetes is a disease resulting from a relative, rather than an absolute, insulin deficiency with an underlying insulin resistance. Type 2 diabetes is associated with obesity, age, and physical inactivity [2, 3]. It is more common as compare to type-1 diabetes, usually 90 to 95%. It is diagnosed in both adults and young people. In this type pancreas does not produce enough insulin to control keeping blood sugar level within normal ranges. Actually it is serious type of diabetes where mostly people are not aware they are suffering from it. Three major causes of diabetes type 2 are lifelong bad diet, inactive or sedentary lifestyle, and overweight [4].

Actually, in the domain of medical treatment by controlling patient food (healthy diet) there are numerous variables that affect the decision process of selecting

© Springer International Publishing AG, part of Springer Nature 2018
M. Ben Ahmed and A. A. Boudhir (Eds.): SCAMS 2017, LNNS 37, pp. 747–758, 2018.
https://doi.org/10.1007/978-3-319-74500-8_68

interesting food list from the patient point of view and efficient list in treatment from the doctor's point of view. These numerous variables causing the differences in the opinions of the practitioners. Also, there are many uncertain risk factors resulted from eating certain types of food with certain amount. Therefore, an accurate tool will be of a great help for an expert to consider all these risk factors and show certain results.

On the other hand the research in developing intelligence knowledge base systems in diabetic domain is important for both health industry and diabetes. Expert system is a computer program that provides expert advice as if a real person had been consulted where this advice can be decisions, recommendations or solutions. A few numbers of expert systems are utilized in diabetic health research where each of these systems attempts solving part or whole of a significant problem to reduce the essential need for human experts and facilitates the effort of new graduates [5].

The paper is organized as follows. Section 2 presents major risk factors Diabetic Diet and Diabetic Food Pyramid. Section 3 describes the related work. Section 4 present the knowledge acquisition and the representation process. Section 5 screening of diabetics. Section 6 reasoning techniques in diabetic expert systems. Section 7 ends up with Conclusion.

2   Related Work

Beulah et al. (2007) [6] introduced the ability to access diabetic expert system from any part of the world. They collect, organize, and distribute relevant knowledge and service information to the individuals. The project was designed and programmed via the dot net framework. The system allows the availability to detect and give early diagnosis of three types of diabetes namely type 1, 2, gestational diabetes for both adult and children.

Szajnar and Setlak (2011) [7] proposed a concept of building an intelligence system of support diabetes diagnostics, where they implemented start-of-art method based on artificial intelligence for constructing a tool to model and analyze knowledge acquired from various sources. The initial target of their system was to function as a medical expert diagnosing diabetes and replacing the doctor in the first phase of illness. Diagnostics the sequence of dealing with their system were as flow: (1) getting patient information and symptoms (2) competing basic medical examination in details (3) based on previous information the system find out whether the patient has diabetes and decides whether it is type 1 or type 2. The systems used decision tree as a model for classification.

Kumar and Bhimrao (2012) [8] developed a natural therapy system for healing diabetic, they aim to help people's health and wellness, which don't cost the earth. Their main goal was to integrate all the natural treatment information of diabetes in one place using ESTA (Expert System Shell for Text Animation) as knowledge based system. ESTA has all facilities to write the rules that will make up a knowledge base. Further, ESTA has an inference engine which can use the rules in the knowledge base to determine which advice is to be given to the user. Their system begins with Consultation asking the users to select the disease (Diabetes) for which they want different type of natural treatment solution then describes the diabetes diseases and their symptoms. After that describes the Natural Care (Herbal/Proper Nutrition) treatment solution of diabetes disease.

Bayu et al. (2011) [9] proposed and boosted algorithm acquires information from historical data of patient's medical records of Mohammad Hoesin public hospital in Southern Sumatera. Rules are extracted from Decision tree to offer decision-making support through early detection of Type-2 diabetes for clinicians, Table 1.

Table 1. Expert systems for diabetes

Authors	System purpose	ML technique	User interface / Application
Kumar and Bhimrao 2012 [8]	Integrate all the natural treatment information of diabetes in one place	rule based	Interactive / Pc
Szajnar and Setlak 2011 [7]	Model and analyze knowledge acquired from various sources	decision tree	Interactive / Pc
Bayu et al. 2011 [9]	An Early Detection Method of Type-2 Diabetes Mellitus in Public Hospital	decision tree	Request/Response / Pc
Beulah et al. 2007 [6]	Detect and give early diagnosis of three types of diabetes for both adult and children	Rule based	Request/Response / Pc

3   Diabetic Diet and Food Groups

3.1   Diabetic Diet

Diabetic Diet for diabetics is simply a balanced healthy diet which is vital for diabetic treatment. The regulation of blood sugar in the non-diabetic is automatic, adjusting to whatever foods are eaten. But, for the diabetic, extra caution is needed to balance food intake with exercise, insulin injections and any other glucose altering activity. This helps diabetic patient to maintain the desirable weight and control their glucose level in their blood. It also helps to prevent diabetes patient from heart and blood vessel related diseases [10].

Research shows that regardless of the makeup of the diet, eating just enough calories to maintain an ideal weight is the most effective dietary strategy to prevent the onset of diabetic. Recommendations of diabetic diet differ for person to person, based on their nutritional needs, lifestyle, and the action and timing of medications [11].

In Type 2 diabetic, the concern may be more oriented to weight loss in order to improve the body's ability to utilize the insulin it does produce. Thus, learning about the basic of food nutrition will be able to help in adjusting diet to suite the particular condition. Recommended daily food portion contains carbohydrates, protein and fat. A Registered Dietitian assesses the nutritional needs of a person with diabetes and calculates the amounts of carbohydrate, fat, protein, and total calories needed per day. He will then convert this information into a recommended list of food for daily diet [11]. See Table 2.

Table 2. Recommended daily food portion

Nutrition	Daily calories
Carbohydrates	(50..55)%
Protein	(15..20)%
Fat	Not more than 30%

3.2   Diabetic Food Pyramid

The Diabetes Food Guide Pyramid is a tool that shows how much you should eat each day from each food group for a healthy diet. The Diabetes Food Guide Pyramid is the best food guide for people with diabetes. The Diabetes Food Guide Pyramid places starchy vegetables such as peas, corn, potatoes, sweet potatoes, winter squash, and beans at the bottom of the pyramid, with grains. These foods are similar in carbohydrate content to grains. Cheese is in the Meat and others group instead of the Milk group because cheese has little carbohydrate content and is similar in protein and fat content to meat [12] (Fig. 1).

Fig. 1. Food Pyramid

Choosing foods from the Diabetes Food Guide Pyramid can help you get the nutrients you need while keeping your blood glucose under control [12].

Foods that are high in carbohydrates increase blood glucose levels and are in the Grains, Beans, and Starchy Vegetables group, the Fruits group, and the Milk group. Other foods that raise blood glucose are Sweets, found in the top of the Pyramid. Starchy foods, sweet foods, fruits and milk are high in carbohydrate. Foods lows in carbohydrates are found in the Vegetables group, Meat and Others group and Fats. Diabetes patient should eat 6 to 11 servings Grains, 2 to 5 servings Group Vegetable, 2 to 4 servings Group Fruit, 2 to 3 servings Group Milk, 2 to 3 servings group protein, Group sugars and oils should rarely be eaten [12].

3.3   Food Groups

Food groups are exchange lists of foods that contain roughly the same mix of carbohydrates, protein, fat, and calories, serving sizes are defined so that each will have the same amount of carbohydrate, fat, and protein as any other. Foods can be "exchanged" with others in a category while still meeting the desired overall nutrition requirements. Food groups can be applied to almost any eating situation and make it easier to follow a prescribed diet. There are six food groups [13]:

- Vegetables
- Starches and Breads
- Fruits
- Milk
- Fat
- Meats and Meat Substitutes

The food groups are based on principles of good nutrition that apply to everyone. The reason for dividing food into six different groups is that foods vary in their carbohydrate, protein, fat, and calorie content. Each group contains foods that are alike; each food choice on a group contains about the same amount of carbohydrate, protein, fat, and calories as the other choices on that group [14].

4   Knowledge Acquisition and Representation

4.1   Knowledge Acquisition

Knowledge acquisition is a very important phase in developing expert systems [4]. Our knowledge has been gained by consultation of nutritionist.

Actually, knowledge acquisition required time of three months form major Ibtehal and Nasik nutritionist of diabetes in the military hospital in Khartoum, in addition to some related books and internet medical web sites. In addition we determine Sudanese food groups in Fig. 2 and analyse the amount of each item in the food groups in Table 3.

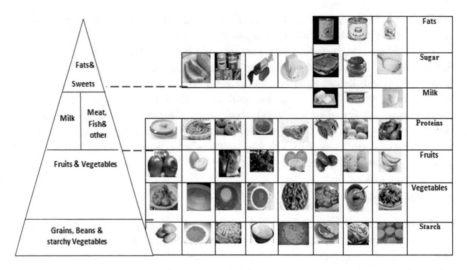

Fig. 2. Sudanese food servings according to the diabetes food guide pyramid

Table 3. Standards of items

Fat and Milk		Sugar		Proteins	
Name	Amount	Name	Amount	Name	Amount
Oil	Spoon (20 g)	Sugar	Spoon(20 g)	Chicken	1/4 piece(250 g)
Shortening	Spoon (20 g)	Jam	Spoon (20 g)	Egg	1 piece
Synths	Spoon (20 g)	Cake	1 piece	Fish	125 g
Milk	1 cup	Tahnia	Spoon (20 g)	Meat	Kumsha (100 g)
Yogurt	100 g	Sweet	1 piece	Tamiea	4 pieces (40 g)
Cheese	50 g	S_drinks	75 ml	Bean	Kumsha (100 g)
–	–	Basta	Small piece	Lentils	Kumsha (100 g)
–	–	–	–	Fual	Kumsha (100 g)
Fruits		Vegetables		Starch	
Name	Amount	Name	Amount	Name	Amount
Banana	piece (100 g)	Salad	Free	Custer	1 cup
Orange	Small piece (100 g)	Molokhia	Kumsha	Kissra	2 pieces (100 g)
Mango	Small piece(100 g)	Bazenjan	Kumsha	Gorasa	1/2 piece (100)
Dates	3 pieces (24 g)	Okra	Kumsha	Bread	1 piece (120 g)
Grapes	10 pieces (120 g)	Potatoes	2 Kumsha	Rice	1 cup
W_melon	2 slice (120)	Regala	2 Kumsha	Pasta	1 cup
Apple	Small piece (100 g)	Taglia	Kumsha	Potato	Big piece
Guava	Small piece (100 g)	Roub	2 Kumsha	Noodles	1 cup

4.2   Knowledge Representation

Knowledge representation allows one to specify and emulate systems of a growing complexity. Knowledge representation schemes indeed have known an important evolution, from basic schemes supporting a rather heuristic approach, to advanced schemes involving a deeper consideration of the various dependencies between knowledge elements [15]. The main Types of diabetes are Type 1, Type 2 and Gestational [16]. Figure 3 describes Knowledge representation of the diabetic serving.

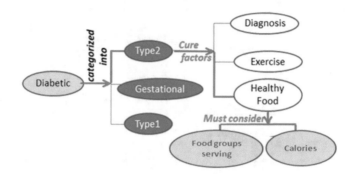

Fig. 3. Knowledge representation

4.3   Food Groups Servings

Some diseases increase the risk of diabetic disease and affect the number of serving in the food groups, the major diseases we get from our Knowledge acquisition are Anorexia, Surgery, Blood pressure, Typhoid, Bitter, Liver problems, Heart disease and Gout. Other factors affect the serving are the patient activity, and weight see Fig. 4. Figure 5 shows a sample of this frame based representation.

4.4   Knowledge Analysis

Serving base. The following is the algorithm to specify the numbers of serving to each patient according to Fig. 4.

- Determine whether the patient is slim or moderate or obese.
- Determine whether the patient activity is high or moderate or little.
- Determine whether the patient infected with (Anorexia, Surgery, Blood pressure, Typhoid, Bitter, Liver problems, Heart disease, Gout).
- Calculate number of servings as follows:
    Vegetable- servings = 3
    If (anorexia = 1) or (surgery = 1) or (age > 65) then fruit- servings = 4 else fruit- servings = 2
    If activity = "normal" then crabs-servings = 6
    Else if activity = "high" then crabs-servings = 8
    If the patient underweight then crabs-servings = 10

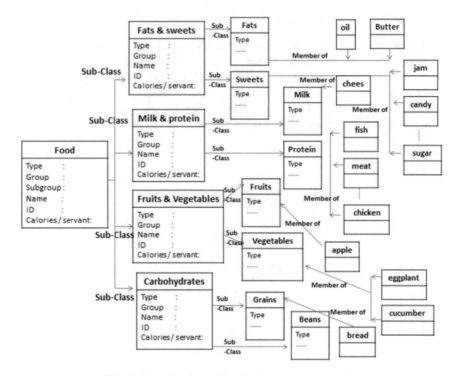

Fig. 4. Sample of diabetics food frame representation

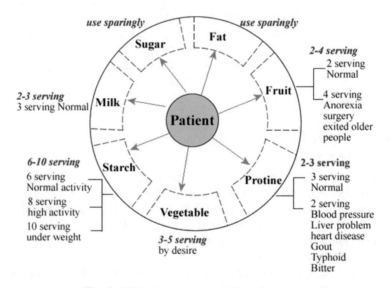

Fig. 5. Diabetics numbers of allowed servings

If ((gout = 1) or (Heart disease = 1) or (Bitter = 1) or (liver problems = 1) or (Blood pressure = 1) or (Typhoid = 1)) then protein-servings = 2 else protein-servings = 3

If ((gout = 1) or (Heart disease = 1) or (Bitter = 1) or (liver problems = 1) or (Blood pressure = 1) or (Typhoid = 1)) then milk-servings = 2 else milk-servings = 3.

Calories base. The following rules are samples of the calculated total calories:

If (MBI > 30) then (patient is obese) Else if (MBI < 18.5) then (patient is slim) Else (patient is normal).

If ((patient is slim) and (activity is very high)) then Total calories = weight * 40.

If ((patient is slim) and (activity is moderate)) then Total calories = weight * 35.

If ((patient is slim) and (activity is little activity)) then Total calories = weight * 30.

If ((patient is normal) and (activity is very high)) then Total calories = weight * 35.

If ((patient is normal) and (activity is moderate)) then Total calories = weight * 30.

If ((patient is normal) and (activity is little activity)) then Total calories = weight * 25.

If ((patient is obese) and (activity is very high)) then Total calories = weight * 30.

If ((patient is obese) and (activity is moderate)) then Total calories = weight * 25.

If ((patient is obese) and (activity is little activity)) then Total calories = weight * 20.

5   Screening of Diabetics

Early Warning Signs for Type 2 Diabetes a blood glucose level should be checked. The criteria testing for Type 2 diabetes in children and adolescents is, overweight (BMI $\geq$ 85th percentile for age and gender, weight for height $\geq$ 85th percentile or weight $\geq$ 120% of ideal for height). And frequency test should be every 2 years and fasting plasma glucose is the preferred method for screening. Diabetes may be diagnosed based on A1C criteria or plasma glucose criteria, either the fasting plasma glucose (FPG) or the 2-h plasma glucose (2-h PG) value after a 75-g oral glucose tolerance test (OGTT), the same tests are used to n diabetes every 3 month to assess the meal planning that If the patient used the meal plan before and his BGL still above 140 or A1C above 6,5, it recommend to visit the doctor [17].

5.1   Laboratory Evaluation

A1C, if results not available within the past 3 months If not performed/available within the past year:

- Fasting lipid profile, including total, LDL, and HDL cholesterol and triglycerides, as needed.
- Liver function tests
- Spot urinary albumin–to–creatinine ratio
- Serum creatinine and eGFR [21].

6 Reasoning Techniques in Diabetic Expert Systems

The abilities of inference, reasoning, and learning are the main features of any expert system. The research area in this field covers a variety of reasoning methodologies, e.g.; automated reasoning, case-based reasoning, commonsense reasoning, multi-model reasoning, fuzzy reasoning, geometric reasoning, non-monotonic reasoning, model-based reasoning, probabilistic reasoning, causal reasoning, qualitative reasoning, spatial reasoning and temporal reasoning [18]. In this section we focus our discussion about the main characteristics of three of the reasoning methodologies which are commonly used in developing diabetic expert systems, namely; reasoning with production rules, fuzzy-rules, and case-based reasoning.

6.1 Reasoning with Production Rules

Production rules are the most commonly technique used in developing the inference engine of expert system. Forward chaining can be used to produce new facts (hence the term "production" rules), and backward chaining can deduce whether statements are true or not. Rule-based systems were one of the first large-scale commercial successes of artificial intelligence research [19].

6.2 Reasoning with Cases

Case-Based Reasoning (CBR) means reasoning from experiences (old cases) in an effort to solve problems, critique solutions and explain anomalous situations. The CBR systems' expertise is embodied in a collection (library) of past cases rather, than being encoded in classical rules. CBR allows the case-library to be developed incrementally, while its maintenance is relatively easy and can be carried out by domain experts [20].

6.3 Reasoning with Fuzzy Rules

In the rich history of rule-based reasoning in AI, the inference engines almost without exception were based on Boolean or binary logic. However, in the same way that neural networks have enriched the AI landscape by providing an alternative to symbol processing techniques, fuzzy logic has provided an alternative to Boolean logic-based systems. Fuzzy logic deals with truth values which range continuously from 0 to 1. Thus something could be half true 0.5 or very likely true 0.9 or probably not true 0.1. The use of fuzzy logic in reasoning systems impacts not only the inference engine but the knowledge representation itself [18].

7 Conclusions

Type-2 diabetes is the most common form of diabetes. This paper presents the first phase of developing an efficient expert system for diabetic Type-2 diet. The structure of the system contains three steps. First calculate total needs of calories, second determines the amount calories of the items and finally determines the proper diet.

Self-monitor for patient of type 2 diabetes is possible by getting proper amount of daily proper diet satisfy the amount of calories. The servings of meals calculate according to Body Mass Index (MBI) and the type of activity for the patient and the additional patient diseases.

The food groups contain the same amount of carbohydrate, protein, fat, and calories Sudanese food groups contains different meals so you don't have to eat the same foods all the time. After collecting knowledge and perform the necessary analysis semantic network and food serving representation, Currently we are working on developing mobile-based expert system in Arabic language interface for diabetes diet that intended to be used in Sudan and Arab countries.

The research field covered a variety of reasoning methodologies. Case based reasoning is the more efficient, powerful and less cost. Our research was motivated by the need of such techniques, therefore the reasoning techniques for diabetics expert system has been presented in this paper as platform towards designing and implementation expert systems for diabetes.

References

1. Yang, H.H., Miller, S.: A PHP-CLIPS Based Intelligent System for Diabetic Self-diagnosis. Department of Math & Computer Science, Virginia State University Petersburg (2006)
2. Shortliffe, E.H., Perreault, L.E. (eds.): Medical Informatics: Computer Applications in Health Care and Biomedicine. Springer, New York (2001)
3. Federal Bureau of Prisons Management of Diabetes Clinical Practice Guidelines June (2012)
4. Forbes, D., Wongthongtham, P., Singh, J.: Development of Patient-Practitioner Assistive Communications (PPAC) Ontology for Type 2 Diabetes Management. Curtin University, Perth (2013)
5. Song, B.-H., Park, K.-W., Kim, T.Y.: U-health expert system with statistical neural network. Adv. Inform. Sci. Serv. Sci. 3(1), 54–61 (2011)
6. Beulah Devamalar, P.M., Bai, T., Srivatsa, S.K.: An Architecture for a Fully Automated Real-Time Web-Centric Expert System. World Academy of Science, Engineering and Technology (2007)
7. Szajnar, W., Setlak, G.: A concept of building an intelligence system to support diabetes diagnostics. Studia Informatica (2011)
8. Kumar, S., Bhimrao, B.: Development of knowledge Base Expert System for Natural treatment of Diabetes disease. (IJACSA) Int. J. Adv. Comput. Sci. Appl. 3(3) (2012)
9. Bayu, A.T., et al.: An early detection method of type-2 diabetes mellitus in Public Hospital. TELKOMNIKA 9(2), 287–294 (2011)
10. The Diabetic Exchange List. http://www.glycemic.com/DiabeticExchange
11. Diet for diabetes patient. http://www.medmint.com/CONTENT/Diabetics/Diabetics_7.html
12. Igbal, A., Nagwa, M.: Health Guide for Diabetics. Sudan Federal Ministry of Health (2010)
13. Garcia, M.A., Gandhi, A.J., Singh, T., Duarte, L., Shen, R., Ponder, M.D.S., Ramirez, H.: Esdiabetes (an expert system in diabetes). JCSC 16, 166–175 (2001)
14. Diabetes Education and Prevention World Diabetes Day. http://www.diabetesdiabetic-diet.com
15. Grimm, S., Hitzler, P., Abecker, A.: Knowledge Representation and Ontologies Logic, Ontologies and SemanticWeb Languages, pp. 37–87. University of Karlsruhe, Germany (2007)

16. Al-Ghamdi, A.A.-M., et al.: An expert system of determining diabetes treatment based on cloud computing platforms. Int. J. Comput. Sci. Inform. Technol. **2**(5), 1982–1987 (2011)
17. Sue Kirkman, M., Briscoe, V.J.: Diabetes in Older Adults: A Consensus Report, American Diabetes Association and the American Geriatrics Society (2012)
18. Zadeh, L.A.: Fuzzy sets. Inf. Control **8**, 338–353 (1965)
19. Salem, A-B.M., Roushdy, M., HodHod, R.A.: A rule-base expert system for diagnosis of heart diseases. In: Proceedings of 8th International Conference on Soft Computing MENDEL, Brno, Czech Republic, 5–7 June 2002, pp. 258–263 (2002)
20. Salem, A.-B.M., Voskoglou, M.G.: Applications of CBR methodology to medicine. Egypt. Comput. Sci. J. **37**(7), 68–77 (2013). ISSN 1110-2586, Special Issue for EMMIT 9th International Conference for Scientific and Social Development in Mediterranean Countries, Nador, Morocco, 21–23 October 2013
21. Shubrook, J., et.al,: Standards of Medical Care in Diabetes—2017 Abridged for Primary Care Providers. American Diabetes Association, 15 December 2016

Quality Control Results for Linear Accelerator at Oncology Center in Nouakchott

Ahmed El Mouna Ould Mohamed Yeslem[1][(✉)] [iD], Moussa Ould Cheibetta[2], Jilali Ghassoun[3], Oum Keltoum Hakam[1], Slimane Semghouli[4], and Abdelmajid Choukri[1] [iD]

[1] Science Faculty, Kenitra, Morocco
ahmedelmouna@yahoo.fr, choukrimajid@yahoo.com
[2] National Oncology Centre, Nouakchott, Mauritania
[3] Sciences Faculty of Semlalia, Marrakech, Morocco
[4] Higher Institute of Nursing Professions and Health Techniques, Agadir, Morocco

Abstract. Radiation therapy plays an important role in the treatment of malignant tumour. The quality control for linear accelerator is one of the keys to ensure the correct and safe implementation of accurate radiotherapy.

The National Center of Oncology in Nouakchott is equipped with a linear accelerator which provides two energies in photon regime 6 MV and 18 MV.

The aim of this work is to measure the percentage depth dose (PDD) by two different ionization chambers: PTW 0.125 cm^3 and PTW 0.6 cm^3 and to compare the measured results with the results calculated by Treatment Planning Systems (TPS).

For the energy of 6 MV beam photon and for different dimensions of the field size, we have measured the percentage depth dose by two ionization chambers PTW 0,125 cm^3 and PTW 0,6 cm^3.

From these results, we have plotted the curves of the PDD and the relative deviation.

We compared the measurements of the percentage depth dose by the two ionization chambers.

We have found that the results obtained are consistent with those calculated by TPS.

There was note a slight difference between the depths measured by the two ionization chambers (PTW 0,125 cm^3 and PTW 0, 6 cm^3 chamber) and also that the latter is related to the field size.

Keywords: Quality control · PTW 0.125 cm^3 and PTW 0.6 cm^3 · (TPS) Linear accelerator

1 Introduction

The use of nuclear techniques are in speed extend in medical field in Africa.

The National Center of Oncology (CNO) at Nouakchott is equipped with equipment and skilled personnel able to work on sophisticated medical devices.

© Springer International Publishing AG, part of Springer Nature 2018
M. Ben Ahmed and A. A. Boudhir (Eds.): SCAMS 2017, LNNS 37, pp. 759–765, 2018.
https://doi.org/10.1007/978-3-319-74500-8_69

In addition, the Center is equipped with radiotherapy, nuclear medicine and chemotherapy equipment that comply with the international standards and are controlled by the International Atomic Energy Agency (IAEA).

In general, external radiotherapy uses ionizing radiation for the tumor treatment. The use of these radiations requires the utmost vigilance on the part of the medical physicist and the personnel who use them. However, the results of the treatment depend a lot on the precision of the dose delivered to the tumor [1–3].

The main objective of radiotherapy is the treatment of cancerous tumors. All cells are sensitive to radiation and all can be destroyed by high dose. The objective of radiotherapy is to deliver a dose in order to destroy the tumor without producing significant side effects (complications) in the healthy tissues [4, 5].

In order to check the quality of the accelerator and related equipment, we have performed measurements (percentage depth dose and dose profile) in water phantom by an ionization chamber for different energies and field sizes and at different Skin- Source-Distance (SSD).

To compare the measured results we have calculated the same parameters by Treatment Planning Systems (TPS), in order to compare measurements with calculation following the (IAEA) recommendations.

The (TPS) is a treatment planning software allowing to predict, according to a given ballistics, an established medical prescription, a chosen energy, an anatomical configuration, the dose at all points of the space [6–8].

2   Materials and Methods

Measurements of percentage depth dose and dose profile were carried out using a water phantom, connected to a PC. The system is controlled for the acquisition of the dosimetric data by MEPHYSTO mc^2 software. The dosietric measurements were realized using an ionization chamber associated with an electrometer and the chamber used for acquisition can move in three directions [9, 10].

The material used in this work is:

(1) Linear accelerator CLINAC 2100DHX, developed by the constrictor VARIAN MEDICAL SYSTEM, of two energies of photons of 6 MV and 18 MV (Fig. 1(a)).
(2) Mini water tank MP3-P (water phantom): The phantom used in this work is a cubic tank with a length of 60 cm (Fig. 1(b)).
(3) Cylindrical ionization chambers: TM31010 Semiflex chamber of 0.125 cm^3 and PTW 0.6 cm^3 (Fig. 1(c)).
(4) PTW electrometer: The collected charge (or intensity) produced in an ionization chamber is extremely low, its measurement requires a very sensitive device called electrometer (Fig. 1(d)).
(5) Medical Physics Control Center MEPHYSTO mc^2: MEPHYSTO is a software for the acquisition of therapeutic beam data and data analysis in radiotherapy (Fig. 1(e)).

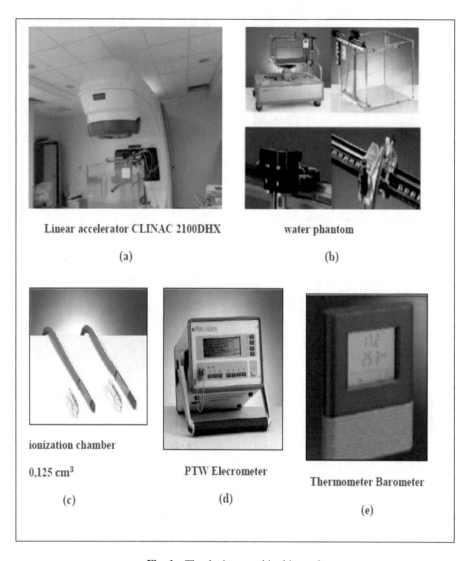

Fig. 1. The devices used in this work

3 Results

The results of the measurements are compared with the results calculated by TPS, Fig. 2 show a comparison between the measured and the TPS calculated for 6 MV photon beam and a (SSD) = 100 cm, and the field sizes of: $(10 \times 10, 20 \times 20, 30 \times 30, 40 \times 40)$ cm^2.

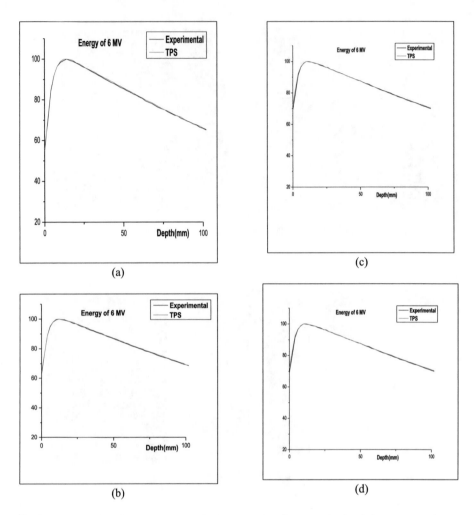

Fig. 2. Comparison of measured and calculated results of percentage depth dose curves of the 6 MV photon beam. (a) is for field size 10 cm × 10 cm, (b) is for 20 cm × 20 cm, (c) is for 30 cm × 30 cm, (d) is for 40 cm × 40 cm.

According to Fig. 3, there is a difference between the percentage depth doses measured by the two detectors, this difference is mainly and clearly observed in the first zone (the buil-up zone).

$$\text{relative deviation} = \left| \frac{(A - B)}{A} \right| \times 100 \tag{1}$$

A: measure obtained by ion chamber PTW 0,125 cm^3; B: measure obtained by ion chamber PTW 0,6 cm^3.

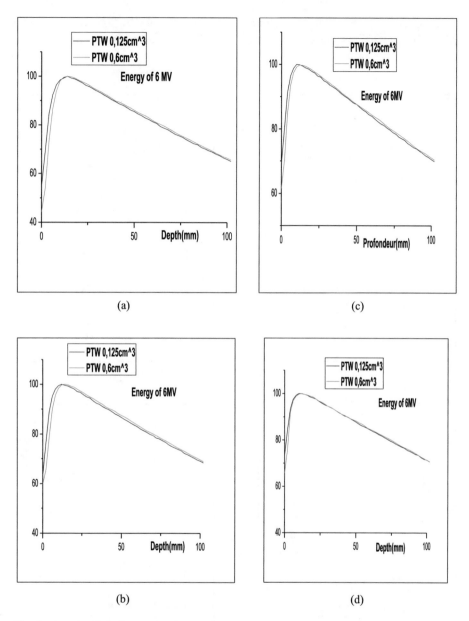

Fig. 3. For photon beam energy 6 MV, (PDD) acquired by two different ionization chambers PTW 0,125 cm³ and PTW 0,6 cm³, (a) is for field size 10 cm × 10 cm, (b) is for 20 cm × 20 cm, (c) is for field size 30 cm × 30 cm and (d) is for 40 cm × 40 cm.

According to these figures, it is noted that the maximum of the relative deviation between the measurements obtained by the ionization chambers that we have used, for the energy of 6 MV, is 22.27%, for the field (10 × 10) cm². 14, 24%, is for field size

(20×20) cm^2. 12, 35%, is for field size (30×30) cm^2 (Fig. 4). 10, 36%, is for field size (40×40) cm^2. These results show that the difference between the curves of the

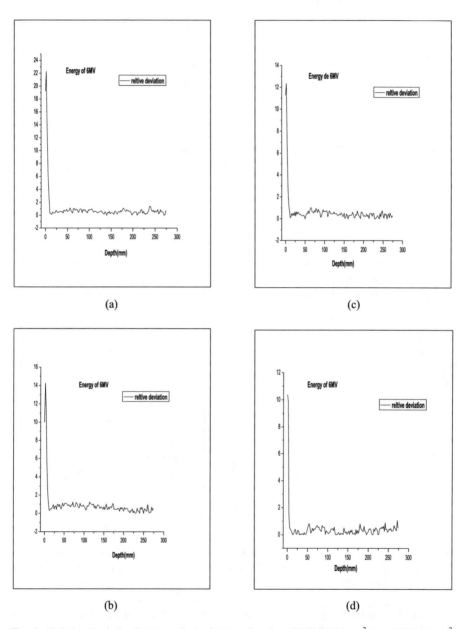

(a)

(c)

(b)

(d)

Fig. 4. Relative deviation between the ionization chambers PTW 0,125 cm^3 and PTW 0,6 cm^3, for 6 MV photon beam, (a) for field size 10 cm × 10 cm, (b) 20 cm × 20 cm, (c) 30 cm × 30 cm, (d) 40 cm × 40 cm.

percentage depth doses measured by the two detectors is bigger if the size of the field is lower.

4   Conclusion

In this work we performed a general quality control, based on the depth dose measurement (PDD) by two different ionization chambers: PTW 0.125 cm^3 and PTW 0.6 cm^3 and compared the results depth dose with the results calculated by TPS. We found that the results obtained are consistent with the results calculated by TPS. The comparison of the figures show differences between the curves of the percentage depth dose measured by the PTW 0.125 cm^3 and the curves of the percentage depth dose measured by the PTW 0.6 cm^3. We note that the difference between these curves is higher in the first zone, whereas it is low in the second zone. In appearance, we have remarked that the difference between the percentage depth doses measured by the detectors PTW 0.125 cm^3 and PTW 0.6 cm^3 is related to the field size of the photon beam.

References

1. Roy, M.L.: Study of national dosimetric standards for external beam radiotherapy: application to conformal irradiations. University of Nice Sophia-Antipolis (2012)
2. Blazy-Aubignac, L.: Quality control of the treatment planning systems dose calculations in external radiation therapy using the penelope Monte-Carlo code. University of Toulouse III – Paul Sabatier (2007)
3. Rosenwald, J.C., Bonvalet, L., Mazurier, J., Métayer, C.: Recommandations pour la mise en service et l'utilisation d'un système de planification de traitement en radiothérapie (TPS), RAPPORT SFPM N° 27 (2010)
4. Ding, G.X.: Energy spectra, angular spread, fluence profiles and dose distributions of 6 and 18 MV photon beams: results of Monte Carlo simulations for a Varian 2100EX accelerator. Phys. Med. Biol. **47**, 1025–1046 (2002)
5. Purdy, J.A.: Buildup/surface dose and exit dose measurements for 6 MV linear accelerator. Med. Phys. **13**, 259–262 (1986)
6. Engler, M.J., Jones, G.L.: Small beam calibration by 0.6 and 0.2 cm3 ionization chambers. Med. Phys. **11**, 822–826 (1984)
7. Villani, N., Gérard, K., Marchesi, V., Huger, S., François, P., Noël, A.: Statistical process control applied to intensity modulated radiotherapy pretreatment controls with portal dosimetry. Cancer/Radiothérapie **14**, 189–197 (2010)
8. Coffey II, C.W., Beach, J.L., Thompson, D.J., Mendiondo, M.: Xray beam characteristics of the Varian Clinac 6100 linear accelerator. Med. Phys. **7**, 716–722 (1980)
9. McKerracher, C., Thwaites, D.I.: Assessment of new small-field detectors against standard-field detectors for practical stereotactic beam data acquisition. Phys. Med. Biol. **44**, 2143–2160 (1999)
10. Krithivas, G., Rao, S.N.: A study of the characteristics of radiation contaminants within a clinically useful photon beam. Med. Phys. **12**, 764–768 (1985)

A Hybrid Approach for French Medical
Entity Recognition and Normalization

Allaouzi Imane[✉] and Mohamed Ben Ahmed

LIST/FSTT, Abdelmalek Essaadi University, Tangier, Morocco
imane.allaouzi@gmail.com, med.benahmed@gmail.com

Abstract. Medical document written in natural language is available in electronic form, and it constitutes an invaluable source for medical research. This paper describes our system based on hybrid approach for the task of Named Entity Recognition and Normalization of French medical documents using QUAERO corpus [1]. To evaluate our system, we took part in three subtasks: Entity Normalization, Named Entity Extraction and Classification which involved 10 categories including Anatomy, Chemicals & Drugs, Devices, Disorders, Geographic Areas, Living Beings, Objects, Phenomena, Physiology and Procedures. The results on both tasks, Named Entity Recognition and Normalization, demonstrate high performance as compared to other methods for French Medical Entity Recognition and Normalization.

Keywords: Medical entity recognition · Automatic categorization
Normalization · UMLS · Machine learning · Knowledge-based · NLP

1 Introduction

Over the past decades, a vast amount of medical document written in natural language is available in electronic form, and it constitutes an invaluable source for medical research. Therefore, Information Extraction techniques have been used to automatically extracting terms and concepts to obtain an organized and a structured representation of free-text documents.

Information extraction (IE) is an area of natural language processing that deals with finding factual information in a specific domain, from one or several documents that are written in a natural language in order to get structured information.

The task of information extraction has a particular interest in named entity recognition (NER), where one tries to recognize (single or subsequent) tokens in text that together constitute a rigid designator phrase, and to determine the category type to which these phrases belong [2]. Within the task, it is often necessary to perform so called Named Entity Normalization (NEN). One can define Normalization as standardizing named entity into a unique form or attributing it to a unique identifier in a terminology.

We focus on the challenge of Named Entity Recognition and Normalization in the medical domain and specifically textual biomedical document written in French. It can be divided in three subtasks: the recognition of biomedical terms and concepts from documents, then classification of these entities into ten categories: Anatomy, Chemicals

© Springer International Publishing AG, part of Springer Nature 2018
M. Ben Ahmed and A. A. Boudhir (Eds.): SCAMS 2017, LNNS 37, pp. 766–777, 2018.
https://doi.org/10.1007/978-3-319-74500-8_70

& Drugs, Devices, Disorders, Geographic Areas, Living Beings, Objects, Phenomena, Physiology and Procedures. And finally Normalization consists of assigning entities to their Unique Concept Identifiers (CUIs).

The aim of this study is fourfold: First, we provide a brief overview of various Named Entity Recognition approaches; second, we describe adopted approach and methodology used to build our system of Entity Recognition and Normalization; third, we present evaluation metrics used to assess the performance of our system followed by presentation, analysis and discussion of results; and finally, we conclude with remarks and some suggestions for future work.

2 Named Entity Recognition Task

The Named Entity Recognition task appeared for the first time in 1995, on the occasion of the Sixth Message Understanding Conference (MUC-6), which basically involves identifying entity names (people and organizations), place names, temporal and numerical expressions.

This task has been applied on a journalistic corpus which included 7 categories divided into 3 types:

- Enamex (Person, Organization and Location).
- Timex (Date, Time).
- Numex (Money and Percent).

By changing domain, organizers realized that some important proper nouns are not included in the MUC definition. To remedy these shortcomings, many researchers have proposed to extend the covered named entity types. For example we find another classification of named entity comprising approximately 150 different types [3].

Medical entity recognition is one of the most important areas in informatics research, as the medical domain contains a very rich and complex vocabulary. To represent medical language different techniques (Thesaurus, Dictionaries, Terminologies, etc.) have been created for a variety of medical domains including genetics, symptoms, anatomy, drugs, diseases, and medical codes.

2.1 Named Entity Recognition Approaches

In the following, we present different approaches for named entity recognition with a number of previous studies applied into French medical documents:

Knowledge-Based Approach. This approach is based primarily on lexical sources (medical terminologies and ontologies) and on a set of handcrafted rules which are produced manually by experts, and primarily based on linguistic descriptions, indices and dictionaries. It is simple to set up, but it takes a lot of time to construct knowledge base because of the high terminological variation and emerging of many new terms and new abbreviations of existing terms in medical domain.

Many existing clinical NLP systems use knowledge-based method to identify medical concepts, such as: a system of automatically retrieve medical problems from free-text documents using MMTx (MetaMap Transfer) with a negation detection algorithm (NeEx), which achieved 89.9% recall and 75.5% precision [4].

Machine Learning (ML) Approach. It is based on learning techniques that aim to give computers the ability to learn without being explicitly programmed. There are three types of ML model: supervised, semi-supervised and unsupervised machine learning model. Supervised machine learning model depends highly on the need of labelled training data, it is most frequently used and has achieved the best performance in medical NER task. However, one of the most important obstacles of this model is that labelled data are not always available. Therefore, a number of research studies have included the use of semi-supervised learning, which aims to improve performance by combining the labelled and unlabelled data. Unlike the previous models, unsupervised machine learning model does not require any training data, its objective is to create the possible annotation from the data. However, this learning technique is not popular among the ML models as does not produce good results without any supervised methods [5].

Among systems that are based on learning techniques for NER task, we found: Structural Support Vector Machines (SSVMs), an algorithm that combines the advantages of both CRFs and SVMs, developed for the Concept Extraction task of the 2010 i2b2 clinical NLP challenge, which was to recognize entities of medical problems, treatments, and tests from hospital discharge summaries achieved a highest F-measure of 85.74% on the test set of 2010 i2b2 NLP challenge [6].

Hybrid Approach. It tends to combine the advantages of knowledge-based and those of ML approach, while eliminating some of their weaknesses. For this approach, the rules are either written manually by an expert and then corrected and improved automatically, or they are automatically learned and then manually reviewed [7]. In the 2010 i2b2/VA Workshop on Natural Language Processing challenge, the authors used hybrid approach for the task of extracting clinical entities including medical, problems, tests and treatments have achieved a maximum overall F-score of 0.8391 [8].

3   System Description

In this section, we describe resources and methodology used to build our system of Biomedical Entity Recognition and Normalization in French narratives.

3.1   Resource

The QUAERO French Medical Corpus. Is a French annotated resource; it has been developed for the task of medical entity recognition and normalization [1]. The QUAERO corpus consists of two sub-corpora: titles from French MEDLINE, which contains journal citations and abstracts for biomedical literature from around the world, and some drug inserts published by the European Medicines Agency (EMEA). It comprises annotations for 10 types of biomedical entities corresponding to UMLS

Semantic Group: Anatomy, Chemical and Drugs, Devices, Disorders, Geographic Areas, Living Beings, Objects, Phenomena, Physiology and Procedures.

The annotation process was guided by concepts in the Unified Medical Language System (UMLS) in such a way that an entity was only annotated if the concept belonged to one of the 10 types of biomedical entities. Entity annotated is supplied in the BRAT standoff format and includes for each entity an id, the relevant UMLS category, the offset position (start and end) and the textual content of the annotation [9].

The Fig. 1 belows illustrate an example of the manual annotation obtained with the following sentence: "Le risque de réactions d' hypersensibilité a été plus important au cours des premières perfusions ainsi que chez les patients recevant de nouveau TYSABRI après une exposition initiale courte (une ou deux perfusions) suivie d'une période prolongée sans traitement (trois mois ou plus)".

T135	DISO 4746 4762	hypersensibilité
#135	AnnotatorNotes T135 C0020517	
T136	PROC 4807 4817	perfusions
#136	AnnotatorNotes T136 C0031001	
T137	LIVB 4837 4845	patients
#137	AnnotatorNotes T137 C0030705	
T138	CHEM 4866 4873	TYSABRI
#138	AnnotatorNotes T138 C1529600	
T139	DISO 4884 4894	exposition
#139	AnnotatorNotes T139 C2905623	
T140	PROC 4925 4935	perfusions
#140	AnnotatorNotes T140 C0031001	
T141	PROC 4976 4986	traitement
#141	AnnotatorNotes T141 C0087111	

Fig. 1. Example of manual annotation in the BRAT standoff format.

Unified Medical Language System (UMLS). Is developed by the US National Library of Medicine (NLM) in 1986 [10], with the aim of constituting a unified language from existing thesauri, classifications and nomenclatures to promote automatic indexing of biomedical documents. There are three main UMLS knowledge sources:

– Metathesaurus: This serves as the core database for the UMLS. It is the raw collections of concepts and terms from various controlled vocabularies and their relationships. It contains approximately 3.44 million concepts and 13.7 million unique concept names from 199 source vocabularies. The Metathesaurus is intended to be used mainly by developers of systems in medical informatics, after having signed a license agreement because it includes vocabulary content produced by different copyright holders as well as content produced by NLM [11].
– Semantic Network: Consists of a set of broad subject categories (Semantic Types) that provide a consistent categorization of all concepts represented in the UMLS Metathesaurus, and a set of useful and important relationships (Semantic Relations) that exist between Semantic Types. Each concept in the Metathesaurus is assigned one or more semantic types (categories) [12].

– Specialist Lexicon: Contains syntactic, morphological, and orthographic information for both biomedical vocabulary and commonly occurring English words.

Google Translation API

Scikit-Learn Package. We have implemented the proposed System in Python language using scikit-learn package; a set of python modules for machine learning and data mining.

3.2   Methodology

Biomedical Entity Recognition and Normalization system is designed to automatically recognize and normalize interest entities in French biomedical documents written in natural language. As shown in the Fig. 2, it consists of two major parts:

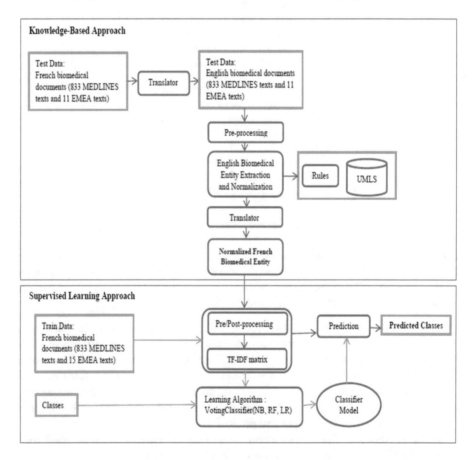

Fig. 2. Our hybrid approach pipeline.

Biomedical Entity Recognition. We proposed a hybrid approach to automatically extract biomedical entity (single or compound) from French document, and classify them into 10 categories corresponding to UMLS Semantic Group: Anatomy (ANAT), Chemicals and drugs (CHEM), Devices (DEVI), Disorders (DISO), Geographic areas (GEOG), Living beings (LIVB), Objects (OBJC), Phenomena (PHEN), Physiology (PHYS) and Procedures (PROC).

The hybrid approach integrates both Knowledge-Based and Machine Learning approach; Knowledge-Based approach was applied to extract biomedical entity from French document. The Unified Medical Language System (UMLS) was used as unique lexical sources with a set of handcrafted rules (string matching), which tries to match single or compound entity from biomedical text with UMLS concepts. To build this model, biomedical text was translated into English; using Google translation API; due to a lack of French version of the UMLS (The UMLS already contains a number of French vocabularies, but their coverage is rather limited), and then a set of pre-processing was applied including tokenization, Stop word removal, Part-Of-Speech tagging and chunking.

Once all entities were extracted, a supervised machine learning method was adopted to classify them into 10 categories: ANAT, CHEM, DEVI, DISO, GEOG, LIVB, OBJC, PHEN, PHYS and PROC. These entities were translated into French and undergo post-processing steps to reduce the number of false positive detection, which include: Tokenization, Remove French ligature, Remove diacritic marks and Stemming. In order to use supervised learning algorithms, a training set was built to train classification model, it consists of 833 MEDLINE titles and 15 EMEA documents. This data was converted into a TF-IDF matrix after having applied the previous post-processing to deal with imperfect data.

To build our classifier model, we have adopted an ensemble classification technique based on Weighted Majority Vote, which is a combination of three base classifiers including Naive Bayes, Random Forests and Logistic Regression. The main idea of Weighted Majority Vote is to assign a specific weight to each classifier in order to work with the weights; we collect the predicted class probabilities for each classifier, multiply it by the classifier weight, and take the average. Based on these weighted average probabilities, we can then assign the class label. The description of each classifier is given in the following Table 1.

Table 1. Description of the three classifiers

Classifier	Main idea	Advantages	Weaknesses
Naïve Bayes (NB)	The method is considered naive due to its assumption that every word in the document is conditionally independent from the position of the other words given the category of the document The classifier learns a set of probabilities from the training data during the learning phase. It then uses these probabilities and the Bayes theorem to classify any new documents [13]	– Fast to train/to classify – Requires a small amount of training data to estimate the parameters – Good results obtained in the most of the cases – Can be used for both binary and multiclass classification problems	– Its main disadvantage is that it can't learn Interactions between features
Random Forests (RF)	Are an ensemble classifier. That consists of many decision trees and outputs the class that is the mode of the classes output by individual trees [14]	– It is unexcelled in accuracy among current algorithms – It runs efficiently in large databases – It gives estimates of what variables are important in the classification	– Random forests have been observed to overfit for some datasets with noisy Classification/ Regression tasks [15]
Logistic Regression (LR)	Is a discriminative classifier. It consists of analyzing a dataset in which there are one or more independent variables that determine an outcome. The outcome is measured with a dichotomous variable (in which there are only two possible outcomes)	– Easily extended to multiple classes (multinomial regression) – Quick to train – Good accuracy for many simple data sets – Resistant to overfitting	– Independent Observations Required – Linear decision boundary

Normalization. Normalization Seeks to map extracted entities to UMLS Concept Unique Identifiers (CUIs). The system obtains normalized information from UMLS Metathesaurus directly after identifying biomedical entities.

4   Results and Discussion

To evaluate our solution, there are many standard performance measures to choose from. The standard metrics used to evaluate our system are: Precision (P), Recall (R), and their weighted mean F1-measure. They are very successful and popular, since the main issue in these metrics lies in their binary decision process: whether a predicted entity element is correct or not [16].

Let's, True Positive (TP): The number of cases which were correctly predicted (entity predicted exists in the training data), False Positive (FP): The number of cases which were predicted entity is correct and does not exist in the training data, and False Negative (FN): The number of cases which were incorrectly predicted (entity not predicted and exists in the training data). Metrics are formulated as:

$$\text{Recall}(R) = TP/(TP + FN) \tag{1}$$

$$\text{Precision}(P) = TP/(TP + FP) \tag{2}$$

$$F1 - measure = (2 * R * P)/(R + P) \tag{3}$$

The evaluation task was done using QUAERO corpus, it consists of two annotated subsets: training set (833 MEDLINE texts and 15 EMEA texts) used to train classifier, and test set (833 MEDLINES texts and 11 EMEA texts) used as reference to determine entity extraction and normalization performance, as well as, to test our classifier model against unseen data.

The results obtained for French biomedical entity extraction and normalization with knowledge-based approach are represented in Table 2. Compared with results on MEDLINE set, our system demonstrated better performance on EMEA texts in terms of F-measure for both entity extraction and normalization task, maybe this is due to the nature of data (EMEA is a long text and MEDLINE titles is very short text). F-measure is under the overage because of two reasons: Firstly, the high number of False Positive (FP) due to UMLS, which contains non specialized terms, such as: Octobre, Academie, Nationale, Couleur, etc., and secondly, the high number of False Negative (FN),which is justified by the inexact translation problem.

Table 2.   French biomedical entity extraction and normalization results on the QUAREO corpus.

Entity extraction						
Corpus	TP	FP	FN	Recall	Precision	F-measure
EMEA	1732	2892	963	0.6427	0.3745	0.4733
MEDLINE	1678	3081	1316	0.5604	0.3526	0.4328
Entity normalization						
Corpus	TP	FP	FN	Recall	Precision	F-measure
EMEA	1630	3101	1065	0.6048	0.3445	0.4390
MEDLINE	1521	3245	1473	0.50801	0.3191	0.3920

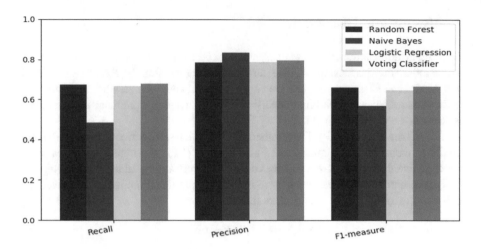

Fig. 3. Classification performance results, in terms of Precision, Recall, and F1-measure of biomedical entity by different classifiers.

Figure 3 shows a graphical summary of classification performance results, in terms of Precision, Recall and F1-measure, of biomedical entity using three different classifiers and averaged by the Voting Classifier. The three classifiers (RandomForestClassifier, GaussianNB and LogisticRegression) are trained and used to train a hard-voting Voting Classifier with weights [1, 2, 4], which means that the predicted probabilities is not the same for the three classifiers when the averaged probability is calculated. As can be seen, Voting Classifier outperforms other classifiers in term of Recall (67.94%) and F1-measure (66.71%), and come second with Precision of 79.64% preceded by GaussianNB classifier with Precision of 83.49%.

Table 3 shows that voting classifier had good performance for most of the 10 categories, especially "LIVB" with 96.64% of F1-measure, while some others categories

Table 3. Voting classifier results across the different categories.

	Precision	Recall	F1-measure
ANAT	0.8081	0.6478	0.7191
CHEM	0.9395	0.3205	0.4779
DEVI	0.8889	0.1667	0.2807
DISO	0.4744	0.9185	0.6256
GEOG	1.0000	0.5909	0.7429
LIVB	0.9848	0.9487	0.9664
OBJC	0.5091	0.3944	0.4444
PHEN	0.7143	0.2632	0.3846
PHYS	0.6863	0.5882	0.6335
PROC	0.9707	0.8406	0.9010
Avg/Total	**0.7855**	**0.6738**	**0.6618**

had low recall, such as: DEVI with 16.67% and PHEN with 26.32%. This is generally characterised by the rare class problem.

In the last year, CLEF eHealth [17] offered challenge of named entity recognition and normalization of biomedical entities in French narratives, using QUAERO corpus, Named entity recognition involved the same ten types invoked in our study. Participants adopted different approach, some of them relied on knowledge-based approach and others relied on machine-learning approach. The best performances in terms of F-measure are reported in the following Table 4.

Table 4. Comparison of the best French biomedical entity recognition and normalization systems, using QUAERO corpus [17].

Authors	Techniques	Task	F1-measure	
			EMEA	MEDLINE
Erasmus team [18]	The system is based on knowledge-based approach, Central in this approach is indexing with French terminologies from the UMLS supplemented with automatically translated English UMLS terms, followed by several post-processing steps to reduce the number of false-positive detections	Entity recognition	**0.749**	**0.698**
		Entity normalization	**0.666**	**0.605**
SIBM team [19]	They approached entity extraction from the provided QUAERO dataset as an indexing task relying on multiple knowledge Organization Systems (KOS) partially or totally translated into French. The extraction method, ECMT (Extracting Concepts with Multiple Terminologies), performs bag of words concept matching at the sentence level [17]	Entity recognition	0.444	0.520
		Entity normalization	0.315	0.388

Comparing our system results with those of previous studies, we can conclude that our proposed approach gives very good result, with second best performance after Erasmus team approach.

5   Conclusion and Future Work

In this paper, we propose a system for French biomedical Entity Normalization and Recognition including ten categories corresponding to UMLS Semantic Group:

Anatomy, Chemicals & Drugs, Devices, Disorders, Geographic Areas, Living Beings, Objects, Phenomena, Physiology, and Procedures.

Our system achieved good results in terms of F1-measure, but still under the average because of the high number of false positive. This can be improved by avoiding the identification of commonly occurring English words. Our classifier can also be improved by adding into training data more examples about rare classes.

References

1. Névéol, A., Grouin, C., Leixa, J., Rosset, S., Zweigenbaum, P.: The QUAERO French medical corpus: a resource for medical entity recognition and normalization. In: Fourth Workshop on Building and Evaluating Resources for Health and Biomedical Text Processing – BioTxtM, pp. 24–30 (2014)
2. György, S.: Feature engineering for domain independent named entity recognition and biomedical text mining applications. In: Research Group on Artificial Intelligence of the Hungarian Academy of Sciences and the University of Szeged, pp. 5–6 (2008)
3. Sekine, S., Sudo, K., Nobata, C.: Extended named entity hierarchy. In: LREC-2002 (2002)
4. Meystre, S.M., Haug, P.J.: Comparing natural language processing tools to extract medical problems from narrative text. In: AMIA Annual Symposium Proceedings, pp. 525–529 (2005)
5. Alfred, R., Leong, L.C., On, C.K., Anthony, P.: Malay named entity recognition based on rule-based approach. Int. J. Mach. Learn. Comput. 4(3), 301–302 (2014)
6. Tang, B., Cao, H., Wu, Y., Jiang, M., Xu, H.: Clinical entity recognition using structural support vector machines with rich features. In: Proceedings of the ACM Sixth International Workshop on Data and Text Mining in Biomedical Informatics, p. 1 (2012)
7. Abacha, A.B., Zweigenbaum, P.: Medical entity recognition: a comparison of semantic and statistical methods. In: Proceedings of BioNLP Workshop, pp. 57–58 (2011)
8. Jiang, M., Chen, Y., Liu, M., Rosenbloom, S.T., Mani, S., Denny, J.C., Xu, H.: A study of machine-learning-based approaches to extract clinical entities and their assertions from discharge summaries. J. Am. Med. Inform. Assoc. 18, 601–606 (2011)
9. Ho-Dac, L.M, Tanguy, L., Grauby, C., Hnub, N., Mby A.H., Malosse, J., Rivière, L., Veltz-Mauclair, A., Wauquier, M.: LITL at CLEF eHealth2016: recognizing entities in French biomedical documents. In: CLEF 2016 Online Working Notes, p. 3. CEUR-WS (2016)
10. McCray, A.T.: The scope and structure of the first version of the UMLS semantic network. In: Proceedings of the Annual Symposium on Computer Application in Medical Care, pp. 126–130 (1990)
11. National Library of Medicine: Semantic networks. In: UMLS Reference Manual. U.S. National Library of Medicine, National Institutes of Health, Bethesda (2009). Chapter 1
12. National Library of Medicine: Semantic networks. In: UMLS Reference Manual. U.S. National Library of Medicine, National Institutes of Health, Bethesda (2009). Chapter 5
13. Elnahrawy, E.M.: Log-based chat room monitoring using text categorization: a comparative study. In: International Conference on Information and Knowledge Sharing, Acta Press Series, pp. 1–2 (2002)
14. Breiman, L.: Mach. Learn. 45, 5 (2001). https://doi.org/10.1023/A:1010933404324
15. Segal, M.R.: Machine Learning Benchmarks and Random Forest Regression. Center for Bioinformatics & Molecular Biostatistics (2014)
16. Galibert, O., Rosset, S., Grouin, C., Zweigenbaum, P., Quintard, L.: Structured and extended named entity evaluation in automatic speech transcriptions. In: Proceedings of 5th International Joint Conference on Natural Language Processing, p. 521 (2011)

17. Névéol, A., Cohen, K.B., Grouin, C., Hamon, T., Lavergne, T., Kelly, L., Goeuriot, L., Rey, G., Robert, A., Tannier, X., Zweigenbaum, P.: Clinical information extraction at the CLEF eHealth evaluation lab 2016. In: CLEF 2016 Online Working Notes. CEUR-WS (2016)
18. Van Mulligen, E., Afzal, Z., Akhondi, S.A., Vo, D., Kors, J.A.: Erasmus MC at CLEF eHealth 2016: concept recognition and coding in French texts. In: CLEF 2016 Online Working Notes. CEUR-WS (2016)
19. Cabot, C., Soualmia, L.F., Dahamna, B., Darmoni, S.J.: SIBM at CLEF eHealth evaluation lab 2016: extracting concepts in french medical texts with ECMT and CIMIND. In: CLEF 2016 Online Working Notes. CEUR-WS (2016)

Analysis of Evolutionary Trends of Incidence and Mortality by Cancers

Hajar Saoud[1]([⊠]), Abderrahim Ghadi[1], and Mohamed Ghailani[2]

[1] LIST Laboratory, University of Abdelmalek Essaadi (UAE), Tangier, Morocco
saoudhajar1994@gmail.com, ghadi05@gmail.com
[2] LabTIC Laboratory, University of Abdelmalek Essaadi (UAE),
Tangier, Morocco
ghalamed@gmail.com

Abstract. Cancer has become the disease of the century, it knew a great evolution in recent years and it reaches several patients each year, it becomes necessary to find solutions to fight against this disease. Then our thesis comes in this direction, it will provide an analysis of the evolutionary trends of incidence and mortality by cancers in Morocco over a period of time.

This paper presents the state of art of the existing methods of analysis, projection and prediction of incidence and mortality by cancers.

At first, we will give a vision of the research carried out by the ministry of health that can be considered as the starting point of our subject. Then we will explain the three projection models and we will compare the existing prediction methods: Classical approach and Bayesian approach.

Also we will give a vision about the Material and methods that we will use.

Keywords: Cancer · Incidence · Mortality · Projection
Bayesian method · Forecasting

1 Introduction

Cancer has become the major cause of mortality in the word with approximately 14.1 million new cases (7 427 000 men and 6 663 000 women) and 8.2 million cancer-related deaths (4 653 000 men and 3 548 000 women) in 2012 [1].

Among men, the five most types of cancer diagnosed according to the statistics of 2012 were the lung (16.7% of the total), prostate (15.0%), colorectum (10.0%), stomach (8.5%), and liver (7.5%). Among women, the five most types of cancers diagnosed were the breast (25.2% of the total), colorectum (9.2%), lung (8.7%), cervix (7.9%), and stomach (4.8%) [1].

Among men, lung cancer occurs the first place in incidence with (34.2 per 100 000) followed by prostate cancer (31.1 per 100 000). Among women, breast cancer had a higher number of incidence (43.3 per 100 000) than the others types of cancers, followed by colorectal cancer (14.3 per 100 000) [1].

In our study we will concentrate on cancers in Morocco because cancers occupy an increasingly important place in Moroccan health preoccupations.

The number of deaths according to the statistics of 2012 is 22900 cases (12500 men and 10400 women) [2], and according to the statistics of the WHO (World Health

M. Ben Ahmed and A. A. Boudhir (Eds.): SCAMS 2017, LNNS 37, pp. 778–788, 2018.
https://doi.org/10.1007/978-3-319-74500-8_71

Organization) about the incidence of cancer in Morocco in 2012. Among women, breast cancer had the highest incidence (6.650 cases), followed in second place by Cervical cancer (2.258 cases), cancer of the colon and rectum (1.126), thyroid cancer (929) and ovarian cancer (735).

Among men, lung cancer occurs the first place with (3.497 cases), followed by prostate cancer (2.332), bladder cancer (1.429), cancer of the colon and rectum (1.358) and non-Hodgkin's lymphoma (1.089).

And according to the gravity of the cancer and its frequency in Morocco, cancer is becoming an object of preoccupation more and more frequent in the Moroccan medical community. Therefore, it was necessary to estimate and forecast the rates of mortality and incidence by cancers over the next few years in order to control the disease and to find solutions for the fight.

The work presented in this article is the result of bibliographic studies on the methods of estimating of the incidence and mortality by cancers in the past years and the prediction for the following years, conducted as part of a current thesis on analysis of evolutionary trends of incidence and mortality by cancers.

2 Material and Methods

The data required in this research are the number of cancer incidence and deaths in Morocco and the corresponding population size.

Populations sizes (past and future) were estimated by the High Commission for Planning in Morocco (HCP) [3], based on the results of five censuses (1960, 1971, 1982, 1994 and 2004). The data, available from 1960 to 2050 for ages 0 to 75 years and over, were aggregated by age groups of five year (16 classes of age from 00–04 years to age 75 and over).

Cancer incidence data were provided by the Cancer Registry of the Region Greater Casablanca [4–6] and the Cancer Registry of Rabat [7, 8], the officially published data in Morocco of incidence by cancer are of the region Greater Casablanca and Rabat.

Firstly, a comparison must be made between the different projection models: the complete age-period-cohort model, as well as the age-cohort and age-period models [9, 10], and choose the most suitable for our study.

Then, we must make a comparison between the prediction methods existing to choose the most appropriate, there are two approaches: classical approach and Bayesian approach.

The most used method was the Bayesian approach [11] where the past and future mortality and incidence parameters are estimated using Bayesian inference by assigning a probability distribution a priori to the parameters and calculating their distribution a posterior from the information provided by the data (observation).

3   Cancers Registries

The incidence data of cancers that are officially published in Morocco, by cancers registries, are data of Rabat and the region of Greater Casablanca. In the register of Rabat, new cases of cancer diagnosed are registered in from the year 2005, for the register of the region Greater Casablanca new cases of cancer diagnosed are registered in from the year 2004.

For the Rabat register of 2005 [7], and the three registers of Greater Casablanca [4–6], they calculated the incidence rates without making predictions for the following years.

The Rabat Cancer Registry of the period 2006–2008 [8], carried out by (la Direction de l'Epidemiologie et de Lutte contre les Maladies), contains a study similar to our research topic, they have made estimates for past periods (2006–2008) to make a projection up to 2020, it can be considered as a starting point of our subject.

To calculate the number of incidences by cancers they are based on the estimate of the population of Rabat, they treated all the types of cancers (by age and sex) and for forecasts of incidences for the year 2020 they are based on the incidence rates between 2006–2008 (by age and sex) and the evolution of the total number of the Moroccan population, the forecasts were calculated for all Morocco they assumed that the rates of the incidence of Rabat are similar among the rest of the Moroccan population.

Then in their study they are based only on the demographic evolution of the Moroccan population which corresponds to a hypothesis of stability of the risks, they did not take in consideration the evolution of other factors like for example the factor cohort, this factor plays an important role in the projection of the disease, it shows the evolution of the disease for each generation and this can help to know the causes of the disease and their evolutions.

Studies have been carried out in France on lung cancer mortality among French women [12, 13], the taking of consideration of the birth cohort effect has clearly shown the effect of the evolution of female smoking in France for each generation.

Also another study was carried out on breast cancer among French women [14], they took into account the cohort effect which showed well the increase of incidence rates for each generation.

4   The Projection Models

To project the incidence and mortality by cancers there are three models: the age-period-cohort model (APC) the completed model or the models age-period (AP), age-cohort (AC) partial models.

Those models are the most used in the projection of the evolution of the events, they give the possibility to study the evolution over time of a phenomenon or a measure, through the temporal effects of the cohort (birth cohort) and the period (observation period), adjusted for the age of the individuals at the time of the phenomenon. They are initially created for demographic use, but they are quickly adapted to epidemiology to palliate the limitations of methods of standardization and to track the progression of the diseases over time.

The age variable represents the age of the person at the moment of his confrontation with the phenomenon, generally it is the time elapsed between the birth of the person and the appearance of the phenomenon.

The period variable is the period of observation of the phenomenon, it brings together the events that are likely to affect simultaneously and in the same way, for all cohorts of birth regardless of the age of the individuals of these cohorts.

Finally, the cohort variable represents the birth cohort. It brings together all individuals who are born in the same periods of time (precise time intervals), with the idea that individuals of the same cohort will be treated in the same way over time.

Those models will allow to us to project the incidences and mortalities by cancers in future and past.

The choice of one of these three models is based on the evolution of the risk factors of the disease, the satisfaction and the precision of the results.

Age-period model was used in the study that was carried out in the region of Birmingham (U.K.) on bladder cancer incidence among men [9], it was the more adapted and it has given satisfactory results.

Age-period model was used in the study that has been carried out on lung cancer mortality among French women [12, 13] they chose the age-cohort model. The choice of this model is based on two findings. First, the age-cohort model seemed well adapted to the nature of the evolution of the risk factor of smoking if we compared it with the age-period model. Second, aggregation of data in five-year periods in women does not leave enough degrees of freedom to correctly estimate the period effect [15].

Age-period-cohort model was used in the study that was carried out in France on lung cancer mortality among men [15] and the study of the lung cancer incidence in the Bas-Rhin [16], the age-cohort-period complete model was the more adapted because there is a stabilization of lung cancer mortality rates on the contrary to women, because of a decrease in tobacco consumption.

5 Forecasting Methods

The methods used to make the predictions are based either on a classical approach or on a Bayesian approach of projection models.

5.1 Classical Approach

The classical approach of projection models consists to make a linear combination between cohort and period effects in order to conclude the future unknown effects that will be obtained by generalized linear models [17].

Generalized linear models. Generalized linear models, as it was explained in [22], give us the possibility to study and search the connection between the variable of response Y and a set of predictor variables X1...Xk. A GLM is a generalization of the classical linear models.

Generalized linear models are composed of three components:

- Y, is the response variable, a random component with which a probability distribution is associated.
- X1....XK set of the explanatory variables (predictors), used as predictors in the model, they are written as a linear combination that is named the deterministic component.
- The link describes the relationship between the mathematical expectation of the response variable Y and the linear combination of the variables X1....Xk.

The probability distribution of the response variable belongs to exponential family which includes the Poisson, normal, binomial, exponential and gamma distributions.

The deterministic component, that is also called a linear predictor, is written as a linear combination $\beta0 + \beta1\ X1 + \ldots + \beta k\ Xk$, specifies the predictors.

The component link is the relation between the deterministic component $\beta0 + \beta1\ X1 + \ldots + \beta k\ Xk$ and the response variable Y.

It specifies how the mathematical expectation of Y denoted λ is related to the linear predictor constructed from the explanatory variables.

The expectation λ can be modeled directly (usual linear regression) or model a function $f(\lambda)$ of the expectation:

$$f(\lambda) = \beta0 + \beta1\ X1 + \ldots + \beta kXk$$

This function $f(\lambda)$, that is called the link function, gives the possibility to model the logarithm of the expectation $f(\lambda) = \log(\lambda)$. The Models using this link function are log-linear models.

Poisson distribution. We know that the cancers data are non-negative, so they will be modeled as the log-scale. The Poisson distribution, one of the generalized linear models, will be the more appropriate because it is written as a log-scale and count data of cancers are not proportions [17], this method was explained in [24].

Poisson law. Let λ a real and Y is a real random variable, we said that Y follows a Poisson law of parameter λ, $Y \sim P(\lambda)$, if and only if for any natural integer k,

$$P(Y = k) = e^{-\lambda}\frac{\lambda^k}{k!}$$

As a result, $E(Y) = V(Y) = \lambda$.

Poisson regression model. The log of the parameter λ is modeled as a linear combination of the explanatory variables (x) and their parameters β:

$$\ln(y) = \alpha + \beta_1 x_1 + \ldots + \beta_i x_i + \ldots + \beta_k x_k$$

- y is a realization of the variable Y according to a Poisson law.
- α ordered at origin.
- β_i Coefficient associated with the ith explanatory variable x_i.

Estimation of the model. The goal is to estimate α and the vector β of the coefficients β_i with the method of the maximum likelihood. And as result of Poisson law the regression will be written like that:

$$\ln(E(Y)) = \ln(\lambda) = \alpha + \beta_1 x_1 + .. + \beta_i x_i + .. + \beta_k x_k$$

Likelihood:

$$L = \prod_{i=1}^{n} P(Y_i = k_i) = \prod_{i=1}^{n} e^{-\lambda_i} \frac{\lambda_i^{k_i}}{k_i!}$$

With n the number of observations.

$$\lambda_i = e^{\alpha + \beta x_i}$$

$$x_i = (x_{i1} \ldots x_{ij}) \beta = (\beta_1 \ldots \beta_j)$$

Logarithm of Likelihood:

$$\ln(L) = \sum_{i=1}^{n} [k_i \ln(k_i) - \lambda_i] - c$$

Maximization because of the derivative:

$$s(\alpha, \beta) = \begin{array}{c} \frac{\partial \ln(L)}{\partial \alpha} \\ \frac{\partial \ln(L)}{\partial \beta} \end{array} = \begin{array}{c} \sum_{i=1}^{n} (y_i - \lambda_i) \\ \sum_{i=1}^{n} (y_i - \lambda_i) \end{array}$$

$$= \sum_{i=1}^{n} (y_i - \lambda_i) \frac{1}{x_i}$$

Newton-Raphson method: Algorithm:

$$\begin{array}{c} \alpha_{k+1} \\ \beta_{k+1} \end{array} = \begin{array}{c} \alpha_k \\ \beta_k \end{array} + I^{-1}(\alpha_k, \beta_k) s(\alpha_k, \beta_k)$$

Where I^{-1} is the variance-covariance matrix.

We stop when $\begin{array}{c} \alpha_{k+1} \\ \beta_{k+1} \end{array} \approx \begin{array}{c} \alpha_k \\ \beta_k \end{array}$.

The deviance. The deviance allows to evaluate the fit quality of the model based on differences between observations and estimates

$$\text{Deviance} = 2[L_{sat} - L]$$

- L_{sat} value of log-likelihood of saturated model.
- L value of log-likelihood of estimated model.

Related work. In the research that was carried out in England and Wales [18], they used this classical model to project and forecast lung cancer mortality rates. So, to estimate the future effects they are based on the nearest past effects estimated and linear regression. In [19] they mentioned that this method has drawbacks because the choice of past values and the regression used is arbitrary.

In the research [21] they used three methods to project the mortality by cancer in Swiss as it was mentioned in [19]. In the first method, they projected the mortality by caner with hypothesis that the period effect does not change. In the second method they used linear regression which is used in [18] applied on the period effect. In the third method they integrated the evolution of the risk factors to determine the futures effects as for example the risk factor smoking. Despite those methods have produced results that can lead to qualitative conclusions but those methods do not always produce the same projections [19].

5.2 Bayesian Approach

Bayesian analysis [11] is a statistical descriptive analysis method among others. And because of the limitations of the classical methods they are elaborated as part of the Bayesian approach to simplify the use of all the mathematics resources. This method can be used in several fields of application of the usual methods of analysis because it brings appreciable changes on the existing methods which lead to more completed conclusion.

Bayesian analysis is used in forecasting and estimation, past and future parameters are estimated using this Bayesian method by giving a probability distribution a priori to the parameters and calculating their distribution a posterior from the information provided by the data.

The principle is to estimate the parameters effect of age, effect of the period of observation and effect of the birth cohort from the data and then to project the effects period and cohort to estimate the specific incidences and mortalities.

Principles of the Bayesian Approach. This approach was explained in [15].

Bayesian inference. Let a set of observation noted $y = y_1,.....,y_n$ and let θ a parameter and other unobserved quantities.

Law a priori: the information a priori on the parameter θ signifies all information available on θ outside that brought by the observations, its probability noted: $p(\theta)$.

Law a posteriori: it is the conditional law of θ knowing y. Its probability is noted $p(\theta \mid y)$. Under the Bayes formula, we have:

$$p(\theta|y) = \frac{p(y|\theta)\,p(\theta)}{\int p(y|\theta)\,p(\theta)\,d\theta}$$

The density of the joint law of (θ, y) is

$$p(\theta, y) = p(y|\theta)\, p(\theta)$$

The marginal law density of y is

$$p(y) = \int p(\theta, y)\, d\theta$$
$$= \int p(y|\theta)\, p(\theta)\, d\theta$$

Applied to the conditional likelihood $l(\theta|y)$, the expression of the Bayes theorem becomes:

$$p(\theta|y) = \frac{l(\theta|y)\, p(\theta)}{\int l(\theta|y)\, p(\theta)\, d\theta}$$

Since $l(\theta|y) = p\,(y|\theta)$

From the expression of $p\,(\theta \mid y))$, it is possible to calculate the a posteriori marginal probability density of θ, the expectation (conditional) a posteriori and the covariance matrices a posterior of θ by integration in the continuous case and by summation in the discrete case [15].

Hope a posteriori:

$$E(\theta|y) = \int \theta\, p(\theta|y)\, d\theta$$

The a posterior covariance matrix:

$$cov(\theta|y) = \int (\theta - E(\theta|y))(\theta - E(\theta|y))'\, p(\theta|y)\, d\theta$$

Principle of projection. This Bayesian inference is adapted to estimate future values in [11] as it was mentioned and explained in [15].

If y_P represents the vector of the past values of the variable y, θ is the vector of the distribution parameters of y, $p\,(y_P, \theta)$ the density of the joint distribution of y_p and θ, p (θ) the density of the distribution of θ, $p(y_P|\theta)$ the distribution of y_p conditionally to θ, $p(\theta \mid y_p)$ the distribution of θ conditionally to y_p, the Bayesian model is written:

$$p\!\left(y_{p},\theta\right) = p(\theta)p\!\left(y_{p}|\theta\right) = p\!\left(y_{p}\right)p\!\left(\theta|y_{p}\right)$$

From where:

$$p\!\left(\theta|y_{p}\right) = \frac{p\!\left(y_{p}|\theta\right)\, p(\theta)}{\int p\!\left(y_{p}|\theta\right)\, p(\theta)\, d\theta}$$

$p(\theta)$ is said to be a priori density, $p\,(\theta_P \mid y)$ is the density of the a posterior law of θ.

If, y_F represents the future values of y. If, in addition, the model is based on a regression with as explanatory variable W and if w_P and w_F are the values taken by W, respectively for past and future times and if w is the set of past and future values of W:

$$p\left(y_F|y_p, w\right) = \int p\left(\theta|y_p, w_p\right)p\left(y_F|y_p, w_F, \theta\right)d\theta$$

$p(y_F|y_P, w)$ is the density of the a posteriori predictive distribution of y_F, $p(\theta \mid y_P, w_F)$ is the a posteriori distribution of θ [15].

Prior distributions for cohort, period and age effects. We will take the example of the completed model age-period-cohort as it was explained in [15].

To calculate the posterior we should estimate the prior correctly so the model imposes a priori constraints between the successive parameters of the covariates age, period and cohort.

For The age, the general dependency relation between the successive effects is expressed by the conditional distribution of relative parameters to the age and the relation between the expectation of an effect and the values taken by the neighboring effects. The same thing for the cohort and period effects.

So the conditional distribution for the age will be written as follows:

$$\alpha_i|\alpha_{i'}, i' \neq i \sim N\left(\mu_{\alpha_i}, \frac{1}{\omega\tau_\alpha}\right) \text{ and } \mu_{\alpha_i} = E(\alpha_i|\alpha_{i'}, i' \neq i)$$

For the period:

$$\beta_j|\beta_{j'}, j' < j \sim N\left(\mu_{\beta_j}, \frac{1}{\tau_\beta}\right) \text{ and } \mu_{\beta_j} = E(\beta_j|\beta_{j'}, j' < j)$$

For the cohort:

$$\gamma_k|\gamma_{k'}, k' < k \sim N\left(\mu_{\gamma_k}, \frac{1}{\tau_\gamma}\right) \text{ and } \mu_{\gamma_k} = E(\gamma_k|\gamma_{k'}, k' < k)$$

This writing means that the age effect α_i, cohort effect β_j and period effect γ_k follow a normal law conditionally to other age effects, cohort effect and period effect, respectively.

τ_α, τ_β and τ_γ are the precision of α_i, β_j and γ_k (inverse of the variance), respectively. ω is the number of neighboring parameters on which the age effect that will be estimated depends.

Related work. For the Bayesian analysis, it is used in all these researches [12, 13, 15, 16], these analysis was carried out with the software Bugs [20], a tool that is dedicated to Bayesian analysis and implemented by Gibbs sampling. Gibbs sampling is a special case of Monte Carlo methods by Markov chain (MCMC) [19].

5.3   Comparison

It seems that Bayesian approach is more appropriate than the classical approach for estimating and forecasting the rates of future incidence and mortality by cancer, because in classical approaches the choice of past values is arbitrary which may lead to not precise results so why they were elaborated as part of the Bayesian approach which leads to more accurate estimates than other models and also allows to improve existing procedures sometimes poorly adapted to particular situations.

In the research [23] they mentioned that the a priori that has been used in the Bayesian approach gives more accurate estimates than classical approaches and also the rates estimated by linear models do not last for long periods of time [19].

And from the comparison carried out in research [19] for classical age-period-cohort that is used in [18] they mentioned that model give a satisfied results but it is not perforce the appropriate model because before applying the regression it is necessary to make several tests and several analysis to know the past values on which they will be based and the most appropriate regression, in the Bayesian approach all this can be concluded from the data.

6   Conclusion

Cancers occupy an increasingly important place in health preoccupations in Morocco, they represent the second cause of death in Morocco and burden the resources of the healthcare system.

This thesis consists to do an analysis of the evolutionary trends of the incidence and mortality by cancers in Morocco in a given period.

In this article, we have given a vision on the three projection models: age-period-cohort, age-period and age-cohort, and we have compared the two methods of forecasting: classical approach and Bayesian approach.

The next step of this research is the choosing of one type of cancer, recovering its data of incidence and mortality and applying the methods that were described in this paper.

This work can be distributed in three steps, first the recovery of incidence and mortality data by cancers, then the choice of projection model between the three models of projections that have already been defined. And finally, the calculation of future statistics by the Bayesian method.

References

1. World Cancer Report (2014). www.iarc.fr/en/publications/books/wcr/IARC-COVER.pdf
2. Rapport mortalite Maroc (2014). http://www.who.int/cancer/country-profiles/mar_fr.pdf
3. Maaroufi, Y.: Projections de la population totale par groupe d'âge et sexe (en milliers et au milieu de l'année): 1960-2050. http://www.hcp.ma/Projections-de-la-population-totale-par-groupe-d-age-et-sexe-en-milliers-et-au-milieu-de-l-annee-1960-2050_a676.html
4. Registre des cancers de la Région du Grand Casablanca: Année e (2004). www.contrelecancer.ma/fr/documents/registre-des-cancers-de-la-region-du-grand-casabla/

5. Registre des cancers de la Région du Grand Casablanca 2005–2006–2007. http://www.contrelecancer.ma/site_media/uploaded_files/RCRC_-_28_mai_2012.pdf
6. Registre des cancers de la Region du Grand Casablanca pour la période 2008–2012. http://www.contrelecancer.ma/site_media/uploaded_files/RCRGC.pdf
7. Registre des Cancers de Rabat (2005). http://biblio.medramo.ac.ma/bib/RECRAB_2005.pdf
8. Registre des Cancers de Rabat 2006–2008. http://biblio.medramo.ac.ma/bib/Registre-Cancer-Rabat%202006-2008.pdf
9. Clayton, D., Schifflers, E.: Models for temporal variation in cancer rates. I: age–period and age–cohort models. Stat. Med. **6**(4), 449–467 (1987)
10. Clayton, D., Schifflers, E.: Models for temporal variation in cancer rates. II: Age-period-cohort models. Stat. Med. **6**(4), 469–481 (1987)
11. Mouchart, M.L.: Inference bayesienne: principles generaux. In: Droesbecke, J., Fine, J., Saporta, G. (eds.) Methodes bayesiennes en statistique, pp. 101–102. Technip, Paris (2002)
12. Eilstein, D., Uhry, Z., Chérié-Challine, L., Isnard, H.: Mortalité par cancer du poumon chez les femmes françaises. Analyse de tendance et projection à l'aide d'un modèle âge-cohorte bayésien, de 1975 à 2014. Rev. Dépidémiologie Santé Publique **53**(2), 167–181 (2005)
13. Eilstein, D., Uhry, Z., Cherie-Challine, I., Isnard, H.: Mortalite par cancer du poumon chez les femmes en France, analyse de tendance et projection de 1975 a 2019
14. Bouée, S., Grosclaude, P., Alfonsi, A., Florentin, V., Clavel-Chapelon, F., Fagnani, F.: Projection de l'incidence du cancer du sein en 2018 en France. Bull. Cancer (Paris) **97**(3), 293–299 (2010)
15. Eilstein, D., Uhry, Z., Chérié-Challine, L., Bloch, J.: Mortalité par cancer du poumon en France métropolitaine Analyse de tendance et projection de 1975 à 2014. In: Tavel, P. (ed.) Modeling and Simulation Design. AK Peters Ltd., Natick (2007). H. I. I. de veille sanitaire
16. Eilstein, D., Quoix, E., Hédelin, G.: Incidence du cancer du poumon dans le Bas-Rhin : tendance et projections en 2014. Rev. Mal. Respir. **23**(2), 117–125 (2006)
17. McCullagh, P., Nelder, J.A.: Generalized Linear Models, 2nd edn. Chapman and Hall, London (1989)
18. Osmond, C.: Using age, period and cohort models to estimate future mortality rates. Int. J. Epidemiol. **14**(1), 124–129 (1985)
19. Bray, I.: Application of Markov chain Monte Carlo methods to projecting cancer incidence and mortality. Appl. Stat. **51**, 151–164 (2002)
20. Spiegelhalter, D., Thomas, A., Best, N., Gilks, W.: Bugs 0.5 bayesian inference using Gibbs sampling manual (version ii) (1996)
21. Negri, E., La Vecchia, C., Levi, F., Randriamharisoa, A., Boyle, P.: The application of age, period and cohort models to predict Swiss cancer mortality. J. Cancer Res. Clin. Oncol. **166**, 207–214 (1990)
22. Lejeune, M., Sprta, G., Droesbeke, J.: Modèles statiques pour données qualitatives (2005)
23. Berzuini, C., Clayton, D.: Bayesian analysis of survival on multiple time scals. Statist. Med. **13**, 823–838 (1994)
24. Poisson regression. https://perso.univ-rennes1.fr/valerie.monbet/ExposesM2/2013/RegressionPoissonAR.pdf

Simplified and Efficient Framework for Securing Medical Image Processing Over Cloud Computing

Mbarek Marwan[(✉)] , Ali Kartit, and Hassan Ouahmane

LTI Laboratory, ENSA, Chouaïb Doukkali University, Avenue Jabran Khalil Jabran,
BP 299, El Jadida, Morocco
marwan.mbarek@gmail.com, alikartit@gmail.com,
hassan.ouahmane@yahoo.fr

Abstract. As the utilization of imaging tools increases in healthcare domain, cloud applications are becoming a popular approach for processing medical data. In addition to being the most affordable solution, this approach offers simple and efficient mechanisms to manage clients' data. In this regard, this concept aims at completely replacing traditional on-premises data centers and moving computations to the cloud. These advantages would inevitably accelerate the adoption of cloud-based data processing in the healthcare industry. Despite many advantages of using cloud services, outsourcing data processing to an external provider could jeopardize the duty to preserve the confidentiality of patients' data. In fact, security and privacy issues obstruct this paradigm from achieving greater success in medical sector. In this context, there are several approaches and techniques to protect clients' data against potentially malicious cloud providers. A survey of existing methods shows that computations on outsourcing data are either not supported, or will pose more challenges. Moreover, they have been shown to be poorly suited to digital records because they process each pixel separately. The eventual goal is to propose a simple and efficient technique for handling clients' data safely. The proposal provides also a mechanism to ensure integrity checking of outsourced data. The novelty of our work is using K-means algorithm and watermarking method to secure cloud services. The implementation results prove that this methodology ensures security and QoS requirements.

Keywords: Image processing · Cloud computing · Security · K-means

1 Introduction

Cloud computing is a parallel and distributed system consisting of a collection of data centers to offer easily accessing IT services. Moreover, it introduces a new way that enables users to quickly store and compute their data. This model is in general the result of many developments in computer science like smart grid, virtualization, High Performance Computing (HPC), Service-Oriented Architecture (SOA). Basically, there are four principal methods for deploying and accessing cloud resources, i.e., private, public, hybrid and community. In this environment, three major services are offered by cloud providers and tailored to better meet clients' needs. The main features and characteristics of this technology are defined by the National Institute of Standards and Technology

© Springer International Publishing AG, part of Springer Nature 2018
M. Ben Ahmed and A. A. Boudhir (Eds.): SCAMS 2017, LNNS 37, pp. 789–802, 2018.
https://doi.org/10.1007/978-3-319-74500-8_72

(NIST) [1]. Clearly, this paradigm has the potential to facilitate the utilization of IT services in medical sector. This study explores the feasibility of using cloud services to process images remotely. In this scenario, on-demand imaging tools are delivered to consumers to process health records without the need for traditional applications. In other words, data processing takes place on powerful cloud servers to improve flexibility, efficiency and performance. After the processing stage, the response is sent back to healthcare professionals using a specific web interface. In this regard, Fig. 1 presents a general overview of the use of cloud computing for remote image processing.

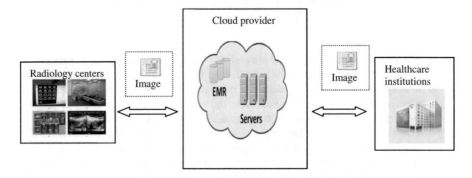

Fig. 1. Basic idea for the medical image processing using cloud

Even with these advantages, however, the utilization of cloud may expose personal data to serious security problems. This is caused by many risk factors, including those related to cloud technology, such as virtualization [2], data location [3], web technology [4] and interoperability [5, 6]. Besides these well-known problems, handling and analyzing medical images over cloud requires additional security measures. The results of the literature review reveal that many techniques are used to protect clients' privacy, especially SOA, homomorphic encryption, secret share scheme. However, these methods still have some drawbacks regarding performance and the inability to comply with data privacy requirements, which limits its applications in cloud computing. Contrary to this, our novel technique to secure image processing is simple and efficient.

The present paper is organized as follows. Section 2 and 3 discuss existing methods for protecting image processing when using cloud computing. Section 4 presents the proposed solution to maintain privacy of patients and meet security requirements. In Sect. 5, we discuss simulation results to highlight the proposed method. We end this paper in Sect. 6 by concluding remarks and a discussion of future work.

2   Related Work

In [7], a novel homomorphic technique is used for processing digital records remotely. To deal with security challenges, the authors use Learning With Error (LWE) scheme to allow a user to perform operations on encrypted data without need of decryption. The purpose of this work is to develop a cryptosystem that helps consumers to execute addition and multiplication on encrypted images. These above reasons make the proposed technique of interest in cloud security and privacy. Because of the large size of medical images, using homomorphic encryption has negative side effects on performance due to its heavy computational costs, which limits its utilization in imaging tools.

Mohanty et al. [8] present a solution to process data using cloud tools. Clearly, the proposed framework provides a mechanism to visualize data effectively. In this respect, a user encrypts medical data using Shamir Secret Share (SSS) in such a way that each image can be analyzed on multiple nodes. In addition, the authors apply pre-classification volume ray-casting method to reinforce the protection and privacy of medical data. In this scheme, n shares are created from the original image using Shamir's (k, n) threshold. This ensures data confidentiality because less than k centers can never reconstruct the secret image. Furthermore, secured volume ray-casting is performed in a cluster system, thereby workload will be balanced across all nodes.

In [9], Ali Mirarab et al. use an infrastructure composed of Eucalyptus system and ImageJ tools to outsource data processing. In this approach, Eucalyptus is an open source that offers a cluster with multiple nodes for high performance and availability. This framework involves the integration of genetic algorithm (GA) and particle swarm optimization (PSO) to manage computational resources more efficiently. Besides, this solution has two main components to handle digital data, namely ImageJ plug-ins to facilitate the implementation of this framework and ImageJ Macro to process input data. Since the measurement of security is based only on ImageJ, the proposal cannot be used to protect confidentiality and integrity of digital data.

In [10], it is possible to process encrypted data using the proposed cryptosystem. This scheme is based principally on Residue Number System (RNS) that allows computations to be carried out directly on data stored in the cloud. In this concept, users need to apply homomorphic encryption before uploading them to the cloud service. This new method has the main objective to perform addition, subtraction and multiplication on ciphertext. Mainly, the authors develop a mechanism that enables users to apply Sobel filter for edge detection on images in cloud. The key goal is to reduce cyber threats and to perform data analysis safely. However, performance is the major disadvantage of using this technique as a data security measure.

Huang et al. [11] implement a new approach to secure image processing when using distributed systems. In this solution, an image is processed in many nodes using MapReduce function which is implemented in Hadoop system. This concept aims at enhancing data protection by dividing data into multiple portions, thereby avoiding disclosure of sensitive information. However, data are not totally protected against untrusted service providers because they are processed in a plaintext format. In addition, this framework uses a simple mechanism that is based only on login and password to ensure access control.

In [12], Bednarz et al. present a cloud solution for medical images analysis purpose. To this aim, the authors develop a framework that is composed mainly on NetCTAR, a runtime environment and toolbox called CSIRO. Typically, the proposed framework is implemented on OpenStack system to enhance availability. The objective is to provide the application with two main services, i.e., HCA-Vision to quantify cell features, MILXView to process 3D images, and X-TRACT dedicated to X-ray images. The proposed solution, however, suffers from some limitations because it does not implement security measures related to integrity and authentication.

Todica et al. [13] propose a new distributed computing system to protect cloud services, especially image processing. Typically in this approach, Service-Oriented Architecture (SOA) is used to guarantee availability and reliability. In particular, cloud providers leverage SOA to manage interoperability issues, which the major barriers that health systems are facing. Moreover, this method is the simplest way to spread workloads among clusters to avoid resources restrictions. Basically, the authors use XML standard to develop this framework that helps clients to process their data remotely. However, the proposal does not offer necessary security mechanisms to ensure data confidentiality.

In [14], Hadoop system is used in this study to perform complex image processing. Essentially, Vemula et al. implement MapReduce functions to distribute requests among clusters for reasons related to the availability of cloud services. Moreover, a clustering technique is used to split an image into multiple portions, thereby processing each part on different nodes. This allows parallel extraction of an image to enhance system performance and reliability. Since the secret is divided into various regions, it is hard to extract information about the plaintext. Consequently, this solution enables a user to process digital records remotely using Hadoop platform. For example, Laplacien filter and Canny Edge Detection (CED) are applied on cipher data to prove the correctness of this proposed model.

Sathish et al. [15] introduce Dynamic Switch of Reduce Function (DSRF) as an effective way to improve MapReduce functions. Basically, this technique is used for decreasing idle time and improving performance. This is achieved by means of a variety of components to process images more efficiently in cloud environment, including the Master module that divides an image, and MapReduce function which processes digital records. Although it processes data, this framework does not provide any security mechanism during data analysis over cloud.

In [16], Chiang et al. describe the promise and the potential of Service-Oriented Architecture (SOA) in cloud computing. By effectively using SOA, many nodes of the distributed system can participate in data processing. Meanwhile, the authors use ImageJ software that offers necessary image processing operations. To efficiently access and process data, the framework is composed mainly of four components: the presentation layer, service layer, business logic and ImageJ tool. Despite its potential to process data, security measures need more improvements to meet privacy requirements.

Moulick et al. [17] illustrate an easy and effective solution for data processing. In this regard, SOA technique is used to implement image processing as a service. The DIPE system offers numerous algorithms that have been proposed to handle digital records. Essentially, the proposed framework has three models, including Programming, Services and Messaging. The need to protect sensitive data when using cloud services is the major barrier to accepting this proposed approach.

In [18], Kagadis et al. propose a simple system that detect brain tumor efficiently. The proposed framework uses cloud resources to remotely process medical images. In this study, the proposal is divided into four modules, namely the front-end, intelligent load balancer as a dispatcher mechanism, universal storage, and processing VMs dedicated to image processing. Additionally, the authors use role-based authentication model for privacy-preservation purposes.

A case study presented by Hu et al. [19], where a novel method based on Secure Multiparty Computation (SMC) is used for data protection. In this concept, distributed linear image filtering is applied for a secure image processing. Basically, the proposed solution uses a combination of rank reduction and random permutation to process data in an encrypted format. Additionally, the secure inner product protocol is implemented to perform linear filtering safely. In this environment, the proposed scheme is composed typically on two entities: one party holds the secret data and the other party offers required tools to process data without revealing any confidential information. In general, this technique is used in relatively simple cases where data processing requires just basic operations.

Similarly, the authors in [20] focus on SMP and investigate the utilization of this technique in cloud based image processing. Technically, feature vectors are used to represent each digital data, especially significant portions. Hence, SMP is applied directly on generated vectors. By using this method, users can perform face detection in a secure manner, and hence, can be implemented in cloud environment. The main disadvantage of this solution is that it is impossible to reconstruct the original image from the generated vectors.

3   Analysis of Existing Approaches

In the era of cloud computing, data processing as a service has continued to grow in popularity and application among healthcare organizations. In this concept, consumers use on-demand, cost-efficient tools to analyze digital records. The main problem with this technique is that it exposes clients' data to security risks. In response to this issue, many techniques are used to avoid accidental data disclosure that results from using cloud computing. What all these techniques have in common is the potential to process data remotely. However, these techniques still have some serious disadvantages that limit their usefulness. For example, SOA is an easy way to increase performance by dividing the workload among nodes in a cluster. However, it does not guarantee the privacy of medical information because data are typically processed in a plaintext format. In contrast, homomorphic approach is an efficient method to operate encrypted data and to avoid malicious utilization of confidential data. Despite their great success, applying these techniques will unfortunately have a negative impact on running speed. Although the SSS technique ensures the confidentiality of data, it is difficult to directly process shares created by SSS technique. In fact, it is necessary to adopt image processing algorithms before applying them on shares. In general, SMP promotes collaborative work in order to outsource computation to an external party. The utilization of this concept in image processing involves complex tasks for designing secure protocols.

To summarize, current available solutions necessitate additional enhancement to satisfy the demands of security and performance. This is basically due to the fact that they require pixel-by-pixel processing approach, which affects running time. In this respect, a simple method based on segmentation is proposed to address security issues in data processing over cloud computing. In addition, we propose a mechanism to keep up data integrity for our proposal in the most efficient way possible.

4   Proposed Framework

The utilization of cloud services exposes data to increasing security risks despite its relatively low cost and wide availability. In most cases, clients need to encode their sensitive data before sending them to the cloud computing. Mainly, it keeps private information secure, thereby reducing the risk of insider threat. Generally, with this approach it is difficult to process data stored on cloud and, simultaneously, protect patient's data privacy. In this regard, the proposed framework aims to address this issue by providing simple and efficient mechanisms to use cloud services safely. The novelty of this solution is designing an architecture composed of three main elements: consumers, third party and service providers. In this case, we introduce a new entity that acts as an intermediary between the client and cloud providers to enhance data protection. Furthermore, we use segmentation technique to avoid data disclosure. Meanwhile, we rely on watermarking method to maintain data integrity.

4.1   The Fundamentals of the Proposed Framework

As aforementioned, the classical architecture of cloud computing poses many security problems which have not been well addressed. In this respect, we design a framework to process data securely using cloud computing. For this reason, we add a new module called CloudSec dedicated to alleviating security and privacy concerns inherent to cloud technology. Basically, this module relies on security measures to prevent unauthorized use of confidential data. Additionally, it ensures authentication to check that the consumer has permission to access to a specific cloud service. A brief overview of the proposed solution is depicted in Fig. 2.

The proposed solution is based on distributed data processing approach to significantly improve performance. To this aim, a multi-cloud model is typically used to process data on different cloud providers. So, a digital record is firstly split into many parts. As a result, only a small part of the original data is analyzed on a cloud provider. This would dramatically improve data confidentiality; thereby minimize the impact of insider threats.

Normally, healthcare organizations communicate with CloudSec using SSL protocol for authentication and data encryption. At the same time, this module stores metadata in a secure local database. This information is often necessary to identify clients and locate their data. In doing so, cloud providers cannot reveal personal information related to cloud consumers; hence, it is the easiest way to ensure anonymity. As outlined above, CloudSec is the primary element in securing the proposed architecture. In addition to enforcing access control policy, CloudSec offers security tools to protect digital records during cloud services utilization. In this study, CloudSec converts the secret image into small parts composed of the subsets of pixels with similar colour. Consequently, many

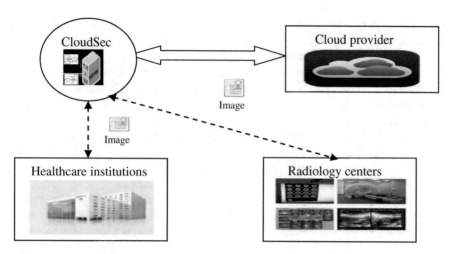

Fig. 2. Architecture of the proposed framework

regions are generated form an image by using a clustering technique, i.e., K-means. In this classification technique, pixels are arranged in random position in order to change the image content. The primary objective of clustering approach is to transform a meaningful image into scrambled versions, essentially because of substantial changes in pixel position. Interestingly, each service provider analyzes only small piece of original image to make sure that health information is protected. In this environment, K-means algorithm is usually implemented at CloudSec module level. In this respect, CloudSec divides the secret image into various small parts, thereby acting as a proxy server to ensure that communication between clients and cloud providers is safe. The main function of each module of the proposed system is illustrated in Fig. 3.

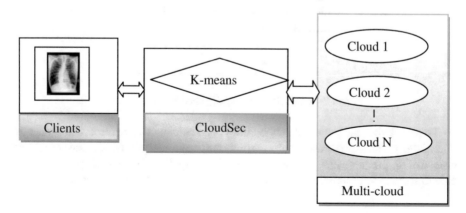

Fig. 3. The principle of our proposed approach

Besides, using multi-region segmentation would help cloud providers analyze medical records easily. Indeed, this technique consists in the grouping of pixels on the basis of their gray values, which facilitates data processing and transmission. Another mechanism of security to be considered is integrity issue. To this end, we propose a mechanism for comparing the current state of the stored shares to a previously recorded state in order to detect any malicious modification. In this framework, we propose a hybrid solution for integrity checking to enhance security in cloud computing. Basically, we use watermarking technique as an efficient approach to prevent accidental changes due to attacks against the integrity of the original images. Specifically, we insert a digital signature in each part to maintain the integrity of the digital records. Since medical images are sensitive, we select a reversible watermarking method based on the Thodi algorithm [21, 22] to avoid data degradation. Typically, the Thodi algorithm is an efficient mechanism to survive normal image processing operations.

Accordingly, we create the digital signature for each portion, relying on Secure Hash Algorithm SHA-256 such as Message Code Authentication (MCA). Basically, we use the Thodi algorithm for watermark embedding to easily verify that images are not changed during data processing. So, clients rely on a digital fingerprint of an image contents which were calculated before and after cloud utilization to ensure the integrity of medical images, as illustrated in Fig. 4.

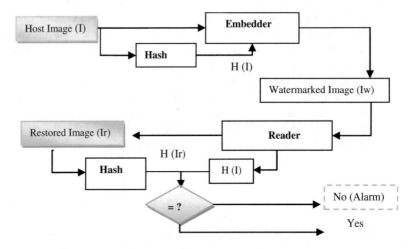

Fig. 4. The principle of integrity checking mechanism

To summarize, the proposed methodology is an efficient way to meet security needs. Therefore, it is an appropriate solution to promote image analysis using cloud computing instead of on-premises imaging tools and healthcare applications. The proposal aims not only to secure cloud-based image processing, but also to enhance system performance.

4.2   An Overview of the Main Used Methods

There are a broad range of issues that must be addressed before implementing cloud services, issues related to storage, software, and virtual machines. The most common encryption standards are widely applied to enforce data security and privacy. There are two major types of encryption methods. The first group only protects data against unauthorized users, while the second type is designed to enable users to perform mathematical operations securely. In this study, we present the most essential security measures for data security, which are typically implemented in the CloudSec module.

Reversible Watermarking. Image-quality degradation caused by digital watermark insertion is the major consideration in our choice of watermarking algorithm. Typically, the proposed method would allow a user to easily insert and remove a message from digital image without a noticeable degradation. In addition, it is important to select a scheme that can provide higher embedding capacity. For integrity of electronic records, this technique seeks to insert a digital signature in a lossless manner, thereby avoiding damage to images. More precisely, the embedded hash function is extracted to verify the integrity of the original. Since there are various schemes, it is necessary to evaluate and compare existing models in order to choose to the appropriate one. In this study, we rely on the study presented in [23] to achieve this objective. Clearly, reversible watermarking algorithms that use expansion approaches are usually the simplest and most effective methods. Besides, this study shows that the Thodi [21, 22] algorithm has a remarkable features in terms of performance and robustness. Accordingly, we select this scheme to insert digital watermark for integrity checking purpose.

Multi-region Segmentation. In data processing, this process plays a major role in identifying and delineating objects in images. Therefore, this technique is widely used in medical bioinformatics to improve disease diagnosis and treatment. Basically, an image is partitioned into a number of logical segments in order to simplify data processing. In general, there are two main approaches to detect meaningful discontinuities in an image: edge and region based methods. The region-based algorithms take into consideration pixel values to group similar pixels in a specific region. In the second approach, we focus on identifying boundaries that separate one region from another. Technically, this can be achieved by classifying pixels on edge or non-edge. In this study, we suggest clustering technique to generate many regions based on the similarity of pixel's features. Specifically, we choose K-means algorithm to affect image's pixels to a specific cluster, thereby creating portions with homogeneous visual content. Consequently, this technique would use a medical image to generate several scrambled regions. Meanwhile, this mechanism will enable distributed data processing by spreading data across multiple servers. In other words, the principal concept of segmentation approach is to provide a simple method of processing data without revealing confidential information. As a result, this technique allows CloudSec module to divide digital records into predefined number of regions before sending them to the multicloud system to enforce data security.

5   Simulation Results

The proposed framework employs the concept of clustering technique to divide a digital image into multiple segments. The key idea behind this technique is that cloud providers cannot view patients' medical records or gain unauthorized access to any medical information. As outlined above, we suggest K-means as a segmentation method to classify each pixel to the closest cluster. In this study, we will firstly discuss the principle of K-means scheme, and then we implement this solution to demonstrate its correctness. The algorithm 1 provides the pseudo code of a generalized version of the conventional K-means clustering algorithm [24]. In this scheme, m[i] is usually used to denote the membership of point xi.

Algorithm 1. K-means

Input: $X = \{x_1, x_2,..., x_N\} \in \mathbb{R}^D$ (N×D input data set)
Output: $C=\{c_1, c_2,..., c_k\} \in \mathbb{R}^D$ (K cluster centers)
We randomly choose a subset C of X as the initial set of cluster centers;
  1: while termination criterion is not satisfied do
  2:      for (for (i = 1; i ≤ N; i = i +1) do
  3:         Each element x_i is assigned to the nearest cluster;
  4:         $m[i] = \text{argmin} \| x_i - c_k \|^2$
              $k \in \{1, 2,..., K\}$
  5:    end for
  6:    Compute the cluster centers;
  7:      For (k = 1; k ≤ K; k = k + 1) do
  8:         Cluster S_k contains the set of points x_i, which are basically nearest to the center c_k;
  9:             $S_k = \{x_i \mid m[i] = k\}$;
Determine the new center c_k as the mean of the points that belong to S_k;
 10:    $c_k = \frac{1}{|S_k|}\sum_{x_i \in S_k} x_i$
 11:   end for
 12: end for
 13: return $<c_1, c_2,..., c_k>$

In most cases, clustering techniques are a very attractive and promising solution to extract the area of interest from a background area. The entire procedure involves multiple steps in order to accomplish this goal. In such a processing model, partial stretching enhancement is firstly performed on medical images to improve the quality level of digital data. Second, subtractive clustering technique is performed to create initial centroids, which are the cornerstone of K-means algorithm. Generally, we rely on the density of surrounding data points to accomplish this task efficiently. Finally, it is ever recommended to use a median filter to create the final image without any unwanted objects and regions. Interestingly, we apply common image processing only on each generated region instead of the original image.

In this study, we seek to enhance the quality of encrypted image using cloud services. To this end, we process these generated shares in order to get the final result. After postprocessing, we reconstruct the final image using the same methodology. In this section, we present some experimental results to simulate our solution. In this context, we use two colours medical images to evaluate the effectiveness of the proposed approach. First, we apply Gaussian filter to reduce the level of noise in the secret image, as illustrated in Fig. 5.

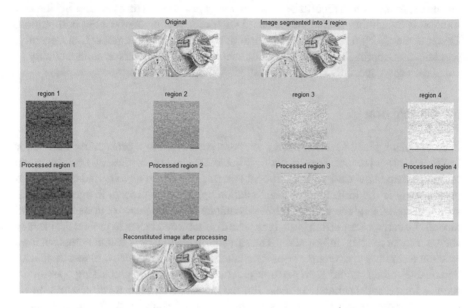

Fig. 5. The experimental results using Gaussian filter

In the second case, image enhancement was performed using K-means clustering technique. Normally, this method is widely used for increasing medical image's quality. For this reason, we increase the value of pixels' intensity to enhance the contrast and brightness of the output image. In this work, we apply our method on a RGB image to simulate the proposed approach, as shown in Fig. 6.

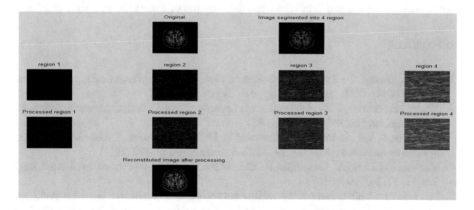

Fig. 6. The experimental results for intensity enhancement

The present work proves the feasibility and utility of applications of the clustering technique in cloud security. In particular, the simulation result shows that this technique protects digital records against potential internal threats. In fact, this method allows clients to process health records on different cloud providers. The basic idea behind this solution is processing data in public cloud without revealing confidential information. This is achieved by using K-means algorithm to creating multi regions. Consequently, segmentation approach is a simple and efficient method to guarantee confidentiality in cloud services, especially image processing.

6 Conclusion

In recent years cloud services have been perceived as a more prospective choice for healthcare organizations due to better adjustment and utilization of computational resources than is the case with traditional systems. The main novelty and contribution of this study is the utilization of segmentation approach to process images safely. In such a model, CloudSec uses clustering technique to create many regions from health records. Usually, each portion is analyzed in a distinct cloud server to protect the privacy and security of medical information. The simulation results show that the proposal is an efficient technique to overcome security challenges in cloud services. In fact, this solution enables cloud providers to perform practically basic image processing operations, such as image enhancement. Technically, the multi-region segmentation based on K-means algorithm is the core element of our proposed framework. To meet security requirements, we suggest a simple methodology to check the integrity of outsourced data. Basically, we use a hybrid approach based on watermarking techniques and digital signature to detect loss of data or any accidental modifications in image content. Consequently, this solution can help address the issue of trust in cloud services, and hence, increase cloud adoption in the healthcare domain. As for the future works, we intend to extend this framework to integrate complex image processing operations. Additionally, we will use some indices to evaluate the robustness of our proposed security mechanisms, such as correlation coefficient and the Number of Pixels Change Rate (NPCR).

References

1. Mell, P., Grance, T.: The NIST definition of cloud computing. Technical report, National Institute of Standards and Technology, vol. 15, pp. 1–3 (2009)
2. Mazhar, A., Samee, U.K., Athanasios, V.: Security in cloud computing: opportunities and challenges. Inf. Sci. **305**, 357–383 (2015)
3. Fernandes, D.A.B., Soares, L.F.B., Gomes, J.V., Freire, M.M., Inácio, P.R.M.: Security issues in cloud environments: a survey. Int. J. Inf. Secur. **13**(2), 113–170 (2013)
4. Marwan, M., Kartit, A., Ouahmane, H.: Cloud-based medical image issues. Int. J. Appl. Eng. Res. **11**, 3713–3719 (2016)
5. Petcu, D.: Portability and interoperability between clouds: challenges and case study. In: Abramowicz, W., Llorente, I.M., Surridge, M., Zisman, A., Vayssière, J. (eds.) ServiceWave 2011. LNCS, vol. 6994, pp. 62–74. Springer, Heidelberg (2011)

6. Ahuja, S.P., Maniand, S., Zambrano, J.: A survey of the state of cloud computing in healthcare. Network Commun. Technol. **1**(2), 12–19 (2012)
7. Challa, R.K., Kakinada, J., Vijaya Kumari, G., Sunny, B.: Secure image processing using LWE based homomorphic encryption. In: Proceedings of the IEEE International Conference on Electrical, Computer and Communication Technologies (ICECCT), pp. 1–6 (2015)
8. Mohanty, M., Atrey, P.K., Ooi, W.-T.: Secure cloud-based medical data visualization. In: Proceedings of the ACM Conference on Multimedia (ACMMM 2012), Japan, pp. 1105–1108 (2012)
9. Mirarab, A., Ghasemi Fard, N., Shamsi, M.: A cloud solution for medical image processing. Int. J. Eng. Res. Appl. **4**(7), 74–82 (2014)
10. Gomathisankaran, M., Yuan, X., Kamongi, P.: Ensure privacy and security in the process of medical image analysis. In: Proceedings of the IEEE International Conference on Granular Computing (GrC), pp. 120–125 (2013)
11. Huang, Q., Ye, L., Yu, M., Wu, F., Liang, R.: Medical information integration based cloud computing. In: Proceedings of the IEEE International Conference on Network Computing and Information Security (NCIS), pp. 79–83 (2011). https://doi.org/10.1109/ncis.2011.24
12. Bednarz, T., Wang, D., Arzhaeva, Y., Szul, P., Chen, S., Khassapov, A., Budett, N., Gureyev, T., Taylor, J.: Cloud-based image analysis and processing toolbox for biomedical applications. In: Proceedings of the 8th IEEE International Conference on eScience, Chicago, pp. 8–12 (2012)
13. Todica, V., Vaida, M.F.: SOA-based medical image processing platform. In: Proceedings of the of IEEE International Conference on Automation, Quality and Testing, Robotics (AQTR), pp. 398–403 (2008). https://doi.org/10.1109/aqtr.2008.4588775
14. Vemula, S., Crick, C.: Hadoop image processing framework. In: Proceedings of the of IEEE International Congress on Big Data (BigData Congress), New York, pp. 506–513 (2015). https://doi.org/10.1109/bigdatacongress.2015.80
15. Sathish, V., Sangeetha, T.A.: Cloud-based image processing with data priority distribution mechanism. J. Comput. Appl. **06**(1), 6–8 (2013)
16. Chiang, W., Lin, H., Wu, T., Chen, C.: Building a cloud service for medical image processing based on service-orient architecture. In: Proceedings of the 4th International Conference on Biomedical Engineering and Informatics (BMEI), pp. 1459–1465 (2011). https://doi.org/10.1109/bmei.2011.6098638
17. Moulick, H.N., Ghosh, M.: Medical image processing using a service oriented architecture and distributed environment. Am. J. Eng. Res. (AJER) **02**(10), 52–62 (2013)
18. Kagadis, G., Alexakos, C., Papadimitroulas, P., Papanikolaou, N., Megalooikonomou, V., Karnabatidis, D.: Cloud computing application for brain tumor detection. European Society of Radiology, Poster C-1851, pp. 1–16 (2015)
19. Hu, N., Cheung, S.C.: Secure image filtering. In: Proceedings of the of IEEE International Conference on Image Processing (ICIP 2006), pp. 1553–1556 (2006)
20. Avidan, S., Butman M.: Blind vision. In: Leonardis, A., Bischof, H., Pinz, A. (eds.) Computer Vision – ECCV 2006. Lecture Notes in Computer Science, vol. 3953, pp. 1–13. Springer, Heidelberg (2006). https://doi.org/10.1007/11744078_1
21. Thodi, D.M., Rodriguez, J.J.: Prediction-error based reversible watermarking. In: Proceedings of the International Conference on Image Processing (ICIP 2004), vol. 3, pp. 1549–1552 (2004)

22. Thodi, D.M., Rodriguez, J.J.: Expansion embedding techniques for reversible watermarking. IEEE Trans. Image Process. **16**(3), 721–730 (2007)
23. Khan, A., Siddiqa, A., Munib, S., Malik, S.A.: A recent survey of reversible watermarking techniques. Inform. Sci. **279**, 251–272 (2014)
24. Gan, G., Ma, C., Wu, J.: Data Clustering: Theory, Algorithms and Applications. Society for Industrial and Applied Mathematics. SIAM, Philadelphia (2007)

AWS for Health Care System

Abdelilah Bouslama[(✉)] and Yassin Laaziz

LabTic, ENSA of Tangier, Abdelmalek Essaadi University, Tétouan, Morocco
bo.abdelilah@gmail.com

Abstract. In the last decade information technology is considered not just as basic support tool but the core of all modern system such as healthcare, transportation, pollution … etc. that why this this paper come to provide a practical reference in information technology (IT) and healthcare industry which become under pressure to improve safety, efficiently, and to reduce human errors, and provide secure access to timely information. Also this document explain the reason behind implications of cloud computing in particular Amazon cloud as one of leaders of cloud services to handle IOT real time data of heterogeneous data sources and with revolutionel protocols suc MQTT. This documents includes description of few services that we can use in order to handle huge traffics of data coming from different data sources [7]. Also we provide short guidance and strategies (detection, processing, and saving) and demo designed to describe the communications possibilities between aws services, and to help researchers in their future projects.

Keywords: AWS · S3 · Cloud computing · MQTT

1 Introduction

In most developed countries, the aging problem has existed for many decades, while, in developing countries, population aging has taken place relatively recently due to the demographic changes and strong gains in life expectancy. A UN study in 2013 said the number of older persons is 841 million in 2013 and the older population will almost triple by 2050, when it is expected to surpass the two billion mark. The share of the working-age population will continue to fall during the next four decades, to about 51% in 2050 [2]. In contrast the old-age dependency ratio will grow rapidly, as a result a society's capacity for taking care of its elderly members could be overwhelmed.

In this Paper we describe the utility of developing healthcare system in cloud environment especially in amazon cloud as one of the leader of cloud services.

2 Related Work

IoT innovation incorporated twisting in the keen power matrix accompanies a cost of putting away and handling the volume of information consistently. This information incorporates end clients stack request, control lines blames, system's parts status, booking vitality utilization, gauge conditions, progressed metering records, blackout

© Springer International Publishing AG, part of Springer Nature 2018
M. Ben Ahmed and A. A. Boudhir (Eds.): SCAMS 2017, LNNS 37, pp. 803–809, 2018.
https://doi.org/10.1007/978-3-319-74500-8_73

administration records, endeavor resources, and numerous all the more Hence, service organizations must have the product and equipment abilities store, oversee, and handle the generated data effectively [5]. Authors [5] proposed cloud support data management smart cities and according to the authors the expanding significance of conveying smart city advances and applications, [14] which brought about an expanding measure of information produced by these applications, the requirement for a comprehensive information administration framework expanded because of the need for effectively gathering and preparing these informations.

3   Cloud Computing

Cloud computing is the on-demand delivery of compute power, database storage, applications, and other IT resources through a cloud services platform via the internet with pay-as-you-go pricing. It provides a simple way to access servers, storage, databases and a broad set of application services over the Internet. A Cloud services platform such as Amazon Web Services owns and maintains the network-connected hardware required for these application services, while you provision and use what you need via a web application [1].

4   Amazon Cloud Services

Amazon Web Services provides online services for other websites or client-side applications. [1] Most of these services are not exposed directly to end users, but instead offer functionality that other developers can use in their applications. Amazon Web Services' offerings are accessed over HTTP, using the REST architectural style and SOAP protocol. All services are billed based on usage.

In the next paragraphs we give short introduction of services that amazon offer and we can use it in healthcare System:

4.1   Amazon S3 (Simple Storage)

Is object storage with a simple web service interface to store and retrieve any amount of data from anywhere on the web. It is designed to deliver 99.999999999% durability, and scale past trillions of objects worldwide. It stores arbitrary objects (computer files) up to 5 terabytes in size, each accompanied by up to 2 kilobytes of metadata. Objects are organized into buckets (each owned by an Amazon Web Services account), and identified within each bucket by a unique, user-assigned key. Amazon Machine Images (AMIs) which are used in the Elastic Compute Cloud (EC2) can be exported to S3 as bundles [3].

S3 have multiple advantages:

- SIMPLE: Amazon S3 is simple to use with a web-based management console and mobile app. Amazon S3 also provides full REST APIs and SDKs for easy integration with third-party technologies [3].

- DURABLE: Amazon S3 provides durable infrastructure to store important data and is designed for durability of 99.999999999% of objects. Your data is redundantly stored across multiple facilities and multiple devices in each facility [3].
- SCALABLE: With Amazon S3, you can store as much data as you want and access it when needed. You can stop guessing your future storage needs and scale up and down as required, dramatically increasing business agility [3].
- SECURE: Amazon S3 supports data transfer over SSL and automatic encryption of your data once it is uploaded. You can also configure bucket policies to manage object permissions and control access to your data using IAM [3].
- AVAILABLE: Amazon S3 Standard is designed for up to 99.99% availability of objects over a given year and is backed by the Amazon S3 Service Level Agreement, 27 ensuring that you can rely on it when needed. You can also choose an AWS Region to optimize for latency, minimize costs, or address regulatory requirements [3].

4.2   Amazon EC2

A web service that provides secure, resizable compute capacity in the cloud. It is designed to make web-scale computing easier for developers. The Amazon EC2 simple web service interface allows you to obtain and configure capacity with minimal friction. It provides you with complete control of your computing resources and lets you run on Amazon's proven computing environment. Amazon EC2 reduces the time required to obtain and boot new server instances (called Amazon EC2 instances) to minutes, allowing you to quickly scale capacity, both up and down, as your computing requirements change [4].

Amazon EC2 changes the economics of computing by allowing you to pay only for capacity that you actually use. Amazon EC2 provides developers and system administrators the tools to build failure resilient applications and isolate themselves from common failure scenarios [4].

4.3   Amazon DynamoDB

Since Big Data leads to huge traffics and a high processing time during data analysis, NoSQL solutions are much more efficient than traditional databases. They are also more flexible in the integration of new data, than data models, which are the basis of relational databases [4].

A second reason to consider a NoSQL solution is that modern systems need to be continuously available. Note that this is different from just "high availability" [4], where unplanned downtime, although not desired, is still expected. Continuous availability describes a feature of systems that can't go down.

The benefits of NoSQL solutions explain why these technologies are gaining market acceptance. Google even predicted that NoSQL will completely replace the relational databases in the next few years [6].

Amazon provide Amazon DynamoDB as a fully managed NoSQL database service that provides fast and predictable performance with seamless scalability. Amazon DynamoDB offers the following benefits:

- NO ADMINISTRATIVE OVERHEAD: Amazon DynamoDB manages the burdens of hardware provisioning, setup and configuration, replication, cluster scaling, hardware and software updates, and monitoring and handling of hardware failures. [5]
- UNLIMITED SCALE: The provisioned throughput model of Amazon DynamoDB allows you to specify throughput capacity to serve nearly any level of request traffic. With Amazon DynamoDB, there is virtually no limit to the amount of data that can be stored and retrieved [5].
- ELASTICITY AND FLEXIBILITY: Amazon DynamoDB can handle unpredictable workloads with predictable performance and still maintain a stable latency profile that shows no latency increase or throughput decrease as the data volume rises with increased [5].

4.4   Amazon IOT

AWS IoT provides secure, bidirectional communication between Internet-connected things (such as sensors, actuators, embedded devices, or smart appliances) and the AWS cloud. This enables you to collect telemetry data from multiple devices and store and analyze the data. You can also create applications that enable your users to control these devices from their phones or tablets [7].

AWS IoT Components AWS IoT consists of the following components: Device gateway Enables devices to securely and efficiently communicate with AWS IoT. Message broker Provides a secure mechanism for things and AWS IoT applications to publish and receive messages from each other. You can use either the MQTT protocol directly or MQTT over WebSocket to publish and subscribe. You can use the HTTP REST interface to publish. Rules engine Provides message processing and integration with other AWS services. You can use a SQL-based language to select data from message payloads, process and send the data to other services, such as Amazon S3, Amazon DynamoDB, and AWS Lambda. [12] You can also use the message broker to publish messages to other subscribers. Security and Identity service Provides shared responsibility for security in the AWS cloud. Your things must keep their credentials safe in order to securely send data to the message broker. The message broker and rules engine use AWS security features to send data securely to devices or other AWS services. Thing registry Sometimes referred to as the device registry. Organizes the resources associated with each thing. You register your things and associate up to three custom attributes with each thing. You can also associate certificates and MQTT client IDs with each thing to improve your ability to manage and troubleshoot your things [7].

MQTT Protocol. After the 2000' many organisation start using IoT protocol in many implementation. The first one was MQTT from IBM as a lightweight asynchronous publish-subscribe messaging protocol that relies on the MQTT broker to facilitate the messages between the publisher and subscriber. MQTT employs Transmission Control Protocol (TCP) to provide a reliable communication channel between the IoT devices. The small header of MQTT protocol (2 bytes only) allows the message delivery with minimal bandwidth and yet reliable connection [10]. In addition, a study of MQTT performance in comparison with HTTPS in 3G and Wifi environments has been made in [10] that shows that MQTT can handle more data with lower energy consumption (cf. Table 1).

Table 1. Comparison between HTTPS and MQTT protocol

	3G		WIFI	
	HTTPS	MQTT	HTTPS	MQTT
%Battery/Hours	18.79%	17.8%	5.44%	3.66%
Message/Hour	1926	21685	5229	23184
%Baterry/ Message	0.00975	0.00082	0.00104	0.00016

4.5 Amazon Elastic Load Balancing (ELB)

Automatically distributes incoming application traffic across multiple EC2 instances. It enables you to achieve greater levels of fault tolerance in your applications, seamlessly providing the required amount of load balancing capacity needed to distribute application traffic. Elastic Load Balancing offers two types of load balancers that both feature high availability, automatic scaling, and robust security. These include the Classic Load Balancer that routes traffic based on either application or network level information, and the Application Load Balancer that routes traffic based on advanced application-level information that includes the content of the request. The Classic Load Balancer is ideal for simple load balancing of traffic across multiple EC2 instances.

4.6 Amazon SNS (Simple Notification Service)

Is a web service that coordinates and manages the delivery or sending of messages to subscribing endpoints or clients. In Amazon SNS, there are two types of clients— publishers and subscribers—also referred to as producers and consumers. Publishers communicate asynchronously with subscribers by producing and sending a message to a topic, which is a logical access point and communication channel. Subscriber. consume or receive the message or notification over one of the supported protocols when they are subscribed to the topic. When using Amazon SNS, you (as the owner) create a topic and control access to it by defining policies that determine which publishers and subscribers can communicate with the topic. A publisher sends messages to topics that they have created or to topics they have permission to publish to. Instead of including a specific destination address in each message, a publisher sends a message to the topic. Amazon SNS matches the topic to a list of subscribers who have subscribed to that topic, and delivers the message to each of those subscribers. Each topic has a unique name that identifies the Amazon SNS endpoint for publishers to post messages and subscribers to register for notifications. Subscribers receive all messages published to the topics to which they subscribe, and all subscribers to a topic receive the same messages.

5 System Architecture

The system of our demo base on three main parts [13]:

- DATA PRODUCER: this is the sender of data. This producer if represent by a Raspberry pi(in real environment it collects all the signals from the sensors and forwards them to the Internet [8]) with programming tool called NOD-RED. This tool is used for wiring together hardware devices, APIs and online services in new and interesting ways. It provides a browser-based editor that makes it easy to wire together flows using the wide range of nodes in the palette that can be deployed to its runtime in a single-click.
- Gateway: is represented By IOT aws. Where we have create our objects, rules and X502 certificate to secure our communication [11].
- DESTINATION: we'll use DynamoDb to store incoming data from IOT gateway.
- EVENT: represent the alert which the system will send to physician as notification on emergency situation, this alert part use SNS service in order to push notification Android Application connect to another service called MobileHub.
- DATA EXPLORATION: We use Elastic MapReduce as service based on hadoop for quickly & cost-effectively process of vast amounts of data.

The next figure shows the communication between different (Fig. 1).

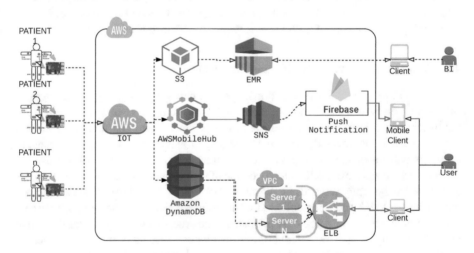

Fig. 1. Components of our system.

6 Conclusion

In this article, we have studied different software technologies which amazon offer that can be implemented for transfer, storage processing of data resulting from an IoT application in health care system and its combinations. This combination is constituted of

Raspberry pi, MQTT in aws IOT, S3 for saving data, SNS and MobileHub for notifications, and DynamoDB for storing data in NOSQL environnement.

Acknowledgments. I would like to express my special thanks of gratitude to my teachers Mr. Laaziz Yassin and Mr. Abdelhak Tali as well as My research colleague Mr. Eddabah who gave me the golden opportunity to do this wonderful practical research on the topic aws for health care system, which also helped me in doing a lot of Research and i came to know about so many new things I am really thankful to them.

References

1. Aws team, Overview of Amazon Web Services, Abstract. https://d0.awsstatic.com/whitepapers/aws-overview.pdf
2. Joe, W., Rudra, S., Subramanian, S.V.: Horizontal inequity in elderly, health care utilization: evidence from India. J. Korean Med. Sci. **30**(Suppl 2), 155–166 (2015)
3. Aws Team, Amazon S3. https://www.amazonaws.cn/en/s3/
4. Datastax, "Why NoSQL" - 10-2012.pdf", pp. 1–2 (2012). datastax.com/wp-content/uploads/2012/10/WP-DataStax-WhyNoSQ.pdf?2
5. Marina, B., Nabi, M.: Integrating cloud computing in IoT (Internet of Things) Applications. http://dx.doi.org/10.21474/IJAR01/2660
6. Gubbi, J.: Internet of Things (IoT): A Vision, Architectural Elements, and Future Directions. Department of Electrical and Electronic Engineering, The University of Melbourne, VIC - 3010, Australia (2013)
7. Aws Team. http://docs.aws.amazon.com/iot/latest/developerguide/what-is-aws-iot.html
8. Doukas, C.: Bringing IoT and Cloud Computing towards Pervasive Healthcare, Dept. of Information & Communication Systems Engineering, University of the Aegean Samos, Greece (2012)
9. Lam, J.-H., Lee, S.-G., Lee, H.-J., Oktian, Y.E.: Securing SDN southbound and data plane communication with IBC. Mob. Inf. Syst. Article ID 1708970, 12 p. (2016)
10. Nicholas, S.: Power Profiling: HTTPS Long Polling vs. MQTT with SSL, on Android. http://stephendnicholas.com/archives/1217
11. Internet of Things (IoT) Meets Big Data and Analytics: A Survey of IoT Stakeholders. http://cdn2.hubspot.net/hub/328935/file-2561965259-pdf/
12. Gigli, M., Koo, S.: Internet of Things: services and applications categorization. Adv. Internet Things **1**, 27–31 (2011)
13. Gómez, J.: Patient monitoring system based on Internet of Things. Procedia Comput. Sci. **83**, 90–97 (2016)
14. Nagib, A.M., et al.: SIGHTED: a framework for semantic integration of heterogeneous sensor data on the Internet of Things. Procedia Comput. Sci. **83**, 529–536 (2016)

Comparative Study of Different Open-Source Hospital Information Systems in Order to Develop an Application for the Healthcare Institution's Needs in Morocco

Youssef Bouidi[✉], Mostafa Azzouzi Idrissi, and Noureddine Rais

Laboratory of Informatics, Modeling and Systems (LIMS),
Sidi Mohamed Ben Abdallah University (USMBA), Fez, Morocco
y.bouidi@gmail.com, azzouziidrissi@gmail.com,
noureddine.rais@usmba.ac.ma

Abstract. The information system (IS) was becoming an indispensable management tool in various fields of human activity. In the health sector, it has an important role in improving patient management and quality of healthcare services, in optimizing resources and in effectiveness of biomedical research. The hospital information system (HIS) began to emerge in recent decades in many developing countries. Our study aims at first for examining many different open source applications and, then, selecting the most appropriate applications to use and adapt for healthcare structures in countries where the public management allocated budgets, in the different healthcare structures, does not allow the purchase of a commercial solution and a necessary after-sales services. A bibliographical study does not allow us to discover related works. From were, the great interest of our comparative study.

We have studied nine open-source hospital information systems: OpenMRS, OpenEMR, MediBoard, HospitalOs, HOSxP, PatientOS, Care2x, MedinTux and OpenHospital.

We have used the SQALE method to evaluate the quality of these applications. We have got the source code using the "sourceforge.net" website. We have used the Sonar platform to measure the source code's quality, and the "OpenHub.net" website to get the evaluation of the activity of Open-Source communities of each HIS.

The obtained results allowed us to select MediBoard and OpenEMR as a basis to develop an application for the healthcare institution's needs, in particular for the first and second levels, in Morocco.

Keywords: Hospital information system · Health system · Open-Source
SQALE method · Sonar platform · Sourceforge.net · openHub.net
Technical debt

© Springer International Publishing AG, part of Springer Nature 2018
M. Ben Ahmed and A. A. Boudhir (Eds.): SCAMS 2017, LNNS 37, pp. 810–823, 2018.
https://doi.org/10.1007/978-3-319-74500-8_74

1 Introduction

Actually, information is the most important resource of every organization. In fact, any structure has its data that describe organizational activities. To manage this data we need to use an information system. It is a set of peoples, services, events, features, and relationships, which define organizations. Therefore, it is a set of information about system's elements. In other words, an information system is an integrated company tool dedicated to data management, process, storage and accessibility. It is defined as a socio-technical subsystem of the company, the term "Socio" refers to the population concerned by the procedure's information. The term "Technique" refers to the ways of this processing [1]. According to this definition, Hospital Information System is the set of technological tools that ensure the management of hospital data, such as the patient information, the service information, doctors, nurses and logistic data. According to [1], a HIS is an application developed to manage the medical, administrative, legal and financial aspects of the hospital.

The first HIS returns to 1965 when Lockheed Martin started a project of collecting information concerning the usability of such systems [2]. Lockheed built a prototype of hospital information system called the Medical Information System (MIS), which was tested by El Camino Hospital [2, 3]. In the 1970s and 1980s, other hospitals around the world has used hospital information systems [4] with the integrated features such as planning, recording and instrument automation [5]. In the 1990s, Japan used HISs at the most of hospital activities, including the preparation of health's reclamation assurance. In 1991, 81.6% of all Japanese hospitals used health informatics technology [6, 7]. The integration of HIS with assurance association systems has also been prototyped, which creates the potential for more comprehensive databases on patients' medical records [6]. During this period, hospitals and developers also focused on two specific objectives for the HIS: to adapt the system to the clinical environment and to establish communication links between the HIS and other external entities, such as Laboratories and pharmacies [6]. The achievement of these integration goals was in the mid-1990s, with many health systems linked through interfaces. The way to integrate HIS with other health care systems was still difficult because of the diversity of tasks, technical limitations, preference for departmental systems and the philosophy of developers. Another key issue in this integration process was the execution of synchronous updates between heterogeneous systems and the management of communication servers [8–10]. Through the 2000s, component-based technologies, communication systems, distributed systems and network architectures gave the possibility to homogenize and share data at a new level. Also, in this period, communication standards was improved, such as Health Level 7 and XML packaging structures like the Simple Object Access Protocol (SOAP), which have contributed to insure an Interoperable environment around Electronic Health Records (EHR) [11].

HISs have been much developed to better serve different needs of health care institutions and services. There are different types of HISs, commercial, non-commercial, closed source and open source. The major problem in HISs adoption for health institutions is the difficulty of adaptation to its specific and local needs. In developing countries, the health sector suffers from different management problems linked to the lack of resources.

Adopting a solution in these countries was never that simple. Indeed, solutions that exist have a difference in their costs and access to sources code. Developing countries, because of its low budget allocated to the health sector, cannot use HISs or they adopt limited ones as low-cost closed solutions, these solutions can never cover all of health care institution's needs. Generally, these solutions are incomplete or too complex to configure.

Open source HISs proposed in this search work as a solution that has a free and open access to source code. This kind of solutions present an advantage summarized in the ease of adoption despite of limited resources, it ensures a better adaptation due to the permitted modification in source codes. However, in this moment, there are various HISs Open Source with different qualities, which allowed us to introduce this comparative study to evaluate and examine different open source solutions and choose a flexible one that has a better quality and can be adapted to health care institution's needs.

Therefore, this work consists in examining, evaluating and comparing open source HISs qualities in order to select the one that can be adopted and adapted to health care institutions specifications.

2   Materials

In this section, we will detail the chosen HISs in our study, and then we will describe the platforms used to recover the source code and to implement the method adopted in this study.

2.1   Hospital Information Systems

We looked for HISs that have active contributions. In fact, we have nine solutions:

OpenMRS is a collaborative project designed to manage healthcare, especially, in developing countries. OpenMRS was a response to many challenges like the serious diseases (HIV, Malaria). This system was created in 2004 by Paul Biondich and Burke Mamlin from the Indiana University School of Medicine after a visit to the Academic Model Providing Access to Healthcare Project in Kenya. It has many collaborators in different specialties as volunteers. OpenMRS exist actually in many countries around the world for research, clinical use, development, evaluation and other uses [12].

OpenEMR Is an Electronic Health Record for a medical practice management. It is ONC Complete Ambulatory EHR Certified with international uses. It was originally developed by Syntech organisation in 2001 as Medical Practice Professional like version 1.0. Actually, the system is on version 5 since 2017 and its code repository was migrated to the GitHub [13].

MediBoard Is an open source web application designed to manage health establishments. It is a Hospital Information System created by Thomas Despoix and Romain Ollivier. This HIS is actually until version 0.4.0 developed by the OpenXtrem organization. It is a modular system based on web technologies to handles all patient files, workflows and planning of all health establishment activities [14].

HospitalOS is a HIS and a research /development project designed to every small-sized hospital. Hospital OS was created and developed for Thailand community. It is actually until version 3.9 and it is featured especially by the treatment of patients [15].

Care2x, Is an integrated HIS started as "Care 2002"project in 2002. The first official release was until version 1.1 in 2004. In 2003, the project name was changed to Care2x. The last stable release was in 2012 until version 2.6.29. Its design can handle both of medical and non-medical services. Care2x has many features that include especially the smart search and the multiple custom languages. [16]

OpenHospital was developed by Informatici Senza Frontiere, in collaboration with students of Volterra San Donà di Piave Technical highschool, in 2005. It was implemented at St Luke's Hospital, Angal, in Nebbi District, Northern Uganda. It was used also in Kenya, Afghanistan, Benin and Congo. Actually, this HIS is at its seventh release which is multi-user, has an extended patient database, has a historical integrated patient and gives an internal communication, reports and statistics. [17]

MedinTux was initiated by the doctor Roland Sevin since ten years. It is distributed under the CeCiLL V2 license which is equivalent to the GPL license adapted to the French legislation. It was originally written for French emergency services, it can be used in a multi users environment and offers many features like consultations, prescriptions, real time visualization and statistics [18]

HosxP is a HIS used in over 70 hospitals in Thailand. It was called KSK-HDBMS. Its development started in 1999 by Suratemekul to be continued by employers of its company Bangkok Medical Software. It is distributed under a GPL license and free only for its Primary Health Care Unit version. [19]

PatientOS is a HIS for small hospitals and clinics. It is a web-based application under the GPL license. [20]

2.2   Source Forge

This platform was our repository to get the source codes for each HIS. Source Forge is a software forge, a website hosting the development of free and open-source software, operated by the company Geeknet and which uses the Apache-Allura forge [21]. It allows developers to host software projects by offering tools for their management as well as their versions. It also offers a wiki space for documentation, as well as a subdomain for each project and gives to the community an opportunity to judge projects by selecting specific indicators that are useful in providing project's statistics.

2.3   SonarQube

We used this platform to evaluate the source code quality of each HIS. It is an open source platform for continuous inspection of code quality and an implementation of the method used in this study [22]. It is a web-based application that analyzes the project source code following a set of quality characteristics. It gives the indicators that evaluate projects by providing numerous statistics. It has many features like the complete dashboard of indicators and the identification of project weaknesses [22].

We used SourceForge to get the source code for the nine HIS that will be analyzed and evaluated using SonarQube and OpenHub to compare their qualities. To evaluate the quality of source codes we adopt the SQALE method that will be developed in the following chapter.

3   Method

After collecting the source code of open source HISs, we arrive to the evaluation phase of their qualities. In order to examine them, we used the McClain model which defines the quality of IS on three levels. On the other hand, we used the SQALE method to evaluate the quality of source codes by applying the concept of technical debt.

Our objective is focused on quality evaluation of HIS. To achieve this we have adopted the concept of quality evaluation of an information system. This concept consists of evaluating IS components according to a model that respects IS definition in its organism. In 1992 the quality of an IS was defined by DeLone and McLean model which splits the quality into two parts, system and information quality [23] (Fig. 1) .

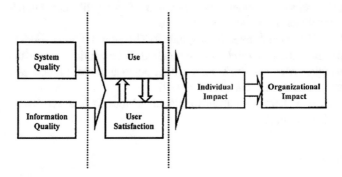

Fig. 1.   Information system success model [23].

The quality of the technical part of the system is determined by ease access, short response time, a practical tool for user, which helps in a more efficient work. The quality of the information produced is determined by the accuracy of the information, its accessibility, its exhaustiveness and its reliability.

This model was improved in 2003 by adding a third dimension which concerns the quality of the service [23]. According to this dimension, the IS quality is also measured by its activity, security, updating and maintenance (Fig. 2).

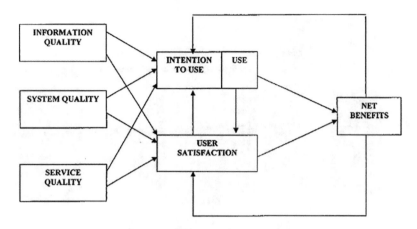

Fig. 2. Updated IS success model [23].

We used the SQALE method to evaluate the technical quality of the system and information for each IS. This method is based on the concept of the technical debt of the source code. According to Jean-Louis Letouzey in the official documentation published in 2016 [24], this method, in its latest version, respects nine fundamental principles:

1. The quality of the source code is a non-functional requirement.
2. The requirements in relation to the quality of the source code have to be formalized according to the same quality criteria such as any other functional requirement.
3. Assessing the quality of a source code is in essence assessing the distance between its state and its expected quality objective.
4. The SQALE Method assesses the distance to the conformity with the requirements by considering the necessary remediation cost of bringing the source code to conformity.
5. The SQALE Method assesses the importance, the impact of a non-conformity by considering the resulting costs of delivering the source code with this non-conformity.
6. The SQALE Method respects the representation condition.
7. The SQALE Method uses addition for aggregating the remediation costs, the non-remediation costs and for calculating its indicators.
8. The SQALE Method's Quality Model is orthogonal.
9. The SQALE Method's Quality Model takes the software's lifecycle into account.

The SQALE method is based on four concepts presented as successive stages by Jean-Louis Letouzey in 2016 [24]. Indeed, a quality model, an analysis model, indices and finally indicators define the SQALE structure (Fig. 3).

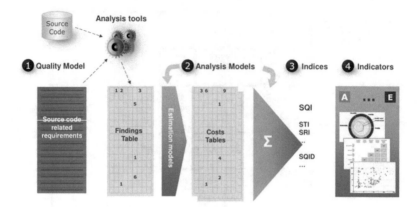

Fig. 3. The SQALE structure [25]

The quality model proposed by this method aims to organize the non-functional requirements related to the quality of the code. It is organized into three hierarchical levels. The first level is composed of characteristics derived from the standards as key factors in describing the quality of the code such as stability, reliability, variability, efficiency, safety, maintenance, portability and reuse. The second level named sub-characteristics used to combine groups of requirements into two types: those corresponding to life cycle activities and other results of generally recognized taxonomies. The third level consists of requirements for the internal attributes of the source code.

According to Jean-Louis Letouzey [24], the SQALE analysis model accomplish two main tasks: first, it applies rules to normalize the measurements by transforming them into costs, and the second defines rules to aggregate these normalized values. The SQALE method defines the cost aggregation rules either in the quality model tree or in the source code artifact hierarchy (Fig. 4).

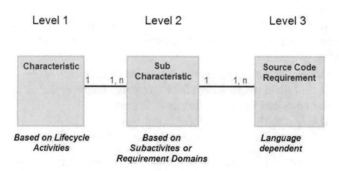

Fig. 4. Levels of SQALE quality model [24]

The indices in this method represent the costs and are measured on the same scale in order to manipulate all the operations permitted for a scale of this type. Jean-Louis Letouzey confirmed that the characteristic indices of SQALE are the Testability Index (STI), the reliability index (SRI), the index of changeability (SCI), the efficiency index

(SEI), Security Index (SSI), Maintainability Index (SMI), Portability Index (SPI) and Reuse Index (SRuI) [24] (Fig. 5).

Fig. 5. SQALE quality's characteristics [24]

By summing all remediation costs related to all quality model requirements, the cost of remediation for all characteristics of the quality model can be estimated. This measure is the quality index SQALE: SQI. The SQALE quality index is a precise implementation of the concept of "technical debt" associated with the source code.

The SQALE method defines four indicators related to the quality characteristics, allowing a highly synthesized representation of the quality of an application [24]. The SQALE Rating consists of producing a derived measure or an ordinary scale subdivided on five levels from A (Green) to E (Red).

Kivat consists of presenting the SQALE Rating in concentric areas and targeting the quality of each project according to its values. The Kiviat presents both the Rating of all the projects compared according to the characteristics of the quality model (Fig. 6).

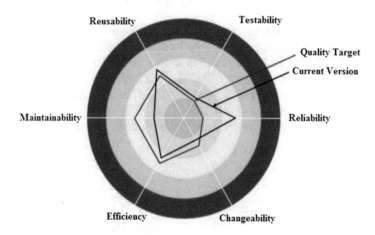

Fig. 6. The Kiviat indicator [25]

The SQALE pyramid helps to make appropriate decisions, taking into account the dependence of the characteristics of the quality model on the life cycle.

The fourth indicator is the Debt Map graph, which represents the artefacts of the assessment scope drawn on two dimensions: the first is Technical Debt (SQI) and the second is the Business Impact (SBII) (Fig. 7).

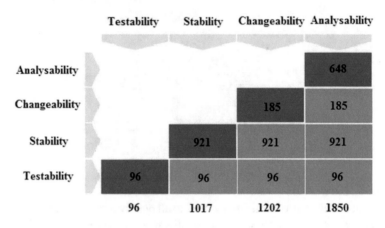

Fig. 7. The SQALE pyramid [25]

This method evaluates the quality of the information system by evaluating the first two dimensions defined by W. DeLone and E. McLean [23]. Then we evaluated the quality of the service, which presents the third dimension of the proposed model, by evaluating activity, maintenance, security and updating rate of the IS.

This evaluation is done by combining characteristics given by the SQALE method and confirmed information by the Open Source community. Being Open Source projects, each HIS has its own activity indices which are summarized in the contribution rate, the validation rate, the date of the last contribution and the date of the last stable version (Fig. 8).

To carry out this evaluation, we called our materials mentioned in the previous chapter. We used SonarQube to implement the SQALE method by installing the client platform and connecting to the Sonar server to scan the source code. We also retrieved the indices of HIS activities using the OpenHub platform.

In this section, we have stated the method adopted in this comparative study. After defining the quality evaluation dimensions of the HISs and the SQALE method, we implemented this method using the SonarQube platform to obtain and visualize the evaluation results.

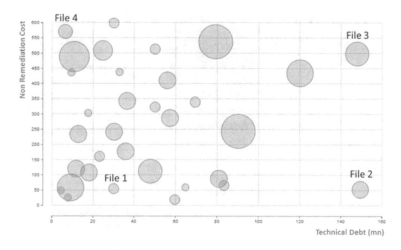

Fig. 8. SQALE Debt Map Graph [26]

4 Results and Discussions

Applying this method, allowed us to have specific criteria of quality that present our referential of comparison of HIS. Indeed, the SonarQube platform is used to evaluate the quality of characteristics of the source code applying the concept of technical debt (Table 1).

Table 1. The used indicators for characteristics evaluation

Evaluation dimension	Characteristic	Used indicators
System & information	Reliability	SPR
	Complexity	Rate
	Rule Compliance	SPR
	Documentation	Rate
Service	Activity	Size & Rate & Date
	Security	SPR
	Maintainability	SPR & Debt Map

The measures obtained was about Reliability, Security, Maintainability, Complexity, Documentation and Rules Compliance.

After having implemented the SQALE method and collected statistics in the open source community, we have put two sets of characteristics that was the basis of evaluation of the studied HISs. The first one was by the test-based selection. However, the second one was by the verification-based selection.

In this table, we mentioned the used indicators to evaluate each characteristic. These indicators are the SQALE Pyramid Rating, Rate, Size, Debt Map Graph and Date (Table 2).

Table 2. The sub Characteristics of the Evaluation

Evaluation dimension	Characteristic	Sub characteristic
System	Complexity	
	Rule Compliance	RCR
	Documentation	
	Internationalization	
	Modularity	
Information	Interoperability	
	tractability	
	Reliability	
	Activity	Contributions
		Date of last activity
		Commits
	Security	
	Maintainability	
	Functional coverage	

On this table, we present the verified-characteristics. This verification is done for the sonar selected HISs in order to be based on the SQALE method results.

Indeed, the sonar test was done using Sonar Scanner on each HIS, and it was applied on the following characteristics:

- Reliability
- Complexity
- Security
- Maintainability
- Documentation
- Rules Compliance

The results given by the Sonar server use the SQALE indicators to evaluate the quality of the HIS source code according to the technical debt of each defined characteristic. Then, we obtained the following results of Characteristic's Rating distribution according to each HIS's source code:

These results show that MediBoard is technically better distributed. Indeed, the evaluation of source codes has proved that MediBoard is technically more satisfying according to Model quality characteristics (Table 3).

Table 3. Comparative study of HISs activities in 2016

HISs	Commits	Contribution
OpenEMR	1566	58
OpenMRS	522	81
MediBoard	2381	12
Open Hospital	70	2

According to OpenHub platform's results, we got the following statistics concerning HISs actives:

However, after evaluating the HIS we observed that MediBoard, despite of its activity fall, has proved its technical performance (Fig. 9).

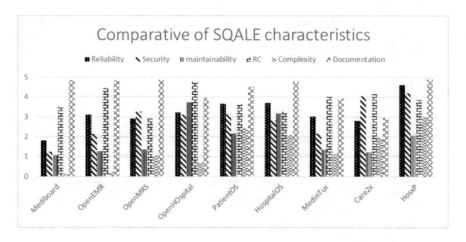

Fig. 9. The Characteristic Rating of HISs

5 Conclusion

According to importance of information systems in the business management, Hospitals have presented a need for IS adoption to facilitate and control their complex structures and organizations. This need was relative to the country's categories. Therefore, it was an adoption lag and a difference in the functional priorities ensured by the HISs. The adoption of a HIS, especially in a developing country, requires an open source solution due to lack of resources. For this reason, our study was a comparison for open source solutions (Table 4).

Table 4.   : the Virified Characteristics of HISs

HIS	Indicators	
	Interoperability	Traceability
MediBoard	+	+
OpenEMR	+	+
OpenMRS	−	+
Care2x	−	−
Hospital OS	−	−
Open Hospital	+	−
MedinTux	−	−
HOSxP	−	−
PatientOS	−	−

In this study we used the Source Forge platform to recover the source codes of nine HISs. We evaluated these HISs by the SQALE method by adopting its implementation presented in the SonarQube platform. This platform allowed us to evaluate the quality of the HIS source code. We used the OpenHub referential to evaluate the HIS's activities as Open Sources.

The result obtained gave us two HISs to be a basis for developing a solution that will be adapted to the levels 1 and 2 of the Moroccan health system. These HISs were in the order MediBoard and OpenEMR.

References

1. Hannah, K.J., Ball, M.J.: Strategic Information Management in Hospitals An Introduction to Hospital Information Systems. Springer Science Business Media, New York (2004)
2. Lockheed Aircraft Corporation, Lockheed Hospital Information System, p. 82 (1965)
3. Gall, J.E., Norwood, D.D., Hospital, E.C.: Demonstration and Evaluation of a Total Hospital Information System. NCHSR research summary series, Education, and Welfare, Public Health Service, Health Resources Administration, National Center for Health Services Research, U.S. Dept. of Health, p. 38 (1977)
4. Krobock, J.R.: A taxonomy: hospital information systems evaluation methodologies. J. Med. Syst. **8**, 419–429 (1984)
5. Rubinoff, M., Yovits, M.M.C.: Perspectives in clinical computing. Adv. Comput. **16**, 128–177 (1977)
6. Yoshikawa, A., Ishikawa, K.B.: Information Systems in Japanese Healthcare, H. I. T. D. o. I. I. f. t. H. Industry. Greenwood Publishing Group, pp. 82–95 (1995)
7. Miyake, H.: Hospital information systems in Japan: their trends of past, present and future. Jpn Hosp. **6**, 35–44 (1987)
8. Smith, J.: Acute care health information management systems. In: Health Management Information Systems: A Handbook for Decision Makers, McGraw-Hill International, pp. 167–210 (1999)
9. Lenz, R., Blaser R., Kuhn, K.A.: Hospital information systems: chances and obstacles on the way to integration. In: Kokol, P., Zupan, B., Stare, J. (eds.) Medical Informatics Europe, 1 edn., pp. 25–30. IOS Press (1999)
10. Stuewe, S.: Interface tools for healthcare information technology. In: Beaver, K. (ed.) Healthcare Information Systems, 2 edn., pp. 31–46. CRC Press (2002)
11. Van de Velde, R., Degoulet, P.: Introduction: the evolution of health information systems. In: Clinical Information Systems: A Component-Based Approach, pp. 1–14. Springer (2003)
12. OpenMRS LLC, Introducing OpenMRS 2.0 (2014)
13. OpenEMR Project: OpenEMR Project (2012)
14. MediBoard 2014
15. Webster, P.C.: The rise of open-source electronic health records (2011)
16. care2x 2013
17. Informatici Senza Frontiere, Open Hospital - The project (2014)
18. MedinTux, Documentation MedinTux (2012)
19. SourceForge, HOSxP (2002)
20. SourceForge, PatientOS (2007)
21. Source Forge, About (2016)
22. SonarSource 2017

23. Delone, W.H., Mclean, E.R.: The DeLone and McLean model of information systems success: a ten-year update (2003)
24. Letouzey, J.-L.: The SQALE method for managing Technical Debt (2016)
25. Letouzey, J.L.: Meaningful insights into your Technical Debt (2012)
26. Letouzey, J.L.: The SQALE method (2014)

Using Homomorphic Encryption
in Cloud-Based Medical Image Processing:
Opportunities and Challenges

Mbarek Marwan$^{(\boxtimes)}$ ⓘ, Ali Kartit, and Hassan Ouahmane

LTI Laboratory, ENSA, Chouaïb Doukkali University, Avenue Jabran Khalil Jabran,
BP 299 El Jadida, Morocco
marwan.mbarek@gmail.com, alikartit@gmail.com,
hassan.ouahmane@yahoo.fr

Abstract. The modern radiologists and doctors often use imaging software for a better diagnosis, and hence, to deliver improved patient care. Obviously, since digital records contain useful biological information, they are now being massively produced. Unfortunately, the maintenance and licenses required to mange an on-premise solution is a costly and time-consuming approach. To overcome this challenge, healthcare organizations implement cloud services to process medical images with the help of remote tools and facilities. In fact, cloud computing has, since its emergence, changed the way we build and develop IT services. In this new paradigm, computational resources are accessed and downloaded as needed through the Internet. The aim of this concept is to outsource computations to an external entity in order to reduce costs. In this context, the billing system of this model is based mainly on software utilization and Service-Level Agreement (SLA) contract. Despite its promising features, the adoption of this technology in healthcare domain gives rise to different problems regarding privacy and security. In this respect, we are interested in highlighting opportunities and challenges of image processing using cloud, as well as suggestions for future developments. We are particularly focused on homomorphic encryption techniques and their applications in data processing. There are only a few studies in the literature that deal with this subject. The results of the present study reveal that homomorphic encryption is a promising technique to ensure data privacy when using cloud services. However, existing schemes still have some limitations, especially computational costs.

Keywords: Medical image · Cloud computing · Homomorphic encryption
Security

1 Introduction

Currently, a wide range of solutions have been suggested to help healthcare professionals process their data. In particular, advanced imaging tools have a huge potential to support the diagnostic process. Naturally, this implies additional costs related to IT needs, i.e., platform and software. For this reason, healthcare organizations prefer to use cloud

© Springer International Publishing AG, part of Springer Nature 2018
M. Ben Ahmed and A. A. Boudhir (Eds.): SCAMS 2017, LNNS 37, pp. 824–835, 2018.
https://doi.org/10.1007/978-3-319-74500-8_75

computing rather than building local applications; hence, eliminating the need to build in-house data centers. The primary goal of this model is to increase efficiency and reduce costs by sharing resources. In addition, clients are charged based on cloud solutions utilization and Quality of service (QoS) requirements. Technically, numerous leading edge technologies are the core elements of cloud computing, such as Service-Oriented Architecture (SOA), Parallel Distributed Systems (PDS), virtualization and data dedu-plication. Essentially, they seek to simplify the storage and analysis of medical data by means of multi-tenancy and virtualization techniques. Furthermore, they allow ubiqui-tous access to cloud services, which are delivered and released with minimal manage-ment effort [1]. Yet, various types of cloud deployment models have been implemented in practice to comply with security requirements and clients' needs. For example, public, private, hybrid and community cloud represent possible cloud deployment models. This study aims to investigate the opportunities and barriers of the implementation of the software as a service (SaaS) in the healthcare domain. In this scenario, healthcare professionals use cloud applications in order to analyze medical images. The cloud providers perform complex image processing using cloud technologies, and then, return the result to the client. The principle of cloud-based image processing is presented in Fig. 1.

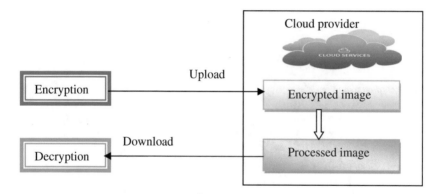

Fig. 1. The basic concepts of cloud-based medical image processing

Despite the significant benefits of cloud services, security and privacy are some of the principal barriers limiting the effectiveness of this model [2–8]. Besides, it is manda-tory to protect medical records because they are sensitive and personal data. In spite of its demonstrated economic advantages, the migration to cloud-based medical image processing faces many challenges. Indeed, security and privacy are the major barriers that obstruct the acceptance of this approach. This is due to the fact that cloud computing inherits security risks of its preceding technologies. Additionally, a medical image is crucial data that need to be intact during cloud utilization. In this work, we examine the feasibility of homomorphic approach to secure data processing, particularly medical images.

The rest of this paper is organized as follows. Sections 2 and 3 present the funda-mentals of homomorphic encryption and its applications in cloud-based medical image processing. Sections 4 and 5 present and discuss the existing implementations of homo-morphic approach to secure data processing. Finally, we conclude this paper in Sect. 6 by remarks and future work.

2   Homomorphic Encryption

In the case of cloud-based services such as image processing, both data and computations are outsourced to an external provider. Accordingly, organizations are more exposed to many security problems associated with cloud as discussed in [2–8]. In addition, medical images are sensitive data because they are used in diagnosis process. For these reasons, it is necessary to implement precautionary measures to secure cloud-based image processing. The classical approach attempts to maintain data privacy by encrypting medical records before uploading them to the cloud. Basically, clients use conventional cryptography techniques such as RSA, DES and AES to prevent unauthorized disclosure of confidential data. In such a scenario, the data need to be decrypted in order to process them. A detailed description of a workflow representing this approach is illustrated in Fig. 2.

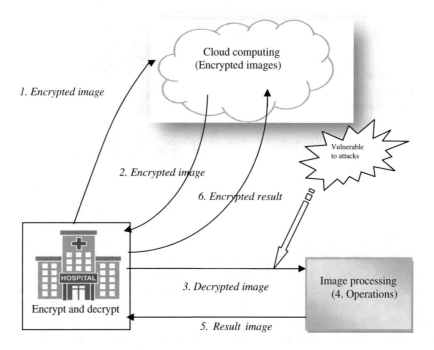

Fig. 2.   Classical approach for image processing over cloud computing

Obviously, since medical images are sent to an external entity in plaintext format, they are continuously exposed to various risks and threats. As an effective alternative, the homomorphic approach is used to securely process data using cloud resources. The main benefit of this approach is that computation operations are directly executed on the encrypted data. More formally, the decryption of the operation performed on an encrypted image is identical to the result that takes as its input the encrypted data of the same operation on the original image. Consequently, this technique is an appropriate method to handle data processing matters in cloud computing. In this regard, Fig. 3 represents the principle of homomorphic encryption and its application in cloud-based medical image processing.

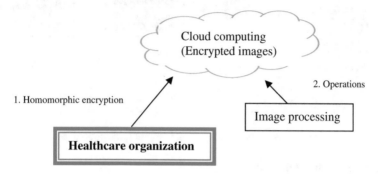

Fig. 3. Image processing using homomorphic encryption approach

3   Homomorphic Schemes

The fundamental goal of homomorphic encryption is to allow a user to execute certain types of arithmetic operations on encrypted data. This implies that clients have to encrypt their medical images before sending them to the cloud computing. In this case, we use homomorphic algorithms to protect sensitive data because they often support both addition and multiplication operations. Consequently, this approach is suitable to perform computations over cloud-based applications as shown in Fig. 4. So, the original image is transformed into encrypted form $E(x)$. Next, we apply basic image processing operations, like linear image filtering h, to get $F_2(E(x), h)$. Finally, we obtain the result after decryption process $y = D\big(F_2(E(x), h)\big)$.

There are several schemes in the area of homomorphic encryption. Mainly, two types are available to process encrypted data, i.e., Partially Homomorphic Encryption (PHE) and Fully Homomorphic Encryption (FHE). In essence, the first one serves to execute only one arithmetic operation, while the second one can simultaneously support two operations. For instance, in the PHE category, we typically use RSA and Paillier to execute multiplication and addition respectively. In contrast, both operations can be carried out using a FHE algorithm, for example, Enhanced Homomorphic Encryption Scheme (EHES).

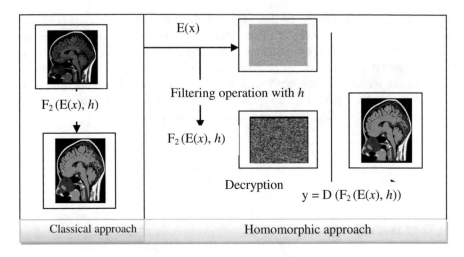

Fig. 4. Image processing using homomorphic encryption

3.1 Homomorphic Addition [9–11]

Paillier's cryptosystem is typically an encryption mechanism with respect to addition. In cryptography, this homomorphic scheme uses three functions, i.e., key generation, encryption and decryption. Assume that p and q are two large primes, so that the composite is evaluated as $n = p \cdot q$. In the key generation process, we select g in such a way that n and $L(g^{\lambda} \bmod n^2)$ are coprimes. Particularly, L denotes the function $L(u) = (u-1)/n$ and λ denotes the Carmichael function $\lambda(p \cdot q) = 1 \, cm \, (p-1, q-1)$, where lcm refers to the least common multiple. In this context, Table 1 gives a detailed description of the main functions of the Paillier algorithm.

Table 1. Fundamental of the Paillier cryptosystem

Function	Input	Computation	Output
Key generation	$p, q \in \mathbb{P}$	Compute $n = p \cdot q$ Choose $g \in \mathbb{Z}_{n_2}^*$ such that $\gcd(L(L(g^{\lambda} \bmod n^2)), n) = 1$ with $L(u) = \dfrac{u-1}{n}$	Public key: $pk = (n, g)$ Secret key: $sk = (p, q)$
Encryption	$m \in \mathbb{Z}_n$	Choose $r \in \mathbb{Z}_n^*$ Compute $c = g^m r^n \bmod n^2$	$c \in \mathbb{Z}_n^2$
Decryption	$c \in \mathbb{Z}_n^2$	Compute $m = \dfrac{L(c^{\lambda} \bmod n^2)}{L(g^{\lambda} \bmod n^2)} \bmod n$	$m \in \mathbb{Z}_n$

Beside data security, using Paillier algorithm allows one to perform basic mathematical operations on encrypted data. Characteristics and main features of this algorithm are demonstrated throughout the following example. In this regard, we encode two values m_1 and m_2 using the public key pk to obtain C_1 and C_2, respectively. Furthermore, E is the encryption function and D stands for decryption.

$$C_1 = E(m_1, pk) \tag{1}$$

$$C_1 = g^{m_1} r_1^n (\bmod n^2) \tag{2}$$

$$C_2 = E(m_2, pk) \tag{3}$$

$$C_2 = g^{m_2} r_2^n (\bmod n^2) \tag{4}$$

Then,

$$C_1 \cdot C_2 = E(m_1, pk) \cdot E(m_2, pk) \tag{5}$$

$$C_1 \cdot C_2 = (g^{m_1} r_1^n)(g^{m_2} r_2^n)(\bmod n^2) \tag{6}$$

$$C_1 \cdot C_2 = g^{m_1 + m_2} (r_1 r_2)^n (\bmod n^2) \tag{7}$$

$$C_1 \cdot C_2 = E(m_1 + m_2, pk) \tag{8}$$

$$E(m_1, pk) \cdot E(m_2, pk) = E(m_1 + m_2, pk) \tag{9}$$

Therefore, we get the following Equation [12].

$$D(E(m_1, pk) \cdot E(m_2, pk)(\bmod n^2)) = m_1 + m_2(\bmod n) \tag{10}$$

As illustrated above, Paillier algorithm offers a helpful solution to perform addition operation on encrypted data. More precisely, the product of two ciphertexts is equal to the decryption of the sum of their corresponding. Thus, it is a promising technique in security of image processing. Consequently, it is a suitable approach to outsource computations to an external cloud provider.

3.2 Homomorphic Multiplication [13, 14]

Another alternative is to use the RSA algorithm, which processes encrypted data without having to decrypt them first. In cryptography, the RSA cryptosystem is a public-key scheme, and is widely used for executing multiplication on ciphertext. In fact, this algorithm is designed to prevent data disclosure when processing outsourced data. In this scheme, we arbitrarily select the two large primes p and q, and then calculate $n = p \cdot q$. We also choose e which needs to satisfy the following condition gcd $(e, \varphi(p \cdot q)) = 1$, where φ is Euler's totient function and $\varphi(n) = (p-1)(q-1)$. The RSA scheme typically consists of three parts, namely key generation, encryption algorithm (E), and decryption algorithm (D), as detailed in Table 2.

Table 2. Fundamentals of the RSA algorithm

Function	Input	Computation	Output
Key generation	$p, q \in \mathbb{P}$	Compute $n = p \cdot q$; $\varphi(n) = (p - 1)(q - 1)$ Choose e such that $\gcd(e, \varphi(n)) = 1$ Determine d such that $e.d \equiv 1 \bmod \varphi(n)$	Public key: $pk = (e, n)$ Secret key: $sk = (d)$
Encryption	$m \in \mathbb{Z}_p$	Compute $c = m^e \bmod n$	$c \in \mathbb{Z}_n$
Decryption	$c \in \mathbb{Z}_n^2$	Compute $m = c^d \bmod n$	$m \in \mathbb{Z}_n$

To explore this scheme, we encode two m_1 and m_2 via the algorithm RSA and its public key $pk = (n, e)$ to obtain two encrypted values C_1 and C_2, respectively.

$$C_1 \cdot C_2 = E(m_1, pk) \cdot E(m_2, pk) \tag{11}$$

$$= m_1^e \cdot m_2^e (\bmod n) \tag{12}$$

$$= (m_1 \cdot m_2)^e (\bmod n) \tag{13}$$

$$= E(m_1 \cdot m_2, pk) \tag{14}$$

As seen in the Eq. 14, the RSA algorithm is an efficient privacy protection mechanism and also homomorphic with respect to multiplication. In other words, this method enables an external entity to handle ciphertext without revealing the secret data. Consequently, it is an appropriate approach to outsource computations to a cloud provider because the multiplication operation can be done publicly.

3.3 Fully Homomorphic

In this category, a single algorithm offers the ability to carry out addition and multiplication operations on encrypted data. Recently, a wide variety of schemes have been developed because of the growing use of fully homomorphic encryption. In this work, we present the most commonly used technique to perform mathematic operations on encrypted data, i.e., the Enhanced Homomorphic Encryption Scheme (EHES) suggested by Gorti et al. [15]. In parallel, this algorithm prevents an unauthorized disclosure of data being stored on cloud computing. Table 3 illustrates the fundamentals of the EHES algorithm.

Table 3. Fundamentals of the EHES cryptosystem

Function	Input	Computation	Output
Key generation	$p, q \in \mathbb{P}$	Compute $n = p \cdot q$	Public key: $pk = (n)$ Secret key: $sk = (p, q)$
Encryption	$m \in \mathbb{Z}_p$	Generate a random number r Compute $c = m + r \times p^q \ (\bmod n)$	$c \in \mathbb{Z}_n$
Decryption	$c \in \mathbb{Z}_n$	Compute $m = c \bmod n$	$m \in \mathbb{Z}p$

To illustrate the main features of this protocol, we choose two values x and y that belong to Z_p. In this case, Enc and Dec stand for encryption and decryption functions respectively. In the encryption process, we define the public key pk = (n), and also the private key sk = (p, q).

In this scheme,

$$\text{Enc}(x \cdot y) \equiv \text{Enc}(x)\,\text{Enc}(y)\,(\text{mod } n), \text{ or } x \cdot y = \text{Dec}(\text{Enc}(x)\,\text{Enc}(y)) \tag{15}$$

$$\equiv \text{Enc}(x)\,\text{Enc}(y)(\text{mod } p) \tag{16}$$

Similarly,

$$\text{Enc}(x + y) \equiv \text{Enc}(x) + \text{Enc}(y)\,(\text{mod } n), \text{ or}$$
$$x + y = \text{Dec}(\text{Enc}(x) + \text{Enc}(y)) \tag{17}$$

$$\equiv \text{Enc}(x) + \text{Enc}(y)\,(\text{mod } p) \tag{18}$$

As seen from the Eqs. 16 and 18, the EHES cryptosystem ensures data security and supports both addition and multiplication operations.

4 Related Work

In [16], the authors use an efficient method to handle encrypted data. The main goal of this technique is to prevent unauthorized disclosure of medical information in cloud environment, where all data are being stored on external servers. In this scheme, R. Challa et al. propose the homomorphic encryption technique to process data securely. More functionally, the authors use Learning With Error (LWE) concept to carry out basic mathematical operations. In other words, this method helps cloud consumers to execute both addition and multiplication operations on encrypted images hosted in the cloud computing. In such a scenario, clients' digital data need to be encoded using the homomorphic algorithms before sending them to a cloud provider. Based on these considerations, the proposed framework is an efficient solution in order to develop cloud applications that offer image processing as a service. Indeed, the proposal contains necessary security measures to help protect data privacy when using remote cloud tools. Consequently, healthcare organizations can rely on this solution to delegate image analysis to a third party in a secure manner. Since homomorphic encryption is a time-consuming technique, it is often not practical for cloud applications involving large data such as medical records. Hence, it is important to address this issue to meet the quality of services requirements.

In [17], the authors suggest a novel approach to improve homomorphic encryption. Essentially, this function uses Residue Number System (RNS) to process encrypted data. In other words, the proposal is homomorphic with respect to addition, subtraction and multiplication. Of course, the main reason to implement this technique is to maintain security and safety during cloud usage. Besides helping clients to outsource computations, the proposed approaches allow secure sharing of the Electronic Patients Record

(EPR). For example, this solution offers the possibility to apply Sobel filter on encrypted data without decrypting the original medical image.

Ramulu et al. [18] illustrate a framework to process medical image in encrypted domain. This system is useful to address security problems that occur when using cloud computing. Typically, it provides safety measures to prevent internal manipulations of data by an untrusted cloud provider. The proposed method uses homomorphic techniques during the encryption process. This method is a mixture of two advanced techniques, i.e., Discrete Wavelet Transform (DWT) and Paillier cryptosystem. Basically, the Paillier algorithm is used to carry out addition operation on ciphertext without the need to decrypt it for privacy protection reasons. This is the primary reason behind using this algorithm to encode the approximation coefficients used in data processing. For instance, the proposed solution is able to securely perform the 2-D Haar wavelet transform to demonstrate the feasibility of this approach. Similarly, the original image can be accurately reconstructed by means of Paillier and IDWT (Inverse Discrete Wavelet Transform).

In [19], Habeep et al. use an efficient mechanism to solve the problem of data expansion occurring when using homomorphic encryption. The proposed technique intends to enhance system performance, which is the major drawback of homomorphic approach. In essence, this solution uses multilevel DWT/IDWT concept in the last step to handle to data expansion issue. In the same line, the Multiplicative Inverse Method (MIM) is implemented in this framework in order to limit the quantization factor during decryption process. To accomplish this, the authors rely particularly on the Paillier scheme as it is an additive homomorphic cryptosystem. Consequently, this model allows users to delegate basic image processing techniques to an external cloud provider.

Yang et al. [20] illustrate a solution for online image processing services. The primary goal of this study is to develop an efficient way to outsource computations securely. To this aim, the authors extend the classical Gentry's homomorphic encryption to perform floating-point arithmetic operations. Furthermore, the proposal is a symmetric encryption instead of public key encryption in order to ease the process of encryption. Results of the experiment show that the proposed technique can resist the statistical analysis attacks. Hence, it is an appropriate solution to process images using cloud technology.

5   Discussion

In general, the fundamental goal of cryptography is to allow secure storage of sensitive data and prevent information disclosure. Beside security reasons, homomorphic encryption schemes offer the possibility to effectively perform mathematical operations on encrypted data without revealing the contents to the data processor. Hence, the application of these types of encryption techniques is continuing to gain popularity, especially in image processing. The main emphasis of this work is to explore the possibilities offered by this new approach to secure cloud-based applications. Today, homomorphic encryption has become a promising solution to process data in encrypted data. For this reason, various secure frameworks are suggested to outsource image processing to an external cloud provider. In this context, there are two methods to achieve this objective. The first one uses Partial Homomorphic Encryption (PHE) to perform only one operation, while the second one supports many operations by using Fully Homomorphic Encryption (FHE) algorithms. By using these techniques, users can protect privacy in online image processing services offered by a cloud provider. However, the majority of existing schemes encrypt each pixel separately, which increases the runtime. Additionally, this approach still faces several barriers that obstruct its adoption. Although, the PHE schemes protect data against common attacks, they often consist of only one mathematical operation. Hence, its utilization in real applications involving more than one operation is not entirely satisfactory. To mitigate this issue, a hybrid solution based on different schemes is suggested to meet the operational needs. Recently, the FHE scheme is introduced to allow a user to perform both addition and multiplication operation. Unfortunately, both schemes are still far from being practical to process on the encrypted medical images efficiently because these cryptosystems are usually too slow [21]. Therefore, using homomorphic cryptosystems in cloud computing has often a negative effect on system performance, especially execution time. Obviously, this has serious consequences regarding the provider's reputation in spite of the potential of fully homomorphic encryption schemes to improve efficiency compared to partial ones. For these considerations, tremendous efforts have been made to extend the classical homomorphic encryption algorithms to meet cloud services requirements. For example, the work in [20] intends to develop Gentry's homomorphic encryption in such a way that it operates on encrypted floating numbers efficiently. Meanwhile, it seeks to improve system performance.

6   Conclusion

In cloud-based applications, healthcare organizations just need to upload their digital data to cloud computing and then wait for the result of post-processing. To outsource image processing is actually the primary reason for adopting cloud computing instead of in-house data centers. Additionally, this model delivers only needed resource to consumers in order to cut costs. Although this concept offers many advantages, potential data disclosure risks are the main preoccupations when installing cloud computing. For these objectives, several security mechanisms are suggested to process medical images securely. In this regard, using homomorphic encryption becomes a promising solution for addressing security problems in the Software as a Service (SaaS) model. These classes of algorithms can operate on encrypted data and maintain privacy during data processing. Functionally, it encodes data in such a way that an authorized user can execute basic mathematical operations, especially addition and multiplication. To accomplish this, there are two categories of algorithms for encryption data, namely Partial Homomorphic Encryption (PHE) and Fully Homomorphic Encryption (FHE) algorithms. Globally, PHE offers the possibility to do one operation, while the second type can perform several operations. So far this technique has been applied to perform only some basic operations. It appears from the present study that it does not fully satisfy the necessary quality requirements of imaging tools, particularly in terms of computational costs. For these objectives, considerable efforts have been made to develop and improve existing schemes to analyze data efficiently and accurately. In the same line, the proposed architecture in [20] allows users to perform arithmetic operations on floating point numbers to meet image processing requirements. In summary, the results of the present study indicate that homomorphic encryption is a promising technique, and hence, can be implemented in practical image processing using cloud applications. In our future work, we will propose an improved scheme that uses homomorphic encryption to enhance running time. First, we intend to implement and compare different existing schemes. Second, we will extend the selected algorithm to meet healthcare requirements regarding efficiency, security and privacy. In addition, we plan to introduce segmentations techniques to split an image into two regions: Region of Interest (RoI) and Region of Non Interest (RoNI). Accordingly, only the RoI region will be encrypted for enhancing the system performance.

References

1. Mell, P., Grance, T.: The NIST definition of cloud computing, vol. 15, pp. 1–3. Technical report, National Institute of Standards and Technology (2009)
2. Mazhar, A., Samee, U.K., Athanasios, V.: Security in cloud computing: opportunities and challenges. Inf. Sci. **305**, 357–383 (2015)
3. Marwan, M., Kartit, A., Ouahmane, H.: Cloud-based medical image issues. Int. J. Appl. Eng. Res. **11**, 3713–3719 (2016)
4. Fernandes, D.A.B., Soares, L.F.B., Gomes, J.V., Freire, M.M., Inácio, P.R.M.: Security issues in cloud environments: a survey. Int. J. Inf. Secur. **13**(2), 113–170 (2013)
5. Abbas, A., Khan, S.U.: E-health cloud: privacy concerns and mitigation strategies. In: Gkoulalas-Divanis, A., Loukides, G. (eds.) Medical Data Privacy Handbook, pp. 389–421. Springer, Cham (2015). https://doi.org/10.1007/978-3-319-23633-9_15
6. Pearson, S., Benameur, A.: Privacy, security and trust issues arising from cloud computing. In: Proceedings of the 2010 IEEE Second International Conference on Cloud Computing Technology and Science, CLOUDCOM 2010, pp. 693–702. IEEE Computer Society, Washington, DC (2010)
7. Al Nuaimi, N., AlShamsi, A., Mohamed, N., Al-Jaroodi, J.: E-health cloud implementation issues and efforts. In: Proceedings of the International Conference on Industrial Engineering and Operations Management (IEOM), pp. 1–10 (2015)
8. Raykova, M., Zhao H., Bellovin, S.M.: Privacy enhanced access control for outsourced data sharing. In: Keromytis A.D. (ed.) Financial Cryptography and Data Security, FC 2012. LNCS, vol. 7397. Springer, Heidelberg (2012)
9. Gomathikrishnan, M., Tyagi, A.: HORNS-a homomorphic encryption scheme for cloud computing using residue number system. IEEE Trans. Parallel Distrib. Syst. **23**(6), 995–1003 (2011)
10. Paillier, P.: Public-key cryptosystems based on composite degree residuosity classes. In: Stern, J. (ed.) EUROCRYPT 1999. LNCS, vol. 1592, pp. 223–238. Springer, Heidelberg (1999). https://doi.org/10.1007/3-540-48910-x_16
11. Bhabendu, K.M., Debasis, G.: Fully homomorphic encryption equating to cloud security: an approach. J. Comput. Eng. (IOSRJCE) **9**, 46–50 (2013)
12. Yi, X., Paulet, R., Bertino, E.: Homomorphic Encryption and Applications. SpringerBriefs in Computer Science. Springer, Cham (2014). https://doi.org/10.1007/978-3-319-12229-8

13. Rivest, R., Shamir, A., Adleman, L.: A method for obtaining digital signatures and public-key cryptosystems. Commun. ACM **21**(2), 120–126 (1978)
14. Kota, C.M., Aissi, C.: Implementation of the RSA algorithm and its cryptanalysis. In: Proceedings of the ASEE Gulf-Southwest Annual Conference. University of Louisiana at Lafayette (2002)
15. Gorti, V.S., Garimella, U.: An efficient secure message transmission in mobile ad hoc networks using enhanced homomorphic encryption scheme. Global J. Comput. Sci. Technol. Netw. Web Secur. **3**, 20–33 (2013). version 1.0
16. Challa, R.K., Kakinada, J., Vijaya Kumari, G., Sunny, B.: Secure image processing using LWE based homomorphic encryption. In: Proceedings of the IEEE International Conference on Electrical, Computer and Communication Technologies (ICECCT), pp. 1–6 (2015)
17. Gomathisankaran, M., Yuan, X., Kamongi, P.: Ensure privacy and security in the process of medical image analysis. In: Proceedings of the IEEE International Conference on Granular Computing (GrC), pp. 120–125 (2013)
18. Kanithi, S.R., Latha, A.G.: Secure image processing using discrete wavelet transform and Paillier cryptosystem. Int. J. Mag. Eng. Technol. Manag. Res. **2**(10), 1270–1276 (2015)
19. Habeep, N., Dayal Raj, R.: Homomorphic encrypted domain with DWT methods. Int. J. Invent. Comput. Sci. Eng. **1**(3), 2348–3431 (2014)
20. Yang, P., Gui, X., An, J., Tian, F.: An efficient secret key homomorphic encryption used in image processing service. Int. J. Secur. Commun. Netw. Hindawi, 2017, article ID 7695751, 11 p. (2017). https://doi.org/10.1155/2017/7695751
21. El Makkaoui, K., Beni Hassane, A.: Challenges of using homomorphic encryption to secure cloud computing. In: Proceedings of the International Conference on Cloud Technologies and Applications (CloudTech) (2015)

The 1st International Workshop on Industry 4.0 and Smart Manufacturing: WI4SM

Impact of the Multi-criteria Methods in Supporting

Innovation "Application of the AHP Method Within an Automotive Firm, Morocco"

Jihane Abdessadak, Houda Youssouf$^{(\boxtimes)}$, Akram El Hachimi,
Kamal Reklaoui, and Abdelatif Benabdellah

Lab: Engineering, Innovation and Management of Industrial Systems,
FSTT, B.P. 416 Tangier, Morocco
hyoussou@gmail.com

Abstract. The international automotive market has for a long time been dominated by a few classical industries, which today increasingly fear the arrival of competition from Asian countries, notably China and India, which promise to "break prices".

To cope with this competition, accentuated by globalization, companies must always think about adapting their products while maintaining good quality and preserving their brand images. To this end, they always use innovative methods to create new ideas to differentiate themselves from competitors.

This paper comes to emphasize the importance of the multi-criteria methods and more precisely the AHP method to support innovation.

Keywords: AHP method · Decrease of diversity · Multi-criteria methods
Innovation · Monozukuri

1 Introduction

Despite the euphoria that reigns in Silicon Valley - the birthplace of American innovation - many economists are pessimistic. According to them, if for a decade, the most industrialized countries struggle to maintain growth and employment, it is because of the lack of major innovations. For others, however, there must be optimism. Several recent studies estimate the potential gain from the most promising innovations at tens of thousands of billions of dollars.

Today, the automotive industry devotes all its energy to innovation. This is indicated by the 2014 ranking of patent applicants from the National Institute of Intellectual Property (Inpi). If the automotive is at the forefront of innovation, it is because it faces many challenges that force builders to heighten their innovation efforts. [1] But how can they lead to innovative ideas without getting stuck with the fateful phrase "It has already been done", trying to understand the reasons that made the "what has already been done" into "a good idea ", or studying the mistakes of others and knowing how to choose the right solution for a given problem. It is for this purpose that come the multi-criteria analysis

© Springer International Publishing AG, part of Springer Nature 2018
M. Ben Ahmed and A. A. Boudhir (Eds.): SCAMS 2017, LNNS 37, pp. 839–851, 2018.
https://doi.org/10.1007/978-3-319-74500-8_76

methods. These methods allow the user to analyze and pinpoint the problem and thus innovate.

In this paper, we are going to explain and expose how the AHP method –which is a multi-criteria method-, can help the firm X (A multinational automotive company) to create and to take the plunge to innovation.

2   Monozukuri and Decrease of Diversity

2.1   Monozukuri

Monozukuri is a Japanese concept designing art, science and craft to design and produce technical objects. It is generally used to designate manufacturing activities.

The concept of Monozukuri appeared around 900 AD in Kyoto (Japan), during the Edo period. But since the early 2000 s, Monozukuri has been given a place of honor by public institutions and major Japanese industrial companies such as NEC, Nissan, Toshiba, Sharp and Toyota. The reason is to maintain the competitiveness of the country's industry, on the one hand threatened by the American and European dominance of the world market, the rise of China, and put back on the other hand because of wars and conflicts. [2, 3]

The philosophy of Monozukuri has a broad meaning that encompasses virtually all business activities such as process engineering, supply chain management and, to a lesser extent, purchasing activities.

With its modern sense, the Monozukuri approach can optimize the entire value chain to improve quality and reduce costs: from the design of the product to what the customer really needs, detailing of each component, up to delivery to final customer, suppliers, packaging of components, transport, stock supply and on the production line … etc. The Monozukuri is therefore a complete approach seeking to reduce the total cost of the product with irreproachable quality and customer value [3].

Monozukuri is a global production approach, used by different automotive ventures such as Nissan, Denso for several years with very good results and is now deployed since February 2011 in all the Renault Group's industrial sites [4].

This approach aims to reduce the total cost of the car while improving its quality. This involves working on the entire value chain of the vehicle or mechanical component from conception to delivery to the customer. This means that the firm teams will all follow the same indicator: the cost of the vehicle delivered to the dealers or the TdC, even if these teams work on different criteria and are not always compatible. For example, the purchasing department has an interest in finding the least expensive component, while engineering is concerned with the quality of the items and processes [5].

The Monozukuri is a global organization that includes:

- Co-operation of all business lines through a transversal approach.
- Customer expectations.
- Benchmarking and continuous improvement.

The indicator for measuring the competitiveness of a mechanical member or a vehicle id then the Total Delivered Cost: the full cost. This indicator is common to all trades. For the firm X, the goal of the Monozukuri is to reduce the TdC of its products by 12% [6].

The Total Delivered Cost (TdC) is the largest perimeter of operating costs that can be allocated to the Product. It is the total cost of designing, manufacturing and transporting vehicles or organs to their place of consumption.

This makes it possible to provide a key indicator in the economic management of the projects, in particular the Monozukuri Project, to improve profitability and increase sales by optimizing the total cost.

Monozukuri is an efficient approach since it determines the total cost of the vehicle between the start of production and the time of delivery. The Monozukuri takes into account all the costs incurred by the various elements of production - logistics, design, manufacturing ... It was found that at the interface of these trades were productivity deposits and that the sum of Optima did not correspond to an overall optimum. The Monozukuri makes it possible to optimize the interfaces so that the overall optimum cost is the only decision factor [2, 7].

The Monozukuri principles on which the firm X is based in the application of the Japanese approach can be summarized in the diagram and detailed below (Fig. 1):

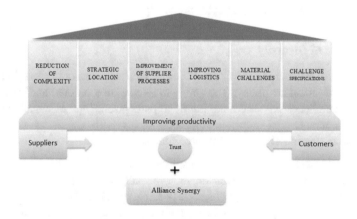

Fig. 1. Application pillars of Monozukuri in the firm X

REDUCTION OF COMPLEXITY:

- Reduction of the diversity of references
- Adjust product to customer requirements (Right content, Right sizing)

STRATEGIC LOCATION:

- Reduction of capital expenditure
- Reduction of customs fees
- Optimization Sourcing, especially in the country/ region of the plant

IMPROVEMENT OF SUPPLIER PROCESSES:

- Integration of suppliers in the perimeter of the plant
- Optimizing the supplier value chain
- Commercial actions

IMPROVING LOGISTICS:

- Technical actions to improve the cost of purchasing parts
- Arbitration between internal production and external procurement
- Reduction of costs and inventories

CHALLENGES MATERIAL:

- Reduction of raw material costs

SPECIFICATIONS CHALLENGE:

- Adjust the workforce of the plant to its activity

2.2   Decrease of Diversity

The reduction of diversity is one of the pillars on which the firm x is based in the application of its Monozukuri policy. The aim is to reduce, in a rational and cost-effective manner, the number of references per family of parts present and managed in the factories of the company.

This diversity comes mainly from three sources:

- Technical constraints related to the functioning of the parts in their system.
- Optional requirements of the customer, especially as the car market trend in its development to give as many choices as possible to the buyer.
- Variations from design and engineering offices, although these errors are rare given the strict approach taken by these institutions.

The diversity of references translates into additional costs related to the management of these articles, the greater the diversity, the more costly its management is, and the greater the need to control the nomenclature [6].

The reduction of the diversity takes place on several aspects of the manufacture of the finished product; its effects are summarized in:

- Optimizing inventory management: Reducing diversity contributes directly to the optimization of stock management by reducing the storage area for parts, since each reference is stored separately in order to avoid Risk of confusion, which increases the occupied area and complicates the logistical maneuver within the plant.
- Improvement of working conditions: Reduction of the risk of operator errors, which subsequently leads to an over-quality, the cost of which is not amortized in the sale price of the vehicle or a non-compliance with customer requirements.
- Reduction in the number of references ordered from suppliers: It is known that in all customer-supplier relationships the greater the number of parts ordered, the more the price is conditioned, as opposed to when the range of orders is diversified because the supplier must adapt to the requirements.
- Improvement of the cycle time: Reduction of the cycle time by eliminating the congestion at the edge of the assembly line and thus promoting the application of the 5S, which leads to improved operator efficiency thanks to the reduction in the number of steps By post and the ease of selection of the parts to be assembled.
- Participation in the standardization of packaging: Reducing the diversity of the number of references per piece contributes to the unification of packaging, which has a positive impact on logistical costs [6, 7].

3   AHP: Analytic Hierarchy Process

The multi-criteria analysis methods are decision-support tools developed since the 1960 s. Many methods have been proposed for the evaluation of efficiency choices. The basic idea is to consider all the criteria taking into account, to attribute to them a weight related to their relative importance, to rate each action against all the criteria and finally to aggregate these results [8].

The most well-known methods of multi-criteria analysis are: the PROMETHEE method, which allows to visualize conflicts and synergies between criteria, the family of ELECTRE methods which is based on the elimination of the worst options and the choice of the best complexity) And the AHP (analytical hierarchical process) which allows to structure the problem of the simple method and to evaluate the criteria and the choice to be made.

The Analytical Hierarchical Process was designed, explained and demonstrated mathematics in 1971 by the American theorist of Iraqi origin Thomas Lorie SAATY, currently Professor at the University of Pittsburgh in the United States. The method allows managers to structure the complexity problems they face by making judgments based on their experience and available informational data. Its application is simple; it can be done by an individual alone or in a group [9, 10].

The AHP method is subjective since it appeals to the judgment of the individuals who apply it. Its unique weighting scale makes it possible to consider the measurable and non-measurable criteria; it does not correspond to the unit of measurement [11].

Among the many applications of AHP discussed in the literature:

- Decide on the location of offshore industrial project in 2002 (University of Cambridge)
- Decide on the best way to manage US watersheds in 2003 (US Department of Agriculture)
- Quantifying the overall quality of software systems in 2005 (Microsoft Corporation)
- Decide on the best way to reduce the impact of global warming in 2007 (Fondazione Eni Enrico Mattei)
- The main characteristics of AHP compared to other methods are:
- Consideration of qualitative and quantitative aspects
- Determination of weights by binary comparison
- Amount of data required by the limited analyst
- Opportunity to consider inconsistencies of policy makers
- Ability to use dedicated software to perform Expert Choice analysis [13].

4   Application of AHP in an Automotive Industry

4.1   Analysis Phase

The aim objective of this study is to reduce the diversity of references by investing a minimum of resources but the inefficiency of the distribution by section and the absence of any other common indicator, between the pieces leads to a single possible solution: to rely on the judgment of the experts for the ranking of these in order of priority, taking into account all the factors revealing the susceptibility of presence of unwanted diversity and gain related to suppression of the references.

The complication of this approach lies in the dual aspect of the objective Choice of intervening factors and the difficulty of their price being taken into account jointly The heterogeneity, or even the contradiction, existing between the latter, hence the need for a tool Help with multi-criteria decision price.

This tool is used to enable the returns to target the most diversity It thus has an economizer of time and effort while focusing actions and Increasing profitability into proposals.

The decision on which families of parts are most likely to present harmful diversity and at the same time the most cost-effective in terms of cost if the diversity is reduced.

The phases of application of the multi-criteria help tool for the classification of the parts families are:

- Consideration of the criteria influencing the decision.
- Evaluation of the weights relative to these criteria.
- Establishing a weighting scale for each criterion.

The criteria were established at a Brainstorming meeting with the pilot, the facilitator and the diversity reduction project team.

A Benchmarking with the firm X was carried out in order to exploit upstream the criteria considered for downstream judgment of the validity of the proposals. Thus, six evaluation criteria were imported and consolidated by two additional criteria: Impact on Design and Impact on Customer Interest [8, 10].

- Impact customer interest: The customer is the king, his judgment of the product and the options offered by the dealer are red lines not to be exceeded during the processing of the part. Its implicit and explicit requirements must not be subject to any.
- Price: The price of the piece reflects its importance and the complexity of packaging during storage and handling.
- Supplier location: Delivery time is a very important concept not to be overlooked.
- Impact design: Each part in the vehicle operates in a complete system, so any changes in the neighboring rooms (functional or dimensional) must be taken into consideration. The more complex the piece, the less useful it is to target it.
- Surface: The surface that the product or the piece takes over the stock or on the edge of a chain directly influences the TdC.
- Reference number: the aim is to reduce the number of references by parts.
- Rate of use: The use of the part provides a key data on the profitability of the part and also on the costs related to the logistics.
- Number of sites concerned: The interest is more important, within the framework of the firm X policy of optimization of the "CARRY OVER", if the part concerns several sites.

While the eight criteria used in the Decision Support Tool for Diversity Reduction are all influential in determining which pieces of interest to address, their order of importance is not similar. In order to determine their relative importance levels, a relative weight evaluation matrix is developed and distributed distinctly for objectivity on 8 experts from different trades and members of the Monozukuri team within the firm (so that all Aspects of the product value chain are taken into consideration) [8, 14].

The calculation of the eigenvector relative to the greatest eigenvalue of this matrix will allow obtaining the relative weights of each of the eight criteria leading to the optimal judgment. This calculation was carried out using the "Expert Choice" software as shown in the Fig. 2 below:

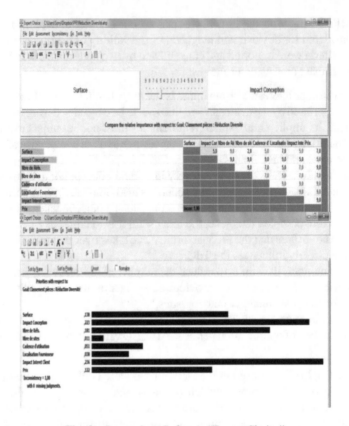

Fig. 2. Screenshots Software "Expert Choice"

A coherence study is then carried out with the aim of eliminating non-consistent opinions. The method of verification consists in calculating the coherence indicator of each matrix using the following equation:

$$IC = \frac{\lambda max - N}{N - 1}$$

Where λmax is the highest eigenvalue of the matrix and N dimension of the matrix, and then divide this index by the index ratio relative to the size of the matrix (Number of criteria)

$$RC = \frac{IC}{RI}$$

The RI for N = 8 is 1.41

The result of the division is called a Coherence Ratio, it is that which makes it possible to judge clearly the coherence of the matrix, if RC ≤ 0.1, the judgment is coherent, the matrix is admitted, if not, if RC > 0.1, the matrix is inconsistent and therefore not considered in the study [15].

After averaging each criteria. The final weights for each criterion are (Fig. 3):

Surface	Impact design	Number of Ref.	Number of sites	Rate of use	Supplier location	Impact Customer interest	Price
0,10	0,13	0,09	0,02	0,15	0,04	0,26	0,17

Fig. 3. Weight of criteria

The rating scale (odd from 1 to 9) differs from criterion to criterion due to the difference in size and the nature of the criterion (measurable or estimable). Scales proportional to measurable criteria such as the surface are scored to take into account all existing values and to split the domain adequately. Note 9 is affected when the piece has a maximum interest, and so on up to 1 for those that are of no interest, and also for the estimable criteria (design impact and customer interest) except that the ratings in this Cases represent whether the impact is very low, low, medium, strong or extreme [14, 16].

The table below summarizes the weighting scales approved by the experts and used for the eight criteria (Fig. 4):

Paramètres

Score	Surface (m^2)	Niveau d'impact	Nbre Références	Nbre Sites	Cadence Annuelle	Localisation Fournisseur (Km)	Interet Client	Coût Pièce (€)
1	Inférieur à 1	Très Fort	Egal à 2	1	Inférieur à 25%	Inférieur à 50	Très Fort	Inférieur à 5
3	Compris entre 1 et 3	Fort	Entre 2 et 4	2	Entre 25% et 50%	Entre 50 et 500	Fort	Compris entre 5 et 25
5	Compris entre 3 et 6	Moyen	Entre 4 et 6	3	Entre 50% et 75%	Entre 500 et 1000	Moyen	Compris entre 25 et 50
7	Supérieur à 6	Faible	Entre 6 et 10	4	Entre 75% et 100%	Entre 1000 et 5000	Faible	Compris entre 50 et 100
9	Bord de Chaine	Très Faible	Supérieur à 10	5	Egale à 100%	Supérieur à 5000	Très Faible	Supérieur à 100

Fig. 4. Weight scale

4.2 Results

The nomenclature contains 2461 families of parts, a first filtration consisting of eliminating single-reference parts (presenting no diversity), which reduces the number of large parts families to 607 families.

The analysis included 50 of the most important pieces in diversity chosen and proposed by each of the preparers and the comity of the engineering department (Fig. 5).

	Désignation	Surface (m²)	Impact Conception	Nbre Références	Nbre de sites	Cadence (an)	Local. Fournisseur (km)	Impact intérêt client	Prix (€)
5	ACCOUDOIR SUR PASSAGE DE ROUE	5,00	Très Fort	4	1	91%	45	Très Fort	6,07 €
6	AERATEUR	1,55	Faible	3	3	80%	4500	Très Fort	1,30 €
7	ALTERNATEUR	21,45	Très Fort	7	5	29%	3500	Très Fort	47,85 €
8	AUTORADIO	0,74	Moyen	4	2	55%	1700	Fort	55,48 €
9	BATTERIE	14,30	Fort	3	4	40%	650	Très Fort	24,41 €
10	BOITIER CHAUFFAGE	44,29	Fort	4	1	52%	45	Très Fort	70,51 €
11	BOITIER ELCQ CALIBRE DIES VHCP CTR MOT	0,48	Très Fort	6	2	53%	1700	Très Fort	63,98 €
12	BOITIER ELCQ PROGRAMMABLE CONTROLE MOT	0,72	Très Fort	2	2	46%	1700	Très Fort	66,30 €
13	BOITIER ELCQ PROGRAMME ESS CTR MOTEUR	0,96	Très Fort	5	2	49%	1700	Très Fort	31,89 €
14	CALCULATEUR AIDE PARKING	1,12	Très Fort	16	2	47%	4500	Très Fort	2,16 €
15	CALCULATEUR UNITE CTL HABITACLE	1,42	Très Fort	2	2	18%	4500	Très Fort	5,31 €
16	CALE GARNITURE LAT G COFFRE	1,30	Très Faible	2	1	63%	46	Très Faible	12,32 €
17	CARENAGE SOUS MOTEUR	Bord de Chaine	Très Faible	4	1	71%	45	Faible	3,28 €
18	COMMANDE SOUS VOLANT	3,99	Fort	2	2	36%	1700	Très Fort	84,33 €
19	COMMANDE VITESSE ASS	1,10	Très Fort	2	2	83%	4500	Très Fort	18,00 €

Fig. 5. Pieces Data

The data collection included for each of the pieces:

- Calculation of the number of references and determination of the number of sites concerned.
- Contact the purchasing department for prices and suppliers.
- Logistics contact for usage rates.
- Manual measurement of the areas occupied by the packaging.
- Decision on the impact on the design and on the customer.

The score for the different pieces is calculated by summing the scores relative to each criterion multiplied by the weight of this one. The scores thus obtained, the pieces are classified and processed in priority order (Fig. 6).

	Désignation	Surface	Impact Conception	Nbre Références	Nbre Sites	Cadence	Local. Fournisseur	Impact Intérêt client	Prix	Score	Classement
5	GARNITURE DE PORTE AV	9	9	7	1	5	1	9	3	6,548	1
6	TAPIS HABITACLE	9	9	7	3	5	7	7	3	6,362	2
7	PARE-SOLEIL D	7	9	9	1	5	1	9	1	6,170	3
8	ECRAN PASSAGE ROUE AV D/G	7	9	5	1	5	1	9	3	6,143	4
9	GARNITURE DE PAVILLON X92	7	9	7	1	3	3	9	3	6,127	5
10	PLANCHE BORD	9	9	9	1	3	1	5	5	5,737	6
11	CALE GARNITURE LAT G COFFRE	3	9	3	1	5	1	9	3	5,524	7
12	VOLANT DIRECTION	9	7	3	1	5	1	7	3	5,376	8
46	BOITIER ELCQ PROGRAMMABLE CONTROLE MOT	1	1	3	3	3	7	1	7	2,897	42
47	TUYAU CONDENSEUR DETENDEUR	1	3	3	1	5	7	3	1	2,874	43
48	MOTEUR	7	1	5	1	5	5	1	1	2,821	44
49	BOITIER ELCQ PROGRAMME ESS CTR MOTEUR	1	1	5	3	3	7	1	5	2,733	45
50	DEMARREUR	3	3	3	5	3	7	1	3	2,718	46
51	TABLEAU BORD	1	5	3	1	3	7	1	3	2,663	47
52	CALCULATEUR AIDE PARKING	3	1	9	3	3	7	1	1	2,620	48
53	CALCULATEUR UNITE CTL HABITACLE	3	1	3	3	1	7	1	3	2,099	49
54	ECRAN THERMIQUE BOITIER DIRECTION	3	1	3	1	3	7	1	1	1,995	50

Fig. 6. Pieces notes

In addition to the targeted channels and Brainstorming carried out by the project team, the involvement of a maximum number of speakers will enrich the flow of proposals. After ranking the sample of the pieces, searches for proposals were carried out by means of complete technical studies of each family of parts in order of priority.

To this end, tools and techniques have been used to structure and facilitate the investigation of applicable and cost-effective ideas among these techniques:

- Physical examinations of the parts on the ground during the weekly chain returns.
- Dimensional and functional studies on CATIA and CAD software.
- Brainstorming organized using the process of methodical questioning.
- Collecting ideas through the guide pamphlet and fiche created and distributed for this purpose.

Of the 50 pieces selected and classified by the multi-criteria decision support tool, 34 were studied. The results of these studies resulted in 14 proposals and 2 others thanks to the fact sheets distributed (Fig. 7).

	Decided	Ongoing	Abandoned
Suggestions	2	11	3
Ref No. to remove	24	20	4
Surface Area gain (m²)	23,79	51,03	12,07
Annual Gain in Euro	21280	46130	10920
Gain / vehicle in Euro	0,304	0,659	0,156

Fig. 7. Synthesis

The top innovative idea we got was the car sun visor (Fig. 8):

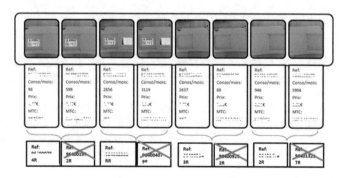

Fig. 8. Sun visor references

We went from 8 references to 4, we optimized in terms of space and we have allowed the company an annual gain of about 4 K €

5   Conclusion

In a rapidly evolving economic system, companies always seek to maximize their profits by adapting to market needs. That's why they use innovation.

The AHP method that we applied in this automotive company, allowed us to push managers to have new avenues of improvement or even innovation.

The reduction of diversity allows us to broaden the strategies of work within the company as well as to increase its competitiveness in the market.

Acknowledgment. We would like to thank the company that allowed us to do this study and provided us with all the resources and information we needed.

References

1. lll.pdf
2. Saito, K.: Development of the University of Kentucky – Toyota Research Partnership: Monozukuri: PART II, vol. 17, no. 5 (2006)
3. Ranky, P.G.: Eighteen 'monozukuri-focused' assembly line design and visual factory management principles with DENSO industrial examples. Assem. Autom. **27**(1), 12–16 (2007)
4. Rse, L., Ong, L., Daiichi, F.: La responsabilité sociale des entreprises au Japon, de l'époque d'Edo à la norme ISO 26 000 et à l'accident nucléaire de Fukushima, pp. 33–37 (2003)
5. Shiro, F., Kaoru, Y.: Is Japanese manufacturing style (so-called Monozukuri) really robust? - Causal loop diagram and modeling analysis. In: Shiro, F. (ed.) Proceedings 27th International Conference System Dynamics Society, no. 5, pp. 1–26 (2009)
6. Fujimoto, T.: The Japanese Manufacturing Industries - Its Capabilities and Challenges - Evolutionary Analysis of Capability and Architecture (2011)
7. Aoki, K., Staeblein, T., Tomino, T.: Monozukuri capability to address product variety: a comparison between Japanese and German automotive makers. Int. J. Prod. Econ. **147**(PART B), 373–384 (2014)
8. Hsu, Y.L., Lee, C.H., Kreng, V.B.: The application of Fuzzy Delphi Method and Fuzzy AHP in lubricant regenerative technology selection. Expert Syst. Appl. **37**(1), 419–425 (2010)
9. Sivilevičius, H., Maskeliūnaite, L.: The criteria for identifying the quality of passengers' transportation by railway and their ranking using AHP method. Transport **25**(4), 368–381 (2010)
10. Saaty, T.L.: Decision making with the analytic hierarchy process. Int. J. Serv. Sci. **1**(1), 83 (2008)
11. Danner, M., et al.: Integrating patients' views into health technology assessment: Analytic hierarchy process (AHP) as a method to elicit patient preferences. Int. J. Technol. Assess. Health Care **27**(4), 369–375 (2011)
12. Vučijak, B., Kurtagić, S.M., Silajdžić, I.: Multicriteria decision making in selecting best solid waste management scenario: a municipal case study from Bosnia and Herzegovina. J. Clean. Prod. **130**, 166–174 (2016)
13. Yang, T., Hsieh, C.H.: Six-Sigma project selection using national quality award criteria and Delphi fuzzy multiple criteria decision-making method. Expert Syst. Appl. **36**(4), 7594–7603 (2009)

14. Stefanović, G., Milutinović, B., Vučićević, B., Denčić-Mihajlov, K., Turanjanin, V.: A comparison of the Analytic Hierarchy Process and the Analysis and Synthesis of Parameters under Information Deficiency method for assessing the sustainability of waste management scenarios. J. Clean. Prod. **130**, 155–165 (2016)
15. Dong, Y., Zhang, G., Hong, G., Xu, Y.: Consensus models for AHP group decision making under row geometric mean prioritization method. Decis. Support Syst. **49**(3), 281–289 (2010)
16. Benaïm, C., Perennou, D.-A., Pelissier, J.Y., Daures, J.-P.: Using an analytical hierarchy process (AHP) for weighting items of a measurement scale: a pilot study. Rev. Epidemiol. Sante Publique **58**, 59–63 (2010)

Computer Vision Control System for Food Industry

Hadj Baraka Ibrahim[(✉)] ⓘ, Oussama Aiadi, Yassir Zardoua,
Mohamed Jbilou, and Benaissa Amami

Faculty of Sciences and Techniques, Abdelmalek Essaadi University, Tangier, Morocco
i.baraka@gmail.com, oussama.aiadi@gmail.com,
yassirzardoua@gmail.com, mjbilou@hotmail.fr,
b_benaissa@hotmail.com

Abstract. Defects in production can appear as a result of material and human errors. A food product with a defect may be the cause of direct and indirect losses caused by product recalls, logistic problems, destruction of defective products, reputation issues, possible penalties, etc. The purpose of this work is to minimize the human intervention and substitute it to the maximum extent with an automatic inspection system based on the artificial vision technology that will carry out the inspection task. To do so, the main functions of the system had to be elaborated in a specification. Examples of the inspection functions are barcode reading, label verification, content level checking, etc. Each function is performed and tested independently. The tests carried out made it possible to record the conditions necessary for the execution of each inspection, which is an indispensable factor in achieving the appropriate physical design. These functions are grouped in a single application with three interfaces (login, inspection and configuration). It allows the user to inspect the products according to a configuration that he can define. This work was developed based on National Instrument platform, the software code was made with LabVIEW software, which resources and libraries are adequate for such a work.

Keywords: Image processing · Pattern matching · Score · Edge detection
Image binarization · Particle analysis

1 Introduction

Machine vision consist of using devices for optical noncontact sensing to automatically receive and interpret an image of a real scene in order to obtain information and/or control machines or processes.

Inspection by machine vision allows to increase the performances and the rates of production, improve the quality of products and ensure the safety. The advantage of a vision system is manifested in the possibility of systematically carrying out several different controls on all products at the same time.

The applications of the vision machine can be technically divided into four types: localization, measurement, inspection and identification. The applications of the vision machine can also be classified according to the type of sector, namely Automotive, Electronics, Pharmaceutical, Food, etc.

© Springer International Publishing AG, part of Springer Nature 2018
M. Ben Ahmed and A. A. Boudhir (Eds.): SCAMS 2017, LNNS 37, pp. 852–872, 2018.
https://doi.org/10.1007/978-3-319-74500-8_77

2 Problem Statement

Demand for food products continues to rise, leading manufacturers to increase production rates. However, this is not sufficient to meet the demand, defective products may appear as a result of systematic or random errors.

Production defects is one of the major reasons for product recall. Because of these recalls, the manufacturer is obliged to bear additional costs in terms of transport, storage and sometimes safe destruction of the returned products. Indirect costs may also be incurred in the event of damage to reputation. The accumulation of these charges during a single year can generate a large loss.

One of the methods developed in the past to overcome this problem is to mandate an operator to perform a manual inspection. This is not very practical since a human being may lose quickly the concentration, especially if several criteria have to be verified on the same product. Given today's high production rates, human inspection will not help.

3 Purpose

The work therefore consists of designing a prototype of a system to automate the inspection of products. Below are the inspections that this system will be able to perform:

- Labeling
  - Presence
  - Appropriate affixing
  - Correspondence to the right product
  - Readability of information
- Bottles
  - Correct filling
  - Cap
    - Present
    - Bottle correctly closed
- Barcodes
  - Readability
  - Matching
- Container shape
  - Container intact
- Surface quality
  - Intact surface

The application to be designed will be flexible, meaning that the system will not be designed to inspect a specific food product, but will be dedicated for any food product chosen by the user who can therefore prepare the food products to be inspected through a configuration interface.

4   General Architecture of the System

As shown in Fig. 1, the inspection system will use artificial vision. The image of the scene taken by the camera [1, 2] will be analyzed by a processing unit and a decision will be made to approve or reject the product.

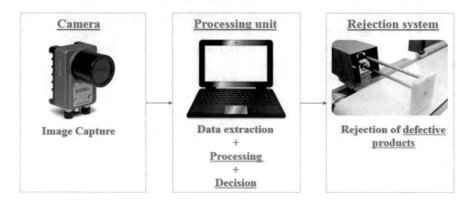

Camera	Processing unit	Rejection system
Image Capture	Data extraction + Processing + Decision	Rejection of defective products

Fig. 1.   General architecture of the vision machine

5   The Software

The system is developed with LabVIEW software. The "Vision & Development" module enables the development and deployment of industrial vision applications. It includes hundreds of image acquisition functions from a large number of cameras, image processing and analysis to meet the various challenges presented by vision applications.

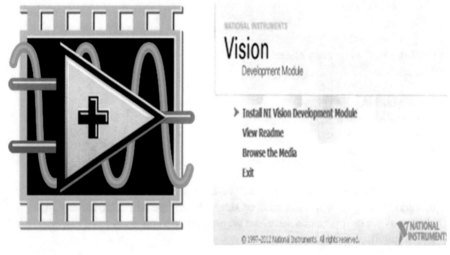

5.1 Preliminary Design

To meet the already set objectives, a group of functions has been prepared (called SubVI's). One or more of these functions contribute to performing one or more inspections such as inspection of labels, shapes, caps, etc. The aim of this step is to allow a modular programming, which will facilitate the development of the final application. These functions are explained in detail here below.

Pattern Matcher Function. Pattern matching is a method for finding regions in a grayscale image that corresponds to an image representing the reference model. All the features provided by LabVIEW for model comparison can be accessed and used with the *Pattern Matcher* function. As shown in Fig. 2, the various entries to this function allow the user to define several options for the search.

For example, the Match Mode entry in Fig. 2-[1] makes it possible to define the comparison mode, it is therefore possible to specify whether the inclined models with respect to the reference one shall be taken into account in the final result. It is even possible to define the comparison accuracy through the Minimum Match Score input (Fig. 2-[2]), to limit the search area by defining the Search Area entry (Fig. 2-[3]), as well as several other options.

When a matching model is found, the *Pattern Matcher* function provides important information through the Matches output (Fig. 2-[4]), i.e. the position, the angle, the scale, the score, and so on.

Pattern Matcher is an indispensable base for all other functions (SubVIs) that will be used, because before operating on any scene, it will be necessary to find the model on which a given function will operate. An example where the *Pattern Matcher* function is applied will be illustrated.

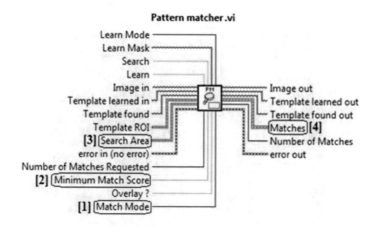

Fig. 2. Pattern Matcher function

Figure 3 shows the application of the *Pattern Matcher* function.

Fig. 3. Pattern Matcher function "application"

Dimensions Calculator Function. The role of this function is to ensure that the shape of a container is intact by taking samples of the dimensions of the object under inspections [3, 4]. In Fig. 5, 40 samples were selected (Fig. 4).

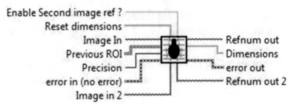

Fig. 4. Dimensions Calculator function

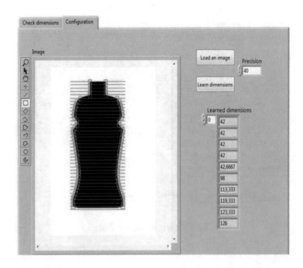

Fig. 5. Dimension Calculator function "application" (configuration)

The *Dimensions Calculator* function is based on edge detection and begins by first drawing lines inside the rectangle drawn by the user to determine the edges of the object. An edge is determined by detecting sudden changes in the values of the pixels on a given line. The next step is to calculate the distance between each edge on the same line and then store them in a reference table.

During the inspection, it will be necessary to determine the edges of the object under verification, then compare them with the values contained in the reference table. This case is one of the cases where the Pattern Matcher function plays an essential role, because before starting the process of drawing lines and determining the edges of the object under inspection, it is necessary to identify the region or rectangle within which this process will be performed, this region is determined by the Pattern Matcher function.

A deformed or damaged object will have dimensions different from those of reference, which makes it possible to deduce whether the shape of the container is good or not (Fig. 6):

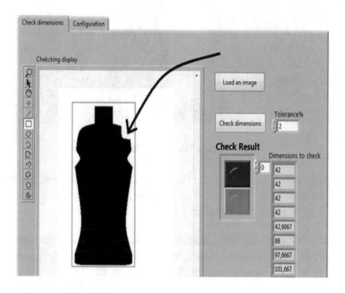

Fig. 6. Dimension Calculator function "application" (inspection)

However, this function only works if the background and the object have different and unique colors, such as white and black. To do this, the function *Image Binarizator* has been created.

Image Binarizator Function. This function operates on 8-bit grayscale images, so each pixel takes a value between 0 and 255. Thanks to the Threshold input (Fig. 7-[1]), the binarization threshold can be set between 0 and 255 [1]. The binarized image in black and white (Fig. 8) is obtained through the binarized Image output (Fig. 7-[2]).

Image binarizator.vi

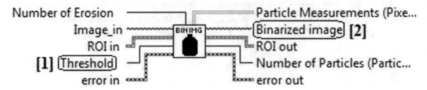

Fig. 7. Image binarizator function

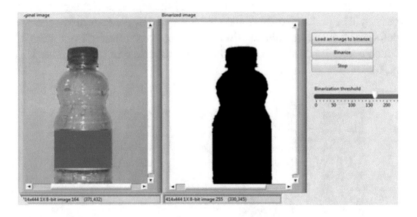

Fig. 8. Image binarizator "application"

The Binarized Image function is used in conjunction with the Dimensions Calculator function to determine edges, it's used also with the *Filling Checker 1* function to check the level of filling of transparent bottles.

Filling Checker 1 Function. This function is designed to check the level of filling of bottles. The image first passes through a binarization process, then through the determination of the air level by edge detection (Figs. 9 and 10).

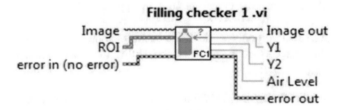

Fig. 9. Filling checker 1 function

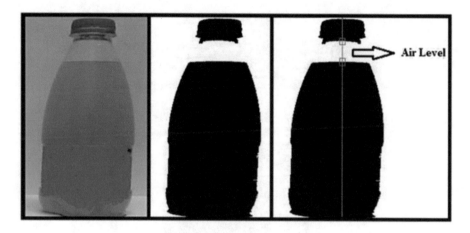

Fig. 10. Air level detection process

The variation of the level of liquid will obviously change the level of detected air, which allows us to deduce the correct filling of the bottles (Fig. 11).

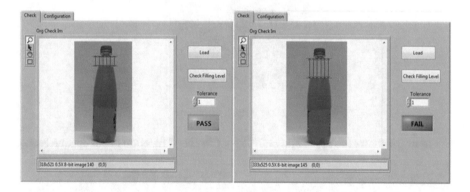

Fig. 11. Filling checker 1 "application"

This function is however not applicable on opaque bottles because the binarization process will not be able to make appear air level. For this reason, another function called *Filling Checker 2* has been created.

Filling Checker 2 Function. For this function to work properly, it is necessary to have the appropriate illumination, which reveals the level of liquid (Fig. 13). This function uses the Pattern Matcher function, it looks for a shape similar to the liquid level, but it's limited to the area where the appearance of the liquid level represents the correct filling, if no level form is found, the bottle is declared badly filled (Fig. 12).

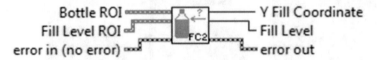

Filling checker 2.vi

Bottle ROI ▭▭▭▭▭▭ ▭▭▭▭▭ Y Fill Coordinate
Fill Level ROI ▭ ▭▭▭▭▭ Fill Level
error in (no error) ▭▭▭ ▭▭ error out

Fig. 12. Filling checker 2 function

Fig. 13. Liquid level form

Figure 14 shows the application of *Filling Checker 2* function on an opaque bottle.

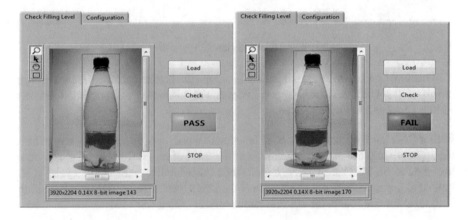

Fig. 14. Filling Checker 2 "application"

Barcode Checker Function. *LabVIEW* already provides a function called *IMAQ READ BARCODE* for reading the barcode [4]. Yet *Barcode Checker* function, which is based on this same function, allows a quick and easy programming for the reading of the barcode.

The *IMAQ READ BARCODE* function can be used to read the barcode after selecting it and indicates its type. What distinguishes *Barcode Checker* is its ability to automatically recognize the barcode selected by the user without the need to indicate the type of barcode it wants to inspect. In addition, *Barcode Checker* allows a barcode learning process and then verifies whether the bar codes under inspection are in accordance with the learned codes. The *Barcode Checker* function can recognize eight types of barcodes, which are MSI, UPCA, EAN8, EAN13, CODE 128, CODE 93, CODE 39, CODABAR (Fig. 15).

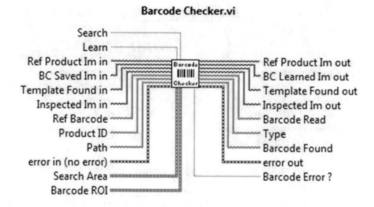

Barcode Checker.vi

Fig. 15. *Barcode Checker* function

Figure 16 shows the application of *Barcode Checker*:

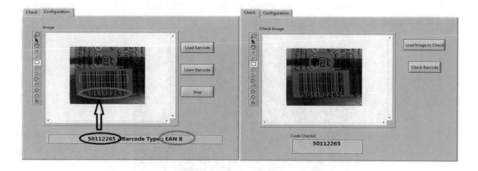

Fig. 16. *Barcode Checker* "application"

Surface Defect Checker Function. The *Surface Defect Checker* function allows you to find defects on the inspected surface, whether it is folds, tears or dirt. This function, like most of those performed, requires an intact image of the inspected surface to be used as a reference. The *Surface Defect Surface* function proceeds by comparing the pixels of the image under inspection with the homologous pixels of the reference image. The comparison is made according to a tolerance that can be specified since it is almost impossible for the pixels of the image under verification to be exactly equal to the homologous pixels of the reference image. A defect will cause a large change of the pixels in the place where it appears, which allows us to conclude that the surface is damaged. Defects are classified into two types (Fig. 17):

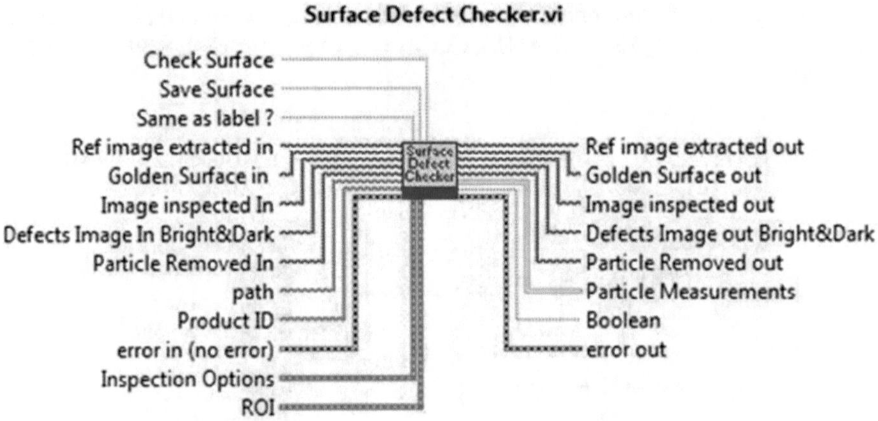

Surface Defect Checker.vi

Check Surface
Save Surface
Same as label ?
Ref image extracted in — Ref image extracted out
Golden Surface in — Golden Surface out
Image inspected In — Image inspected out
Defects Image In Bright&Dark — Defects Image out Bright&Dark
Particle Removed In — Particle Removed out
path — Particle Measurements
Product ID — Boolean
error in (no error) — error out
Inspection Options
ROI

Fig. 17. Surface defect checker function

- Bright Defect, it occurs when the pixel under inspection has a lower value than its reference counterpart.
- Dark Defect, it occurs when the pixel under inspection has a value greater than its reference peer.

To illustrate the functioning of this function, a comparison of the two labels of Fig. 18 is made:

Fig. 18. Intact and defective surface

The comparison result is an image with a black background, representing the light and dark faults respectively in green and red (Fig. 19). Some defects appear as small particles, these are not real defects, as has been said, this is due to the fact that the pixels of an image are not exactly equal to the pixels of the reference image, even if the comparison is made with tolerance.

Fig. 19. Preliminary result of comparison

Fig. 20. Comparison result after removal of small particles

For this reason, small particles are removed because the actual defect causes a large particle. Figure 20 shows the final comparison result after the suppression of the parasitic particles.

To indicate the error, the coordinates of the center of gravity of the area representing the defect are determined, which allows the user to see it (Fig. 21).

Fig. 21. Surface defect checker "application"

Label Tester and Plug Tester Function The *Label Tester* function allows multiple inspections on product labels. It can verify that a label is present, correct (the correct label is affixed to the corresponding product), well oriented and well positioned. The *Plug Tester* function verifies that the plug is present and affixed appropriately. These

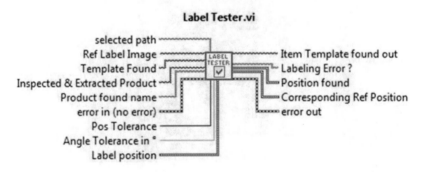

Fig. 22. Label Tester function

two functions use the *Pattern Matcher* function, especially the output information such as angle, position, etc. (Figs. 22, 23, 24 and 25).

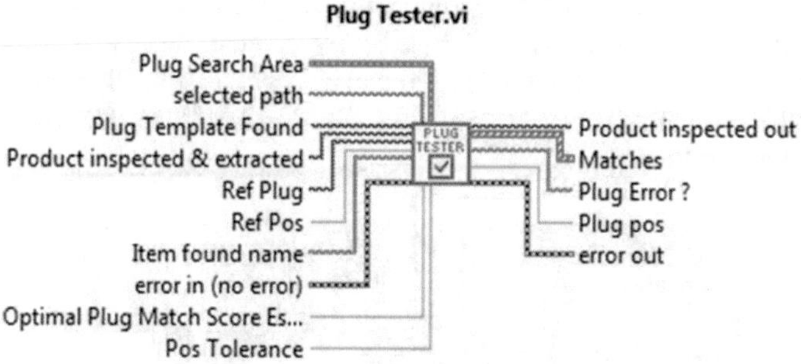

Plug Tester.vi

Plug Search Area
selected path
Plug Template Found
Product inspected & extracted
Ref Plug
Ref Pos
Item found name
error in (no error)
Optimal Plug Match Score Es...
Pos Tolerance

PLUG TESTER

Product inspected out
Matches
Plug Error ?
Plug pos
error out

Fig. 23. Plug Tester function

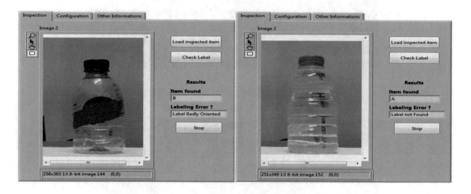

Fig. 24. Label Tester "application"

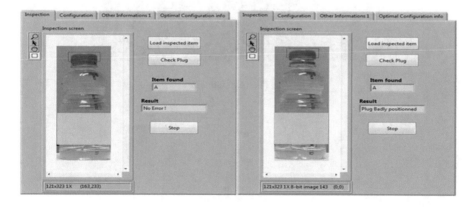

Fig. 25. Plug Tester "application"

5.2   Design of the Final Solution

Realized Prototype.   Figure 26 shows the hardware part of the system, it comprises all
the elements necessary for taking the image, this is done by a position sensor (Fig. 26-
[1]) which triggers the camera (Fig. 27) [1, 5], which takes an image of the product to
be inspected. An internal lighting (Fig. 28) makes it possible to illuminate the product
suitably for the treatment. The ejection system (Fig. 26-[2]) allows the defective products
to be pushed out of line.

Fig. 26.   Global view of the prototype

Fig. 27.   NI smart camera 1742

Fig. 28. Internal lighting

Developed Application. The developed application consists of two main interfaces: configuration interface (Fig. 29) and inspection interface (Fig. 30). The configuration interface (Fig. 29) allows the user to configure multiple products, hence the need for proper identification of the product being inspected before loading the appropriate reference data.

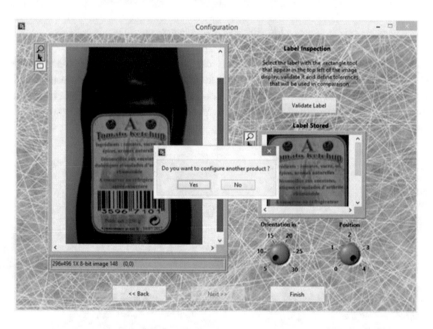

Fig. 29. Configuration interface

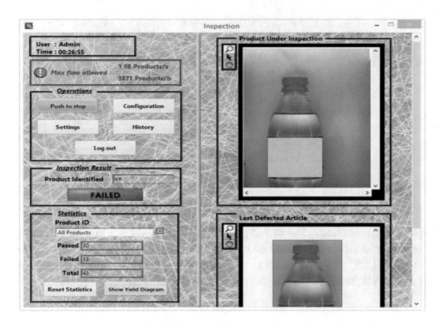

Fig. 30. Inspection interface

Identification Troubleshooting. The correct identification of a product is not a simple task and presents some problems. These problems are due to defective products whose score falls considerably, this may lead to non-identification of the product if the scored score is below the threshold or to confuse it with another if one of the configured products attained a qualifying score.

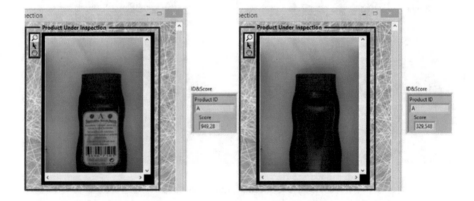

Fig. 31. Influence of a label defect on the score "before implementing the solution"

The types of defects that seriously lower the score are labeling errors. A label usually takes up a large space on the face of the product and therefore, if a label is absent for example, the score will decrease inversely proportional to the size of the missing label.

Figure 31 shows that the absence of a label decreased the score by 620, and that causes a problem.

The solution adopted to overcome this problem is to make a search by mask. For labeling inspection, the user is requested (through the configuration interface) to select the product label, the selected area (which represents the label) is automatically set as a mask. This is a region that will be excluded when comparing the inspected item with the reference one. Thus, an absent or incorrect label will not influence the score significantly. A good result was obtained by this solution, Fig. 32 shows that the absence of a label made the score decrease by a value of 60 instead of 600.

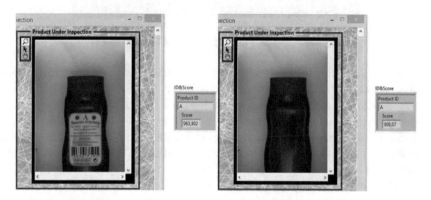

Fig. 32. Influence of a lack of labeling on the score "after implementing the solution"

The designed system has the ability to recognize products of the same brand and having the same size and shape. The products in Fig. 33 are easily identifiable by their unique color. However, the product can not recognize them by color since the camera is monochrome. Figure 34 shows how the monochrome camera sees the image of these bottles.

Fig. 33. RGB image of products of the same brand

Fig. 34. Grayscale image of the same brand products

After the transformation of the images of the bottles into grayscale, it's noticeable that there is a certain resemblance especially between the bottles (1) and (2) in Fig. 34. Initially, the search process consisted of loading the reference products one by one and comparing them with the inspected image. If a product scores above the threshold, the comparison is stopped even if there are other reference products that have not yet been compared, this process was followed because it is assumed that the product that does not correspond to the one inspected will never score above the threshold but this assumption turned out to be wrong.

The remark that made it possible to develop the solution is that even if more than one product presents a score above the threshold, the product that actually corresponds to the one inspected always marks the highest score. For this reason, the identification process has been changed so that comparisons of the reference products with the object in front of the camera stops only when the comparison of all the configured products

Fig. 35. Score marked by different products of the same brand

would have finished. The name of each product compared and its marked score are recorded in a table, the product name that really matches the one under inspection is none other than the one that scored the highest value.

Figure 35 shows that there are 3 products that scored above the threshold. The real product inspected is Ice Orange, since it has the highest score. According to the old research process, the identified product would have been Ice Ananas, as it is the first product that has a qualifying score (above 600).

System Specifications and Features.

Features	Description
Inspections provided	Labeling, barcode, surface quality, level of filling, caps, container shape
Number of faces inspected	1
Inspection conditions	• Aligned and oriented products • Maximum size allowed: 19×8 cm^2
Several inspections on the same product	Yes
Inclusion of multiple products in the same configuration	
Cadence	75 item/min(Depends on the configuration complexity)
Vision technology	PC based system
Power supply	220 V (single phase)/50 Hz

5.3   Prospects and Improvements

The achieved work meets the objectives set at the beginning of the project, however, this system is likely to be improved and may be the basis for other research projects. The following axes are proposed:

- Inspection angles: A multi-camera system can be considered to cover the product from several angles. Another technique based on a single camera can be used to cover the entire lateral surface of a product. This method consists in rolling the product in front of the camera, which will allow it to detect the entire lateral surface, however, this technique requires a special conveyor for rolling the product which must be of cylindrical nature.
- System automation: Several functions can be added to the vision system, such as automatically adjusting the speed of the conveyor to the maximum cadence allowed, or else stop the conveyor if the rate of defects exceeds a certain threshold.
- Prospecting and exploring 3D techniques.

Acknowledgements. This research was partly supported by MASTER TEC and PRODEV SYSTEM Company. The authors thank Abdelhamid Elwahabi for participation in developing the vision system.

References

1. National Instrument: Getting Started with the NI 17xx Smart Camera
2. National Instrument: NI Vision NI 17xx Smart Camera User Manual For NI [5]1722/1742/1744/1762/1764 Smart Cameras
3. Kwon, K.-S., Ready, S.: Practical Guide to Machine Vision Software
4. Cristopher, G.: Relf. Image Acquisition and Processing with LabVIEW
5. National Instrument: USER GUIDE NI Smart Camera I/O Accessory

Improvement of Scheduling, Assignment of Tools on Assembly Machines Connected to a Supervision Interface

El Fekri Yassine[1(✉)] ⓘ, Ouardouz Mustapha[1], Bel Fekih Abdelaziz[1], and Bernoussi Abdessamad[2]

[1] MMC, Faculty of Sciences and Techniques, B.P.416, Tangier, Morocco
elfekri@gmail.com
[2] GAT Faculty of Sciences and Techniques, B.P.416, Tangier, Morocco

Abstract. In this work we consider a scheduling problem in an assembly area composed of several machines of various characteristics and used to resist welding different parts using specific tools.

The area is equipped with a monitoring interface installed on remote computer representing in real time the state of the machines, the progress data of the schedule are extracted from this application.

The problem consists in optimizing the efficiency of the zone by minimizing the time of the logistic shutdown (change of tools, overproduction processing time, disruptions due to stock shortages…)

The results will be illustrated through an application (case study) carried out on this area.

Keywords: Assignment · Load balancing · Optimization

1 Introduction

Consumers less and less covet standard products made in very large quantities. Indeed, the latter prefer to buy a product of excellent quality, corresponding perfectly to their needs at an affordable price.

This new reality forces companies to shift from mass production, at low prices, aimed at a standard customer, to an increasingly diversified quality production. In addition, the globalization of markets requires these same companies to optimize their manufacturing costs in order to offer a competitive selling price and accelerate the marketing of their new products [1]. Thus, to meet customer requirements, companies try to make their production methods more flexible and efficient.

To achieve this, equipment utilization rates must be increased, productivity and process capacity must be increased, production must be balanced with economical batch sizes by distributing them over time and thus be able to produce several Typologies of products within a defined timeframe and on the same production chain [2], to arrive at this stage requires in-depth work on the performance of the production system "In general, when trying to apply the Toyota production system it is necessary to start by distributing, or smoothing the production. This responsibility rests primarily

© Springer International Publishing AG, part of Springer Nature 2018
M. Ben Ahmed and A. A. Boudhir (Eds.): SCAMS 2017, LNNS 37, pp. 873–883, 2018.
https://doi.org/10.1007/978-3-319-74500-8_78

with the people in charge of controlling or managing production." F. Cho, president of Toyota Motor Corporation.

Unbalanced manufacturing programs on a production line increase the frequency of series changes and duration, which penalizes the productive performance of machines, we propose in our research a method that will minimize the changes all by ensuring a balancing of the capacity load.

2 Problem Statement

The 150 items manufactured in the assembly area (our study sector) are managed by the SAP integrated management software, these references generate large volumes of information, the value of which varies considerably over time.

Within these increasingly complex environments, the optimization and management of production resources becomes a priority within the company.

If its functions are not managed correctly, the productivity is penalized for several reasons:

- Unbalance of the load distribution on the machines.
- Full in the overall equipment effectiveness (OEE)
- The risks of stopping the chain.
- Risk of penalization from delay to delivery.

The major risk for the company is to pursue a strategy of overinvestment or underinvestment in terms of the number of resources used [3].

2.1 Performance Evaluation

The measure of the area performance indicator overall equipements effectiveness shows that there is a significant drop that started as early as week 09 (Fig. 1).

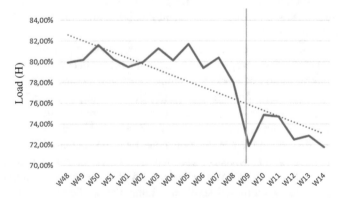

Fig. 1. Overall Equipment Effectiveness (OEE)

The production managers justify this by the integration of the new X52 project and its coincidence with the rise in pace.

2.2   Measurement of Load Distribution

The distribution of the load per machine recorded during a given week confirms that the distribution of the load is not balanced on the machines (Figs. 2 and 3).

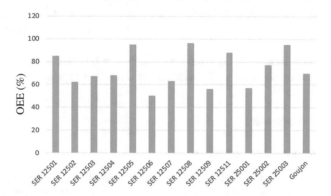

Fig. 2. Graphical representation of load distribution over week 9 per machine

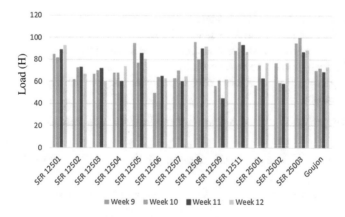

Fig. 3. Graphical representation of the load distribution over 4 weeks.

The distribution of the load recorded over several weeks shows that the problem of imbalance in the distribution of the load persists and varies from one week to another, The problem of unbalancing the distribution of the load is linked to the following constraints:

2.2.1   Technological Constraints

The area contains 41 tools and 14 machines, these machines are not the same, each one only has specific tools, and each tool requires different technical characteristics (machine power, height of the table compared to that of the machine, tool).

These constraints depend on the technology of the machine and the tool (power required to carry out the welding, welding program required by the tool, etc.)

2.2.2   The Constraints of Variation of the Customer Needs

Customer needs are not stable, there are always peak demands that are due to the customer's rising pace, or falls in need due to production delays in assembly lines or leave.

The problem is then to find dynamic assignments of the tools to the machines, with the aim of balancing the load all by adapting to the technological constraints.

3   Problem Approach

3.1   Formalization of the Problem

For the formulation of our problem, we would first need to describe the data for the study area and the constraints imposed. The formulation of the problem will then consist in formulating the objectives to be achieved (optimally) according to the data and constraints of the problem [4, 5, 6].

Data for the study area

- Each item is assembled by a single tool.
- A tool can assemble from 1 to several items.
- A tool is assigned to a single machine for the same batch.
- A machine supports 1 to several tools.
- The changeover time of a tool series depends on the tool setting at the table height of the machine.

This real problem is translated in mathematical form as follows:

$$(P) \begin{cases} \min f(x) = \sum_{i=1}^{n}\sum_{j=1}^{m} X_{ij}\tau_{ij} \quad \text{with} \quad X_{ij} = \{0; 1\} \quad i = \{1,\ldots,n\} \, et \, j = \{1,\ldots,m\} \\[2mm] \text{Under the constraints} \quad \sum_{j=1}^{m} X_{ij} = 1 \, , \, i = \{1,\ldots\ldots,n\} \\[2mm] \sum_{i=1}^{n} X_{ij}C_{ij}\frac{Ct}{m} + \varepsilon \, , \, j = \{1,\ldots\ldots,m\} \\[2mm] \sum_{i=1}^{n}\sum_{j=1}^{m} X_{ij}C_{ij} = Ct \end{cases}$$

i : the number of tools

j : the number of machines

X_{ij} : The binary matrix of variables, it takes the values 0 or 1

$$X_{ij} = \begin{cases} 0 \text{ if the tool i is not assigned to the machine j} \\ 1 \text{ if the tool i is assigned to the machine j} \end{cases}.$$

τ_{ij} : The time matrixis the time required for the series change of tool i in the machine j

C_{ij} : The matrix of the load of tool i on the machine j

$C_{ij} = Ch_i \times a_{ij}$

With:

a_{ij} : The matrix of adaptation of the tool i to the machine j

$$a_{ij} = \begin{cases} 1, \text{ If the tool i can be mounted on the machine} \\ 0, \text{ If not} \end{cases}$$

Ch_i : The Load of the tool i

4 Problem Resolution

To better illustrate our approach, without loss of generality, we consider the case of three machines (m = 3). Indeed the case where m = 1 or 2 the problem is simple to formulate and to solve and the case where m = 3 constitutes a generalizable case even in the case where m is greater than three. Indeed, in this case (m greater than three), the different machines can be divided into three different classes depending on the properties of each machine and all the machines of the same class have the same properties.

With m = 3 machines and n = 8 tools, we have n x m = 24 variables and n + m + 1 = 12 constraints.

All the simulations are done on the open solver; this software is available on the website http://opensolver.org/, and which quickly integrates on the Excel office.

In this case:

$$f(x) = \sum_{i=1}^{8} \sum_{j=1}^{3} X_{ij} t_{ij}$$

It will be the objective function

$$\min f(x) , \quad Xij \in \{0; 1\}$$

The solution should check the following constraints

- Each tool can only be adapted to one machine

$$\sum_{j=1}^{3} X_{ij} = 1$$ (C1)

$$i = 1, \ldots \ldots, 8$$

- Each machine j has a maximum capacity dj

$$\sum_{i=1}^{8} X_{ij} C_{ij} \leq d_j$$ (C2)

$$j = 1, \ldots \ldots, 3$$

The sum of all loads assigned to the machines must coincide with the total load to be treated:

$$\sum_{i=1}^{8} \sum_{j=1}^{3} X_{ij} C_{ij} = Ct$$ (C3)

- For the maximum capacity of the machine j the average capacity Ct/3 was taken with an Epsilon tolerance:

$$d_j = \frac{Ct}{3} + \varepsilon$$

The problem of minimization will be resolved as follow:

$$a = (a_{ij})_{i,j} = \begin{pmatrix} 1 & 0 & 1 \\ 1 & 1 & 1 \\ 0 & 1 & 0 \\ 1 & 1 & 0 \\ 1 & 0 & 0 \\ 1 & 1 & 1 \\ 1 & 0 & 0 \\ 1 & 1 & 1 \\ 1 & 1 & 1 \end{pmatrix} \quad Ch = (Ch_i)_i = \begin{pmatrix} 5.8 \\ 6 \\ 9.5 \\ 10 \\ 8 \\ 7.2 \\ 9.5 \\ 5 \end{pmatrix}$$

Which gives:

$$C = (C_{ij})_{i,j} = (Ch_i a_{ij})_{i,j} = \begin{pmatrix} 5.8 & 0 & 5.8 \\ 6 & 6 & 6 \\ 0 & 9.5 & 0 \\ 10 & 10 & 0 \\ 8 & 8 & 8 \\ 7.2 & 0 & 0 \\ 9.5 & 9.5 & 9.5 \\ 5 & 5 & 5 \end{pmatrix}$$

We suppose the tools are grouped into 3 families according to their HOF (height of tool).

One wants to assign the tools of the same family on the same machine to minimize the times of change.

If we put the tools in HOF identical in the same machine, it will be easy to make changes of tools, and it takes only half an hour, on the other hand if one tool is changed by another of different HOF Achieve a full hour of series change.

So, for each machine, we define the corresponding HOF, so that our assignment matrix stores the tools of the same HOF on the same machine.

In our example, it is assumed that there are three families of HOF: Type 1, Type 2 and Type 3 (Table 1).

Table 1. Distribution of tools by family

HOF		
Family 1	Family 2	Family3
O1	O3	O7
O2	O4	O8
O6	O5	

So the matrix τ_{ij} Defines in a time unit (hours) the tool change time i on the machine j,

$$\tau = (\tau_{ij})_{ij} = \begin{pmatrix} 0.5 & 1 & 1 \\ 0.5 & 1 & 1 \\ 1 & 0.5 & 1 \\ 1 & 0.5 & 1 \\ 1 & 0.5 & 1 \\ 0.5 & 1 & 1 \\ 1 & 1 & 0.5 \\ 1 & 1 & 0.5 \end{pmatrix} \qquad X_{ij} = \begin{pmatrix} X_{11} & X_{12} & X_{13} \\ X_{21} & X_{22} & X_{23} \\ X_{31} & X_{32} & X_{33} \\ X_{41} & X_{42} & X_{43} \\ X_{51} & X_{52} & X_{53} \\ X_{61} & X_{62} & X_{63} \\ X_{71} & X_{72} & X_{73} \\ X_{81} & X_{82} & X_{83} \end{pmatrix}$$

$$Ct = 61h \;\; ==> \;\; dj = \frac{61}{3} + \varepsilon = 20.33 + 0,5 \quad j = 1\ldots3$$

The solution obtained by the software is then:

$$X^{opt} = \begin{pmatrix} 1 & 0 & 0 \\ 0 & 0 & 1 \\ 0 & 1 & 0 \\ 0 & 1 & 0 \\ 1 & 0 & 0 \\ 1 & 0 & 0 \\ 0 & 0 & 1 \\ 0 & 0 & 1 \end{pmatrix}$$

The assignment matrix calculated by the balance gives us an optimized load balancing (Fig. 4).

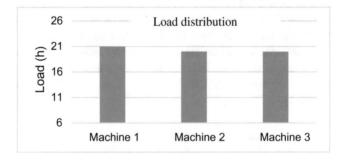

Fig. 4. Result of the load distribution per machine

5   Generalization of the Solution

5.1   General Description

The results obtained below are applied subsequently to the PRP assembly area considered to be the core of the production line, the assembly area produces 150 references (semi-finished and finished product)

The finished products are sent directly to the shipment so that they can be shipped directly to the customer, a poor scheduling can be very expensive, so this zone directly feeds the MAG and Robotic area and a semi-finished product delay or poor coordination directly influences the downstream zone.

5.2   Data for the Study Area

The PRP assembly area is composed of 14 machines; these small resistance welding presses (PRP) are SERRA brand and divided into three families according to their power. So we have machines of 125KVA, 170KVA and 250KVA. Some of these machines are equipped with automatic small parts launchers (nut screws, etc.) and the other hand where the operator assembles the parts under the machine and the machine performs the welding.

The tool is the element that carries the electrodes and also supports the mask. There are 41 PRP tools. Each PRP machine takes a defined number of tools; we speak of machine-tool flexibility.

5.3   Modeling, Simulation and Interpretation

The table below compares the load distribution during weeks 15 to 18. First, based on a fixed assignment (old method used), and second places use the assignment results obtained by discounter.

The feasibility of the work carried out is based on the evolution of the results of the distribution of the load resulting from the tool that has been realized in relation to the fixed allocation that the production managers follow [7].

5.3.1   Results Interpretation

a. Load distribution

After a first planning based on a fixed assignment, we note that the distribution of the load is not balanced, which leads to the overloading of machines, below we have the number of machines overloaded per week (Table 2).

Table 2. Number of machines overloaded

	Number of machines overloaded	
	Planning with fixed assignment.	Planning using the mathematic model
Week 15	3	0
Week 16	3	0
Week 17	2	0
Week 18	3	0

After the simulation of the distribution of the load using the tool that was realized, we find that the results of assignments obtained have eliminated the number of machines with the load exceeds the capacity in 2 teams (machines overloaded).

This tool will facilitate the task of smoothing the load from a first planning.

b. Tools set-up time

The number of tool changes can be calculated from the results obtained by the load balancing tool developed, assuming that the tool will only be affected once during the week.

In the example discussed above, the total sum of the change times (see below) of the tools is found during each week (Table 3).

Table 3. Set-up time

Number of the week	Tool change time
Week 15	21.5 h
Week 16	22 h
Week 17	21 h
Week 18	23 h

5.4   Estimated Time Savings of Tool Changes

Measurement times of the tools measured during the weeks up to 18 were measured above and it was found that if no possible aliases were maintained which prevented the tool from being mounted once a week, we will arrive at a probable gain of 45.5% in time of tool changes, obtaining this gain requires the ability to manage production contingencies.

6   Conclusion

Establish mathematical modeling of assignment problems by operational research, once simulated and integrated, it was a step towards more efficient system, this work made it possible to reduce the series change stops by 45% as they contributed to the elimination of overstocking and stock-out situations.

The results obtained demonstrated the failure of the working tool used, which did not take into account external constraints and other uncertainties of production. As a result, and through the mathematical modeling of the problem, a dynamic tool has been put in place to balance the loads taking into account technical constraints and fluctuations in customer demand.

Through the simulation of the load distribution over different weeks and by different scenarios, it was possible to determine the critical points to be dealt with, which helped to integrate other functions to improve the balancing of the load distribution, and to develop an action plan to increase the flexibility of production.

References

1. MARTIN, Alain and Chantal BONNEFOUS,Gestion de la production, Chapitre 9 - «La gestion d'atelier par les contraintes», 5th edn. (2011)
2. Prins, C., Sevaux, M.: Programmation linéaire avec Excel: 55 problèmes d'optimisation modélisés pas à pas et résolus avec Excel, chapitre 13 - Emplois du temps et gestion de personnel, pp. 215–350, 3 Mars 2011
3. Huq, F., Cutright, K., Martin, C.: Employee scheduling and makespan minimization in a flow shop with multi-processor work stations: a case study. Omega **32**(2), 121–129 (2004)
4. Yalaoui, F.: Optimisation et gestion de la production: problème de conception et problème d'ordonnancement. Habilitation à diriger les recherches, Université de Technologie de Compiègne (2006)

5. Lalande, J.F.: Equilibrage des lignes de montage, Rapport de Stage ISIMA (2000)
6. Amen, M.: An exact method for cost-oriented assembly line balancing. Int. J. Prod. Econ. **64**, 187–195 (2000)
7. Som, H.: Modèle Mathématique pour le problème de l'Équilibrage de la Charge sur une ligne de Production, Rapport de projet CUST (2001)

Contribution of Industrial Information Systems to Industrial Performance: Case of Industrial Supervision

Derboul Ahmed[1], Hadj Baraka Ibrahim[2(✉)], and Chafik Khalid[1]

[1] ENCG, Tangier, Morocco
derboul@gmail.com, khchafik@yahoo.fr
[2] LIST, Faculty of Sciences and Technologies, Tangier, Morocco
i.baraka@gmail.com

Abstract. The industrial information system (SII) as a tool for decision-making and management of production systems in real time is one of the main areas of research in industrial systems management.

Industrial supervision is today among the most developed IIS in the field of industrial workshops management. All the studies carried out have shown that the supervisory systems have a positive impact on the process of steering production systems but this is function of several constraints and variables. Our research problem is part of this context, which is about the evaluation of the contribution of supervisory systems to the performance of production processes.

In this work, we selected the company Fromagerie Bel Maroc (FBM) as a field of investigation to assess the impact of the SCADA system on improving performance indicators for the production function.

To achieve this objective, we tracked trends in the historical performance indicators for manufacturing facilities managed by industrial supervision.

The results of this study allowed us to confirm the positive and direct impact of industrial supervision on the performance of the production system.

Keywords: Information system · Production system · Industrial supervision
Control system

1 Introduction

Globalization requires increased competitiveness. Multinational corporations have considerable means to mitigate their consequences by seeking more strategic rather than operational alternatives. On the other hand, SMEs companies (Small and Medium-sized Entreprised) are obliged to invest more in operational processes by seeking simple solutions to optimize physical and information flows.

At present, a very attractive strategy begins to occupy a prominent place in the strategic pyramid of companies, it's Industry 4.0. The ultimate goal of this strategy is to digitize all the processes, thus creating value, reducing paperwork and increasing the rate of customer service in order to remain competitive.

© Springer International Publishing AG, part of Springer Nature 2018
M. Ben Ahmed and A. A. Boudhir (Eds.): SCAMS 2017, LNNS 37, pp. 884–901, 2018.
https://doi.org/10.1007/978-3-319-74500-8_79

Digitalizing processes, based on the Internet of Things (IoT), systematically leads to propose new technologies, new methods and approaches ranging from strategic processes to operational ones.

Industrial information systems are currently a major challenge and a real opportunity to implement the industry approach 4.0, enabling companies to significantly reduce decision-making and information flows and consequently the physical flow. Industrial information systems are mainly located at both strategic and tactical levels and interact continuously with the operational level.

We can distinguish 3 families of application from industrial information systems:

1. Control-command applications: such as a computer system which performs the acquisition of data via sensors and elaborates commands sent to the physical process by means of actuators. A control-command system receives information on the state of the external process, processes the data and, depending on the result, evaluates a decision that acts on that external environment to ensure a stable state[1].
2. SCADA applications (Supervisory Control and Data Acquisition) or industrial supervision system which consists in monitoring the operating state of a process in order to bring it to its optimum operating point and maintain it there[2].
3. MES applications, which constitute an information and communication system for industrial production, enable management control and monitoring of the work in progress in the workshop[3].

Our research problematic falls within this context, which is concerned with the evaluation of the contribution of SCADA systems to the performance of the production function. Our main research question can be formulated as follows:

How do SCADA systems contribute to the performance of the production function? To deal with this problem, this work will be structured in two axes:

1. The first axis is reserved for the presentation of fundamental concepts and research trends on subjects directly and indirectly related to our research problem.
2. The second axis is devoted entirely to the empirical study which seeks to evaluate the contribution of SCADA systems to the improvement of the performance of the production system. The company FBM based in Tangier is selected to constitute our theoretical analysis and experimentation area.

2 Conceptual Framework of the Study

Currently, Industrial Supervision is one of the main functions developed within the framework of control and command of production workshops. Indeed, several research projects propose different methods adapted to the design and implementation of the

---

[1] See Cottet and Grolleau (2005).
[2] See Chartres (1997).
[3] Logica CMG (Référence ***2006).

supervision of production systems[4] On the other hand, various implementation tools are proposed on the market under the term of Industrial Supervisors.

The objective of this theoretical part is to present the state of the art of industrial supervision by trying to answer different questions concerning the definition of the concept of supervision, the functionalities and the architecture of supervisory systems.

First, we analyze the main definitions of the concept and recall the history of supervision in the industrial context, we then study the characteristics and functionalities of the main industrial supervisors. Finally, we will discuss the problem of integrating a supervisor into the control structure of a production system.

2.1   Industrial Supervision: Definitions and History

Industrial supervision consists in monitoring the operating state of a process in order to bring it to and keep it at its optimum operating point[5].

It responds to the need of industrial companies to visualize their industrial processes, in an economic context of productivity and flexibility, supervision has benefited from an exceptional technological advance.

The literature provides several definitions of supervision. Here we have the three definitions that seem to be close to the research and software used in the industry:

Definition of EXERA6 (Association des EXploitants d'Equipement de mesure, de Régulation et d'Automatisme):
« Supervision is essentially a function of centralizing information, pre-processing information for the level of management and their visualization. The supervision allows real-time visualization of the state of evolution of an automated installation, so that the operator can take decisions as quickly as possible to achieve the production objectives (cadences, quality, safety, etc.) »

Definition of CETIM (CEentre Technique des Industries Mécaniques):
«It is the set of tasks that aim to monitor the operating state of a process, in order to bring it as close as possible to its nominal operating point and maintain it there. »

Definition of GRP-SPSF (Groupement de Recherche en Productique):
« Supervision aims to control the execution of an operation or work carried out by others without going into the details of this execution. It covers the normal and abnormal functioning aspects. »

It is clear from these three definitions that supervision has a decision-making role based on the collection and processing of production data in real time. But it seems to us that the first definition is more complete.

As for the history, the supervision of the processes was originally assured by human operators who had real-time knowledge of the data of their installation. In turn, the foreman was informed within an hour of the events that had occurred on his team. The

---

[4]   Artigues C., Roubelat F.: A polynomial activity insertion algorithm in multi-resource schedule with cumulative constraints and resource flexibility.

[5]   See Chartres (1997).

managers of the company received only aggregated information on the functioning of the workshop. Afterwards, it was useful to centralize the data. Thus, control rooms have appeared and were equipped with a large wall display representing the vision of the operators of the industrial process. Quickly, with the computer boom, the lights were replaced by screens and keyboards. The goal remained the same: monitoring and controlling an industrial production process.

With the complexity of the processes, the synthetic and global operator view has been replaced by a multitude of screens with animated mimics represented on smaller surfaces and with a multitude of views accessible at will. Starting from a simplified global view, it is possible to browse a sort of tree displaying the content of the processes down to the smallest details[6]. Each view is assigned certain functionalities allowing the operator to carry out specific commands. Subsequently, other features appeared such as error logging, alarm management, log management, etc.

Currently, and with automation, industrial supervision has become a tool for steering continuous and manufacturing processes necessary to ensure exhaustive traceability and control of the quality of products requested, hence the proliferation of these systems. Features have also evolved: synthesis reports, process and product parameters curves can be edited, preventive maintenance plans can also be integrated, etc. Its scope is extended to functions of a production control system such as the case of MES systems. In addition, the supervisory system has now become an integral part of the company's information system.

Compared to the market supply, many manufacturers and facilities makers have developed their own supervisory systems. At the same time, publishers of IT solutions have started marketing supervisory software.

2.2   Industrial Supervision: Functional and Technical Architecture

According to a recent survey of supervisory practitioners, most of the solutions are based on supervisory software packages. The current market trend is the harmonization of supervision applications functionalities. The basic functions are: database, communication, recipes, archiving, visualization (synoptics), alarm manager and calculation task (see Fig. 1).

Other advanced functions such as diagnosis and reconfiguration have been developed at the academic level but remain limited in current commercial products.

Compared to the current market offer, most supervisors consist of a central engine with functional modules for the acquisition of data from equipment, display and processing of this data, communication with other applications. They offer functional modules such as:

- The operator's assistance in its actions of control of the production process (dynamic HMI interface …);
- The visualization of the state and evolution of an automated process control system, with the identification of anomalies (alarms) in the form of several synoptics;

---

[6] See Millot (1988).

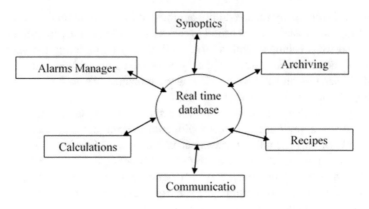

Fig. 1. Features of an Industrial Supervisor

- The graphical representation of different process data and the historized variables analysis tools;
- The collection of real-time information on processes from remote sites (machines, workshops, factories …) and their archiving;
- Production Tracking Reports;
- Alarms handler (Alarm management system);
- Business functions (recipe manager, batch tracking, maintenance…);
- Statistical Process Controls (SPC);
- Scripts that provide easy access to all supervisor features (animations, files, mathematical calculations, logical command sequences …). They integrate the principles of object-oriented programming and are open to standard databases (ACCESS, ORACLE, DBM, SYBASE).

The main industrial supervision software on the market are:

- PcVue32 (Arc Informatique)
- InTouch (Factory Systems)
- Fix 32 (Emerson)
- WinCC (Siemens)
- Panorama (Europ Supervision)
- Monitor Pro (Schneider)

For the Supervisory Control and Data Acquisition (SCADA) concept used by Americans instead of the French understanding of supervision, there are two functions: Data Acquisition, which means a data acquisition function et Supervisory Control for driving, managing modes and processing alarms. The following diagram shows a functional architecture of the SCADA system or supervision (Fig. 2).

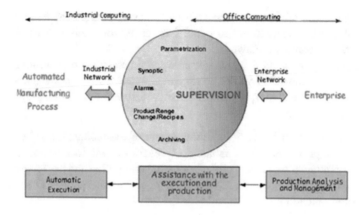

Fig. 2. SCADA System Features

As regards the hardware architecture of the supervisory system, we have retained the scheme proposed by Pierre BONNET[7] which shows that the supervisory stations are located between the automation systems and the MES application and are linked together by Ethernet networks and field networks (Fig. 3).

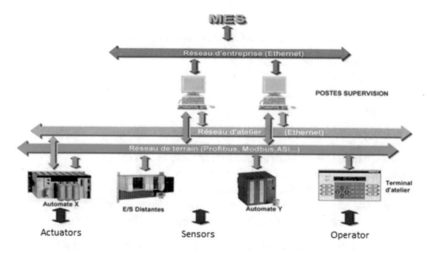

Fig. 3. The hardware architecture of the supervisory system

In the literature, we distinguish several hardware architectures with the software packages of supervision[8]:

- Single-station architecture where a supervisor manages the data coming from one or more devices and all functions are concentrated on a single station.

---

[7] See Bonnet (2010).
[8] See Pujo and Kieffer (2002).

- Multi-stations parallel architecture where several workstations share supervision, and identical applications can run in parallel on each workstation. All information is shared via the computer network.
- Client-Server Architecture, which is used when several operator stations are required with a single connection to the control equipment. In this case, the server is in communication with the equipment and distributes the data to the supervisory stations which are customers.

The communication between the components of the technical architecture of the supervision is a primordial function and decisive for the efficiency of the system. The modes of communication in the world of industrial computing and especially in the field of automatism are multiple and pose the problem of interoperability. There are several types of communication:

- Local communications to supervisory station
- External communications with industrial equipment
- Communications with databases
- Communication between different components of supervision

2.3 Contribution and Impact of Industrial Supervision on Production Performance

The supervisory system provides the following main functions:

- Continuous process monitoring.
- Real-time control.
- Automation and Protection.
- Remote control and operation.

These functions allow the SCADA to:

- Acquire quantitative measurements immediately and over time
- Detect, diagnose and correct problems as they arise
- Measuring trends over time and preparing reports and tables
- Discover and eliminate bottlenecks over time and improve efficiency
- Ability to control complex processes with a few specialized employees.

In relation to its contribution to the performance of the production system, Aditya Bagri and Al identified the main contributions and positive impacts on the following criteria of production performance:

1. Security of operators and equipment through predefined security processes managed by the SCADA system.
2. Maintenance costs through centralized control and monitoring to minimize downtime.
3. Production costs through the optimization of human and material resources.
4. Productivity through analysis of processes used to improve the efficiency of production facilities and with integration of other company systems.

5. The quality of the product produced by statistical analysis of process and product data using the standard SCADA functionality.
6. Regulatory compliance for traceability using the basic function "data acquisition" of industrial supervision.

3   Empirical Study

This part is completely reserved for the empirical study which seeks to evaluate the contribution of the SCADA system to the improvement of the performance of the production system. The FBM company located in Tangier is selected to constitute our theoretical analysis and practical exploitation area. This part is composed of the context of the study, the methodological framework of research adopted and the results obtained.

3.1   Presentation of the Field of Investigation

Established in 1979 in MOROCCO, FBM is one of the most important subsidiaries of the French Groupe Bel and the best performers in MOROCCO. It makes a strong contribution to strengthening the Group's position as one of the world's leading cheeses brands in terms of turnover and profitability.

With the start of export in 1997, the company implemented a development strategy that affected its organization, its production tool and its information system. As for the production system, the company set as a strategic objective the modernization of its production line. The operational objectives of this strategy are determined and monitored:

- The upgrading and automation of the existing UHT line for the production of cheese paste.
- The implementation of a second UHT line to increase its production capacity in order to meet the growing local demand and the launch of the export.
- The deployment of the SIEMENS WinCC industrial supervision solution to improve the management of these two lines, to optimize production resources and to control food safety.

3.2   Description of the Production Process

The UHT production line is composed by the following installations:

- A plant for the preparation of melting salts
- A reconstituted milk preparation plant
- A mixer of the MP mix
- A UHT sterilization installation

The following Fig. 4 shows the flowchart of the production line.

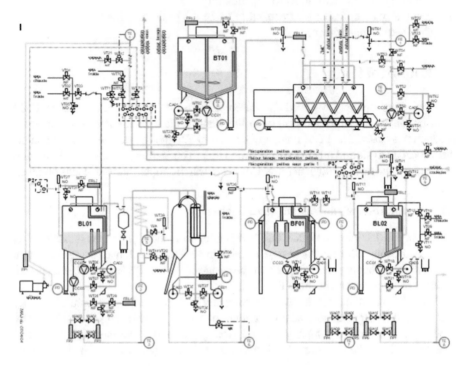

Fig. 4. Flowchart of the production line

3.3   Overview of the Selected Monitoring Application

SIMATIC WinCC is a modular process control system that offers powerful automation monitoring functions. WinCC offers full Windows-based SCADA functionality for all sectors, from single-user configuration to distributed multi-user configurations with redundant servers and multi-site solutions with Web clients (Figs. 5, 6, 7).

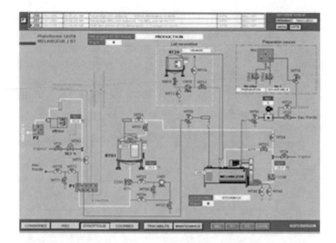

Fig. 5. Sample of vizualization - WinCC

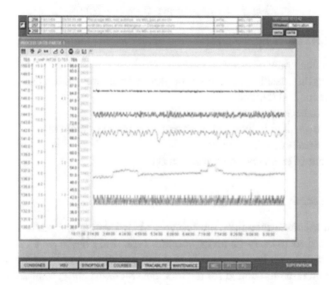

Fig. 6. Graphical tools view - WinCC

Fig. 7. Alarm management view - WinCC

Industrial HMI functions are part of the basic system configuration:

- Fully graphical view of process
- Conducting the machine or installation through a customizable user interface with its own menus and toolbars
- Signaling and acknowledgment of events
- Archiving measured values and messages in a process database
- Process and archive data logging
- Management of users and their access rights

3.4 The Research Methodology

The nature of the research object, our research objectives and our status as a researcher practitioner lead us to carry on an exploratory type study based on a case study since the level of knowledge on our specific problem and our research object which is the SCADA system remains limited for the moment (Stebbins 2001, Stier Adler and Clark 2011)

Three arguments led us to choose the approach of the clinical study:

- This method is often considered the most appropriate approach to address and analyze contemporary phenomena within a real context[9].

---

[9] « *A case study is an empirical investigation that analyzes a contemporary phenomenon within a real context, especially when the boundaries between phenomena and context are not clearly evident* » Cited on p 60 of Rantz Rowe « Faire de la recherche en systèmes d'information » Vuibert, Paris, 2002.

- The clinical study is essentially privileged by the Logistics Higher Studies Institute (Institut des Hautes Études Logistiques) to deal with the problems of research in logistics and information systems[10]
- On a personal level, my presence in the company as production manager, allowed me to master the stakes, processes and mechanisms of the production system of "Fromagrie Bel Maroc".

As for data collection techniques, four methods were used to evaluate the impact of WinCC on the performance of the production function:

- Participant observation
- Analysis of company's internal documents related to the project
- Interview with drivers of the UH line, main users of the new SCADA system
- Historical Trends in Performance Indicators of the Production System

3.5 Project Overview

The project involves setting up an industrial supervision system for the UHTA, UHTB and UHTC production lines with ancillary facilities. It will regroup the complete chain of each installation (see Fig. 8), specifically:

- The Powdering section "Powder and reconstituted milk dosing"
- The Sauces section "Preparation and dosage of sauces"
- The Pre-Cooking section "Mixers/BT"
- The Sterilization section "Part 1"
- The Creaming section and pooring machines feeding "Part 2"
- The section "Recovery of the cream"
- The section "Pooring machines feeding"

---

[10] See Dornier (1996).

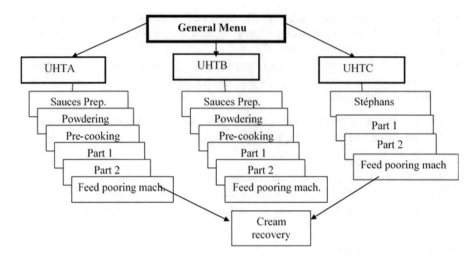

Fig. 8. Perimeter and general architecture of the project

The system shows the synoptics of the installations described above in accordance with the Actual circuit diagrams with all the actuators, sensors, tanks in synoptic formats that reflect reality while having dynamic displays of process data (Levels, T °, Flows etc.) with real-time animations. It will mainly consist of:

- Graphical parts « Synoptics »
- "Process Alarms and Defaults" parts
- "Trends and Recordings" parts
- "Archiving" parts

All controls will be centralized on the PC which will be placed in the control room with the possibility of remote controls (case of Mixers).

Compared to the material aspect, the system is composed of:

- Software WINCC SIEMENS Version 6.1 + SP3
- WINCC Runtime (Licence pour 64 K Power Tags)
- 02 Computers including 1 redundant
- 1 ETHERNET Network Coupler

For communication, the central PC will be connected to the network via ETHERNET port. Some of the PLCs communicate via PROFIBUS network (UHTA, UHTB, UHTC). PLC CPUs will communicate via ETHERNET network through CP ports (see Fig. 9).

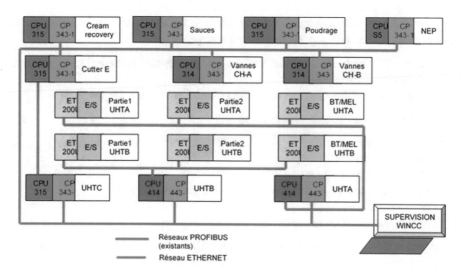

Fig. 9. Network architecture

With regard to the technical data of the installations to be managed by the system, the project envisages about 2000 variables. They are presented in the following Table 1:

Table 1. Synthesis of the numbers of variables to be processed

INSTALLATION	CPU	COUPLER	Network	Number of variables
UHTA	414	CP443-1	ETHERNET	512
UHTB	414	CP443-1	ETHERNET	512
UHTC	315	CP343-1	ETHERNET	256
Preparation of SAUCES	315	CP343-1	ETHERNET	64
POWDERING	315	CP343-1	ETHERNET	64
Cream recovery	315	CP343-1	ETHERNET	64
Chain A valves	314	CP343-1	ETHERNET	128
Chain B valves	314	CP343-1	ETHERNET	128

The expected benefits from the implementation of the WinCC solution are as they are declared by the factory management in an official announcement to launch the project:

- Improvement of UHT process management.
- Ensuring the traceability of the UHT production and cleaning process.
- Increase Productivity.
- Reduce manufacturing time and costs.

On the other hand, the production manager added other objectives for setting up the supervision system during a meeting with the project team:

- Improve the management of the workshop by configuring and editing the synthesis reports of the production.
- Integrate preventive maintenance of the production line into the solution.

As for the project team, it is composed of:

- Technical Manager
- Production Manager
- UHT Workshop Manager
- Electrical and automation manager who is the project manager
- The IT coordinator
- The external automation expert

A project schedule that covers a period of 6 months has been established and validated by the project team. The main stages of the project are:

1. Establishment of the specifications
2. Determining the need for supervisory measurement instruments
3. Risk analysis on the UHT process
4. Training of operators
5. Writing the user manual
6. Establishment of summary statements
7. Integration of preventive maintenance and UHT supervisor manual tasks
8. Launching of supervision in industrial test phase on UHT B as pilot line
9. Deployment on other UHT lines

3.6   Main Results

This section is devoted to the presentation of the results of our empirical study. First, we present the results of the interviews with the end users who are mainly the UHT supervisors, about the contributions of the new system. Second, we present the impact of the system on the evolution of certain indicators of the production process.

The interview with the system supervisors
The interviews with the 6 UHT supervisors reflected a unanimity on the positive and direct contributions of the WINCC system. The interviewees' responses are summarized in the following table:

The contributions of the system
▪ Protection against accidents at work ▪ Protection of installations ▪ Diagnosis of abnormalities ▪ Planning of maintenance operations ▪ Troubleshooting support ▪ Continuous process monitoring. ▪ Improvement in installations reliability and availability ▪ Operator Efficiency ▪ Viewing events and history ▪ Reduction of material losses

Historical Trends in Performance Indicators of the Production System

The objective of this pertinent and original method is to analyze the impact of the system on the evolution of these indicators of the production process over time since its introduction in 2007. The performance indicators directly related to the selected system are:

- The rate of material losses
- The centrality and dispersion of product parameters
- The overall effectiveness

The rate of material losses

Process	2005	2006	2007	2008	2009	2010	2011	2012
UHT	3.75	3.58	2.75	2.35	2.01	1.49	1.32	1.14

The overall effectiveness

Process	2005	2006	2007	2008	2009	2010	2011	2012
UHT	97.5	97.8	98.6	99.2	99.5	99.5	99.6	99.6

Average values and dispersion of product parameters

Parameter	m/σ[a]	Goal	2005	2006	2007	2008	2009	2010	2011	2012
ES	m	**46,4**	46,68	46,63	46,54	46,38	46,42	46,40	46,40	46,40
Dry extract	σ		0,29	0,30	0,25	0,20	0,15	0,12	0,08	0,08
G/S	m	**56,5**	56,75	56,70	56,55	56,53	56,51	56,50	50,52	56,51
Fat	σ		0,48	0,51	0,45	0,40	0,37	0,33	0,29	0,25
pH	m	**5,95**	5,93	5,94	5,94	5,95	5,95	5,95	5,95	5,95
Acidity	σ		0,05	0,05	0,03	0,01	0,01	0,01	0,01	0,01
TX	m	**20**	23,12	22,24	21,35	20,81	20,25	20,10	20,04	20,10
Texture	σ		3,45	2,81	2,50	1,50	1,43	1,18	1,05	0,09

[a]m (Average value)/σ(Standard deviation)

Historical trends show that all of the selected indicators improved with the introduction of the WinCC system in 2007.

4   Conclusion

The objective of this work was to evaluate the impact, through a case study, of the SCADA system on the performance of the production process. Two techniques were used to achieve this objective: Interviews and tracking of trends in indicator history in direct relation to the system. The results of this study allowed us to confirm and measure the positive and direct impact of industrial supervision on the performance of the production system.

The approach adopted for research cannot provide a universal answer to our problem. The results obtained cannot be generalized to other cases or to other sectors. Nevertheless, this approach remains very useful for framing subsequent studies concerning more extensive fields of investigation (comparative study, sectoral survey, survey questionnaire, etc.).

References

Cottet, F., Grolleau, E.: Systèmes temps réel de contrôle commande Conception et implémentation, Dunod (2005)

Chartres, J.-M.: Supervision: outil de mesure de la production. Techniques d'ingénieurs, R7630 (1997)

Artigues, C., Roubelat, F.: A polynomial activity insertion algorithm in multiressource schedule with cumulative constraints and resource flexibility. Eur. J. Oper. Res. **127**(2), 179–198 (2000)

Rowe, R.: Faire de la recherche en systèmes d'information. Vuibert, Paris (2002)

Dornier, P.P.: Thèse doctorale, les différents modèles de la révolution logistique (1996)

Stebbins, R.A.: Exploratory Research in the Social Sciences. Sage Publication, Thousand Oaks (2001)

Stier Adler, E., Clark, R.: An Invitation to Social Research: How It's Done, 4th edn. Cengage Learning, Wadsworth (2011)

Millot, P.: la supervision des procédés automatisés et ergonomie. HERMES, Paris (1988)

Site de l'association MESA Internationale (Manufacturins Exection system Association) (2012). www.mesa.org

Pujo, P., Kieffer, J.P.: Fondements de pilotage des systèmes de production. HERMES, Paris (2002)

Bonnet, P.: Introduction à la supervision, université LILIE 1 (2010)

CIMAX: Introduction à la supervision, Edition applicatif, no. 4 (1998)

Bailey, D., Wrihgt, E.: Practical SCADA for Industry. Elsevier, Amsterdam (2003). IDC Technologies

Vieille, J.: Industrial information systems – ISA88/95 based functional definition. In: WBF European Conference Mechelen, Belgium, 13–15 November 2006

Kirti, : SCADA: supervisory control and data acquisition. Int. J. Eng. Comput. Sci. **3**(1), 3743–3751 (2014). ISSN: 2319-7242

Author, F., Author, S.: Title of a proceedings paper. In: Editor, F., Editor, S. (eds.) CONFERENCE 2016, LNCS, vol. 9999, pp. 1–13. Springer, Heidelberg (2016)

Author, F., Author, S., Author, T.: Book title. 2nd edn. Publisher, Location (1999)

Author, F.: Contribution title. In: 9th International Proceedings on Proceedings, pp. 1–2. Publisher, Location (2010)

LNCS Homepage. http://www.springer.com/lncs. Accessed 21 Nov 2016

Prediction of Temperature Gradient on Selective Laser Melting (SLM) Part Using 3-Dimensional Finite Element Method

Mohammed Abattouy[1]([✉]) [iD], Mustapha Ouardouz[1],
and Abdes-Samed Bernoussi[2]

[1] MMC, Faculté des Sciences et Technique de Tanger,
Université Abdelamalik Essadi, Tangier, Morocco
abattouy.mohammed@yahoo.com
[2] GAT, Faculty of Sciences and Techniques, Tangier, Morocco
a.samed.bernoussi@gmail.com

Abstract. Additive manufacturing (AM) or known as 3D printing is a direct digital manufacturing process where a 3D part can be produced, layer by layer from 3D digital data with no use of conventional machining and casting. AM has developed over the last 10 years and has showed significant improvement in cost reduction of critical component. This can be demonstrated through reduced material waste, improved design freedom and reduced post processing.

Modeling the AM process provides an important insight into physical phenomena that lead to improve final material properties and product quality and predict the final workpiece characteristics.

It's very challenging to measure the temperature gradient due to the transient nature and small size of molten pool on SLM. A 3-dimensional finite element model has been developed to simulate multilayer deposition to predict temperature gradient on melting pool of stainless steel, as well as a review of different models used to simulate the selective laser melting is given.

Keywords: Selective Laser Melting (SLM) · Finite Element Model (FEM)
Temperature gradient · Molten pool

1 Introduction

As a first step, Rapid Prototyping was used to create prototypes from CAD for communication and inspection purposes in a short time, whereas nowadays Rapid prototyping (RP) is used to produce directly and quickly end-use parts with complex shapes. Selective laser melting (SLM), laser metal deposition (LMD), Warm and arc additive layer manufacturing (WAALM), [1], etc., are emerging technologies allowing manufacturing in a single step parts from their 3D CAD models. The main advantages of the additive Manufacturing are its ability to create complex geometry difficult to obtain with other conventional techniques, and the reduction of masse and manufacturing process. This work deals only with the selective laser melting (SLM) process where the part is first cut numerically into thin layers to get a multilayered CAD model. Then, each metallic powder layer is successively laid down on a horizontal bed-plate and

© Springer International Publishing AG, part of Springer Nature 2018
M. Ben Ahmed and A. A. Boudhir (Eds.): SCAMS 2017, LNNS 37, pp. 902–909, 2018.
https://doi.org/10.1007/978-3-319-74500-8_80

converted in a solid material by melting with a high intensity laser beam on the previous solidified layers. The path of the laser beam is controlled in agreement with the cross-sections generated from the 3D CAD model (Fig. 1). The material density of the manufactured parts is close to the density of molded parts. For example, this process is very versatile to produce parts used in aerospace, medical and automotive industry [2]. The laser beam generates heating cycles in the vicinity of its radiation, so violent cooling takes place with strong temperature gradients in the solidified layer. As cyclic thermal expansions and contractions far exceed the maximum elastic strain of the material, heterogeneous plastic strains are cumulated in the manufactured part generating internal stresses. Their level can reach the strength of the material and cracks may appear during the process or reduce the fatigue life of the part, so it is necessary to analyze the temperature distribution in order to reduce material defects.

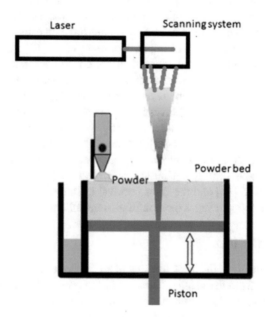

Fig. 1. Selective laser melting process.

2   Related Work

Researchers have employed finite element analysis (FEA), one of the most widely used numerical methods in use today, for temperature field analysis of various additive layer manufacturing processes, Abattouy et al. used a 3D finite element model to simulate laser cutting of metallic workpiece [3]. In early papers, Childs [4] explored the influence of process parameters on the mass of melted single layers in SLM and concluded that an increase in scanning speed resulted in a larger melted mass. Matsumoto et al. [5] suggested a method for calculating the distribution of temperature and stress in the SLM of single metallic layers; they concluded that the solid layer on the powder bed warped because of heating and cooling while the laser scanned on the track. When the neighboring track began to solidify, large tensile stresses take place at the side end of the solid part. Dai and Shaw [6] searched the effect of the laser scanning strategy on residual stress and distortion and confirmed that a scanning pattern with frequent 90° changes in the scanning direction at every turn could lead to a reduction of concave upwards and downwards distortions.

Hussein et al. [7] proposed a transient finite element model for the analysis of temperature and stress fields in single layers built without support in SLM. They concluded that the predicted length of the molten pool increased at higher scanning speeds, while both the width and depth of the molten pool decreased. Li and Gu [8] explored the thermal behavior during the SLM of aluminum alloy powders and obtained the optimum molten pool width (111.4 μm) and depth (67.5 μm) for a specific combination of parameters (laser power 250 W and scan speed 200 mm/s). Roberts et al. [9] used an element birth-and-death strategy to analyse the 3D temperature field in multiple layers within a powder bed. Similar studies have investigated the behaviour of other materials during SLM. For example, Kolossov et al. [10] suggested a 3D finite element model to predict the temperature distribution on the top surface of a titanium powder bed during the laser sintering process; their work demonstrated that changes in thermal conductivity determined the behaviour and development of thermal processes. In addition, Patil and Yadava [11] investigated the temperature distribution in a single metallic powder layer during metal laser sintering and found that temperature increased with increases in laser power and laser on-time, but decreased with increases in laser off-time and hatch spacing. Residual stresses have been investigated with experimental methods like the hole drilling technique presented in [12]. The bridge curvature technique presented in [13] and the layer removal technique presented in [14] are also used to analyze the residual stress distributions in metallic parts manufactured by SLM.

2.1   Governing Equation

In selective laser melting, the heat transfer in the material dominated by conductive heat transfer results from the localized heating of the powder bed by the laser beam. In such a case, the spatial and temporal distribution of temperature is governed by the heat conduction equation, which can be expressed as:

$$\rho C_p \frac{\partial T}{\partial t} = k_{xx} \frac{\partial^2 T}{\partial x} + k_{yy} \frac{\partial^2 T}{\partial y} + k_{zz} \frac{\partial^2 T}{\partial z} + \ddot{Q}(r) \tag{1}$$

where T is the temperature, t is the time, (x, y, z) are the spatial co-ordinates, k_{xx}, k_{yy} and k_{zz} are the thermal conductivities, ρ is the density, C_p is the specific heat and $\ddot{Q}$ is the heat source term. The selective laser melting process is usually carried out in a chemically inert gaseous environment. The thermal interaction at the boundaries between the material and the surroundings can be given as

$$-k \frac{\partial T}{\partial t} = h(T_{amb} - T) + \sigma\varepsilon(T^4 - T_{amb}^4) \tag{2}$$

Where h is the heat transfer coefficient, T_{amb} is the temperature of the gaseous environment, ε is the emissivity of the material and σ is the Stefan-Boltzman constant. While the above equations holds true in general, appropriate values of the physical quantities (e.g. thermal conductivities, specific heat, etc.) must be used to get an accurate and realistic output.

2.2   Laser Model

The heat source term $\dddot{Q}(r)$ is used to indirectly model the laser beam in the heat conduction equation (Eq. 1) as a volumetric internal heat generation.

The heat flux $\dddot{Q}(r)$ follows a Gaussian distribution and can be expressed as:

$$\dddot{Q}(r) = \frac{2AP}{\pi r_0} \exp\left(\frac{-2r^2}{r_0^2}\right) \tag{2}$$

The irradiance of a simple laser beam with a Gaussian distribution profile is defined by Eq. (2), where P is the power of the laser beam.

2.3   Establishment of Finite Element Model

A full three-dimensional FEM of the specimen is developed and the necessary correlations are implemented to make the model as accurately as possible, Fig. 2 shows the initial grid employed for thermal analysis. A variable spacing grid system with a fine grid near the heat source and a course grid away from the heat source has been used for model the heat transferring. The finite element model of the material is stainless steel and its size is 10 mm × 10 mm × 0.3 mm. In order to ensure the accuracy of the

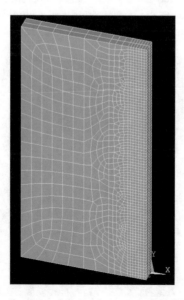

Fig. 2. Three dimensional finite element model of laser sintering.

calculation, the sintering powder is meshed by Solid70 with hexahedral and eight-node and the grid size is 0.1 mm × 0.1 mm × 0.1 mm while the rest of the material is meshed by Solid90.

3 Thermal Physical Properties of Material

The thermal physical properties of stainless steel are shown in Table 1 for different temperature (from 20 °C to 2500 °C). However, the effective thermal conductivity of powder bed is the most important properties which can significantly affect the accuracy of the final result.

Table 1. Thermal physical properties of stainless steel.

	Density	Specific heat (J/K$_g$.K)	Thermal conductivity (W/m.K)
$T_1 = 20\ ^\circ C$	7820	460	50
$T_2 = 200\ ^\circ C$	7700	480	47
$T_3 = 500\ ^\circ C$	7610	530	40
$T_4 = 750\ ^\circ C$	7550	675	27
$T_5 = 1000\ ^\circ C$	7490	670	30
$T_6 = 1500\ ^\circ C$	7350	660	35
$T_7 = 1700\ ^\circ C$	7300	780	40
$T_8 = 2500\ ^\circ C$	7090	820	55

4 Simulation Results and Discussion

Figures 3, 4 and 5 shows the temperature and temperature gradient contours in the cross-section perpendicular to the laser scanning direction. In center zone the highest temperature is observed from the lateral surface, the Gaussian beam is responsible for

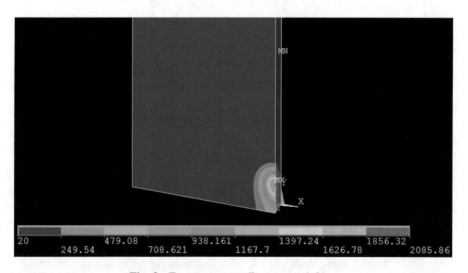

Fig. 3. Temperature gradient, at t = 1.64 s.

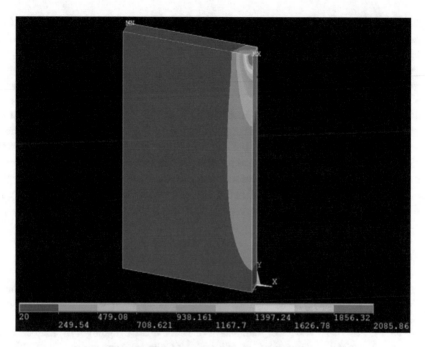

Fig. 4. Temperature gradient, at t = 5.64 s.

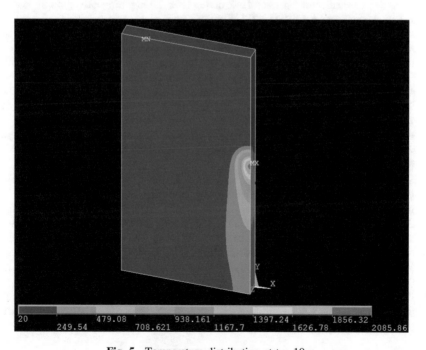

Fig. 5. Temperature distribution at t = 10 s.

higher heat flux. The relative heat generation and dissipation at the contact surfaces causes the temperature contour in the center zone to make a "V" shape of distribution.

5 Conclusion

A three-dimensional modeling and measurement of the temperature evolution in the stainless steel is conducted and the experimental values found in related works validate the efficiency of the proposed model. The prediction and measurement shows that the maximum temperature gradients in longitudinal and lateral directions are located center zone of laser beam. The prediction implies that the higher stress is located in the region from top surface of laser scan zones and the developed model had good capability for predicting the temperature cycles throughout selective laser melting.

References

1. Abattouy, M., Ouardouz, M., Bernousi, A.: Additive manufacturing laboratory inside Moroccan universities: a case study. In: Proceedings of the Conference on 6th International Conference on Additive Technologies, ICAT 2016, Nurnberg, Germany (2016)
2. Abattouy, M., Ouardouz, M., Bernousi, A.: la fabrication additive au Maroc: Etude prospective en milieu universitaire et industrial. In: Colloque international TELECOP'2017 & 10 éme Journée Franco-Maghrébines des microondes et Applications (2017)
3. Yang, L., Ye, D., Bai, C., Abattouy, M., Yang, W.: Numerical simulation and experimental research on reduction of taper and HAZ during laser drilling using moving focal point. Int. J. Adv. Manuf. Technol. 91(1), 1171–1180 (2017)
4. Childs, T.: Selective laser sintering (melting) of stainless and tool steel powders: experiments and modelling. Proc. Inst. Mech. Eng. Part B J. Eng. Manuf. 219(4), 339–357 (2005)
5. Dai, K., Shaw, L.: Distortion minimization of laser-processed components through control of laser scanning patterns. Rapid Prototyp. J. 8(5), 270–276 (2002)
6. Matsumoto, M., Shiomi, M., Osakada, K., Abe, F.: Finite element analysis of single layer forming on metallic powder bed in rapid prototyping by selective laser processing. Int. J. Mach. Tools Manuf. 42(1), 61–67 (2002)
7. Hussein, A., Hao, L., Yan, C., Everson, R.: Finite element simulation of the temperature and stress fields in single layers built without-support in selective laser melting. Mater. Des. 52, 638–647 (2013)
8. Li, Y., Gu, D.: Parametric analysis of thermal behavior during selective laser melting additive manufacturing of aluminum alloy powder. Mater. Des. 63, 856–867 (2014)
9. Roberts, I.A., Wang, C.J., Esterlein, R., Stanford, M., Mynors, D.J.: A three dimensional finite element analysis of the temperature field during laser melting of metal powders in additive layer manufacturing. Int. J. Mach. Tools Manuf. 49(12–13), 916–923 (2009)
10. Kolossov, S., Boillat, E., Glardon, R., Fischer, P., Locher, M.: 3D FE simulation for temperature evolution in the selective laser sintering process. Int. J. Mach. Tools Manuf. 44(2–3), 117–123 (2004)
11. Patil, R.B., Yadava, V.: Finite element analysis of temperature distribution in single metallic powder layer during metal laser sintering. Int. J. Mach. Tools Manuf. 47(7–8), 1069–1080 (2007)

12. Dadbakhsh, S., Hao, L.: Effect of Al alloys on selective laser melting behaviour and microstructure of in-situ formed particle reinforced composites. J. Alloys Compd. **541**, 328–334 (2012)
13. Thijs, L., Kempen, K., Kruth, J.P., Van Humbeeck, J.: Fine-structured aluminium products with controllable texture by selective laser melting of pre-alloyed AlSi10Mg powder. Acta Mater. **61**(5), 1809–1819 (2013)
14. Wong, M., Tsopanos, S., Sutcliffe, C.J., Owen, I.: Selective laser melting of heat transfer devices. Rapid Prototyp. J. **13**(5), 291–297 (2007)

The 1st International Workshop on Mathematics for Smart City: MCS'17

Interaction Between the VaR of Cash Flow and the Interest Rate Using the ALM

Mostafa El Hachloufi[✉], Driss Ezouine[✉],
and Mohammed El Haddad[✉]

Department of Management-Faculty of Juridical,
Economical and Social Sciences, University of Mohamed V, Rabat, Morocco
elhachloufi@yahoo.fr, ezouinedriss@gmail.com,
mo.haddad@hotmail.com

Abstract. In this paper, we propose an approach to study the impact of the interest rate on the risk of variation in cash flows measured by the value at risk (VaR) using stochastic processes and ALM technics.

This approach provides a decision-making tool for manage asset, liability funds to bankers insurers and all companies operating in the financial sector.

Keywords: Interest rates · VaR · Cash flow · ALM technics
Stochastic processes

1 Introduction

ALM is one of the main tools used to help solve rate variation problems in financial institutions such as banks and insurance companies. He plays a very important role in managing the various activities of the financial institute.

Appropriate liquidity and balance sheet management are a key factor in ensuring the activity of financial institutions and are a tool for managers to make decisions about risk management with variations in the interest rate.

The activities of companies, whether banks, insurance companies or for-profit corporations, generate cash flows affecting the balance sheet as assets or liabilities. This financial flow and its risk are influenced by the variation in the interest rate. In the case of a positive or negative variation, it will have an impact on the assets, liabilities or both at the same time.

The objective of this paper is to study this influence by treating the impact of the change in the interest rate on the risk of this flow.

2 Interest Rate

The interest rate is defined as the economic remuneration of time. This is the amount a borrower is willing to pay to his lender in addition to the capital, and it is based on the credit risk of that borrower.

© Springer International Publishing AG, part of Springer Nature 2018
M. Ben Ahmed and A. A. Boudhir (Eds.): SCAMS 2017, LNNS 37, pp. 913–918, 2018.
https://doi.org/10.1007/978-3-319-74500-8_81

The evolution of the interest rate can be modeled by several stochastic processes whose Vasicek process or model is the most popular. This stochastic process is called the process of return to the mean.

The Vasicek model assumes that the current short-term interest rate is known, while the future values of this rate follow the following equation:

$$dr_t = \eta(\bar{r} - r_t)dt + \sigma dz_t \tag{1}$$

Where:

η : is the rate of return of the interest rate to the average.
$\bar{r}$: is the average interest rate.
σ : is the volatility of the interest rate which is assumed to be independent of r_t.
z_t : is a Brownian movement such as $dz_t = \varepsilon_t \sqrt{dt}$ with $\varepsilon_t \sim N(0, 1)$.

3   The VaR of Financial Flow

The VaR is a measure of risk most widely used in financial markets to quantify the maximum loss on a portfolio for a given horizon and confidence level. It depends on three elements:

The distribution of the portfolio's profits and losses for the holding period.
The level of confidence.
The holding period of the asset.

Analytically, the VaR with time horizon t and the threshold probability α the number such as:

$$P[\Delta R \leq VaR(t, \alpha)] = \alpha \tag{2}$$

where:

- t: Horizon associated with VaR which is: 1 day or more than one day.
- α: The probability level is typically 95%, 98% or 99%.

The financial flow $F_i, i = 1, \ldots, n$ from a company E is a set of n amounts received or paid by it at different times $t_i, i = 1, \ldots, n$ to the future whose interest rate corresponds to the t_i is r_{t_i}. Interest rates $r_{t_i}, i = 1, \ldots, n$ are assumed to be independent.

The present value of this financial flow is given by:

$$V_r = \sum_{i=1}^n F_i r_{t_i} = F_1 r_{t_1} + F_2 r_{t_2} + \ldots + F_n r_{t_n}$$

Consider two future flows: assets A and liabilities P given by:

$$A = \{(A_i, r_{t_i}), \ i = 1, \ldots, n\}), \qquad P = \{(P_i, r_{t_i}), \ i = 1, \ldots, n\}$$

The surplus relative to the couple of flow $\{A, P\}$ in relation to the interest rate r_{t_i}, $i = 1, \ldots, n$, noted S_{r_t}, is given by:

$$S_{r_t} = \sum_{i=1}^{n} (A_i - P_i) \times r_{t_i}$$

$$\text{or} \quad S_{r_t} = \sum_{i=1}^{n} AP_i \times r_{t_i} = \sum_{i=1}^{n} F_i \times r_{t_i}$$

$$\text{where} \quad F_i = AP_i = (A_i - P_i)$$

Assuming that the interest rate follows a Vasicek process, i.e.:

$$dr_t = \eta(\bar{r} - r_t)dt + \sigma dz_t$$

Thus, we can develop this model which is used to describe the interest rate as follows:

$$d(e^{\eta t} r_t) = e^{\eta t} dr_t + r_t \eta e^{\eta t} \Rightarrow e^{\eta t} dr_t = d(e^{\eta t} r_t) - r_t \eta e^{\eta t}$$

In other,

$$dr_t = \eta(\bar{r} - r_t)dt + \sigma dz_t \Rightarrow e^{\eta t} dr_t = e^{\eta t}\eta(\bar{r} - r_t)dt + e^{\eta t}\sigma dz_t$$

Then we get:

$$d(e^{\eta t} r_t) = \eta \bar{r} e^{\eta t} dt + e^{\eta t}\sigma dz_t \Rightarrow r_t = r_0 e^{-\eta t} + \int_0^t \eta e^{-\eta(t-s)}\bar{r} ds + \sigma \int_0^t e^{-\eta(t-s)} dz_s$$

So the interest rate r_t can be expressed as follows:

$$r_t = \bar{r} + (r_0 - \bar{r})e^{-\eta t} + \sigma \int_0^t e^{-\eta(t-s)} dz_s \tag{3}$$

Knowing that z_t is a process such as $dz_t = \varepsilon_t \sqrt{dt}$ with $\varepsilon_t \rightarrow N(0, 1)$, so dz_t follows the normal distribution and the variable $\left(\sigma \int_0^t \eta e^{-\eta(t-s)} dz_s\right)$ also follows the normal distribution.

The term $\bar{r} + (r_0 - \bar{r})e^{-\eta t}$ is not a random term then $r_t | r_0$ is a random variable that follows the normal distribution.

Using isometry we find that:

$$E\left[\left(\sigma \int_0^t \eta e^{-\eta(t-s)} dz_s\right)\right] = 0 \text{ and}$$

$$E\left[\left(\sigma \int_0^t \eta e^{-\eta(t-s)} dz_s\right)^2\right] = \int_0^t \left(\sigma e^{-\eta(t-s)}\right) ds = \frac{\sigma^2}{2\eta}(1 - e^{-2\eta t})$$

Then the mean and the variance are respectively:

$$E(r_t|r_0) = \bar{r} + (r_0 - \bar{r})e^{-\eta t}$$

$$V(r_t|r_0) = \frac{\sigma^2}{2\eta}\left(1 - e^{-2\eta t}\right)$$

So the random variable $r_t|r_0$ follows the normal distribution of mean and variance respectively $\bar{r} + (r_0 - \bar{r})e^{-\eta t}$ and $\frac{\sigma^2}{2\eta}\left(1 - e^{-2\eta t}\right)$, i.e.:

$$r_t \rightarrow N\left(\bar{r} + (r_0 - \bar{r})e^{-\eta t}, \sigma\sqrt{\frac{1}{2\eta}\left(1 - e^{-2\eta t}\right)}\right)$$

Knowing that:

$$V_{r_t} = \sum_{i=1}^{n} F_i r_{t_i} = F_1 r_{t_1} + F_2 r_{t_2} + \ldots + F_n r_{t_n} \quad \text{so} \quad dV_{r_t} = \sum_{i=1}^{n} F_i dr_{t_i}$$

and

$$dr_t = \eta(\bar{r} - r_t)dt + \sigma dz_t$$

$$\text{so} \quad dV_{r_t} = \sum_{i=1}^{n} F_i dr_{t_i} = \sum_{i=1}^{n} F_i [\eta(\bar{r} - r_{t_i})dt_i + \sigma dz_{t_i}]$$

$$\Rightarrow dV_{r_t} = \sum_{i=1}^{n} F_i \eta(\bar{r} - r_{t_i})dt_i + F_i \sigma dz_{t_i}$$

$$\Rightarrow \Delta V_{r_t} = \sum_{i=1}^{n} F_i \eta(\bar{r} - r_{t_i})\Delta t_i + F_i \sigma \sqrt{\Delta t_i}\varepsilon_{t_i}$$

So $\quad E(\Delta V_{r_t}) = \sum_{i=1}^{n} F_i \eta(\bar{r} - r_{t_i})\Delta t_i \quad$ and $\quad V(\Delta V_{r_t}) = \sum_{i=1}^{n} (\sigma F_i)^2 \Delta t_i$

Let $\quad \alpha_t = \sum_{i=1}^{n} F_i \eta(\bar{r} - r_{t_i})\Delta t_i \quad$ and $\quad \beta_t = \sum_{i=1}^{n} (\sigma F_i)^2 \Delta t_i \quad$ then

$$\Delta V_{r_t} \rightarrow N\left(\alpha_t, \sqrt{\beta_t}\right)$$

The calculation of the VaR at α is given by the following equation:

$$P(\Delta V_{r_t} \leq VaR_\alpha) = \alpha$$

Knowing that $\Delta V_{r_t} \rightarrow N\left(\alpha_t, \sqrt{\beta_t}\right)$ therefore

$$P\left(\frac{\Delta V_{r_t} - \alpha_t}{\sqrt{\beta_t}} \leq \frac{VaR_\alpha - \alpha_t}{\sqrt{\beta_t}}\right) = \alpha$$

$$\Rightarrow \quad \frac{VaR_\alpha - \alpha_t}{\sqrt{\beta_t}} = \tau_\alpha.$$

Thus

$$VaR_\alpha = \sum_{i=1}^{n} \eta F_i(\bar{r} - r_{t_i})\Delta t_i + \tau_\alpha \sigma^2 \sum_{i=1}^{n} F_i^2 \Delta t_i \tag{4}$$

4 Conclusion

In this paper, we have developed an approach based on a mathematical formula to study the impact of the interest rate on the risk of the financial flow using value at risk as a risk measure and ALM to express the variability of a company's financial flows.

This approach allows us to evaluate the risk of financial flows in relation to the variation in the interest rate which gives rise to a decision-making tool for the management of funds, whether at the level of the assets or the liabilities of the company.

References

1. Adam, A.: Handbook of Asset and Liability Management: From Models to Optimal Return Strategies. Wiley, Hoboken (2007)
2. Chang, H.: Dynamic mean-variance portfolio selection with liability and stochastic interest rate. Econ. Model. **51**, 172–182 (2015)
3. Zenios, S.A., Ziemba, W.: Handbook of Asset Liability Management, vol. 1, 1st edn, pp. 1689–1699. North Holland, Amsterdam (2015)
4. Dempster, M.A.H., Germano, M., Medova, E.A., Villaverde, M.: Global Asset Liability Management (2002)
5. Elhachloufi, M., Guennoun, Z., Hamza, F.: Stocks portfolio optimization using classification and genetic algorithms. Appl. Math. Sci. **6**(94), 4673–4684 (2012)
6. Elhachloufi, M., Guennoun, Z., Hamza, F.: Minimizing risk measure semi-variance using neural networks and genetic algorithms. J. Comput. Optim. Econ. Financ. **4** (2012)
7. Elhachloufi, M., Guennoun, Z., Hamza, F.: Optimization of stocks portfolio using genetic algorithms and value at risk. Int. J. Math. Comput. **20**(3) (2012)

8. Elhachloufi, M., Guennoun, Z., Hamza, F.: Optimization stock portfolio optimization using neural network and genetic algorithm. Int. Res. J. Financ. Econ. (104) (2013)
9. Jorion, P.: Value at Risk the New Benchmark for Managing Financial Risk, 3rd edn. McGraw-Hill Education, New York City (2007)
10. Blomvall, J.: Measurement of interest rates using a convex optimization model. Eur. J. Oper. Res. **256**, 308–316 (2017)
11. Zhang, Y., Chen, Z., Li, Y.: Bayesian testing for short term interest rate models. Financ. Res. Lett. (2016)
12. Wang, X., Xie, D., Jiang, J., Wu, X., He, J.: Value-at-risk estimation with stochastic interest rate models for option-bond portfolios. Financ. Res. Lett. (2016)

Optimal Reinsurance Under CTV Risk Measure

Abderrahim El Attar[1]($\boxtimes$), Mostafa El Hachloufi[2],
and Zine El Abidine Guennoun[1]

[1] Department of Mathematics, Faculty of Sciences-Rabat,
Mohamed V University, Rabat, Morocco
estimabd@hotmail.com, guenoun@fsr.ac.ma
[2] Department of Statistics and Mathematics, Faculty of Juridical Sciences,
Economic and Social-Ain Sebaa, Casablanca, Morocco
elhachloufi@yahoo.fr

Abstract. In this paper we proposed a new model for optimizing reinsurance which acts on the Conditional Tail Expectation (CTV) and the technical benefit. In this model, we have determined optimal reinsurance treaty parameters that minimize (CTV) under the constraint of technical benefit which must also be maximal. The minimization procedure is based on augmented Lagrangian method and genetic algorithms in order to solve the optimization program of this model.

Keywords: Augmented Lagrangian · Conditional Tail Variance
Genetic algorithms · Optimization · Premium principle · Reinsurance
Technical benefit

1 Introduction

The search for an optimal reinsurance plan is always an important part of actuarial mathematics. The main objective of an insurer is undoubtedly the maximization of the expected technical benefits and the minimization of risk measures under certain constraints.

In this context, several criteria were proposed to determine an optimal choice of reinsurance, including the Mean-variance criterion, criterion of minimization of the probability of ruin and the criterion of minimization of risk measures such as Value at Risk (VaR) and Conditional Tail Expectation (CTE).

The recent generation of reinsurance optimization research is based on the criterion of insurance risk minimization. This criterion was introduced for the first time in the work of Cai and Tan [5] who proposed reinsurance optimization models based on the minimization of risk measures such as Value at Risk (VaR) and Conditional Tail Expectation.

(CTE) by demonstrating the existence of explicit optimal retention in the case of "surplus-loss" treaties.

The Cai and Tan result was then improved by Balbas et al. [3] and Tan et al. [8].

© Springer International Publishing AG, part of Springer Nature 2018
M. Ben Ahmed and A. A. Boudhir (Eds.): SCAMS 2017, LNNS 37, pp. 919–930, 2018.
https://doi.org/10.1007/978-3-319-74500-8_82

However, the risk minimization criterion focuses solely on risk and does not take into consideration the yield (technical benefit) of the ceding company. In addition, the insurer is motivated by the purchase of reinsurance only if it hopes to have a positive gain expectancy after the recourse of reinsurance.

It should be noted that the risk measures used in these models are not all consistent measures.

Then in this context we proposed a new reinsurance optimization model using a new coherent risk measure based on the principle of expected value and Conditional Tail Variance (CTV). We also have developed an optimization procedure based on the augmented Lagrangian and the genetic algorithms, in order to solve the optimization program of this model.

Thus, we have developed a resolution procedure based on the augmented Lagrangian and the genetic algorithms in order to apply it in agreement with the different mathematical approaches and statistical tools to solve the problem of optimization of this model. It is a powerful optimization algorithm capable of both maximizing technical profit and minimizing the risk of an insurance company under certain constraints. This has made it possible to solve, in very short times, increasingly complex optimization problems with several constraints of equality and/or inequality, notably thanks to computers.

This approach can be seen as a decision support tool that can be used by managers to minimize the risk and maximize the return of an insurance company.

The organization of this chapter is as follows, the Sect. 2 presents new coherent risk measures, then in the Sect. 3 we formulate our optimization problem, then in the Sect. 4 we propose the optimization procedure by the Augmented Lagrangian method and the Genetic Algorithms. Finally, in the Sect. 5, we illustrate our model of optimization by a sample application.

2 Presentation of New Coherent Risk Measure

Definition 1. The Conditional Tail Variance corresponds to the risk X with a level of probability $\theta \in \,]0, 1[$ noted by $CTV_\theta(X)$ is defined by:

$$CTV_\theta(X) = Var[X|X \geq VaR_\theta(X)] \tag{1}$$

We know that (CTV) is not a coherent risk measure. (See Valdez [7]).

Proposition 1. Consider the following risk measurement:

$$\Pi_\theta(X) = \mu_X + q\sqrt{CTV_\theta(X)}, q > 0 \tag{2}$$

With μ_X is the average of X and q is a positive constant not null.
Then $\Pi_\theta(X)$ is a coherent risk measure.

That is to say, it satisfies the coherence properties of Artzner [2].
Demonstration: Verification the coherence properties of $\Pi_\theta(X)$:

P1: Not negativity:
Because $CTV_\theta(X) = Var[X|X \geq VaR_\theta(X)] \geq 0$

P2: Homogeneous positive:
We consider a random variable $Y = \beta X$, for any positive real β.
Either the following θ-th quantile:
Pose

$$x_\theta = VaR_\theta(X) \text{ and } y_\theta = VaR_\theta(Y)$$
$$y_\theta = VaR_\theta(Y) = VaR_\theta(\beta X) = \beta VaR_\theta(X) = \beta x_\theta$$

With the average $\mu_Y = \beta \mu_X$
The Conditional Tail Variance of Y corresponds to the level of probability θ is equal:

$$CTV_\theta(Y) = Var[Y - \mu_Y|X > y_\theta] = Var[\beta(X - \mu_X)|X > x_\theta]$$
$$= \frac{1}{1-\theta} \int_{x_\theta}^{\infty} \beta^2(x - \mu_X)^2 f_X(x)dx = \beta^2 CTV_\theta(X)$$

Then

$$\Pi_\theta(Y) = \mu_Y + q\sqrt{CTV_\theta(Y)}$$
$$= \beta\left(\mu_X + q\sqrt{CTV_\theta(X)}\right) = \beta \Pi_\theta(X)$$

P3: Translational invariant:
We consider a random variable $Y = X + \beta$, for any positive real β.
Let the θ-th quantile

$$y_\theta = VaR_\theta(Y) = VaR_\theta(X + \beta) = VaR_\theta(X) + \beta = x_\theta + \beta$$

With the average $\mu_Y = \mu_X + \beta$
The Conditional Tail Variance de Y Corresponds to the level of probability θ is equal

$$CTV_\theta(Y) = Var[Y - \mu_Y|Y > y_\theta]$$
$$= Var[(X + \beta - \mu_X - \beta)|X + \beta > x_\theta + \beta] = CTV_\theta(X)$$

Then

$$\Pi_\theta(Y) = \mu_Y + q\sqrt{CTV_\theta(Y)} = \mu_X + \beta + q\sqrt{CTV_\theta(X)} = \beta + \Pi_\theta(X)$$

P4: Sub-additivity

Let two risks X and Y we then have:

$$\Pi_\theta(X+Y) \leq \Pi_\theta(X) + \Pi_\theta(Y)$$

Let $S = X + Y$

Then the mean of S is equal $\mu_S = \mu_X + \mu_Y$

We notice s_θ the θ-th quantile of S such as,

$$s_\theta = VaR_\theta(S) = VaR_\theta(X+Y)$$

The sub-additivity of $CTV_\theta(S)$ has led to the sub-additivity of $\Pi_\theta(.)$.

$$CTV_\theta(S) = Var[S - \mu_S | S > s_\theta] = Var[X + Y - \mu_X - \mu_Y | S > s_\theta]$$

$$= Var[(X - \mu_X) + (Y - \mu)_Y | S > s_\theta]$$

$$= Var[(X - \mu_X) | S > s_\theta] + Var[(Y - \mu_Y) | S > s_\theta]$$

We have
$$+ 2Cov[(Y - \mu_Y) | S > s_\theta, [(Y - \mu_Y) | S > s_\theta]$$

$$\leq Var[(X - \mu_X) | S > s_\theta] + Var[(Y - \mu_Y) | S > s_\theta]$$

$$+ 2Var[(Y - \mu_Y) | S > s_\theta] Var[(Y - \mu_Y) | S > s_\theta]$$

$$= (Var[(X - \mu_X) | S > s_\theta] + Var[(Y - \mu_Y) | S > s_\theta])^2$$

From where

$$CTV_\theta(S) \leq (CTV_\theta(X) + CTV_\theta(Y))^2$$

We obtain immediately the sub-additivity of Π_θ.

3   Formulation of the Optimization Problem

Let a portfolio of claims expenses be represented by the random variables $X_1, \ldots, X_N$, continuous and positive, of the distribution function $F_{X_1}, \ldots, F_{X_N}$ and the density function $f_{X_1}, \ldots, f_{X_N}$, in exchange for the respective premiums $P_1, \ldots, P_N$, with $P = \sum_{i=1}^{N} P_i$.

The risks are considered independent and identically distributed, and independent of N.

With, the sum of $(X_i)_{i=1,\ldots,N}$ is zero, if $N = 0$.

Let X is a random variable that designates the total amount of claims, of the distribution function F_X and of the survival function $S_X(x)$.

The reinsurance contract is defined as follows:

$$X = \sum_{i=1}^{N} X_i = \sum_{i=1}^{N} X_i^A + \sum_{i=1}^{N} X_i^R = X^A + X^R \tag{3}$$

Or X^A is the insurer's claims burden and X^R is the claims burden transferred to the reinsurer.

The reinsurer's charge shall in no case exceed the total claims burden, as it should not be negative, i.e. $0 \leq X^R \leq X$.

Assume that the reinsurance premium uses the principle of premium based on mathematical expectation with a safety load η, i.e.

$$\pi(X^R) = (1 + \eta)E(X^R) \tag{4}$$

Let λ is the reinsurance treaty parameter applied to cover the risk of loss. This parameter represents the transfer rate in the case of proportional reinsurance, and the retention limit in the case of non-proportional reinsurance.

Such as:

$$X^A = f(X, \lambda) \text{ and } X^A = g(X, \lambda)$$

Where f and g are two measurable random functions with values in $\mathbb{R}^+$.

The technical benefit of the ceding company is defined by:

$$B(X, \lambda) = \sum_{i=1}^{N} \left[P_i - \pi(X_i^R) - X_i^A \right] = P - \pi(g(X, \lambda)) - f(X, \lambda) \tag{5}$$

Then the mathematical expectation of B is given by:

$$E(B(X, \lambda)) = P - (1 + \eta)\pi(g(X, \lambda)) - E(f(X, \lambda)) \tag{6}$$

We assume that the insurer is seeking optimal reinsurance that minimizes a risk measure for its technical benefit.

According to the coherence properties, the translational invariant and the positive homogeneous, the

We note

$$\Pi_\theta(X) = \mu_X + \sqrt{CTV_\theta(X)} \tag{7}$$

Based on the properties of the coherence (The invariant by translation and the positive homogeneous), the Π_θ-measure of $B(X, \lambda)$ is presented by:

$$\Pi_\theta(B(X, \lambda)) = \Pi_\theta(P - \pi(g(X, \lambda)) - f(X, \lambda)) = P - \pi(g(X, \lambda)) + \Pi_\theta(-f(X, \lambda)) \tag{8}$$

Then

$$\min_\lambda \{\Pi_\theta(B(X, \lambda))\} \equiv \min_\lambda \{\Pi_\theta(-f(X, \lambda)) - \pi(g(X, \lambda))\} \tag{9}$$

We construct the following optimization program:
(CTV)-minimization:

$$\begin{cases} \min_\lambda \{Z(\lambda) = \Pi_\theta(-f(X,\lambda)) - \pi(g(X,\lambda)) = -E(f(X,\lambda)) + \sqrt{CTV_\theta(f(X,\lambda))} - \pi(g(X,\lambda))\} \\ sc\langle P - \pi(g(X,\lambda)) - E(f(X,\lambda)) \geq k \end{cases} \quad (10)$$

Such as k is the expectation of minimal benefit, which he set by the insurance company to be protected from bankruptcy.

We will treat the *(CTV)-minimization* for a form of proportional reinsurance of the "quote part" type using the Pareto probabilistic law.

So we will solve this problem of optimization by a dynamic method based on Augmented Lagrangian and Genetic Algorithms.

In addition, the Augmented Lagrangian method consists in replacing a constrained optimization problem with an unconstrained problem by adding a penalty term to the objective.

We have thus obtained a problem of optimization without constraints which is easy to solve with the Lagrangian method Augmented in convention with the Genetic Algorithms which are developed by the software Matlab which contains programs more efficient and easy to execute.

Case of a "quote part" reinsurance treaty

In the case of proportional reinsurance of the "quote part" type with a proportionality factor $0 < \alpha \leq 1$ constant. The insurer supports the portion $X^A = (1 - \alpha)X$ and the portion $X^R = \alpha X$ transferred to the reinsurer.

Let $\pi(.)$ the principle of premium applied by the insurance company to cover the part of reinsurance.

With

$$\pi(X^R) = \pi\left(\sum_{i=1}^{N} X_i^R\right) = (1+\eta)E(X^R) = \alpha(1+\eta)E(X) \quad (11)$$

From the property of translational invariance and positive homogeneity (because $\Pi_\theta(X, \alpha)$ is coherent) we have:

$$\Pi_\theta(X^A) = \Pi_\theta(f(X,\alpha)) = (1-\alpha)\Pi_\theta(X) = (1-\alpha)\left(E(X) + \sqrt{CTV_\theta(X)}\right)$$
$$\Rightarrow \Pi_\theta(f(X,\alpha)) = (1-\alpha)\left(-E(X) + \sqrt{CTV_\theta(X)}\right) \quad (12)$$

We calculate the mathematical expectation of the technical benefit:
We have

$$E(B(X,\alpha)) = P - (1+\eta)\pi(g(X,\alpha)) - E(f(X,\alpha)) = P - (1+\eta\alpha)E(X) \quad (13)$$

We get the following optimization problem:

(CTV)-minimization:

$$\begin{cases} \min_\alpha \{Z(\alpha) = (1-\alpha)\left(-E(X) + \sqrt{CTV_\theta(X)}\right) - \alpha(1+\eta)E(X)\} \\ s.t \begin{cases} P - (1+\eta\alpha)E(X) \geq k \\ \alpha \in \,]0,1] \end{cases} \end{cases} \tag{14}$$

We calculate now $CTV_\theta(X)$ by the following theorem

Theorem 1. Let X is a random variable that designates an insurance risk, of mathematical expectation $E(X)$. Then

$$CTV_\theta(X) = (VaR_\theta(X) - E(X))^2$$
$$+ 2E[X - VaR_\theta(X)|X \geq VaR_\theta(X)].(E(X^*) - E(X))$$

With X^* is a new random variable with the following density function:

$$f_{X^*}(x) = \frac{S_X(x)}{E\left((X - VaR_\theta(X))_+\right)}, \quad x > VaR_\theta(X)$$

(The proof of this theorem is detailed in Valdez [7]).
We have

$$S_X(VaR_\theta(X)) = 1 - \theta \text{ then } VaR_\theta(X) = S_X^{-1}(1-\theta)$$

We also have

$$E[X - VaR_\theta(X)|X \geq VaR_\theta(X)] = E\left[X - S_X^{-1}(1-\theta)|X \geq S_X^{-1}(1-\theta)\right.$$
$$= \frac{1}{1-\theta} E\left((X - S_X^{-1}(1-\theta))_+\right) \tag{15}$$

Then

$$CTV_\theta(X) = \left(S_X^{-1}(1-\theta) - E(X)\right)^2 + \frac{2}{1-\theta}E\left((X - S_X^{-1}(1-\theta))_+\right).(E(X^*) - E(X))$$

We obtain the following optimization problem:

(CTV)-minimization:

$$\begin{cases} \min_\alpha \left\{Z(\alpha) = (1-\alpha)\left(-E(X) + \sqrt{(S_X^{-1}(1-\theta) - E(X))^2 + \frac{2}{1-\theta}E[(X - S_X^{-1}(1-\theta))_+].(E(X^*) - E(X))}\right) - \alpha(1+\eta)E(X)\right\} \\ s.t \begin{cases} P - (1+\eta\alpha)E(X) \geq k \\ \alpha \in \,]0,1] \end{cases} \end{cases}$$

$$\tag{16}$$

4 Procedure for Optimization by Augmented Lagrangian and Genetic Algorithms

To solve the previously reformulated optimization problems, we use the Augmented Lagrangian Algorithm and the Genetic Algorithms. This is a simple concept that transforms an optimization problem with constraint into a problem or a sequence of optimization problems without constraint. This approach is very often used; it allows to obtain a solution of sufficient quality quickly without having to enter the sophisticated Algorithm of optimization with constraints. The Augmented Lagrangian Algorithm consists in replacing a constrained optimization problem by a series of problems without constraint by adding a penalty term to the objective. In the Augmented Lagrangian, if the problem is subject to inequality constraints, it is sufficient to introduce a deviation variable to transform these constraints into equality constraints and to add a positivity constraint to the deviation variable.

Then the optimization problem is rewritten in the following equivalent form:

$$\begin{cases} \min_\lambda \{Z(\lambda)\} \\ sc \quad E(B(\lambda)) \geq k \end{cases} \Leftrightarrow \begin{cases} \min_\lambda \{Z(\lambda)\}, \quad \varepsilon \geq 0 \\ E(B(\lambda)) - k + \varepsilon = 0 \end{cases} \tag{17}$$

Let the Augmented Lagrangian function $L(\lambda, \rho, r, \varepsilon)$ defined as follows:

$$L(\lambda, \rho, r, \varepsilon) = Z(\lambda) - \rho(h(\lambda) - \varepsilon) - r(h(\lambda) - \varepsilon)^2 \tag{18}$$

With

- $Z(\lambda)$: the objective function to optimize;
- λ: is the reinsurance treaty parameter;
- $h(\lambda) = E(B(\lambda)) - k$: constraint of inequality;
- ρ: is the Lagrange multiplier;
- $r > 0$ is the penalty parameter.

The principle of the method consists in resolving, iteratively, the problem without constraints which minimizes the Augmented Lagrangian function $L(\lambda, \rho, r, \varepsilon)$.

$L(\lambda, \rho, r, \varepsilon)$ should be minimized in relation to λ and to $\varepsilon > 0$, or ε is the deviation variable.

The minimization with respect to ε, for λ fixed, can be carried out analytically to ε. We have more

$$\frac{\partial L}{\partial \varepsilon} = \rho + 2r(h(\lambda) - \varepsilon) = 0 \text{ with } r > 0$$

Thus, the minimum is met for $\varepsilon^* = h(\lambda) + \frac{\rho}{2r}$

We thus reduce to a problem dependent only on variable (λ, ρ, r):

$$L(\lambda, \rho, r) = Z(\lambda) + \frac{\rho^2}{4r} \text{ with } r > 0 \text{ and } \lambda > 0$$

However, the approach of Lagrangian Augmented alone does not generally allow to optimize explicitly in practice and especially in the cases of the most complex problems mathematically and which are difficult to solve algebraically.

This is how we have developed it in this work through a dynamic Algorithm that combines Augmented Lagrangian and Genetic Algorithms.

---

1. Create the Augmented Lagrangian function L ;

2. Initialize the multiplier and the penalty factor $(\rho, r) = (\rho_0, r_0)$;

3. Initialize the reinsurance treaty parameter $\lambda = \lambda_0$;

4. i=0;

5. As long as the stopping criterion is not verified, i.e. $\nabla L(\lambda, \rho, r) > \varepsilon'$, with $\varepsilon' > 0$;

- $i \leftarrow i+1$;

- Create the objective function $L_i(\lambda_i, \rho_i, r_i)$;

6. Create an inner population {of size n };

7. For $j=1$ to n do

- Calculate the probability of selection $P_{ij} \leftarrow \frac{1}{n-1}\left(1 - \frac{L_{ij}}{\sum_{i \in n} L_{ij}}\right)$;

end for

8. Run Classic GA for the objective function $L_i(\lambda_i, \rho_i, r_i)$;

9. Return the result λ^*

---

5 Example

Assume that the total claims burden X Follows a parameter Pareto law $X \sim Pareto(a, x_0)\, a, x_0 \in \mathbb{R}_+^*$.

$$f_X(x, a, x_0) = \begin{cases} \frac{a x_0^a}{x^{a+1}} & if \quad x \geq x_0 \\ 0 & if \quad x_0 < x \end{cases}$$

The distribution function and the survival function are respectively given by:

$$F_X(x, x_0, a) = 1 - \left(\frac{x_0}{x}\right)^a \text{ and } S_X(x, x_0, a) = \left(\frac{x_0}{x}\right)^a$$

with $x \geq x_0 > 0$.

The expectation and variance of X are given respectively by:

$$E[X] = \frac{x_0 a}{a - 1}, \text{if } a > 1 \text{ and } \sigma^2[X] = \frac{x_0^2 a}{(a - 2)(a - 1)}, \text{if } a > 2.$$

We have according to Theorem 1.

$$CTV_\theta(X) = (x_\theta - E(X))^2 + \frac{2}{1 - \theta} E((X - x_\theta)_+).(E(X^*) - E(X))$$

With X^* is a new random variable with the following density function:

$$f_{X^*}(x) = \frac{S_X(x)}{E((X - x_\theta)_+)}, \quad x > x_\theta$$

Let the θ-th quantile such as $x_\theta = VaR_\theta(X)$

$$S_X(x_\theta) = S_X(VaR_\theta(X)) = S_X(S_X^{-1}(1 - \theta)) = 1 - \theta$$

Which implies

$$\left(\frac{x_0}{x_\theta}\right)^a = 1 - \theta$$

Then

$$E((X - x_\theta)_+) = \int_{x_\theta}^\infty S_X(x)dx = \int_{x_\theta}^\infty \left(\frac{x_0}{x}\right)^a dx = \frac{x_\theta}{a - 1}\left(\frac{x_0}{x_\theta}\right)^a$$

From where

$$\frac{2}{1 - q} E((X - x_\theta)_+) = 2\frac{1}{1 - \theta}\frac{x_\theta}{a - 1}\left(\frac{x_0}{x_\theta}\right)^a = \frac{2x_\theta}{a - 1}$$

On the other hand, we compute the density function of the new random variable X^*:

$$f_{X^*}(x) = \frac{S_X(x)}{E((X - x_\theta)_+)} = \frac{\left(\frac{x_0}{x}\right)^a}{\frac{x_\theta}{a - 1}\left(\frac{x_0}{x_\theta}\right)^a} = \frac{(a - 1)x_\theta^{a-1}}{x^{(a-1)+1}}, \quad x > x_\theta$$

This is the Pareto law density of the parameters $(a - 1, x_q) \in \mathbb{R}_+^* \times \mathbb{R}_+^*$.

With

$$E[X^*] = \frac{x_q(a-1)}{a-2}, \quad if \quad a > 2$$

It follows that

$$CTV_\theta(X) = \left(x_\theta - \frac{x_0 a}{a-1}\right)^2 + \frac{2x_\theta}{a-1} \cdot \left(\frac{x_\theta(a-1)}{a-2} - \frac{x_0 a}{a-1}\right)$$

The *(CTV)-minimization* is the next:

$$\begin{cases} \min_\alpha \left\{ Z(\alpha) = -(1-\alpha)\frac{x_0 a}{a-1} + (1-\alpha)\sqrt{\left(x_\theta - \frac{x_0 a}{a-1}\right)^2 + \frac{2x_\theta}{a-1} \cdot \left(\frac{x_\theta(a-1)}{a-2} - \frac{x_0 a}{a-1}\right)} - (1+\eta)\alpha\frac{x_0 a}{a-1} \right\} \\ sc \left/ \begin{array}{l} P - (1+\eta\alpha)\frac{x_0 a}{a-1} \geq k \\ \alpha \in \,]0, 1] \end{array} \right. \end{cases} \tag{19}$$

Numerical application

Consider the following data:

$\theta = 0.05$, $\eta = 2$, $a = 3$, $x_0 = 1$, $P = 57$ et $k = 85$.

Similarly, we applied the Augmented Lagrangian program in agreement with the Genetic Algorithm, we obtained the following result:

De même, nous avons obtenu le résultat suivant:

- The best choice for session rate, $\alpha^* = 0.667$;
- Optimal objective function, $Z(\alpha^*) = -0.146$;
- The minimum Π_θ-measure of technical benefit, $\Pi_\theta(B(\alpha^*)) = P + Z(\alpha^*) = 56.853$;
- The expected technical benefit, $E(B(\alpha^*)) = 53.499$.

6   Conclusion

In this paper, we proposed a reinsurance optimization model, based on the minimization of the Conditional Tail Variance (CTV) risk measures under the constraint of technical benefit which must also be maximal.

This model of optimization depends on the distribution of the claims expenses (insurance risk), the probability threshold, and the safety load factor.

We have created an optimization procedure based on the Augmented Lagrangian method and the Genetic Algorithms to solve the optimization problem of this model.

The result found in this model has shown its efficiency, either in terms of accuracy (improved optimization that acts on both risk and technical benefit), or in terms of its simplicity (effective for optimization problems enormously difficult to solve algebraically).

References

1. Andreani, R., Birgin, E.G., Martínez, J.M., Schuverdt, M.L.: On augmented Lagrangian methods with general lower-level constraint. SIAM J. Optim. **18**, 1286–1302 (2007)
2. Artzner, P.: Application of coherent risk measures to capital requirements in insurance. North Am. Actuarial J. **3**(2), 11–25 (1999)
3. Balbas, A., et al.: Optimal reinsurance with general risk measures. Insur. Math. Econ. **44**, 374–384 (2009)
4. Dbabis, B.: Modèles et méthodes actuarielles pour l'évaluation quantitative des risques en environnement Solvabilité II. Thèse de doctorat, Université Paris Dauphine (2013)
5. Cai, J., Tan, K.S.: Optimal Retention for a stop loss reinsurance under the VaR and CTE risk measures. ASTIN Bull. **37**, 93–112 (2007)
6. Denault, M.: Coherent allocation of risk capital. J. Risk **4**, 1–34 (2001)
7. Valdez, E.: On tail conditional variance and tail covariances. Faculty of Commerce and Economics, University of New South Wales Sydney, Australia (2004)
8. Tan, K.S., Weng, C., Zhang, Y.: Optimality of general reinsurance contracts under CTE risk measure. Insur. Math. Econ. **49**(2), 175–187 (2011)
9. Kaas, R., et al.: Modern Actuarial Risk Theory. Kluwer Academic Publishers, Dordrecht (2001)
10. Hong, L., Elshahat, A.: Conditionnal tail variance and conditionnal tail skewness in finance and insurance. Bradley University (2011)
11. Boudreault, M.: Mathématiques du risque, Document de référence, Département de mathématiques, Université du Québec à Montréal (2010)

Influence of Wall Deformation on a Slip Length

Redouane Assoudi$^{(\boxtimes)}$ ⓘ, Khalid Lamzoud ⓘ,
and Mohamed Chaoui ⓘ

Faculty of Science, Moulay Ismail University, 11201 Meknes, Morocco
r.assoudi@gmail.com

Abstract. This paper presents the effect of a wall deformation on the boundaries conditions of a shear flow of the viscous fluid over a deformable wall which has a periodic deformation and small amplitude. The Reynolds number for the flow over a wall is low and the creeping flow equations apply. The no-slip boundary condition on the deformable wall applies. By using an asymptotic expansion, the analytic expression is obtained for the slip length.

Keywords: Creeping flow · Shear flow · Deformable wall
Slip length · Amplitude

1 Introduction

Modeling fluid flows past a deformable surface is important for the optimal design and fabrication of microfluidic devices whose applications range from medicine to biotechnology [2, 8], and requires some assumption about the nature of the fluid motion (the boundary condition) at the solid interface. One of the simplest boundary conditions is the no-slip condition [1, 5], which dictates that a liquid element adjacent to the surface assumes the velocity of the surface. There is a large literature on surface deformation influence in fluid mechanics. The most of this literature concerns turbulent flow, or of an experimental nature. The question often how microscopic deformation parameters influence macroscopic flow. Many studies showed that deformation surfaces limited a flow, with a no slip boundary condition, behaves as plan surface with a slip condition [3, 7, 9, 10]. Jansons [4] showed that very small amounts of roughness can well approximate a no-slip boundary condition macroscopically. Priezjev [6] by using the (MDS), he investigated the influence of molecularscale surface roughness on the slip behavior in thin liquid films. He has showed that the slip length increases almost linearly with the shear rate for atomically smooth rigid walls and incommensurate structures of the liquid/solid interface.

In this paper, we consider the steady shear flow over a periodically deformable surface. We will use the asymptotic approach to derive the analytical expression of the stream function of the flow generated by the deformation, then we derive the analytic expression of the slip length.

© Springer International Publishing AG, part of Springer Nature 2018
M. Ben Ahmed and A. A. Boudhir (Eds.): SCAMS 2017, LNNS 37, pp. 931–936, 2018.
https://doi.org/10.1007/978-3-319-74500-8_83

2 Materials and Methods

We consider a right-handed system of rectangular Cartesian coordinates $(\tilde{X}, \tilde{Y}, \tilde{Z})$ with unit vectors $(\mathbf{i}_x, \mathbf{i}_y, \mathbf{i}_z)$ is attached to a deformable wall. The wall deformation is defined by the profile $\tilde{Z}_p = a\varepsilon\mathcal{R}(\tilde{X})$, with ε is a dimensionless parameter, it is assumed to be very low to unit $(\varepsilon \ll 1)$ and $\mathcal{R}(\tilde{X})$ is a normalized function, assumed to be periodic with a period $\tilde{L}$, then it is expanded in the $(\tilde{X}, \tilde{Y}, \tilde{Z})$ frame in Fourier series:

$$\mathcal{R}(\tilde{X}) = c_0 + \sum_{n=1}^{\infty} \left(c_n \cos(n\omega\tilde{X}) + s_n \sin(n\omega\tilde{X}) \right) \tag{1}$$

The c_0, c_n and s_n are the Fourier coefficients, and $\omega = 2\pi/\tilde{L}$. We consider a shear flow above this deformable wall as shown in Fig. 1. The flow away from the wall is defined by the linear velocity field $\tilde{\mathbf{V}}_s^{\infty} = k_s\tilde{Z}\mathbf{i}_x$, but near the rough wall their form is no longer linear, that due to the perturbation generated by the roughness. Let $(\tilde{\mathbf{V}}_s, \tilde{P}_s)$ the velocity and pressure field of the flow in the vicinity of the wall.

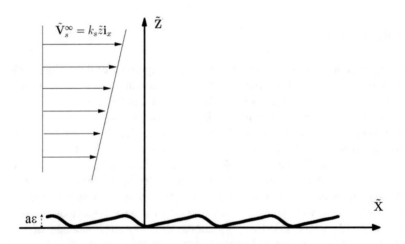

Fig. 1. Shear flow over a deformable wall.

In the right-handed system of rectangular Cartesian coordinates $(\tilde{X}, \tilde{Y}, \tilde{Z})$ the velocity field can be expressed as:

$$\tilde{\mathbf{V}}_s = \tilde{U}_s\mathbf{i}_x + \tilde{w}_s\mathbf{i}_z \tag{2}$$

The field flow $(\tilde{\mathbf{V}}_s, \tilde{P}_s)$ is governed by the Stokes equations by assuming that their Reynolds number is small.

$$\mu_f \tilde{\nabla}^2 \tilde{\mathbf{V}}_s = \tilde{\nabla} \tilde{P}_s$$
$$\tilde{\nabla}.\tilde{\mathbf{V}}_s = 0 \tag{3}$$

With the boundaries conditions:

$$\tilde{\mathbf{V}}_s = 0, \quad \text{on the plan wall } \tilde{Z} = 0$$

$$\tilde{\mathbf{V}}_s = \tilde{\mathbf{V}}_s^\infty, \quad \text{at infinity from the plan wall} \tag{4}$$

In the case of small deformation amplitude $\varepsilon \ll 1$, the velocity and pressure fields can be developed asymptotically as follow:

$$\tilde{\mathbf{V}}_s = \tilde{\mathbf{V}}_s^{(0)} + \varepsilon \tilde{\mathbf{V}}_s^{(1)} + O(\varepsilon^2) \tag{5a}$$

$$\tilde{P}_s = \tilde{P}_s^{(0)} + \varepsilon \tilde{P}_s^{(1)} + O(\varepsilon^2) \tag{5b}$$

With

- $\left(\tilde{\mathbf{V}}_s^{(0)}, \tilde{P}_s^{(0)}\right)$ are the velocity and pressure field of a linear shear flow limited by a smooth (virtual) wall located in $\tilde{Z} = 0$ with a no-slip condition.
- $\varepsilon\left(\tilde{\mathbf{V}}_s^{(1)}, \tilde{P}_s^{(1)}\right)$ are the velocity and pressure of the field ε generated by the wall roughness.

The flows $\left(\tilde{\mathbf{V}}_s^{(0)}, \tilde{P}_s^{(0)}\right)$ and $\left(\tilde{\mathbf{V}}_s^{(1)}, \tilde{P}_s^{(1)}\right)$ are governed by the Stokes equations. By substituting the expressions (5a) and (5b) of velocity and pressure fields of the global flow in the Stokes Eq. (3), by linearity of the Stokes equations, then the separation of terms of order 1 and of order ε, we obtain:
The flow of order 1:

$$\mu_f \tilde{\nabla}^2 \tilde{\mathbf{V}}_s^{(0)} = \tilde{\nabla} \tilde{P}_s^{(0)} \tag{6}$$

$$\tilde{\nabla}.\tilde{\mathbf{V}}_s^{(0)} = 0$$

With the boundary conditions:

$$\tilde{\mathbf{V}}_s^{(0)} = 0, \quad \text{on the plan wall } \tilde{Z} = 0$$

$$\tilde{\mathbf{V}}_s^{(0)} = \tilde{\mathbf{V}}_s^\infty, \quad \text{at infinity from the plan wall} \tag{7}$$

The solutions of the velocity and pressure fields of the order 1, according to the boundary conditions, are trivial:

$$\tilde{\mathbf{V}}_s^{(0)} = \tilde{\mathbf{V}}_s^\infty = \tilde{k}_s \tilde{Z} \mathbf{i}_x \tag{8a}$$

$$\tilde{P}_s^{(0)} = 0 \tag{8b}$$

For the flow of order ε, we must first find the boundary conditions. On the deformable wall, whose profile is described by the equation $\tilde{Z}_p = a\varepsilon\mathcal{R}(\tilde{X})$, we use the Taylor expansion in $\tilde{Z} = 0$ to express the velocity vector $\tilde{\mathbf{V}}_s$. By using the boundary conditions (7) of the flow $(\tilde{\mathbf{V}}_s^{(0)}, \tilde{P}_s^{(0)})$, we find the expression of the flow $(\tilde{\mathbf{V}}_s^{(1)}, \tilde{P}_s^{(1)})$ on a virtual plane $\tilde{Z} = 0$:

$$\left[\tilde{\mathbf{V}}_s^{(1)}\right]_{\tilde{Z}=0} = -a\mathcal{R}(\tilde{X})\left[\frac{\partial \tilde{\mathbf{V}}_s^{(0)}}{\partial \tilde{Z}}\right]_{\tilde{Z}=0} = -a\mathcal{R}(\tilde{X})\tilde{k}_s \mathbf{i}_x \tag{9}$$

The boundary conditions at infinity from the wall of the perturbed velocity field $\tilde{\mathbf{V}}_s$ and unperturbed velocity field $\tilde{\mathbf{V}}_s^{(0)}$ give the expression of the velocity field of order $\varepsilon\tilde{\mathbf{V}}_s^{(1)}$ at infinity $(\tilde{Z} \to \infty)$;

$$\tilde{\mathbf{V}}_s^{(1)} \to 0 \qquad \tilde{Z} \to \infty \tag{10}$$

The condition (9) appears as a tangential velocity which varies with the position $\tilde{X}$ on the virtual plane wall localized in $\tilde{Z} = 0$, and it is analogous to a Navier slip condition with length slip.

3   Stream Function of Order ε and Slip Length

The order ε solution with conditions (9) and (10) is searched for in terms of a stream function $\tilde{\psi}_s^{(1)}$. Following [3], an appropriate solution is written in the form:

$$\tilde{\psi}_s^{(1)} = d_0 Z + f_0 + \sum_{n=1}^{\infty}\left[(d_n Z + f_n)\cos(n\omega\tilde{X}) + (e_n Z + g_n)\sin(n\omega\tilde{X})\right]\exp(-n\omega\tilde{Z}) \tag{11}$$

By using the boundary conditions, we find the expressions of the coefficients d_0, f_0, d_n, f_n, e_n and g_n:

$$d_0 = -c_0 a\tilde{k}_s$$

$$d_n = -a\tilde{k}_s c_n$$

$$f_n = 0$$

$$e_n = -a\tilde{k}_s s_n$$

$$g_n = 0$$

According to Hocking [3], the value $\beta = -\varepsilon d_0/ak_s = \varepsilon c_0$ determines the dimensionless slip length, which means that the rough wall behaves like a plan wall located in $\tilde{Z} = 0$ with a slip condition

$$\tilde{U}_s = -a\beta \frac{\partial \tilde{U}_s}{\partial \tilde{Z}} \tag{12}$$

And then, the flow away from the rough surface can be written as follow

$$\tilde{U}_s \sim (\tilde{Z} - a\beta)k_s \tag{13}$$

The above expression (12) shows the minus sign, this sign depends on the reference plane $\tilde{Z} = 0$ where it is located with respect to the wall deformation. In our model, it is at the bottom of the deformation, so the minus sign indicates that the equivalent plane surface with adhesion condition is located above the plane wall $\tilde{Z} = 0$, between peaks and valleys, with a dimensional slip length $\beta = \varepsilon c_0$.

This value of dimensionless slip length can be calculated also by using the expression giving by Tuck and Kouzoubov [9] for the case of a shear flow near a deformable wall with a periodic deformation:

$$\beta_{TK} = -\varepsilon^2 k \sum_{n=1}^{\infty} n(a_n^2 + b_n^2) \tag{14}$$

With a_n, b_n are the Fourier series coefficients and k is the wavenumber. The reference plane $\tilde{Z} = 0$ chosen by Tuck and Kouzoubov [9] is located at the roughness average position (Table 1).

Table 1. Slip length in terms the amplitude of the wall deformation.

ε	β	β_{Tk}
0.01	0.005	0.005
0.03	0.015	0.015
0.05	0.025	0.025
0.07	0.035	0.035
0.09	0.045	0.045
0.1	0.05	0.049
0.2	0.1	0.097

4 Conclusions

In this paper the effect of the wall deformation on the boundaries conditions, in the case of the shear flow limited by the deformable wall, is investigated by solving the Stokes equations. Using the asymptotic expansion, the analytical expression of the slip length is given. The results show that the no slip velocity condition on the deformable wall turn into the slip velocity condition on the virtual plan wall located at $\tilde{Z} = 0$ with the slip length that is calculated for the no-symmetric deformation. The results show also that slip length is dependent on the deformation amplitude ε of the wall.

References

1. Batchelor, G.K.: An Introduction to Fluid Dynamics. Cambridge University Press, Cambridge (2000)
2. Beebe, D.J., Mensing, G.A., Walker, G.M.: Physics and applications of microfluidics in biology. Ann. Rev. Biomed. Eng. **4**(1), 261–286 (2002)
3. Hocking, L.: A moving fluid interface on a rough surface. J. Fluid Mech. **76**(4), 801–817 (1976)
4. Jansons, K.M.: Determination of the macroscopic (partial) slip boundary condition for a viscous flow over a randomly rough surface with a perfect slip microscopic boundary condition. Phys. Fluids **31**(1), 15–17 (1988)
5. Lamb, H.: Hydro Dynamics. 6th ed. 738. Dover, New York (1932)
6. Priezjev, N.V.: Effect of surface roughness on rate-dependent slip in simple fluids. J. Chem. Phys. **127**(14), 144708 (2007)
7. Priezjev, N.V., Darhuber, A.A., Troian, S.M.: Slip behavior in liquid films on surfaces of patterned wettability: comparison between continuum and molecular dynamics simulations. Phys. Rev. E **71**(4), 041608 (2005)
8. Squires, T.M., Quake, S.R.: Microfluidics: fluid physics at the nanoliter scale. Rev. Mod. Phys. **77**(3), 977 (2005)
9. Tuck, E., Kouzoubov, A.: A laminar roughness boundary condition. J. Fluid Mech. **300**, 59–70 (1995)
10. Vinogradova, O.I., Yakubov, G.E.: Surface roughness and hydrodynamic boundary conditions. Phys. Rev. E **73**(4), 045302 (2006)

Existence of Sign Changing Radial Solutions for Elliptic Equation Involving the p-Laplacian on Exterior Domains

Boubker Azeroual[1]([⊠])[iD] and Abderrahim Zertiti[2]

[1] Faculty of Science, Abedelmalek Esaadi University,
BP 2121, 93000 Tetouan, Morocco
azeroualaboubaker@gmail.com

[2] Laboratoire d'Analyse Fonctionnel non Linèaire Application à la physique thèorique et la thèorie de la dynamique des populations,
Université Abedelmalek Essaadi, BP 2121, 93000 Tetouan, Morocco

Abstract. In this paper we prove the existence of radial solutions having a prescribed number of sign change to the p-Laplacian $\Delta_p u + f(u) = 0$ on exterior domain of the ball of radius $R > 0$ centred at the origin in $\mathbb{R}^N$. The nonlinearity f is odd and behaves like $|u|^{q-1}u$ when u is large with $1 < p < q+1$ and $f < 0$ on $(0, \beta)$, $f > 0$ on (β, ∞) where $\beta > 0$. The method is based on a shooting approach, together with a scaling argument.

Keywords: Exterior domain · p-Laplacian
Sign changing radial solution

1 Introduction

In this paper we deal with the existence and multiplicity of classical radial sign-changing solutions to the Dirichlet boundary problem involving the p-Laplacian

$$\Delta_p u + f(u) = 0 \quad \Omega, \tag{1}$$

$$u = 0 \, \partial\Omega, \tag{2}$$

$$\lim_{|x|\to\infty} u(x) = 0. \tag{3}$$

Where $\Omega = \{x \in \mathbb{R}^N| \ |x| > R\}$ is the complement of the ball of radius $R > 0$ centred at the origin with $|x| = \sqrt{x_1^2 + x_2^2 + \ldots + x_N^2}$ is the standard norm of $\mathbb{R}^N$. Also, $\Delta_p u$ is the p-Laplacian of the function u with $\Delta_p u = div\left(|\nabla u|^{p-2}\nabla u\right)$. We will assume henceforth that the function f satisfies the following hypotheses:

(H1) $f : \mathbb{R} \to \mathbb{R}$ is odd and locally Lipschitzian,
(H2) $f(u) = |u|^{q-1}u + g(u)$ with $1 < p < q+1$ and

$$\lim_{|u|\to\infty} \frac{|g(u)|}{|u|^q} = 0,$$

© Springer International Publishing AG, part of Springer Nature 2018
M. Ben Ahmed and A. A. Boudhir (Eds.): SCAMS 2017, LNNS 37, pp. 937–960, 2018.
https://doi.org/10.1007/978-3-319-74500-8_84

(H3) There exists $\beta > 0$ such that

$$f(0) = f(\beta) = 0 \text{ where } f < 0 \text{ on } (0, \beta), f > 0 \text{ on } (\beta, \infty),$$

(H4) If $p > 2$ we also assume for some $\eta > 0$

$$\int_0^\eta \frac{1}{|F(u)|^{\frac{1}{p}}} \, du = \infty,$$

where $F(u) = \int_0^u f(s) \, ds$.

As a consequence of the previous assumptions we have:

(i) $F(u) \to \infty$ as $|u| \to \infty$. F is even and bounded below by some $-F_0 < 0$ on $\mathbb{R}$, i.e.

$$F(u) \geq -F_0 \quad \forall u \in \mathbb{R}. \tag{4}$$

(ii) F is strictly increasing in (β, ∞) and decreasing in $(0, \beta)$.

(iii) F has a unique positive zero, $\gamma > \beta$ and $F < 0$ on $(0, \gamma)$, $F > 0$ on (γ, ∞).

Remark 1. If $1 < p \leq 2$ it follows from the fact that f is locally Lipschitzian the assumption (H4) also holds.

Indeed, by (H1), f is locally Lipschitzian and $f(0) = 0$, then there exists $L > 0$ and $\alpha > 0$ such that $|f(u)| \leq L|u|$, for all $|u| \leq \alpha \leq \eta$. As F is even we deduce that $|F(u)| \leq \frac{L}{2}u^2$, then it follows that

$$\int_0^\alpha \frac{1}{|F(u)|^{\frac{1}{p}}} \, du \geq \left(\frac{2}{L}\right)^{\frac{2}{p}} \int_0^\alpha \frac{1}{u^{\frac{2}{p}}} \, du = \infty.$$

Radial symmetric solutions to (1)–(3) satisfy the problem

$$\left(r^{N-1}\Phi_p(u')\right)' + r^{N-1}f(u) = 0 \quad R < r, \tag{5}$$

$$u(R) = 0 \qquad u(r) \to 0 \qquad r \to \infty, \tag{6}$$

where $\Phi_p(s) = |s|^{p-2}s$. Also $'$ denotes the derivative with respect to $r = |x| \geq 0$ with $x \in \mathbb{R}^N$ and for radial functions as it is usual we shall write $u(x) = u(r)$. We note that Φ_p is odd and differentiable on $\mathbb{R} \setminus \{0\}$ with $\Phi_p'(s) = (p-1)|s|^{p-2}$ and $\Phi_p^{-1} = \Phi_{p'}$, where p' the Hölder conjugate exponent of p.

We will be interested only in classical solutions of (5)–(6) i.e., $u \in C^1([R, \infty), \mathbb{R})$ and $\Phi_p(u') \in C^1([R, \infty), \mathbb{R})$.

The research of radial solution of elliptic equations with zero Dirichlet boundary conditions (1)–(3) with the usual Laplace operator ($p = 2$) has widely studied by many authors via variational methods when Ω is bounded domain or the whole space $\mathbb{R}^N$, under different regularity and growth assumptions of nonlinearity of f for instance see [1–4], exploring the symmetry of the problem (1)–(3) to prove the existence of infinity radial solution by means the Mountain pass theorem. In particular when Ω is a ball and is f non increasing by an other argument well-know plane method to prove the existence and multiplicity of radial solution to this problem see [5]. However, these arguments are quite difficult and provide no specific information of qualitative properties. Then it was an open question as to whether solutions exist with prescribed number of zeros. Jones and Küpper in [7] addressed this question using a dynamical systems approach and an application of the Conley index [9]. In [10] Mcleod et al. established the existence of sign changing bound state solutions by using the shooting techniques and a scaling argument when f satisfies appropriate sign conditions and is of subcritical growth. The author Pudipeddi [11] extended the previous result for the p-Laplacian where $1 < p < N$ and $\Omega = \mathbb{R}^N$ using the same approach when f is locally Lipschitz and odd and behaves like $|u|^{q-1}u$ for u sufficiently large with $p < q + 1 < \frac{Np}{N-p}$.

Recently on exterior domain, there has been an interest in studying this question if $p = 2$ we mention as instances [6,8]. Here we use the shooting argument and a ' simple ' ordinary differential equation proofs to etablisch that (1)–(3) has an infinite number of radial solutions with a prescribed number of zeros.

Our paper is organized as follows: in Sect. 2 we begin to establish some preliminary results concerning the existence and proprieties of radial solutions. In Sect. 3 we show that there are solutions with arbitrarily number large of zeros by using a scaling argument and finally, we shall prove the following our main theorem.

Theorem 1. *Assuming (H1)–(H4) and $N \geq 2$. Then for each nonnegative integer k, there exist two radially symmetric solutions u_k and v_k of problem (1)–(3) which have exactly k zeros on (R, ∞) such that $v'_k(R) < 0 < u'_k(R)$.*

2   Preliminaries

To deal with the problem (5)–(6), we will use a shooting method and consider the initial value problem

$$\left(r^{N-1}\Phi_p(u')\right)' + r^{N-1}f(u) = 0 \quad \text{if} \quad r > R, \tag{7}$$

$$u(R) = 0 \quad \text{and} \quad u'(R) = a > 0. \tag{8}$$

To emphasize the dependence of the solution to (7)–(8) in the shooting parameter a we will denote it u_a.

Lemma 1. *Assume (H1) and (H2) hold. Then (7)–(8) has a unique solution u_a defined on interval $[R, \infty)$. Moreover, $a \to u_a$, $a \to u'_a$ are continuous on $(0, \infty)$.*

Proof. Let u be a solution of (7)–(8) and integrating (7) on $[R, r]$ we obtain

$$r^{N-1}\Phi_p(u') = R^{N-1}a^{p-1} - \int_R^r t^{N-1}f(u)\, dt. \tag{9}$$

We rewrite this as

$$u'(r) = \left(\frac{R}{r}\right)^k \Phi_{p'}\left(a^{p-1} - \int_R^r \left(\frac{t}{R}\right)^{N-1} f(u)\, dt\right), \tag{10}$$

where $k = \frac{N-1}{p-1}$. Integrating (10) on $[R, r]$ we obtain

$$u(r) = \int_R^r \left(\frac{R}{t}\right)^k \Phi_{p'}\left(a^{p-1} - \int_R^t \left(\frac{s}{R}\right)^{N-1} f(u)\, ds\right) dt.$$

Let $\epsilon > 0$. Denote $C^0[R, R + \epsilon]$ the Banach space of real continuous functions on $[R, R + \varepsilon]$ endowed with the sup norm $\| \ \|$. Let $a, \delta_0 > 0$ are fixed such that $\delta_0 < a^{p-1}$ and we define the complete metric space by

$$E := \{(u, v) \in \left(C^0[R, R + \epsilon]\right)^2 \ | \ u(R) = 0, \ v(R) = a^{p-1}\},$$

endowed with the distance $d(x_1, x_2) = max(\|u_1 - u_2\|, \|v_1 - v_2\|)$ where $x_i = (u_i, v_i)$ for $i = 1, 2$.

Denote $B_\delta^\epsilon(a) := \{(u, v) \in E \ | \ d\left((u, v), (0, a^{p-1})\right) \leq \delta\}$ the closed ball of (E, d).

Existence and uniqueness for(7)–(8) result from the study of fixed point of the application $\mathfrak{F}_a : (u, v) \in E \to (\tilde{u}, \tilde{v}) \in E$ where $\tilde{u}$ and $\tilde{v}$ are defined by

$$\tilde{u}(r) = \int_R^r \left(\frac{R}{t}\right)^k \Phi_{p'}(v)\, dt,$$

$$\tilde{v}(r) = a^{p-1} - \int_R^r \left(\frac{t}{R}\right)^{N-1} f(u)\, dt.$$

We will show that $\mathfrak{F}_a$ is a contraction mapping of $B_\delta^\epsilon(a)$ into itself for ϵ, δ small enough.

For all $(u, v) \in B_\delta^\epsilon(a)$ and $r \in [R, R + \epsilon]$ we have

$$|\tilde{u}(r)| \leq \int_R^r \left(\frac{R}{t}\right)^k |\Phi_{p'}(v)|\, dt,$$

$$\leq \epsilon \|v\|^{p'-1},$$

$$\leq \epsilon \left(\delta_0 + a^{p-1}\right)^{p'-1}.$$

Therefore for ϵ small enough we have

$$\|\tilde{u}\| \leq \epsilon \left(\delta_0 + a^{p-1}\right)^{p'-1} \leq \delta.$$

Furthermore therefore

$$|\tilde{v}(r) - a^{p-1}| \leq \int_R^r \left(\frac{t}{R}\right)^{N-1} |f(u(t))|\, dt,$$

$$\leq M \int_R^{R+\epsilon} \left(\frac{t}{R}\right)^{N-1} dt,$$

$$\leq \frac{MR}{N}\left((1 + \frac{\epsilon}{R})^N - 1\right) \leq \delta,$$

where $M = \sup\limits_{|s| \le \delta_0} |f(s)|$ and for ϵ small enough. Then it follows that $\|\tilde{v} - a^{p-1}\| \le \delta$, which implies that $(\tilde{u}, \tilde{v}) \in B_\delta^\epsilon(a)$, for ϵ small enough.

Now, let $x_i = (u_i, v_i) \in B_\delta^\epsilon(a)$ for $i = 1, 2$ then

$$d\Big(\mathfrak{F}_a(x_1), \mathfrak{F}_a(x_2)\Big) = max\Big(\|\tilde{u}_1 - \tilde{u}_2\|; \|\tilde{v}_1 - \tilde{v}_2\|\Big).$$

For $r \in [R, R+\epsilon]$ fixed, thanks to the mean value theorem we obtain

$$\Phi_{p'}(v_1(r)) - \Phi_{p'}(v_2(r)) = \Phi'_{p'}(w)(v_1(r) - v_2(r)), \tag{11}$$

where $w = \alpha\, v_1(r) + (1-\alpha)\, v_2(r)$ for some $0 < \alpha < 1$ and $\Phi'_{p'}(w) = (p'-1)\,|w|^{p'-2}$. As $\|v_i - a^{p-1}\| \le \delta \le \delta_0$, then for each $i = 1, 2$ we have

$$a^{p-1} - \delta_0 \le v_i(t) \le a^{p-1} + \delta_0 \quad \forall t \in [R, R+\epsilon],$$

therefore there exists two constants $c_1 = a^{p-1} - \delta_0 > 0$ and $c_2 = a^{p-1} + \delta_0 > 0$ and for all $i = 1, 2$ we see that

$$c_1 \le v_i(t) \le c_2, \quad \forall t \in [R, R+\epsilon].$$

Then it follows that

$$(p'-1)\,|w|^{p'-2} \le C_p,$$

where $C_p = (p'-1)\,c_2^{p'-2}$ if $1 < p \le 2$ and $C_p = (p'-1)\,c_1^{p'-2}$ if $p > 2$. Further from (11) we have

$$\|\Phi_{p'}(v_1) - \Phi_{p'}(v_2)\| \le C_p\|v_1 - v_2\|.$$

Therefore

$$\|\tilde{u}_1 - \tilde{u}_2\| \le \epsilon\, C_p\|v_1 - v_2\|. \tag{12}$$

On the other hand we see that

$$|\tilde{v}_1(r) - \tilde{v}_2(r)| \le \int_R^r \left(\frac{t}{R}\right)^{N-1} |f(u_1) - f(u_2)|\, \mathrm{d}t.$$

By virtue of (H1), then there exists a constant $K > 0$ and for each $|s|$ and $|t| \le \delta_0$ we have
$$|f(s) - f(t)| \le K|s - t|.$$

Since $\|u_i\| \le \delta \le \delta_0$ for $i = 1, 2$ then it follows that

$$\|\tilde{v}_1 - \tilde{v}_2\| \le \lambda(\epsilon)\|u_1 - u_2\|,$$

where $\lambda(\epsilon) = \frac{KR}{N}\Big((1 + \frac{\epsilon}{R})^N - 1\Big)$. Thus, from (12) we have

$$d\Big(\mathfrak{F}_a(x_1), \mathfrak{F}_a(x_2)\Big) \le max\Big(\epsilon\, C_p, \lambda(\epsilon)\Big)\, d(x_1, x_2).$$

Since $\lambda(\epsilon) \to 0$ as $\epsilon \to 0$. Thus by the contraction mapping principle it follows that for ϵ small enough $\mathfrak{F}_a$ has a unique fixed point denoted $x_a = (u_a, v_a) \in B_\delta^\epsilon(a)$ such that

$$u_a(r) = \int_R^r \left(\frac{R}{t}\right)^k \Phi_{p'}(v_a) \, \mathrm{d}t$$

$$v_a(r) = a^{p-1} - \int_R^r \left(\frac{t}{R}\right)^{N-1} f(u_a) \, \mathrm{d}t.$$

Therefore $u_a, v_a \in C^1([R, R+\varepsilon], \mathbb{R})$, in addition

$$u_a' = \left(\frac{R}{r}\right)^k \Phi_{p'}(v_a) \quad \text{and} \quad \Phi_p(u_a') = \left(\frac{R}{r}\right)^{N-1} v_a$$

then it follows that $\Phi_p(u_a') \in C^1([R, R+\varepsilon], \mathbb{R})$. Hence for some $\epsilon > 0$, the problem (7)–(8) has a unique classical solution u_a defined on the interval $[R, R+\varepsilon]$.

Next, we will show that the solution u_a can be extended on $[R, +\infty)$, we define the energy of solution to (7)–(8) as

$$E_a(r) = \frac{|u_a'|^p}{p'} + F(u_a) \quad \forall r \geq R. \tag{13}$$

Differentiating E_a and using (7) gives

$$E_a'(r) = -\frac{N-1}{r} |u_a'|^p \leq 0. \tag{14}$$

Which implies E_a is non-increasing on $[R, +\infty)$, so

$$\frac{|u_a'|^p}{p'} + F(u_a) \leq \frac{a^p}{p'}.$$

From (4) we have

$$|u_a'| \leq \left(a^{p-1} + p' F_0\right)^{\frac{1}{p}} = M_a. \tag{15}$$

Then we have $|u_a'|$ is uniformly bounded on wherever it is defined, as $u_a(R) = 0$ thus it follows that $|u_a|$ is uniformly bounded on wherever it is defined. Hence the existence on all of $[R, +\infty)$ follows.

Now, we will show the continuous dependence of solutions on initial conditions. For this let $a > 0$ and $a_j \to a$ as $j \to \infty$ and denote $u_j(r) = u_{a_j}(r)$ for all j. In the following we shall prove that $u_j \to u_a$ and $u_j' \to u_a'$ as $j \to \infty$ on compact subsets of $[R, \infty)$.

Indeed, as the sequence (a_j) is bounded by some $A > 0$ and from (15) for all j, we get

$$|u_j'(r)| \leq \left(A^{p-1} + p' F_0\right)^{\frac{1}{p}} = M_2 \quad \forall r \geq R.$$

Therefore $u_j'(r)$ is uniformly bounded. Next, we will show that $u_j(r)$ is uniformly bounded.

Suppose by the way of contradiction that there exists a sequence $r_j \geq R$ such that $|u_j(r_j)| \to \infty$ as $j \to \infty$. Since $F(u) \to \infty$ as $|u| \to \infty$ then $F(u_j(r_j)) \to \infty$ as $j \to \infty$. As

$$E_{a_j}(r_j) = E_j(r_j) \geq F(u_j(r_j)).$$

Then $E_j(r_j) \to \infty$ as $j \to \infty$. Since E_j is non-increasing and (a_j) is bounded then we have

$$E_j(r_j) \leq E_j(R) = \frac{a_j^p}{p'} \leq \frac{A^p}{p'} < \infty.$$

Which contradict to $E_j(r_j) \to \infty$ as $j \to \infty$. Thus there exists $M_1 > 0$ and $M_2 > 0$ for all j such that

$$|u_j(r)| \leq M_1 \quad \text{and} \quad |u_j'(r)| \leq M_2 \quad \forall r \geq R.$$

Which implies that u_j and u_j' are uniformly bounded and equicontinuous. Then by Arzela-Ascoli's theorem there exists subsequence still denoted u_j such that $u_j(r) \to u(r)$ as $j \to \infty$ uniformly on compact subsets of $[R, \infty)$. Therefore it follows from (H1) that $f(u_j) \to f(u)$ as $j \to \infty$ uniformly on compact subsets of $[R, \infty)$ and since $a_j \to a$, then we get

$$w_j(r) = a_j^{p-1} - \int_R^r \left(\frac{t}{R}\right)^{N-1} f(u_j(t))\, dt$$

Then

$$w_j(r) \to a^{p-1} - \int_R^r \left(\frac{t}{R}\right)^{N-1} f(u(t))\, dt = w(r),$$

uniformly on compact subsets of $[R, \infty)$ as $j \to \infty$. Which implies that $\Phi_{p'}(w_j')$ converges to $\Phi_{p'}(w)$ uniformly on compact subsets of $[R, \infty)$. By virtue of (10) we obtain

$$u_j'(r) = \left(\frac{R}{r}\right)^k \Phi_{p'}(w_j'(r)) \to \left(\frac{R}{r}\right)^k \Phi_{p'}(w(r)) = v(r) \quad \text{as } j \to \infty,$$

uniformly on compact subsets of $[R, \infty)$.

Furthermore from

$$u_j(r) = \int_R^r u_j'(t)\, dt \to \int_R^r v(t)\, dt \quad as \ j \to \infty,$$

pointwise on $[R, \infty)$ and $u_j(r) \to u(r)$, therefore it follows that u is differentiable and $u' = v$. Hence $u_j \to u$ and $u_j' \to u'$ uniformly on compact subsets of $[R, \infty)$ and finally, $a \to u_a$, $a \to u_a'$ are continuous on $(0, \infty)$. This completes the proof of Lemma 1. □

Remark 2. It is immediate that $u_a \in C^2([R, R+\varepsilon])$ for $1 < p \leq 2$ and u is C^2 only at the point $r \in [R, R+\varepsilon]$ such that $u_a'(r) \neq 0$ for $p > 2$.

Proposition 1. *Assume (H1)–(H3) hold and u_a be solution of (7)–(8).*

(i) *Then $r \to E_a(r)$ is non-increasing and $\lim_{r\to\infty} E_a(r)$ is finite. In addition, if there exists $r_0 > R$ such that $E_a(r_0) < 0$ then $u_a > 0$ on $[r_0, \infty)$.*

(ii) *If $\lim_{r\to\infty} u_a(r) = \ell$ exists, then ℓ is a zero of f. Moreover*

$$\lim_{r\to\infty} u'_a(r) = 0 \quad \text{and} \quad \lim_{r\to\infty} E_a(r) = F(\ell).$$

(iii) *Then $u'_a > 0$ on a maximal nonempty open interval (R, M_a) where either*

   (a) $M_a = \infty$, $\lim_{r\to\infty} u_a(r) = \beta$ *and* $0 \le u_a(r) < \beta$ *for all* $r \ge R$.

   or

   (b) M_a *is finite and* $u_a(M_a) \ge \beta$.

Proof. By virtue of (14) we have E_a is non-increasing and since $E_a(r) \ge -F_0$ implying the existence of $\lim_{a\to\infty} E_a = \zeta_a$ and is finite. Now, if there exists $r_0 > R$ such that $E_a(r_0) < 0$ by monotonicity of E_a we have $E_a(r) \le E_a(r_0) < 0$ for $r \ge r_0$. By contradiction we suppose that there exists $r_1 > r_0$ zero of u_a, then it follows that $E_a(r_1) = \frac{|u'_a|^p}{p'} \ge 0$, which contradicts to $E_a(r_1) < 0$. Hence $u_a > 0$ on $[r_0, \infty)$.

For (ii) we suppose that $\lim_{r\to\infty} u_a(r) = \ell$, then by continuity of F we have

$\lim_{r\to\infty} F(u_a(r)) = F(\ell)$, so from (13) we have $\lim_{r\to\infty} |u'_a(r)| = \left(p'(\xi_a - F(\ell))\right)^{\frac{1}{p}} = m$. Assume to the contrary that $m > 0$, then for $0 < \epsilon < m$ there exists $r_0 > R$ such that $|u'_a| \ge m - \epsilon > 0$ for each $r \ge r_0$. Thus it follows that $|u_a(r) - u_a(r_0)| \ge (m-\epsilon)(r-r_0)$ which implies that $\lim_{r\to\infty} u_a(r) = \infty$. This is a contradiction. Hence $\lim_{r\to\infty} u'_a(r) = 0$ and $\zeta_a = F(\ell)$.

By (10) and applying L'Hôpital's rule we obtain

$$
\begin{aligned}
0 &= \lim_{r\to\infty} \frac{\Phi_p(u'_a(r))}{r} \\
&= \lim_{r\to\infty} \frac{R^{N-1}a^{p-1}}{r^N} - \frac{\int_R^r t^{N-1} f(u_a)\,dt}{r^N} \\
&= -\lim_{r\to\infty} \frac{r^{N-1} f(u_a(r))}{N r^{N-1}} \\
&= -\frac{f(\ell)}{N}.
\end{aligned}
$$

Therefore $f(\ell) = 0$.

Next for (iii), As $u'_a(R) = a > 0$ and by continuity then, there exists $\epsilon > 0$ such that $u'_a > 0$ on $(R, R + \epsilon)$. Denote (R, M_a) a maximal nonempty open interval where $u'_a > 0$. If $M_a = \infty$ then $u_a > 0$ is increasing and bounded above on $[R, \infty)$, therefore it follows that $\lim_{r\to\infty} u_a(r) = \ell$ exists. By virtue of (ii) we have $f(\ell) = 0$ and $\lim_{r\to\infty} u'_a(r) = 0$. Thus $\ell = \beta$ and $0 \le u_a < \beta$ on $[R, \infty)$. Hence (a) is proven.

For (b), if $M_a < \infty$ we must have $u'_a(M_a) = 0$ and $u'_a > 0$ on (r, M_a) for $R < r < M_a$. Assume to contrary that $0 < u_a(M_a) < \beta$ then by (H3) it follows

$f(u_a) < 0$ on (r, M_a). Integrating (7) on (r, M_a) and using the fact $u_a'(M_a) = 0$ we get

$$r^{N-1}\Phi_p(u_a'(r)) = \int_r^{M_a} t^{N-1} f(u_a(t))\,dt < 0.$$

Which implies that $u_a' < 0$ on (r, M_a), this is a contradiction. Thus $u_a(M_a) \geq \beta$. Which completes the proof of Proposition 1. $\qquad\square$

Lemma 2. *Assume* (H1)–(H3) *hold. Then*

1. *u_a has a maximum at $M_a > R$ for a sufficiently large. Moreover $|u_a|$ has a global maximum at M_a.*
2. *$\lim_{a\to\infty} u_a(M_a) = \infty$ and $\lim_{a\to\infty} M_a = R$.*

Proof. For (1), we suppose by the way of contradiction that $M_a = \infty$. Then $u_a' > 0$ on $[R, \infty)$ and by *(ii)–(iii)* of Proposition 1 we have $0 \leq u_a(r) < \beta$, $u_a(r) \to \beta$ and $u_a'(r) \to 0$ as $r \to \infty$.

Let $y_a = \frac{u_a(r)}{a}$. It is straightforward using (7)–(8) to show

$$\left(r^{N-1}\Phi_p(y_a')\right)' + r^{N-1}\frac{f(ay_a)}{a^{p-1}} = 0 \quad \forall r > R, \tag{16}$$

$$y_a(R) = 0 \quad \text{and} \quad y_a'(R) = 1. \tag{17}$$

Then it follows that

$$\left(\frac{|y_a'|^p}{p'} + \frac{F(ay_a)}{a^{p-1}}\right)' = -\frac{N-1}{r}|y_a'|^p \leq 0.$$

Therefore

$$\frac{|y_a'|^p}{p'} + \frac{F(ay_a)}{a^{p-1}} \leq \frac{1}{p'}.$$

By virtue of (4) and for a sufficiently large we obtain

$$\frac{|y_a'|^p}{p'} \leq \frac{1}{p'} + \frac{F_0}{a^{p-1}} \leq \frac{1}{p'} + \frac{1}{p} = 1.$$

As $0 < y_a < \frac{\beta}{a}$ then for a sufficiently large we get $0 < y_a < 1$. It follows that $|y_a|$ and $|y_a'|$ are uniformly bounded if a is sufficiently large. Then by the Arzela-Ascoli theorem we deduce that $y_a \to y$ uniformly as $a \to \infty$ on the compact sets of $[R, \infty)$, for some subsequence denoted the same by y_a with y is a continuous function on $[R, \infty)$ where $y(R) = 0$.

Integrating (16) on $[R, r]$ gives

$$\Phi_p(y_a') = \left(\frac{R}{r}\right)^{N-1} - \int_R^r \left(\frac{t}{r}\right)^{N-1} \frac{f(ay_a)}{a^{p-1}}\,dt.$$

Since ay_a is bounded and f is continuous then $\frac{f(ay_a)}{a^{p-1}} \to 0$ uniformly on $[R, \infty)$ as $a \to \infty$. Therefore it follows that $\Phi_p(y_a'(r))$ converges to $\left(\frac{R}{r}\right)^{N-1}$ uniformly on compact subsets of $[R, \infty)$. Which implies that y_a' converges uniformly as $a \to \infty$ on compact subsets of $[R, \infty)$ to a continuous function denoted $z(r) = \left(\frac{R}{r}\right)^k$ with $k = \frac{N-1}{p-1}$. Moreover z' exists and is continuous.

Furthermore from

$$y_a(r) = \int_R^r y_a'(t) \, dt$$

Letting $a \to \infty$, we obtain that

$$y(r) = \int_R^r z(t) \, dt.$$

Then y_a is continuously differentiable and $y' = z$, thus $y_a' \to y'$ uniformly as $a \to \infty$ on the compact subsets of $[R, \infty)$. Since $0 < y_a < \frac{\beta}{a}$ so $y_a \to 0$ as $a \to \infty$, then $y \equiv 0$ would imply that $y' \equiv 0$. Which contradicts to $y'(R) = 1$. Thus u_a has a local maximum at $M_a > R$.

Next, we will show that $|u_a|$ has a global maximum at M_a. Otherwise, suppose that there exists $r_1 > M_a$ such that $|u_a(r_1)| > |u_a(M_a)|$. From (iii) of Proposition 1 we have $u_a(M_a) \geq \beta$, since F is even and increasing on (β, ∞) therefore it follows

$$E_a(r_1) = F(u_a(r_1)) = F(|u_a(r_1)|) > F(u_a(M_a)) = E_a(M_a).$$

By the monotonicity of E_a we have $r_1 \leq M_a$, which contradicts to $r_1 > M_a$.

For (2), we begin to claim that $\lim_{a \to \infty} u_a(M_a) = \infty$.

To proof this, assume to contrary that for a sufficiently large, there exists a constant $C > 0$ independent of a such that $|u_a(M_a)| \leq C$. As $|u_a|$ has a global maximum at M_a then $|u_a(r)| \leq C$ for all $r \geq R$.

Let $y_a = \frac{u_a}{a}$. We proceed in the same way as (1), we show that $y_a \to y$ with $y \equiv 0$ and $y'(R) = 1$ this is a contradiction. Hence $u_a(M_a) \to \infty$ as $a \to \infty$. In the following we want to show that $M_a \to R$, as $a \to \infty$.

From (1) and by monotonicity of E_a, for a sufficiently large we see that $E_a(r) \geq E_a(M_a) = F(u_a(M_a))$ on $[R, M_a]$.

Denote $x_a = u_a(M_a)$, so $|u_a'| = u_a'(r) \geq (p')^{\frac{1}{p}} \left(F(x_a) - F(u_a(r))\right)^{\frac{1}{p}}$ for all $r \in [R, M_a]$. Integrating this on $[R, M_a]$ gives

$$\int_0^{x_a} \frac{ds}{\left(F(x_a) - F(s)\right)^{\frac{1}{p}}} = \int_R^{M_a} \frac{u_a'(r)}{\left(F(x_a) - F(u_a(r))\right)^{\frac{1}{p}}} \, dr,$$

then

$$\int_0^{x_a} \frac{ds}{\left(F(x_a) - F(s)\right)^{\frac{1}{p}}} \geq (p')^{\frac{1}{p}} (M_a - R) > 0. \tag{18}$$

First we estimate the integral on $[\frac{x_a}{2}, x_a]$ for a sufficiently large. From (H2) we have for x large enough that $f(x) \geq \frac{1}{2} x^q$ and since $u_a(M_a) = x_a \to \infty$ as $a \to \infty$

we therefore have for a large enough that

$$Q_a = \min_{[\frac{x_a}{2}, x_a]} f \geq \frac{1}{2^{q+1}} (x_a)^q.$$

As $p < q + 1$ then it follows that

$$\frac{x_a^{1-\frac{1}{p}}}{Q_a^{\frac{1}{p}}} \leq 2^{\frac{q+1}{p}} (x_a)^{\frac{p-q-1}{p}} \to 0, \quad \text{as} \quad a \to \infty.$$

Thus

$$\lim_{a \to \infty} \frac{(x_a)^{1-\frac{1}{p}}}{Q_a^{\frac{1}{p}}} = 0. \tag{19}$$

Since $F(u)$ is increasing for u large enough, it follows that for $\frac{x_a}{2} \leq t \leq x_a$ we have by the mean value theorem for some c such that $\frac{x_a}{2} \leq t < c < x_a$:

$$F(x_a) - F(t) = f(c)(x_a - t) \geq Q_a(x_a - t). \tag{20}$$

Thus

$$\int_{\frac{x_a}{2}}^{x_a} \frac{dt}{\left(F(x_a) - F(t)\right)^{\frac{1}{p}}} \leq \left(\frac{1}{Q_a}\right)^{\frac{1}{p}} \int_{\frac{x_a}{2}}^{x_a} \frac{dt}{\left(x_a - t\right)^{\frac{1}{p}}}$$

$$\leq \left(\frac{p}{(p-1)2^{1-\frac{1}{p}}}\right) \frac{(x_a)^{1-\frac{1}{p}}}{Q_a^{\frac{1}{p}}}.$$

From (19) we therefore have

$$\lim_{a \to \infty} \int_{\frac{x_a}{2}}^{x_a} \frac{dt}{\left(F(x_a) - F(t)\right)^{\frac{1}{p}}} = 0. \tag{21}$$

Next we estimate the integral of left-hand side of (18) on $[0, \frac{x_a}{2}]$. Then we have $F(t) \leq F(\frac{x_a}{2})$ for all $t \in [0, \frac{x_a}{2}]$ and a sufficiently large. Thus by (20) we have

$$F(x_a) - F(t) \geq F(x_a) - F(\frac{x_a}{2}) \geq Q_a \frac{x_a}{2}.$$

Then it follows that

$$\int_0^{\frac{x_a}{2}} \frac{dt}{\left(F(x_a) - F(t)\right)^{\frac{1}{p}}} \leq 2^{\frac{1}{p}-1} \frac{(x_a)^{1-\frac{1}{p}}}{Q_a^{\frac{1}{p}}}.$$

Therefore by (19) we have

$$\lim_{a \to \infty} \int_0^{\frac{x_a}{2}} \frac{dt}{\left(F(x_a) - F(t)\right)^{\frac{1}{p}}} = 0. \tag{22}$$

Combining (21) and (22) we conclude that

$$\lim_{a \to \infty} \int_0^{x_a} \frac{dt}{\left(F(x_a) - F(t)\right)^{\frac{1}{p}}} = 0.$$

Hence from (18) we see that $M_a \to R$ as $a \to \infty$. Which completes the proof of Lemma 2. ☐

Lemma 3. *Assume* (H1)–(H4) *hold. Then* $u_a > 0$ *on* (R, ∞) *for* $a > 0$ *sufficiently small.*

Proof. If $M_a = \infty$ we have $u_a(r) > 0$ for each $r > R$ and so we are done in this case. If $M_a < \infty$, there are two cases:

1. If $u_a(M_a) < \gamma$, since $E_a(M_a) = F(u_a(M_a)) < 0$ then by virtue of *(i)* in Proposition 1 we have $u_a > 0$ on $[M_a, \infty)$, as $u_a' > 0$ on $[R, M_a)$ so $u_a > 0$ on (R, ∞) then it follows we are done in this case as well.
2. If $u_a(M_a) \geq \gamma$, so there exists two real r_a and s_a with $R < r_a < s_a < M_a$ such that $u_a(r_a) = \beta$ and $u_a(s_a) = \frac{\beta+\gamma}{2}$. By the monotonicity of E_a we get

$$E_a(r) = \frac{|u_a'|^p}{p'} + F(u_a) \leq \frac{a^p}{p'}, \quad \forall r \geq R.$$

Therefore

$$\frac{|u_a'|}{\left(a^p - p'F(u_a)\right)^{\frac{1}{p}}} \leq 1 \quad \forall r \geq R. \tag{23}$$

As $u_a' > 0$ on $[R, r_a]$ and by integrating (23) on $[R, r_a]$, we obtain

$$\int_0^\beta \frac{dt}{\left(a^p - p'F(t)\right)^{\frac{1}{p}}} = \int_R^{r_a} \frac{u_a'(r)}{\left(a^p - p'F(u_a(r))\right)^{\frac{1}{p}}} dr$$

$$\leq r_a - R.$$

Using (H4) and Remark 1, then we see that

$$\int_0^\beta \frac{dt}{\left(a^p - p'F(t)\right)^{\frac{1}{p}}} \to \left(\frac{p-1}{p}\right)^{\frac{1}{p}} \int_0^\beta \frac{dt}{|F(t)|^{\frac{1}{p}}} = \infty \quad , \text{ as } \quad a \to 0^+.$$

Thus it follows that $\lim_{a \to 0^+} r_a = \infty$, also $\lim_{a \to 0^+} s_a = \infty$.
  Next, by virtue of (13) and (14) we have

$$\left(r^\alpha E_a(r)\right)' = \left(r^\alpha\right)' F(u_a(r)), \tag{24}$$

where $\alpha = p'(N-1) > 1$ because $N \geq 2$ and $p' > 1$. Integration (24) on $[r_a, s_a]$ and using (H3), we obtain

$$s_a^\alpha E_a(s_a) - r_a^\alpha E_a(r_a) = \int_{r_a}^{s_a} \left(r^\alpha\right)' F(u_a(r)) \, dr$$

$$\leq F(\frac{\beta+\gamma}{2}) \int_{r_a}^{s_a} \left(r^\alpha\right)' dr$$

$$\leq F(\frac{\beta+\gamma}{2}) \left(s_a^\alpha - r_a^\alpha\right).$$

As $F(u_a(r)) \leq 0$ on $[R, r_a]$ and by (24) we have $r \to r^\alpha E_a(r)$ is decreasing, by integrating again (24) on $[R, r_a]$ we obtain,

$$r_a^\alpha E_a(r_a) = R^\alpha E_a(r_a) + \int_R^{r_a} \left(r^\alpha\right)' F(u_a(r)) \, dr \leq R^\alpha \frac{a^p}{p'}.$$

Hence it follows that

$$s_a^\alpha E_a(s_a) \leq R^\alpha \frac{a^p}{p'} + F(\frac{\beta+\gamma}{2}) \left(s_a^\alpha - r_a^\alpha\right). \tag{25}$$

Now, by means value theorem and since $\alpha > 1$ we therefore have,

$$s_a^\alpha - r_a^\alpha \geq \alpha \, r_a^{\alpha-1} (s_a - r_a). \tag{26}$$

By integrating (23) on $[r_a, s_a]$ we see that

$$\int_\beta^{\frac{\beta+\gamma}{2}} \frac{dt}{\left(a^p - p'F(t)\right)^{\frac{1}{p}}} = \int_{r_a}^{s_a} \frac{u_a'(r)}{\left(a^p - p'F(u_a(r))\right)^{\frac{1}{p}}} \, dr$$

$$\leq s_a - r_a.$$

Using (4), taking $0 < a < 1$ and for each $t \in [\beta, \frac{\beta+\gamma}{2}]$ we get

$$a^p - p'F(t) \leq 1 + p'F_0.$$

Therefore

$$s_a - r_a \geq \frac{\gamma-\beta}{2}\left(1 + p'F_0\right)^{-\frac{1}{p}}.$$

From (26) we deduce that

$$s_a^\alpha - r_a^\alpha \geq \frac{\gamma-\beta}{2}\left(1 + p'F_0\right)^{-\frac{1}{p}} \alpha \, r_a^{\alpha-1}.$$

As $\lim_{a \to 0+} r_a = \infty$ it follows that

$$\lim_{a \to 0+} s_a^\alpha - r_a^\alpha = \infty.$$

Then by virtue of (25) and since $F(\frac{\beta+\gamma}{2}) < 0$ we have

$$\lim_{a \to 0+} s_a^\alpha E_a(s_a) = -\infty.$$

Hence, for small enough positive a we get $E_a(s_a) < 0$. Thus by *(i)* in Proposition 1 it follows that $u_a > 0$ on $[s_a, \infty)$ if a is sufficiently small positive. As $u_a > 0$ and increasing on $[R, s_a)$ so, $u_a > 0$ on $(R, s_a]$. Hence $u_a > 0$ on (R, ∞) for small enough positive a. This completes the proof of Lemma 3.   □

Lemma 4. *Assume* (H1)–(H4) *hold. Then* u_a *has only simple zeros on* $[R, \infty)$.

Proof. The proof of this lemma is similar to [11, Lemma 2.6]. Assume to the contrary that u_a reaches a double zero at some point $r_0 > R$. Let

$$r_1 = \inf_{r > R} \{r \mid u_a(r) = u_a'(r) = 0\}.$$

As $u_a(r_0) = u_a'(r_0) = 0$ then $R \le r_1 < \infty$. The first we want to show that $r_1 = R$.

By contradiction we assume that $r_1 > R$ and let $r \in (\frac{R+r_1}{2}, r_1)$ fixed. Integrating (14) on (r, r_1) we see that

$$E_a(r_1) - E_a(r) = - \int_r^{r_1} \frac{(N-1)|u_a'|^p}{t} \, dt.$$

From (13) and $E_a(r_1) = 0$ we see that

$$E_a(r) = \frac{|u_a'|^p}{p'} + F(u_a) = \int_r^{r_1} \frac{(N-1)|u_a'|^p}{t} \, dt. \tag{27}$$

Denote $w = \int_r^{r_1} \frac{(N-1)|u_a'|^p}{t} \, dt$ and differentiating gives

$$-\frac{N-1}{r} |u_a'|^p = w'.$$

Substituting in (27), we obtain

$$w' + \frac{\alpha}{r} w = \frac{\alpha}{r} F(u_a), \tag{28}$$

where $\alpha = p'(N-1) > 1$. Multiplying both sides by r^α we have

$$(r^\alpha w)' = \alpha r^{\alpha-1} F(u_a).$$

Integrating the above on $[r, r_1]$ we obtain

$$r_1^\alpha w(r_1) - r^\alpha w(r) = \alpha \int_r^{r_1} t^{\alpha-1} F(u_a) \, dt.$$

As $u(r_1) = 0$, so for r sufficiently close to r_1 we have $|u_a| \le \beta$ and by (H3) implies that $F(u_a) < 0$ on (r, r_1). Since $w(r_1) = 0$, then it follows for r sufficiently close to r_1

$$w = \frac{\alpha}{r^\alpha} \int_r^{r_1} t^{\alpha-1} |F(u_a)| \, dt.$$

Combining the equation above and (27), we therefore have for r sufficiently close to r_1

$$|u_a'(r)|^p = p'\Big(|F(u_a(r))| + \frac{\alpha}{r^\alpha}\int_r^{r_1} t^{\alpha-1}|F(u_a)|\,dt\Big). \qquad (29)$$

Notice that for $r < r_1$ and r sufficiently close to r_1 then $u_a'(r) \neq 0$.

Otherwise there exists $r_2 < r_1$ such that $u_a'(r_2) = 0$ then by (29) we deduce that $u_a \equiv 0$ on (r_2, r_1) and by continuity we have $u_a(r_2) = u_a'(r_2) = 0$. Which contradict the definition of r_1. Thus without loss of generality we assume that $u_a' < 0$ for $r < r_1$ with r is sufficiently close to r_1. Thus we have $0 < u_a \leq \beta$ on (r, r_1). Since F is decreasing on $[0, \beta]$, therefore it follows that $|F(u_a(r))| > |F(u_a(t))| > 0$ for each $r < t < r_1$.

From (29) we therefore have

$$|u_a'|^p \leq p'\Big(|F(u_a(r))| + \frac{|F(u_a(r))|}{r^\alpha}(r_1^\alpha - r^\alpha)\Big) = p'|F(u_a(r))|\Big(\frac{r_1}{r}\Big)^\alpha$$

$$\leq p'2^\alpha|F(u_a(r))| \quad (\ \frac{r_1}{r} < 2 - \frac{R}{r} < 2).$$

Thus

$$|u_a'| \leq C_{p,\alpha}|F(u_a(r))|^{\frac{1}{p}},$$

where $C_{p,\alpha} = (p'2^\alpha)^{\frac{1}{p}}$. Dividing by $|F(u_a(r))|^{\frac{1}{p}}$, integrating the above on $[r, r_1]$ and using (H4) we have

$$\infty = \int_0^{u_a(r)} \frac{1}{|F(s)|^{\frac{1}{p}}}\,ds = \int_r^{r_1} \frac{|u_a'(t)|}{|F(u_a(t))|^{\frac{1}{p}}}\,dt \leq C_{p,\alpha}(r_1 - r) < \infty.$$

This is impossible, therefore $r_1 = R$ implying $u_a'(R) = 0$. This is a contradiction with $u_a'(R) = a > 0$. Hence u_a has only simple zeros on $[R, \infty)$. This completes the proof of Lemma 4. $\qquad \square$

3   Solutions with a Prescribed Number of Sign Changin to (7)–(8)

In this section we are interested to study the behaviour of zeros of solution to (7)–(8) assuming (H1)–(H4) hypotheses. By virtue of Lemma 3, if a is small enough we saw that u_a has no zeros on (R, ∞) and by Lemma 4, we get u_a has only simple zeros. In the following as in [10, 11] we want to show that u_a has an arbitrary large number of zeros on (R, ∞) for a large enough. Which leads to the existence of sign changing solution.

Let

$$\lambda_a^{\frac{p}{q-p+1}} = u_a(M_a)$$

and we define,

$$v_{\lambda_a}(r) = \lambda_a^{\frac{-p}{q-p+1}} u_a(M_a + \frac{r}{\lambda_a}) \quad \forall r \geq 0.$$

By (7), it is straightforward to show that

$$\left((\lambda_a M_a + r)^{N-1} \Phi_p(v'_{\lambda_a})\right)' + \lambda_a^{\frac{-pq}{q-p+1}} \left(\lambda_a M_a + r^{N-1}\right)^{N-1} f(\lambda_a^{\frac{p}{q-p+1}} v_{\lambda_a}) \quad (30)$$

$$v_{\lambda_a}(0) = 1 \quad \text{and} \quad v'_{\lambda_a}(0) = 0. \quad (31)$$

By Lemma 2 and $q - p + 1 > 0$ we have that

$$\lim_{a \to \infty} \lambda_a = \infty. \quad (32)$$

Lemma 5. *As $a \to \infty$, $v_{\lambda_a} \to v$, uniformly on compact subsets of $[0, \infty)$, and $v$satisfies*

$$\left(\Phi_p(v')\right)' + |v|^{q-1} v = 0, \quad (33)$$

$$v(0) = 1, \quad v'(0) = 0. \quad (34)$$

Proof. By virtue of (H2) we have

$$F(u) = \frac{1}{q+1} |u|^{q+1} + G(u),$$

where $G(u) = \int_0^u g(s) \, \mathrm{d}s$. Then it follows that

$$\lim_{|u| \to \infty} \frac{F(u)}{|u|^{q+1}} = \frac{1}{q+1} + \lim_{|u| \to \infty} \frac{G(u)}{|u|^{q+1}}.$$

By using the L'Hospital's rule we have,

$$\lim_{u \to \infty} \frac{G(u)}{u^{q+1}} = 0.$$

Since G is even then

$$\lim_{|u| \to \infty} \frac{G(u)}{u^{q+1}} = 0.$$

Thus

$$\lim_{|u| \to \infty} \frac{F(u)}{|u|^{q+1}} = \frac{1}{q+1}. \quad (35)$$

Let

$$E_{\lambda_a}(r) = \frac{|v'_{\lambda_a}|^p}{p'} + \lambda_a^{\frac{-p(q+1)}{q-p+1}} F(\lambda_a^{\frac{p}{q-p+1}} v_{\lambda_a}).$$

By differentiating we therefore have

$$E'_{\lambda_a}(r) = -\frac{(N-1)|v'_{\lambda_a}|^p}{r} \le 0.$$

Then E_{λ_a} is decreasing on $[0, \infty)$ which implies that

$$E_{\lambda_a}(r) \le E_{\lambda_a}(0) = \lambda_a^{\frac{-p(q+1)}{q-p+1}} F(\lambda_a^{\frac{p}{q-p+1}}).$$

By (32) and (35) we see that

$$\lambda_a^{\frac{-p(q+1)}{q-p+1}} F(\lambda_a^{\frac{p}{q-p+1}}) \to \frac{1}{q+1} \text{ as } \to \infty.$$

Thus for a large enough we see that

$$E_{\lambda_a}(r) \le \frac{2}{q+1} \quad \forall r \ge 0.$$

By (4) and (32) for a large enough, it follows that

$$\frac{|v'_{\lambda_a}|^p}{p'} \le \frac{2}{q+1} + \lambda_a^{\frac{-p(q+1)}{q-p+1}} F_0 \le \frac{3}{q+1} \quad \forall r \ge 0.$$

Hence we have $|v'_{\lambda_a}|$ is uniformly bounded on $[0, \infty)$ by $M_{p,q} = \left(\frac{3p'}{q+1}\right)^{\frac{1}{p}}$, for a sufficiently large. From Lemma 2 we obtain

$$|v_{\lambda_a}| \le \lambda_a^{\frac{-p}{q-p+1}} u_a(M_a) = 1.$$

Thus $|v_{\lambda_a}|$ and $|v'_{\lambda_a}|$ are uniformly bounded. By Arzela-Ascoli's theorem, for some subsequence still denoted v_{λ_a}, we have $v_{\lambda_a} \to v$ uniformly on compact subsets of $[0, \infty)$ and v is continuous.

On the other hand, by integrating (30) on $[0, r]$ and using (31), we obtain

$$\left(\lambda_a M_a + r\right)^{N-1} \Phi_p(v'_{\lambda_a})$$
$$= -\int_0^r \left(\lambda_a M_a + t\right)^{N-1} \lambda_a^{\frac{-pq}{q-p+1}} f(\lambda_a^{\frac{p}{q-p+1}} v_{\lambda_a}) \, dt$$
$$= -\int_0^r \left(\lambda_a M_a + t\right)^{N-1} \left(|v_{\lambda_a}|^{q-1} v_{\lambda_a} + \lambda_a^{\frac{-pq}{q-p+1}} g(\lambda_a^{\frac{p}{q-p+1}} v_{\lambda_a})\right) \, dt.$$

Therefore

$$v'_{\lambda_a} = -\Phi_{p'}\left(\int_0^r \left(\frac{\lambda_a M_a + t}{\lambda_a M_a + r}\right)^{N-1} \left(|v_{\lambda_a}|^{q-1} v_{\lambda_a} + \lambda_a^{\frac{-pq}{q-p+1}} g(\lambda_a^{\frac{p}{q-p+1}} v_{\lambda_a})\right) \, dt\right) \quad (36)$$

From (H2) and since g is continuous, then for all $\epsilon > 0$ there exists $C > 0$ such that

$$|g(u)| \le C + \epsilon |u|^q \quad \forall u \in \mathbb{R},$$

it follows that

$$|g(\lambda_a^{\frac{p}{q-p+1}} v_{\lambda_a})| \le C + \epsilon \lambda_a^{\frac{pq}{q-p+1}} |v_{\lambda_a}|$$
$$\le C + \epsilon \lambda_a^{\frac{pq}{q-p+1}} \quad \left(|v_{\lambda_a}| \le 1\right).$$

Thus

$$\lambda_a^{\frac{-pq}{q-p+1}} |g(\lambda_a^{\frac{p}{q-p+1}} v_{\lambda_a})| \leq \lambda_a^{\frac{-pq}{q-p+1}} C + \epsilon.$$

By virtue of (32) we have $\lim_{a \to \infty} \lambda_a^{\frac{-pq}{q-p+1}} = 0$. Then it follows that, for a sufficiently large

$$\lambda_a^{\frac{-pq}{q-p+1}} |g(\lambda_a^{\frac{p}{q-p+1}} v_{\lambda_a})| \leq 2\epsilon,$$

which implies that $\lambda_a^{\frac{-pq}{q-p+1}} g(\lambda_a^{\frac{p}{q-p+1}} v_{\lambda_a}) \to 0$ as $a \to \infty$ uniformly on $[0, \infty)$.

From (32) we have $\lambda_a M_a \to \infty$ as $a \to \infty$, we therefore have $\left(\frac{\lambda_a M_a + t}{\lambda_a M_a + r} \right)^{N-1} \to 1$, as $a \to \infty$ uniformly on compact subsets of $[0, \infty)$. Since $v_{\lambda_a} \to v$ as $a \to \infty$ uniformly on compact subsets of $[0, \infty)$ and by (36) we deduce that

$$v'_{\lambda_a} \to -\Phi_{p'}\left(\int_0^r |v|^{q-1} v \, dt \right) \equiv w \quad \text{on } [0, \infty),$$

w is continuous. Furthermore from

$$v_{\lambda_a} = 1 + \int_0^r v'_{\lambda_a} \, dt,$$

since $v_{\lambda_a} \to v$ as $a \to \infty$ uniformly on compact subsets of $[0, \infty)$, $v'_{\lambda_a} \to w$ as $a \to \infty$ (pointwise) and $|v'_{\lambda_a}|$ is uniformly bounded by $M_{p,q}$, therefore by applying the dominated convergence theorem we have

$$v = 1 + \int_0^r w(t) \, dt.$$

Thus $v' = w$. Then it follows from (36) that

$$v' = -\Phi_{p'}\left(\int_0^r |v|^{q-1} v \, dt \right).$$

Hence $v \in C^1[0, \infty)$ and v satisfies (33)–(34). End of the proof of Lemma 5. □

Lemma 6. *Let v as in Lemma 5. Then v has a zero on $[0, \infty)$.*

Proof. Suppose by the way of contradiction that $v > 0$ on $[0, \infty)$. By integrating (33) on $[0, r]$ we obtain

$$-\Phi_p(v') = \int_0^r v^q \, dt > 0.$$

So $v' < 0$ and v is decreasing on $[0, \infty)$. It follows that

$$|v'(r)|^{p-1} = \int_0^r v^q \, dt \geq r \, v^q(r),$$

which implies that

$$\frac{-v'(r)}{v^{\frac{q}{p-1}}(r)} \geq r^{\frac{1}{p-1}}.$$

By integrating the above on $[0, r]$ we get

$$\frac{p-1}{q-p+1}\left(v^{-\frac{q-p+1}{p-1}}(r) - 1\right) \geq \frac{p-1}{p} r^{\frac{p}{p-1}},$$

as $q - p + 1 > 0$ we see that

$$v^{-\frac{q-p+1}{p-1}}(r) \geq 1 + \frac{q-p+1}{p} r^{\frac{p}{p-1}} \geq \frac{q-p+1}{p} r^{\frac{p}{p-1}},$$

therefore

$$v^{q+1} \leq C_{p,q}\, r^{\frac{-p(q+1)}{q-p+1}},$$

where

$$C_{p,q} = \left(\frac{p}{q-p+1}\right)^{\frac{(q-p+1)(q+1)}{p-1}}.$$

Therefore it follows that

$$\int_1^\infty v^{q+1}(t)\, dt \leq C_{p,q} \int_1^\infty r^{\frac{-p(q+1)}{q-p+1}}\, dt < \infty.$$

By continuity of v we have v is bounded on compact set $[0, 1]$ so,

$$\int_0^\infty v^{q+1}(t)\, dt < \infty. \tag{37}$$

Let for each $r \geq 0$,

$$E(r) = \frac{|v'(r)|^p}{p'} + \frac{1}{q+1}|v(r)|^{q+1}.$$

From (33) we obtain $E'(r) = 0$ on $[0, \infty)$, it implies that $E \equiv E(0) = \frac{1}{q+1}$. Therefore it follows that

$$|v'(r)|^p = \frac{p'}{q+1}\left(1 - |v(r)|^{q+1}\right). \tag{38}$$

Since $v > 0$ and from (33) we see that

$$\left(v\,\Phi_p(v')\right)' = v'\,\Phi_p(v') - \left(\Phi_p(v')\right)' = |v'|^p - v^{q+1}.$$

Integrating this on $[0, r]$ and using (34) we have

$$v(r)\,\Phi_p(v'(r)) = \int_0^r |v'|^p - v^{q+1}\, dt,$$

since $v > 0$ and $v' < 0$, thus it follows that

$$\int_0^\infty |v'|^p\, dt \leq \int_0^\infty v^{q+1}\, dt < \infty. \tag{39}$$

By integrating (38) on $[0, r]$ we obtain

$$\int_0^r |v'|^p dt = \frac{p'}{q+1} \int_0^r (1 - v^{q+1})\, dt = \frac{p'}{q+1}\left(r - \int_0^r v^{q+1}\, dt\right).$$

Letting $r \to \infty$ and using (37) we have

$$\int_0^\infty |v'|^p\, dt = \infty.$$

Which contradicts (39). This completes the proof of Lemma 6. $\square$

Lemma 7. *If a is sufficiently large. Then u_a has an arbitrarily large number of zeros on (R, ∞).*

Proof. We begin to establish the following claim.

Claim: v has an infinite number of zeros on $[0, \infty)$.

Indeed, by Lemma 6 v has a zero $z_1 > 0$ such that $v > 0$ and $v' < 0$ on $(0, z_1)$. Next we want to show that v has a first local minimum on (z_1, ∞) denoted m_1. So, suppose by contradiction that v is decreasing on $[0, \infty)$. Since v is bounded and decreasing then $\lim_{r \to \infty} v(r) = \ell$ exists. As in the proof of *(ii)* Proposition 1 we therefore have, $\lim_{r \to \infty} v'(r) = 0$ and $\ell = 0$. From (38) and letting $r \to \infty$ we obtain

$$\lim_{r \to \infty} |v'(r)|^p = \frac{p'}{q+1} > 0.$$

Which contradicts to $\lim_{r \to \infty} v'(r) = 0$. Hence, v has a first minimum denoted $m_1 > z_1$ and $v_1 = v(m_1) < 0$. Thus v satisfies,

$$\Phi_p(v') = -\int_{m_1}^r |v|^{q-1} v\, dt, \quad \forall r > m_1$$

$$v(m_1) = v_1 \quad \text{and} \quad v'(m_1) = 0.$$

In a similar way of the Lemma 6 we can show that v has a second zero at $z_2 > z_1 > R$ and has a second extrema $m_2 > z_2$. Continuing in this way we can get an arbitrarily large number of zeros for v on $[0, \infty)$. Which is complete the proof of claim.

Now, since $v_{\lambda_a} \to v$ as $a \to \infty$ uniformly on compact subsets of (R, ∞). By virtue of the previous claim and (32), it follows that v_{λ_a} has an arbitrary large number of zeros for a large enough. Finally, as $u_a(M_a + \frac{r}{\lambda_a}) = \lambda_a^{\frac{p}{q-p+1}} v_{\lambda_a}(r)$, hence we can get as many zeros of u_a as desired on (R, ∞) for a sufficiently large. End of the proof of Lemma 7. $\square$

4   Proof of the Main Result

For $k \geq 1$ defined by set

$$S_k = \{\, a > 0 \ \mid \ u_a(r) \ \text{has exactly } k \text{ zeros on } (R, \infty)\,\}$$

and
$$S_0 = \{\, a > 0 \quad | \quad u_a(r) > 0, \forall r > R \,\}.$$

If a is sufficiently small it follows by Lemma 3 that $u_a > 0$ for any $r > R$, so S_0 is nonempty and from Lemma 7 we see that S_0 is bounded from above.

Let
$$a_0 = \sup S_0.$$

Lemma 8. $u_{a_0} > 0$ for all $r > R$.

Proof. Assume to the contrary that there exists a zero $z_0 > R$ of u_{a_0}. Then $u_{a_0}(z_0) = 0$ and $u_{a_0} > 0$ on (R, z_0) and by Lemma 4 we have $u'_{a_0}(z_0) < 0$. Thus $u_{a_0} < 0$ on $(z_0, z_0 + \epsilon)$ for some $\epsilon > 0$. If a close to a_0 with $a < a_0$ and by continuous dependence of solutions on initial conditions, it follows that $u_a \leq 0$ on $(z_0, z_0 + \epsilon)$. Which contradicts to the definition of a_0. □

Lemma 9. $E_{a_0} \geq 0$ on $[R, \infty)$.

Proof. By the definition of a_0 if $a > a_0$ then u_a has a zero denoted $z_a > R$. We begin to show the following result

$$\lim_{a \to a_0^+} z_a = \infty. \tag{40}$$

Indeed, suppose by contradiction there exists a subsequence of z_a still denoted z_a converges to some $z \in [R, \infty)$, as $a \to a_0^+$. As u_a converges uniformly on a compact set $[R, z + 1]$, Therefore it follows that

$$0 = \lim_{a \to a_0^+} u_a(z_a) = u_{a_0}(z).$$

Which contradicts to Lemma 8. Hence (40) is proven.

Now, assume to the contrary that there exists $r_0 > R$ such that $E_{a_0}(r_0) < 0$, then $E_a(r_0) < 0$ for a close to a_0^+. By the definition of a_0 and taking $a > a_0$, there exists a zero z_a of u_a such that $E_a(z_a) \geq 0 > E_a(r_0)$. From the monotonicity of E_a we have $z_a < r_0 < \infty$. Which contradicts (40). End of the proof of Lemma 9. □

Lemma 10. u_{a_0} has a local maximum at $M_{a_0} > R$.

Proof. By the way of contradiction, suppose that $u'_{a_0} > 0$ on $[R, \infty)$. From *(iii)* of the Proposition 1 we see that $\lim_{r \to \infty} u_{a_0}(r) = \beta$ and by the definition of a_0 if $a > a_0$ then u_a has a zero denoted $z_a > R$. Next, from *(ii)* of Proposition 1 therefore it follows that $\lim_{r \to \infty} E_{a_0}(r) = F(\beta) < 0$. From Lemma 9 we see that $E_{a_0}(r) \geq 0$. Thus $0 \leq \lim_{r \to \infty} E_{a_0}(r) = F(\beta) < 0$, this is contradictory. End of the proof of Lemma 10. □

Lemma 11. $u'_{a_0} < 0$ on (M_{a_0}, ∞).

Proof. We argue by the contradiction. We distinguish two cases:

(1) If there exists $r_1 > M_{a_0}$ such that $u'_{a_0}(r_1) = 0$ and $u'_{a_0} < 0$ on (M_{a_0}, r_1),
    or
(2) If $u'_{a_0} = 0$ on $[M_{a_0}, \infty)$.

In the first case we have $0 < u_{a_0}(r_1) < \beta$.

Indeed assume to contrary that $u_{a_0}(r_1) \geq \beta$, then by H3 we see that $f(u_{a_0}(r_1)) \geq 0$. Further from (10) and the fact that $u'_{a_0}(M_{a_0}) = 0$ we have

$$0 = -r^{N-1} \Phi_p(u'_{a_0}(r_1)) = \int_{M_{a_0}}^{r_1} t^{n-1} f(u_{a_0}(t)) \, dt. \tag{41}$$

As u_{a_0} is non-increasing on $[M_{a_0}, r_1]$ then $u_{a_0}(t) \geq u_{a_0}(r_1)$ for all $t \in [M_{a_0}, r_1]$ and by using (H3) it follows that $f(u_{a_0}(t)) \geq f(u_{a_0}(r_1)) \geq 0$. From (41) we have $f(u_{a_0}) \equiv 0$ on $[M_{a_0}, r_1]$ implying that $u \equiv \beta$ on $[M_{a_0}, r_1]$, which contradicts to $u'_{a_0} < 0$ on (M_{a_0}, r_1). Thus $0 < u_{a_0}(r_1) < \beta$ implying $E_{a_0}(r_1) = F(u_{a_0}(r_1)) < 0$. Which contradicts to Lemma 9, then it follows we are done in this case.

In the second case we must have that $u = c$ on $[M_{a_0}, \infty)$. Then by *(ii)* and *(iii)* of Proposition 1 it follows that $f(c) = 0$ and $c = \beta$. Thus $0 < u_{a_0} < \beta$ on (R, M_{a_0}) and by using (H3) we obtain $f(u_{a_0}) < 0$. From (40) and $u'_{a_0}(M_{a_0}) = 0$ we see that

$$r^{N-1} \Phi_p(u'_{a_0}(r)) = \int_r^{M_{a_0}} t^{n-1} f(u_{a_0}(t)) \, dt < 0.$$

Therefore we have $u'_{a_0} < 0$ on (R, M_{a_0}), this a contradiction then it follows we are done in this case as well. End of the proof of Lemma 11.   □

By Lemmas 10 and 11 then it follows that $\lim_{r \to \infty} u_a(r) = \ell$ exists and by *(ii)* of Proposition 1, we have $f(\ell) = 0$ and $\lim_{r \to \infty} E_{a_0}(r) = F(\ell)$. Then either $\ell = 0$ or $\ell = \beta$.

If $\ell = \beta$ again by *(ii)* of Proposition 1 we therefore have $\lim_{r \to \infty} E_{a_0}(r) = F(\beta) < 0$. By Lemma 9 we have $E_{a_0}(r) \geq 0$ for each $r \geq R$, so it follows that $F(\beta) = \lim_{r \to \infty} E_{a_0}(r) \geq 0$. Which contradicts to $F(\beta) < 0$. Hence $\ell = 0$ and finally we have found a non-negative solution of (5)–(6).

Next by [11, Lemma 4.3], if $a > a_0$ and $a \to a_0$ then u_a has at most one zero on (R, ∞). From the definition of a_0 if $a > a_0$ we have u_a has at least one zero. Thus for $a > a_0$ close to a_0 the solution u_a has exactly one zero. Then it follows that S_1 nonempty and by Lemma 7 we see that S_1 is bounded above. Let

$$a_1 = \sup S_1.$$

As above lemmas by using a similar argument, we can show that u_{a_1} has one simple zero and $\lim_{r \to \infty} u_{a_1}(r) = 0$. Hence, it follows that there exists a solution of (5)–(6) which has exactly one sign change in (R, ∞).

Proceeding inductively we can show that, for each $k \in \mathbb{N}$ there exists a solution $u_{a_k} = u_k$ of (5)–(6) which has exactly k zeros on (R, ∞) with $u'_k(R) > 0$.

Now, in the case $a < 0$ we consider the problem

$$\left(r^{N-1}\Phi_p(v')\right)' + r^{N-1}f(v) = 0 \quad \text{if } R < r, \tag{42}$$

$$v(R) = 0 \quad \text{and} \quad v'(R) = a < 0. \tag{43}$$

We denote $w(r) = -v(r)$ on $[R, \infty)$, as f and Φ_p are odd, then it follows the problem (42)–(43) is equivalent to

$$\left(r^{N-1}\Phi_p(w')\right)' + r^{N-1}f(w) = 0 \quad \text{if R < r,}$$

$$w(R) = 0 \quad \text{and} \quad w'(R) = -a > 0.$$

Next, according to the case $a > 0$ we deduce that, for each $k \in \mathbb{N}$, the problem (5)–(6) has a solution w_k which has exactly k zeros on (R, ∞) with $w'_k(R) > 0$. Hence, for each $k \in \mathbb{N}$ integer, the problem (5)–(6) has a solution $v_k = -w_k$ which has k zeros on (R, ∞) and $v'_k(R) < 0$. End of proof of the main Theorem 1.

References

1. Berestycki, H., Lions, P.L.: Non linear scalar fields equations I, existence of a ground state. Arch. Rat. Mech. Anal. **82**, 313–345 (1983). https://doi.org/10.1007/BF00250555
2. Berestycki, H., Lions, P.L.: Non linear scalar fields equations II, existence of a ground state. Arch. Rat. Mech. Anal. **82**(4), 347–375 (1983). https://doi.org/10.1007/BF00250556
3. Berger, M.S., Schechter, M.: Embedding theorems and quasi-linear elliptic boundary value problems for unbounded domain. Trans. Am. Math. Soc. **172**, 261–278 (1972)
4. Berger, M.S.: Nonlinearity and Functionnal Analysis. Academic Free Press, New York (1977)
5. El Hachimi, A., de Thelin, F.: Infinitely many radially symmetric solutions for a quasilinear elliptic problem in a ball. J. Differ. Equ. **128**, 78–102 (1996). https://doi.org/10.1006/jdeq.1996.0090
6. Iaia, J.: Existence of solutions for semilinear problems with prescribed number of zeros on exterior domains. J. Math. Anal. Appl. **446**, 591–604 (2017). https://doi.org/10.10016/j.jmaa.2016.08.063
7. Jones, C.K.R.T., Küpper, T.: On the infinitely many solutions of a semilinear equation. SIAM J. Math. Anal. **17**, 803–835 (1986). https://doi.org/10.1137/0517059
8. Joshi, J., Iaia, J.: Existence of solutions for semilinear problems with prescribed number of zeros on exterior domains. Electron. J. Differ. Equ. **112**, 1–11 (2016)
9. Jiu, Q., Su, J.: Existence and multiplicity results for Dirichlet problems with p-Laplacian. J. Math. Anal. Appl. **281**, 587–601 (2003). https://doi.org/10.1016/S0022-247X(03)00165-3

10. Mcleod, K., Troy, W.C., Weissler, F.B.: Radial solution of $\Delta u + f(u) = 0$ with prescribed numbers of zeros. J. Differ. Equ. **83**, 368–378 (1990). https://doi.org/10.1016/0022-0396(90)90063-U
11. Pudipeddi, S.: Localized radial solutions for nonlinear p-Laplacian equation in $\mathbb{R}^N$. Electron. J. Qual. Theory Differ. Equ. **20**, 1–22 (2008). https://doi.org/10.14232/ejqtde.2008.1.20

On the Linear Essential
Spectrum Operator

Hassan Outouzzalt$^{(\boxtimes)}$

LabSi, FSJES & FSA, University Ibn Zohr, Agadir, Morocco
h.outouzzalt@uiz.ac.com

Abstract. This paper presents: Let A be a unital C^*-algebra of real rank zero and B be a unital semisimple complex Banach algebra. We characterize linear maps from A onto B that compress different essential spectral sets such as the (left, right) essential spectrum, the semi-Fredholm spectrum, and the Weyl spectrum. Essentially spectrally bounded linear mapping from A onto B are also characterized.

Keywords: Fredholm elements · Essential spectrum
Essential spectral radius

1 Introduction

Linear preserver problems is an active research area in matrix and operator theory. These problems involve linear or additive maps that leave invariant certain relations, or subsets, or functions. Over the past decades much work has been done on linear preserver problems on matrix or operator spaces. Often, the characterization of such linear preservers reveal the algebraic structures, in many cases, they are in fact Jordan homomorphisms; see surveys papers [3] and the references therein.

Throughout, A and B will denote infinite dimensional unital semisimple Banach algebras over the field $\mathbb{C}$ of complex number, unless specified otherwise. The unit is denoted by **1**. A linear mapping $\varphi : A \rightarrow B$ is said to be Jordan homomorphism if $\varphi(a^2) = \varphi(a)^2$ for all $a \in A$, or equivalently

$$\varphi(ab + ba) = \varphi(a)\varphi(b) + \varphi(b)\varphi(a)$$

for all $a, b \in A$. Clearly, every homomorphism and every anti-homomorphism is a Jordan homomorphism. For further properties of Jordan homomorphisms, we refer the reader to [3,9–11]. The map φ is said to be essentially spectrally bounded if there exists a positive constant M such that

$$r_e(\varphi(a)) \leq M r_e(a)$$

for all $a \in A$, where $r_e(.)$ stands for the essential spectral radius. In [6], the authors characterized linear maps on the algebra $\mathcal{L}(\mathcal{H})$ of all bounded

© Springer International Publishing AG, part of Springer Nature 2018
M. Ben Ahmed and A. A. Boudhir (Eds.): SCAMS 2017, LNNS 37, pp. 961–965, 2018.
https://doi.org/10.1007/978-3-319-74500-8_85

linear operators on an infinite dimensional Hilbert space $\mathcal{H}$ that are essentially spectrally bounded, extending some recent results obtained in [5] concerning linear essential spectral radius (essential spectrum) preservers. They proved that a linear surjective up to compact operators map from $\mathcal{L}(\mathcal{H})$ into itself is essentially spectrally bounded if and only if it preserves the ideal of compact operators and the induced linear map on the Calkin algebra is either a continuous epimorphism or a continuous anti-epimorphism multiplied by a nonzero scalar. Recently, in [7], as a local version, the authors studied linear maps on the algebra $\mathcal{L}(\mathcal{H})$ of all bounded linear operators on an infinite dimensional Hilbert space $\mathcal{H}$ that compress the local spectrum and the ones that are locally spectrally bounded.

The object of this note is to study essential spectrum compressors and essentially spectrally bounded linear maps between Banach algebras.

2   Essentially Spectrally Bounded Linear Maps

First, let us recall the following useful facts about Fredholm theory in semisimple Banach algebras that will be often used in the sequel.

Let A be a semisimple Banach algebra. The socle of A, $\mathrm{Soc}\,(A)$, is defined to be the sum of all minimal left (or right) ideals of A. The ideal of inessential elements of A is given by

$$I(A) := \bigcap \{P : P \in \Pi_A : \mathrm{Soc}\,(A) \subseteq P\},$$

where Π_A denotes the set of all primitive ideals of A. It is a closed ideal of A. We call $\mathcal{C}(A) := A/I(A)$ the generalized Calkin algebra of A. It should be noted that a semisimple Banach algebra is finite dimensional if and only if it coincides with its socle; see for instance [2, Theorem 5.4.2]. Since our algebras are always supposed to be infinite dimensional, the generalized Calkin algebra introduced above is not trivial.

An element $a \in A$ is called left semi-Fredholm (resp. right semi-Fredholm) if it is left (resp. right) invertible modulo $\mathrm{Soc}\,(A)$, and is called Fredholm if it is invertible modulo $\mathrm{Soc}\,(A)$. The element a is said to be Atkinson if it is left or right semi-Fredholm. It is known that left (resp. right) invertible modulo $\mathrm{Soc}\,(A)$ is equivalent to left (resp. right) invertible modulo $I(A)$.

For every $a \in A$ we set

$$\sigma_e(a) := \{\lambda \in \mathbb{C} : a - \lambda \, isnotFredholm\},$$
$$\sigma_{le}(a) := \{\lambda \in \mathbb{C} : a - \lambda \, isnotleftsemi-Fredholm\},$$
$$\sigma_{re}(a) := \{\lambda \in \mathbb{C} : a - \lambda \, isnotrightsemi-Fredholm\},$$

and

$$\sigma_{SF}(a) := \{\lambda \in \mathbb{C} : a - \lambda \, isnotAtkinson\}.$$

These are called respectively the essential spectrum, the left essential spectrum, the right essential spectrum, and the semi-Fredholm spectrum of a.

For an element $a \in A$, the essential spectral radius is defined by

$$r_e(a) := \max\{|\lambda| : \lambda \in \sigma_e(a)\}.$$

It coincides with the limit of the convergent sequence $(\|a^n\|_e^{\frac{1}{n}})_{n \geq 1}$, where $\|a\|_e := \|\pi(a)\|$ is the essential norm of a and π is the canonical projection from A onto $\mathcal{C}(A)$. We refer the reader to [12,13] and the monographs [1,4] for basic facts concerning Atkinson and Fredholm theory in Banach algebras.

A linear map $\varphi : A \to B$ is said to be surjective up to inessential elements if $B = \varphi(A) + I(B)$. It is called spectrally bounded if there exists a positive constant M such that $r(\varphi(a)) \leq Mr(a))$ for all $a \in A$, where $r(.)$ denotes the classical spectral radius. The following, quoted in [6], is needed in what follows.

Let A be a unital purely infinite C*-algebra with real rank zero and let B be a unital semi-prime Banach algebra. If $\varphi : A \to B$ be a surjective spectrally bounded linear map, then there exist a central invertible element c, viz., $\varphi(\mathbf{1})$, and a Jordan epimorphism $J : A \to B$ such that $\varphi(x) = cJ(x)$ for all $x \in A$.

Proof. See [6, Lemma 1]   □

The problem of characterizing essentially spectrally bounded it seems to be difficult even when A and B are supposed to be C*-algebras of real rank zero. Recall that a C*-algebra A is of real rank zero if the set of all finite real linear combinations of orthogonal projections is dense in the set of all self adjoint elements of A; see [8]. However, We give a positive answer to this problem when A is a purely infinite C*-algebra of real rank zero. A C*-algebras A is purely infinite if it has no characters and if, for every pair of positive elements a, b in $\mathcal{A}$ with $a \in \overline{\mathcal{A}b\mathcal{A}}$, there is a sequence $(x_n)_{n \in \mathbb{N}}$ in $\mathcal{A}$ such that $a = \lim_n x_n^* b x_n$; see [14].

The main result of this section is the following. It characterizes essentially spectrally bounded linear maps.

Theorem 2.1. *Let φ be a linear surjective up to compact operators map from a purely infinite C*-algebras with real rank zero A into semisimple a Banach algebra B. If φ is essentially spectrally bounded, then*

$$\varphi(I(A)) \subseteq I(B)$$

and the induced mapping $\widehat{\varphi} : \mathcal{C}(A) \to \mathcal{C}(B)$ defined by

$$\widehat{\varphi}(\pi(a)) := \pi(\varphi(a)), \ (a \in A),$$

is a continuous Jordan epimorphism multiplied by an invertible central element of $\mathcal{C}(B)$.

Proof. Assume that there is a positive constant M such that $r_e(\varphi(x)) \leq Mr_e(x)$ for all $x \in A$. We first show that φ maps $I(A)$ into $I(B)$. So pick an inessential element $a \in I(A)$, and let us prove that $\varphi(a)$ is inessential as well. Let y be an arbitrary element in B and note that, since φ is surjective up to inessential elements, there exist $x \in A$ and $y_0 \in I(B)$ such that $y = \varphi(x) + y_0$. For every $\lambda \in \mathbb{C}$, we have

$$
\begin{aligned}
r(\lambda \pi(\varphi(a)) + \pi(y)) = r(\pi(\lambda \varphi(a) + y)) &= r_e(\lambda \varphi(a) + y) \\
&= r_e(\varphi(\lambda a + x) + y_0) \\
&= r_e(\varphi(\lambda a + x)) \\
&\leq Mr_e(\lambda a + x) = Mr_e(x) = Mr(\pi(x)).
\end{aligned}
$$

Since $\lambda \mapsto r(\lambda \pi(\varphi(a)) + \pi(y))$ is a subharmonic function on $\mathbb{C}$, Liouville's Theorem implies that

$$
r(\pi(\varphi(a)) + \pi(y)) = r(\pi(y)).
$$

As y is arbitrary in B, it follows from semi-simplicity of $\mathcal{C}(B)$ and the Zemánek's characterization of the radical [2, Theorem 5.3.1] that $\pi(\varphi(a)) = 0$ and $\varphi(a) \in I(B)$.

Therefore $\varphi(I(A)) \subseteq I(B)$, and φ induces a linear map $\widehat{\varphi} : \mathcal{C}(A) \to \mathcal{C}(B)$ defined by $\widehat{\varphi}(\pi(x)) := \pi(\varphi(x))$ for all $x \in A$. The map $\widehat{\varphi}$ is obviously surjective and spectrally bounded. Note that, by [14, Proposition 4.3] the quotient of a C^*-algebra of real rank zero by a closed ideal is a C^*-algebra of real rank zero and the quotient of a purely infinite C^*-algebra by a closed ideal is a purely infinite C^*-algebra. Thus, the desired conclusion follows by applying Lemma 1.   □

For an infinite-dimensional complex Hilbert space $\mathcal{H}$, $\mathrm{Soc}\,(\mathcal{L}(\mathcal{H})) = \mathcal{F}(\mathcal{H})$ is the ideal of all finite rank operators on $\mathcal{H}$, $I(\mathcal{L}(\mathcal{H})) = \mathcal{K}(\mathcal{H})$ is the closed ideal of all compact operators on $\mathcal{H}$, and the generalized Calkin algebra $\mathcal{C}(\mathcal{L}(\mathcal{H}))$ coincides with the usual Calkin algebra $\mathcal{C}(\mathcal{H}) = \mathcal{L}(\mathcal{H})/\mathcal{K}(\mathcal{H})$, and it is prime. Thus, a Jordan homomorphism $\widehat{\varphi} : \mathcal{C}(\mathcal{H}) \to \mathcal{C}(\mathcal{H})$ is either a homomorphism or an anti-homomorphism by [9,10].

More generally, if A is a C^*-algebra, then $\mathrm{Soc}\,(A)$ is the set of all finite rank element in A, and $I(A) = \overline{\mathrm{Soc}\,(A)} = \mathcal{K}(A)$, the set of all compact elements in A; see [4]. Recall that an element x of A is said to be finite (resp. compact) in A if the wedge operator $x \wedge x : A \to A$, given by $x \wedge x(a) = xax$, is a finite rank (resp. compact) operator on A. Note that even if the C^*-algebra A is prime, the generalized Calkin algebra $\mathcal{C}(A) = A/\mathcal{K}(A)$ is not necessary prime. For example, consider A the C^*-algebra generated by $\mathcal{K}(\mathcal{H})$ and two orthogonal infinite dimensional projections on a Hilbert space $\mathcal{H}$, and note that $\mathcal{C}(A) \cong \mathbb{C}^2$ is not prime. However, when A is factor, the ideal $\mathcal{K}(A)$ is the largest ideal of type I, and $\mathcal{C}(A)$ is a prime C^*-algebra.

Let φ be a surjective up to inessential elements linear map from a purely infinite C^*-algebra A with rank real zero into a factor B. Then the following assertions are equivalent.

(i) φ is essentially spectrally bounded.
(ii) $\varphi(I(A)) \subseteq I(B)$ and the induced mapping $\widehat{\varphi} : \mathcal{C}(A) \to \mathcal{C}(B)$ is either a continuous epimorphism or a continuous anti-epimorphism up to a nonzero complex scalar.

Proof. We only need to proof that (i) ⇒ (ii) as (ii) ⇒ (i) follows easily. If φ is essentially spectrally bounded then, by Theorem 2.1 and the fact that the center of $\mathcal{C}(B)$ is trivial, we infer that $\widehat{\varphi}$ is a continuous Jordan epimorphism multiplied by a nonzero complex number c. As the algebra $\mathcal{C}(B)$ is prime, then by [11] the map $\widehat{\varphi}$ is, in fact, either an epimorphism or an anti-epimorphism multiplied by c. □

References

1. Aiena, P.: Fredholm and Local Spectral Theory, with Applications to Multipliers. Kluwer Academic Publishers, New York (2004)
2. Aupetit, B.: A Primer on Spectral Theory. Springer, New York (1991)
3. Aupetit, B.: Spectrum-preserving linear map between Banach algebra or Jordan-Banach algebra. J. Lond. Math. Soc. **62**(3), 917–924 (2000)
4. Barnes, B.A., Murphy, G.J., Smyth, M.R.F., West, T.T.: Riesz and Fredholm Theory in Banach Algebra. Pitman, London (1982)
5. Bendaoud, M., Bourhim, A., Sarih, M.: Linear maps preserving the essential spectral radius. Linear Algebra Appl. **428**, 1041–1045 (2008)
6. Bendaoud, M., Bourhim, A.: Essentially spectrally bounded linear maps. Proc. Amer. Math. soc. **137**(10), 3329–3334 (2009)
7. Bendaoud, M., Sarih, M.: Locally spectrally bounded linear maps. Math. Bohem **136**(1), 81–89 (2011)
8. Brown, L.G., Pedersen, G.K.: C^*-algebras of rank real zero. J. Funct. Anal. **99**, 131–149 (1991)
9. Cui, J., Hou, J.: Linear maps between Banach algebras compressing certains spectral functions. Rocky Mt. J. Math. Soc. **34**(2), 565–584 (2004)
10. Herstein, I.N.: Jordan homomorphisms. Trans. Amer. Math. Soc. **81**, 331–341 (1956)
11. Jacobson, N., Rickart, C.E.: Jordan homomorphism of rings. Trans. Amer. Math. Soc. **69**, 479–502 (1950)
12. Rowell, J.W.: Unilateral Fredholm theory and unilateral spectra. Proc. Roy. Irish. Acad. **84**, 69–85 (1984)
13. Schmoger, C.: Atkinson theory and holomorphic functions in Banach algebras. Proc. Roy. Irish. Acad. **91**, 113–127 (1991)
14. Kirchberg, E., Rørdam, M.: Non-simple purely infinite C^*-algebras. Amer. J. Math. **122**, 637–666 (2000)

Complex Event Processing and Role-Based Access Control Implementation in ESN Middleware

Yassir Rouchdi[1]([✉]), Khalid El Yassini[1], and Kenza Oufaska[2]

[1] IA Laboratory, Faculty of Sciences Meknes, Moulay Ismail University,
Faculty of Sciences - BP 11201 Zitoune, Meknes, Morocco
yassir.rouchdi@gmail.com, khalid.elyassini@gmail.com
[2] TICLab, International University of Rabat, Rabat, Morocco
Kenza.Oufaska@uir.ac.ma

Abstract. This paper presents Radio frequency identification components, functioning and Middleware's role. It discusses ESN middleware architecture and explains its security and privacy issues, including a discussion about resolving these problems by applying Role based access Control model as an authentication tool regulating back-end application's access to data. Moreover, it presents the proposed architecture of our three layers middleware 'UIR-', Explaining how Complex event processing can handle RIFD and WSN data, shows RBAC rules application and gives details on the implementation process.

Keywords: ESN middleware · RFID · WSN · CEP · RBA

1 Introduction

The internet of things holds the promise to offer advanced connectivity of devices, networks, and services that goes beyond machine-to-machine communications and to cover a wide range of protocols, domains, and applications. The interconnection of these embedded devices is expected to marshal in automation in nearly all fields, while also empowering advanced applications and elaborating to areas such as smart cities. In order to do so, IoT assembles both wireless and wired technologies into the same network, using Low-power wide-area networking (LPWAN) for long-range wireless connections, HALOW and LTE-advanced for medium-range, and Radio Frequency identification between many others (Bluetooth Low Energy, NFC, WIFI …) for short-range communication. One of the focuses of scientists nowadays is resolving the issues occurring during the use of combined technologies resulting in unexpected complex events.

WSN and RFID are both very efficient and reliable technologies but aren't any exception of the rule, their combination means dealing with RFID imperfect privacy and security and every WSN complex event issue, during this paper we will be trying to resolve some of the problems, therefore making both RFID and WSN more secure and stable. This paper presents Radio frequency identification components, functioning and Middleware's role. It discusses ESN middleware architecture and explains its security and privacy issues,

© Springer International Publishing AG, part of Springer Nature 2018
M. Ben Ahmed and A. A. Boudhir (Eds.): SCAMS 2017, LNNS 37, pp. 966–975, 2018.
https://doi.org/10.1007/978-3-319-74500-8_86

including a discussion about resolving these problems by applying Role based access Control model as an authentication tool regulating back-end application's access to data. Moreover, it presents the proposed architecture of our three layers middleware 'UIR-', Explaining how Complex event processing can handle RIFD and WSN data, shows RBAC rules application and gives details on the implementation process.

2   RFID Components, Functioning and Middleware

RFID stands for Radio Frequency Identification. Its importance and efficiency are expressed by the vast amount of medical, military and commercial applications using this approach Worldwide [1]. Billions of the RFID systems are operated in transportation (automotive vehicle identification, automatic toll system, electronic license plate, electronic manifest, vehicle routing, vehicle performance monitoring), banking (electronic checkbook, electronic credit card), security (personnel identification, automatic gates, surveillance) and medical (identification, patient history) [2].

2.1   RFID Components

RFID systems are basically composed of three elements: a tag, a reader and a middleware deployed at a host computer. The RFID tag is a data carrier part of the RFID system, which is placed on the objects to be uniquely identified. The RFID reader is a device that transmits and receives data through radio waves using the connected antennas. Its functions include powering the tag, and reading/writing data to the tag [1].

Unique identification or electronic data stored in RFID tags can be consisting of serial numbers, security codes, product codes and other specific data related to the tagged object. The available RFID tags in today's market could be classified with respect to different parameters. For example with respect to powering, tags may be passive, semi-passive, and active. In terms of access to memory, the tags may be read-only, read-write, Electrically Erasable Programmable Read-Only Memory, Static Random Access Memory, and Write-once read-many. Tags have also various sizes, shapes, and may be classified with respect to these geometrical parameters. The RFID reader is a device that transmits and receives data through radio waves using the connected antennas.

RFID reader can read multiple tags simultaneously without line-of-sight requirement, even when tagged objects are embedded inside packaging, or even when the tag is embedded inside an object itself. RFID readers may be either fixed or handheld, and are now equipped with tag collision, reader collision prevention and tag-reader authentication techniques [2–4]. Figure 1 illustrates RFID components.

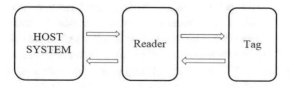

Fig. 1.   RFID components

2.2   RFID Middleware

Radio Frequency Identification (RFID) technology holds the promise to automatically and inexpensively track items as they move through the supply chain. The proliferation of RFID tags and readers will require dedicated middleware solutions that manage readers and process the vast amount of captured data [6]. The efficiency of an RFID application depends on the precision of its hardware components, and the reliability of its middleware. Which is the computer software that provides services to software applications beyond those available from the operating system.

It can be described as "software glue" [7]. Middleware makes it easier for software developers to perform communication and input/output, so they can focus on the specific purpose of their application. Middleware includes Web servers, application servers, content management systems, and similar tools that support application development and delivery. It is especially integral to information technology based on Extensible Markup Language (XML), Simple Object Access Protocol (SOAP), Web services, SOA, Web 2.0 infrastructure, and Lightweight Directory Access Protocol (LDAP) [3, 8].

2.3   Middleware's Basic Functions

The three primary functions of an RFID middleware can be broadly classified as:

- 'Device integration' (that is, connecting to devices, communicating with them in their prescribed protocols and interpreting the data).
- 'Filtering' (the elimination of duplicate or junk data, which can result from a variety of sources, for example: the same tag being read continuously or spikes or phantom reads caused by interference).
- 'Feeding applications', with relevant information based on the information collected from devices after properly performing the appropriate conversions and formatting [7].

2.4   Middleware Architecture

The usual architecture of an RFID middleware is presented in Fig. 2.

Data Processor and Storage: The data processor is responsible for the management and processing of raw data from the readers. This component is also responsible for storing the raw tag data, so that it can be processed. Part of the important processing logic performed here is the Filtering and Grouping of RFID Data. This component also manages the level event data associated with the application. By way of example, when all applications request data captures in the same time interval, processing of the time stamp is performed by this component and the data is then transmitted to the applications.

Application Interface: The application interface component manages the interface of the middleware in dependence on those of the applications. It provides the application with an API (application programming interface) to communicate and evoke the RFID middleware. It accepts application requests and translates them down to the underlying components of the middleware. This component is responsible for integrating enterprise applications with RFID middleware.

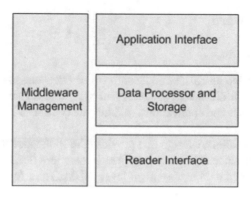

Fig. 2. Middleware architecture [11]

Middleware Management: The middleware management component helps with the management of the RFID middleware conspiracy. It provides information about all the processes running in the middleware. The middleware management provides the administrator with the following features:

1. Add, remove, and modify the RFID readers connected to the system.
2. Change various settings by applications.
3. Enable and disable various functions supported by the middleware.

3 Related Works

In the RFID domain, Savant middleware is a successful implementation of the EPC network. Currently, many of the large IT companies already offer commercial RFID software, such as SUN EPC Network and IBM WebSphere RFID Premises Server. More recently, CEP technology is used in several RFID middleware systems. Event processing language was used to define complex events. In this paper we will be applying complex event processing to combine unions and intersections of both RFID and WSN simple events, defined as complex events.

Concerning the WSN part, several collectors of sensor data and sharing architecture already distinguished [3, 6, 7]. Global Network of sensors (GSN) middleware which is the database Capture of virtual sensors and powerful query tools makes access to the heterogeneous wireless sensor knots easier. Hi-fi architecture [8] is a hierarchical architecture for processing distributed RFID data and network of sensors. It includes many components, such as data receiver, data stream processor, data sender, resource manager, query listener, etc. This article propose a new approach. Instead of building our architecture from scratch, UIR middleware design is built according to already developed RFID standards. it leads to a framework suitable for both RFID and WSN integration applications.

4   UIR Middleware

4.1   UIR Architecture

We propose to develop an RFID middleware called UIR- bearing in mind the design problems discussed in the second chapter. Our system is organized as a three-tiered architecture, with back-end applications, middleware (UIR-) and both RFID and WSN hardware.

UIR- middleware offers a design that provides the application with a neutral device protocol and an independent platform interface. It integrates three hardware abstraction layer (HAL), event and data management layer (EDML) and Application Abstraction Layer (AAL).

The next figure (Fig. 3) illustrates UIR-RFID architecture.

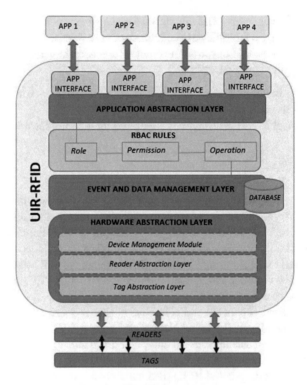

Fig. 3.   UIR-RFID architecture

4.2   Hardware Abstraction Layer

HAL is the lowest layer of (UIR-) and is responsible for interaction with the hardware. It allows access to devices and tags in an independent manner of their various characteristics through layers of tag abstraction and reader.

The reader abstraction provides a common interface for accessing hardware devices with different characteristics such as protocols (ISO 14443, EPC Gen2, ISO 15693), UHF (HF) and host side interface Interface (RS232, USB, Ethernet).

The abstraction of the reader exposes simple functions such as opening, closing, reading, writing, etc. To accomplish complex operations of the readers.

The abstraction of readers and tags in UIR- make it extensible to support various tags, readers and sensors.

The device management module in HAL is responsible for the dynamic loading and unloading of the reader libraries depending on the use of device hardware. This allows the system to be light because only the required libraries are loaded. This layer contains the devices for various operations, as specified by the upper layers. It is also responsible for monitoring and reporting the status of the device. Some of the functions provided by the HAL to access RFID hardware are as follows:

- The Device-opening: function is responsible for opening a connection with the device. The connection parameters are provided as an argument to this function. When a successful connection is made to the reader, a response is returned by this function. This response is then used as a reference to access the device in subsequent calls.
- The Device-reading function: reads data from the internal reader. The read parameters such as the protocol to be read by the reader, the size of the data to be read, are specified as arguments of this function. The function responds successfully if valid data is present in the reader if not with an error code.
- The Device-Writing function writes data to the Tag. Arguments Specified with this function, the unique ID partially or totally, which triggers the data to be written to the tag. The function responds successfully if the data is written to the Tag or returns an error code (for example, when the tag is not identified only).

4.3   Event and Data Management Layer (EDML)

EDML handles various reader-level operations, such as reading tags and informing readers of disconnected notions such as device failure, write failure, and so on. The layer acts as a conduit between the hardware abstraction layer (HAL) and the application abstraction layer (AAL). It accepts commands from AAL, processes them and therefore issues commands to HAL. Similarly, the responses are brought from the HAL, processed and transmitted to the LAA by this layer.

The EDML is the kernel of this middleware. It filters out uninteresting data, formats the remaining useful data and builds complex events according to real-time specifications. The event specification analyzer interprets and transforms event specifications into four processes steps: filtering, grouping, aggregating, and complex construction of events. The volume of event data is very important in the NSE middleware system. The filter selects only those events in the upper layer, thus reducing the reports data dramatically. In the ratio to the upper layer, event data are separated in several groups for a clear demonstration. The aggregation provides statistical information event data. By aggregating, the volume of the declared data may be reduced again. Later, simple events are grouped together to form complex events. Complex events provide more meaningful reporting and improve automation of the system.

Simple and Complex Events in ESN Middleware

In the ESN middleware, two types of event objects exists, RFID event object and sensor event object. the RFID event object is defined as a tuple (ID, L, T) where ID represents the EPC code of the RFID tag, L for Location and T for time. The sensor event object, which is more complex than RFID, is defined hierarchically. The first layer is still a tuple (ID, L, T, D), where ID is the identification of a Sensor node, L Location, T Time and D is the sensor data (Temperature, weight …). The ID includes both the reader ID and the ID of the sensor node. To achieve unique IDs identification, the reader's EPC codes, and the sensor node's can be used as identifiers. Whereas in the second layer of event object in D, a sensing type tuple such as (humidity, temperature, pressure) can be included.

A simple event is the RFID event or sensor with constraint. For example, the RFID event (S1) in an application level (localization constraint) is a simple RFID event.

S1 = (ID, L {L = "Test Location"}, T) and the sensor event (S2) per example, with a temperature higher than $20°$ is also a simple sensor event. S2 = (ID, L, T, D {Value of D. temperature > 20}). A complex event is a combination of simple events or complex events with the following set: **AND** ($\land$): E1 $\land$ E2 represents two events, where E1 and E2 occur together. **OR** ($\lor$): E1 $\lor$ E2 means E1 or E2 occurs. **NOT** (!): ! E1 means that E1 does not occur. **SEQ** ($\rightarrow$): E1 $\rightarrow$ E2 means that E1 is followed by E2. Relative peripheral (Rp): Rp (E1, E2, E3) means that E2 occurs between E1 and E3, perhaps several times.

4.4   Application Abstraction Layer (AAL)

The Application Abstraction Layer (AAL) provides various applications with an independent interface to RFID hardware. The interface is designed as an API by which Applications use UIR-RFID services. All operations at the application level such as reading, writing, etc. Are interpreted and translated into the lower layers of UIR- by the AAL. In order to restrain unauthorized back-end application from getting access to Data, we used Role-Based access control method of regulating access to guarantee data protection from unauthorized back-end applications.

Role-Based Access Control

To clarify the notions presented in the previous section, we give a simple formal description, in terms of sets and relations, of role based access control. No particular implementation mechanism is implied. For each subject, the active role is the one that the subject is currently using:

- AR(s: subject) = {the active role for subject s}.

  Each subject may be authorized to perform one or more roles:

- RA(s: subject) = {authorized roles for subject s}.

Each role may be authorized to perform one or more transactions:

- TA(r: role) = {transactions authorized for role r}.

    Subjects may execute transactions.

- The predicate exec(s, t) is true if subject 's' can execute transaction 't' at the current time, otherwise it is false:

- Exec(s: subject, t: tran) = true if subject s can execute transaction t.

RBAC Primary Rules

1. Role assignment: A subject can exercise a permission only if the subject has selected or been assigned a role. $\forall$s: subject, t: tran (, exec(s, t) $\Rightarrow$ AR(s) $\neq$ O/).
2. Role authorization: A subject's active role must be authorized for the subject. With rule 1 above, this rule ensures that users can take on only roles for which they are authorized. $\forall$s: subject (, AR(s) $\subseteq$ RA(s)).
3. Permission authorization: A subject can exercise a permission only if the permission is authorized for the subject's active role. With rules 1 and 2, this rule ensures that users can exercise only permissions for which they are authorized. $\forall$s: subject, t: tran (, exec(s, t) $\Rightarrow$ t $\in$ TA(AR(s))).

4.5 UIR-RFID Implementation

For the UIR-RFID implementation, we propose the use of Microsoft Visual Studio .Net 2010 as Framework and development tool. The reasons for this choice are the powerful utilities for Application Development that this framework provides. The code to use to develop the Project is C Sharp (C#). We propose the use of graphical user interface features provided by the .Net Framework and Microsoft Visio 2013 to develop conceptual models and middleware architecture. For the purpose of data management and storage, we offer Microsoft SQL Server 2008, It is a cohesive set of tools, utilities and interfaces collaborating to provide excellent data management. The database schema generated by this DBMS provides a comprehensive view of the data and its relationships. To view the database, retrieve, modify, delete, and store data, we propose the use of the SQL language (Structured Query Language).

5 Application Example

A smart medicament transportation application, can illustrate the use of our ESN middleware. Medicaments put into small boxes are tagged and localized by an Alien reader installed in the transportation vehicle. Meanwhile, a sensor node equipped with a microcontroller, a transceiver, a temperature sensor and a humidity sensor senses the temperature and humidity of the Medicaments environment. Sensor nodes in different Vehicles compose a WSN and transfer data to reader using. As an example, one of the complex events in the above application is generated when there is Medicaments on the truck and temperature is above 13°. For this complex event, the corresponding event rules are as follows: S1 = (ID{ID = 'MedsID'}, L{L = "reader A"}, T) S2 = (ID, L{L = "Track A"}, T, D{D.temperature.value > 13}) C1 = S1 ∧ S2. When this complex event is generated, EPCIS gets a report, and action would be token.

6 Conclusion

Our proposed middleware (UIR-) architecture offered a solution to many issues discussed in earlier chapters. Resolving the Multiple Hardware Support issue on The reader abstraction layer that provides a common interface for accessing RFID hardware devices with different characteristics, Resolving Synchronization and Scheduling in the middleware on The EDML that manages data flow between the other layer handling various reader-level operations, Servicing Multiple Applications and offering a Device Neutral Interface to the applications on The Application Abstraction Layer (AAL) that provides various applications with an independent interface to RFID hardware, Resolving Scalability problems on The Hardware Abstraction Layer that allows access to devices and tags in an independent manner of their various characteristics through layers of tag and reader abstraction. Moreover, we explain the use of CEP technology in the ESN middleware, the integration architecture between RFID and WSN. The events of RFID, WSN and their interactions were also analyzed. By adopting CEP technology, we built a middleware system which has the functions of filtering, grouping and aggregation of real-time event data. And Regulating Access to data by using The RBAC Mechanism that makes sure only authorized users (applications) access the needed data depending on the permissions allowed and the role assigned.

References

1. Ajana, M., Boulmalf, M., Harroud, H., Hamam, H.: A policy based event management middleware for implementing RFID applications. In: IEEE International Conference on Wireless and Mobile Computing, Networking and Communications (2009)
2. United States Government Accountability Office: INFORMATON SECURITY Radio Frequency Identification Technology in the Federal Government. United States Government Accountability Office, May 2005. http://epic.org/privacy/surveillance/spotlight/0806/gao05551.pdf
3. Sheng, Q., Li, X., Zeadally, S.: Enabling next-generation RFID applications: solutions and challenges. Computer **41**(9) (2008)
4. Al-Mousawi, H.: Performance and reliability of Radio Frequency Identification (RFID)", in Agder University College, June 2004. http://student.grm.hia.no/master/ikt04/ikt6400/g28/Document/Master_Thesis.pdf
5. Kefalakis, N., Leontiadis, N., Soldatos, J., Donsez, D.: Middleware building blocks for architecting RFID systems. Mob. Lightweight Wirel. Syst. **13**, 325–336 (2009)
6. Su, X., Chu, C.-C., Prabhu, B.S., Gadh, R.: On the creation of automatic identification and data capture infrastructure via RFID and other technologies. In: Yan, L., Zhang, Y., Yang, L.T., Ning, H. (eds.) The Internet of Things: From RFID to the Next-Generation Pervasive Networked Systems, p. 24. Auerbach Publications, Taylor & Francis Group (2007)

7. Bornhövd, M.C., Lin, T., Haller, S., Schaper, J.: Integrating automatic data acquisition with business processes - experiences with SAP's auto-ID infrastructure. In: Proceedings of the 30th International Conference on Very Large Data Bases (VLDB), Toronto (2004)
8. Bell, S.: RFID Technology and Applications, pp. 6–8. Cambridge University Press, London (2011)
9. Catherine O'Connor, M.: RFID is the key to car clubs success. RFID J. (2011)
10. Russell, R.: Manufacturing execution systems: moving to the next level. Pharm. Technol. **28**, 38–50 (2004)
11. Darwish, M.: Analysis of ANSI RBAC support in commercial middleware. Ph.D. thesis, University of British Columbia, Vancouver, Canada, April 2009
12. Sandhu, R., Ferraiolo, D.F., Kuhn, D.R.: The NIST model for role-based access control: toward a unified standard. In: 5th ACM Workshop Role-Based Access Control, pp. 47–63, July 2000
13. Ferraiolo, D.F., Kuhn, D.R., Sandhu, R.: RBAC standard rationale: comments on a critique of the ANSI standard on role-based access control. IEEE Secur. Priv. **5**(6), 51–53 (2007)
14. Thiell, M., Zuluaga, J., Montanez, J., van Hoof, B.: Green logistics – global practices and their implementation in emerging markets, p. 2, Colombia (2011)
15. Sandhu, R., Coynek, E.J., Feinsteink, H.L., Youman, C.E.: Role-based access control models. IEEE Comput. **29**(2), 38–47 (1996)
16. Sandhu, R.: Role-Based Access Control (RBAC). CS 6393 Lecture 3, 29 January 2016
17. Zhang, T., Ouyang, Y., He, Y.: Traceable air baggage handling system based on RFID tags in the airport. J. Theoret. Appl. Electron. Commer. Res. **3**(1), 106–115 (2008). School of Computer Science and Engineering, Beijing University of Aeronautics and Astronautics, China
18. Weil, R., Coyne, E.: ABAC and RBAC: scalable, flexible, and auditable access management. IT Prof. **15**, 14–16 (2013)
19. Jin, C., Shen, A., Yu, W.: The RBAC system based on role risk and user trust. Int. J. Comput. Commun. Eng. **5**, 374 (2016)

Big Data Analytics for Supply Chain Management

Mariam Moufaddal[1]([⊠]) [iD], Asmaa Benghabrit[2],
and Imane Bouhaddou[1]

[1] LM2I Laboratory, ENSAM Meknes, Marjane 2, P.B. 15250, Al-Mansor,
Meknes, Morocco
moufaddalm@gmail.com, b_imane@yahoo.fr
[2] LMAID Laboratory, ENSMR Rabat, Avenue Hadj Ahmed Cherkaoui,
P.B. 753, Agdal, Rabat, Morocco
a.benghabrit@gmail.com

Abstract. All our daily digital actions generate data at an alarming velocity, volume and variety. To extract meaningful value from big data, we need optimal processing power, analytics capabilities and skills. Nowadays, big data solutions are widely applied in different types of organizations. Such solutions bring multiple benefits in managing supply chains. The aim of this paper is to give an overview of big data analytic techniques used in supply chain management based on the latest version of the SCOR model.

Keywords: Big data · Analytic techniques · Supply chain
Supply chain management · SCOR model

1 Introduction

The sustained success of Internet powerhouses such as Amazon, Google, Facebook, and eBay provides evidence of a fourth production factor in today hyper-connected world [1]. Besides resources, labor, and capital, there is no doubt that information has become an essential element of competitive differentiation. Even though companies have relied on tools and techniques for many years to make decisions based on relevant information, with the rise of the digital consumer, these tools become obsolete, and new technologies are needed. Matthias Winkenbach, Director MIT Megacity Logistics Lab, has reported that more and more companies are sitting on tons of data, but they do not know what to do with it, or how to understand it. These massive data are called nowadays «Big Data».

Many think that the appearance of Big Data allowed only to deal with large volumes of data in near real-time with reliability, but it is in dealing with a multiplicity of data from various sources that big data gave the way to new methods of analysis and use of data in various domains. Thus, supply chains are ideally placed to benefit from the technological and methodological advancements of Big Data. It brings a new source of competitive advantages for managers to carry out supply chain management so as to obtain enhanced visibility, the ability to adjust under demand and capacity fluctuations in a real-time basis [5].

© Springer International Publishing AG, part of Springer Nature 2018
M. Ben Ahmed and A. A. Boudhir (Eds.): SCAMS 2017, LNNS 37, pp. 976–986, 2018.
https://doi.org/10.1007/978-3-319-74500-8_87

The rest of this paper is organized as follow: in the next section, we explain the relation between Supply Chain Management (SCM) and big data. Then we describe the Supply Chain Operations Reference (SCOR) model as the standard diagnostic tool for SCM. The third section assigns, to each SCOR area, the adequate type of analytic techniques, algorithms and big data technologies. Concluding remarks and some potential-work suggestions are given in the last section.

2 Supply Chain and Big Data are a Perfect Match

2.1 «Big» Data

While the term "big data" is relatively new, the act of gathering and storing large amounts of information for eventual analysis is ages old. Organizations collect data from a variety of sources, including business transactions, social media and information from sensors or machine-to-machine data [6]. Data streams in at an unprecedented speed and must be dealt with in a timely manner. Big Data can be described as datasets which could not be captured, managed, and processed by general computers within an acceptable scope [7].

The name has come to be the technology improving the storage, management, processing, interpretation, analysis and visualization of the huge flood of data [2]. The initial technical definition of Big Data, which is the 3Vs – Velocity, Variety and Volume, was provided in a research report of Meta Group (now Gartner) [3]. Later as the experience with Big Data growth, the definition was extended to 5Vs with Veracity and Visualization then to 7Vs with Variability and Value. We explicit these characteristics as follow:

Velocity is dealing with streaming data in an unprecedented speed and a timely manner. RFID tags, sensors and smart metering are driving the need to deal with torrents of data in near-real time. Variety is examining data coming in all types of formats; from structured, numeric data in traditional databases to unstructured emails, videos, audios and financial transactions. Volume is how much data we possess; what used to be measured in Gigabytes is now measured in Zettabytes or even Yottabytes. Veracity is all about making sure the data is accurate, which requires processes to keep the bad data from accumulating in systems. Visualization means that data must be read and understood at first sight by users and decision makers. Variability is different from variety. A coffee shop may offer 6 different blends of coffee, but if we get the same blend every day and it tastes different every day, that is variability. The same is true of data, if the meaning is constantly changing it can have a huge impact on data homogenization. Value is the end of the game. After addressing all characteristics – which takes a lot of time, effort and resources – we need to be sure the organization is getting value from the data.

Now credit reference agency Experian – in a recently published white paper [4], are proposing to add another "V" to the checklist which is Vulnerability and it discusses the privacy issue with Big Data.

Today, humanity produces each year a volume of digital information in the order of Zettabyte. That is almost as many bytes as there are stars in the universe [8]. Already in 2010, Eric Schmidt, Google boss, announced that "each two days, we produce as much information as we have generated since the dawn of civilization until 2003". These astronomical amounts of data that are exchanged constantly have great strategic interest for any organization that will be able to sort, analyze and extract actionable information [9].

2.2   SCM: Latest Version of the SCOR Model

The supply chain for a product is the network of firms and facilities involved in the transformation process from raw materials to a product and in the distribution of that product to customers [10]. The needs and capabilities of material suppliers, service suppliers, and especially customers are incorporated into strategic planning, as firms view operations in terms of supply chain interactions and strategies [11]. The challenge of coordinating operations across all facets of a business has become known as Supply Chain Management (SCM). Copacino highlights the importance of integration in his definition of SCM: "The new vision of Supply Chain Management links all the players and activities involved in converting raw materials into products and delivering those products to consumers at the right time and at the right place in the most efficient manner." [12].

The Supply Chain Operations Reference (SCOR) model has been developed to describe the business activities associated with all phases of satisfying a customer demand. The model itself contains several sections and is organized around the five primary management processes of Plan, Source, Make, Deliver and Return. The Plan process balances aggregate demand and supply to develop a course of action which best meets sourcing, production, and delivery requirements. The Source process procures goods and services to meet planned or actual demands. Whereas the Make process transforms product to a finished state, to meet planned or actual demand. The Deliver process provides finished goods and services to customers. Finally, the Return process is associated with moving material from a customer back through the supply chain to address defects in product, ordering, or manufacturing, or to perform upkeep activities[1].

In the latest version of the SCOR model, another management process was added. It is the Enable process which is behind all the other processes and it supports the realization and governance of the planning and execution processes of supply chains.

By describing supply chains using these processes building blocks, the model can be used to describe supply chains that are very simple or very complex using a common set of definitions. As a result, disparate industries can be linked to describe the depth and breadth of virtually any supply chain. Thus, leading logistical practice has shifted from an exclusively internal focus to collaboration across the full range of supply chain participants [13].

3   Technologies and Algorithms for SCA

Along the supply chain, many data sources (GPS, QR codes, sensors, social network…) are providing us with huge amounts of data. The majority of raw data does not offer a lot of value in its unprocessed state. Of course, by applying the right set of tools and techniques, we can pull powerful insights from this stockpile of bits.

---

[1] From http://docs.huihoo.com/scm/supply-chain-operations-reference-model-r11.0.pdf.

3.1   Analytic Techniques Types: Where There is Data, There is Analytics

As digitization, has become an integral part of everyday life, data collection has resulted in the accumulation of huge amounts of data that can be used in various beneficial application domains [14] mainly supply chain management.

Supply chain analytics focuses on the use of information and analytical techniques to drive better decisions regarding flows in the supply chain [10]. Put differently, supply chain analytics focuses on analytical approaches to make decisions that better match supply and demand. Following our literature review, several advanced analytic techniques were found that can be categorized into three types: descriptive, prescriptive and predictive [15].

The descriptive analytics aims to identify problems and opportunities within existing processes and functions. It derives information from significant amounts of data and provides information regarding "what has happened?" and "what is happening at the moment?". The GPS, RFID technologies and sensors collect data based on a real-time which will be summarized and converted into information relative to location and quantity of goods in a supply chain [16, 17]. This information provides managers with tools to make the necessary adjustments.

The predictive analytics is the use of algorithms and techniques such as machine learning, social network analysis, data mining (i.e., regression analysis) [15, 19, 20] to make predictions concerning the future intended from Big Data. Emotions, opinions, and behaviors are all subjective information that is extracted from vast amounts of data and is intended to answer the question "What might happen?".

The prescriptive analytics are often referred to as advanced analytics. It derives decision recommendations based on descriptive and predictive analytics models, mathematical optimization models and simulation techniques [16, 17, 21]. It can be used to drive important, complex and time-sensitive decisions as it addresses the question "What should we do?".

3.2   Panorama of Analytic Techniques

In any big data setup, the first step is to capture lots of digital information. With data in hand, we can begin doing analytics. But where do we begin? What are Big Data analytics techniques which are used in each supply chain process? And which type of analytics is most appropriate for our big data environment?

To address these questions, Table 1 summarizes technologies and outlines algorithms and analytic techniques used to process data along the supply chain based on a literature review. We decided to go through all the supply chain processes as each process presents constraints and need special analytic techniques to get insights from the data generated from it.

The purpose of this literature review is to identify SCOR areas which are influenced by the big data analytics. So, following this study, we noticed that there are some SCOR areas that researchers focused on more than others which are the Plan, Make and Deliver & Return processes.

For supply chain planning, companies use mostly predictive approaches to determine planned capabilities and gaps in demand or resources and identify actions to

Table 1. Classification of analytics techniques for each SCOR area.

	Predictive analytics	Prescriptive analytics	Descriptive analytics	Big Data technologies
Plan	• Causal forecasting methods (non-linear, and logistic regression) [10] • Simulation methods (Monte Carlo) [22, 23] • Time-series analysis (exponential smoothing: used in short-term and intermediate range forecasting, auto-regressive model) [10, 24–27] • Data/text/web mining (sentiment analysis clustering, association rules, classification) and forecasting [10, 15, 16, 20, 28, 29] • Machine learning (supervised and unsupervised learning) [16] • Social network analysis [16, 25] • Correlation analysis [30] • Predictive modeling (KNN, neural networks) [25]	• Optimization methods: strategic planning [15, 24], operational planning [10, 15, 22]		• SAP SB for advanced planning and optimization • Hadoop ecosystem • Citus DB
Source	• Time series methods (moving average, exponential smoothing, autoregressive models) [10] • Data-mining techniques (cluster analysis, market basket analysis, sentiment analysis) [15, 22, 28, 33] • Causal forecasting methods (linear, non-linear, and logistic regression) [10] • Social network analysis • Statistical forecasting techniques (for inventory needs) [34, 35]	• Multi-criteria decision-making techniques (Analytic Hierarchy Process – AHP) [36, 37] • Enterprise social networking [38] • Game theory (auction design, contract design) [39] • Optimization methods (linear and non-linear programming) [10, 15, 17] • Cost modeling [40, 41]	• Supply chain mapping (tracking every flow in the SC) [38] • Statistics (false nearest neighbour algorithm, Gauss–Newton algorithm) [10, 15–17, 42]	• Sourcemap.com: open source tool • SAP invoice and goods receipt reconciliation • SAP SB for inventory management • SAP SB for extended warehouse management

(continued)

Table 1. (continued)

	Predictive analytics	Prescriptive analytics	Descriptive analytics	Big Data technologies
Make	• Time series methods (moving average, exponential smoothing, autoregressive models) [10] • Data-mining techniques (sentiment analysis, clustering, association rules) [10, 15, 19, 22, 25, 44] • Machine learning algorithms (supervised and unsupervised learning, by reinforcement) [19] • Associative forecasting methods (executive opinion) [52, 53] • Predictive maintenance analytics (deep learning) [54]	• Network design (graph algorithm, cutset saturation method) [45] • Mixed-Integer Linear Program (MILP) [46] • Genetic algorithms [49] • Manufacturing scheduling (synchro PRO, visual ERP) [49] • Workforce scheduling [50] • Optimization methods (linear and non-linear programming) [10, 15, 17, 42] • The burbidge connectance concept [51] • Deduction graph [47]	• Supply chain visualization (data visualization approach) [10, 17, 18, 47] • Induction graph [48]	• SAP ERP: applied planning and optimization modules • Internet of things • Systema big data real time analysis for manufacturing • Tool for Action Plan Selection (TAPS) [55] • LINGO [47] • Apache Mahout ML • TableauSoftware
Deliver & return	• Data-mining techniques (sentiment analysis, regression analysis) [10, 15, 16, 22, 56] • Machine learning (supervised and unsupervised learning) [16] • Social network analysis [16] • Time series forecasting (Naïve, simple moving average, seasonal indexes) [56] • Transport analytics (the bottleneck approach) [57] • Predictive modeling (KNN, neural networks) [25]	• Stochastic dynamic programming [66] • Multi-commodity network flow model: linear programming formulation [67] • Network simplex method [68] • Optimization methods (linear and non-linear programming) [10, 15, 17, 18] • Mathematical models [17, 21, 42] • MapReduce programming model [60]	• Supply chain visualization [18] • Cognitive mapping [59]	• The P&G business sphere • SAP supply chain control tower • WasteRoute vehicle-routing software • Internet of things (i.e., GPS, RFID tags) • SAP SB for transportation management • Hadoop ecosystem • Decision explorer [47] • IBM PureData
Enable	• Scenario and risk profile analysis, Monte Carlo and discrete-event simulation methods [22, 32]	• Heuristic/meta-heuristic approaches: bid-price controls (bid-prices are approximated using linear programming), branch-and-bound, branch-and-cut [57] • Price-optimization algorithms [57] • Mathematical models and simulation techniques [10, 18] • Risk assessment [61]	• Data visualization techniques [18] • Influence diagram [58]	• In-memory database: OLAP & OLTP • SAP SB for product lifecycle management • Infosys SC risk management • Infosys SC performance simulator • Analytica [47]

correct these gaps. They begin with data mining techniques such as clustering or market basket analysis for analyzing purchase models, knowing customer's perceptions regarding products and services and finding the factors which determine the demand [62]. These factors will then be analyzed using time-series approaches to predict product demand [26, 27]. This demand prediction constitutes the main input of planning in supply chain. It is used at the strategic, tactical and operational levels to plan operations (procurement, production and distribution) and sales to synchronize demand with offer [10, 16].

In Make process, supply chain analytics enable manufacturers to understand the different production costs involved and how they influence the bottom line. For that, companies use the prescriptive analytics at the tactical level such as genetic algorithms [49] to determine the capacity of plants. They use predictive analytics to schedule workforce and perform preventive maintenance tasks. Finally, they can use descriptive analytics to get insights regarding the production capacity levels and decide whether improvements are needed to maximize productivity [63, 64].

In Deliver & Return processes, companies use mostly the prescriptive and predictive approach to plan distribution and transport [10]. According to the council of supply chain management professionals [43], global logistics generates massive amounts of data as – shippers, logistics service providers and carriers – manage logistics operations. This data is generated from different sources and each portion of it is intended for making a different decision. For instance, location-based data that is obtained from RFID tags, GPS chips in mobile devices, EDI transactions [5] and GPS-enabled big data telematics can be harnessed for logistics planning purposes. This deals with the distribution of products from supply points (i.e., warehouses) to demand points (i.e., retailers' sites). Fuel consumption, usury, and therefore maintenance and vehicles substitution generate important costs that can be limited by analyzing these location-based data using data mining techniques and time-series methods. However, the benefits are not limited only to the economic profitability. They are also environmental and human. Collecting data, analyzing it, making it relevant and workable enables an intelligent driving. Driving better means reducing greenhouse gas emissions and consequently carbon footprint. Driving better means also reducing stress at the wheel, yet increasing safety of drivers.

Finally, there is growing interest in the use of data and advanced analytics for supply chain management. Well-planned and implemented decisions contribute directly to the bottom line by lowering sourcing, manufacturing, maintenance, transportation, storage, and disposal costs.

4   Conclusion and Future Research Directions

Driven by the development of new technologies and the pressure of environmental issues, the way supply chains are managed is about to undergo profound changes. The challenge will be to adapt as quickly as possible or risk being irremediably outdated. Adopting new technologies that may improve efficiency and profitability in supply chains is a way to differentiate from competitors. To achieve this goal, managers need to have an overview of their customers and their needs. Thus, Big Data is "the revolutionary technology" that gathers data from multiple sources and extracts value from it using tools and methods designed for this purpose.

It is important to remember that the primary value from Big Data comes not from the data in its raw form, but from the processing and analysis of it and the insights, products, and services that emerge from analysis [31]. No need to be Amazon to benefit from the advantages of big data.

In this paper, we highlighted the importance of Big data analytics for supply chains. We have presented a panorama of algorithms, analytic techniques and technologies of big data used in supply chain management through a literature review. In our analysis, we have identified the main processes modeling the global supply chains based on SCOR model. For each process, we have defined the type of analytic technique to be performed and the big data technology that goes with it.

We identified three types of analytic techniques: descriptive analytics that answers the question "what is going on?", predictive analytics that answers the question "what is going to happen?" and prescriptive analytics which answers the question "What should we do?". The question that still does not have an answer is "why did something happened at a certain moment in the past?". This type of analytics is used to validate or to reject the different business hypotheses. So as future research direction, we propose to cover inquisitive analytics known as diagnostic analytics which answer this question.

Furthermore, we have particularly noticed that most studies concern only three supply chain processes out of six. This is probably because big data analytics is still on its early stages of development in supply chains. We estimate that all processes are important and must benefit from big data technologies, and it is also an almost virgin field in which one can work and participate in its development.

We also noticed that the less treated SCOR area was the Enable process. Admittedly it is a new process that was added in the last version of the SCOR model, but it is the SCOR area that coordinates between supply chain partners and support processes in executing operations. Indeed, a supply chain consists of many parts or elements of various types, which are linked to each other directly or indirectly. These various elements and their interrelationships are significant for complexity occurring in a system [65]. We can consider as future research focusing the study on this process and figuring out the theories which can be mobilized for resolving the complexity of the supply chain using big data analytics.

References

1. Martin, J., et al.: Big data in logistics a DHL perspective on how to move beyond the hype. DHL Customer Solutions & Innovation, pp. 1–30 (2013)
2. Kaisler, S., et al.: Big data: issues and challenges moving forward. In: 2013 46th Hawaii International Conference on System Sciences (HICSS), Wailea, Maui, HI, 7–10 January, pp. 995–1004. IEEE (2013)
3. Laney, D.: Meta group (now Gartner) (2001)
4. White paper - A Data Powered Future, Experian 2016

5. Swaminathan, S.: The effects of big data on the logistics industry: PROFIT ORACLE Technology Powered. Business Driven (2012). http://www.oracle.com/us/corporate/profit/archives/opinion/021512sswaminathan-1523937.html
6. SAS Institue: What is big data. http://www.sas.com/en_us/insights/big-data/what-is-big-data.html
7. Chen, M., et al.: Big data: a survey. Mobile Netw. Appl. **19**(2), 171–209 (2014)
8. IDC: International Data Corporation (2012)
9. Acteaos (2015). http://acteos.fr/actualite-and-evenements/big-data-ameliore-supply-chain-management
10. Souza, G.C.: Supply chain analytics. Bus. Horiz. **57**, 595–605 (2014)
11. Stank, T.P., et al.: Supply chain collaboration and logistical service performance. J. Bus. Logist. **22**(1), 29–48 (2001)
12. Copacino, W.C.: Supply Chain Management: The Basics and Beyond. APICS Series on Resource Management, p. 5. St. Lucie Press, Boca Raton (1997)
13. Stein, M., Voehl, F.: Macrologistics Management. St. Lucie Press, Boca Raton (1998)
14. Al Nuaimi, E., et al.: Application of big data to smart cities. J. Internet Serv. Appl. **6**, 25 (2015)
15. Hahn, G.J., et al.: A perspective on applications of in-memory analytics in supply chain management. Decis. Support Syst. **76**, 45–52 (2015)
16. Chae, B., Yang, C., et al.: The impact of advanced analytics and data accuracy on operational performance: a contingent resource based theory (RBT) perspective. Decis. Support Syst. **59**, 119–126 (2014)
17. Groves, W., et al.: Agent-assisted supply chain management: analysis and lessons learned. Decis. Support Syst. **57**, 274–284 (2014)
18. Duta, D., Bose, I.: Managing a big data project: the case of Ramco Cements Limited. Int. J. Prod. Econ. **165**, 293–306 (2015)
19. Zhong, R.Y., et al.: A big data approach for logistics trajectory discovery from RFID-enabled production data. Int. J. Prod. Econ. **165**, 260–272 (2015)
20. Kahn, K.B.: Solving the problems of new product forecasting. Bus. Horiz. **57**, 607–615 (2014)
21. Hazen, H.T., et al.: Data quality for data science, predictive analytics, and big data in supply chain management: an introduction to the problem and suggestions for research and applications. Int. J. Prod. Econ. **154**, 72–80 (2014)
22. Waller, M.A., Fawcett, S.E.: Data science, predictive analytics, and big data: a revolution that will transform supply chain design and management. J. Bus. Logist. **34**, 77–84 (2013)
23. O'Dwyer, J., Renner, R.: The promise of advanced supply chain analytics. Supply Chain Manag. Rev. **15**, 32–37 (2011)
24. Stadtler, H.: Supply chain management and advanced planning: basics, overview and challenges. Eur. J. Oper. Res. **163**, 575–588 (2005)
25. Manyika, J., et al.: Big Data: The Next Frontier for Innovation, Competition, and Productivity. McKinsey Global Institute, New York City (2011)
26. Cheikhrouhou, N., et al.: A collaborative demand forecasting process with event-based fuzzy judgements. Comput. Ind. Eng. **61**(2), 409–421 (2011)
27. Li, B., Li, J., Li, W., Shirodkar, S.A.: Demand forecasting for production planning decision-making based on the new optimized fuzzy short time-series clustering. Prod. Plan. Control **23**(9), 663–673 (2012)
28. Cohen, J., et al.: MAD skills: new analysis practices for big data. J. VLDB Endow. **2**(2), 1481–1492 (2009)
29. Demirkan, H., et al.: Leveraging the capabilities of service-oriented decision support systems: putting analytics and big data in cloud. Decis. Support Syst. **55**(1), 412–421 (2013)

30. Sanders, N.R.: Big Data Driven Supply Chain Management: A Framework for Implementing Analytics and Turning Information into Intelligence. 1st edn. Pearson Education, Upper Saddle River (2014). 26 p. ISBN 10:0133801284

31. Goetschalckx, M., et al.: Strategic network design. In: Stadtler, H., Kilger, C. (eds.) Supply Chain Management and Advanced Planning: Concepts, Models, Software, and Case Studies, pp. 117–132. Springer, Berlin (2008)

32. Dickersbach, J.T.: Supply Chain Management with APO: Structures, Modelling Approaches and Implementation of SAP SCM 2008, 3rd edn. Springer, Berlin (2009)

33. Rey, T., Kordon, A., Wells, C.: Applied Data Mining for Forecasting Using SAS. SAS Institute, Cary (2012)

34. Downing, M., Chipulu, M., Ojiako, U., Kaparis, D.: Advanced inventory planning and forecasting solutions: a case study of the UKTLCS Chinook maintenance programme. Prod. Plan. Control **25**(1), 73–90 (2014)

35. Wei, C., Li, Y., Cai, X.: Robust optimal policies of production and inventory with uncertain returns and demand. Int. J. Prod. Econ. **134**(2), 357–367 (2011)

36. Ho, W., Xu, X., Dey, P.: Multi-criteria decision making approaches for supplier evaluation and selection: a literature review. Eur. J. Oper. Res. **202**(1), 16–24 (2010)

37. Ekici, A.: An improved model for supplier selection under capacity constraint and multiple criteria. Int. J. Prod. Econ. **141**(2), 574–581 (2013)

38. Cisco: Building a collaboration architecture for a global supply chain (2013). Accessed 14 June 2013. http://www.cisco.com/en/US/solutions/collateral/ns340/ns858/c22-613680_sOview.html

39. Simchi-Levi, D., et al. (eds.) Handbook of Quantitative Supply Chain Analysis: Modeling in the E-Business Era. Kluwer Academic Publishers, Boston (2004)

40. Jain, S., Lindskog, E., Andersson, J., Johansson, B.: A hierarchical approach for evaluating energy trade-offs in supply chains. Int. J. Prod. Econ. **146**(2), 411–422 (2013)

41. Apte, A.U., Rendon, R.G., Salmeron, J.: An optimization approach to strategic sourcing: a case study of the United States Air Force. J. Purch. Supply Manag. **17**(4), 222–230 (2011)

42. Chae, B., et al.: The impact of advanced analytics and data accuracy on operational performance: a contingent resource based theory (RBT) perspective. Decis. Support Syst. **59**, 119–126 (2014)

43. Stroh, M.B.: What is Logistics. Logistics Network (2002). http://www.logisticsnetwork.net/articles/What%20is%20Logistics.pdf

44. Kwon, K., et al.: A real-time process management system using RFID data mining. Comput. Ind. **65**, 721–732 (2014)

45. Funaki, K.: State of the art survey of commercial software for supply chain design (2009). Accessed 18 June 2013. http://scl.gatech.edu/research/supply-chain/GTSCL_scdesign_software_survey.pdf

46. Paksoy, T., et al.: A multi-objective mixed-integer programming model for multi echelon supply chain network design and optimization. System Research and Information Technologies, METU (2009)

47. Tan, K.H., et al.: Harvesting big data to enhance supply chain innovation capabilities: an analytic infrastructure based on deduction graph. Int. J. Prod. Econ. **165**, 223–233 (2015)

48. Zighed, D.A., Rakotomalala, R.: Graphes d'induction. Hermes Science publications, Paris (2000)

49. Pinedo, M.: Scheduling Theory, Algorithms and Systems, 3rd edn. Springer, New York (2008)

50. Campbell, G.: Overview of workforce scheduling software. Prod. Invent. Manag. J. **45**(2), 7–22 (2009)

51. Burbidge, J.L.: A Production System Variable Connectance Model. Cranfield Institute of Technology, London (1984)
52. Lu, C., Wang, Y.: Combining independent component analysis and growing hierarchical self-organizing maps with support vector regression in product demand forecasting. Int. J. Prod. Econ. **128**(2), 603–661 (2010)
53. Beutel, A.L., et al.: Safety stock planning under causal demand forecasting. Int. J. Prod. Econ. **140**(2), 637–639 (2012)
54. IBM: White paper 2014, Big data and analytics in travel and transportation, beyond the hype: solutions that deliver big value (2014)
55. Tan, K.H., Platts, K.: Linking objectives to action plans: a decision support approach based on the connectance concept. Decis. Sci. **34**(3), 569–593 (2003)
56. Wang, G., et al.: Big data analytics in logistics and supply chain management: certain investigations for research and applications. Int. J. Prod. Econ. **176**, 98–110 (2016)
57. Melo, M.T., et al.: Facility location and supply chain management: a review. Eur. J. Oper. Res. **196**(2), 401–412 (2009)
58. Shachter, R.D.: Evaluating influence diagrams. Oper. Res. **34**(6), 871–882 (1986)
59. Buzan, T.: Use Your Head. BBC/Ariel Books, London (1982)
60. Ayed, B., et al.: Big data analytics for logistics and transportation. In: 2015 4th IEEE International Conference on Advanced logistics and Transport (ICALT), pp. 311–316 (2015)
61. Shen, Y., Willems, S.P.: Strategic sourcing for the short-lifecycle products. Int. J. Prod. Econ. **139**(2), 575–585 (2012)
62. Feki, M., Boughzala, I., Wamba, S.F.: Big data analytics-enabled supply chain transformation: a literature review (2016)
63. Noyes, A., Godavarti, R., Titchener-Hooker, N., Coffman, J., Mukhopadhyay, T.: Quantitative high throughput analytics to support polysaccharide production process development. Vaccine **32**(4), 2819–2828 (2014)
64. Jodlbauer, H.: A time-continuous analytic production model for service level, work in process, lead time and utilization. Int. J. Prod. Res. **46**(7), 1723–1744 (2008)
65. Lockamy III, A., McCormack, K.: Linking SCOR planning practices to supply chain performance, an explorative study. Int. J. Oper. Prod. Manag. **24**(12), 1192–1218 (2004)
66. Winston, W.L.: Operations Research: Applications and Algorithms, 7th edn. Duxbury Press, Belmont (2003). Chapter 19, Example 3
67. Karakostas, G.: Faster approximation schemes for fractional multicommodity flow problems. In: Proceedings of the Thirteenth Annual ACM-SIAM Symposium on Discrete Algorithms, pp. 166–173 (2002)
68. Orlin, J.B.: A polynomial time primal network simplex algorithm for minimum cost flows. Math. Program. **78**(2), 109–129 (1997)

Modeling and Resolution of Consumer Behavior Problem in both Periods Active/Retired

Badreddine El Goumi[✉][ID], Mohammed El Khomssi, and Jamali Alaoui Amine

Laboratory of Modeling and Scientific Computation,
Sciences and Technologies Faculty, Sidi Mohamed Ben Abdellah University,
B.P. 2202 Route d'Imouzzer, Fez, Morocco
badreddinegoumi@gmail.com, khomsixmath@yahoo.fr, aminejamali@gmail.com

Abstract. Pension fund is an issue for social as well as economic, the political decision is based on a strategic choice to answering the economic, demographic evolution and social stability. The complexity of the issue obliges economists and researchers to solve this problematic by a mathematical models. In this paper, we present an optimization model applied to the pension problem while integrating the behavior of the agent during the two periods active/retired for optimal consumption and budget constraints. We propose in the numerical simulation a comparison between the pension funds functioned with distribution and capitalization to guide the agent in terms of its optimal choice of pension fund.

Keywords: Optimization · Equilibrium · Active/retired
Consumption · Budget constraints · Pension · Decision

1 Introduction

Retirement is one of the fundamental pillars of social protection, to provide a replacement income to the elderly after years of activity. It is one of the ways to fight against poverty and the preservation of social cohesion. Demographic trends and labor market situation now constitute a major challenge for the future of retirement. Indeed, the socioeconomic and demographic factors, including lower mortality rates, the economic crisis, the aging population and longer life expectancy, upset the balance of pension systems and weigh heavily on their functioning [13,15,18]. For these reasons, the debates on the future of pensions are, increasingly, an important place in the political and union circles, and even in public opinion in Morocco [1,3]. Uncertainties about the future of pensions and the fear of non-sustainability of schemes and preservation of their financial stability constitute therefore a challenge for the coming years. However, although

B. El Goumi—This author is the one who did all the really hard work.
M. El Khomssi—This work is under the supervision of this author.
J. A. Amine—This author is involve in this work.

© Springer International Publishing AG, part of Springer Nature 2018
M. Ben Ahmed and A. A. Boudhir (Eds.): SCAMS 2017, LNNS 37, pp. 987–996, 2018.
https://doi.org/10.1007/978-3-319-74500-8_88

the reform of pension systems ensures current and future affiliates a decent replacement income, the study of this issue is still a difficult problem in future. In fact, this reform often requires a change in the functioning, these results in judgments and reforms in that system. Mathematical models are a necessary tool to present key issues in choosing a mode of operation and financing [5–12, 14, 17]. The methodology of Dupuis and El Moudden [4] presents an analytical model that can account realistically the conditions in which is realized the balance of a plan distribution and the relationships between demographic and economic variables at issue of retirement. Moroccan pension funds managed by pay-as-you-go and by capitalization represent our object of modeling. In this work we also present a mathematical modeling in the form of an optimization model of the behavior of the agent which represents the consumer choosing the optimal consumption quantity between the active and retired periods, to maximize the utility function under constraints the two periods. The application of this model allows guiding the agent in the optimal choice of his optimal consumption in both periods with knowledge and clarity.

2   Mathematical Modeling of Pension Problem Based on the Consumption of an Agent

Each agent can live two periods, he works during his first period of life and he is retired during the second. The age of retirement is thus fixed by construction. We introduce a model of pensions and examine how it affects the consumption and savings choices of agents. This individual behavior will serve as a basis for studying the macroeconomic reference model of Diamond (1965) [4]. The behavior of an agent is represented by the following mathematical program (P):

$$
\begin{cases}
\max U(c_t, c_{t+1}) \\
under.Constraints \\
c_t + s_t \leq (1 - \theta_t)w_t & (1) \\
c_{t+1} \leq R_{t+1}s_t + \lambda_{t+1}w_{t+1} & (2) \\
c_t \geq 0, c_{t+1} \geq 0, s_t \geq 0
\end{cases}
$$

Parameter of the model:
w_t: Average real wage of the agent in the first period.
θ_t: Contribution rate in the first period.
R_{t+1}: Return of the savings of the first period.
λ_{t+1}: Replacement rate in the second period.

Variables of the model:
c_t: Quantity consumed during the first period of life when the agent is active.
s_t: Savings made by the asset during the first period.
c_{t+1}: The consumption of the agent during its second period of Life.

Objective function:
Represents the consumption preferences of the agent by a utility function to maximize for optimal consumption.

Constraints:

(1) This constraint ensures the consumption of the agent during his first period of life which must not exceed his real wage reduced by a rate of retirement contribution in order to save an amount during this period.

(2) Ensures the behavior of consumption during the second period which must not also exceed the replacement wage including the return on savings of the first period of life.

We remark from the model that the two consumptions c_t and c_{t+1} are like opposite alternatives, so when the agent increases its consumption c_t this leads to the decrease of c_{t+1}, thus we can model the utility function $U(c_t, c_{t+1})$ as a weighted sum where the consumer must choose the optimum quantity level between c_t and c_t to maximize $U(c_t, c_{t+1})$ under the budget constraints of the two periods.

We adopt the constant elasticity of substitution (CES) instantaneous utility function [2], which takes into account the different financial and economic hazards and is written as follows:

$$U(c_t, c_{t+1}) = \frac{1}{1-\sigma} \left[\left[(w_t c_t^{-\rho} + (1-w_t) c_{t+1}^{-\rho})^{\frac{-1}{\rho}} \right]^{1-\sigma} - 1 \right]$$

With

σ: Risk aversion coefficient.

w_t: The consumption parameter (proportion) or the weight of consumption in relation to the preferences in the utility of the agent.

ρ: Substitution rate.

2.1   Pension Fund by Capitalization

In this case, the pension received is equal to the sum of placement made capitalized and the amount reported by the placement of the contribution:

$$\lambda_{t+1} w_{t+1} = R_{t+1} w_t \theta_t$$

In this system, forced savings replaces voluntary savings then the interest rate will remunerate the investment instead of a pension fund, and the contribution rate will have a sense of consumer desire, therefore the regime by capitalization will have no a macroeconomic impact, because the accumulation of capital will be made by the contribution effort of the employee without the intervention of a pension fund or the state. The budget constraint discounted of the whole life gives:

$$c_t + \frac{c_{t+1}}{R_{t+1}} \leq (1-\theta_t) w_t + \frac{\lambda_{t+1} w_{t+1}}{R_{t+1}}$$

Then, in the case of a regime by capitalization, we will have:

$$c_t + \frac{c_{t+1}}{R_{t+1}} \leq w_t$$

The model (P) of the agent in the regime by capitalization becomes:

$$\begin{cases} \max U(c_t, c_{t+1}) = \frac{1}{1-\sigma}\left[\left[(w_t c_t^{-\rho} + (1-w_t)c_{t+1}^{-\rho})^{\frac{-1}{\rho}}\right]^{1-\sigma} - 1\right] \\ under.Constraints \\ c_t + \frac{c_{t+1}}{R_{t+1}} \leq w_t \\ c_t \geq 0 \\ c_{t+1} \geq 0 \end{cases} \quad (1)$$

Then we conclude that the wealth of the agent is reduced to the wage, it is not affected by the pension fund, the only account for the agent its total saving: $(s_t + w_t\theta_t)$. As a result, the pension fund by capitalization does not totally affected by changes in demographic and economic factors.

2.2   Pension Fund by Pay-As-You-Go (PAYG)

The sums collected on wages or contributions represent immediately and in full the benefits distributed to retirees. Thus, the equation of equilibrium is:

$$N_t w_t \theta_t = N_{t-1}\lambda_t w_t$$

And we have: $N_t = (1 + n_t)N_{t-1}$ With:
N_t: The size of the active population.

N_{t-1}: the size of the population of liabilities.

n_t: Growth rate of the labor force.

We thus obtain a simple form of the equation of equilibrium:

$$\theta_t = \frac{1}{1+n_t}\lambda_t$$

Thus, according to (2) we have the following relation:

$$c_{t+1} \leq R_{t+1}s_t + (1 + n_{t+1})\theta_{t+1}w_{t+1}$$

Then the budget constraint discounted becomes:

$$c_t + \frac{c_{t+1}}{R_{t+1}} \leq (1 - \theta_t)w_t + \frac{(1 + n_{t+1})\theta_{t+1}w_{t+1}}{R_{t+1}}$$

Thus, according to this constraint, we see an increase in the size of the population directly leads to an improvement in the pension.

We then obtain the model (P) of the agent in the pension fund by PAYG as follow:

$$
\begin{cases}
\max U(c_t, c_{t+1}) = \frac{1}{1-\sigma}\left[\left[(w_t c_t^{-\rho} + (1-w_t)c_{t+1}^{-\rho})^{\frac{-1}{\rho}}\right]^{1-\sigma} - 1\right] \\
under.Constraints \\
c_t + \frac{c_{t+1}}{R_{t+1}} \leq (1-\theta_t)w_t + \frac{(1+n_{t+1})\theta_{t+1}w_{t+1}}{R_{t+1}} \quad\quad (1) \\
c_{t+1} \leq R_{t+1}s_t + \lambda_{t+1}w_{t+1} \quad\quad (2) \\
c_t \geq 0, c_{t+1} \geq 0, s_t \geq 0
\end{cases}
$$

We have found a nonlinear optimization model, we use the internal point algorithm as a programming tool via the Matlab software, and the solutions are the optimal consumption of the agent in both periods.

3   Application, Numerical Simulation and Analysis of Results

3.1   Presentation of Data

We present in this section an application in the analysis of data of Moroccan pension funds for a practical exploitation of the mathematical model that we proposed. Throughout this numerical application, we compare the results of the PAYG pension systems and of those by capitalization. We assume that the return on savings is fixed at 0.05 *i.e.* $R_{t+1} = 0.05$ and the wages of active agents w_t and retirees w_{t+1} are assumed bounded such that $w_t = [1000DH, 40600]$ and $w_{t+1} = [1000DH, 40600]$. After the resolution of this model which will be donne via MATLAB software using algorithm: 'interior-point'. Indeed, our objective in this part is to show the impact on the consumption of the agent throughout his life according to the evolution of his salary, the return of his savings and the calculation parameters of the pension funds (Table 1).

Table 1. Data of each Moroccan pension fund (data given [16])

Pension funds	Contribution rate θ_t and θ_{t+1}	Replacement rate λ_{t+1}	Dependency ratio $(1 + N_{t+1})$
CMR[a]	0.2	0.875	1.03
CNSS[b]	0.1389	0.7	1.095
RCAR[c]	0.18	0.76	1.03

[a]Moroccan pension fund.
[b]National fund of social security.
[c]Collective fund of allocation of pension.

3.2 Result of the Comparison Between the Moroccan Pension Funds Operating by PAYG and the Pension Funds Operating by Capitalization Where the Agent is in Asset Period

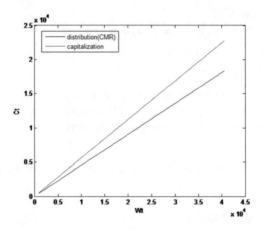

Fig. 1. Comparison between CMR and pension fund by capitalization in asset period.

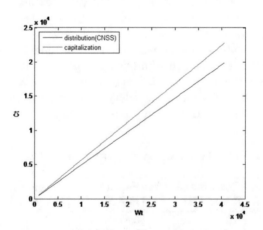

Fig. 2. Comparison between CNSS and pension fund by capitalization in asset period.

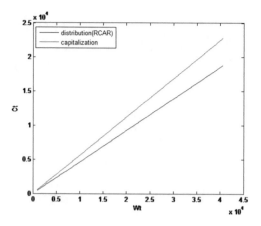

Fig. 3. Comparison between RCAR and pension fund by capitalization in asset period.

3.3 The Result of the Comparison Between the Moroccan Pension Funds Operating by Pay-As-You-Go and the Pension Fund by Capitalization Where the Agent is in Liability Period

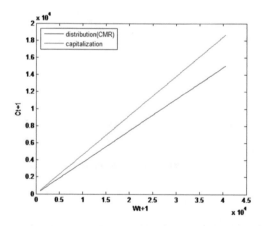

Fig. 4. Comparison between CMR and pension fund by capitalization in liability period

3.4 Interpretation and Discussion of Result

In Figs. 1, 2 and 3, it can be seen that the consumption of the active agent according to the evolution of his salary in the pension fund operate with capitalization is more preferable than all the Moroccan pension funds operating on a PAYG, in addition, it can be seen in Fig. 2 that there is no major difference between the CNSS pension fund and pension fund by capitalization thanks to the high

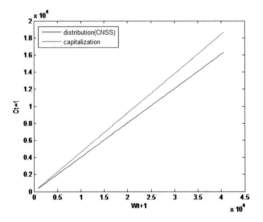

Fig. 5. Comparison between CNSS and pension fund by capitalization in liability period.

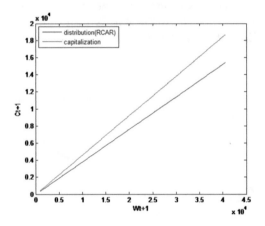

Fig. 6. Comparison between RCAR and pension fund by capitalization in liability period.

independence ratio and the reduced contribution rate compared to other PAYG pension funds. So the agent is advised to join the pension fund by capitalization instead of a pay PAYG pension funds because the consumption of the pension fund by capitalization has always been superior in all the salaries considered in our study. This shows the inability of these PAYG pension funds to meet future commitments to its affiliates. This situation is imputed to many criteria due to demographic, economic, financial and social factors that the parametric construction of these pension funds is not able to take care of.

In Figs. 4, 5 and 6, it can be seen that the consumption of the retired agent according to the evolution of his salary in the pension fund by capitalization is always more preferable than PAYG pension funds. We also observe that there is not a large gap between the CNSS pension fund and pension fund by capitalization thanks to the high independence ratio and the reduced contribution rate compared to the other PAYG pension funds, so we go out with the same conclusion in both figures that recommends the active or retired agent to adhere to a pension fund by capitalization that always marks superiority in our study.

4 Conclusion

In the present work, we proposed a general approach allowing the implementation of a mathematical model based on the determination of optimal consumption of the agent in the two active/retired periods. This model allows guiding the consumer under his budgetary constraints including the return on savings in the first period. Our model will be solved via the MATLAB software using the interior point algorithm. The results obtained after the numerical simulation show the superiority of the pension fund by capitalization, and these results will have a great impact on the agent's consumption throughout his life according to the evolution of his salary, the return of his savings and the calculation parameters of pension funds.

References

1. Benjelloun, S.: Une première évaluation des réformes des retraites au Maroc: Redistribution, inégalités et pauvreté, 4èmes Journees Internationales du Developpement du Gretha-Gres. Inégalités et Développement: Nouveaux Enjeux, Nouvelles Mesures. UNIVERSITE BORDEAUX IV, MONSTESQUIEU, 13–15 June 2012
2. Borsch, A., Ludwig, A., Winter, J.: Ageing, pension reform and capital flows: a multi country simulation model. Economica **73**, 625–658 (2006)
3. Cherkaoui, M.: Vieillissement, transition démographique et crise des systèmes de retraite: cas du Maroc. Université Pierre Mendes, France (2009)
4. Dupuis, J., El Moudden, C.: A Contribution to the Empirics of Economic Growth, Economie des Retraites. Economica, Paris (2002)
5. El Goumi, B., El Khomssi, M., Fikri, M.: Model for the management of pension fund with deterministic and stochastic parameters. IEEE Spectr. J. (2016). ISBN 978-1-4673-8571-8
6. Faleh, A.: Un modèle de programmation stochastique pour l'allocation stratégique d'actifs d'un régime de retraite partiellement provisionné, Caisse des Dépôts et Consignations. ISFA- Université Lyon, 2 Feb 2011. hal-00561965
7. Faleh, A., Planchet, F., Rullire, D.: Les générateurs de scénarios économiques: de la conception la mesure de la qualité. Assurances et gestion des risques **78**(1/2), 1–30 (2010)
8. Frauendorfer, K., Jacoby, U., Schwendener, A.: Regimes witching based portfolio selection for pension funds. J. Bank. Financ. **31**(2007), 2265–2280 (2007)
9. Geyer, A., Ziemba, W.T.: The innovest Austrian pension fund financial planning model InnoALM. Oper. Res. **56**(2008), 797–810 (2008)
10. Hilli, P., Koivu, M., Pennanen, T., Ranne, A.: A stochastic programming model for asset and liability management of a Finnish pension company. Ann. Oper. Res. **152**(2007), 115–139 (2007)

11. Kouwenberg, R.: Scenario generation and stochastic programming models for asset liability management. Eur. J. Oper. Res. **134**(2001), 51–64 (2001)
12. Kusy, M.I., Ziemba, W.T.: A bank asset and liability management model. Oper. Res. **34**(3), 356–376 (1986)
13. Legros, F.: Allongement de l'esprance de vie et choix du systme de retraite. Revue Conomique **53**, 809–823 (2002)
14. Pennacchi, G., Rastad, M.: Portfolio allocation for public pension funds. J. Pension Econ. Financ. **17**, 221–245 (2011)
15. Planchet, F., Thrond, P.E., Kamega, A.: Scnarios conomiques en assurance- Modlisation et simulation. Economica, Paris (2009)
16. Rapport sur le Systme de retraite au Maroc: diagnostic et propositions de rformes. Cour des comptes Royaume du Maroc, Morocco (2013)
17. Talfi, M.: Organisation des systmes de retraites et modlisation des fonds de pension. University Claude Bernard, Lyon (2007)
18. Verniere, L: Les indicateurs de rendement et de rentabilit de la retraite. Branche Retraites de la Caisse des dpts et consignation, pp. 98–107 (1998)

The Pricing of Interest Rate Derivatives: Caps/Floors and the Construction of the Yield Curve

Ghizlane Kouaiba[1](✉) and Moulay-Driss Saikak[2]

[1] Faculty of Science, Ibn Tofail University, Kenitra, Morocco
kouaiba.ghizlane@gmail.com
[2] Faculty of Law, Economics and Social Sciences, Mohamed V University,
Rabat, Morocco
http://fs.uit.ac.ma/

Abstract. The aim of this paper is to highlight the theoretical foundations of caps and floors that are distinguished among derivatives meeting the requirements of most investors in the financial sphere. Then, the key elements of the calculation were defined: the LIBOR rate and the forward rate. This paper focuses on the construction of the yield curve, a fundamental approach in the analysis of derivatives. After a description of the characteristics of these products, a keen interest was devoted to their valorisation by referring to the model of Black (1976).

Keywords: LIBOR rate · Forward rate · Black · Cap · Floor
Caplet · Floorlet · Yield curve

1 Introduction

Derivatives have been traded since the 1980s. They are therefore important instruments of investment for the various financial market participants, reflecting their daily transactions, which quickly evolved into trillions of US dollars and particularly swaptions and caps/floors, which are supposed to be traded on the OTC market. Indeed, caps and floors emanating from the family of interest rate derivatives have emerged remarkably in the high-frequency trading platforms and will thus represent the framework of this work. All the techniques given through the financial market to price interest rate derivatives in 1980s and 1973s based respectively on the original work of Heath, Jarrow and Morton (HJM) and the Black and Scholes paper show that the major aim of investors was to look for an optimal opportunities to hedge and manage their portfolios against a rise or fall in interest rate levels. The modeling of the yield curve still at this last ten years used by the academics and traders (Vasicek model in 1977s and Cox, Ingersoll and Ross (CIR) in the 1985s). Faced with the multiplicity of the market's instruments (caplets, floorlets, swaptions.....), Vasicek and Cox et al. made a specific formulas depending on the assumption of the dynamic of the instantaneous of short rate. However, the evolution of interest rates is expressed as a stochastic process resulting from a partial differential equation.

© Springer International Publishing AG, part of Springer Nature 2018
M. Ben Ahmed and A. A. Boudhir (Eds.): SCAMS 2017, LNNS 37, pp. 997–1005, 2018.
https://doi.org/10.1007/978-3-319-74500-8_89

2 Principles of Operation

The interest rate caps are called on the financial market to hedge against a rise in rates above a cap rate. Assume that the ceiling interest rate will be 3% and the notional amount associated with this transaction will be 1 million of dollars with a 3-month tenor and a 5-year term. If the LIBOR rate reaches 4%, the bound must pay the following interest at the end of the quarter: 0.25 * 0.04 * 1 000 000 = 10 000$. In this case, the heading that we are considering will generate a payoff of 10 000-0.25 * 0.03 * 1 000 000$ = 10 000-7 500 = 2 500$. In practice, the tenor is often expressed in exact number of days. In addition, the LIBOR rate [8,11] value is consulted on each adjustment date. If this rate is lower than that of the ceiling, the heading will not generate any flow in this case. The functioning of the floors is not as different as the caps. Just this time, the derivative generates a non-zero flow if the LIBOR rate on each readjustment date falls below an already set threshold called the floor rate.

- N: The notional
- fl_{t_i}: The LIBOR rate observed at each moment t_i
- R_{k_1} and R_{k_2} respectively represent the ceiling and the floor rates

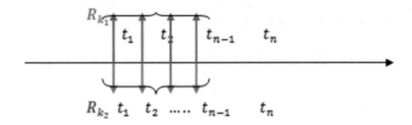

Fig. 1. The operation of caps and floors

if $fl_{t_i} > R_{k_1}$: the cap generates a non-zero flow at time i equal to $N\delta_i(f_{t_i} - R_{k_1})$.
if $fl_{t_i} < R_{k_2}$: the floor generates a non-zero flow at time i equal to $N\delta_i(R_{k_2} - f_{t_i})$.
Where $\delta_i = [t_i, t_{i+1}]$ for $\forall i$.

By the union of two floorlets, we can obtain a floor composed by two periods of readjustment of the rate, a floor that starts in t_i and which ends in t_{i+2}:

$$floor = \{fl^{t_i}, fl^{t_{i+1}}\} \, [5] \tag{1}$$

By analogy, two caplets grouped simultaneously give a heading on two periods of readjustment:

$$cap = \{cl^{t_i}, cl^{t_{i+1}}\} \, [5] \tag{2}$$

There is theoretically a relation between the cap and the floor similar to that of the put-call parity:

$$Cap\ Price = Floor\ Price + Swap\ Value \tag{3}$$

3   LIBOR Rates and Forward Rates

3.1   LIBOR Rates

Internationally, banks negotiate loans and borrowings for all currencies and maturities generally less than one year, ranging from overnight to 12 months. The rate at which highly rated institutions lend is called the London Interbank Offer Rate (LIBOR) [11], which is a benchmark for the world's major banks that deal with short-term loans. The LIBOR rate is always identified by a maturity called tenor and a currency such as the dollar (USD), euro (EUR), pound sterling (GBP), etc (Table 1).

Table 1. The values of the LIBOR rate, source: global-rates.com [7]

Maturity	LIBOR$ 14/09/2017	LIBOR$ 13/09/2017	LIBOR$ 12/09/2017
Overnight	1,17889%	1,17722%	1,17778%
1 Week	1,19667%	1,19556%	1,19667%
1 Month	1,23444%	1,23444%	1,23667%
2 Months	1,27167%	1,27167%	1,27222%
3 Months	1,32111%	1,32000%	1,31917%
6 Months	1,45861%	1,45830%	1,45444%
12 Months	1,71956%	1,71233%	1,70956%

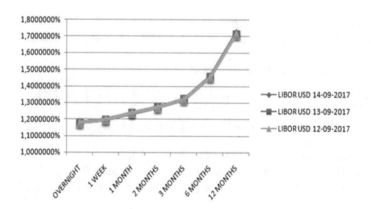

Fig. 2. The graphical representation of LIBOR rates

3.2   Forward Rates

A Forward Rate Agreement (FRA) [11] is a contract in which two parties today agree (t = 0) on an interest rate $F(t_1, t_2)$ for a period t_1 to t_2. More generally, if two zero-coupon rates, denoted R_1 and R_2, correspond to maturities t_1 and t_2 The forward rate for the period between t_1 and t_2 is defined by:

$$F(t_1, t_2) = \frac{(R_2 t_2 - R_1 t_1)}{(t_2 - t_1)} \, [4] \tag{4}$$

4   The Construction of the Yield Curve

Often there is a lack of the continuing curve of imported rates, often for precise maturities, by software applied in the field of market finance: Bloomberg..., etc. In order to value the different structured products. As a result, several methods have been put in place to guarantee its construction.

4.1   The Bootstrapping Method

This approach is based on the calculation of the zero-coupon yield curve from market data.

For maturities of less than one year:
The zero-coupon rate is calculated from bond prices. Subsequently, we can proceed by linear interpolation or cubic interpolation to extract the value of the rate for each instant less than one year.

$$N = P(0, t)(1 + R(0, t))^{\frac{1}{t}} \, [10] \tag{5}$$

Where:

N: notional of the zero-coupon bound.

$P(0, t)$: Price of zero-coupon in date 0 and the maturity t is in $0 < t \leq 1$ year.

$R(0, t)$: Interest rate covered period $[0, t]$.

For maturities between n_i years and $n_{(i+1)}$ years for i from 1 to N :

$$P(0, t_i) = C \times P_{ZC}(0, t_i) + (N + C) \times P_{ZC}(0, t_{i+1}) \, [10] \tag{6}$$

- C: coupon of the bound which price is $P(0, t_i)$.
- $P_{ZC}(0, t_i)$: price of zero-coupon bound with the maturity n_i.
- $P_{ZC}(0, t_{i+1})$: price of zero-coupon bound with the maturity n_{i+1}.

Where $t_i \leq n_i$ and $n_i < t_i \leq n_{i+1}$.

Linear and Cubic Interpolation. Linear interpolation is the easiest way to calculate a value of a continuous function (the curve of the rates, noted $R(0, t)$ between two moments t_1 and t_2 different:

$$R(0, t) = R(0, t_1) + (t - t_1) \frac{R(0, t_2) - R(0, t_1)}{t_2 - t_1} \tag{7}$$

As for the cubic interpolation, we connect between two moments t_1 and t_2 by a polynomial of order 3. In other expression we have:

$$\begin{cases} R_1(0,t) = a_1t^3 + b_1t^2 + c_1t + d_1 \\ R_2(0,t) = a_2t^3 + b_2t^2 + c_2t + d_2 \\ R_3(0,t) = a_3t^3 + b_3t^2 + c_3t + d_3 \end{cases} \tag{8}$$

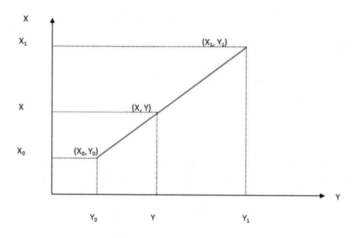

Fig. 3. Illustration of linear interpolation

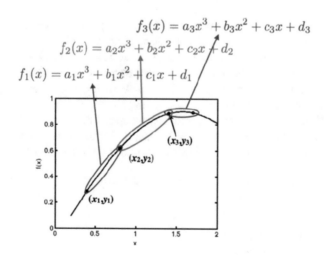

Fig. 4. Illustration of cubic interpolation: n polynomials of degree 3

4.2   Nelson-Siegel Method

This method obviously allows describing the dynamics of the instantaneous rates. Such an approach calculates zero-coupon rates $R(0,T)$ through the following formula:

$$R(0,T) = \beta_0 + \beta_1 \frac{(1 - exp(-\frac{T}{\alpha}))}{\frac{T}{\alpha}} + \beta_2 \left(\frac{(1 - exp(-\frac{T}{\alpha}))}{\frac{T}{\alpha}} - exp(-\frac{T}{\alpha}) \right) \ [2, 10] \quad (9)$$

Where:

- β_0: is interpreted as the long run levels of interest rates.
- β_1: is the short-term component.
- β_2: is the medium-term component.
- α: is the decay factor.

4.3   Stochastic Method

The Vasicek Model and the CIR Model. In practice, the term structure of interest rates measures the relationship among the yields on default-free securities that differ only in their term to maturity. To estimate it, we use the stochastic models like Vasicek model as a one-factor model which considered the short interest rate as a risk. All models of one-factor start by specifying the stochastic differential equation. This equation can be written:

$$dr = f(t,r)dt + \rho(t,r)dw(t) \ [9] \quad (10)$$

Where $w(t)$ is a Wiener process, and f(t,r) represents the drift coefficient, while $\rho(t,r)$ is the diffusion term. By specifying those functions, many researchers have proposed their own term structure interest rates. Vasicek gives an explicit characterization of the term structure of interest rates in an efficient market. The model is widely used for pricing the bonds. Additionally, it uses the Ornstein Uhlenbeck process to compute the spot interest rate. This model is a one-factor model which means that rates depend on the spot interest rate. Thus, the spot rate defines the whole term structure. Vasicek illustrates the general model by assuming that: the spot rate $r(t)$ follows the Ornstein-Uhlenbeck process:

$$dr = \alpha(\gamma - r(t))dt + \sigma dz \ [3, 9] \quad (11)$$

Where:

- α: Speed of reversion.
- γ: Long term mean level.
- σ: Instantaneous volatility.

The solution of the differential equation depending on Vasicek model is given by:

$$r(t) = r_0 e^{-\alpha t} + \gamma(1 - e^{-\alpha t}) + \sigma e^{-\alpha t} \int_0^t e^{\alpha s} dw(s) \quad (12)$$

The Ornstein-Uhlenbeck process with $\alpha > 0$ is sometimes called the elastic random walk. It is a Markov process with normally distributed increments. The instantaneous drift $(\gamma - r(t))$ presents a force that keep pulling the process towards it's long term mean γ with magnitude proportional to the deviation of the process from the mean. The stochastic element, which has a constant instantaneous variance σ^2, causes the process to fluctuate around the level γ in an erratic, but continuous. Based on the model of one-factor, the CIR model is defined as follows:

$$f(t, r) = a(b - r)$$

and

$$\rho(t, r) = \sigma \sqrt{r} \ [6]$$

Where a, b and σ are positive constants.

5   The Pricing of Caps/Floors

Let a cap of maturity T subdivided into regular readjustment sub-intervals $\delta_i = [t_i, t_{i+1}]$ a notional L and a cap rate R_K. The payoff of a caplet is given by:

$$N\delta_i max(f_{t_i} - R_K, 0) \ [4, 5] \tag{13}$$

The two rates used in the predefined formula follow the same space δ_i. The payoff of a floorlet is nothing else than:

$$N\delta_i max(R_K - f_{t_i}, 0) \ [4, 5] \tag{14}$$

This brings us to note that the floor concept is purely considered as a portfolio of European puts. Unlike the cap that can be seen as call payments on the underlying rates. The current financial market assumes that the underlying rates admit a representation of their partial differential equation in the form of the Black model founded in 1976 by Fischer Black:

$$df_t = \sigma f_t dw_t \ [4, 5] \tag{15}$$

Where σ and w_t respectively express the market volatility assumed to be constant and the Brownian motion associated with the process f_t. The solution of the model, which can be obtained by using the Ito formula, expresses the relation between the rate underlying the instant 0 and that observed in t_i:

$$f_{t_i} = f_0 exp(-\frac{\sigma^2}{2} + \sigma w_{t_i}) \tag{16}$$

Where $f_{t_0} = f_0$.

Lemma 1. *Itô's lemma If we consider that $(X_t)_{t \in [0,T]}$ is the Itô's process defined by: $X_t = X_0 + \int_0^t G_m dm + \int_0^t H_m dB_m$ and F is is a twice-differentiable continuous function. So, we have the following expression:*

$$F(X_t) = F(X_0) + \int_0^t F'(X_m)dm + \frac{1}{2}\int_0^t F''(X_m)d\langle X, X\rangle_m \qquad (17)$$

where

$$\langle X, X\rangle_t = \int_0^t |H_m|^2 dm \qquad (18)$$

Under the risk-neutral probability Q and under the assumption of the log-normality of the underlying rates, the caplet pays a flow at time 0 equal to:

$$E^Q\left(P(0, t_{t_{i+1}})N\delta_i max(f_{t_i} - R_K, 0)\right) \qquad (19)$$

This can be calculated otherwise:

$$N\delta_i P(0, t_{t_{i+1}})[F_{t_i}N(d_1) - R_K N(d_2)]\,[4,5] \qquad (20)$$

Where:

– $N(d)$: represents the density function of the Gaussian law given by:

$$N(d) = \int_{-\infty}^d \frac{1}{\sqrt{2\Pi}\sigma}\exp(-\frac{x^2}{2\sigma^2})dx\,[5]$$

– $d_1 = \frac{ln(\frac{F_{t_i}}{R_K}) + \sigma^2\frac{t_i}{2}}{\sigma\sqrt{t_i}}\,[4,5]$ and $d_2 = d_1 - \sigma\sqrt{t_i}\,[4,5]$
– $P(0, t_{i+1})$ is the zero-coupon price that covers the period $[0, t_{i+1}]$.
– F_{t_i} the forward rate observed between t_i and t_{i+1}.

By instruction, the price of a floorlet is given by:

$$N\delta_i P(0, t_{t_{i+1}})[R_K N(-d_2) - F_{t_i}N(-d_1)]\,[4] \qquad (21)$$

6   Implied Volatility

In the most frequent cases, all the elements necessary to calculate the price of a cap or a floor are available. In the contrary case, it is desired to calculate the value of one of these inputs knowing the price given by the market. The Newton-Raphson algorithm allows calculating the volatility of a derivative knowing well its market price by using the recurring formula:

$$\sigma_{n+1} = \sigma_n + \frac{P^{market} - P^{Black}}{\frac{dP^{Black}}{d\sigma}}\,[1] \qquad (22)$$

The algorithm stops at iteration i once $|\sigma_{i+1} - \sigma_i|$ is less than a certain precision that it is already fixed and finally retains the value σ_{i+1}.

7   Calculating the Greeks

the Greek parameters are obviously used to quantify the price derivative (P), in this case for caps and floors, with respect to the market parameters (volatility σ, maturity T, underlying asset S ......). this quantity is effectively linked to the optimal hedging strategy. For example, the parameter Delta: $\Delta = \frac{dP}{dS}$: [4], is used to estimate the sensitivity of the price following a fluctuation of the price of the underlying, The delta of a call is positive whereas the delta of a put is negative since the price of a call (put) is an increasing (decreasing) function of the price of the underlying stock. Therefore, the variation of this strategy with respect to the price of its underlying is called Gamma: $\Gamma = \frac{d\Delta}{dS} = \frac{d^2P}{dS^2}$: [4]. The dependence of the price on the maturity and its sigma risk parameter leads us to define two other Greek parameters: the parameter Vega and the parameter rho. the first is defined by the rate of variation in the value of a portfolio of derivatives with the respect to the volatility of the underlying asset, denoted by Vega: $\nu = \frac{dP}{d\sigma}$: [4], the said parameter is already used in the initiated section "implied volatility", whereas the Theta: $\Theta = \frac{dP}{dT}$: [4] parameter measures the variation of the portfolio's value with the respect to the lifetime of the option.

References

1. Coqueret, G.: La volatilité implicite, Dalloz (2009)
2. Gbongue, F., Planchet, F.: Analyse comparative des modèles de construction d'une courbe des taux sans risque dans la zone CIPRES, ISFA : Laboratoire De Science Actuarielle Et Financiere, Université Lyon 1 (2013)
3. Gupta, A., Subrahmanyam, M.G.: Pricing and hedging interest rate options: evidence from cap–floor markets. J. Bank. Finance **29**(2005), 701–733 (2005). https://doi.org/10.1016/j.jbankfin.2004.05.025
4. Hénot, C., Hull, J., Deville, L., Roger, P.: Options, futures et autres actifs dérivés. Pearson, 6 ème édition (2007). ISBN 978-2-7 440-7179-9
5. Kosowski, R.L., Neftci, S.N.: Principles of Financial Engineering, 3rd edn. Academic Press, Elsevier, San Diego (2015). ISBN 978-0-12-386968-5
6. Poulsen, R.: Working with the Cox-Ingersoll-Ross Model. AMS, IMF, UK (2003)
7. LIBOR rates (USD) (2017). http://fr.global-rates.com
8. Ribonato, R., McKay, K., White, R.: The SABR/LIBOR Market Model: Pricing, Calibration and Hedging for Complex Interest-rate Derivatives. Wiley, Chichester (2006). ISBN 978-0-470-74005-7
9. Saikak, M., Raouf, M.: Estimation of the term structure of interest rates for Moroccan financial market using vasicek model. Monetary Research Center, Bulgaria, pp. 3–6 (2016). ISSN 2534–9600
10. Thérond, P.: Génération de scénarios économiques: Modélisation des taux d'intérêt. ISFA, Université Lyon 1 (2013)
11. Economic and financial website. www.Investopedia.com

Stochastic Model of Economic Growth with Heterogeneous Technology and Technological Upgrading

Amine Jamali Alaoui[(⊠)][iD], Mohammed El Khomssi[(⊠)],
and Badreddine El Goumi[(⊠)]

Laboratory of Modeling and Scientific Computation,
Sciences and technologies Faculty,
Sidi Mohamed Ben Abdellah University,
B.P. 2202 Route d'Imouzzer, Fez, Morocco
aminejamali@gmail.com, khomsixmath@yahoo.fr, badreddinegoumi@gmail.com

Abstract. In our paper, we present in the first place a stochastic model of optimal growth with heterogeneous technology with job qualifications. We are based on the models of Pesaran, Binder and Romer with technological transfer, we find the optimal fraction for the qualification of the employment in order to obtain a more favorable income. Second, we establish the conditions on the coefficient of technological diffusion, for which the country partially upgrading with their technological level.

Keywords: Economic growth · Heterogeneous technology
Stochastic model · Technological upgrading

1 Introduction

Economic growth is a pillar of macroeconomics, for this reason it is essential to understand the main mechanisms responsible for this growth, since even small deviations in growth rates can influence the quality of life of several generations. For this reason several researchers have proposed mathematical models of economic growth to analyze on the one hand and to explain on the other hand the impact of several factors on the economies of the countries. Thus to forecast and estimate this responsible growth of the riches between the countries.

Among the researchers we quote for example the models of: Samuelson [1], Ramsey [2], Solow [3], Domar [4], Cass and Koopmans [5], Romer (Mankin, Romer, Weil) [6], Aghion and Howitt [7], Harrod [8], Acemoglu [9], Novales, Alfonso [10], Novales, Ruiz [11], Asfiji et al. [12], Saby and Dominique [13] etc.

On the other hand among the development indicators is the growth rates of technical progress in the countries. The hypothesis of a homogeneous growth

---

A. Jamali Alaoui— This author is the one who did all the really hard work.

M. El Khomssi—This work is under the supervision of this author.

B. El Goumi—This author is involve in this work.

© Springer International Publishing AG, part of Springer Nature 2018
M. Ben Ahmed and A. A. Boudhir (Eds.): SCAMS 2017, LNNS 37, pp. 1006–1014, 2018.
https://doi.org/10.1007/978-3-319-74500-8_90

rate of technology for all countries is largely rejected by empirical studies (see, for example, Lee et al. [14] or Liberto et al. [15]). Since the theoretical work of Nelson and Phelps [16] and the empirical work of Coe and Helpman [17], the concept of imperfect diffusion of technical progress at the international level has become crucial for understanding technological disparities in performance between countries.

Our paper is organized as follows, in the second paragraph we discuss a heterogeneous growth model. We establish the optimum quantity of physical and human capital, and thus the fraction of the appropriate time for the qualification of employment, in order to attain a more favorable income, and consequently a sustained growth. in the third paragraph we draw on the work of Benhabib and Spiegel, we show the conditions on which a country will partially catch up to their level of technological growth.

2   Optimal Stochastic Growth Model with Heterogeneous Technology

The major assumptions of the convergence result in the Solow model, are the factorial decline in capital returns and a homogeneity of technological diffusion at the international level. This second hypothesis implies an equality of growth rates of technical progress at the international level (Phillips and Sul [18]). According to the empirical work of El Ghak [19] and Liberto et al., the hypothesis of homogeneity is largely rejected and there is a strong heterogeneity of technological levels between countries. According to Faberger, the disparities in growth rates between countries are mainly due to differences in their levels of technology.[1] According to the work of Seyma and Alev [20], Ethier [21], McDonald [22], Benhabib [23], these disparities in growth rates come from human capital and invasion in the research and development.

In this section, we propose an optimal growth model to determine the optimal amount of effective human capital per employment that gives us a maximum effective GNI per employment, and consequently an improvement in per capita income of the country to consider.

2.1   Notations and Model Macroeconomic Assumptions

Notations
   In the remainder of this article, and for a country i, we denote by:

t is The time.
A_i Is the level of technology of the country i.
A_l Represents the global technological frontier, it corresponds to the index of the most advanced capital goods.

---

[1] The test of the homogeneity hypothesis on a sample of 102 non-oil countries over the period 1960–1989 gives a rate of growth of technical progress between −3.1 and 7.4 between countries.

τ_i Is the fraction of time spent by the economy on the accumulation of know-how.

K_i Is physical capital, H_i is human capital.

L_i Is the work, which represents the quantity of manpower involved in the productions.

Y_i Gross national income (GNI).

$\tilde{a}_i = \frac{H_i}{A_i}$ Is the proportion of learning in relation to the global technological frontier.

$N_{i,t}$; Is the total number of inhabitants of the country i, at time t.

$k_i = \frac{K_i}{L_i}$, $h_i = \frac{H_i}{L_i}$ et $y_i = \frac{Y_i}{L_i}$ Are, respectively, capital per manpower, human capital per manpower and GNI per manpower.

For any aggregate F_i, we note and define by:

For any vector of real inputs variables $(x_1, x_2, \ldots, x_j)$, $F_{i,x_1,x_2,\ldots,x_j} = F_i(x_1, x_2, \ldots, x_j)$.

$\gamma_{F_i} = \frac{\frac{\partial F_i}{\partial t}}{F_i}$ Is the growth rate of the aggregate F_i.

$f_i = \frac{F_i}{L_i}$ Is the aggregate per manpower.

$\hat{f}_i = \frac{F_i}{A_i L_i}$ Is the effective aggregate per manpower.

We pose $\Omega = [0,1] \times [0, +\infty[$. For reasons of regularity, it is assumed that the functions K_i et H_i are in $\mathcal{C}^2(\Omega) \bigcap \mathcal{L}^\infty(\Omega)$.

We denote respectively by s_i, s_{K_i}, s_{H_i} and s_{R_i}: The total investment rate, the investment rate in production, the investment rates in the accumulation of know-how and the investment rates for research and development sector.

Macroeconomic assumptions of the model

1. The income Y_i is divided into three; the consumption C_i, the expenses G_i and investment I_i.

$$Y_i = S_i + C_i + G_i$$

2. The investment I_i is also split into three: investment for production, it's noted I_{K_i}, the second denoted I_{H_i}, is the one for the accumulation of know-how and the third denoted I_{R_i}, is the one for research and development sector. So we have:

$$I_i = s_i Y_i$$

$$I_{K_i} = s_{K_i} Y_i$$

$$I_{H_i} = s_{H_i} Y_i$$

and

$$s_i = s_{K_i} + s_{H_i} + s_{R_i}$$

3. The employment and technology of the country, which are random variables.
4. Technology of the country grows logistically, and the global technological frontier grows Malthusian.

5. Human capital is dependent on; the level of education of manpower, the fraction of time devoted to the accumulation of know-how and the global technological frontier.
6. The income Y_i is assumed to be a Cobb Douglas function.
7. Physical and human capital, are believed respectively by rates s_{K_i}, s_{H_i} and decrease by constant depreciation rates δ_i and η_i. Where δ_i and η_i, are respectively the depreciations of physical capital and human capital.

2.2   The Model

In our modeling, we assume that human capital is dependent on the level of schooling and the qualification of manpower.

We base ourselves on the empirical work of Samuelson, Nelson, Phelps and Romer with technological transfer, we propose a new dynamic equation of accumulation of know-how:

$$\frac{\partial H_{i,\tau,t}}{\partial t} = \chi_{i,t} e^{\tau_i \theta} H_{i,\tau_i,t}^{1-\lambda_i} A_t^{\lambda_i} \tag{1}$$

Where:

$\chi_{i,t}$ Is the level of schooling of labor.

λ_i Is the coefficient of technological diffusion.

θ Is a strictly positive constant.

According to the second hypothesis, the employment and technology of the country, which are random variables, we put:

$$A_{i,t} = A_{i,t_0} e^{\int_{t_0}^{t} \gamma_{A_{i,t}} dt + \varepsilon_{i,t}} \tag{2}$$

and

$$L_{i,t} = L_{i,t_0} e^{\int_{t_0}^{t} \gamma_{L_{i,t}} dt + \nu_{i,t}} \tag{3}$$

Where

$\varepsilon_{i,t}$ and $\nu_{i,t}$ are respectively stochastic variables of technology and employment, which satisfy:

$$\frac{d\varepsilon_{i,t}}{dt} = -(1 - \mu_{ia})\varepsilon_{i,t} + \rho_{ia,t} \tag{4}$$

And

$$\frac{d\nu_{i,t}}{dt} = -(1 - \mu_{il})\nu_{i,t} + \rho_{il,t} \tag{5}$$

Such us:

(μ_{ia}, μ_{il}) are constants.

$\rho_{ia,t} \sim iid(0, \sigma_{ia}^2)$ and $\rho_{il,t} \sim iid(0, \sigma_{il}^2)$.

σ_{ia} and σ_{il} are the standard deviations of technology and employment.

The income Y_i is assumed to be a Cobb Douglas function, it's defined by:

$$Y_{i,\tau_i,t} = K_{i,\tau_i,t}^{\alpha_i} H_{i,\tau_i,t}^{\beta_i} (A_{i,t} L_{i,t})^{1-\alpha_i-\beta_i} \tag{6}$$

Where:

$(\alpha, \beta) \in [0,1]^2$ are respectively. The elasticity's of physical and human capital. In terms of effective aggregate per manpower, we have:

$$\hat{y}_{i,\tau_i,t} = \frac{Y_{i,\tau_i,t}}{A_{i,t}L_{i,t}} = \hat{k}_{i,\tau_i,t}^{\alpha_i} \hat{h}_{i,\tau_i,t}^{\beta_i} \tag{7}$$

We consider that the dynamic equations of physical capital and human capital are respectively given by:

$$\frac{\partial K_{i,\tau_i,t}}{\partial t} = s_{K_i} Y_{i,\tau_i,t} - \delta_i K_{i,\tau_i,t} \tag{8}$$

$$\frac{\partial H_{i,\tau_i,t}}{\partial t} = s_{H_i} Y_{i,\tau_i,t} - \eta_i H_{i,\tau_i,t} \tag{9}$$

We note by Γ_{t_0}, the information set at time t_0.

In terms of effective aggregate per manpower, and by using the Eqs. (3) to (9), our growth model (denoted (P)) is the following:

$$\begin{cases} \max_{\hat{k}_i, \hat{h}_i} E\left[\int_{t_0}^T \varrho_i^t \hat{y}_{i,\tau,t} dt | \Gamma_{t_0}\right] \\ S.C \\ \frac{\partial \hat{k}_{i,\tau_i,t}}{\partial t} = s_{K_i} \hat{y}_{i,\tau_i,t} - (\gamma_{A_{i,t}} + \gamma_{L_{i,t}} + \frac{d}{dt}(\varepsilon_{i,t} + \nu_{i,t}) + \delta_i)\hat{k}_{i,\tau_i,t} \\ \frac{\partial \hat{h}_{i,\tau,t}}{\partial t} = s_{H_i} \hat{y}_{i,\tau_i,t} - (\gamma_{A_{i,t}} + \gamma_{L_{i,t}} + \frac{d}{dt}(\varepsilon_{i,t} + \nu_{i,t}) + \eta_i)\hat{h}_{i,\tau_i,t} \\ \hat{y}_{i,\tau_i,t_0} = \hat{y}_{i0} \\ \hat{y}_{i,\tau_i,T} = q\hat{y}_{i0} \\ q > 1 \end{cases} \tag{10}$$

ϱ_i^t Is discount factor.

3   Model Resolution for Developing Economies

The world technological level grows with the rate:

$$\gamma_{A_{l,t}} = \frac{\frac{dA_{l,t}}{dt}}{A_{l,t}} \tag{11}$$

in developing economies, this rate is constant:

$$\gamma_{A_{l,t}} = g = constante$$

Using Eq. (1), for all $t_0 \in [0, +\infty[$ and $\forall(\tau, t) \in \Omega$ we have:

$$\int_{t_0}^t \frac{\frac{\partial H_{i,\tau,s}}{\partial s}}{H_{i,\tau,s}^{1-\lambda_i}} ds = e^{\tau_i \theta} \int_{t_0}^t \chi_{i,s} A_{l,s}^{\lambda_i} ds$$

The resolution of Eq. (1) gives:

$$H_{i,\tau,t} = \left[\lambda_i A_{l,t_0}^{\lambda_i} e^{\tau_i \theta} \int_{t_0}^t \chi_{i,s} e^{g\lambda_i(s-t_0)} ds - H_{i,\tau_i,t_0}^{\lambda_i}\right]^{\frac{1}{\lambda_i}} \tag{12}$$

in case where $(\delta_i = \eta_i)$, by using the Eqs. (8) and (9) we have:

$$s_{H_i}\frac{\partial K_{i,\tau_i,t}}{\partial t} - s_{K_i}\frac{\partial H_{i,\tau_i,t}}{\partial t} = -\delta_i(s_{H_i}K_{i,\tau_i,t} - s_{K_i}H_{i,\tau_i,t}) \qquad (13)$$

We put:

$$X_{i,\tau_i,t} = s_{H_i}K_{i,\tau_i,t} - s_{K_i}H_{i,\tau_i,t} \qquad (14)$$

Then we get:

$$\frac{\partial X_{i,\tau_i,t}}{\partial t} = -\delta_i X_{i,\tau_i,t} \qquad (15)$$

The solution of the last equation is as follows:

$$X_{i,\tau_i,t} = X_{i,\tau_i,t_0}e^{-\delta_i(t-t_0)} \qquad (16)$$

So, the physical capital is given by:

$$K_{i,\tau_i,t} = \frac{1}{s_{H_i}}[s_{K_i}H_{i,\tau_i,t} + (s_{H_i}K_{i,\tau_i,t_0} - s_{K_i}H_{i,\tau_i,t_0})e^{-\delta_i(t-t_0)}] \qquad (17)$$

K_{i,τ_i,t_0} and H_{i,τ_i,t_0}, are respectively the initial physical capital and human capital, at the moment t_0.

Return to our model, under the optimality condition, and in case where $\delta_i = \eta_i$, we have:

$$\hat{k}_{i,\tau_i,t} = \frac{\alpha}{\beta}\hat{h}_{i,\tau_i,t} \qquad (18)$$

We put:

$$\rho_{i,t} = \rho_{ia,t} + \rho_{il,t} \qquad (19)$$

and

$$\varphi_{i,t} = -(1 - \mu_{ia})\varepsilon_{i,t} - (1 - \mu_{il})\nu_{i,t} + \rho_{i,t} \qquad (20)$$

We have: $\rho_{i,t} \sim iid(0, \sigma_i^2)$,
where:

$$\sigma_i^2 = \sigma_{ia}^2 + \sigma_{il}^2 + cov(\rho_{ia,t}, \rho_{il,t})$$

Then we have the following theorem.

Theorem 1. *We assume that* $\delta_i = \eta_i$ *and*

$$q > \max\left(1, (\alpha + \beta)\left(\int_{t_0}^{T}\gamma_{L_{i,t}} + \gamma_{A_{i,t}}dt + \nu_{i,T} + \varepsilon_{i,T}\right)\right)$$

Then the solution of the problem (P) *is as follows:*

$$\begin{pmatrix}\hat{h}_{i,\tau_i,t} \\ \hat{k}_{i,\tau_i,t}\end{pmatrix} = (\frac{\alpha}{\beta})^{\frac{\alpha}{1-\alpha-\beta}}\left(\frac{s_{H_i}}{\gamma_{A_{i,t}} + \gamma_{L_{i,t}} + \varphi_{i,t} + \delta_i}\right)^{\frac{1}{1-\alpha-\beta}}\begin{pmatrix}\frac{\alpha}{\beta} \\ 1\end{pmatrix} \qquad (21)$$

We put:

$$Q(T) = \ln \left[\frac{\frac{q}{\alpha+\beta} - \left(\int_{t_0}^{T} \gamma_{L_{i,t}} + \gamma_{A_{i,t}} dt + \nu_{i,T} + \varepsilon_{i,T} \right)}{\int_{t_0}^{T} \chi_{i,t} dt} \right] \tag{22}$$

And we assume that and the proportion of learning in relation to the global technological frontier of country i is constant, i.e.:

$$\hat{a}_{i,\tau_i,t} = \hat{a}_i = constant$$

Then, we have the below corollary:

Corollary 1. *The optimal fraction of the qualification of the employment to attain an income level* $\hat{y}_{i,\tau_i,T} = q\hat{y}_{i0}$, *where q verifies condition to mention in theorem, is as follows:*

$$\tau_{i,max} = Q(T) + \lambda_i \ln(\hat{a}_i) \tag{23}$$

4   Technological Upgrading Conditions

By retaining a simplified version of the Benhabib and Spiegel model [23], the function of disseminating technical progress would be expressed the following form:

$$\gamma_{A_{i,t}} = s_{iR} + \lambda_i \left(1 + \frac{A_{i,t}}{A_{l,t}} \right) \tag{24}$$

Where A_l measures the technological level of the leading country, and s_{iR} the country-specific innovation rate i.

We base on the work of Benhabib and Spiegel, we show that the convergence of technological levels will depend on value of the speed of diffusion of technical progress. Then we have the following theorem.

Theorem 2. *Under the hypotheses of Theorem 1 and the Corollary 1 above, we have the following convergence result:*

$$\lim_{x \mapsto +\infty} = \begin{cases} \frac{s_{iR}+\lambda_i-\gamma_{A_l}}{\lambda_i}, & if \lambda_i > \gamma_{A_l} - s_{iR} \\ 0, & if \lambda_i < \gamma_{A_l} - s_{iR} \end{cases} \tag{25}$$

Corollary 2. *The country i to a partial catch up of its technological level, if the technological diffusion coefficient* λ_i *is sufficiently high and verified the following relation:*

$$\lambda_i > \gamma_{A_l} - s_{iR} \tag{26}$$

5   Conclusion and Perspectives

In this paper we discussed a heterogeneous growth model in which technology and employment are stochastic variables. Afterwards, we have established that sustained growth depends on the qualification of manpower, in other words the apprenticeship of new technology and the rate of technological diffusion of the country.

For the perspective, we will work on some empirical test of the model, so we will test the result of convergence. We will apply this model for countries in the MENA region (Middle East and North Africa).

References

1. Stolper, W., Samuelson, P.: Protection and real wages. Econ. J. **38**(152), 543–559 (1941). JSTOR 2224098
2. Ramsey, F.P.: A mathematical theory of saving. Appl. Anal. **46**(3–4), 219–239 (1928, 1992)
3. Solow, R.M.: A contribution to the theory of economic growth. Q. J. Econ. **70**(1), 65–94 (1956)
4. Domar, E.: Capital expansion, rate of growth, and employment. Econometrica **14**(2), 137–147 (1946). https://doi.org/10.2307/1905364
5. Koopmans, T.C.: On the concept of optimal economic growth. In: The Economic Approach to Development Planning, pp. 225–287. Rand McNally, Chicago (1965)
6. Mankiw, N.G., Romer, D., Weil, D.N.: A contribution to the empirics of economic growth. Q. J. Econ. **107**(2), 407–437 (1992)
7. Aghion, P., Howitt, P.: A model of growth through creative destruction. Econometrica **60**(2), 323–351 (1992). https://doi.org/10.2307/2951599
8. Harrod, R.F.: An essay in dynamic theory. Econ. J. **49**(193), 14–33 (1939). https://doi.org/10.2307/2225181
9. Acemoglu, D., Robinson, J.A.: Persistence of power, elites and institutions. Am. Econ. Rev. **98**(1), 267–293 (2008)
10. Novales, A., Fernández, E., Ruiz, J.: Endogenous growth with accumulation of human capital. In: Economic Growth: Theory and Numerical Solution Methods, pp. 342–376. Berlin, Springer (2009). ISBN 978-3-540-68665-1
11. Novales, A., Fernandez, E., Ruiz, J.: Theory and Numerical Solution Methods, 558 p. Springer, Heidelberg (2014)
12. Asfiji, N.S., Isfahane, R.D., Bakhshi Dastjerdi, R., Fakhar, M.: Mathematical model of Solow economic growth model Modern Sientific Press Chernivtsi. Int. J. Mod. Math. Sci. **4**(3), 118–125 (2012)
13. Saby, B., Dominique, S.: Les Grandes Thories Conomiques. Dunod, Paris (2003). ISBN 2 10 008153 5
14. Lee, K., Pesaran, M., Smith, R.: Growth and convergence in multi-country empirical stochastic Solow model. J. Appl. Econom. **12**(4), 357–392, 319 pp. (1997, 2004)
15. Liberto, D., et al.: Past dominations, current institutions and the Italian regional economic performance. Intal. J. Pure Appl. Math **3**(3), 255–273 (2002, 2008)
16. Nelson, R., Phelps, E.: Investment in humans, technological diffusion, and economic growth. Am. Econ. Rev. Pap. Proc. **61**, 69–75 (1966)
17. Coe, D., Helpman, E.: International R&D spillovers. Eur. Econ. Rev. **39**(5), 859–887 (1995)
18. Kong, J., Phillips, P.C.B., Sul, D.: Weak s- convergence: theory and applications. Cowles Foundation Discussion Papers 2072, Cowles Foundation for Research in Economics, Yale University (2017)
19. El Ghak, T.: Productivite et clubs de convergence productivity and convergence clubs. **2**(3), 19–46 (2009). https://doi.org/10.1016/S1999-7620(09)70015-1
20. Cavdar, S.C., Aydin, A.D.: Empirical analysis about technological development and innovation indicators. J. Procedia - Soc. Behav. Sci. **195**, 1486–1495 (2015). https://doi.org/10.1016/j.sbspro.2015.06.449
21. Ethier, W.J., Markusen, J.R.: Multinational firms, technology diffusion and trade. J. Int. Econ. **41**, 1–28 (1996)

22. McDonald, S., Robert, J.: Growth and multiple forms of human capital in an augmented Solow model: a panel data investigation. Econ. Lett. **74**(2002), 271–276 (2002)
23. Benhabib, J., Spiegel, M.: Human capital and technology diffusion. In: Handbook of Economic Growth, chap. 13, vol. 1, Part A, pp. 935-966. Elsevier (2005)

Optimization of Design of Wind Farm Layout for Maximum Wind Energy Capture: A New Constructive Approach

Mohamed Tifroute$^{(\boxtimes)}$ and Hassane Bouzahir

LISTI, ENSA, Ibn Zohr University, PO Box 1136, Agadir, Morocco
`mohamed.tifroute@gmail.com`

Abstract. Wind energy is becoming an attractive source of clean energy. However, this type of power source is subject to power reductions due to losses in wind energy conversion system and to frequent changes in wind velocity. For that reason, the important phase of a wind farm design is solving the Wind Farm Layout Optimization Problem (WFLOP), which consists in optimally positioning the turbines within the wind farm so that the wake effects are minimized and therefore the expected power production is maximized. This problem has been receiving increasing attention from the scientific community. In this paper, a mathematical optimization scheme is employed to optimize the locations of wind turbines with respect to maximizing the wind farm power production. To formulate the mathematical optimization problem, we used Jensen's wake model. We calculate the wake loss and we express the expected wind farm power as a differentiable function in terms of the locations of the wind turbines. In this paper Furthermore, we develop a New Constructive Approach (NCA) to find the best solution to the wind turbines placement problem. Lastly, our results are compared with those in some other ealier studies.

Keywords: Wind power · Wind farm layout optimization problem
Wake effect · New Constructive Approach

1 Introduction

The wind energy has gained great attention because it represents an important option for reducing the reliance on hydrocarbons for energy production, especially for electricity. But like all current technologies, wind energy poses challenges such as the reduction of the wind speed due to the wake (turbulence) effect of other turbines. Normally, if a turbine is within the area of the wake caused by another turbine, or just in the close area behind it, the wind speed suffers a reduction, and therefore there is a decrease in the production of electricity. Accurately predicting and reducing these losses may improve the feasibility of wind farm installations. A way of reducing and assessing these losses is to find the best possible way of positioning the turbines in the wind farm.

© Springer International Publishing AG, part of Springer Nature 2018
M. Ben Ahmed and A. A. Boudhir (Eds.): SCAMS 2017, LNNS 37, pp. 1015–1027, 2018.
https://doi.org/10.1007/978-3-319-74500-8_91

We call this problem the Turbine Positioning Problem (TPP). It considers the impact of turbines on the others and takes into account the terrain and wind conditions in the region. The existing works that have tackled this problem are limited. In addition, most of these works have been carried out by the wind engineering and wind energy communities, whereas the effort has been done by the optimization community. Existing algorithms include only genetic algorithms and simulated annealing. There is, therefore, potential for improvement by using other optimization techniques, such as mixed-integer programming, dynamic programming, and stochastic programming. As it will be clear what follows, the main reasons why this problem has been largely disregarded by the operations research community are its nonlinearity and the difficulty in obtaining data about the problem instances. Few articles have been published to date that apply mathematical optimization to the problem of positioning turbines inside a wind farm. Mosetti et al. [1] first approached the problem proposing a position optimization scheme based on genetic algorithms. Grady et al. [2] expanded this approach predicting optimal wind farm configurations for simple cases. Ozturk and Norman [3] used a greedy heuristic method which consists in trying different operations recursively (add, remove and move a turbine) in order to maximize the profit. It is only very recently that Park and Law [4] have optimized the layout of a wind farm with 80 turbines using sequential convex programming. In [4], the author's main focus was on demonstrating the usability of a novel continuous wake model that the authors proposed based on Jensen's model [5], and which the authors calibrated with CFD simulation data. Although the authors solved a continuous variable WFLO problem using a mathematical programming method, the numerical experiments were not sufficient to fully document the effectiveness of the proposed optimization approach. In particular, the authors considered only one wind regime, a fixed number of turbines, and a single starting point for the optimization, focusing instead on characterizing the influence of the wake decay constant on the attainable wind farm efficiency, and the wind direction was not fully discussed in their optimization models. This paper extends the followed procedure in [4] by using a New Constructive Approach (NCA) to maximize the wind farm power function and optimize the wind farm performance.

2   Problem Formulation and Methodology

Firstly, a comprehensive model is set up. Both the wake effect impact from all upstream wind turbines as well as the impact of the wind speed variation on wake effect itself is included in this model. Then the energy yields calculation model is described in Sect. 2. The wind speed U (at a given location, height, and direction) follows a Weibull distribution

$$p_u(u, k, c) = \frac{k}{c}(\frac{u}{c})^{k-1} e^{(-\frac{u}{c})^k} \tag{1}$$

where:

- k (non-dimensional) is a parameter related to the shape of the function: high values are related to distribution concentrated around a given value; low values are related to a distribution very spread in the different values.
- $c(m/s)$ is the scale parameter that fixes the position of the curve, with higher values for the sites with strong wind and lower values for still sites.

Statistical characteristics of Dakhla site according to the Weibull c and k parameters, identified according to the least square (LS) and the standard deviation (SD) method. The annual mean speed V_m for the site is also shown, computed according to the data, to the Weibull as identified by SD method (WSD), and according to the data as available in the WEB (Table 1).

Table 1. Weibull distribution parameters and mean wind speed [11].

Parameters	Dakhla
$k(LS)$	1.86
$c(LS)$ (m/s)	9.40
$K(SD)$	2.1
$c(SD)$ (m/s)	9.45
V_m (m/s)(data)	8,4
V_m (m/s) (WSD)	8,36
V_m (m/s) (WSD)	8,36

The k and c parameters must be identified for each given location, and $p_u(.)$ is the probability density function. Note that Assumption 4 may not hold for short-time horizons; however, a great number of sites have shown Weibull distribution of the wind speed [8].

Assumption. At a given height, wind speed (a parameter of the Weibull distribution function) v is a continuous function of the wind direction θ, i.e., $k = k(\theta), c = c(\theta), 0° \leq \theta < 360°$. In other words, wind speeds at different locations with the same direction share the same Weibull distribution across a wind farm. The parameter θ is also assumed to have a continuous probability distribution $p_\theta(\theta)$.

In a relatively flat terrain, Assumption 5 is a reasonable one. Moreover, if the wind farm does not cover a wide range of terrain, the wind speeds at a fixed direction should share a similar distribution. Future research could consider the wind speed distributions changing with directions and locations. Wind direction is an important parameter in this paper. Figure 1 illustrates wind direction for a wind farm, where North is defined as $90°$ and East is defined as $0°$. Though wind turbines may follow different layout patterns, here it is assumed that all turbines are placed within a circular boundary of a wind farm.

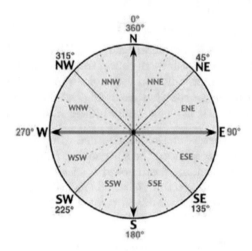

Fig. 1. Wind-direction classification in term of states[10].

The Wind direction is a type of directional data. It has unique characteristics that are different from those of linear data. As mentioned above, let θ be a random variable that measures the directional data and takes on values in the range from $0°$ to $360°$. An analysis of θ depends on the selection of a starting point as the "zero-direction" and the direction of rotation, either clockwise or counterclockwise. The "beginning" of the directional data always coincides with the "end", i.e., $0° = 360°$, and the measurement is also periodic, with θ being the same as $\theta + p \times 2\Pi$ for any integer p [9]. In fact, wind direction data can be classified into several intervals based on the natural geographical directions-namely, North (N), North-North- East (NNE), North-East (NE), East-North-East (ENE), East (E), East- South-East (ESE), South-East (SE), South-South-East (SSE), South (S), South-South-West (SSW), South-West (SW), West-South-West (WSW), West (W), West-North-West (WNW), North-West (NW) and North-North-West (NNW)-as shown in Fig. 2. All of these geographical directions are very useful, particularly for technologies involving compass measurements. This study tries to define the states of the Markov chain based on the geographical directions before a detailed analysis is made. However, only a main geographical direction will be defined as a state: namely, North ($337.5° < \theta < 22.5°$), North-East ($22.5° < \theta < 67.5°$), East ($67.5° < \theta < 112.5°$), South-East ($112.5° < \theta < 157.5°$), South ($157.5° < \theta < 202.5°$), South-West ($202.5° < \theta < 247.5°$), West ($247.5° < \theta < 292.5°$) and North-West ($292.5° < \theta < 337.5°$). In total, eight states of the state space of the processus are determined, which can be written as $S = \{E, N, NE, NW, S, SE, SW, W\}$.

The behaviors of the wind direction can be classified as a stochastic process $\theta = \{\theta_t, t = 0, 1, 2, \cdots, T\}$, where θ_t denotes the directional state of the wind blowing at time t. The random variable θ_t takes on values in the state space S, where $S = \{E, N, NE, NW, S, SE, SW, W\}$; equivalently, the state space can also be indexed in such a way, $S = \{1, 2, 3, 4, 5, 6, 7, 8\}$ that it indicates a total

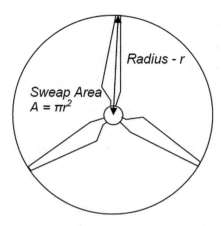

Fig. 2. The swept area of the turbine

of eight state spaces for wind direction. For example, if the process is in state NE at time t, then $\theta_t = NE$ or equivalently $\theta_t = 3$.

2.1 Power Function of a Wind Turbine

A German physicist Albert Betz concluded in 1919 that no wind turbine can convert more than 16/27 of the kinetic energy of the wind into mechanical energy turning a rotor. To this day, this is known as the Betz Limit or Betz' Law. The function of induction factor is defined as:

$$2\alpha(1 - \alpha)^2 \tag{2}$$

The theoretical maximum power efficiency of any design of wind turbine is 0.59 (i.e. no more than per cent of the energy carried by the wind can be extracted by a wind turbine). This is called the "power coefficient" and is defined as: Also, wind turbines cannot operate at this maximum limit. The value is unique to each turbine type and is a function of the wind speed that the turbine is operating in. Once we incorporate various engineering requirements of a wind turbine - strength and durability in particular - the real world limit is well below the Betz Limit with values of 35–45 common even in the best-designed wind turbines. By the time we take into account the other factors in a complete wind turbine system - e.g. the gearbox, bearings, generator and so on - only 10–30 of the power of the wind is ever actually converted into usable electricity. Hence, the extractable power from the wind is given [4] by:

$$P_{avail} = \begin{cases} 0; & U \leq U_{in} \\ \frac{1}{2}AU^3C_p^*; & U_{in} \leq U < U_{out} \\ \frac{1}{2}AU_{out}^3C_p^*; & U \geq U_{out} \end{cases}$$

The values for the cut in wind speed U_{in} and the cut out wind speed U_{out} are $3\,\text{m/s}$ and $12\,\text{m/s}$, respectively. When $U \geq U_{out}$; C_p is adjusted to keep the clipped wind turbine power $\frac{1}{2}AU_{out}^3C_p^*$, generally referred as the rated power.

2.2   Jensen's Wake Model

The Jensen wake model is used to generate wake speed of downstream wind turbines. This model was first developed by N.O. Jensen in 1983 which is a simple analytical model with a short calculation time. This paper adopts the Jensen model for its simplicity. However, any wake model could be used with our new approach proposed in this paper. The wake model is, thus, derived by conserving the momentum downstream of the wind turbine. The velocity in the wake is given as a function of downstream distance from the turbine hub and it is assumed that the wake expands linearly downstream. If the near field behind a wind turbine is neglected, the resulting wake behind the wind generator can be treated as a turbulent wake. This model is based on the assumption that the wake is a turbulent and the contribution of tip vortices is neglected. Thus, this means that this wake model is strictly applicable only in the far wake region. Based on the findings from his study, Jensen 1983 recommends the Jensen's model to be used for the energy predictions in offshore wind farms, as it gives a good trade-off between prediction errors.

To get the relationship of downstream velocity, Eq. (1) can be solved with respect to velocity. The resulting equation gives the downstream velocity as a function of upstream wind velocity. According to Betz theory the value of U_0 in terms of U will be given as in the following equation:

$$U = (1 - 2a)U_0 \tag{3}$$

where a is the axial flow induction coefficient or induction factor, it is assumed that the wake downstream expands linearly. Thus, the path traced by the wind downstream follows the conical shape of disturbance. The radius of the cone can be estimated by using the following equation:

$$R_W = d\frac{(1 + 2\alpha X)}{2} \tag{4}$$

Given that D is wind turbine diameter and R_w is radius of expanding wake. The determination of α is sensitive to factors including ambient turbulence, turbine induced turbulence and atmospheric stability. This can be calculated by using an analytical expression, as below Eq. (4):

$$\alpha = \frac{0,5}{\ln\left(\frac{z}{z_0}\right)} \tag{5}$$

The realistic model of the velocity profile inside the wake region, a modulation term is recommended by Jensen [5], where a Gaussian profile for the velocity deficit is used as an alternative to the uniform profile. The modulation term is defined mathematically as:

$$f(\theta) = \begin{cases} \dfrac{1 + \cos(9\theta')}{2}; & \theta' \leq \frac{\pi}{9} \\ 0; & \theta' > \frac{\pi}{9}, \end{cases}$$

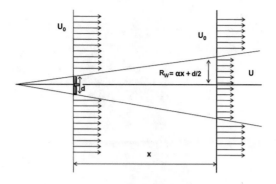

Fig. 3. Wind turbine wake model.

where θ' is the angular (polar) location of where the velocity is calculated with respect to the central axis of the wake. Given the wind direction θ_t and the locations of two wind turbines i and j, denoted as $l_i(x_i; y_i)$ and $l_j(x_j; y_j)$ as shown in Fig. 4, the downstream wake inter distance d_{ij} and the radial wake inter distance r_{ij} between the hubs of wind turbines i and j can be determined, respectively, as

$$d_{ij} = \sqrt{(x_i - x_j)^2 + (y_i - y_j)^2} \cos(\theta') \tag{6}$$

$$r_{ij} = \sqrt{(x_i - x_j)^2 + (y_i - y_j)^2} \sin(\theta') \tag{7}$$

where $\theta' = |\theta_{ij} - \theta_t|$ and $\theta_{ij} = tan^{-1} \frac{|y_j - y_i|}{|x_j - x_i|}$ are the angle between the two wind turbines i and j.

With this modification, the wind velocity is a function of x and y coordinates, as below:

$$U_k(x, y) = U_0[1 - \frac{2}{3}f(\theta')(\frac{d}{d + 2\alpha X})^2] \tag{8}$$

When a wind turbine faces multiple wake effect from upstream wind turbines, the resulting velocity U_i can be calculated by equating the sum of the kinetic energy deficits of each wake to the kinetic energy deficit of the mixed wake at that point.

$$U = U_0[1 - \sqrt{\sum_{k=1}^{n}(1 - \frac{U_k}{U_0})^2}] \tag{9}$$

2.3   Wind Farm Layout Optimization Using Jensen's Wake Model

The wind farm optimization can be executed by applying specific objective function. The most extensive method is by maximizing the total power of the wind farm that is produced. The power produced is contingent up the total number

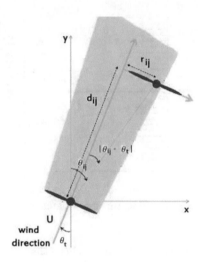

Fig. 4.

of wind turbines on a farm and their positions with respect to one another in order to reduce the wake effect. Wind farm layout optimization is referred as the optimization task that chooses the optimal turbine positions. Optimization does not necessarily mean finding the optimum solution to a problem since it may be unfeasible due to the characteristics of the problem, which in many cases are classed NP-hard problem [7]. The first step is to define the space of feasible solutions: In the case of a Wind Farm, these feasible solutions will be all the possible layouts. That is, in all the possible combinations of wind turbines, no two turbines are closer than a certain minimum distance. If $l_i = (x_i, y_i)$ is the position of turbine i and $l_j = (x_j, y_j)$ the position of turbine j, we can express this constraint as:

$$\sqrt{(x_i - x_j)^2 + (y_i - y_j)^2} \geq 5D$$

Given $5D$ is the minimum turbine distance (normally expressed in terms of the rotor diameter D), the nonlinear proximity constraint can be defined as:

$$h(x, y) = 5D - \sqrt{(x_i - x_j)^2 + (y_i - y_j)^2} \leq 0 \tag{10}$$

The configuration suggested by Mosetti and others, 1994, was selected for simulation and comparison of results, where the speed and the direction of wind are constants. Table 2 gives the values of the used parameters of entry.

And a constant thrust coefficient $C_T = 0.88$. The total power of N_T turbines is expressed as:

$$P_T = \sum_{k=1}^{N^T} (0, 3U_k^3) \tag{11}$$

Table 2. Input parameters used by Mosseti et al.

Input parameters	Value
Roughness	0.3
Wind velocity in free flow	12 m/s
Hub height (Z)	60 m
Rotor diameter (D)	40 m
Wind farm dimension	2000 m × 2000 m
C_T	0.88

The objectify function $f(x, y)$ is the expected wind farm power:

$$f(x, y) = E[P(U)] = \sum_{t=1}^{T} \sum_{j=1}^{N_U} \sum_{k=1}^{N_T} (0, 3U^3) Pr(U_j, \theta_t) \tag{12}$$

Where N_T is the number of wind turbines in a wind farm. The expectation is expressed in terms of the joint probability distribution $Pr(U, \theta_t)$ of the wind speed U and the wind direction θ_t. Here, the expected wind farm power is approximated as the sum of the power produced by the wind turbines weighted by the joint probability $Pr(U_j, \theta_t) = P(\theta_i)P(U_t|\theta_t)$ for the discrete wind speed U_j and wind direction θ_t. Note that T and N_U are the numbers of bins for the discretization of wind direction θ and wind direction U, respectively.
The optimization problem is defined for each position $l = (x, y)$ as:

$$\max_{l \in \Omega} f(x, y)$$
Subject to: $h(x, y)_{l \in \Omega} \leq 0$;
with $\Omega = \{x | \underline{x}_i \leq x_i \leq \overline{x}_i, i = 1, \cdots, N_T\}$

Where the objective function $f_0 : \mathbb{R}^2 \longrightarrow \mathbb{R}$ is supposed to be twice differentiable continuous at x for all $x \in \mathbb{R}^2$, and the $h_j(x)$ are M inequality constraint functions. $\underline{x}_i$ and $\overline{x}_i$ respectively indicate lower and upper bounds of continuous real variable x_i that usually reflect the numerical considerations.

2.4   A New Constructive Approach (NCA)

In this paper, we develop a new approach to find the optimal location of the turbines within the wind farm so that the wake effects are minimized and therefore the expected power production is maximized. The Step zero of our approach consist to fixed an initial block of turbines, in step 1, the positions of the turbines of block 1 are the algorithm determinate using our algorithm, so that no turbines of Block 1 is to influence by the turbines of the preceding block, therefore all the turbines of block 1 will be out of fields wake to generate by the turbines of the preceding block, and for this reason it is necessary that the distance between each two turbines of the initial block and the following block are higher than 5D, and the angle θ' is higher than $\pi/9$, the angle θ' and the distance are defined as follows: where $\theta' = |\theta_{ij} - \theta_t|$ and $\theta_{ij} = tan^{-1}\frac{|y_j-y_i|}{|x_j-x_i|}$ are the angle between the two wind turbines i and j, and $X_{ij} = \sqrt{(x_i - x_j)^2 + (y_i - y_j)^2}$.

We proceeded in the same manner to determine the blocks until determining the positions of the last block (Table 3).

Table 3. Optimal locations of 26 turbines.

Coordinate X	Coordinate Y	Velocity U_d (m/s)	Power P_k (KW)
99	149	12	518.4
101	341	12	518.4
101	543	12	518.4
96	945	12	518.4
201	859	12	518.4
240	643	12	518.4
237	445	12	518.4
237	246	12	518.4
223	1063	12	518.4
243	34	12	518.4
362	351	12	518.4
370	550	12	518.4
362	750	10,9249	391,18
374	143	12	518.4
355	969	12	518.4
593	22	12	518.4
514	248	11,6913	479,41
504	456	12	518.4
509	653	11,8304	496,73
520	861	12	518.4
525	1082	12	518.4
681	979	12	518.4
669	761	12	518.4
645	559	12	518.4
659	354	11,99	517,74
731	128	11,5041	456,76

2.5   Results and Discussion

In a way to verify the NCA performance, some simulations were done and the results were compared with other author's works [1,2,10]. These works were

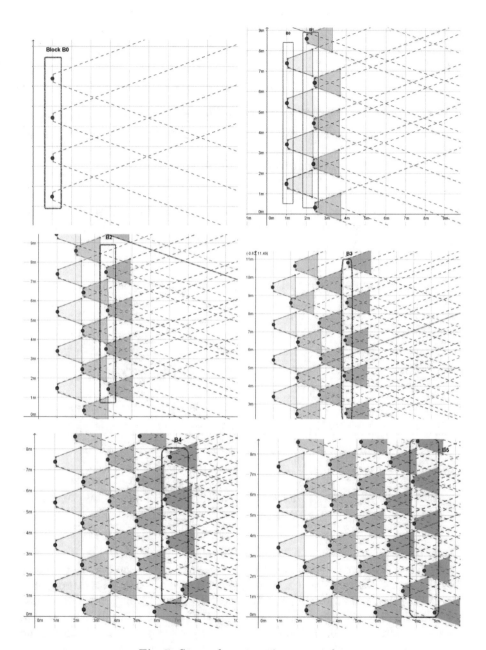

Fig. 5. Steps of constructive approach.

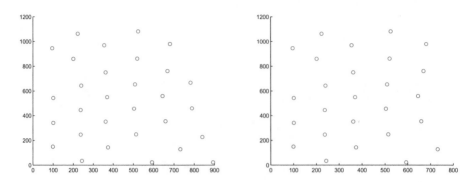

Fig. 6. Optimal locations of 26 and 30 turbines.

chosen because they use the same parameters used by Mosetti et al. model as well as Jensen's wake model [5] and probabilistic algorithms as optimization tool, guaranteeing control and reliability of results (Table 4).

Table 4. Results for comparison with previous studies.

Mosetti et al. [1]	Grady et al. [2]	Marmidis et al. [10]
Number of turbines: 26	Number of turbines: 30	Number of turbines: 32
Total power [kW]: 12921	Total power [kW]: 14764	Total power [kW]: 13467
Efficiency: 95, 5	Efficiency: 94, 6	Efficiency: 80, 9
WindFarmer	PCOA	PCOA
Number of turbines: 26	Number of turbines: 30	Number of turbines: 32
Total power [kW]: 13127	Total power [kW]: 14790	Total power [kW]: 15574
Efficiency: 97	Efficiency: 94.8	Efficiency: 93.6
NCA	NCA	NCA
Number of turbines: 26	Number of turbines: 30	Number of turbines: 32
Total power [kW]: 13228	Total power [kW]: 15285	Total power [kW]: 16330
Efficiency: 98	Efficiency: 98	Efficiency: 98.4

From the results presented in the previous table, some remarks can be made. The first is the quality of our methodology, as our results are close by contribution to the other works. In addition, the effectiveness of the algorithm developed in this work. This has been verified because of the proximity Results compared to the work of Mosetti et al. [1] and Grady et al. [2], do Couto et al. [11] considering that both of them used Genetic algorithm for optimization. Finally, comparing the results of the NCA with Those obtained by Marmidis et al. [10], There is a significant increase in the efficiency and production of the wind farm. This difference is related to the optimization method used because Marmidis et al. Which uses the Monte method Carlo and our method which uses a constructive app-roach block by block, proved more efficient. Moreover, in our configuration we occupied that 50 percent of the ground of park whose surface is 2000 m × 2000 m.

3   Conclusion

This paper focuses on the existing literature on the wake effect modeling, a New Constructive Approach (NCA) for optimizing in onshore wind farm layout was presented. The optimization model considered wind farm radius and turbine dis-tance constraints. However, other constraints can be easily incorporated in this model. The model maximizes the energy production by placing wind turbines in such a way that the wake loss is minimized.

The optimal solution maximized energy production while satisfying all con-straints, the algorithm finds not the global optimum but one of local optimums. Therefore, it is important to ensure that a local optimum value is close to the global maximum (although it is unknown). For the NCA method, the optimal solution obtained depends on the initial Block.

References

1. Mosetti, G., Poloni, C., Diviacco, B.: Optimization of wind turbine positioning in large wind farms by means of a genetic algorithm. J. Wind Eng. Ind. Aerodyn. **51**, 105–116 (1994)
2. Grady, S.A., Hussaini, M.Y., Abdullah, M.M.: Placement of wind turbines using genetic algorithms. Renew. Energy **30**, 259–270 (2005)
3. Ozturk, U.A., Norman, B.A.: Heuristic methods for wind energy conversion system positioning. Electr. Power Syst. Res. **70**, 179–185 (2003)
4. Park, J., Law, K.H.: Layout optimization for maximizing wind farm power pro-duction using sequential convex programming. Appl. Energy **151**, 320–334 (2015)
5. Jensen, N.O.: A note on wind generator interaction. RisoNational Laboratory (1983)
6. Katic, I., Hotrup, J., Jensen, N.A.: Simple model for cluster efficiency. In: European Wind Energy Association Conference and Exhibition, pp. 407–410 (1986)
7. Gary, M., Johnson, D.: Computers and Intractability a Guide to the Theory of NP- Completeness. W. H. Freeman & Co., San Francisco (1979)
8. Manwell, J.F., McGowan, J.G., Rogers, A.L.: Wind Energy Explained: Theory, Design and Application, 1st edn. Wiley, London (2002)
9. Jammalamadaka, S.R., SenGupta, A.: Topics in Circular Statistics. World Scien-tific Publishing, Singapore (2001)
10. Marmidis, G., Lazarou, S., Pyrgioti, E.: Optimal placement of wind turbines in a wind park using Monte Carlo simulation. Renew. Energy **33**, 1455–1460 (2008)
11. do Couto, T.G., Farias, B., Diniz, A.C.G.C., de Morais, M.V.G.: Optimization of wind farm layout using genetic algorithm. In: 10th World Congress on Structural and Multidisciplinary Optimization, Orlando, Florida, USA, May 2013

Some Properties of Pettis Integrable Multivalued Martingales

M'hamed El-Louh$^{(\boxtimes)}$ and Fatima Ezzaki

Laboratoire Modélisation et Calcul Scientifique, Département de Mathématiques,
Faculté des Sciences et Techniques, Fès, Morocco
ellouh.mhamed@gmail.com

Abstract. We present a new result of Pettis integrable multivalued martingale. A result presented in this paper is a new version of uniformly integrable martingale in Pettis integration. A classical theorem of vector uniformly integrable martingale in Bochner integration is stated by Egghe [11]. A multivalued version of this result is proved by Hiai and Umegaki in [20].

Keywords: Conditional expectation · Martingale
Multivalued martingale · Pettis integral · Uniformly Pettis integrable
Pettis integrable multivalued martingale

1 Introduction

The notion of multivalued martingales extends those of real and vector martingale; in fact the values of multivalued martingale are closed and convex subsets of a separable Banach space E. The theory of Pettis integration in infinite dimensional space is more general concept than Bochner integrability one. The multivalued Pettis integrability theory has been massively developed by Castaing and Valadier [4], Musial [22–25], Egghe [11,12], Hiai and Umegagi [20] and others. This theory has recently attracted the attention of several authors, for example Akhiat et al. [1], Ezzaki and El Harami [17], Musial [22], El Amri [13], Godet-Thobie and Satco [19], Hiai and Umegaki [20] and others. Our paper is organized as follows: In the second section we present some Properties of Pettis integrable function and some related definitions. In the third section we prove that every uniformly Pettis integrable multivalued martingale is a Pettis regular martingale. Some Properties of Pettis integrable multifunctions are also presented.

2 Notations and Definitions

Let $(\Omega, \mathcal{A}, P)$ be a probability space and $\mathcal{B}$ a sub σ-algebra of $\mathcal{A}$, let X a random variable defined on Ω with values in a separable Banach space E. If a conditional expectation of X exists we denote it by $E^{\mathcal{B}}X$.

© Springer International Publishing AG, part of Springer Nature 2018
M. Ben Ahmed and A. A. Boudhir (Eds.): SCAMS 2017, LNNS 37, pp. 1028–1034, 2018.
https://doi.org/10.1007/978-3-319-74500-8_92

Definition 1. *Let $(\mathcal{A}_n)_{n\geq 1}$ be an increasing sequence of $\mathcal{A}$ such that $\sigma(\cup_n\mathcal{A}_n) = \mathcal{A}$. A sequence $(X_n)_{n\geq 1}$ of random variables with values in E is said to be adapted sequence to $(\mathcal{A}_n)_{n\geq 1}$ if X_n is $\mathcal{A}_n$-measurable for all $n \geq 1$.*

Definition 2. *An adapted sequence $(X_n, \mathcal{A}_n)_{n\geq 1}$ is a martingale if for all $n \geq 1$, $X_n = E^{\mathcal{A}_n} X_{n+1}$ a.s.*

Definition 3. *An adapted sequence $(X_n, \mathcal{A}_n)_{n\geq 1}$ is a regular martingale if there exists a random variable X such that $X_n = E^{\bar{\mathcal{A}}_n} X$ a.s. $\forall n \geq 1$.*

Definition 4. *A function $X : \Omega \to E$ is said to be Pettis integrable if:*

(1) X is scalarly measurable; for each $x^* \in E^*$, the real function $< x^*, X(.) >$ is measurable,
(2) X is scalarly integrable; for all $x^* \in E^*$, $< x^*, X(.) > \in L^1$,
(3) for each $A \in \mathcal{A}$, there exists $X_A \in E$ such that $< x^*, X_A > = \int_A < x^*, X > dP$, for every $x^* \in E^*$. We denote X_A by $\int_A X dP$ and we call it the Pettis integral of X on A.

– A multifunction $X : \Omega \to cwk(E)$ is $\mathcal{A}$-measurable if for every open set $U \in E$, the set

$$X^-U = \{\omega \in \Omega : X(\omega) \cap U \neq \emptyset\}$$

is in $\mathcal{A}$ (*see,* [4]).

A sequence $(X_n)_{n\geq 1}$ of measurable multifunctions with values in a set of closed and convex subset of E $(cc(E))$ is said to be adapted to $(\mathcal{A}_n)$ if, for any $n \geq 1$, X_n is $\mathcal{A}_n$-measurable.

– Measurable multifunctions are also called random set. Let X be a random set $\Omega \to cwk(E)$. A measurable function $f : \Omega \to E$ is said to be selector of X if $f(\omega) \in X(\omega)$ a.s. The support function of X denoted by $\delta^*(., X)$ and for every $X \in cwk(E)$ defined on E^* by

$$\delta^*(x^*, X) = \sup\{< x^*, x > : x \in X\}, \; for \; each \; x^* \in E^*$$

– A multifunction $\mathcal{A}$-measurable $X : \Omega \to cwk(E)$ is scalary integrable if $\delta^*(x^*, X(.))$ is integrable for every $x^* \in E^*$.
– A measurable multifunction X is said to be integrable if it admits one Bochner integrable selection.
– A multifunction X is integrable if the distance function $d(0, X(.))$ is integrable.
On the other hand X said to be integrably bounded (or strongly integrable) if the function $h(X) \in L^1$, where $h(X)(\omega) = \sup_{x\in X(\omega)} \|x\|_E$. A scalarely integrable random set $X : \Omega \to cwk(E)$ is Pettis integrable if for all $A \in \mathcal{A}$ there exists $C_A \in cwk(E)$ such that

$$\int_A \delta^*(x^*, X) dP = \delta^*(x^*, C_A), \; for \; every \; x^* \in E^*.$$

A multivalued Pettis sequence $(X_n)_{n\geq 1}$ is said to be Pettis uniformly integrable if the set $\{\delta^*(x^*, X_n), n \geq 1, x^* \in B^*\}$ is uniformly integrable in L^1. We denote by:

- $P_E^1(\mathcal{A})$ the space of all $\mathcal{A}$-measurable and Pettis-integrable E-valued functions defined on $(\Omega, \mathcal{A}, P)$.
- $P_{cwk(E)}^1(\mathcal{A})$ (resp. $P_{cc(E)}^1(\mathcal{A})$) the set of all Pettis-integrable $cwk(E)$-valued (resp. $cc(E)$-valued) multifunctions.
- $S_X^{Pe}(\mathcal{A})$ the set of all $\mathcal{A}$-measurable and Pettis-integrable selections of X.
- The multivalued Aumann-Pettis integral $\int_\Omega X dP$ of $cwk(E)$-valued Pettis-integrable multifunction X is defined by

$$\int_\Omega X dP = \{\int_\Omega f dP : f \in S_X^{Pe}(\mathcal{A})\}$$

where $\int_\Omega f dP$ denote the Pettis integral of the Pettis selector mapping $f : \Omega \to E$, of X. A random set X is Pettis Aumann-integrable if $S_X^{Pe} \neq \emptyset$ then $\int_\Omega X dP \neq \emptyset$ too. Given a measurable multifunction $X : \Omega \to cwk(E)$, we denote by $S_X^1(\mathcal{A})$, or simply by S_X^1, the subset of $L_X^1(\mathcal{A})$ defined by

$$S_X^1(\mathcal{A}) = \{f \in L_X^1(\mathcal{A}) : f(\omega) \in X(\omega) \ a.s.\}$$

and we denote by $S_X^{Pe}(\mathcal{A})$, the subset of $P_X^1(\mathcal{A})$ defined by

$$S_X^{Pe}(\mathcal{A}) = \{f \in P_X^1(\mathcal{A}) : f(\omega) \in X(\omega) \ a.s.\}$$

A sequence $(X_n)_{n\geq 1}$ in $P_{cwk(E)}^1(\mathcal{A})$ is adapted if for each $n \geq 0$, $X_n \in P_{cwk(E)}^1(\mathcal{A}_n)$. Given a sub-$\sigma$-algebra $\mathcal{A}_n$ of $\mathcal{A}$. The conditional expectation with respect to $\mathcal{A}_n$ of $cwk(E)$ valued Pettis integrable multifunction X denoted by $E^{\mathcal{A}_n} X$ is the unique $\mathcal{A}_n$-measurable function such that, $\forall A \in \mathcal{A}_n$

$$\int_A E^{\mathcal{A}_n} X dP = \int_A X dP, \quad \forall A \in \mathcal{B}.$$

See Akhiat et al. [1].

A $cwk(E)$-valued sequence $(X_n)_{n\geq 1}$ Mosco-converges to a Weakly convex compact set X_∞ if

$$X_\infty = s - liX_n = w - lsX_n,$$

where

$$s - liX_n = \{x \in E : \|x_n - x\| \to_{n\to\infty} 0; \ x_n \in X_n\}$$

and

$$w - lsX_n = \{x \in E : x = w - \lim_{j\to\infty} x_j; \ x_j \in X_{n_j}\}$$

and s (resp. w) is the strong (resp. weak) topology in E.

3   Uniformly Pettis Integrable Multivalued Martingale

In this section we present a new version of uniformly integrable martingale in Pettis integration.

Theorem 1. [1] *Assume that E^* is a separable Banach space.*
Let $\mathcal{B}$ be a sub σ-algebra of $\mathcal{A}$ and let X be a $cwk(E)$-valued Pettis-integrable multifunction such that $s\,E^{\mathcal{B}}|X| < \infty$. Then there exists a unique $\mathcal{B}$ -measurable, $cwk(E)$-Pettis-integrable multifunction, denoted by $E^{\mathcal{B}}X$, which enjoys the following property: For every $h \in L^{\infty}(\mathcal{B})$, one has

$$\int_{\Omega} hE^{\mathcal{B}}XdP = \int_{\Omega} hXdP.$$

Where $\int_{\Omega} hE^{\mathcal{B}}XdP$ and $\int_{\Omega} hXdP$ denote the $cwk(E)$-valued Aumann Pettis integral of $hE^{\mathcal{B}}X$ and hX respectively.

Lemma 1. *Assume that E^* is separable. Let X be a $cwk(E)$-valued Pettis-integrable multifunction such that*

$$E^{\mathcal{A}_n}|X| < \infty$$

for each $n \geq 1$.
Then we have

$$M - \lim_{n\to\infty} E^{\mathcal{A}_n}X = X \ a.s.$$

Theorem 2. [13] *Assume that E is separable. Let $X : \Omega \to cwk(E)$ be a scalarely integrable multifunction. Then the following properties are equivalent:*

 (i) X is Pettis integrable,
 (ii) $\{\delta^*(x^*, X(.)),\ x^* \in B^*\}$ is uniformly integrable,
 (iii) $\{<x^*, f(.)>,\ x^* \in B^*, f \in S_X^1\}$ is uniformly integrable,
 (iv) every measurable selection of X is Pettis integrable,
  (v) $\forall A \in \mathcal{A},\ \int_A XdP \in cwk(E).$

Theorem 3. [25] *Let $X : \Omega \to cwk(E)$ be scalarly integrable and let $\{X_n : \Omega \to cwk(E) :\ n \geq 0\}$ be a sequence of multifunctions Pettis integrable in $cwk(E)$ and satisfying the following two conditions:*

 (i) the set $\{\delta^*(x^*, X_n(.)) : \|x^*\| \leq 1, n \geq 0\}$ is uniformly integrable;
 (ii) $\lim_n \delta^*(x^*, X_n) = \delta^*(x^*, X),\quad a.s.\ \forall\, x^* \in E^*.$

Then X is Pettis integrable in $cwk(E)$ and,

$$\lim_n \delta^*(x^*, \int_A X_n dP) = \delta^*(x^*, \int_A XdP)$$

for every $x^* \in E^*$ and $A \in \mathcal{A}$.

Theorem 4. [1] *Assume that E^* is separable. Let $\mathcal{B}$ be a sub-σ -algebra of $\mathcal{A}$ and let X be a $cwk(E)$-valued Pettis-integrable multifunction such that $E^{\mathcal{A}_n}|X| \in [0, \infty[$. Then there exists a unique $\mathcal{B}$-measurable, $cwk(E)$- valued Pettis integrable multifunction, denoted by $Pe - E^{\mathcal{B}}X$, which enjoys the following property: For every $h \in L^\infty(\mathcal{B})$, one has*

$$Pe \int_\Omega hPe - E^{\mathcal{B}}XdP = Pe - \int_\Omega hXdP,$$

where

$$Pe - \int_\Omega hPe - E^{\mathcal{B}}XdP$$

and

$$Pe - \int_\Omega hXdP$$

denote the $cwk(E)$-valued Aumann Pettis integral of $hPe - E^{\mathcal{B}}X$ and hX respectively

Lemma 2. *Let $(C_n)_{n\geq1}$ be a Pettis random sets with values in $cwk(E)$ satisfying the following conditions:*

(i) there exists L Pettis integrable random set such that $C_n \subset L \ \forall n \geq 1$ and $0 \in L$,

(ii) $\forall x^* \in D^* \lim_n \delta^*(x^*, C_n(\omega))$ exists for all $\omega \in \Omega$.

Then there exists a Pettis integrable random set C such that

$$\lim_n \delta^*(x^*, C_n(\omega)) = \delta^*(x^*, C(\omega)), \quad a.s. \ \forall \ x^* \in E^*$$

Theorem 5. *Let E be a separable Banach space with strongly separable dual E^*. Let $(X_n, \mathcal{A}_n)$ be a Pettis integrable martingale with values in $cwk(E)$, satisfying the following conditions:*

(i) There is a Pettis integrable multifunction

$$K : \Omega \to cwk(E)$$

such that $X_n(\omega) \subset K(\omega)$ for all $n \geq 0$ and for all $\omega \in \Omega$ and $0 \in K(\omega)$ $\forall \omega \in \Omega$.

(ii) $\forall n \geq 1$, $E^{\mathcal{A}_n}|K| < \infty$.

Then there exists a Pettis integrable and measurable multifunction X with values in $cwk(E)$ such that $X_n = E^{\mathcal{A}_n}X$ a.s $\forall n \geq 1$.
and

$$M - \lim_n X_n = X \ a.s$$

References

1. Akhiat, F., Castaing, C., Ezzaki, F.: Some variational convergence results for multivalued martingales. Adv. Math. Econ. **13**, 1–33 (2010)
2. Amrani, A., Castaing, C.: Weak compactness in Pettis integration. Bull. Polish Acad. Sci. **44**(2), 67–79 (1996)
3. Auman, R.J.: Integrals of set valued functions. J. Math. Anal. Appl. **12**, 1–12 (1965)
4. Castaing, A., Valadier, M.: Convex Analysis and measurable multifunctions. University des sciences et techniques de languedoc U.E.R de, Mathématiques (1975)
5. Castaing, C., Saadoune, M.: Dunford Pettis types theorem and convergences in set valued integration. J. Nonlinear Convex Anal. **1**(1), 37–71 (1999)
6. Castaing, C., Ezzaki, F., Lavie, M., Saadoune, M.: Weak star convergence of martingales in a dual space. Working paper, Département de Mathématiques, Université Montpellier II, September 2009
7. Castaing, C., Ezzaki, F.: SLLN for convex random sets and random lower semicontinuous integrands. Atti. Sem. Mat. Fis. Univ. Modena **XLV**, 527–553 (1997)
8. Castaing, C., Ezzaki, F.: Some convergence results for multivalued martingales in the limit. In: Séminaire d Analyse Convexe, Montpellier, exposé No 1 (1990)
9. Castaing, C., Ezzaki, F.: Mosco convergence for multivalued martingales in the limit. C. R. Acad. Sci. Paris Ser. **I**(312), 695–698 (1991)
10. Castaing, C., Ezzaki, F.: Convergence of conditional expectation for unbounded closed convex random sets. Stud. Math. **124**(2), 133–148 (1997)
11. Egghe, L.: Stopping Time Techniques for Analysts and Probabilists. London Mathematical Society Lecture Note Series, no. 100. Cambridge University Press, Cambridge (1984)
12. Egghe, L.: Convergence of adapted sequences of Pettis-integral of closed valued multifunctions. Pacific J. Math. **114**(2), 345–366 (1984)
13. El Amri, K.: intégral de Pettis et convergence, Thèse de Doctorat Univérsité Mohamed V Faculté des Sciences Rabat (2000)
14. El Harami, M.: Contribution aux espérences conditionnelles des ensembles aléatoires Pettis intégrables et Applications à la convergence des classes des martingales, Thèse de Doctorat Univérsité Sidi Mohamed Ben Abdellah Faculté des Sciences et Techniques Fès (2012)
15. Ezzaki, F.: A general dominated convergence theorem for unbounded random sets. Bull. Polish Acad. Sci. Math. **44**(3), 353–361 (1996)
16. Ezzaki,F.: Contributions aux problèmes de convergence des suites adaptées et des ensembles aléatoires. Thèse de Doctorat, Université de Montpellier II (1993)
17. Ezzaki, F., El Harami, M.: General Pettis conditional expectation and convergence theorems. Int. J. Math. Statist. **11**, 91–111 (2012)
18. Geitz, R.: Pettis integration. Proc. Am. Math. Soc. **8**, 81–86 (1981)
19. Godet-Thobie, C., Satco, B.: Decomposability and uniform integrability in Pettis integration. Quaestiones Math. **29**, 39–58 (2006)
20. Hiai, F., Umegaki, H.: Integrals, conditional expectations, and martingales of multivalued functions. J. Multivar. Anal. **7**, 149–182 (1977)
21. Marraffa, V.: Stochastic process of vector valued Pettis and MacShane integrable functions. Folia Math. **12**(1), 25–37 (2005)
22. Musial, K.: Vitali and Lebesgue convergence theorems for Pettis integral in locally convex spaces. Atti Sem. Mat. Fis. Univ. Modena **XXXV**, 159–166 (1987)

23. Musial, K.: The weak Radonn-Nikodym property in Banach space. Studia Math. **64**(2), 151–173 (1979)
24. Musial, K.: Martingales of Pettis integrable functions. In: Proceedings of the Conference on Measure Theory. Oberwolfach Lecture Notes in Mathematics, vol. 794, pp. 324–339. Springer, Berlin (1980)
25. Musial, K.: Pettis integrability of multifunctions with values in arbitrary Banach space. J. Convex Anal. **18**(3), 769–810 (2011)

Optimization of Actuarial Calculation Processes by Application of Stochastic Methods

Jamal Zahi and Merieme Samaoui[(⊠)]

Hassan 1st University, Settat, Morocco
mersamaoui@yahoo.fr

Abstract. The approach presented in this paper is a contribution to research on the optimization of actuarial calculation processes.

The social insurance companies are required by law to integrate technical provisions into their liabilities and to take them into consideration, insofar as they can guarantee future commitments vis-a-vis their members and/or subscribers. The calculation of its provisions is a major issue for hedge funds.

This work has the following objectives: to evaluate the costs and to properly manage the technical provisions taking into account the rates of regulations, the future constraints, and their random nature.

Keywords: Risk management · Actuarial calculations · Solvency
Technical provisions · Deterministic methods · Stochastic methods

1 Introduction

Social hedge funds are subject to very long-term risks from one year to the next in terms of their financial obligations such as real income growth, structural changes in the economy, disability, medical cover… [1].

To analyze the financial impact of these risks, actuaries use techniques in mathematics, economics and statistics to model future events. Usually, their work involves the quantification of amounts that represents a sum of money or future financial liability on a given date.

The problem of better adapting the funds required of insurance and reinsurance companies with the risks that they incur in their business is the objective of the European regulatory reform of the world of insurance Solvency II.

The calculation of technical provisions is one of the pillars to guarantee a solvency margin and to ensure the sustainability of the insurance system [2]. Technical provisions are generally defined by two types:

1. Mathematical reserves representing the present value of the insurer's future benefits for events occurring prior to the date of the inventory [3].
2. Allowances for claims payable (ACP), which represent the estimated.

© Springer International Publishing AG, part of Springer Nature 2018
M. Ben Ahmed and A. A. Boudhir (Eds.): SCAMS 2017, LNNS 37, pp. 1035–1041, 2018.
https://doi.org/10.1007/978-3-319-74500-8_93

2 Methods of Calculation

In the insurance industry, there are several methods for calculating technical provisions that can be grouped into two categories:

a. Deterministic methods,
b. Stochastic methods.

The deterministic methods (Chain Ladder, London Chain, Least Squares of De VYFLDER, etc.) showed good results on the calculation.
On the other hand, they have limits since:

   i. They are based on the last amount of charge, which may distort the amount of the technical provision calculated,
  ii. they work only for regular and stable triangles,
 iii. they do not take into account non-constant inflation.

Hence the use of a stochastic approach that can respond to the need to quantify the uncertainty presented in the results obtained via deterministic methods. Uncertainty analysis meets the requirements of Solvency II with the determination of Risk Margin. This margin therefore rests on the construction of confidence intervals around the hope of predicted recovery.

Stochastic models are used to determine the degree of uncertainty in the reserve, which can be crucial information for the company's financial strategies. The stochastic approach assumes that the data used (Amount, rate) can be considered as random variables. We present in this section some stochastic methods of calculation in provisioning, on the other hand we will present the results obtained by the application of a chosen method [4].

2.1   Model of Mack (Thomas Mack)

Thomas MACK proposed a parametric method which corresponds to the stochastic version of the deterministic method of CHAIN LADDER. It provides an estimate of the mean and the standard deviation for the estimator R of the variable R which represents the technical provision to be constituted. This allows us to model it using a normal or log-normal law.
The deterministic model of CHAIN LADDER:

$$C_{i,j+1} = \lambda_j \times C_{i,j} \quad for \; i \in \{1,\ldots,n\}, j \in \{1,\ldots,n-1\}$$

becomes stochastic:

$$E[C_{i,j+1}] = \lambda_j \times E[C_{i,j}] \quad for \; i \in \{1,\ldots,n\}, j \in \{1,\ldots,n-1\}$$

For each year of occurrence, a factor is used to quantify the increase in claims from year k to year k + 1.

Mack's model is based on the same assumptions as the CHAIN LADDER method, namely the independence of years of occurrence and regularity of payments.

Generally, short-term risks (health) are modeled by a normal law and long risks by a log-normal law.

2.2   Generalized Linear Models

The generalized linear model was originally developed in 1972 by Nelder and Weddeburn. It was widely repeated subsequently by Mc Cullagh and Nelder in 1983, Aithkin and Al in 1990 and Lindset in 1989.

The generalized linear model is an extension of the Gaussian linear model because it allows to consider a law of probability other than the Gaussian law and a function of link other the identity. Linear models are characterized by three components:

- Random component: non-cumulative overlap amounts allow an exponential structure issue.
- Systematic component: deterministic component of the model.
- Function of the link: there exists a functional relation between the random components and the systematic component.

2.3   Model of Bootstrap

The BOOTSTRAP method was introduced by Efron in 1979 to estimate the bias and variability of an estimator in a non-parametric context. The principle consists in simulating a large number of samples of size N, by randomly drawing N observations from an initial sample of N independent and identically distributed random variables.

We can say that the Bootstrap technique is a particular method of resampling. It replaces the theoretical inferences of statistical analysis by repeating the resampling of the initial data and making statistical inference on these bootstrap samples.
The application of BOOTSTRAP is based on two assumptions:

  i. The independence of the observations (hence the draw with delivery)
 ii. The uniqueness of the distribution law of each element that composes the initial sample.

Although the first approach is more robust than the Residual Bootstrap, only the latter can be implemented in terms of provisioning.

3   Results (Practical Case)

3.1   Application Domain

Health insurance is a branch of social security that aims to provide access to care for all members, it allows several people to share risks.

In health insurance or health insurance, we can talk about two groups of risks:
Major risks: This category includes heavy and serious illnesses that involve significant expenses such as hospitalization, surgery and other specialized procedures. The likelihood of occurrence of these events is low, on the other hand, the financial effort required is greater for families both for provident organizations.

Small risks: They concern the mildest cases which require less expenditure but, on the contrary, the frequency is higher. It is ambulatory care.

The management of large risks by the provident bodies is done by issuing the assumption of all or part of the health expenses of its beneficiaries according to a list of precisely defined benefits. The main source of expenditure for the insurance system is the cost of care.

In this section, we will try to calculate the technical provisions applied to the field of health insurance in Morocco, and Thomas Mack's method was chosen as the method of calculation.

3.2   Application

The hypotheses of Mack's model are:

$$E[C_{i,k+1}|C_{i,j}, \ldots, C_{i,k}] = E[C_{i,k+1}|C_{i,k}] = f_k C_{i,k}.$$

Years of occurrence are independent

$$Var(C_{i,k+1}|C_{i,j}, \ldots, C_{i,k}) = \sigma_k^2 C_{i,k}$$

The loss ratio is represented by the triangles of the regulations

Année de survenance	Année de développement				
	1	2	$\cdots$	n-1	n
1	$C_{1,1}$	$C_{1,2}$	$\cdots$	$C_{1,n-1}$	$C_{1,n}$
2	$C_{2,1}$	$C_{2,2}$	$\cdots$	$C_{2,n-1}$	
3	$\vdots$	$\vdots$	$\cdot\cdot\cdot$		
n-1	$C_{n-1,1}$	$C_{n-1,2}$			
n	$C_{n,1}$				

Assumptions Verification:

- 1st hypothesis: The correlation on the ranks is tested.
- 2nd hypothesis: To verify independence between years of occurrence, one must test the existence of a diagonal effect.
- 3rd hypothesis: We will study the graph $C_{(i,k+1)}$ as a function of $C_{(i,k)}$ to validate the existence of a linear relation.

In addition, it must study the graph of residuals r_i, k as a function of C_i, k to check their randomness.

$$r_{i,k} = \frac{C_{i,k+1} - \lambda_k C_{i,k}}{\sqrt{C_{i,k}}}$$

Consider the following non-cumulative settlement triangle:

"i"	k = 1	k = 2	k = 3	k = 4	k = 5	k = 6	k = 7	k = 8
			Thomas Mack - "Distribution-Free" Calculation of Standard Error					
1	315 426	393 800	733 402	1 458 665	1 605 035	1 804 733	1 815 540	1 834 111
2	19 038	87 853	650 990	883 708	1 641 203	1 823 599	2 017 056	
3	8 512	437 420	734 915	1 490 600	1 930 169	2 015 292		
4	44 731	171 565	1 111 424	1 466 413	1 680 488			
5	294 564	403 854	434 833	543 214				
6	248 956	601 902	893 359					
7	58 527	277 245						
8	105 946							

The Process variance reflects the randomness of the realization of a variable following a certain stochastic model

Process Variance Multiplier

Annual	7855819	1435346,92	100315,277	110030,848	3279,12579	8691,80324	3316,36345
Reserve	9516799,33	1660980,34	225633,417	125318,141	15287,2925	12008,1667	3316,36345

Parameter Variance Multiplier

Annual	0,63428529	0,13121753	0,01140574	0,01379277	0,0005445	0,0022449	0,00085654
Reserve	0,79434727	0,16006198	0,02884445	0,01743871	0,00364594	0,00310144	0,00085654

by the application of the model, we have

	Matrix of Estimation Error Factors						
	2	3	4	5	6	7	8
	0,00085654	0,00310144	0,00364594	0,01743871	0,02884445	0,01743871	0,79434727
2	0,00085654	0,00085654	0,00085654	0,00085654	0,00085654	0,00085654	0,00085654
3	0,00310144	0,00085654	0,00310144	0,00310144	0,00310144	0,00310144	0,00310144
4	0,00364594	0,00085654	0,00310144	0,00364594	0,00364594	0,00364594	0,00364594
5	0,01743871	0,00085654	0,00310144	0,00364594	0,01743871	0,01743871	0,01743871
6	0,02884445	0,00085654	0,00310144	0,00364594	0,01743871	0,02884445	0,02884445
7	0,01743871	0,00085654	0,00310144	0,00364594	0,01743871	0,02884445	0,01743871
8	0,79434727	0,00085654	0,00310144	0,00364594	0,01743871	0,02884445	0,01743871

For the covariance matrix:

		Covariance Matrix - Estimation Error Only					
	2	**3**	**4**	**5**	**6**	**7**	**8**
	2037688	2150522	1955109	817728	2143526	1446626	1325762
2	2037688	3556515182	3753451715	3412383908	1427235541	3741241395	2524893778	2313942260
3	2150522	3753451715	1,4343E+10	1,304E+10	5454015986	1,4297E+10	9648590320	8842463426
4	1955109	3412383908	1,304E+10	1,3936E+10	5828931653	1,5279E+10	1,0312E+10	9450305075
5	817728	1427235541	5454015986	5828931653	1,1661E+10	3,0567E+10	2,0629E+10	1,8906E+10
6	2143526	3741241395	1,4297E+10	1,5279E+10	3,0567E+10	1,3253E+11	8,9443E+10	8,197E+10
7	1446626	2524893778	9648590320	1,0312E+10	2,0629E+10	8,9443E+10	3,6494E+10	3,3445E+10
8	1325762	2313942260	8842463426	9450305075	1,8906E+10	8,197E+10	3,3445E+10	1,3962E+12
9								
10								

		Matrix of Estimation Error Correlation Coefficients					
		2	**3**	**4**	**5**	**6**	**7**	**8**		
		59637	119764	118053	107986	364049	191035	1181601	0	0
2	59637	1	0,52552418	0,48469658	0,22162436	0,17232311	0,22162436	0,03283745
3	119764	0,52552418	1	0,9223107	0,42172057	0,3279071	0,42172057	0,06248514
4	118053	0,48469658	0,9223107	1	0,4572435	0,35552781	0,4572435	0,06774847
5	107986	0,22162436	0,42172057	0,4572435	1	0,77754591	1	0,14816716
6	364049	0,17232311	0,3279071	0,35552781	0,77754591	1	1,2860977	0,19055745
7	191035	0,22162436	0,42172057	0,4572435	1	1,2860977	1	0,14816716
8	1181601	0,03283745	0,06248514	0,06774847	0,14816716	0,19055745	0,14816716	1
9								
10								

Finally, we will have the following results:

Diagonal	LDF	Reserve	Ultimate	Proc Mult	Proc SD	CV	Param SD	CV	Total SD	CV	Published
1834 111	1,000	0	1834111								
2017 056	1,010	20632	2037688	3316	82205	398%	59 637	289%	101 559	492%	492%
2015 292	1,067	135230	2150522	12008	160698	119%	119 764	89%	200 418	148%	148%
1680 488	1,163	274621	1955109	15287	172882	63%	118 053	43%	209 344	76%	78%
543 214	1,505	817728	817728	125318	320119	117%	107 986	39%	337 842	123%	123%
893 359	2,399	1250167	2143526	225633	695450	56%	364 049	29%	784 973	63%	63%
277 245	5,218	1169380	1446626	1660980	1550102	133%	191 035	16%	1 561 829	134%	134%
105 946	12,514	1219815	1325762	9516799	3552043	291%	1181 601	97%	3 743 420	307%	307%

Mack's method therefore gives us a total amount of the technical provision of 11559343,05 DHS.

The deterministic method of CHAIN LADDER, based on the use of "link ratios" which are the coefficients of passage between the different years of development, gave us a provision of 5408971,13 DHS.

The Mack model yields standard deviations from estimates on reserves. And mainly it is based on a reduced number of hypotheses that its application requires.

In our case and by the application of the Mack model, the log-normal law was used in order to avoid negative provisions.

Mack's method has an advantage in the calculation of the provision, which can be felt in the case where the health insurance body expects to expand benefits in the benefit mode of care.

4 Conclusion

This work deals with the problem of optimizing actuarial processes through the calculation of technical provisions. We presented the results of the application of Mack's stochastic method applied to the field of health insurance. This method has shown good results on the calculation of the provision since it integrates the variance in the calculation and gives a risk representation in addition to the confidence intervals.
On the other hand, it has limits since:

- The first hypothesis is no longer verified in the case of a change in claims management,

$$E[C_{i,k+1}|C_{i,j}, \ldots, C_{i,k}] = E[C_{i,k+1}|C_{i,k}] = f_k C_{i,k}$$

- It does not take into account non-constant inflation,
- It works only for regular and stable triangles,
- For the recent years of the triangle, it creates considerable uncertainty.

Indeed, with regard to the results obtained and taking into account the evolutions and the constraints which the health insurance bodies undergo, one is obliged to pass to the other models of calculation.

References

1. Provisions for adverse deviations for actuarial valuations of defined benefit pension plans, January 2013
2. Dbabis, M.B.: Actuarial models and methods for quantitative risk assessment in Solvency II environment, October 2012, Version 1, 13 June 2013
3. Favre, J.P.: Financial and actuarial mathematics (2014)
4. Croguennec, J.-B.: Methods of calculating actuarial liabilities in occupational retirement provision, public service cases (2009)

Author Index

© Springer International Publishing AG, part of Springer Nature 2018
M. Ben Ahmed and A. A. Boudhir (Eds.): SCAMS 2017, LNNS 37, pp. 1043–1046, 2018.
https://doi.org/10.1007/978-3-319-74500-8